Mathematics for Economists with Applications

Mathematics for Economists with Applications provides detailed coverage of the mathematical techniques essential for undergraduate and introductory graduate work in economics, business and finance.

Beginning with linear algebra and matrix theory, the book develops the techniques of univariate and multivariate calculus used in economics, proceeding to discuss the theory of optimization in detail. Integration, differential and difference equations are considered in subsequent chapters. Uniquely, the book also features a discussion of statistics and probability, including a study of the key distributions and their role in hypothesis testing. Throughout the text, large numbers of new and insightful examples and an extensive use of graphs explain and motivate the material. Each chapter develops from an elementary level and builds to more advanced topics, providing logical progression for the student, and enabling instructors to prescribe material to the required level of the course.

With coverage substantial in depth as well as breadth, and including a companion website at www.routledge.com/cw/bergin, containing exercises related to the worked examples from each chapter of the book, *Mathematics for Economists with Applications* contains everything needed to understand and apply the mathematical methods and practices fundamental to the study of economics.

James Bergin is Professor of Economics at Queen's University, Canada.

Mathematics for Economists with Applications

James Bergin

Routledge
Taylor & Francis Group
LONDON AND NEW YORK

First published 2015
by Routledge
2 Park Square, Milton Park, Abingdon, Oxon, OX14 4RN

and by Routledge
711 Third Avenue, New York, NY 10017

Routledge is an imprint of the Taylor & Francis Group, an informa business

British Library Cataloguing in Publication Data
A catalogue record for this book is available from the British Library

Library of Congress Cataloging-in-Publication Data
Bergin, James.
Mathematics for economists with applications / by James Bergin. –
1 Edition. pages cm
1. Economics, Mathematical. I. Title.
HB135.B47 2015
510–dc23
2014020372

ISBN: 978-0-415-63827-2 (hbk)
ISBN: 978-0-415-63828-9 (pbk)
ISBN: 978-1-315-81953-2 (ebk)

Typeset in NewCenturySchlbk
by Cenveo Publisher Services

MIX
Paper from responsible sources
FSC® C013604
FSC www.fsc.org

Printed and bound by CPI Group (UK) Ltd, Croydon, CR0 4YY

Contents

Figures

Preface

Courses in undergraduate mathematical economics are taught at various levels, from the second to fourth year, and as preparatory material for graduate students. As a result, the diversity in the backgrounds of students taking such courses is substantial with large variations in training and knowledge. This diverse readership makes it difficult to present the mathematics used in economics at a suitable level. Nevertheless, there is a basic set of mathematical techniques considered prerequisite for many third- and fourth-year courses and which should be discussed in any course on introductory mathematical economics. While the material in this book varies substantially in terms of level and difficulty, it is possible to identify a subset of the material that may be used for a first course on mathematical techniques in economics. The following outline suggests a suitable collection of essential material at the undergraduate level.

Chapter 2 on matrices and systems of equations provides a standard discussion of linear equation systems and rules on matrix manipulation. In particular, the material from Section 2.1 to Section 2.3.5 provides these rules, along with various techniques for solving such systems. Following this, Chapter 3 describes a variety of applications of matrix algebra, including input–output models, adjacency matrices and invariant distributions. These are discussed in Section 3.1 to Section 3.6. Proceeding to functions and calculus, Chapter 5 considers functions of a single variable. This chapter introduces continuity and differentiability of a function, and develops the basic techniques and rules of calculus and differentiation. In addition, the standard functions, including log and exponential function, are discussed in detail.

Chapter 6 provides a range of calculus applications, including optimization and the use of calculus techniques to determine elasticities and taxation incidence. In Chapter 7, the study of functions of many variables begins. Section 7.1 to Section 7.6 develop partial derivatives along with a variety of applications, including the study of contours of functions, the elasticity of substitution and comparative statics calculations. In Chapter 8, Section 8.1 and Section 8.2 cover a broad range of applications of Taylor series expansions, including finding the roots of an equation, the development of numerical optimization routines and the calculation of measures of risk aversion.

Multivariate optimization is discussed in Chapter 11 for the unconstrained case. Sections 11.1 and 11.2 examine the two-variable problem, motivate the first- and second-order conditions and describe the envelope theorem. Chapter 12 develops optimization with equality constraints. Section 12.2 considers the case of two variables, while motivations for the necessary and sufficient conditions for an optimum are given in Section 12.3. Integration is introduced in Chapter 14. Section 14.2 defines the integral, and some basic rules of integration are presented in Section 14.3. Common applications, such as the measurement of welfare surplus and the calculation of present value, are described in Sections 14.4 through 14.7.

In Chapter 16, differential equations are studied. Section 16.1 introduces the topic and Section 16.2 develops the basic analysis for the first-order linear differential equation. Chapter 17 considers difference equations. Section 17.2 provides some motivating examples and Section 17.3 provides the mathematical techniques for solving these equations. Summarizing this suggested course outline:

Chapter	*Section*
2	2.1–2.3.5
3	3.1–3.6
5	5.1–5.6
6	6.1–6.8
7	7.1–7.6
8	8.1–8.2
11	11.1–11.2
12	12.1–12.3
14	14.1–14.7
16	16.1–16.2
17	17.1–17.3

1 Introduction

1.1 Introduction

This introduction provides a brief overview of the contents of the book, briefly describing the focus of each chapter.

1.2 A brief summary by chapter

Chapter 2 introduces matrices and the corresponding linear algebra. A matrix is a rectangular array of numbers, which may be used to succinctly represent sets of equations, network connections, evolutionary dynamic systems and so on. Manipulation of matrices becomes essential to the efficient study of large systems. The following example illustrates the benefit of matrices when moving from single equation problems to systems of equations.

In the supply-and-demand model, demand (D) for some good is negatively related to price, and supply (S) positively related. Thus, $D = \alpha - \beta p$, where p is the price of the good. The letters α and β denote *parameters* of the model with $\alpha > 0$, denoting that an increase in price reduces demand. The formulation of supply is similar – except that supply ought to be positively related to price, so that we postulate that $\delta > 0$ in the relation $S = \gamma + \delta p$. Combining these two relationships gives us the supply-and-demand model:

$$\begin{aligned} D &= \alpha - \beta p,\ \alpha > 0,\ \beta > 0 \qquad (1.1)\\ S &= \gamma + \delta p,\ \gamma < 0,\ \delta > 0. \end{aligned}$$

At what price does the good sell and how much of the good is sold? The market clearing assumption implies that the market clearing price will prevail. Denote by

p_e the price that equates supply and demand. Then $D_e = \alpha - \beta p_e = \gamma + \delta p_e = S_e$. Rearranging, $\alpha - \beta p_e = \gamma + \delta p_e$, and again, $(\alpha - \gamma) = (\beta + \delta)p_e$. Divide both sides by $(\beta + \delta)$ to get the *equilibrium* price $p_e = \frac{(\alpha-\gamma)}{(\beta+\delta)}$. Since at this price the quantity demanded is equal to the quantity supplied, we can determine the equilibrium quantity, denoted Q_e. We find Q_e by substituting p_e into either the supply or demand equation: $Q_e = \alpha - \beta p_e = \alpha - \beta\left[\frac{(\alpha-\gamma)}{(\beta+\delta)}\right] = \frac{(\delta\alpha+\beta\gamma)}{(\beta+\delta)}$. Summarizing:

$$p_e = \frac{(\alpha - \gamma)}{(\beta + \delta)}, \qquad Q_e = \frac{(\delta\alpha + \beta\gamma)}{(\beta + \delta)}. \tag{1.2}$$

With multiple markets, the computations become more complex and motivate the use of matrices.

With two (or more) markets, there may be cross-price effects: the price of bread may affect the demand for margarine; the price of tea may affect the demand for coffee. As a result, the computations to determine the market clearing prices and quantities become more complex. To capture these cross-price effects, write the demand and supply in both markets:

$$\begin{aligned} &\text{Market 1:} \quad D_1 = a_0 + a_1 p_1 + a_2 p_2, \quad S_1 = b_0 + b_1 p_1 + b_2 p_2 \\ &\text{Market 2:} \quad D_2 = \alpha_0 + \alpha_1 p_1 + \alpha_2 p_2, \quad S_2 = \beta_0 + \beta_1 p_1 + \beta_2 p_2. \end{aligned}$$

Equilibrium occurs when prices equate supply and demand in both markets ($D_1 = S_1$ and $D_2 = S_2$). To find the equilibrium level of quantity and price in each market, equate supply and demand in each market.

$$\begin{aligned} &\text{Market 1:} \quad D_1 = S_1 \quad \text{or} \quad a_0 + a_1 p_1 + a_2 p_2 = b_0 + b_1 p_1 + b_2 p_2 \\ &\text{Market 2:} \quad D_2 = S_2 \quad \text{or} \quad \alpha_0 + \alpha_1 p_1 + \alpha_2 p_2 = \beta_0 + \beta_1 p_1 + \beta_2 p_2. \end{aligned}$$

Rearranging these equations gives:

$$\begin{aligned} (a_1 - b_1)p_1 + (a_2 - b_2)p_2 &= (b_0 - a_0) \\ (\alpha_1 - \beta_1)p_1 + (\alpha_2 - \beta_2)p_2 &= (\beta_0 - \alpha_0). \end{aligned}$$

These equations can be solved by substitution. However, the process is tedious. From the first equation

$$p_1 = \frac{(b_0 - a_0)}{(a_1 - b_1)} - \frac{(a_2 - b_2)}{(a_1 - b_1)} p_2.$$

Substitute into the second equation to get:

$$(\alpha_1 - \beta_1)\left[\frac{(b_0 - a_0)}{(a_1 - b_1)} - \frac{(a_2 - b_2)}{(a_1 - b_1)} p_2\right] + (\alpha_2 - \beta_2)p_2 = (\beta_0 - \alpha_0).$$

Collecting terms in p_2 gives:

$$\left[(\alpha_2-\beta_2)-(\alpha_1-\beta_1)\frac{(a_2-b_2)}{(a_1-b_1)}\right]p_2=(\beta_o-\alpha_o)-(\alpha_1-\beta_1)\frac{(b_o-a_o)}{(a_1-b_1)}.$$

Multiplying both sides by (a_1-b_1) gives:

$$[(\alpha_2-\beta_2)(a_1-b_1)-(\alpha_1-\beta_1)(a_2-b_2)]p_2=(\beta_o-\alpha_o)(a_1-b_1)-(\alpha_1-\beta_1)(b_o-a_o).$$

Thus,

$$p_2=\frac{[(\beta_o-\alpha_o)(a_1-b_1)-(\alpha_1-\beta_1)(b_o-a_o)]}{[(\alpha_2-\beta_2)(a_1-b_1)-(\alpha_1-\beta_1)(a_2-b_2)]}.$$

A similar calculation gives p_1. With three or more equations, the procedure becomes progressively more complicated. The use of matrix algebra simplifies problems of this sort significantly. In matrix form:

$$\begin{pmatrix}(a_1-b_1) & (a_2-b_2)\\(\alpha_1-\beta_1) & (\alpha_2-\beta_2)\end{pmatrix}\begin{pmatrix}p_1\\p_2\end{pmatrix}=\begin{pmatrix}b_0-a_0\\\beta_0-\alpha_0\end{pmatrix}$$

or

$$\begin{pmatrix}p_1\\p_2\end{pmatrix}=\begin{pmatrix}(a_1-b_1) & (a_2-b_2)\\(\alpha_1-\beta_1) & (\alpha_2-\beta_2)\end{pmatrix}^{-1}\begin{pmatrix}b_0-a_0\\\beta_0-\alpha_0\end{pmatrix}.$$

The matrix

$$\begin{pmatrix}(a_1-b_1) & (a_2-b_2)\\(\alpha_1-\beta_1) & (\alpha_2-\beta_2)\end{pmatrix}^{-1}$$

is called the inverse of

$$\begin{pmatrix}(a_1-b_1) & (a_2-b_2)\\(\alpha_1-\beta_1) & (\alpha_2-\beta_2)\end{pmatrix}.$$

The inverse can be written (*details later*):

$$\begin{pmatrix}(a_1-b_1) & (a_2-b_2)\\(\alpha_1-\beta_1) & (\alpha_2-\beta_2)\end{pmatrix}^{-1}=\frac{1}{d}\begin{pmatrix}(\alpha_2-\beta_2) & -(a_2-b_2)\\-(\alpha_1-\beta_1) & (a_1-b_1)\end{pmatrix},$$

where $d=(a_1-b_1)(\alpha_2-\beta_2)-(\alpha_1-\beta_1)(a_2-b_2)$. Thus,

$$\begin{aligned}\begin{pmatrix}p_1\\p_2\end{pmatrix}&=\frac{1}{d}\begin{pmatrix}(\alpha_2-\beta_2) & -(a_2-b_2)\\-(\alpha_1-\beta_1) & (a_1-b_1)\end{pmatrix}\begin{pmatrix}(b_0-a_0)\\(\beta_0-\alpha_0)\end{pmatrix}\\&=\frac{1}{d}\begin{pmatrix}(\alpha_2-\beta_2)(b_0-a_0)-(a_2-b_2)(\beta_0-\alpha_0)\\-(\alpha_1-\beta_1)(b_0-a_0)+(a_1-b_1)(\beta_0-\alpha_0)\end{pmatrix}.\end{aligned}$$

So,

$$p_1 = \frac{[(\alpha_2 - \beta_2)(b_0 - a_0) - (a_2 - b_2)(\beta_0 - \alpha_0)]}{[(a_1 - b_1)(\alpha_2 - \beta_2) - (\alpha_1 - \beta_1)(a_2 - b_2)]}$$

$$p_2 = \frac{[-(\alpha_1 - \beta_1)(b_0 - a_0) + (a_1 - b_1)(\beta_0 - \alpha_0)]}{[(a_1 - b_1)(\alpha_2 - \beta_2) - (\alpha_1 - \beta_1)(a_2 - b_2)]}.$$

The early part of Chapter 2 describes the basic rules for matrix manipulation and for solving systems of equations. Not all systems have a solution, and some equation systems may have many solutions. This leads to the task of identifying when equation systems are soluble and the circumstances under which there is a unique solution. The key ideas of rank and linear independence are introduced to study this matter. The chapter concludes with some additional observations on matrix algebra.

In Chapter 3, matrix techniques are used to illustrate the many uses of matrix algebra. These include linear demand systems, input–output techniques, Markov processes and invariant distributions, and least squares estimation. To illustrate, consider input–output analysis, where the economy is viewed as a number of interlinked industries (n industries). The output of an industry is used for final consumption, and in the production of other goods. The total output of an industry is allocated between that used in production in other industries (interindustry demand) and consumption or final demand. Planning to achieve some (target) level of consumption or final demand indirectly implies production levels in each industry. Input–output analysis is concerned with determining the industry production levels required to sustain or achieve target levels across the entire range of final demands. A simple model is developed next to illustrate the procedure. The discussion illustrates the method and the use of matrix methods to determine the necessary level of industry output.

Let a_{ij} = amount of i required to produce 1 unit of j; let x_i = output of i and f_i = final demand for i. Thus, if x_j is the amount of good j produced, $a_{ij}x_j$ is the amount of i required in the production of j. (If one kilo of good j is produced, a_{ij} is the fraction of one kilo of i used in the production of the kilo of j.) The output of industry i is allocated to production in other industries ($a_{ij}x_j$) and final demand (f_i). Thus,

$$x_i = a_{i1}x_1 + a_{i2}x_2 + \cdots + a_{in}x_n + f_i, i = 1, \ldots, n.$$

This can be rearranged:

$$-a_{i1}x_1 - a_{i2}x_2 \cdots + (1 - a_{ii})x_i \cdots - a_{in}x_n = f_i, i = 1, \ldots, n,$$

or, in matrix notation:

$$\begin{pmatrix} (1-a_{11}) & -a_{12} & \cdots & -a_{1n} \\ -a_{21} & (1-a_{22}) & \cdots & -a_{2n} \\ \vdots & \vdots & \ddots & \vdots \\ -a_{n1} & -a_{n2} & \cdots & (1-a_{nn}) \end{pmatrix} \begin{pmatrix} x_1 \\ x_2 \\ \vdots \\ x_n \end{pmatrix} = \begin{pmatrix} f_1 \\ f_2 \\ \vdots \\ f_n \end{pmatrix}.$$

This expression can be written succinctly: $(\boldsymbol{I}-\boldsymbol{A})x = f$, where the matrices are:

$$\boldsymbol{A} = \begin{pmatrix} a_{11} & a_{12} & \cdots & a_{1n} \\ a_{21} & a_{22} & \cdots & a_{2n} \\ \vdots & \vdots & \ddots & \vdots \\ a_{n1} & a_{n2} & \cdots & a_{nn} \end{pmatrix}, \quad \boldsymbol{I} = \begin{pmatrix} 1 & 0 & \cdots & 0 \\ 0 & 1 & \cdots & 0 \\ \vdots & \vdots & \ddots & \vdots \\ 0 & 0 & \cdots & 1 \end{pmatrix}.$$

Chapter 4 develops the theory of linear programming. Optimization arises in many contexts. For example, utility maximization involves determining the consumption bundle that maximizes utility subject to a linear constraint, the budget constraint. While the objective function, the utility function, is non-linear, the set of feasible choices is defined by a linear constraint. Chapter 4 introduces constrained optimization in the linear context: both the objective and the constraints are linear. Linear programming is widely used in industry and so has immediate practical usefulness.

The following transportation problem shows how a linear programming problem arises. A firm has two sources of input, two processing plants and three markets to supply. The shipping costs are given in the following table.

	Source 1	*Source 2*	*Market 1*	*Market 2*	*Market 3*
Plant A	2	3	5	4	5
Plant B	3	2	3	4	6

The chart in Figure 1.1 depicts this graphically.

Let x_{iAj} = # tons from source i through plant A to market j and x_{iBj} = # tons from source i through plant B to market j. Inputs are shipped from source to plant, and then to market. The cost of shipping one ton from source 1 through plant A to market 3 is $2+5=7$ (the cost of shipping from source 1 to plant A plus the cost of shipping from plant A to market 3). Thus, the cost of shipping x_{1A3} tons on this route is $7x_{1A3}$. Other route costs are calculated in the same way. Finally, sources 1 and 2 have, respectively, 10 and 20 units of raw material available.

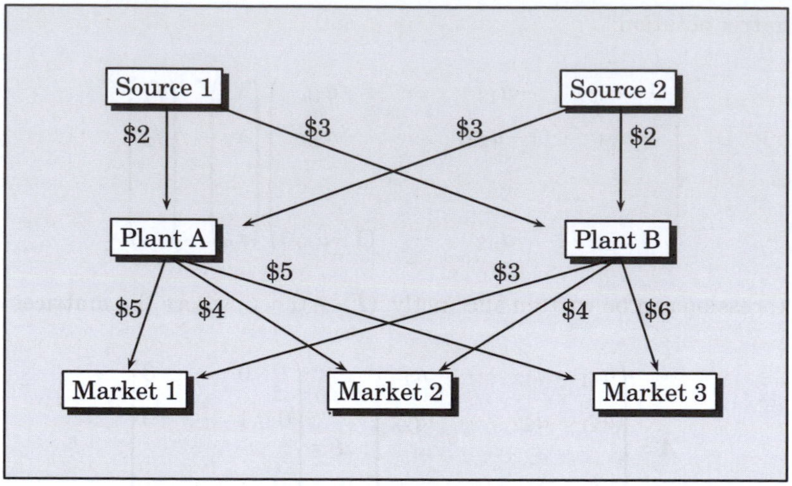

Figure 1.1: Cost flows

Market demand is 10 in market 1, 15 in market 2 and 5 in market 3. This leads to the program:

$$\begin{aligned}\min \quad C = {} & 7x_{1A1} + 6x_{1A2} + 7x_{1A3} + 6x_{1B1} + 7x_{1B2} + 9x_{1B3} + 8x_{2A1} + 7x_{2A2} \\ & + 8x_{2A3} + 5x_{2B1} + 6x_{2B2} + 8x_{2B3},\end{aligned}$$

subject to:

$$x_{1A1} + x_{1A2} + x_{1A3} + x_{1B1} + x_{1B2} + x_{1B3} \leq 10$$

$$x_{2A1} + x_{2A2} + x_{2A3} + x_{2B1} + x_{2B2} + x_{2B3} \leq 20$$

$$x_{1A1} + x_{1B1} + x_{2A1} + x_{2B1} \geq 10$$

$$x_{1A2} + x_{1B2} + x_{2A2} + x_{2B2} \geq 15$$

$$x_{1A3} + x_{1B3} + x_{2A3} + x_{2B3} \geq 5$$

$$x_{iAj}, x_{iBj} \geq 0, \forall i, j.$$

From a theoretical point of view, the subject introduces many key ideas in economics: optimization, constraints and feasible choices, dual or shadow prices of constraints, the notion of a slack variable and corresponding constraint, and so on.

The first part of Chapter 4 introduces the basic concepts used to define a linear program, formulating objectives, constraints and feasible region. Then a graphical approach is used to illustrate the general solution technique. One key idea is that a solution to a linear program can always be found at a corner or vertex of the feasible region. Since there are a finite (although possibly large) number of vertices, this fact greatly simplifies the search for a solution. Examples illustrate this fact and utilize it to locate the optimal solution. In addition, computation of shadow prices may be done directly for simple problems where the graphical technique is used. The second part of the chapter considers the general structure of these problems. Basic solutions are introduced and vertices are identified as basic solutions. The computation of basic solutions is straightforward, so this observation provides a means of locating the vertices of the feasible region. These observations provide a basis for algorithms to determine the solution. The concluding part of the chapter provides a discussion of duality theory and the theory underlying dual or shadow prices.

Chapter 5 is concerned with functions of one variable and begins by showing how various functions, demand functions, present value functions, and so on, arise in economics. A demand function is written $p(q)$, relating price to quantity demanded. To measure price-demand responsiveness, one must consider the price $p(q')$ at which some alternative demand q' would be forthcoming. This naturally leads to the calculation $p(q')-p(q)$, or $\frac{p(q')-p(q)}{q'-q}$, giving the per-unit price variation associated with a quantity variation. Similarly, if a single firm supplies the market, the total revenue to the firm is $r(q) = qp(q)$, and if the cost of output level q is $c(q)$, the firm's profit is $\pi(q) = r(q) - c(q)$. In this way, functions arise routinely in economic problems and calculations such as the increase in cost per unit increase in output $\frac{c(q')-c(q)}{q'-q}$, the marginal cost, lead naturally to calculus and derivatives.

Working with functions requires an understanding of the basic concepts of continuity and differentiability. These are introduced in the early part of the chapter. Following this, a variety of functions are described. The log and exponential functions are discussed in detail since they appear regularly in economic calculations, in particular in calculations concerning growth and present value. Various rules of calculus are described for the functions of functions (sums, products and ratios). Finally, higher-order derivatives are introduced.

In Chapter 6 a number of applications illustrate the use of functions in economics. Optimization for functions of one variable is discussed in detail, by considering the behavior of the first and second derivatives. Subsequently, the determination of optima is related to concavity or convexity of the function, and this in turn to properties of the derivatives. Following the discussion of optimization with a single variable, a range of applications are considered. A model of externalities is developed to describe the free-rider problem and market inefficiency. Then the elasticity function is described and this is later applied to the study of taxation and dead-weight loss. For example, if the pre-tax equilibrium is at Q_e with price p_e and post-tax output and price at Q_t, p_t, the impact of the tax is a quantity reduction of $Q_t - Q_e$ and price increase of $p_t - p_e$. The dead-weight loss is the area between Q_t and Q_e above the supply curve and under the demand curve. In the linear case, this is easy to compute, giving an expression for dead-weight loss. In this chapter, comparative statics are introduced by considering the impact of parameter changes on the equilibrium or the optimal level of a variable. The incidence of a product tax is considered in detail. Finally, the chapter concludes with a discussion of Ramsey pricing.

Chapter 7 introduces functions of many variables and systems of non-linear equations. The chapter begins with the definition of a partial derivative and the notion of a level contour of a function. This is illustrated with the familiar isoquant where curvature is described with the marginal rate of substitution. Partial derivatives are used to develop the notion of a Cournot equilibrium in the context of oligopoly. The market inverse demand function is given by the equation $p = f(Q)$, where p is the market price and Q the aggregate quantity. There are two firms, labeled 1 and 2. If firm i supplies quantity q_i, the aggregate output and supply is $Q = q_1 + q_2$. The cost per unit of output for firm i is given by c_i – so that if firm i produces q_i, its cost of production is $c_i q_i$. When aggregate output is $Q = q_1 + q_2$, the revenue of firm i is $q_i f(Q)$. Thus, the total profit of firm i is

$$\pi_i(q_1, q_2) = q_i f(q_1 + q_2) - c_i q_i.$$

This notation makes explicit the dependence of profit on the output of each firm and illustrates how multivariate functions may arise.

The analysis of firm behavior in the linear demand case $p = a - bQ$ may be taken further here. In this case, the profit of firm i is $\pi_i(q_i, q_j) = pq_i - c_i q_i$, where $i \neq j$. Since price depends on the quantities, $\pi_i(q_i, q_j) = [a - b(q_1 + q_2)]q_i - c_i q_i$, where $i \neq j$. This can be rearranged: $\pi_i(q_i, q_j) = [(a - c_i) - b(q_1 + q_2)]q_i$; again $i \neq j$. How a firm decides on its level of output depends on how the other firm is expected to behave. A myopic assumption for a firm, say firm 1, is that firm 2 will not adjust its output in response to the choice of firm 1. In general, when one firm adjusts it output, the other firm will also wish to do so. In the particular case where both firms' quantity levels are such that neither firm *wants* to adjust its output level, the output levels are equilibrium levels. The fact that a firm does not wish to change its level of output is the relevant issue. The fact that this decision is based on the naive assumption that the other firm would not respond is irrelevant. If firm i takes as given the output of j and maximizes profit with respect to i (by differentiating with respect to q_i), this gives: $(a - c_i) - 2bq_i - bq_j = 0, i \neq j$. Solving for q_i gives $q_i = [(a - c_i) - bq_j]/2b$. In particular, $q_1 = [(a - c_1) - bq_2]/2b$, and $q_2 = [(a - c_2) - bq_1]/2b$. Denote these expressions $q_1(q_2)$ and $q_2(q_1)$, respectively.

These are depicted in Figure 1.2: the interpretation is that when firm 2 supplies the quantity q_2, then firm 1 would wish to choose the output level $q_1(q_2)$ and vice

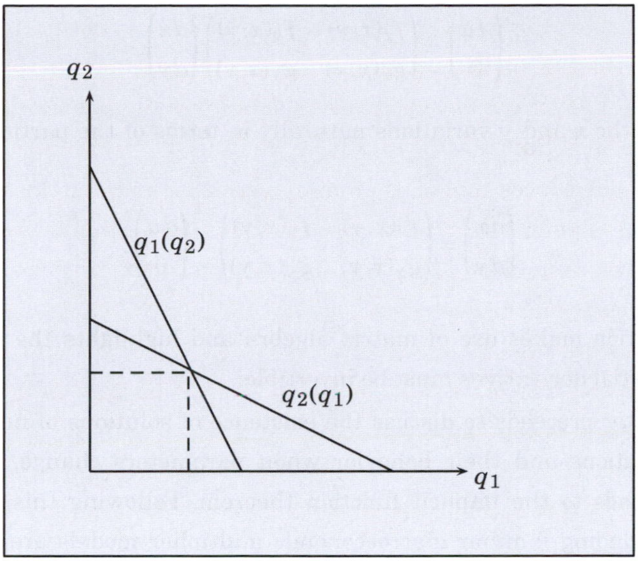

Figure 1.2: Reaction curves

versa. Note that at the quantity pair (q_1^*, q_2^*), firm 1 wishes to choose q_1^* given that firm 2 is supplying q_2^*, and similarly for firm 2.

$$\begin{aligned} q_1 &= \frac{(a-c_1)}{2b} - \frac{1}{2}q_2 = \frac{(a-c_1)}{2b} - \frac{1}{2}\left\{\left[\frac{(a-c_2)}{2b}\right] - \frac{1}{2}q_1\right\} \\ &= \frac{(a-c_1)-(a-c_2)}{2b} + \frac{1}{4}q_1. \end{aligned} \tag{1.3}$$

Thus, the solution values for q_1 and q_2, (q_1^*, q_2^*), are:

$$q_1^* = \frac{4}{6b}\left[(a-c_1) - \frac{1}{2}(a-c_2)\right] \qquad q_2^* = \frac{4}{6b}\left[(a-c_2) - \frac{1}{2}(a-c_1)\right]. \tag{1.4}$$

Further development of the Cournot equilibrium allows a characterization of the price markup over marginal cost in terms of the elasticity. In addition, the Herfindahl–Hirschman index of industry concentration may be defined and related to the equilibrium markup in Cournot competition.

This chapter also develops the techniques of comparative statics and provides a discussion of the implicit function theorem. Given two functions f and g of two variables (x and y), suppose that for some parameter α the functions satisfy $\alpha = f(x,y)$ and $0 = g(x,y)$. Then, at a solution (x,y) the total differentials (with a variation in α, $d\alpha$) give $d\alpha = f_x(x,y)dx + f_y(x,y)dy$ and $0 = g_x(x,y)dx + g_y(x,y)dy$, where, for example, $f_x(x,y)$ is the partial derivative of f at (x,y). Rearranging in matrix form:

$$\begin{pmatrix} d\alpha \\ 0 \end{pmatrix} = \begin{pmatrix} f_x(x,y) & f_y(x,y) \\ g_x(x,y) & g_y(x,y) \end{pmatrix} \begin{pmatrix} dx \\ dy \end{pmatrix}.$$

This sets up the x and y variations naturally in terms of the partial derivatives and $d\alpha$.

$$\begin{pmatrix} dx \\ dy \end{pmatrix} = \begin{pmatrix} f_x(x,y) & f_y(x,y) \\ g_x(x,y) & g_y(x,y) \end{pmatrix}^{-1} \begin{pmatrix} d\alpha \\ 0 \end{pmatrix}.$$

The formulation makes use of matrix algebra and highlights the fact that the matrix of partial derivatives must be invertible.

The chapter proceeds to discuss the existence of solutions of non-linear systems of equations and their behavior when parameters change. An informal discussion leads to the implicit function theorem. Following this, a variety of examples including familiar macroeconomic multiplier models are developed to illustrate the application of the theorem.

Chapter 8 introduces Taylor series expansions. A Taylor series expansion involves approximating a continuous function by a polynomial function, with higher-order polynomials giving better approximations or greater accuracy. The value of this technique is substantial since polynomial functions are relatively simple, and, if the approximation is good, it provides important information on the properties of the original function. The chapter begins with an example showing a quadratic approximation of a function and then proceeds to describe expansions of order n. Some further discussion shows how accuracy improves as the order of the expansion increases. Following this, the chapter develops some applications of Taylor series expansions, beginning with a proof that concavity implies a negative second derivative. This application shows the usefulness of the expansion. A second application concerns numerical optimization. Much of the logic for numerical optimization routines derives from the fact that an arbitrary continuous function may be approximated locally as a quadratic function. This observation forms the basis for many numerical routines. The ideas behind this are developed in the chapter and a variety of examples used to illustrate how the process may be used to locate a maximum or minimum of a function $f(x)$. Exactly the same process may be used to locate a root of a function (a value from x such that the function is equal to 0 at that value ($f(x) = 0$)). Multidimensional generalizations are described. The chapter then discusses risk aversion using Taylor series expansions to provide measures of risk, of risk premia, and of welfare loss resulting from small risks. Finally, a proof of Taylor's theorem is given for both the univariate and multivariate cases.

Vectors are discussed in Chapter 9. A vector can be viewed from a few different perspectives. In one, as a rectangular array of numbers, it is a matrix with either one row or one column. A second interpretation views a vector as a direction, each coordinate representing a length in the corresponding direction. Thus $x = (x_1, x_2)$ is viewed as x_1 units in direction 1 and x_2 units in direction 2. In this view, it makes sense to describe vectors in terms of direction, perpendicularity (or orthogonality), length in a given direction, and so on. In particular, using vectors it is possible to calculate the direction in which a function increases (or decreases) most rapidly. This perspective facilitates the study of many problems, including optimization. Specifically, considering a function $f(x_1, x_2)$, those values of $x = (x_1, x_2)$ that give a constant value for the function f determine a "level contour" of the function f. So, the set $F = \{(x_1, x_2) \mid f(x_1, x_2) = c\}$ is the level contour of the function f at level c. Such curves appear in economics as, for example, iso-cost curves or indifference curves. If f is a utility function, the set F is the set of consumption combinations that give the same level of utility. It is common to consider the slope of the indifference curve by considering the variations of x_1 and x_2 that leave utility constant: $0 = f_{x_1}(x_1, x_2)dx_1 + f_{x_2}(x_1, x_2)dx_2$, with the marginal rate of substitution given by $\frac{dx_2}{dx_1} = -\frac{f_{x_1}(x_1, x_2)}{f_{x_2}(x_1, x_2)}$. But this information is also contained in the vector $(f_{x_1}(x_1, x_2), f_{x_2}(x_1, x_2))$ and, in addition, the absolute magnitudes of the partial derivatives. It turns out that, viewed as a direction $\nabla f = (f_{x_1}(x_1, x_2), f_{x_2}(x_1, x_2))$ points in the direction of steepest increase in the function f. Figure 1.3 illustrates this observation. Starting from x, the value of the function f increases most rapidly in the direction ∇f and ∇f is perpendicular to the level surface of f at x. This fact is used extensively in the discussion of constrained optimization. Finally, Chapter 9 states and proves Farkas' lemma. This useful result may be used to prove the main theorems of linear programming and provides valuable insight into the logic behind important results in constrained optimization theory.

Chapter 10 discusses quadratic forms. A quadratic form is an expression $x'Ax$ where x is a vector and A a matrix. In the univariate case, this just amounts to x^2a. An important question concerns the circumstances in which such an expression is always positive or always negative. For example, what properties of A would imply that $x'Ax > 0$ for all $x \neq 0$. Such a matrix is called a positive definite matrix. When $x'Ax < 0$ for all $x \neq 0$, the matrix is called negative definite. The standard characterization is given in terms of the leading principal minors of the

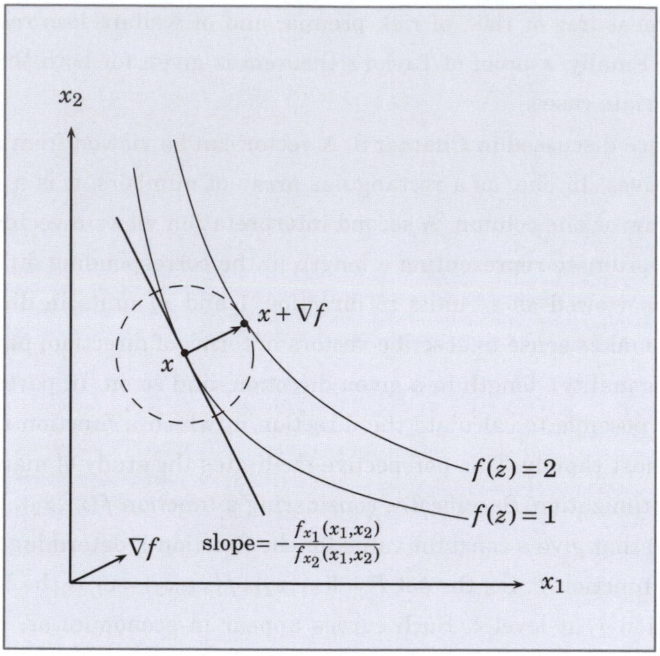

Figure 1.3: Direction of steepest increase

matrix (the leading principal matrix of order k in an $n \times n$ matrix is obtained by deleting the last $n-k$ rows and $n-k$ columns). Positive and negativeness correspond to the cases where $x'Ax \geq 0$ (or $x'Ax \leq 0$) for all x. An important related question concerns positive or negative definiteness when the admissible x are constrained: under what conditions is $x'Ax < 0$ or $x'Ax > 0$ for all non-zero x satisfying a set of linear restrictions: $Bx = 0$? This question arises in constrained optimization. Here, a full characterization is given in terms of an augmented (bordered) matrix and its leading principal minors.

Chapter 11 is the first of three chapters on optimization, and considers multivariate optimization where there are no constraints on the choice variables. The chapter begins with a discussion of the two-variable case, that of maximization or minimization of a function $f(x_1,x_2)$. For example, the standard profit-maximizing problem writes profit $\pi(x_1,x_2) = ph(x_1,x_2) - c_1x_1 - c_2x_2$, where h is a production function, p the price of output and c_i the unit cost of input i.

The first- and second-order conditions for a local maximum or minimum are developed in detail, motivated by using Taylor series expansions, using the Hessian matrix (of second-order partial derivatives). Following this, the envelope theorem is described at length: given a function $f(x,\alpha)$ depending on a choice variable x and parameter α, if $x(\alpha)$ maximizes $f(x,\alpha)$, how does $f^*(\alpha) = f(x(\alpha),\alpha)$ vary with

α? This is the analog of a shadow or dual price in linear programming and can be viewed as a measure of the value of the resource α. The chapter considers the n-variable case, generalizing the initial discussion. The chapter concludes by establishing the connection between concavity, convexity and the properties of the Hessian matrix.

In Chapter 12, maximizing or minimizing a function subject to equality constraints is considered. Familiar problems are the maximization of utility $u(x_1,\ldots,x_n)$ subject to a budget constraint $\sum p_i x_i = y$ and minimization of the cost, $c = \sum w_i x_i$, of achieving a level of output y with a production function $f(x_1,\ldots,x_n)$, $f(x_1,\ldots,x_n) = y$. The following discussion describes utility maximization, to motivate subsequent discussion.

An illustrative example concerns an individual's optimization problem. A consumer's utility depends on the consumption of food, f, and drink, w, according to the formula: $u(f,w) = fw$. The respective prices of food and drink are p_f and p_w and the consumer's income is I. The problem for the consumer is to maximize utility subject to spending no more than I. Thus, the consumer is restricted to choosing those pairs (f,w) that satisfy $p_f f + p_w w = I$. This equation can also be written $w = \frac{I}{p_w} - (\frac{p_f}{p_w})f$ and this equation gives a straight line relation between w and f, with slope $-(\frac{p_f}{p_w})$. See Figure 1.4. The set of pairs (f,w) that yield a given level of utility $\bar{u}$ are given by the equation $\bar{u} = fw$. This determines a relation between f and w, $w = \frac{\bar{u}}{f}$, which gives for any value of f the level of w $(\frac{\bar{u}}{f})$ that achieves utility $\bar{u}$. When $\bar{u} = 1$ then $fw = 1$ so that $w = \frac{1}{f}$; when $\bar{u} = 2$ then $fw = 2$

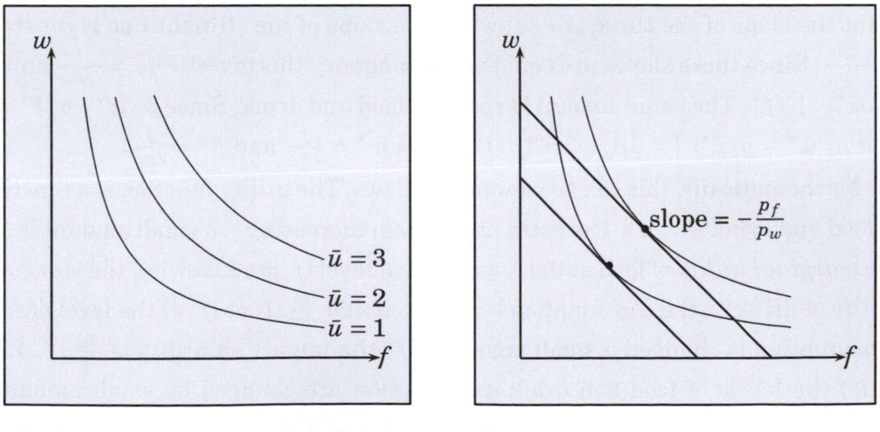

Figure 1.4: Indifference curves and optimal choices

so that $w = \frac{2}{f}$; and when $\bar{u} = 3$ then $fw = 3$ so that $w = \frac{3}{f}$. These calculations indicate that higher levels of utility correspond to higher curves, as in Figure 1.4.

Thus, the highest utility attainable is reached by going to the highest curve having some point in common with the equation $p_f f + p_w w = I$.

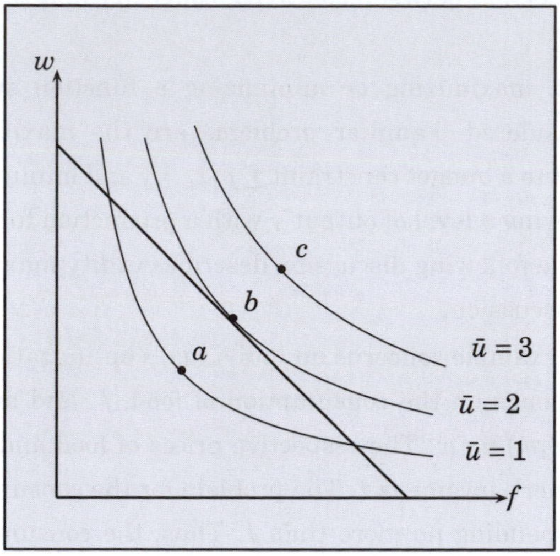

Figure 1.5: Non-optimal, optimal and non-feasible choice

Figure 1.5 illustrates a number of possibilities regarding indifference curves and the budget constraint: (a) feasible points on the indifference curve are not optimal; (c) there are no affordable points giving the utility $\bar{u}$; (b) this is the only possible optimal situation – and there is a unique optimal point $-(f^*, w^*)$. At this point the slope of the curve is $-\frac{w^*}{f^*}$, while the slope of the straight line is constant at $-\frac{p_f}{p_w}$. Since these slopes are equal at the tangency this gives: $-\frac{w^*}{f^*} = -\frac{p_f}{p_w}$ and so $p_w w^* = p_f f^*$. The same amount is spent on food and drink. Since $p_w w^* + p_f f^* = I$ and $p_w w^* = p_f f^*$, $I = 2p_w w^* = 2p_f f^*$. Thus $w^* = \frac{I}{2p_w}$ and $f^* = \frac{I}{2p_f}$.

Mathematically, this can be viewed as follows. The utility function is a function of food and drink $u(f,w)$. The extra utility from increasing f a small amount is $\frac{\partial u}{\partial f}$, the *marginal utility* of food at the consumption level (f,w). Likewise, the *marginal utility* of drink at the consumption level is denoted $\frac{\partial u}{\partial w}$. If at (f,g) the level of food consumption is changed a small amount, df, the impact on utility is $\frac{\partial u}{\partial f}df$. If at (f,w) the levels of food and drink consumption are changed by small amounts, df and dw, the impact on utility is $\frac{\partial u}{\partial f}df + \frac{\partial u}{\partial w}dw$. If the change in the pair (f,w) to $(f+df, w+dw)$ is such that no change in utility occurs, then it must be that $0 = \frac{\partial u}{\partial f}df + \frac{\partial u}{\partial w}dw$. Since the marginal utilities are positive (more food or more drink raises utility), an increase in one must be balanced by a reduction in the other to maintain the same utility level. When f changes and w changes to keep utility constant, it must be that:

$$dw = -\left(\frac{1}{\frac{\partial u}{\partial w}}\right)\frac{\partial u}{\partial f}df \quad \text{or} \quad \frac{dw}{df} = -\frac{\frac{\partial u}{\partial f}}{\frac{\partial u}{\partial w}}.$$

This gives the slope of the indifference curve at (f,g). From the previous discussion, this must equal the slope of the budget constraint at the optimal point where utility is maximized subject to the budget constraint. So, at the optimal point,

$$\frac{\frac{\partial u}{\partial f}}{\frac{\partial u}{\partial w}} = \frac{p_f}{p_w}.$$

The ratio of the marginal utilities is called the marginal rate of substitution, so at the optimal point the marginal rate of substitution is equal to the price ratio. However, this procedure is ad-hoc. The chapter develops systematic procedures for locating the optima.

The initial part of Chapter 12 deals with the two-variable case, giving necessary and sufficient conditions for local maxima and local minima. For the case of just two variables, the constraint (subject to regularity conditions) determines one variable as a function of the other, so that one variable may be eliminated. For example, if the objective is $f(x_1,x_2)$ and the constraint $g(x_1,x_2) = c$, then using the constraint to solve for x_2 as a function of x_1, $x_2(x_1)$, the objective may be maximized as a function of one unconstrained variable, $\max f(x_1,x_2(x_1))$. This observation may be used to justify the second-order conditions. Following this, the n-variable case is considered. Here, the second-order conditions make use of the bordered Hessian matrix and properties derived from the theory of quadratic forms. The value of the Lagrange multiplier is interpreted as a shadow price and a general version of the envelope theorem given for the constrained optimization problem. In utility maximization, these calculations identify the Lagrange multiplier as the marginal value of income; in the cost-minimization case, the Lagrange multiplier is seen to be the marginal cost of additional output. Finally, this chapter discusses optimization in terms of the shape of the objective function and the constraints. The key idea is quasiconcavity (or quasiconvexity) of a function. This property is related to the bordered Hessian and used to provide a sufficient condition for a constrained global maximum.

Chapter 13 considers optimization with inequality constraints. Thus, the task is to maximize or minimize a function $f(x_1,\dots,x_n)$ subject to a constraint $g(x_1,\dots,x_n) \le c$ or a collection of constraints $g_k(x_1,\dots,x_n) \le c_k$, $k = 1,\dots,m$. The chapter begins with some motivating examples. Perhaps the central idea in the chapter is the observation that the direction of improvement in the objective must lie between the gradients of the constraints in the non-feasible direction.

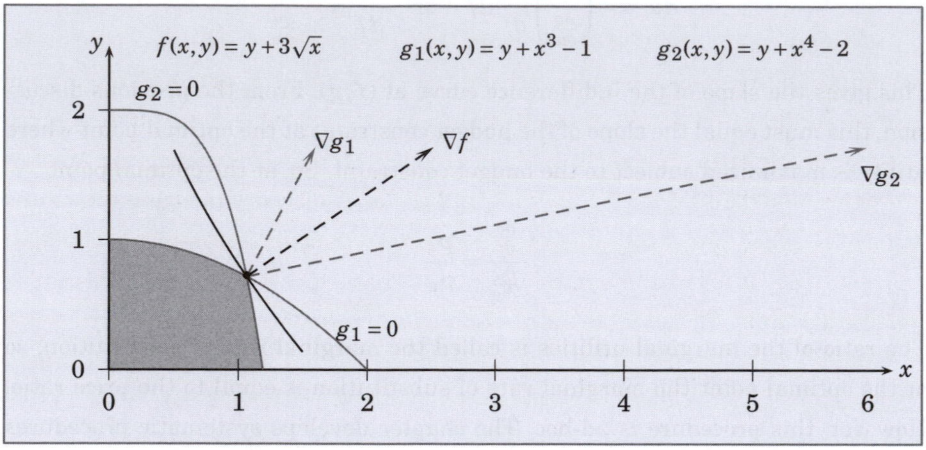

Figure 1.6: Inequality-constrained optimization

Let the feasible region be defined as those pairs (x_1, x_2) satisfying $g_1(x_1, x_2) \le 0$ and $g_2(x_1, x_2) \le 0$. From the figure one may see that there are non-negative numbers λ_1 and λ_2 such that

$$\nabla f = \lambda_1 \nabla g_1 + \lambda_2 \nabla g_2,$$

or, alternatively, $\frac{\partial f}{\partial x_i} = \lambda_1 \frac{\partial g_1}{\partial x_i} + \lambda_2 \frac{\partial g_2}{\partial x_i}$, for $i = 1, 2$. Although non-negativity of the choice variables (x_1, x_2) is not imposed in this example, adding non-negativity constraints is straightforward and done later in the chapter. It turns out that this condition, $\nabla f = \lambda_1 \nabla g_1 + \lambda_2 \nabla g_2$, is, in general, a necessary condition for a local optimum: a candidate solution necessarily satisfies this condition in general. The benefit of this is obvious: it provides a system of equations that helps to locate the solution (but note that there are more unknowns than equations, when the λ variables are included). However, one problem with this insight is that in certain (exceptional) circumstances this equation fails to be a necessary condition for an optimum. This arises, for example, when the constraints have the same gradient at the candidate solution. Recognizing this issue, the constraint qualification is a requirement on the constraints, such that the expression $\nabla f = \lambda_1 \nabla g_1 + \lambda_2 \nabla g_2$ is in fact necessary at a solution. The chapter motivates the constraint qualification with a number of examples. Following this, the non-negativity constraints are introduced, leading to the familiar Kuhn–Tucker conditions.

Chapter 14 discusses integration – the integral of a function. Most of the chapter is spent on applications, beginning with consumer and producer surplus. The chapter works through a range of problems: recovering total cost from marginal costs, calculating dead-weight loss from taxation, inequality measures, Ramsey pricing and welfare measures in consumer theory.

In Chapter 15, the theory of eigenvalues is developed. Many problems involving the use of matrices in economics have a dynamic aspect where a system evolves according to a rule of the form $x_{t+1} = Ax_t$, where A is a square matrix. Systems of this form arise in evolutionary models of population change, in Markov processes, matrix difference equations and so on. The theory of eigenvectors and eigenvalues facilitates the study of such systems. An eigenvector–eigenvalue pair for a matrix A is a vector v and scalar λ, such that $Av = \lambda v$. The vector v is called an eigenvector and the scalar λ an eigenvalue. Typically, an $n \times n$ matrix A has n distinct eigenvalues, $\{\lambda_i\}_{i=1}^n$, and a corresponding set of linearly independent eigenvectors, $\{v_i\}_{i=1}^n$. Arranging the eigenvectors in a matrix $P = [v_1,\dots,v_n]$, the set of equations $Av_i = \lambda_i v_i$ may be written $AP = P\Lambda$, where Λ is a diagonal matrix with $\lambda_1, \lambda_2,\dots,\lambda_n$ on the main diagonal, and 0 elsewhere. Thus, Λ is a diagonal matrix with the eigenvalues on the diagonal. With P non-singular, this may be written $A = P\Lambda P^{-1}$. Considering $A^2 = P\Lambda P^{-1}P\Lambda P^{-1} = P\Lambda^2 P^{-1}$, $A^3 = P\Lambda^3 P^{-1}$, $A^t = P\Lambda^t P^{-1}$, higher powers of A are expressible in terms of powers of Λ. Therefore, the long-run behavior of A^t is determined by Λ. This property has many applications, such as in the study of long-run behavior of Markov processes, in the study of the dynamics of population growth, and so on. In this chapter, eigenvalues are used to study the evolution of different age cohorts in a population and to provide a model of webpage ranking based on the long-run dynamics of a Markov process describing the transition dynamics between webpages.

Dynamic models in economics are frequently developed in terms of differential or difference equations. Chapter 16 develops the theory of differential equations. In the simplest case, a differential equation expresses the derivative of some variable $y'(t)$ as a function of its current value $y(t)$ and the independent variable t: $y'(t) = f(y(t),t)$. (Such an equation is called a first-order differential equation. Regarding terminology, a second-order differential equation has the form $y''(t) = f(y', y, t)$, and so on.) The problem is to determine the function $y(t)$ from this equation. For example, if the value of a loan p grows according to the formula $p'(t) = rp(t) = f(p(t),t)$, where r is the rate of interest at time t, what is $p(t)$? Here, one can see that $p(t) = p(0)e^{rt}$ is the required function, where $p(0)$ is the initial value of the loan. In general, determining the value of y that satisfies $y'(t) = f(y(t),t)$ is very difficult. (For continuous functions f, a solution always exists; the solution is unique if, in addition, the function f satisfies a "Lipschitz" condition.) This difficulty leads to a focus on tractable problems. The chapter begins with (first-order) linear differential equations $y'(t) = b(t)y(t) + c(t)$ or $y'(t) = by(t) + c$ when the coefficients are constant. These equations are solved using standard techniques in the early part of the chapter. Then two useful, but non-linear, first-order differential equations are considered – the logistic and Bernoulli equations. Following this, second- and higher-order differential equations are discussed. A second-order linear differential equation with constant coefficients has the form $ay''(t) + by'(t) + cy(t) + d = 0$. Such a differential equation is again solved by standard techniques but the study of stability (the behavior of the solution $y(t)$ as a function of t) is more involved. The chapter concludes with a discussion of matrix differential equations and the stability of such systems.

In Chapter 17, linear difference equations are discussed. A linear difference equation has the form $y_t = ay_{t-1} + b$. In contrast to the differential equation, in a difference equation variables move in discrete increments. A kth order linear difference equation has the form $y_t = a_1y_{t-1} + a_2y_{t-1} + \cdots + a_ky_{t-k} + b$. The solutions of such systems are relatively straightforward. A slight generalization occurs when the parameter b varies, say as b_t – this is called the variable forcing function case and is discussed for a variety of models relating b to t. The second-order difference equation, $y_t = a_1y_{t-1} + a_2y_{t-1}$, is studied in detail, in particular the stability and dynamic behavior of y_t. The conditions described here apply to autoregressive processes of order 2: $y_t = a_1y_{t-1} + a_2y_{t-1} + \epsilon_t$, where ϵ_t is a random error term. The chapter concludes with a discussion of systems of difference equations.

Chapter 18 introduces some basic material on probability and random variables. The density and distribution function of a random variable are defined. The chapter then develops a few of the fundamental discrete distributions: the Bernoulli, Poisson and binomial distributions. Following this, continuous distributions are introduced. The chapter discusses in detail four important distributions that arise routinely in statistical theory and are the key distributions that appear in hypothesis testing. These are the normal, chi-square, t and F distributions. The relations between these distributions are explored in detail. In particular, the chi-square density is derived from the normal density; the t distribution is derived from the normal and chi-square distributions; and the F distribution is derived from the chi-square distribution.

The concluding chapter, Chapter 19, introduces estimation and some basic properties of estimators. The law of large numbers and the central limit theorem are described. The chapter then proceeds to review confidence intervals and develop the fundamentals of hypothesis testing, including the role of various distributions. Least squares estimation is reviewed and the use of major test statistics (t and F) is explained.

2 Matrices and systems of equations

2.1 Introduction

In linear algebra, a rectangular array of numbers is called a matrix. For many problems in mathematics, economics, and other fields, arranging numbers in such blocks turns out to be very useful – for example, in solving systems of equations, analyzing the evolution of dynamic systems and so on. Working with blocks of numbers requires a methodology for the manipulation of such arrays, so that addition, subtraction, multiplication and other algebraic procedures are defined.

Sections 2.2 and 2.3 introduce arrays of numbers (vectors and matrices) and provide the basic rules of manipulation (addition, subtraction and multiplication). Addition, subtraction and multiplication are natural extensions of the scalar rules. In contrast, matrix inversion is more involved, requiring a number of steps, described in Section 2.3.4. Matrix inversion is defined for square matrices but not all square matrices are invertible. For square matrices, a summary statistic, the determinant, identifies key properties of a matrix. In particular, a square matrix can be inverted if and only if it has a non-zero determinant. The determinant of a matrix is discussed in Section 2.3.5. Inverting a matrix ofter arises in solving systems of equations, but it is not the only way. Sections 2.3.6 and 2.3.7 provide alternative methods. In Section 2.4, linear dependence, a key concept in linear algebra, and independence are discussed. From this, the rank of a matrix is defined and matrix rank is used to discriminate between equation systems with many, one or no solution. Section 2.5 provides a geometric perspective on the task of solving equation systems and proceeds to give necessary and sufficient conditions in terms of rank and linear independence for a system to be soluble. Section 2.6

concludes with an introduction to some special matrices. The remainder of this section provides some elementary motivation for the use of matrices.

2.1.1 Some motivating examples

The following examples illustrate how matrices arise naturally in a variety of areas and where the matrix formulation greatly simplifies subsequent analysis. The first example introduces adjacency matrices, a simple way to represent "closeness" or adjacency between entities – in this case cities, where closeness is represented by the minimum number of flights necessary to travel from one city to another.

EXAMPLE 2.1: Suppose there are four airports, A, B, C and D. The figure shows the flights available from each airport. The only flight from A is to B; from C, flights are available to A and D – and so on. These flight options may be represented in an array, with rows identified with flight origin and columns identified with flight destination. Thus, in the first row (corresponding to A) and the third column (corresponding to B), there is a "1", indicating that there is a flight from A to B. In the third row (corresponding to C), there is a "1" in columns 1 and 4, corresponding to destinations A and D: it is possible to take a flight from C to either A or D.

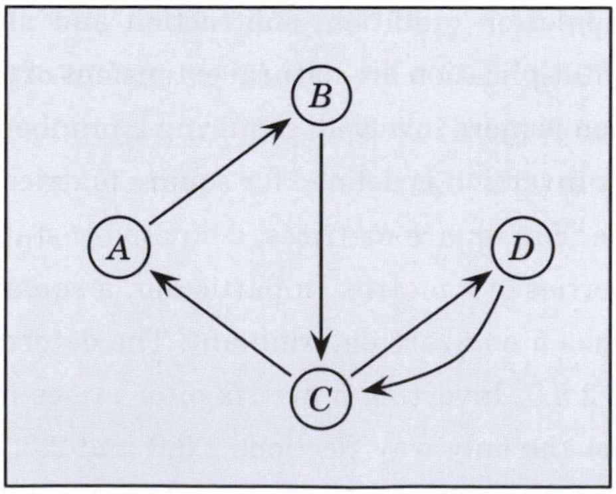

	A	B	C	D
A	0	1	0	0
B	0	0	1	0
C	1	0	0	1
D	0	0	1	0

Write $\boldsymbol{M}$ or $\boldsymbol{M}_1$ for this array or matrix of numbers. Consider now the destinations reached if exactly two flights are taken. From A there is one possible destination, C, reached by a flight to B and then to C. From B there are two possible destinations: A and D with corresponding flight paths $B \to C \to A$ and $B \to C \to D$. In matrix $\boldsymbol{M}_2$, these calculations are represented. Looking at the first row, there is just one possible destination, C, with two flights from A. This is indicated by "1" in column 3 of row 1 of $\boldsymbol{M}_2$. Similarly, from B there are two possible destinations if exactly two flights are taken: A and D. In matrix $\boldsymbol{M}_2$, this is represented by a "1" in row 2 (corresponding to B) column 1 (corresponding to A), and in row 2 column 4 (corresponding to D). Matrix $\boldsymbol{M}_3$ indicates the destinations with exactly three flights. Finally, the matrix $\boldsymbol{M}(2)$ identifies the destinations reachable with two or fewer flights (two or one flights). Simple matrix rules may be used to generate matrices such as $\boldsymbol{M}_3$ or $\boldsymbol{M}(2)$.

$$\boldsymbol{M} = \boldsymbol{M}_1 = \begin{pmatrix} 0 & 1 & 0 & 0 \\ 0 & 0 & 1 & 0 \\ 1 & 0 & 0 & 1 \\ 0 & 0 & 1 & 0 \end{pmatrix}, \quad \boldsymbol{M}_2 = \begin{pmatrix} 0 & 0 & 1 & 0 \\ 1 & 0 & 0 & 1 \\ 0 & 1 & 1 & 0 \\ 1 & 0 & 0 & 1 \end{pmatrix},$$

$$\boldsymbol{M}_3 = \begin{pmatrix} 1 & 0 & 0 & 1 \\ 0 & 1 & 1 & 0 \\ 1 & 0 & 1 & 1 \\ 0 & 1 & 1 & 0 \end{pmatrix}, \quad \boldsymbol{M}(2) = \begin{pmatrix} 0 & 1 & 1 & 0 \\ 1 & 0 & 1 & 1 \\ 1 & 1 & 1 & 1 \\ 1 & 0 & 1 & 1 \end{pmatrix}.$$

◊

The next example describes a standard demand system. Demand for a good typically depends not just on the good's own price, but also on the price of related goods. For example, if two goods are substitutes, such as butter and margarine, one expects the price of one good to affect the demand for the other. But, with such cross-price effects, it becomes more difficult to determine the equilibrium price in each market. The next example illustrates this issue.

EXAMPLE 2.2: This example considers equilibrium price determination. To illustrate, consider the case where there are two markets, 1 and 2, defined by the supply and demand conditions in each market. Suppose the demand and supply in the pair of markets is described by the following equations:

$$D_1 = 20 - 3p_1 + p_2, \qquad S_1 = 2 + 2p_1$$
$$D_2 = 10 + 2p_1 - 4p_2, \qquad S_2 = 4 + 3p_2.$$

So, for example, a unit increase in the price of good 1 reduces the demand for good 1 by 3 units, while an increase in the price of good 2, raises the demand for good 1 (possibly because good 1 is a substitute for good 2, with the higher price for good 2 causing consumers to switch to good 1).

Setting supply equal to demand in each market gives:

$$\begin{aligned} 20-3p_1+\ p_2 &= 2+2p_1 \\ 10+2p_1-4p_2 &= 4+3p_2, \end{aligned}$$

and rearranging,

$$\begin{aligned} 5p_1-\ p_2 &= 18 \\ -2p_1+7p_2 &= 6. \end{aligned}$$

These equations determine the price pair (p_1, p_2) at which both markets clear (supply equals demand in each market). The solution is $(p_1, p_2) = (4,2)$. However, finding such solutions is more difficult when the number of markets is large, with correspondingly large numbers of prices to be determined. This pair of equations is rearranged in matrix form in Example 2.5. The procedure generalizes directly to any number of equations and variables and provides a systematic way of dealing with and solving large systems of equations. ◊

The final example focuses on the dynamic evolution of economic variables. At any point in time, suppose that the current state of the economy evolves to a future state according to a law of motion. Analyzing the long-run behavior of such economic variables is simplified by matrix techniques. The next example illustrates such a dynamic model.

EXAMPLE 2.3: The Phillips curve asserts that the change in inflation is proportional to the gap between the level of unemployment (u) and a natural rate of unemployment ($\bar{u}$). If π denotes the rate of inflation, with subscripts for time periods, then

$$\pi_t - \pi_{t-1} = -\alpha(u_t - \bar{u}).$$

Okun's law connects changes in unemployment to output changes, which are in turn connected to the inflation rate. Assume that this relation has the form

$$u_t - u_{t-1} = \beta(g - \pi_t),$$

where g is the growth in the money supply. Rearranging these equations:

$$\begin{aligned} \pi_t &= \pi_{t-1} - \alpha(u_t - \bar{u}) \\ u_t &= u_{t-1} - \beta(g - \pi_t). \end{aligned}$$

Suppose that $\alpha = \beta = \frac{1}{2}$, so the equations are $\pi_t = \pi_{t-1} - \frac{1}{2}(u_t - \bar{u})$ and $u_t = u_{t-1} - \frac{1}{2}(g - \pi_t)$. Substituting the expression for u_t into the first equation and substituting the expression for π_t into the second equation, some rearrangement gives:

$$\pi_t = \frac{4}{5}\pi_{t-1} - \frac{2}{5}u_{t-1} + \left[\frac{1}{5}g + \frac{2}{5}\bar{u}\right]$$
$$u_t = \frac{2}{5}\pi_{t-1} + \frac{4}{5}u_{t-1} + \left[-\frac{2}{5}g + \frac{1}{5}\bar{u}\right].$$

From these equations, the pair (π_t, u_t) are determined by (π_{t-1}, u_{t-1}) so the evolution of system is determined by the coefficients on π_{t-1} and u_{t-1}. This system can be arranged as a matrix system where the dynamics are determined by the matrix terms (see Example 2.5). ◊

The following discussion focuses on two types of array – vectors, where the array has one row or one column, and matrices, where the number of rows and columns is arbitrary.

2.2 Vectors

Definition 2.1: *A* VECTOR *in $\mathcal{R}^n$ is an ordered collection (row or column) of n real numbers.*

Throughout, lower case bold letters, such as $\boldsymbol{a}$, $\boldsymbol{b}$ and $\boldsymbol{c}$, will denote vectors. To highlight the dimension of a vector, say that $\boldsymbol{a}$ is an n-vector. For example, $\boldsymbol{a} = (1,3)$ and $\boldsymbol{b} = (5,2)$ are 2-vectors, $\boldsymbol{c} = (3,4,1,7)$ is a 4-vector, and so on. If two vectors have the same number of rows or columns they may be multiplied, to form the *inner product*.

Definition 2.2: *Let $\boldsymbol{f}$ and $\boldsymbol{g}$ be two n-vectors with $\boldsymbol{f} = (f_1, f_2, \ldots, f_n)$ and $\boldsymbol{g} = (g_1, g_2, \ldots, g_n)$. The* INNER PRODUCT *of $\boldsymbol{f}$ and $\boldsymbol{g}$ is denoted $\boldsymbol{f} \cdot \boldsymbol{g}$ and defined: $\boldsymbol{f} \cdot \boldsymbol{g} = f_1 g_1 + f_2 g_2 + \cdots + f_n g_n$.*

So the entries of the vectors are multiplied pairwise and added together. For example, with $\boldsymbol{a}$ and $\boldsymbol{b}$ given above, $\boldsymbol{a} \cdot \boldsymbol{b} = 1 \times 5 + 3 \times 2 = 11$. Note that the order does not matter since $\boldsymbol{f} \cdot \boldsymbol{g} = \boldsymbol{g} \cdot \boldsymbol{f}$. Equations may be represented using this notation. The equations $x + y = 5$ and $2x + 3y = 12$ may be written as $(1,1) \cdot (x,y) = 5$ and $(2,3) \cdot (x,y) = 12$.

There is a convenient way to represent $f_1g_1 + f_2g_2 + \cdots + f_ng_n$. Suppose there are 20 numbers, labeled $x_1, x_2, \ldots, x_{20}$. Adding the first ten can be written $x_1 + x_2 + \cdots + x_{10}$, but there is also a shorthand notation for this, $\sum_{i=1}^{10} x_i$, which can be read as saying "add up the numbers labeled from 1 to 10" or "sum from 1 to 10" the numbers x_i. Similarly, to illustrate, $\sum_{i=7}^{15} x_i = \sum_{i=1}^{15} x_i - \sum_{i=1}^{6} x_i$. The sign $\sum$ is called the summation sign. With this notation: $\boldsymbol{f} \cdot \boldsymbol{g} = \sum_{i=1}^{n} f_i g_i = f_1g_1 + f_2g_2 + \cdots + f_ng_n$.

2.3 Matrices

Definition 2.3: *A* MATRIX *is a rectangular array of numbers.*

Throughout, upper case bold letters, such as $\boldsymbol{A}$, $\boldsymbol{B}$ and $\boldsymbol{C}$, will denote matrices. A matrix $\boldsymbol{B}$ is an $m \times n$ matrix if it has m rows and n columns (this is sometimes denoted by $\boldsymbol{B} - m \times n$).

If $\boldsymbol{B}$ is an $m \times n$ matrix we may write it as:

$$\boldsymbol{B} = \begin{pmatrix} b_{11} & b_{12} & \cdots & b_{1n} \\ b_{21} & b_{22} & \cdots & b_{2n} \\ \vdots & \vdots & \ddots & \vdots \\ b_{m1} & b_{m2} & \cdots & b_{mn} \end{pmatrix}.$$

For example, $\boldsymbol{C}$ is a 2×3 matrix:

$$\boldsymbol{C} = \begin{pmatrix} 2 & 11 & 4 \\ 1 & 0 & 6 \end{pmatrix}.$$

Sometimes, the matrix $\boldsymbol{B}$ may be denoted $\{b_{ij}\}_{i \in I, j \in J}$, where I and J are index sets, or just as $\{b_{ij}\}$. An element b_{ij} of the matrix $\boldsymbol{B}$ represents the element in the position given by the ith row and jth column, the (i,j)th element of $\boldsymbol{B}$. There are some special matrices, such as the identity ($\boldsymbol{I}$), null ($\boldsymbol{0}$) and diagonal matrices ($\boldsymbol{D}$):

$$\boldsymbol{I} = \begin{pmatrix} 1 & 0 & \cdots & 0 \\ 0 & 1 & \cdots & 0 \\ \vdots & \vdots & \ddots & \vdots \\ 0 & 0 & \cdots & 1 \end{pmatrix}, \quad \boldsymbol{0} = \begin{pmatrix} 0 & 0 & \cdots & 0 \\ 0 & 0 & \cdots & 0 \\ \vdots & \vdots & \ddots & \vdots \\ 0 & 0 & \cdots & 0 \end{pmatrix}, \quad \boldsymbol{D} = \begin{pmatrix} a_{11} & 0 & \cdots & 0 \\ 0 & a_{22} & \cdots & 0 \\ \vdots & \vdots & \ddots & \vdots \\ 0 & 0 & \cdots & a_{nn} \end{pmatrix}. \tag{2.1}$$

In the diagonal matrix ($\boldsymbol{D}$), the diagonal elements may be arbitrary numbers but the off-diagonal elements must be zero. When a matrix has just one row or one column, it is a vector.

The next sections discuss matrix algebra, in particular addition, subtraction, multiplication and inversion. The rules for addition, subtraction, multiplication are straightforward. Understanding matrix inversion requires some discussion, and the computation of the inverse involves a number of steps.

2.3.1 Matrix addition and subtraction

Let $\boldsymbol{A}$ and $\boldsymbol{B}$ be two $m \times n$ matrices, so that both $\boldsymbol{A}$ and $\boldsymbol{B}$ have the same number of rows and columns.

MATRIX ADDITION: The *sum* of $\boldsymbol{A}$ and $\boldsymbol{B}$ is obtained by adding the entries in the two matrices pairwise:

$$\boldsymbol{A}+\boldsymbol{B} = \begin{pmatrix} a_{11}+b_{11} & a_{12}+b_{12} & \cdots & a_{1n}+b_{1n} \\ a_{21}+b_{21} & a_{22}+b_{22} & \cdots & a_{2n}+b_{2n} \\ \vdots & \vdots & \ddots & \vdots \\ a_{m1}+b_{m1} & a_{m2}+b_{m2} & \cdots & a_{mn}+b_{mn} \end{pmatrix}.$$

MATRIX SUBTRACTION: The *difference* of $\boldsymbol{A}$ and $\boldsymbol{B}$ is obtained by subtracting entries pairwise:

$$\boldsymbol{A}-\boldsymbol{B} = \begin{pmatrix} a_{11}-b_{11} & a_{12}-b_{12} & \cdots & a_{1n}-b_{1n} \\ a_{21}-b_{21} & a_{22}-b_{22} & \cdots & a_{2n}-b_{2n} \\ \vdots & \vdots & \ddots & \vdots \\ a_{m1}-b_{m1} & a_{m2}-b_{m2} & \cdots & a_{mn}-b_{mn} \end{pmatrix}.$$

So the entry in the ith row and jth column of $\boldsymbol{A}+\boldsymbol{B}$ is $a_{ij}+b_{ij}$, and the entry in the ith row and jth column of $\boldsymbol{A}-\boldsymbol{B}$ is $a_{ij}-b_{ij}$. For example, if

$$\boldsymbol{A} = \begin{pmatrix} 1 & 3 & 4 \\ 5 & 8 & 1 \end{pmatrix} \quad \text{and} \quad \boldsymbol{B} = \begin{pmatrix} 2 & 4 & 1 \\ 9 & 0 & 3 \end{pmatrix},$$

then

$$\boldsymbol{A}+\boldsymbol{B} = \begin{pmatrix} 1+2 & 3+4 & 4+1 \\ 5+9 & 8+0 & 1+3 \end{pmatrix} = \begin{pmatrix} 3 & 7 & 5 \\ 14 & 8 & 4 \end{pmatrix};$$

$$\boldsymbol{A}-\boldsymbol{B} = \begin{pmatrix} 1-2 & 3-4 & 4-1 \\ 5-9 & 8-0 & 1-3 \end{pmatrix} = \begin{pmatrix} -1 & -1 & 3 \\ -4 & 8 & -2 \end{pmatrix}.$$

Note from the definition that $\boldsymbol{A}+\boldsymbol{B}=\boldsymbol{B}+\boldsymbol{A}$. Also, $(\boldsymbol{A}+\boldsymbol{B})+\boldsymbol{C}=\boldsymbol{A}+(\boldsymbol{B}+\boldsymbol{C})$. For two matrices $\boldsymbol{A}$ and $\boldsymbol{B}$, the operations of addition and subtraction are defined only if $\boldsymbol{A}$ and $\boldsymbol{B}$ have the same number of rows and columns.

2.3.2 Matrix multiplication

Let $\boldsymbol{A}$ and $\boldsymbol{B}$ be two matrices:

$$\boldsymbol{A}=\begin{pmatrix} a_{11} & a_{12} & a_{13} & \cdots & a_{1n} \\ a_{21} & a_{22} & a_{23} & \cdots & a_{2n} \\ a_{31} & a_{32} & a_{33} & \cdots & a_{3n} \\ \vdots & \vdots & \vdots & \ddots & \vdots \\ a_{m1} & a_{m2} & a_{m3} & \cdots & a_{mn} \end{pmatrix}, \quad \boldsymbol{B}=\begin{pmatrix} b_{11} & b_{12} & b_{13} & \cdots & b_{1p} \\ b_{21} & b_{22} & b_{23} & \cdots & b_{2p} \\ b_{31} & b_{32} & b_{33} & \cdots & b_{3p} \\ \vdots & \vdots & \vdots & \ddots & \vdots \\ b_{n1} & b_{n2} & b_{n3} & \cdots & b_{np} \end{pmatrix}.$$

So $\boldsymbol{A}$ is an $m \times n$ matrix and $\boldsymbol{B}$ is an $n \times p$ matrix. If the number of columns in $\boldsymbol{A}$, n, equals the number of rows in $\boldsymbol{B}$, then the matrices can be multiplied to form the matrix $\boldsymbol{C}=\boldsymbol{AB}$, where $\boldsymbol{C}$ is defined component-wise as

$$c_{ij}=a_{i1}b_{1j}+a_{i2}b_{2j}+\cdots+a_{in}b_{nj}=\sum_{k=1}^{n} a_{ik}b_{kj}.$$

In words, take the ith row of the $\boldsymbol{A}$ matrix and the jth column of the $\boldsymbol{B}$ matrix and form the inner product. Note that this is possible since $\boldsymbol{A}$ has the same number of columns as $\boldsymbol{B}$ has rows so that both the ith row of the $\boldsymbol{A}$ matrix and the jth column of the $\boldsymbol{B}$ matrix are vectors with the same number of elements, n, and the inner product can be formed. For example:

$$\boldsymbol{A}=\begin{pmatrix} 1 & 3 & 2 \\ 4 & 3 & 6 \end{pmatrix}, \quad \boldsymbol{B}=\begin{pmatrix} 1 & 5 & 2 \\ 2 & 5 & 1 \\ 3 & 2 & 2 \end{pmatrix}, \quad \boldsymbol{AB}=\begin{pmatrix} 13 & 24 & 9 \\ 28 & 47 & 23 \end{pmatrix}.$$

EXAMPLE 2.4: Recall the adjacency matrix in Example 2.1. Write $f_{ij}=1$ if there is a flight from city i to city j; and write $f_{ij}=0$ if there is no flight from i to j. For notational convenience, label the cities A, B, C and D as 1, 2, 3 and 4. Then the adjacency matrix may be written:

$$\boldsymbol{M}=\begin{pmatrix} f_{11} & f_{12} & f_{13} & f_{14} \\ f_{21} & f_{22} & f_{23} & f_{24} \\ f_{31} & f_{32} & f_{33} & f_{34} \\ f_{41} & f_{42} & f_{43} & f_{44} \end{pmatrix}=\begin{pmatrix} 0 & 1 & 0 & 0 \\ 0 & 0 & 1 & 0 \\ 1 & 0 & 0 & 1 \\ 0 & 0 & 1 & 0 \end{pmatrix}.$$

Consider the matrices $M^2 = MM$, obtained by multiplying M with itself, and $M^3 = MMM$, obtained by multiplying M by itself three times. These calculations give:

$$M^2 = \begin{pmatrix} 0 & 0 & 1 & 0 \\ 1 & 0 & 0 & 1 \\ 0 & 1 & 1 & 0 \\ 1 & 0 & 0 & 1 \end{pmatrix}, \quad M^3 = \begin{pmatrix} 1 & 0 & 0 & 1 \\ 0 & 1 & 1 & 0 \\ 1 & 0 & 1 & 1 \\ 0 & 1 & 1 & 0 \end{pmatrix}.$$

Let f^2_{ij} be the entry in the ith row and jth column of M^2. From its definition:

$$f^2_{ij} = f_{i1}f_{1j} + f_{i2}f_{2j} + f_{i3}f_{3j} + f_{i4}f_{4j}.$$

If $f_{ik}f_{kj} > 0$, then there is a flight from i to k and a flight from k to j; and if $f^2_{ij} = 0$, then $f_{ik}f_{kj} = 0$ for each k and there is no airport k with connecting flight from i to j. In this example, $f^2_{13} = 1$, indicating that there is exactly one route with two flights from airport 1 to airport 3. (If $f^2_{ij} = 2$ this would mean that there are exactly two distinct routes connecting i and j, each involving two flights.) Thus, f^2_{ij} is the number of distinct routes from i to j involving exactly two flights.

Going further, let f^3_{ij} be the (i,j)th entry in M^3. This gives the number of distinct routes from i to j with exactly three flights. Finally, if we add M and M^2 together:

$$M + M^2 = \begin{pmatrix} 0 & 1 & 0 & 0 \\ 0 & 0 & 1 & 0 \\ 1 & 0 & 0 & 1 \\ 0 & 0 & 1 & 0 \end{pmatrix} + \begin{pmatrix} 0 & 0 & 1 & 0 \\ 1 & 0 & 0 & 1 \\ 0 & 1 & 1 & 0 \\ 1 & 0 & 0 & 1 \end{pmatrix} = \begin{pmatrix} 0 & 1 & 1 & 0 \\ 1 & 0 & 1 & 1 \\ 1 & 1 & 1 & 1 \\ 1 & 0 & 1 & 1 \end{pmatrix} = M(2).$$

Thus, if $f_{ij}(2)$ is the entry in the ith row and jth column of $M + M^2$, then $f_{ij}(2)$ gives the number of distinct routes from i to j with either one or two flights. ◊

With matrix multiplication, there are some basic points of caution to note. Even when the product AB can be formed, this does not imply that BA can be formed. The expression BA is not defined unless $p = m$ (i.e., B has the same number of columns as A has rows). For example:

$$A = \begin{pmatrix} 1 & 3 \\ 5 & 1 \\ 2 & 1 \end{pmatrix}, \quad B = \begin{pmatrix} 2 & 1 \\ 1 & 1 \end{pmatrix} \quad \text{gives} \quad AB = \begin{pmatrix} 1 & 3 \\ 5 & 1 \\ 2 & 1 \end{pmatrix}\begin{pmatrix} 2 & 1 \\ 1 & 1 \end{pmatrix} = \begin{pmatrix} 5 & 4 \\ 11 & 6 \\ 5 & 3 \end{pmatrix},$$

but the expression $\boldsymbol{BA}$ is not defined since $\boldsymbol{B}$ has fewer columns than $\boldsymbol{A}$ has rows. If both $\boldsymbol{AB}$ and $\boldsymbol{BA}$ exist, they are generally not equal and may not be the same dimension. For example:

$$\boldsymbol{B} = \begin{pmatrix} 2 & 1 \\ 1 & 1 \end{pmatrix}, \boldsymbol{C} = \begin{pmatrix} 3 & 0 \\ 1 & 2 \end{pmatrix}, \quad \text{so } \boldsymbol{BC} = \begin{pmatrix} 7 & 2 \\ 4 & 2 \end{pmatrix} \text{ while } \boldsymbol{CB} = \begin{pmatrix} 6 & 3 \\ 4 & 3 \end{pmatrix}$$

and

$$\boldsymbol{A} = \begin{pmatrix} 1 & 3 & 5 \\ 2 & 4 & 6 \end{pmatrix}, \boldsymbol{B} = \begin{pmatrix} 2 & 1 \\ 5 & 7 \\ 3 & 1 \end{pmatrix}, \text{ so } \boldsymbol{AB} = \begin{pmatrix} 32 & 27 \\ 42 & 36 \end{pmatrix}, \text{ and } \boldsymbol{BA} = \begin{pmatrix} 4 & 10 & 16 \\ 19 & 43 & 67 \\ 5 & 13 & 21 \end{pmatrix}.$$

A matrix may be multiplied by a real number by multiplying each element by the number. Take a real number, $\lambda \in \mathscr{R}$ (where $\mathscr{R}$ is the set of real numbers), then $\lambda\boldsymbol{A}$ is defined component-wise. The (i, j)th element of $\lambda\boldsymbol{A}$ is λa_{ij}. For example:

$$\boldsymbol{D} = \begin{pmatrix} 1 & 3 \\ 4 & 2 \end{pmatrix}, \ \lambda\boldsymbol{D} = \begin{pmatrix} \lambda 1 & \lambda 3 \\ \lambda 4 & \lambda 2 \end{pmatrix}, \text{ if } \lambda = 2, \ 2\boldsymbol{D} = \begin{pmatrix} 2 & 6 \\ 8 & 4 \end{pmatrix}.$$

The operations of matrix addition and multiplication have the following properties.

$$\begin{aligned} &\boldsymbol{A}(\boldsymbol{B}+\boldsymbol{C}) = \boldsymbol{AB}+\boldsymbol{AC} \\ &\lambda(\boldsymbol{A}+\boldsymbol{B}) = \lambda\boldsymbol{A}+\lambda\boldsymbol{B}, \ \lambda \text{ a scalar} \\ &(\lambda+\mu)\boldsymbol{A} = \lambda\boldsymbol{A}+\mu\boldsymbol{A}, \ \lambda, \mu \text{ scalars.} \end{aligned}$$

Matrices provide a succinct way of expressing systems of equations. The system of m equations in n unknowns, $\{x_1, x_2, \ldots, x_n\}$:

$$\begin{array}{ccccccccc} a_{11}x_1 & + & a_{12}x_2 & + & \cdots & + & a_{1n}x_n & = & b_1 \\ a_{21}x_1 & + & a_{22}x_2 & + & \ldots & + & a_{2n}x_n & = & b_2 \\ \vdots & + & \vdots & + & \ldots & + & \vdots & = & \vdots \\ a_{m1}x_1 & + & a_{m2}x_2 & + & \ldots & + & a_{mn}x_n & = & b_n \end{array}$$

may be written:

$$\begin{pmatrix} a_{11} & a_{12} & a_{13} & \cdots & a_{1n} \\ a_{21} & a_{22} & a_{23} & \cdots & a_{2n} \\ \vdots & \vdots & \vdots & \ddots & \vdots \\ a_{m1} & a_{m2} & a_{m3} & \cdots & a_{mn} \end{pmatrix} \begin{pmatrix} x_1 \\ x_2 \\ \vdots \\ x_n \end{pmatrix} = \begin{pmatrix} b_1 \\ b_2 \\ \vdots \\ b_m \end{pmatrix}.$$

To illustrate, consider the examples introduced earlier.

EXAMPLE 2.5: Considering Example 2.2 shows how a system of equations may be arranged in matrix form using matrix multiplication:

$$\begin{aligned} 5p_1 - p_2 &= 18 \\ -2p_1 + 7p_2 &= 6 \end{aligned} \qquad \text{or} \qquad \begin{pmatrix} 5 & -1 \\ -2 & 7 \end{pmatrix} \begin{pmatrix} p_1 \\ p_2 \end{pmatrix} = \begin{pmatrix} 18 \\ 6 \end{pmatrix}.$$

Similarly, the Phillips curve equations in Example 2.3 may be written in matrix form:

$$\begin{aligned} \pi_t &= \frac{4}{5}\pi_{t-1} - \frac{2}{5}u_{t-1} + \left[\frac{1}{5}g + \frac{2}{5}\bar{u}\right] \\ u_t &= \frac{2}{5}\pi_{t-1} + \frac{4}{5}u_{t-1} + \left[-\frac{2}{5}g + \frac{1}{5}\bar{u}\right] \end{aligned}$$

or

$$\begin{pmatrix} \pi_t \\ u_t \end{pmatrix} = \begin{pmatrix} \frac{4}{5} & -\frac{2}{5} \\ \frac{2}{5} & \frac{4}{5} \end{pmatrix} \begin{pmatrix} \pi_{t-1} \\ u_{t-1} \end{pmatrix} + \begin{pmatrix} [\frac{1}{5}g + \frac{2}{5}\bar{u}] \\ [-\frac{2}{5}g + \frac{1}{5}\bar{u}] \end{pmatrix}.$$

◊

2.3.3 Matrix transpose

Let $\boldsymbol{A}$ be a matrix, not necessarily square. The transpose of $\boldsymbol{A}$ is obtained from $\boldsymbol{A}$ by "rotating" $\boldsymbol{A}$ so that the ith row becomes the ith column. Write $\boldsymbol{A}'$ for the transpose of $\boldsymbol{A}$. The (i,j)th element of $\boldsymbol{A}'$ is a_{ji}, so the (i,j)th element of the transpose is the (j,i)th element of the original matrix.

In general, if $\boldsymbol{A}$ is an $m \times n$ matrix, then $\boldsymbol{A}'$ is an $n \times m$ matrix. For example,

$$\boldsymbol{B} = \begin{pmatrix} 2 & 4 & 6 \\ 1 & 5 & 3 \end{pmatrix} \Rightarrow \boldsymbol{B}' = \begin{pmatrix} 2 & 1 \\ 4 & 5 \\ 6 & 3 \end{pmatrix}, \quad \boldsymbol{D} = \begin{pmatrix} 3 & 0 \\ 1 & 2 \end{pmatrix} \Rightarrow \boldsymbol{D}' = \begin{pmatrix} 3 & 1 \\ 0 & 2 \end{pmatrix}.$$

When using vectors in matrix computations, dimensions must be assigned – since matrix multiplication depends on correct dimensions. To illustrate, let $\boldsymbol{x}$ and $\boldsymbol{y}$ be the two 3×1 vectors:

$$\boldsymbol{x} = \begin{pmatrix} 1 \\ 2 \\ 3 \end{pmatrix}, \quad \boldsymbol{y} = \begin{pmatrix} 4 \\ 5 \\ 6 \end{pmatrix}.$$

Then, for example, $\boldsymbol{x}' = (1,2,3)$, $\boldsymbol{x}'\boldsymbol{y} = 1 \cdot 4 + 2 \cdot 5 + 3 \cdot 6 = 32 = \boldsymbol{x} \cdot \boldsymbol{y}$. Similarly,

$$\boldsymbol{B}\boldsymbol{x} = \begin{pmatrix} 2 & 4 & 6 \\ 1 & 5 & 3 \end{pmatrix} \begin{pmatrix} 1 \\ 2 \\ 3 \end{pmatrix} = \begin{pmatrix} 28 \\ 20 \end{pmatrix} \quad \text{and} \quad \boldsymbol{y}'\boldsymbol{B}' = (4,\ 5,\ 6) \begin{pmatrix} 2 & 1 \\ 4 & 5 \\ 6 & 3 \end{pmatrix} = \begin{pmatrix} 64 & 47 \end{pmatrix}.$$

However, the expressions $\boldsymbol{xB}$ or $\boldsymbol{B}'\boldsymbol{y}$ are not defined, since the matrices do not conform for matrix multiplication. The transpose operation has the following properties:

$$(\boldsymbol{A}')' = \boldsymbol{A}$$
$$(\boldsymbol{A}+\boldsymbol{B})' = \boldsymbol{A}' + \boldsymbol{B}'$$
$$(\boldsymbol{AB})' = \boldsymbol{B}'\boldsymbol{A}'$$
$$(\boldsymbol{ABC})' = \boldsymbol{C}'\boldsymbol{B}'\boldsymbol{A}'.$$

Note, for example, that if $\boldsymbol{AB}$ exists then $\boldsymbol{A}$ has the same number of columns as $\boldsymbol{B}$ has rows. Say $\boldsymbol{A}$ is an $m \times n$ matrix and $\boldsymbol{B}$ an $n \times p$ matrix: then $\boldsymbol{A}'$ is an $n \times m$ and $\boldsymbol{B}'$ an $p \times n$ matrix so that $\boldsymbol{B}'\boldsymbol{A}'$ exists.

2.3.4 Matrix inversion

When considering a single variable, division is straightforward. Given a number a, provided $a \neq 0$, one can divide 1 by a to get $\frac{1}{a}$, or in alternative notation, a^{-1}. This term, $\frac{1}{a}$ or a^{-1}, is called the inverse of a or "a inverse". It routinely arises when solving equations. If $2x = 4$, then multiplying both sides by $\frac{1}{2}$ gives $(\frac{1}{2})2x = (\frac{1}{2})4$, and since $(\frac{1}{2})2 = 1$ and $(\frac{1}{2})4 = 2$, we get $1x = 2$ or $x = 2$. The same calculation may be expressed as $2^{-1}2x = 2^{-1}4$, and since $2^{-1}2 = 1$, $x = 2^{-1}4 = 2$. For the expression $ax = b$, where a and b are real numbers and $a \neq 0$, exactly the same procedure applies: $a^{-1}ax = a^{-1}b$, and with $a^{-1}a = 1$, $x = a^{-1}b$. We wish to apply exactly the same logic to matrices.

Considering the equation system $\boldsymbol{Az} = \boldsymbol{b}$, if there is a matrix, $\boldsymbol{A}^{-1}$, such that $\boldsymbol{A}^{-1}\boldsymbol{A} = \boldsymbol{I}$, then we can multiply the equation to get $\boldsymbol{A}^{-1}\boldsymbol{Az} = \boldsymbol{A}^{-1}\boldsymbol{b}$ or $\boldsymbol{Iz} = \boldsymbol{A}^{-1}\boldsymbol{b}$ or $\boldsymbol{z} = \boldsymbol{A}^{-1}\boldsymbol{b}$. For example, given the pair of equations

$$x + y = 5 \tag{2.2}$$
$$2x + 3y = 12, \tag{2.3}$$

we wish to find a pair (x^*, y^*) that satisfies both of these equations. One can see that (0,4) satisfies the second equation, but not the first, while (5,0) satisfies the first but not the second. It is straightforward to find a solution by substitution. Take Equation 2.2 and express x in terms of $y : x = 5 - y$. This can be substituted into Equation 2.3 to find: $2(5-y)+3y = 12$ or $10-2y+3y = 12$ or $10+y = 12$, so that $y = 2$. Since $x + y = 5$, this implies that $x = 3$ and so the solution is $(x^*, y^*) = (3, 2)$. However, this solution procedure is not satisfactory for large systems of equations and does not allow one to easily investigate circumstances under which solutions exist or are unique. Writing the pair of equations in matrix form gives:

$$\begin{pmatrix} 1 & 1 \\ 2 & 3 \end{pmatrix}\begin{pmatrix} x \\ y \end{pmatrix} = \begin{pmatrix} 5 \\ 12 \end{pmatrix}.$$

Taking the matrix system on the left, consider multiplying both sides by an appropriate matrix:

$$\begin{pmatrix} 3 & -1 \\ -2 & 1 \end{pmatrix}\begin{pmatrix} 1 & 1 \\ 2 & 3 \end{pmatrix}\begin{pmatrix} x \\ y \end{pmatrix} = \begin{pmatrix} 3 & -1 \\ -2 & 1 \end{pmatrix}\begin{pmatrix} 5 \\ 12 \end{pmatrix}.$$

Carrying out the matrix multiplications on both sides gives:

$$\begin{pmatrix} 1 & 0 \\ 0 & 1 \end{pmatrix}\begin{pmatrix} x \\ y \end{pmatrix} = \begin{pmatrix} 3 \\ 2 \end{pmatrix} \quad \text{or} \quad \begin{pmatrix} x \\ y \end{pmatrix} = \begin{pmatrix} 3 \\ 2 \end{pmatrix}.$$

This "appropriate" matrix is in fact the inverse of $\boldsymbol{A}$:

$$\boldsymbol{A}^{-1} = \begin{pmatrix} 3 & -1 \\ -2 & 1 \end{pmatrix}.$$

Observe that $\boldsymbol{A}^{-1}\boldsymbol{A} = \boldsymbol{I}$:

$$\boldsymbol{A}^{-1}\boldsymbol{A} = \begin{pmatrix} 3 & -1 \\ -2 & 1 \end{pmatrix}\begin{pmatrix} 1 & 1 \\ 2 & 3 \end{pmatrix} = \begin{pmatrix} 1 & 0 \\ 0 & 1 \end{pmatrix}.$$

This following discussion considers matrix inversion, developing the computation of $\boldsymbol{A}^{-1}$.

In general given a system of n equations with n unknowns, $\boldsymbol{Az} = \boldsymbol{b}$, where $\boldsymbol{A}$ is an $n \times n$ square matrix, assuming we can invert $\boldsymbol{A}$, we may solve for $\boldsymbol{z}$ according to $\boldsymbol{z} = \boldsymbol{A}^{-1}\boldsymbol{b}$. However, this is not always the case. Not all square matrices are invertible (have an inverse). Returning to the single equation, $ax = b$, there are three distinct possibilities. The first is that there is a unique solution when $a \neq 0$, so a^{-1} is defined, and the solution is given by $x = a^{-1}b$. A second possibility is that there is no solution – if $a = 0$ and $b \neq 0$ it is impossible to find x satisfying $ax = b$. Finally, there may be an infinite number of solutions – if $a = b = 0$, then any number x satisfies $ax = b$.

The situation is essentially the same in the matrix case. Consider the following three equation systems:

$$\begin{aligned} x + y &= 5 \\ 2x + 3y &= 12 \end{aligned} \qquad \begin{aligned} x + y &= 5 \\ 3x + 3y &= 12 \end{aligned} \qquad \begin{aligned} x + y &= 5 \\ 2x + 2y &= 10. \end{aligned}$$

The first system has a unique solution $(x^*, y^*) = (5, 12)$. The second system requires that $x+y=5$ and $3(x+y)=12$ or $x+y=4$: there is no (x,y) satisfying both equations. The third system requires that $x+y=5$ and $2x+2y=10$ or $x+y=5$ – so any (x,y) satisfying the first equation also satisfies the second, and there are infinitely many solutions.

If the matrix $\boldsymbol{A}$ has an inverse, $\boldsymbol{A}^{-1}$, then $\boldsymbol{A}^{-1}\boldsymbol{A} = \boldsymbol{I}$, the identity matrix and given $\boldsymbol{A}\boldsymbol{z} = \boldsymbol{b}$, we can multiply both sides by $\boldsymbol{A}^{-1}$ to obtain $\boldsymbol{A}^{-1}\boldsymbol{A}\boldsymbol{z} = \boldsymbol{A}^{-1}\boldsymbol{b}$ or $\boldsymbol{z} = \boldsymbol{A}^{-1}\boldsymbol{b}$. In this case, there is a unique solution. When $\boldsymbol{A}$ does not have an inverse, there is either no solution or an infinite number of solutions. That issue is considered in detail in Section 2.5, along with the more general case where the number of equations and the number of variables may differ. A square matrix has an inverse if and only if its determinant (defined next) is non-zero. An alternative criterion for invertibility may be given in terms of matrix rank (developed in Section 2.5.2).

REMARK 2.1: A square matrix has a non-zero determinant if and only if the columns of the matrix are *linearly independent*, in which case the matrix is said to be of *full rank*. Rank and linear independence are introduced in Section 2.4.1. The concept of linear independence provides geometric insight into the circumstances in which a matrix is invertible, but is not required to define the inverse of a matrix. □

When a matrix has an inverse, the inverse can be found in a number of ways. The method described here involves finding in turn a number of matrices, leading to the inverse. Briefly, the steps are to find the determinant along with the minor, cofactor and adjoint matrices; the inverse matrix is the adjoint matrix scaled by the reciprocal of the determinant (for which reason, it must be non-zero). These are developed in turn.

REMARK 2.2: As noted, the method discussed next is not the only way to invert a matrix, but it serves the purpose of introducing determinants, minors and cofactors. Later, alternative methods of inversion are discussed. □

2.3.4.1 The determinant

Let $\boldsymbol{A}$ be a square matrix. The determinant of $\boldsymbol{A}$ is denoted $|\boldsymbol{A}|$. The determinant is defined inductively. If $\boldsymbol{A} = \{a\}$, so that $\boldsymbol{A}$ is a 1×1 matrix, then $|\boldsymbol{A}| = a$ (the determinant is equal to the number, not the absolute value). For example, if

$\boldsymbol{A} = (-2)$, then $|\boldsymbol{A}| = -2$. In the 2×2 case,

$$\boldsymbol{A} = \begin{pmatrix} a_{11} & a_{12} \\ a_{21} & a_{22} \end{pmatrix} \Rightarrow |\boldsymbol{A}| \stackrel{\text{def}}{=} \begin{vmatrix} a_{11} & a_{12} \\ a_{21} & a_{22} \end{vmatrix} = a_{11}a_{22} - a_{21}a_{12}.$$

In words, multiply the pair a_{11} and a_{22} and subtract from that the product of a_{21} and a_{12}. For the 3×3 case the determinant is defined:

$$\boldsymbol{A} = \begin{pmatrix} a_{11} & a_{12} & a_{13} \\ a_{21} & a_{22} & a_{23} \\ a_{31} & a_{32} & a_{33} \end{pmatrix}, \quad |\boldsymbol{A}| = a_{11} \begin{vmatrix} a_{22} & a_{23} \\ a_{32} & a_{33} \end{vmatrix} - a_{12} \begin{vmatrix} a_{21} & a_{23} \\ a_{31} & a_{33} \end{vmatrix} + a_{13} \begin{vmatrix} a_{21} & a_{22} \\ a_{31} & a_{32} \end{vmatrix}. \tag{2.4}$$

For example,

$$\boldsymbol{A} = \begin{pmatrix} 1 & 2 & 3 \\ 1 & 3 & 5 \\ 1 & 5 & 12 \end{pmatrix}, \quad |\boldsymbol{A}| = 1(36-25) - 2(12-5) + 3(5-3) = 11 - 14 + 6 = 3. \tag{2.5}$$

Notice that the definition of the determinant for a 3×3 matrix makes use of determinants of 2×2 matrices. Similarly, the definition of the determinant of a 4×4 matrix utilizes 3×3 determinants, and the definition of the determinant of a $n \times n$ matrix utilizes $(n-1) \times (n-1)$ determinants. So the definition of the determinant is given inductively. The general process is as follows. Given the square matrix $\boldsymbol{A}$, let $\boldsymbol{A}_{ij}$ be the matrix obtained from $\boldsymbol{A}$ by deleting the ith row and jth column. If $\boldsymbol{A}$ is an $n \times n$ matrix, $\boldsymbol{A}_{ij}$ is an $(n-1) \times (n-1)$ matrix. For the $n \times n$ matrix $\boldsymbol{A}$, the determinant may be calculated according to the formula

$$|\boldsymbol{A}| = a_{11} |\boldsymbol{A}_{11}| - a_{12} |\boldsymbol{A}_{12}| + a_{13} |\boldsymbol{A}_{13}| + \cdots + (-1)^{1+n} a_{1n} |\boldsymbol{A}_{1n}|. \tag{2.6}$$

(The pattern may be seen in the computation of the determinant in Equation 2.5.) Observe that in this calculation we are required to compute determinants of the $\boldsymbol{A}_{ij}$ matrices. However, if $\boldsymbol{A}$ is an $n \times n$ matrix, then $\boldsymbol{A}_{ij}$ is an $(n-1) \times (n-1)$ matrix. Therefore, if we have defined the determinant of a 2×2 matrix, we can define the determinant of a 3×3 matrix. In turn, this allows us to define the determinant of a 4×4 matrix, and so on.

Note that the a_{ij} coefficients are drawn from a chosen row, row 1, but in fact any row may be chosen for the calculation. In the example given above, if the second row were chosen, the computation would proceed as:

$$
\begin{aligned}
|A| &= -1\begin{vmatrix} 2 & 3 \\ 5 & 12 \end{vmatrix} + 3\begin{vmatrix} 1 & 3 \\ 1 & 12 \end{vmatrix} - 5\begin{vmatrix} 1 & 2 \\ 1 & 5 \end{vmatrix} \\
&= -1(2\cdot 12 - 3\cdot 5) + 3(1\cdot 12 - 1\cdot 3) - 5(1\cdot 5 - 1\cdot 2) \\
&= -1(9) + 3(9) - 5(3) \\
&= 3.
\end{aligned}
$$

Note that in Equation 2.6, we have a "−" sign before a_{12} while there is a "+" sign before a_{13}. The rule here is that if in a_{ij}, i,j sum to an even number (i.e., $i+j$ is even), then we place a "+" sign before that a_{ij} and if i,j sum to an odd number then we place a "−" sign before that a_{ij}. Finally, note that $(-1)^2 = (-1)(-1) = 1$, $(-1)^3 = (-1)(-1)(-1) = -1$ and, in general, $(-1)^k = 1$ if k is even and $(-1)^k = -1$ if k is odd. Thus

$$
\begin{aligned}
|\,A\,| &= (-1)^{1+1}a_{11}\,|\,A_{11}\,| + (-1)^{1+2}a_{12}\,|\,A_{12}\,| + (-1)^{1+3}a_{13}\,|\,A_{13}\,| + \dots \\
&\quad \dots + (-1)^{1+n}a_{1n}\,|\,A_{1n}\,| \\
&= \sum_{k=1}^{n}(-1)^{1+k}a_{1k}\,|\,A_{1k}\,|. \qquad (2.7)
\end{aligned}
$$

In fact, the determinant can be calculated by expanding along any row. Expanding along the ith row:

$$
\begin{aligned}
|\,A\,| &= \sum_{j=1}^{n}(-1)^{i+j}a_{ij}\,|\,A_{ij}\,| = (-1)^{i+1}a_{i1}\,|\,A_{i1}\,| + (-1)^{i+2}a_{i2}\,|\,A_{i2}\,| \\
&\quad + \dots + (-1)^{i+n}a_{in}\,|\,A_{in}\,|. \qquad (2.8)
\end{aligned}
$$

While these calculations obtain the determinant by expanding along rows, the determinant may also be calculated by expansion along columns. The point is clarified in the discussion of cofactors below.

2.3.4.2 The minor matrix

Given an $n \times n$ matrix A, there is a corresponding matrix, M, called the *matrix of minors*. Recall that A_{ij} is the matrix obtained from A by deleting the ith row and jth column of A. The minor in the ith row and jth column of the minor matrix is the determinant of A_{ij}. That is:

$$m_{ij} = |\,A_{ij}\,|.$$

Let M be the matrix with m_{ij} in the ith row and jth column.

For example, in the 3×3 case:

$$m_{11} = \begin{vmatrix} a_{22} & a_{23} \\ a_{32} & a_{33} \end{vmatrix}.$$

Consider the matrix $\boldsymbol{A}$ in Equation 2.5. The minors are:

$$m_{11} = \begin{vmatrix} 3 & 5 \\ 5 & 12 \end{vmatrix} = 36 - 25 = 11; \quad m_{12} = \begin{vmatrix} 1 & 5 \\ 1 & 12 \end{vmatrix} = 12 - 5 = 7; \quad m_{13} = \begin{vmatrix} 1 & 3 \\ 1 & 5 \end{vmatrix} = 5 - 3 = 2$$

$$m_{21} = \begin{vmatrix} 2 & 3 \\ 5 & 12 \end{vmatrix} = 24 - 15 = 9; \quad m_{22} = \begin{vmatrix} 1 & 3 \\ 1 & 12 \end{vmatrix} = 12 - 3 = 9; \quad m_{23} = \begin{vmatrix} 1 & 2 \\ 1 & 5 \end{vmatrix} = 5 - 2 = 3$$

$$m_{31} = \begin{vmatrix} 2 & 3 \\ 3 & 5 \end{vmatrix} = 10 - 9 = 1; \quad m_{32} = \begin{vmatrix} 1 & 3 \\ 1 & 5 \end{vmatrix} = 5 - 3 = 2; \quad m_{33} = \begin{vmatrix} 1 & 2 \\ 1 & 3 \end{vmatrix} = 3 - 2 = 1.$$

Together, $\boldsymbol{A}$ and its matrix of minors $\boldsymbol{M}$ are:

$$\boldsymbol{A} = \begin{pmatrix} 1 & 2 & 3 \\ 1 & 3 & 5 \\ 1 & 5 & 12 \end{pmatrix} \quad \text{and} \quad \boldsymbol{M} = \begin{pmatrix} 11 & 7 & 2 \\ 9 & 9 & 3 \\ 1 & 2 & 1 \end{pmatrix}.$$

Finally, note that since $m_{ij} = |A_{ij}|$, the determinant may also be written as:

$$|\boldsymbol{A}| = \sum_{j=1}^{n} (-1)^{i+j} a_{ij} m_{ij}. \tag{2.9}$$

2.3.4.3 The cofactor matrix

The cofactor matrix is derived directly from the minor matrix by changing the sign of minors in odd locations on the minor matrix. The ijth cofactor c_{ij} is defined as $c_{ij} = (-1)^{i+j} m_{ij}$. In words, if $i + j$ is even, the cofactor c_{ij} equals the minor m_{ij}; and $c_{ij} = -m_{ij}$ if $i + j$ is odd. More succinctly:

$$c_{ij} = (-1)^{i+j} m_{ij} \tag{2.10}$$

noting that $(-1)^{i+j} = 1$ if $i + j$ is even and $(-1)^{i+j} = -1$ if $i + j$ is odd. Let $\boldsymbol{C}$ be the matrix of cofactors.

For example, in the 3×3 case:

$$\text{If } \boldsymbol{M} = \begin{pmatrix} m_{11} & m_{12} & m_{13} \\ m_{21} & m_{22} & m_{23} \\ m_{31} & m_{32} & m_{33} \end{pmatrix}, \text{ then } \boldsymbol{C} = \begin{pmatrix} c_{11} & c_{12} & c_{13} \\ c_{21} & c_{22} & c_{23} \\ c_{31} & c_{32} & c_{33} \end{pmatrix} = \begin{pmatrix} m_{11} & -m_{12} & m_{13} \\ -m_{21} & m_{22} & -m_{23} \\ m_{31} & -m_{32} & m_{33} \end{pmatrix}.$$

Continuing with the earlier example, the matrix $\boldsymbol{A}$ and its minor and cofactor matrices are given by:

$$\boldsymbol{A} = \begin{pmatrix} 1 & 2 & 3 \\ 1 & 3 & 5 \\ 1 & 5 & 12 \end{pmatrix}, \quad \boldsymbol{M} = \begin{pmatrix} 11 & 7 & 2 \\ 9 & 9 & 3 \\ 1 & 2 & 1 \end{pmatrix}, \quad \boldsymbol{C} = \begin{pmatrix} 11 & -7 & 2 \\ -9 & 9 & -3 \\ 1 & -2 & 1 \end{pmatrix}.$$

Note that, in view of Equations 2.9 and 2.10,

$$|\boldsymbol{A}| = \sum_{j=1}^{n} a_{ij} c_{ij}, \tag{2.11}$$

and this is called cofactor expansion along the ith row. Cofactor expansion will be discussed further.

2.3.4.4 The adjoint matrix

The adjoint matrix of $\boldsymbol{A}$ is denoted adj($\boldsymbol{A}$) and defined as the transpose of the cofactor matrix:

$$\operatorname{adj}(\boldsymbol{A}) = \boldsymbol{C}'.$$

Continuing with the example, the matrices are given by:

$$\boldsymbol{A} = \begin{pmatrix} 1 & 2 & 3 \\ 1 & 3 & 5 \\ 1 & 5 & 12 \end{pmatrix}, \quad \boldsymbol{M} = \begin{pmatrix} 11 & 7 & 2 \\ 9 & 9 & 3 \\ 1 & 2 & 1 \end{pmatrix}, \quad \boldsymbol{C} = \begin{pmatrix} 11 & -7 & 2 \\ -9 & 9 & -3 \\ 1 & -2 & 1 \end{pmatrix}, \quad \boldsymbol{C}' = \begin{pmatrix} 11 & -9 & 1 \\ -7 & 9 & -2 \\ 2 & -3 & 1 \end{pmatrix}.$$

The inverse matrix is defined directly from the adjoint matrix.

2.3.4.5 The inverse matrix

Definition 2.4: *The inverse of* $\mathbf{A}$ *is denoted* $\mathbf{A}^{-1}$ *and defined:*

$$\mathbf{A}^{-1} = \frac{1}{|\mathbf{A}|} \operatorname{adj}(\mathbf{A}).$$

Continuing with the example (recall $|\boldsymbol{A}| = 3$):

$$\operatorname{adj}(\boldsymbol{A}) = \begin{pmatrix} 11 & -9 & 1 \\ -7 & 9 & -2 \\ 2 & -3 & 1 \end{pmatrix}, \quad \boldsymbol{A}^{-1} = \begin{pmatrix} \frac{11}{3} & -\frac{9}{3} & \frac{1}{3} \\ -\frac{7}{3} & \frac{9}{3} & -\frac{2}{3} \\ \frac{2}{3} & -\frac{3}{3} & \frac{1}{3} \end{pmatrix}.$$

The inverse matrix satisfies: $\boldsymbol{A}\boldsymbol{A}^{-1} = \boldsymbol{I} = \boldsymbol{A}^{-1}\boldsymbol{A}$

$$\begin{pmatrix} 1 & 2 & 3 \\ 1 & 3 & 5 \\ 1 & 5 & 12 \end{pmatrix} \begin{pmatrix} \frac{11}{3} & -\frac{9}{3} & \frac{1}{3} \\ -\frac{7}{3} & \frac{9}{3} & -\frac{2}{3} \\ \frac{2}{3} & -\frac{3}{3} & \frac{1}{3} \end{pmatrix} = \begin{pmatrix} 1 & 0 & 0 \\ 0 & 1 & 0 \\ 0 & 0 & 1 \end{pmatrix}.$$

A property of the adjoint (which follows from the cofactor expansion discussed above) is the following:

$$\boldsymbol{A}\,\mathrm{adj}(\boldsymbol{A}) = (\mathrm{adj}\boldsymbol{A})\boldsymbol{A} = \begin{pmatrix} |\boldsymbol{A}| & 0 & \cdots & 0 \\ 0 & |\boldsymbol{A}| & \cdots & 0 \\ \vdots & \vdots & \ddots & \vdots \\ 0 & 0 & \cdots & |\boldsymbol{A}| \end{pmatrix} = |\boldsymbol{A}| \cdot \boldsymbol{I}.$$

So, given a system of n equations in n unknowns, such a system can be written in the form: $\boldsymbol{A}\boldsymbol{x} = \boldsymbol{b}$. Provided $\boldsymbol{A}$ has full rank (is non-singular), we can solve for $\boldsymbol{x}$: $\boldsymbol{x} = \boldsymbol{A}^{-1}\boldsymbol{b}$.

EXAMPLE 2.6: Consider the system $\boldsymbol{A}\boldsymbol{x} = \boldsymbol{b}$, where $\boldsymbol{A}$ and $\boldsymbol{b}$ are given by:

$$\boldsymbol{A} = \begin{pmatrix} 1 & 2 & 4 & 3 \\ 2 & 5 & 9 & 1 \\ 3 & 2 & 1 & 7 \\ 0 & 1 & 6 & 2 \end{pmatrix}, \quad \text{and } \boldsymbol{b} = \begin{pmatrix} 4 \\ 5 \\ 1 \\ 2 \end{pmatrix},$$

and we require the solution $\boldsymbol{x}$. Then

$$\boldsymbol{A}^{-1} = \begin{pmatrix} -\frac{162}{61} & \frac{35}{61} & \frac{51}{61} & \frac{47}{61} \\ \frac{116}{61} & -\frac{10}{61} & -\frac{32}{61} & -\frac{57}{61} \\ -\frac{33}{61} & \frac{6}{61} & \frac{7}{61} & \frac{22}{61} \\ \frac{41}{61} & -\frac{13}{61} & -\frac{5}{61} & -\frac{7}{61} \end{pmatrix} \text{ so that } \boldsymbol{x} = \boldsymbol{A}^{-1}\boldsymbol{b} = \begin{pmatrix} -\frac{328}{61} \\ \frac{268}{61} \\ -\frac{51}{61} \\ \frac{80}{61} \end{pmatrix}.$$

◊

The next example develops the two-market model of Example 2.2 for arbitrary supply and demand parameters.

EXAMPLE 2.7: A two-market model with cross-price effects (the price of butter may affect the demand for margarine; the price of tea may affect the demand for coffee) has the price of both goods enter the demand for each good. This observation leads to the market structure:

$$\begin{aligned} &\text{Market 1:} \quad D_1 = a_0 + a_1 p_1 + a_2 p_2, \quad S_1 = b_0 + b_1 p_1 + b_2 p_2 \\ &\text{Market 2:} \quad D_2 = \alpha_0 + \alpha_1 p_1 + \alpha_2 p_2, \quad S_2 = \beta_0 + \beta_1 p_1 + \beta_2 p_2. \end{aligned}$$

(One might plausibly expect that the supply functions depend on own price ($b_2 = \beta_1 = 0$) but that is unnecessary for the computations.) To find the equilibrium level of quantity and price in each market, equate supply and demand in each market.

$$\begin{aligned} &\text{Market 1:} \quad D_1 = S_1 \quad \text{or} \quad a_0 + a_1 p_1 + a_2 p_2 = b_0 + b_1 p_1 + b_2 p_2 \\ &\text{Market 2:} \quad D_2 = S_2 \quad \text{or} \quad \alpha_0 + \alpha_1 p_1 + \alpha_2 p_2 = \beta_0 + \beta_1 p_1 + \beta_2 p_2. \end{aligned}$$

Rearranging these equations gives:

$$\begin{aligned} (a_1 - b_1)p_1 + (a_2 - b_2)p_2 &= (b_0 - a_0) \\ (\alpha_1 - \beta_1)p_1 + (\alpha_2 - \beta_2)p_2 &= (\beta_0 - \alpha_0). \end{aligned}$$

These equations can be solved by substitution. From the first equation

$$p_1 = \frac{(b_0 - a_0)}{(a_1 - b_1)} - \frac{(a_2 - b_2)}{(a_1 - b_1)} p_2.$$

Substitute into the second equation to get:

$$(\alpha_1 - \beta_1)\left[\frac{(b_0 - a_0)}{(a_1 - b_1)} - \frac{(a_2 - b_2)}{(a_1 - b_1)} p_2\right] + (\alpha_2 - \beta_2)p_2 = (\beta_0 - \alpha_0).$$

Collecting terms in p_2 gives:

$$\left[(\alpha_2 - \beta_2) - (\alpha_1 - \beta_1)\frac{(a_2 - b_2)}{(a_1 - b_1)}\right] p_2 = (\beta_0 - \alpha_0) - (\alpha_1 - \beta_1)\frac{(b_0 - a_0)}{(a_1 - b_1)}.$$

Multiplying both sides by $(a_1 - b_1)$ gives:

$$[(\alpha_2 - \beta_2)(a_1 - b_1) - (\alpha_1 - \beta_1)(a_2 - b_2)]p_2 = (\beta_0 - \alpha_0)(a_1 - b_1) - (\alpha_1 - \beta_1)(b_0 - a_0).$$

Thus,

$$p_2 = \frac{[(\beta_0 - \alpha_0)(a_1 - b_1) - (\alpha_1 - \beta_1)(b_0 - a_0)]}{[(\alpha_2 - \beta_2)(a_1 - b_1) - (\alpha_1 - \beta_1)(a_2 - b_2)]}.$$

A similar calculation gives p_1. ◊

With three or more equations, the procedure in Example 2.7 becomes progressively more complicated and unmanageable. The use of matrix algebra simplifies the problem significantly, and the advantage of matrix algebra becomes more pronounced as the number of equations increases. Example 2.8 reconsiders the problem formulated in matrix terms and shows that a direct generalization to the n-variable case is straightforward.

EXAMPLE 2.8: In matrix form, the system here may be written

$$\begin{pmatrix} (a_1-b_1) & (a_2-b_2) \\ (\alpha_1-\beta_1) & (\alpha_2-\beta_2) \end{pmatrix} \begin{pmatrix} p_1 \\ p_2 \end{pmatrix} = \begin{pmatrix} b_0-a_0 \\ \beta_0-\alpha_0 \end{pmatrix}$$

or

$$\begin{pmatrix} p_1 \\ p_2 \end{pmatrix} = \begin{pmatrix} (a_1-b_1) & (a_2-b_2) \\ (\alpha_1-\beta_1) & (\alpha_2-\beta_2) \end{pmatrix}^{-1} \begin{pmatrix} b_0-a_0 \\ \beta_0-\alpha_0 \end{pmatrix},$$

where the matrix

$$\begin{pmatrix} (a_1-b_1) & (a_2-b_2) \\ (\alpha_1-\beta_1) & (\alpha_2-\beta_2) \end{pmatrix}^{-1}$$

is the inverse of

$$\begin{pmatrix} (a_1-b_1) & (a_2-b_2) \\ (\alpha_1-\beta_1) & (\alpha_2-\beta_2) \end{pmatrix}.$$

Using the procedure above, the inverse is:

$$\begin{pmatrix} (a_1-b_1) & (a_2-b_2) \\ (\alpha_1-\beta_1) & (\alpha_2-\beta_2) \end{pmatrix}^{-1} = \frac{1}{d}\begin{pmatrix} (\alpha_2-\beta_2) & -(a_2-b_2) \\ -(\alpha_1-\beta_1) & (a_1-b_1) \end{pmatrix},$$

where $d = (a_1-b_1)(\alpha_2-\beta_2)-(\alpha_1-\beta_1)(a_2-b_2)$. Thus,

$$\begin{aligned} \begin{pmatrix} p_1 \\ p_2 \end{pmatrix} &= \frac{1}{d}\begin{pmatrix} (\alpha_2-\beta_2) & -(a_2-b_2) \\ -(\alpha_1-\beta_1) & (a_1-b_1) \end{pmatrix}\begin{pmatrix} (b_0-a_0) \\ (\beta_0-\alpha_0) \end{pmatrix} \\ &= \frac{1}{d}\begin{pmatrix} (\alpha_2-\beta_2)(b_0-a_0)-(a_2-b_2)(\beta_0-\alpha_0) \\ -(\alpha_1-\beta_1)(b_0-a_0)+(a_1-b_1)(\beta_0-\alpha_0) \end{pmatrix}. \end{aligned}$$

So

$$\begin{aligned} p_1 &= \frac{[(\alpha_2-\beta_2)(b_0-a_0)-(a_2-b_2)(\beta_0-\alpha_0)]}{[(a_1-b_1)(\alpha_2-\beta_2)-(\alpha_1-\beta_1)(a_2-b_2)]} \\ p_2 &= \frac{[-(\alpha_1-\beta_1)(b_0-a_0)+(a_1-b_1)(\beta_0-\alpha_0)]}{[(a_1-b_1)(\alpha_2-\beta_2)-(\alpha_1-\beta_1)(a_2-b_2)]}. \end{aligned}$$

From this solution one can, for example, calculate the change in the equilibrium price of good 1 resulting from a change in α_0 : $\frac{\Delta p_1}{\Delta \alpha_0} = \frac{(a_2-b_2)}{d}$ $(= \frac{\Delta p_1}{\Delta \beta_0})$. ◊

The following discussion elaborates on various aspects of determinants, cofactors and the inverse.

2.3.5 Additional remarks on determinants

The following sections observe that the calculation of determinants may be obtained by cofactor expansion along any row or column, as described next. In addition, the determinant and inverse of a matrix satisfies a number of important properties, summarized in Sections 2.3.5.2 and 2.3.5.3.

2.3.5.1 Determinants and cofactor expansion

Let

$$A = \begin{pmatrix} a_{11} & a_{12} & \dots & a_{1n} \\ a_{21} & a_{22} & \dots & a_{2n} \\ \vdots & \vdots & \ddots & \vdots \\ a_{n1} & a_{n2} & \dots & a_{nn} \end{pmatrix} \quad \text{and} \quad C = \begin{pmatrix} c_{11} & c_{12} & \dots & c_{1n} \\ c_{21} & c_{22} & \dots & c_{2n} \\ \vdots & \vdots & \ddots & \vdots \\ c_{n1} & c_{n2} & \dots & c_{nn} \end{pmatrix},$$

where C is the cofactor matrix derived from A. Recall that the determinant of A may be calculated:

$$|A| = a_{11}c_{11} + a_{12}c_{12} + \cdots + a_{1n}c_{1n} = \sum_{j=1}^{n} a_{1j}c_{1j},$$

where c_{ij} is the ijth cofactor of A.

From the definition, the determinant is computed by taking the first rows of both A and C, multiplying them pairwise, and summing. Put differently, the determinant of A equals the inner product of row 1 from matrix A with row 1 from matrix C. In fact, the same answer is obtained from the inner product of any matching rows of A and C. Furthermore, the determinant may also be computed by taking the inner product of matching columns: the inner product of the ith column of A and the ith column of C gives the determinant. For this reason, the determinant of $\boldsymbol{A}$ satisfies the following:

$$
\begin{aligned}
|\boldsymbol{A}| &= \sum_{j=1}^{n} a_{ij}c_{ij}, \text{ for any } i \\
&= \sum_{i=1}^{n} a_{ij}c_{ij}, \text{ for any } j.
\end{aligned}
$$

The expression $\sum_{j=1}^{n} a_{ij}c_{ij}$ is called "row expansion" since the summation is along a row (the ith). The expression $\sum_{i=1}^{n} a_{ij}c_{ij}$ is called "column expansion" since the summation is along a column (the jth). The following terms give the expansions associated with each row (for the matrix in Equation 2.5):

$$
\begin{aligned}
a_{11}c_{11} + a_{12}c_{12} + a_{13}c_{13} &= (1)(11) + 2(-7) + 3(2) = 3 \\
a_{21}c_{21} + a_{22}c_{22} + a_{23}c_{23} &= (1)(-9) + 3(9) + (5)(-3) = 3 \\
a_{31}c_{31} + a_{32}c_{32} + a_{33}c_{33} &= (1)(1) + 5(-2) + 12(1) = 3.
\end{aligned}
$$

Similarly, the column expansions are:

$$
\begin{aligned}
a_{11}c_{11} + a_{21}c_{21} + a_{31}c_{31} &= (1)(11) + (1)(-9) + (1)(1) = 3 \\
a_{12}c_{12} + a_{22}c_{22} + a_{32}c_{32} &= (2)(-7) + 3(9) + (5)(-2) = 3 \\
a_{13}c_{13} + a_{23}c_{23} + a_{33}c_{33} &= (3)(2) + (5)(-3) + (12)(1) = 3.
\end{aligned}
$$

So, in the 3×3 case, the determinant is computed by any of the expansions given above. This sometimes provides an easy way to compute a determinant.

EXAMPLE 2.9: In the following example the determinant is computed most easily by cofactor expansion along the row with most 0s.

$$
\boldsymbol{B} = \begin{pmatrix} 3 & 8 & 5 \\ 6 & 3 & 4 \\ 1 & 0 & 0 \end{pmatrix} \Rightarrow |\boldsymbol{B}| = b_{31}c_{31} + b_{32}c_{32} + b_{33}c_{33} = 1 \cdot \begin{vmatrix} 8 & 5 \\ 3 & 4 \end{vmatrix} = 32 - 15 = 17.
$$

◊

The inner product of the ith row of the $\boldsymbol{A}$ matrix and the kth row of the cofactor matrix, with $k \neq i$, or the inner product of the jth column of the $\boldsymbol{A}$ matrix and the kth column of the cofactor matrix, with $k \neq j$, is called *expansion by alien cofactors*. These expansions have the property that these inner products are equal to zero. Thus, $\sum_{j=1}^{n} a_{ij}c_{kj} = 0$, if $k \neq i$ (expansion by ith row of $\boldsymbol{A}$, kth row of $\boldsymbol{C}$). Similarly, $\sum_{i=1}^{n} a_{ij}c_{ik} = 0$, if $j \neq k$ (expansion by jth column of $\boldsymbol{A}$, kth column of $\boldsymbol{C}$).

To illustrate with the example above:

$$\sum_{j=1}^{3} a_{1j}c_{2j} = a_{11}c_{21} + a_{12}c_{22} + a_{13}c_{13} = (1)(-9) + 2(9) + 3(-3) = 0$$

$$\sum_{i=1}^{3} a_{i1}c_{i2} = a_{11}c_{12} + a_{21}c_{22} + a_{31}c_{32} = (1)(-7) + (1)(9) + (1)(-2) = 0.$$

2.3.5.2 Properties of determinants

1. $|\,\boldsymbol{A}\,| = |\,\boldsymbol{A}'\,|$. The determinant of matrix $\boldsymbol{A}$ equals the determinant of the transpose of matrix $\boldsymbol{A}$.

2. Interchange of two rows or two columns reverses the sign of a determinant.

3. Multiplying any *one* row or column by a scalar k alters the determinant k-fold.

4. The addition or subtraction of a multiple of any row (column) to or from any other row (column) leaves the value of the determinant unaltered.

5. If one row or column is a multiple of another, the determinant is 0.

6. If $\boldsymbol{A}$ and $\boldsymbol{B}$ are two square matrices of order n, then $|\,\boldsymbol{AB}\,| = |\,\boldsymbol{A}\,||\,\boldsymbol{B}\,|$. For example,

$$\boldsymbol{A} = \begin{pmatrix} 1 & 3 \\ 2 & 4 \end{pmatrix}, \boldsymbol{B} = \begin{pmatrix} 5 & 7 \\ 6 & 8 \end{pmatrix} \Rightarrow \boldsymbol{AB} = \begin{pmatrix} 23 & 31 \\ 34 & 46 \end{pmatrix},$$

so $|\,\boldsymbol{A}\,| = -2$, $|\,\boldsymbol{B}\,| = -2$ and $|\,\boldsymbol{A}\,||\,\boldsymbol{B}\,| = (-2)(-2) = 4$. Also, $|\,\boldsymbol{AB}\,| = 1058 - 1054 = 4$.

7. In general $|\,\boldsymbol{A}+\boldsymbol{B}\,| \neq |\,\boldsymbol{A}\,| + |\,\boldsymbol{B}\,|$. Any of the following can hold:

(1) $|\,\boldsymbol{A}+\boldsymbol{B}\,| = |\,\boldsymbol{A}\,| + |\,\boldsymbol{B}\,|$; (2) $|\,\boldsymbol{A}+\boldsymbol{B}\,| > |\,\boldsymbol{A}\,| + |\,\boldsymbol{B}\,|$; (3) $|\,\boldsymbol{A}+\boldsymbol{B}\,| < |\,\boldsymbol{A}\,| + |\,\boldsymbol{B}\,|$.

The following matrices illustrate these differences.

$$\boldsymbol{A} = \begin{pmatrix} 1 & 0 \\ 0 & 1 \end{pmatrix}, \boldsymbol{B} = \begin{pmatrix} 1 & 0 \\ 0 & 1 \end{pmatrix}, \boldsymbol{A}+\boldsymbol{B} = \begin{pmatrix} 2 & 0 \\ 0 & 2 \end{pmatrix}.$$

So $|\,\boldsymbol{A}\,| = |\,\boldsymbol{B}\,| = 1$, while $|\,\boldsymbol{A}+\boldsymbol{B}\,| = 4$. Putting $\boldsymbol{C} = -\boldsymbol{B}$, $\boldsymbol{A}+\boldsymbol{C}$ is a matrix of 0s with $|\,\boldsymbol{A}+\boldsymbol{C}\,| = 0$, while $|\,\boldsymbol{A}\,| = |\,\boldsymbol{C}\,| = 1$.

2.3.5.3 Properties of the matrix inverse

1. The inverse is unique: if the matrix $\boldsymbol{A}$ is non-singular (has non-zero determinant) then $\boldsymbol{A}^{-1}$ exists and is unique.

2. The determinant of the inverse is the inverse of the determinant: $|\boldsymbol{A}^{-1}| = \frac{1}{|\boldsymbol{A}|}$.

3. $(\boldsymbol{AB})^{-1} = \boldsymbol{B}^{-1}\boldsymbol{A}^{-1}$. To illustrate,

$$\boldsymbol{A} = \begin{pmatrix} 1 & 0 \\ 2 & 3 \end{pmatrix}, \quad \boldsymbol{B} = \begin{pmatrix} 2 & 5 \\ 2 & 1 \end{pmatrix}, \quad \boldsymbol{A}^{-1} = \begin{pmatrix} 1 & 0 \\ -\frac{2}{3} & \frac{1}{3} \end{pmatrix}, \quad \boldsymbol{B}^{-1} = \begin{pmatrix} -\frac{1}{8} & \frac{5}{8} \\ \frac{1}{4} & -\frac{1}{4} \end{pmatrix},$$

$$\boldsymbol{AB} = \begin{pmatrix} 2 & 5 \\ 10 & 13 \end{pmatrix}, \quad (\boldsymbol{AB})^{-1} = \begin{pmatrix} -\frac{13}{24} & \frac{5}{24} \\ \frac{10}{24} & -\frac{2}{24} \end{pmatrix}, \quad \boldsymbol{B}^{-1}\boldsymbol{A}^{-1} = \begin{pmatrix} -\frac{1}{8} & \frac{5}{8} \\ \frac{1}{4} & -\frac{1}{4} \end{pmatrix} \begin{pmatrix} 1 & 0 \\ -\frac{2}{3} & \frac{1}{3} \end{pmatrix}.$$

4. $(\boldsymbol{A}')^{-1} = (\boldsymbol{A}^{-1})'$. For example,

$$\boldsymbol{A} = \begin{pmatrix} 2 & 1 \\ 5 & 3 \end{pmatrix}, \quad \boldsymbol{A}^{-1} = \begin{pmatrix} 3 & -1 \\ -5 & 2 \end{pmatrix}, \quad (\boldsymbol{A}^{-1})' = \begin{pmatrix} 3 & -5 \\ -1 & 2 \end{pmatrix},$$

while

$$\boldsymbol{A}' = \begin{pmatrix} 2 & 5 \\ 1 & 3 \end{pmatrix}, \quad (\boldsymbol{A}')^{-1} = \begin{pmatrix} 3 & -5 \\ -1 & 2 \end{pmatrix}.$$

5. If $\boldsymbol{A}$ or $\boldsymbol{B}$ is non-singular, then $\boldsymbol{AB} = \boldsymbol{0} \Rightarrow$ either $\boldsymbol{A}$ or $\boldsymbol{B} = \boldsymbol{0}$; $|\boldsymbol{A}| \neq 0 \Rightarrow \boldsymbol{B} = \boldsymbol{0}$, $|\boldsymbol{B}| \neq 0 \Rightarrow \boldsymbol{A} = \boldsymbol{0}$.

2.3.6 Cramer's rule

Let $\boldsymbol{A}$ be an $n \times n$ matrix with non-zero determinant, $|\boldsymbol{A}| \neq 0$, $\boldsymbol{b}$ an $n \times 1$ vector of constants and let $\boldsymbol{x} \in \mathscr{R}^n$ satisfy $\boldsymbol{Ax} = \boldsymbol{b}$. Then $\boldsymbol{A}$ may be inverted to solve for $\boldsymbol{x}$: $\boldsymbol{x} = \boldsymbol{A}^{-1}\boldsymbol{b}$. An alternative method of solving for $\boldsymbol{x}$ that avoids the calculations involved in matrix inversion (computation of minors, cofactors and adjoint) is given by Cramer's rule, which involves the computation of $n+1$ determinants for a system of n equations in n unknowns.

Let $\boldsymbol{A}_j$ be the $\boldsymbol{A}$ matrix with the jth column replaced by $\boldsymbol{b}$. Then the jth element of $\boldsymbol{x}$ satisfies:

$$x_j = \frac{|\boldsymbol{A}_j|}{|\boldsymbol{A}|}, \quad j = 1, \ldots, n.$$

Thus, Cramer's rule provides a relatively simple method for solving equation systems. The following example illustrates.

EXAMPLE 2.10: Consider

$$\begin{pmatrix} 2 & 3 \\ 3 & 6 \end{pmatrix}\begin{pmatrix} x_1 \\ x_2 \end{pmatrix} = \begin{pmatrix} 3 \\ 5 \end{pmatrix}, \text{ then } \boldsymbol{A}_1 = \begin{pmatrix} 3 & 3 \\ 5 & 6 \end{pmatrix}, \boldsymbol{A}_2 = \begin{pmatrix} 2 & 3 \\ 3 & 5 \end{pmatrix}.$$

Thus, $|\boldsymbol{A}| = 3$, $|\boldsymbol{A}_1| = 3$, $|\boldsymbol{A}_2| = 1$ and so $x_1 = \frac{3}{3} = 1$, $x_2 = \frac{1}{3}$. ◊

2.3.7 Gaussian elimination

Gaussian elimination provides an algorithm for the solution of equation systems. The method involves reducing the coefficient matrix to row echelon form and then solving by repeated substitution.

Definition 2.5: *A matrix $\boldsymbol{A}$ is in row echelon form if: (1) all rows consisting entirely of 0s are at the bottom, and (2) in any non-zero row, the first non-zero entry (the leading entry for that row) is in a column to the left of the leading entry of the row below (and hence to the left of all leading entries below it).*

So, the leading entry on a non-zero row is in a column to the right of the leading entry of the row above it.

EXAMPLE 2.11:

$$\boldsymbol{A} = \begin{pmatrix} 1 & 2 & 4 & 3 \\ 0 & 5 & 9 & 1 \\ 0 & 0 & 1 & 7 \\ 0 & 0 & 0 & 2 \end{pmatrix}, \quad \boldsymbol{B} = \begin{pmatrix} 1 & 2 & 4 & 3 \\ 0 & 5 & 9 & 1 \\ 0 & 0 & 0 & 7 \\ 0 & 0 & 0 & 2 \end{pmatrix}, \quad \boldsymbol{C} = \begin{pmatrix} 1 & 2 & 4 & 3 \\ 0 & 5 & 9 & 1 \\ 0 & 0 & 0 & 0 \\ 0 & 0 & 0 & 7 \end{pmatrix}.$$

$\boldsymbol{A}$ is in row echelon form; neither $\boldsymbol{B}$ nor $\boldsymbol{C}$ are in row echelon form. ◊

REMARK 2.3: Some authors define row echelon such that the first non-zero entry in each row equals 1. Others use the term *reduced* row echelon form for the case where the first non-zero entry is one. The latter convention is adopted here. □

The definition applies also to the case where the matrix is not square. Gaussian elimination provides a method for solving systems of equations by reduction to row echelon form. It is an algorithm, based on the augmented matrix of the system.

Given a system of equations $\boldsymbol{Ax} = \boldsymbol{b}$, the *augmented matrix* of the system is defined as $(\boldsymbol{A} : \boldsymbol{b})$, a matrix obtained from $\boldsymbol{A}$ by adding $\boldsymbol{b}$ as a column.

Definition 2.6: *Given a system of equations $\boldsymbol{Ax} = \boldsymbol{b}$, the system is solved by Gaussian elimination as follows:*

1. *Form the augmented matrix.*
2. *Reduce the augmented matrix to row echelon form.*
3. *Solve the reduced system by substitution.*

The point of this procedure is that the reduction to row echelon form greatly simplifies the process of solving by substitution. From the point of view of application, the main difficulty is in reducing the augmented system to row echelon form.

EXAMPLE 2.12: Consider the system $\boldsymbol{Ax} = \boldsymbol{b}$ where:

$$\boldsymbol{A} = \begin{bmatrix} 1 & 2 & 3 \\ 1 & 3 & 5 \\ 1 & 5 & 12 \end{bmatrix}, \quad \boldsymbol{x} = \begin{bmatrix} x_1 \\ x_2 \\ x_3 \end{bmatrix}, \quad \boldsymbol{b} = \begin{bmatrix} 1 \\ 2 \\ 3 \end{bmatrix}.$$

The augmented matrix is:

$$(\boldsymbol{A} : \boldsymbol{b}) = \left[\begin{array}{ccc|c} 1 & 2 & 3 & 1 \\ 1 & 3 & 5 & 2 \\ 1 & 5 & 12 & 3 \end{array}\right].$$

To convert to row-echelon form, subtract the first row from the second, then subtract the first row from the third, and finally, subtract three times the second row from the third:

$$\left[\begin{array}{ccc|c} 1 & 2 & 3 & 1 \\ 1 & 3 & 5 & 2 \\ 1 & 5 & 12 & 3 \end{array}\right] \Rightarrow \left[\begin{array}{ccc|c} 1 & 2 & 3 & 1 \\ 0 & 1 & 2 & 1 \\ 1 & 5 & 12 & 3 \end{array}\right] \Rightarrow \left[\begin{array}{ccc|c} 1 & 2 & 3 & 1 \\ 0 & 1 & 2 & 1 \\ 0 & 3 & 9 & 2 \end{array}\right]$$
$$\Rightarrow \left[\begin{array}{ccc|c} 1 & 2 & 3 & 1 \\ 0 & 1 & 2 & 1 \\ 0 & 0 & 3 & -1 \end{array}\right]. \tag{2.12}$$

The coefficient matrix has now been reduced to row echelon form. This gives the equation system:

$$\begin{pmatrix} 1 & 2 & 3 \\ 0 & 1 & 2 \\ 0 & 0 & 3 \end{pmatrix} \begin{pmatrix} x_1 \\ x_2 \\ x_3 \end{pmatrix} = \begin{pmatrix} 1 \\ 1 \\ -1 \end{pmatrix}. \tag{2.13}$$

Directly, $x_3 = -\frac{1}{3}$. Then, by substitution, $x_2 + 2x_3 = 1$ or $x_2 - 2(\frac{1}{3}) = 1$ or $x_2 = \frac{5}{3}$. Finally, $x_1 + 2x_2 + 3x_3 = 1$, or $x_1 + 2(\frac{5}{3}) - 3(\frac{1}{3}) = 1$. So, $x_1 = 1 + 1 - \frac{10}{3} = -\frac{4}{3}$: $(x_1, x_2, x_3) = (-\frac{4}{3}, \frac{5}{3}, -\frac{1}{3})$.

With direct inversion of $\boldsymbol{A}$ the same result is obtained:

$$\boldsymbol{A}^{-1} = \begin{pmatrix} \frac{11}{3} & -\frac{9}{3} & \frac{1}{3} \\ -\frac{7}{3} & \frac{9}{3} & -\frac{2}{3} \\ \frac{2}{3} & -\frac{3}{3} & \frac{1}{3} \end{pmatrix}, \quad \boldsymbol{A}^{-1}\boldsymbol{b} = \begin{pmatrix} \frac{11}{3} & -\frac{9}{3} & \frac{1}{3} \\ -\frac{7}{3} & \frac{9}{3} & -\frac{2}{3} \\ \frac{2}{3} & -\frac{3}{3} & \frac{1}{3} \end{pmatrix} \begin{pmatrix} 1 \\ 2 \\ 3 \end{pmatrix} = \begin{pmatrix} -\frac{4}{3} \\ \frac{5}{3} \\ -\frac{1}{3} \end{pmatrix}.$$

◊

The Gaussian elimination process can be taken further, by additional reduction of the row echelon form.

Definition 2.7: *A matrix in row echelon form is in reduced row echelon form if: (1) the leading entry in any non-zero row is 1, and (2) each column containing a leading 1 has zeros elsewhere.*

For example, the matrices $\boldsymbol{A}$ and $\boldsymbol{B}$ are in reduced row echelon form:

$$\boldsymbol{A} = \begin{pmatrix} 1 & 3 & 0 & 0 \\ 0 & 0 & 1 & 0 \\ 0 & 0 & 0 & 1 \end{pmatrix}, \quad \boldsymbol{B} = \begin{pmatrix} 1 & 3 & 0 & 0 & 2 \\ 0 & 0 & 1 & 0 & 3 \\ 0 & 0 & 0 & 1 & 4 \\ 0 & 0 & 0 & 0 & 0 \end{pmatrix}.$$

Definition 2.8: *The system $\boldsymbol{A}\boldsymbol{x} = \boldsymbol{b}$ is solved by Gauss–Jordan elimination if the augmented matrix is reduced to reduced row echelon form, and the resulting system solved by substitution.*

EXAMPLE 2.13: Continuing with Example 2.12: (1) subtract twice row 2 from row 1, (2) then add $\frac{1}{3}$ of row 3 to row 1 and (3) subtract $\frac{2}{3}$ of row 3 from row 2.

$$\begin{bmatrix} 1 & 2 & 3 & 1 \\ 0 & 1 & 2 & 1 \\ 0 & 0 & 3 & -1 \end{bmatrix} \Rightarrow \begin{bmatrix} 1 & 0 & -1 & -1 \\ 0 & 1 & 2 & 1 \\ 0 & 0 & 3 & -1 \end{bmatrix} \Rightarrow \begin{bmatrix} 1 & 0 & 0 & -\frac{4}{3} \\ 0 & 1 & 2 & 1 \\ 0 & 0 & 3 & -1 \end{bmatrix}$$

$$\Rightarrow \begin{bmatrix} 1 & 0 & 0 & -\frac{4}{3} \\ 0 & 1 & 0 & \frac{5}{3} \\ 0 & 0 & 3 & -1 \end{bmatrix}.$$

Finally, multiply row 3 by $\frac{1}{3}$ to give:

$$\begin{bmatrix} 1 & 0 & 0 & -\frac{4}{3} \\ 0 & 1 & 0 & \frac{5}{3} \\ 0 & 0 & 1 & -\frac{1}{3} \end{bmatrix} \Rightarrow \begin{pmatrix} 1 & 0 & 0 \\ 0 & 1 & 0 \\ 0 & 0 & 1 \end{pmatrix} \begin{pmatrix} x_1 \\ x_2 \\ x_3 \end{pmatrix} = \begin{pmatrix} -\frac{4}{3} \\ \frac{5}{3} \\ -\frac{1}{3} \end{pmatrix}$$

and the solution for $(x_1, x_2, x_3) = \left(-\frac{4}{3}, \frac{5}{3}, -\frac{1}{3}\right)$ can be read off directly from the equation. ◊

2.3.7.1 Gaussian elimination: further remarks

The following discussion clarifies the relation between Gaussian elimination and matrix operations for exchanging rows. The concluding discussion illustrates how these procedures may be used to find the inverse of a matrix.

Consider the following equation system:

$$\begin{pmatrix} a_{11} & a_{12} & a_{13} \\ 0 & a_{22} & a_{23} \\ 0 & 0 & a_{33} \end{pmatrix} \begin{pmatrix} x_1 \\ x_2 \\ x_3 \end{pmatrix} = \begin{pmatrix} b_1 \\ b_2 \\ b_3 \end{pmatrix}.$$

The coefficient matrix is in row echelon form (in fact a square matrix with 0s below the main diagonal is called an *upper triangular matrix*). Such a system is easy to solve. Directly, $x_3 = \frac{b_3}{a_{33}}$ and then $a_{22}x_2 + a_{23}x_3 = b_2$ gives $a_{22}x_2 + a_{23}\left(\frac{b_3}{a_{33}}\right) = b_2$, so that $x_2 = \frac{1}{a_{22}}\left[b_2 - a_{23}\left(\frac{b_3}{a_{33}}\right)\right]$. Going one step further, substitute the solution values for x_2 and x_3 into $a_{11}x_1 + a_{12}x_2 + a_{13}x_3 = b_1$ to give:

$$a_{11}x_1 + a_{12}\left\{\frac{1}{a_{22}}\left[b_2 - a_{23}\left(\frac{b_3}{a_{33}}\right)\right]\right\} + a_{13}\left\{\frac{b_3}{a_{33}}\right\} = b_1,$$

and rearranging:

$$a_{11}x_1 = b_1 - \left(a_{12}\left\{\frac{1}{a_{22}}\left[b_2 - a_{23}\left(\frac{b_3}{a_{33}}\right)\right]\right\} + a_{13}\left\{\frac{b_3}{a_{33}}\right\}\right).$$

or

$$a_{11}x_1 = \frac{1}{a_{11}}\left[b_1 - \left(a_{12}\left\{\frac{1}{a_{22}}\left[b_2 - a_{23}\left(\frac{b_3}{a_{33}}\right)\right]\right\} + a_{13}\left\{\frac{b_3}{a_{33}}\right\}\right)\right].$$

So if the equation is in upper echelon form, it is easy (albeit tedious) to solve.

Gaussian elimination involves taking a system $\boldsymbol{Ax} = \boldsymbol{b}$ and converting it to row echelon form in a way that leaves the solution(s) unchanged. The general principle is the following. Suppose that $\boldsymbol{Ax} = \boldsymbol{b}$, then multiplying both sides by a square non-singular matrix $\boldsymbol{Q}$ leaves the set of possible solutions unchanged: if $\boldsymbol{Ax}^* = \boldsymbol{b}$, then also $\boldsymbol{QAx}^* = \boldsymbol{Qb}$ and if $\boldsymbol{Ax}^\dagger \neq \boldsymbol{b}$ then $\boldsymbol{QAx}^\dagger \neq \boldsymbol{Qb}$. So, we can solve $\boldsymbol{QAx} = \boldsymbol{Qb}$ to find a solution $\boldsymbol{x}$ to $\boldsymbol{Ax} = \boldsymbol{b}$. To illustrate, in the 2×2 case, let

$$\begin{pmatrix} a_{11} & a_{12} \\ a_{21} & a_{22} \end{pmatrix}\begin{pmatrix} x_1 \\ x_2 \end{pmatrix} = \begin{pmatrix} b_1 \\ b_2 \end{pmatrix}.$$

Assume $a_{11} \neq 0$. (If $a_{11} = 0$, move the second equation to the top. If $a_{11} = a_{21} = 0$, then the equation is insoluble unless $\frac{b_1}{a_{21}} = \frac{b_2}{a_{22}}$ assuming $a_{21}, a_{22} \neq 0$.) So, assume $a_{11} \neq 0$, so that $a_{21} = \alpha_2 a_{11}$ for some number α_2. Then $\boldsymbol{QAx} = \boldsymbol{Qb}$ is:

$$\begin{pmatrix} 1 & 0 \\ -\alpha_2 & 1 \end{pmatrix}\begin{pmatrix} a_{11} & a_{12} \\ a_{21} & a_{22} \end{pmatrix}\begin{pmatrix} x_1 \\ x_2 \end{pmatrix} = \begin{pmatrix} 1 & 0 \\ -\alpha_2 & 1 \end{pmatrix}\begin{pmatrix} b_1 \\ b_2 \end{pmatrix}, \quad \text{where} \quad \boldsymbol{Q} = \begin{pmatrix} 1 & 0 \\ -\alpha_2 & 1 \end{pmatrix}.$$

Multiplying through:

$$\begin{pmatrix} a_{11} & a_{12} \\ a_{21} - \alpha_2 a_{11} & a_{22} - \alpha_2 a_{12} \end{pmatrix}\begin{pmatrix} x_1 \\ x_2 \end{pmatrix} = \begin{pmatrix} b_1 \\ b_2 - \alpha_2 b_1 \end{pmatrix}$$

and using the fact that $a_{21} - \alpha_2 a_{11} = 0$,

$$\begin{pmatrix} a_{11} & a_{12} \\ 0 & a_{22} - \alpha_2 a_{12} \end{pmatrix}\begin{pmatrix} x_1 \\ x_2 \end{pmatrix} = \begin{pmatrix} b_1 \\ b_2 - \alpha_2 b_1 \end{pmatrix},$$

which is a system in upper triangular form. Thus, $x_2 = \frac{b_2 - \alpha_2 b_1}{a_{22} - \alpha_2 a_{12}}$. This assumes that $a_{22} - \alpha_2 a_{12} \neq 0$ (if this were 0, there would be no solution unless $b_2 - \alpha_2 b_1 = 0$). Then x_1 can be solved – either by substitution or one more matrix transformation. To see how the latter proceeds, let β satisfy $a_{12} = \beta(a_{22} - \alpha_2 a_{12})$

and consider the operation, $\boldsymbol{Q}^*\boldsymbol{Q}\boldsymbol{A}\boldsymbol{x} = \boldsymbol{Q}^*\boldsymbol{Q}\boldsymbol{b}$,

$$\begin{pmatrix} 1 & -\beta \\ 0 & 1 \end{pmatrix}\begin{pmatrix} a_{11} & a_{12} \\ 0 & a_{22} - \alpha_2 a_{12} \end{pmatrix}\begin{pmatrix} x_1 \\ x_2 \end{pmatrix} = \begin{pmatrix} 1 & -\beta \\ 0 & 1 \end{pmatrix}\begin{pmatrix} b_1 \\ b_2 - \alpha_2 b_1 \end{pmatrix}, \quad \boldsymbol{Q}^* = \begin{pmatrix} 1 & -\beta \\ 0 & 1 \end{pmatrix}.$$

The equation system is then:

$$\begin{pmatrix} a_{11} & 0 \\ 0 & a_{22} - \alpha_2 a_{12} \end{pmatrix}\begin{pmatrix} x_1 \\ x_2 \end{pmatrix} = \begin{pmatrix} b_1 - \beta[b_2 - \alpha_2 b_1] \\ b_2 - \alpha_2 b_1 \end{pmatrix},$$

with $\beta = \frac{a_{12}}{a_{22} - \alpha_2 a_{12}}$, and x_1, x_2 may be solved directly.

EXAMPLE 2.14: Consider again the system $\boldsymbol{A}\boldsymbol{x} = \boldsymbol{b}$ given by:

$$\begin{pmatrix} 1 & 2 & 3 \\ 1 & 3 & 5 \\ 1 & 5 & 12 \end{pmatrix}\begin{pmatrix} x_1 \\ x_2 \\ x_3 \end{pmatrix} = \begin{pmatrix} 1 \\ 2 \\ 3 \end{pmatrix}.$$

Define the matrices

$$\boldsymbol{Q}_1 = \begin{pmatrix} 1 & 0 & 0 \\ -1 & 1 & 0 \\ 0 & 0 & 1 \end{pmatrix}, \quad \boldsymbol{Q}_2 = \begin{pmatrix} 1 & 0 & 0 \\ 0 & 1 & 0 \\ -1 & 0 & 1 \end{pmatrix}, \quad \boldsymbol{Q}_3 = \begin{pmatrix} 1 & 0 & 0 \\ 0 & 1 & 0 \\ 0 & -3 & 1 \end{pmatrix}.$$

For an arbitrary matrix $\boldsymbol{B}$, $\boldsymbol{Q}_1\boldsymbol{B}$ yields a matrix whose first and third rows are unchanged, when the second row is transformed to the original second row minus the first row.

$$\boldsymbol{B} = \begin{pmatrix} b_{11} & b_{12} & b_{13} \\ b_{21} & b_{22} & b_{23} \\ b_{31} & b_{32} & b_{33} \end{pmatrix}, \quad \boldsymbol{Q}_1\boldsymbol{B} = \begin{pmatrix} b_{11} & b_{12} & b_{13} \\ b_{21} - b_{11} & b_{22} - b_{12} & b_{23} - b_{13} \\ b_{31} & b_{32} & b_{33} \end{pmatrix}.$$

Similarly, $\boldsymbol{Q}_2$ creates a new third row by subtracting the first row from the third leaving the other rows unchanged and $\boldsymbol{Q}_3$ again transforms the third row by subtracting three times the second row from the third. Consider these transformations applied in sequence to $\boldsymbol{A}$.

$$\boldsymbol{Q}_1\boldsymbol{A} = \begin{pmatrix} 1 & 0 & 0 \\ -1 & 1 & 0 \\ 0 & 0 & 1 \end{pmatrix}\begin{pmatrix} 1 & 2 & 3 \\ 1 & 3 & 5 \\ 1 & 5 & 12 \end{pmatrix} = \begin{pmatrix} 1 & 2 & 3 \\ 0 & 1 & 2 \\ 1 & 5 & 12 \end{pmatrix}$$

$$Q_2Q_1A = \begin{pmatrix} 1 & 0 & 0 \\ 0 & 1 & 0 \\ -1 & 0 & 1 \end{pmatrix} \begin{pmatrix} 1 & 2 & 3 \\ 0 & 1 & 2 \\ 1 & 5 & 12 \end{pmatrix} = \begin{pmatrix} 1 & 2 & 3 \\ 0 & 1 & 2 \\ 0 & 3 & 9 \end{pmatrix}$$

$$Q_3Q_2Q_1A = \begin{pmatrix} 1 & 0 & 0 \\ 0 & 1 & 0 \\ 0 & -3 & 1 \end{pmatrix} \begin{pmatrix} 1 & 2 & 3 \\ 0 & 1 & 2 \\ 0 & 3 & 9 \end{pmatrix} = \begin{pmatrix} 1 & 2 & 3 \\ 0 & 1 & 2 \\ 0 & 0 & 3 \end{pmatrix}.$$

This expression parallels the row echelon reduction in Equation 2.12, noting that

$$Q_3Q_2Q_1b = \begin{pmatrix} 1 \\ 1 \\ -1 \end{pmatrix},$$

where b is the 3×1 vector with entries $(1,2,3)$. Thus, $Q_3Q_2Q_1A = Q_3Q_2Q_1b$ yields Equation 2.13. ◊

Finally, notice that the previous example may be carried further to Gauss–Jordan elimination. Define:

$$Q_4 = \begin{pmatrix} 1 & -2 & 0 \\ 0 & 1 & 0 \\ 0 & 0 & 1 \end{pmatrix}, \quad Q_5 = \begin{pmatrix} 1 & 0 & \frac{1}{3} \\ 0 & 1 & 0 \\ 0 & 0 & 1 \end{pmatrix}, \quad Q_6 = \begin{pmatrix} 1 & 0 & 0 \\ 0 & 1 & -\frac{2}{3} \\ 0 & 0 & 1 \end{pmatrix}.$$

Then

$$Q_6Q_5Q_4Q_3Q_2Q_1A = I \Rightarrow Q_6Q_5Q_4Q_3Q_2Q_1Ax = Ix = Q_6Q_5Q_4Q_3Q_2Q_1b$$

and $Q^* \stackrel{\text{def}}{=} Q_6Q_5Q_4Q_3Q_2Q_1$ is equal to A^{-1}. This provides an alternate means of computing the inverse of a square matrix.

2.4 Linear dependence and independence

Consider m n-vectors, $a_1, a_2, \ldots, a_m$ (each vector has n entries or elements), thus, each vector is a list of n real numbers written $a_i \in \mathscr{R}^n = \{(x_1, x_2, \ldots, x_n) \mid x_i \in \mathscr{R}, i = 1, \ldots, n\}$, where $\mathscr{R}$ is the set of real numbers.

Definition 2.9: *The vectors $a_1, a_2, \ldots, a_m$ are said to be linearly dependent if there are numbers $\lambda_1, \lambda_2, \ldots, \lambda_m$, not all zero and*

$$\sum_{i=1}^{m} \lambda_i a_i = \lambda_1 a_1 + \cdots + \lambda_m a_m = 0.$$

Conversely, if the condition $\sum_{i=1}^{m} \lambda_i \boldsymbol{a}_i = 0$*, implies that* $\lambda_1 = \lambda_2 = \cdots = \lambda_m = 0$*, then* $\boldsymbol{a}_1, \boldsymbol{a}_2, \ldots, \boldsymbol{a}_m$ *are said to be linearly independent vectors.*

If $\boldsymbol{a}_1, \boldsymbol{a}_2, \ldots, \boldsymbol{a}_m$ are linearly dependent, then there is a collection of numbers, $\lambda_1, \ldots, \lambda_m$, with at least one not equal to zero and $\sum_{i=1}^{m} \lambda_i \boldsymbol{a}_i = \mathbf{0}$. Suppose that $\lambda_j \neq 0$ in the equation $\sum_{i=1}^{m} \lambda_i \boldsymbol{a}_i = \mathbf{0}$. Rearranging the equation $\lambda_1 \boldsymbol{a}_1 + \cdots + \lambda_j \boldsymbol{a}_j + \cdots + \lambda_m \boldsymbol{a}_m = \mathbf{0}$,

$$\lambda_j \boldsymbol{a}_j = -\lambda_1 \boldsymbol{a}_1 - \cdots - \lambda_{j-1} \boldsymbol{a}_{j-1} - \lambda_{j+1} \boldsymbol{a}_{j+1} - \cdots - \lambda_m \boldsymbol{a}_m = -\sum_{i \neq j}^{m} \lambda_i \boldsymbol{a}_i.$$

Since $\lambda_j \neq 0$, dividing by λ_j gives

$$\boldsymbol{a}_j = -\sum_{i \neq j}^{m} \frac{\lambda_i}{\lambda_j} \boldsymbol{a}_i = \sum_{i \neq j}^{m} \theta_i \boldsymbol{a}_i, \text{ where } \theta_i = -\frac{\lambda_i}{\lambda_j}.$$

Thus, when a set of vectors is linearly dependent, at least one vector in the set can be written as a linear combination of the others: for some j, $\boldsymbol{a}_j = \sum_{i \neq j}^{m} \theta_i \boldsymbol{a}_i$.

Theorem 2.1: *If* $\boldsymbol{a}_1, \ldots, \boldsymbol{a}_m$ *are linearly independent then any subset of* $\boldsymbol{a}_1, \ldots, \boldsymbol{a}_m$ *is linearly independent. Conversely, if a subset of the vectors* $\boldsymbol{a}_1, \ldots, \boldsymbol{a}_m$ *is linearly dependent, the entire set cannot be linearly independent.*

This is easy to see from the definition of linear independence. For example, suppose that $\boldsymbol{a}_1$, $\boldsymbol{a}_2$ and $\boldsymbol{a}_3$ are linearly independent but that $\boldsymbol{a}_1$ and $\boldsymbol{a}_2$ are not. Then there are two numbers, λ_1, λ_2, with at least one not 0, such that $\lambda_1 \boldsymbol{a}_1 + \lambda_2 \boldsymbol{a}_2 = \mathbf{0}$, and in that case $\lambda_1 \boldsymbol{a}_1 + \lambda_2 \boldsymbol{a}_2 + 0 \boldsymbol{a}_3 = \mathbf{0}$, contradicting the assumption that $\boldsymbol{a}_1$, $\boldsymbol{a}_2$ and $\boldsymbol{a}_3$ are linearly independent.

EXAMPLE 2.15 (LINEAR INDEPENDENCE): To illustrate the definition, consider the vectors:

$$\boldsymbol{a}_1 = \begin{pmatrix} 1 \\ 0 \\ 0 \end{pmatrix}, \quad \boldsymbol{a}_2 = \begin{pmatrix} 0 \\ 1 \\ 0 \end{pmatrix}, \quad \boldsymbol{a}_3 = \begin{pmatrix} 0 \\ 0 \\ 1 \end{pmatrix}.$$

Recall that given a number λ and vector $\boldsymbol{a}$, $\lambda \boldsymbol{a}$ is the vector with each element of $\boldsymbol{a}$ multiplied by λ. So,

$$\lambda_1 \boldsymbol{a}_1 + \lambda_2 \boldsymbol{a}_2 + \lambda_3 \boldsymbol{a}_3 = \begin{pmatrix} \lambda_1 \\ 0 \\ 0 \end{pmatrix} + \begin{pmatrix} 0 \\ \lambda_2 \\ 0 \end{pmatrix} + \begin{pmatrix} 0 \\ 0 \\ \lambda_3 \end{pmatrix} = \begin{pmatrix} \lambda_1 \\ \lambda_2 \\ \lambda_3 \end{pmatrix}.$$

This vector can equal **0** only if $\lambda_1 = \lambda_2 = \lambda_3 = 0$. Therefore, the three vectors are linearly independent.

Or consider the vectors:

$$\boldsymbol{a}_1 = \begin{pmatrix} 4 \\ 0 \\ 1 \end{pmatrix} \text{ and } \boldsymbol{a}_2 \begin{pmatrix} 0 \\ 5 \\ 2 \end{pmatrix}. \text{ If } \lambda_1 \boldsymbol{a}_1 + \lambda_2 \boldsymbol{a}_2 = \begin{pmatrix} \lambda_1 4 \\ \lambda_2 5 \\ \lambda_1 + 2\lambda_2 \end{pmatrix} = \begin{pmatrix} 0 \\ 0 \\ 0 \end{pmatrix} \text{ then } \lambda_1 = \lambda_2 = 0.$$

So, $\lambda_1 \boldsymbol{a}_1 + \lambda_2 \boldsymbol{a}_2 = \mathbf{0}$ forces $\lambda_1 = \lambda_2 = 0$, and therefore the two vectors are linearly independent. ◊

EXAMPLE 2.16 (LINEAR DEPENDENCE): To illustrate linearly dependent vectors consider the following:

$$\text{Let } \boldsymbol{a}_1 = \begin{pmatrix} 1 \\ 3 \\ 2 \end{pmatrix}, \boldsymbol{a}_2 = \begin{pmatrix} 2 \\ 4 \\ 6 \end{pmatrix}, \boldsymbol{a}_3 = \begin{pmatrix} 4 \\ 9 \\ 11 \end{pmatrix}, \text{ then } 2\boldsymbol{a}_1 + 3\boldsymbol{a}_2 - 2\boldsymbol{a}_3 = \begin{pmatrix} 0 \\ 0 \\ 0 \end{pmatrix}.$$

So, with $\lambda_1 = 2$, $\lambda_2 = 3$ and $\lambda_3 = -2$, $\lambda_1 \boldsymbol{a}_1 + \lambda_2 \boldsymbol{a}_2 + \lambda_3 \boldsymbol{a}_3 = \mathbf{0}$, and the vectors are linearly dependent.

Or consider the vectors:

$$\boldsymbol{a}_1 = \begin{pmatrix} 2 \\ 4 \end{pmatrix}, \boldsymbol{a}_2 = \begin{pmatrix} 3 \\ 1 \end{pmatrix}, \boldsymbol{a}_3 = \begin{pmatrix} 4 \\ 3 \end{pmatrix}, \text{ then } \boldsymbol{a}_1 + 2\boldsymbol{a}_2 - 2\boldsymbol{a}_3 = \begin{pmatrix} 0 \\ 0 \end{pmatrix},$$

and so the three vectors are linearly dependent. ◊

Definition 2.10: *A set of vectors, $\{v_1, v_2, \ldots, v_m\}$ with $v_i \in \mathcal{R}^n$, is said to span $\mathcal{R}^n$ if every vector in $\mathcal{R}^n$ can be written as a linear combination of elements in this set: given a vector x, there are numbers $\alpha_1, \alpha_2, \ldots, \alpha_m$ such that $\sum_{i=1}^m \alpha_i v_i = x$.*

EXAMPLE 2.17: Let $\boldsymbol{x}$ be an arbitrary vector (x_1 and x_2 arbitrary), and $\boldsymbol{e}_1$, $\boldsymbol{e}_2$ and $\boldsymbol{e}^*$ as given:

$$\boldsymbol{x} = \begin{pmatrix} x_1 \\ x_2 \end{pmatrix}, \boldsymbol{e}_1 = \begin{pmatrix} 1 \\ 0 \end{pmatrix}, \boldsymbol{e}_2 = \begin{pmatrix} 0 \\ 1 \end{pmatrix}, \boldsymbol{e}^* = \begin{pmatrix} \frac{1}{2} \\ \frac{1}{2} \end{pmatrix}.$$

The three vectors $\boldsymbol{e}_1$, $\boldsymbol{e}_2$ and $\boldsymbol{e}^*$ span $\mathcal{R}^2$, because, for any $\boldsymbol{x} \in \mathcal{R}^2$, $\boldsymbol{x} = x_1 \boldsymbol{e}_1 + x_2 \boldsymbol{e}_2$. However, they are not linearly independent since $\boldsymbol{e}^* = \frac{1}{2}\boldsymbol{e}_1 + \frac{1}{2}\boldsymbol{e}_2$. ◊

Definition 2.11: *A basis for $\mathcal{R}^n$ is a linearly independent set of vectors that span the space, $\mathcal{R}^n$.*

EXAMPLE 2.18: The two vectors $\boldsymbol{e}_1$ and $\boldsymbol{e}_2$ span $\mathcal{R}^2$ since for any $\boldsymbol{x} \in \mathcal{R}^2$, $\boldsymbol{x} = x_1\boldsymbol{e}_1 + x_2\boldsymbol{e}_2$. In addition, $\lambda_1\boldsymbol{e}_1 + \lambda_2\boldsymbol{e}_2 = \boldsymbol{0}$ if and only if $\lambda_1 = \lambda_2 = 0$, and so $\boldsymbol{e}_1$ and $\boldsymbol{e}_2$ are linearly independent. These two observations imply that $\boldsymbol{e}_1$ and $\boldsymbol{e}_2$ are a basis for $\mathcal{R}^2$. ◇

Figure 2.1 illustrates combinations of basis vectors. Theorem 2.2 gives one useful fact concerning a basis.

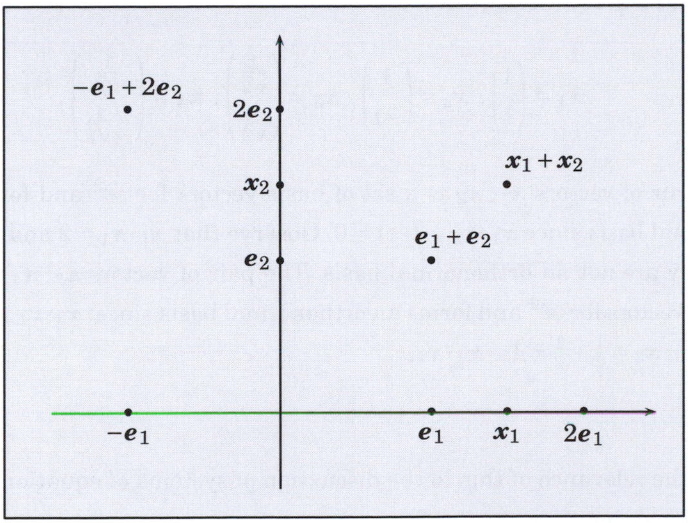

Figure 2.1: Combinations of basis vectors

Theorem 2.2: *Every basis for $\mathcal{R}^n$ contains exactly n vectors.*

Consequently, if any collection of n vectors are linearly independent, then they form a basis for $\mathcal{R}^n$. However, many different collections of vectors can form a basis for $\mathcal{R}^n$, as the next example illustrates.

EXAMPLE 2.19: The pair $\boldsymbol{e}_1$, $\boldsymbol{e}_2$ form a basis for $\mathcal{R}^2$, as does the pair $\boldsymbol{a}_1$, $\boldsymbol{a}_2$, where:

$$\boldsymbol{e}_1 = \begin{pmatrix}1\\0\end{pmatrix},\ \boldsymbol{e}_2 = \begin{pmatrix}0\\1\end{pmatrix},\ \boldsymbol{a}_1 = \begin{pmatrix}2\\1\end{pmatrix},\ \boldsymbol{a}_2 = \begin{pmatrix}1\\3\end{pmatrix}.$$ ◇

Theorem 2.3: *The representation of any vector in terms of a set of basis vectors is unique. If $\boldsymbol{y} \in \mathscr{R}^n$ and the collection of vectors $\boldsymbol{x}_1,\ldots,\boldsymbol{x}_n$ is a basis for $\mathscr{R}^n$ then:*

$$\boldsymbol{y} = \sum_{i=1}^{n} \lambda_i \boldsymbol{x}_i \text{ and } \boldsymbol{y} = \sum_{i=1}^{n} \gamma_i \boldsymbol{x}_i \text{ imply that } \lambda_i = \gamma_i, \text{ for } i = 1,2,\ldots n.$$

This is easy to see because $\mathbf{0} = \boldsymbol{y} - \boldsymbol{y} = \sum_{i=1}^{n}(\lambda_i - \gamma_i)\boldsymbol{x}_i$ and since $\boldsymbol{x}_1,\boldsymbol{x}_2,\ldots,\boldsymbol{x}_n$ are linearly independent, it must be that $\lambda_i = \gamma_i$ for each i.

Definition 2.12: *A set of basis vectors for $\mathscr{R}^n, \boldsymbol{x}_1,\ldots,\boldsymbol{x}_n$ is called an orthogonal basis if $\boldsymbol{x}_i \cdot \boldsymbol{x}_j = 0$ when $i \neq j$. An orthogonal basis is orthonormal if $\boldsymbol{x}_i \cdot \boldsymbol{x}_i = 1, \forall i$.*

EXAMPLE 2.20: In $\mathscr{R}^2$, let:

$$\boldsymbol{x}_1 = \begin{pmatrix} 1 \\ 1 \end{pmatrix}, \ \boldsymbol{x}_2 = \begin{pmatrix} 1 \\ -1 \end{pmatrix}, \ \boldsymbol{x}_3 = \begin{pmatrix} \frac{1}{\sqrt{2}} \\ \frac{1}{\sqrt{2}} \end{pmatrix}, \ \boldsymbol{x}_4 = \begin{pmatrix} \frac{1}{\sqrt{2}} \\ -\frac{1}{\sqrt{2}} \end{pmatrix}.$$

The pair of vectors $\boldsymbol{x}_1$, $\boldsymbol{x}_2$ is a set of basis vectors for $\mathscr{R}^2$ and forms an orthogonal basis since $\boldsymbol{x}_1 \cdot \boldsymbol{x}_2 = 1 - 1 = 0$. Observe that $\boldsymbol{x}_1 \cdot \boldsymbol{x}_1 = 2$ and $\boldsymbol{x}_2 \cdot \boldsymbol{x}_2 = 2$, so they are not an orthonormal basis. The pair of vectors $\boldsymbol{x}_3$, $\boldsymbol{x}_4$ is a set of basis vectors for $\mathscr{R}^2$ and forms an orthonormal basis since: $\boldsymbol{x}_3 \cdot \boldsymbol{x}_4 = \frac{1}{2} - \frac{1}{2} = 0$, and $\boldsymbol{x}_3 \cdot \boldsymbol{x}_3 = \frac{1}{2} + \frac{1}{2} = 1 = \boldsymbol{x}_4 \cdot \boldsymbol{x}_4$. ◊

To see the relevance of this to the discussion of systems of equations, consider the equation system:

$$\begin{pmatrix} a_{11} & a_{12} \\ a_{21} & a_{22} \end{pmatrix} \begin{pmatrix} x_1 \\ x_2 \end{pmatrix} = \begin{pmatrix} b_1 \\ b_2 \end{pmatrix}.$$

In view of the earlier rules for matrix manipulation, this can be rewritten as:

$$\begin{pmatrix} a_{11} \\ a_{21} \end{pmatrix} x_1 + \begin{pmatrix} a_{12} \\ a_{22} \end{pmatrix} x_2 = \begin{pmatrix} b_1 \\ b_2 \end{pmatrix}.$$

So finding a solution involves finding an x_1, x_2 pair that combines the columns of the $\boldsymbol{A}$ matrix to make the vector $\boldsymbol{b}$. If the columns of A are linearly independent then, in view of Theorem 2.2, they form a basis for $\mathscr{R}^2$, and in view of Theorem 2.3, there is a unique pair x_1, x_2 satisfying the equation. More generally, given a square $n \times n$ matrix $\boldsymbol{A}$ and an arbitrary vector $\boldsymbol{b} \in \mathscr{R}^n$, if the columns of $\boldsymbol{A}$ are linearly independent, then there is a unique vector $\boldsymbol{x} \in \mathscr{R}^n$ satisfying $\boldsymbol{A}\boldsymbol{x} = \boldsymbol{b}$. So, linear independence in the columns of $\boldsymbol{A}$ is central to solving the equation system $\boldsymbol{Ax} = \boldsymbol{b}$. The "extent" of linear independence in the columns of a matrix $\boldsymbol{A}$ is measured by the rank of a matrix.

2.4.1 Matrix rank and linear independence

The *column rank* of a matrix is the maximum number of linearly independent columns. The *row rank* of a matrix is the maximum number of linearly independent rows. There are a number of useful facts to know about the rank of a matrix. First, in view of the following result, the *rank* of a matrix is unambiguously defined.

Theorem 2.4: *For any matrix, the row rank equals the column rank.*

So, the term "rank of a matrix" is unambiguous.

EXAMPLE 2.21: Let

$$\boldsymbol{A} = \begin{pmatrix} 1 & 1 \\ 1 & 0 \end{pmatrix}, \quad \text{with columns} \quad a_1 = \begin{pmatrix} 1 \\ 1 \end{pmatrix}, \; a_2 = \begin{pmatrix} 1 \\ 0 \end{pmatrix}.$$

The rank of $\boldsymbol{A} = 2$ since a_1 and a_2 are linearly independent. ◊

EXAMPLE 2.22: Consider:

$$\boldsymbol{B} = \begin{pmatrix} 1 & 4 & 2 \\ 2 & 5 & 1 \\ 3 & 6 & 0 \end{pmatrix} \quad \text{with columns} \quad \boldsymbol{b}_1 = \begin{pmatrix} 1 \\ 2 \\ 3 \end{pmatrix}, \; \boldsymbol{b}_2 = \begin{pmatrix} 4 \\ 5 \\ 6 \end{pmatrix}, \; \boldsymbol{b}_3 = \begin{pmatrix} 2 \\ 1 \\ 0 \end{pmatrix}.$$

The three columns are not linearly independent, since $2\boldsymbol{b}_1 + \boldsymbol{b}_3 = \boldsymbol{b}_2$. But one can find two columns that are independent. Consider the second and third columns and the expression $\lambda_1 \boldsymbol{b}_2 + \lambda_2 \boldsymbol{b}_3 = \boldsymbol{0}$.

$$\lambda_1 \boldsymbol{b}_2 + \lambda_2 \boldsymbol{b}_3 = \lambda_1 \begin{pmatrix} 4 \\ 5 \\ 6 \end{pmatrix} + \lambda_2 \begin{pmatrix} 2 \\ 1 \\ 0 \end{pmatrix} = \begin{pmatrix} \lambda_1 4 + \lambda_2 2 \\ \lambda_1 5 + \lambda_2 1 \\ \lambda_1 6 + \lambda_2 0 \end{pmatrix} = \begin{pmatrix} 0 \\ 0 \\ 0 \end{pmatrix}.$$

The third equation implies that $\lambda_1 6 = 0$ or $\lambda_1 = 0$. Then, the second equation implies that $1\lambda_2 = 0$, so that $\lambda_2 = 0$. Therefore $\boldsymbol{b}_2$ and $\boldsymbol{b}_3$ are linearly independent, so the rank of $\boldsymbol{B}$ is 2. ◊

The fact that the row and column rank are equal can sometimes be useful in determining the rank by observation. Consider the matrix $\boldsymbol{A}$:

$$\boldsymbol{A} = \begin{pmatrix} 1 & 2 & 3 \\ 7 & 3 & 5 \\ 2 & 4 & 6 \end{pmatrix}.$$

Looking at the columns, there is nothing immediately striking, but observe directly that the third row is twice the first row – so the rank is less than three. Denote the rank of a matrix $\boldsymbol{A}$ by $r(\boldsymbol{A})$. The rank of a matrix can be no larger than the number of rows or the number of columns of the matrix.

Theorem 2.5: *Let* $\mathbf{A}$ *be an* $m \times n$ *matrix. Then* $r(\mathbf{A}) \le \min\{m,n\}$*, where* m *is the number of rows, and* n *is the number of columns.*

A square $n \times n$ matrix $\boldsymbol{A}$ is said to be *non-singular* (or "of full rank") if $r(\boldsymbol{A}) = n$ and *singular* if $r(\boldsymbol{A}) < n$. Given an $n \times n$ nonsingular matrix $\boldsymbol{A}$, and vector $\boldsymbol{b} \in \mathscr{R}^n$, the equation system $\boldsymbol{Ax} = \boldsymbol{b}$ has a solution $\boldsymbol{x}^*$. The following fact is important.

Theorem 2.6: *A square* $n \times n$ *matrix* $\mathbf{A}$ *is non-singular if and only if* $|\mathbf{A}| \neq 0$.

Theorem 2.7: *Let* $\mathbf{A}$ *be an* $m \times p$ *matrix,* $\mathbf{B}$ *a* $p \times n$ *matrix, and let* $\mathbf{C} = \boldsymbol{AB}$.

1. *Then* $r(\mathbf{C}) \le \min\{(r(\boldsymbol{A}), r(\boldsymbol{B})\}$.
2. *If* $\boldsymbol{A}$ *is a square non-singular matrix* $r(\mathbf{C}) = r(\boldsymbol{B})$.
3. *If* $\boldsymbol{B}$ *is a square non-singular matrix* $r(\mathbf{C}) = r(\mathbf{A})$.
4. *If* $\boldsymbol{A}$ *is a diagonal matrix* ($a_{ij} = 0$ *if* $i \neq j$) *then* $r(\mathbf{A}) = n$ *if* $a_{ii} \neq 0, \forall i$.

To illustrate the use of Theorem 2.7, consider point 1. Recall an earlier example where:

$$\boldsymbol{A} = \begin{pmatrix} 1 & 3 & 5 \\ 2 & 4 & 6 \end{pmatrix}, \ \boldsymbol{B} = \begin{pmatrix} 2 & 1 \\ 5 & 7 \\ 3 & 1 \end{pmatrix}, \text{ so that } \boldsymbol{C} = \boldsymbol{BA} = \begin{pmatrix} 4 & 10 & 16 \\ 19 & 43 & 67 \\ 5 & 13 & 21 \end{pmatrix}.$$

It is not obvious from inspection that the matrix $\boldsymbol{C}$ has rank less than 3. However, the rank of both $\boldsymbol{A}$ and $\boldsymbol{B}$ can be at most 2, since $\boldsymbol{A}$ has only two rows and $\boldsymbol{B}$ has only two columns, so the rank of $\boldsymbol{C}$ can be 2 at most.

REMARK 2.4: Let $\{\boldsymbol{a}_1, \boldsymbol{a}_2, \dots, \boldsymbol{a}_m\}$ be m vectors with $\boldsymbol{a}_i$ an $n \times 1$ vector. Suppose that the maximum number of linearly independent vectors is p: in this list we can find p linearly independent vectors, but no collection of more than p vectors is linearly independent. Suppose that $\{\boldsymbol{a}_1,\ldots,\boldsymbol{a}_p\}$ are linearly independent and consider any $\boldsymbol{a}_{p+j}$, where $j \geq 1$. Then $\{\boldsymbol{a}_1,\boldsymbol{a}_2,\ldots,\boldsymbol{a}_p,\boldsymbol{a}_{p+j}\}$ are linearly dependent so that there is a collection of scalars $\lambda_1,\ldots,\lambda_p,\lambda_{p+j}$, not all 0 with

$$\lambda_1\boldsymbol{a}_1+\lambda_2\boldsymbol{a}_2+\cdots+\lambda_p\boldsymbol{a}_p+\lambda_{p+j}\boldsymbol{a}_{p+j}=0.$$

Note that $\lambda_{p+j}\neq 0$, since if it were, then $\sum_{i=1}^{p}\lambda_i\boldsymbol{a}_i=0$, with some $\lambda_i\neq 0$ for $i\in\{1,\ldots,p\}$, and this would imply that $\{\boldsymbol{a}_1,\boldsymbol{a}_2,\ldots,\boldsymbol{a}_p\}$ are linearly dependent – a contradiction. So because $\lambda_{p+j}\neq 0$, the equation may be rearranged to give

$$\begin{aligned}\boldsymbol{a}_{p+j}&=-\frac{\lambda_1}{\lambda_{p+j}}\boldsymbol{a}_1-\frac{\lambda_2}{\lambda_{p+j}}\boldsymbol{a}_2-\cdots-\frac{\lambda_p}{\lambda_{p+j}}\boldsymbol{a}_p\\&=\sum_{i=1}^{p}\alpha_i\boldsymbol{a}_i,\quad \alpha_i=-\frac{\lambda_i}{\lambda_{p+j}}.\end{aligned}$$

Therefore, given any collection of vectors and a maximal (sub)set of linearly independent vectors, any vector not in the maximal set may be written as a linear combination of vectors in the maximal subset.

One useful implication of this observation is the following. Let $\boldsymbol{A}$ be a matrix whose columns are $\{\boldsymbol{a}_1,\boldsymbol{a}_2,\ldots,\boldsymbol{a}_m\}$. If the columns are linearly dependent, let $\boldsymbol{A}^*$ be a matrix whose columns consist of any maximal collection of linearly independent columns in $\boldsymbol{A}$ (any maximal collection of linearly independent vectors from $\{\boldsymbol{a}_1,\boldsymbol{a}_2,\ldots,\boldsymbol{a}_m\}$). Then the equation system $\boldsymbol{A}\boldsymbol{x}=\boldsymbol{b}$ has a solution if and only if the system $\boldsymbol{A}^*\boldsymbol{z}=\boldsymbol{b}$ has a solution. □

2.5 Solutions of systems of equations

2.5.1 Solving systems of equations: a geometric view

Consider the equations discussed earlier:

$$\begin{aligned}x+\ y&=5\\2x+3y&=12,\end{aligned}$$

or, in matrix notation:

$$\begin{pmatrix}1&1\\2&3\end{pmatrix}\begin{pmatrix}x\\y\end{pmatrix}=\begin{pmatrix}5\\12\end{pmatrix}.$$

This has the form $\boldsymbol{Az} = \boldsymbol{b}$ where

$$\boldsymbol{A} = \begin{pmatrix} 1 & 1 \\ 2 & 3 \end{pmatrix}, \quad \boldsymbol{z} = \begin{pmatrix} x \\ y \end{pmatrix}, \quad \boldsymbol{b} = \begin{pmatrix} 5 \\ 12 \end{pmatrix}.$$

Note that in the general form: $\boldsymbol{Az} = \boldsymbol{b}$, the matrix $\boldsymbol{A}$ may be of arbitrary size so that this formulation can incorporate large systems of equations. Write $\boldsymbol{a}_i$ for the ith column of A and write $\boldsymbol{A}_i$ for the ith row of $\boldsymbol{A}$. With this notation,

$$\boldsymbol{A} = \begin{pmatrix} \boldsymbol{A}_1 \\ \vdots \\ \boldsymbol{A}_n \end{pmatrix} = \begin{pmatrix} \boldsymbol{a}_1 & \boldsymbol{a}_2 & \cdots & \boldsymbol{a}_n \end{pmatrix}.$$

Then $\boldsymbol{Az} = \boldsymbol{b}$ may be expressed in two ways:

$$(1) \qquad \boldsymbol{Az} = \begin{pmatrix} \boldsymbol{A}_1 \cdot \boldsymbol{z} \\ \vdots \\ \boldsymbol{A}_n \cdot \boldsymbol{z} \end{pmatrix} = \boldsymbol{b} = \begin{pmatrix} b_1 \\ \vdots \\ b_n \end{pmatrix}$$

and

$$(2) \qquad \boldsymbol{Az} = \boldsymbol{a}_1 x_1 + \boldsymbol{a}_2 x_2 + \cdots + \boldsymbol{a}_n x_n = \boldsymbol{b}.$$

Considering (1) first, this says that the solution $\boldsymbol{z}$ satisfies $\boldsymbol{A}_i \cdot \boldsymbol{z} = b_i$, for $i = 1, \ldots, n$. So, there are n equations to be satisfied. For the example discussed above, $\boldsymbol{A}_1 = (1, 1)$, $\boldsymbol{A}_2 = (2, 3)$, $b_1 = 5$, and $b_2 = 12$, so that $\boldsymbol{A}_1 \cdot \boldsymbol{z} = b_1$ is $(1, 1) \cdot (x_1, x_2) = 5$ and $\boldsymbol{A}_2 \cdot \boldsymbol{z} = b_2$ is $(2, 3) \cdot (x_1, x_2) = 12$. These are the original equations $1x_1 + 1x_2 = 5$, $2x_1 + 3x_2 = 12$, and are plotted in Figure 2.2.

The solution to the two equations occurs at the point where both lines intersect – at (3,2). More generally, a set of points of the form $H_i = \{\boldsymbol{z} \mid \boldsymbol{A}_i \cdot \boldsymbol{z} = b_i\}$ is the set of all those $\boldsymbol{z}$ vectors satisfying the ith equation, and is called a hyperplane. So, solving the equation system amounts to finding a point $\boldsymbol{z}$ that is on every hyperplane.

Next, considering (2), the equation system $\boldsymbol{Az} = \boldsymbol{b}$ has the form: $\sum_{i=1}^{n} \boldsymbol{a}_i z_i = \boldsymbol{b}$, where $\boldsymbol{z} = (z_1, z_2, \ldots, z_n)$. In the example, this has the form:

$$\boldsymbol{a}_1 x_1 + \boldsymbol{a}_2 x_2 = \begin{pmatrix} 1 \\ 2 \end{pmatrix} x_1 + \begin{pmatrix} 1 \\ 3 \end{pmatrix} x_2 = \begin{pmatrix} 5 \\ 12 \end{pmatrix}.$$

Now, given a vector $\binom{\alpha}{\beta}$, interpret it as a direction: starting from the point (0,0) go horizontally a distance α, and then go vertically a distance β. Multiplying this vector by a number w scales the vector up or down proportionately: go horizontally

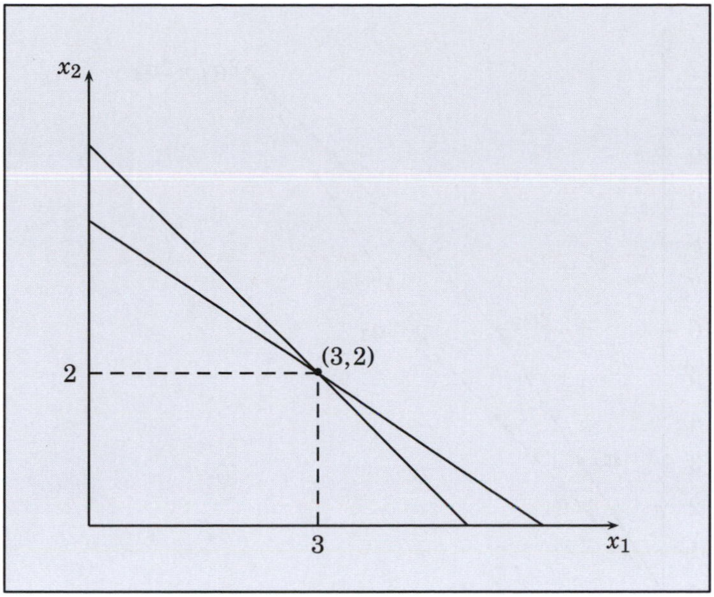

Figure 2.2: Intersecting lines: a unique solution

αw and vertically βw. For this pair of equations, earlier the solution was determined to be $x_1 = 3$ and $x_2 = 2$. So, with $x_1 = 3$, repeat the process "[go horizontally 1 step and vertically 2 steps]" three times; and with $x_2 = 2$, repeat the process "[go horizontally 1 step and vertically 3 steps]" two times. When these are combined, the point (5, 12) is reached – see Figure 2.3.

2.5.1.1 Systems with no solution

From the earlier discussion, in a square system ($\boldsymbol{A}$ an $n \times n$ matrix), $\boldsymbol{Ax} = \boldsymbol{b}$, if $\boldsymbol{A}$ is non-singular, there is a unique solution. So if the system is insoluble, it must be that $\boldsymbol{A}$ is singular – there is linear dependence between rows (or columns). But a singular matrix $\boldsymbol{A}$ does not guarantee that there is no solution. To see this, consider the following sets of equations:

$$(a) \begin{pmatrix} 1 & 1 \\ 2 & 2 \end{pmatrix} \begin{pmatrix} x \\ y \end{pmatrix} = \begin{pmatrix} 2 \\ 4 \end{pmatrix}, \quad (b) \begin{pmatrix} 1 & 1 \\ 2 & 2 \end{pmatrix} \begin{pmatrix} x \\ y \end{pmatrix} = \begin{pmatrix} 2 \\ 3 \end{pmatrix}.$$

Or, writing the equations in terms of columns of the coefficient matrix:

$$(a) \begin{pmatrix} \boldsymbol{a}_1 & \boldsymbol{a}_2 \end{pmatrix} \begin{pmatrix} x \\ y \end{pmatrix} = \boldsymbol{b}, \quad (b) \begin{pmatrix} \boldsymbol{a}_1 & \boldsymbol{a}_2 \end{pmatrix} \begin{pmatrix} x \\ y \end{pmatrix} = \boldsymbol{b}^*.$$

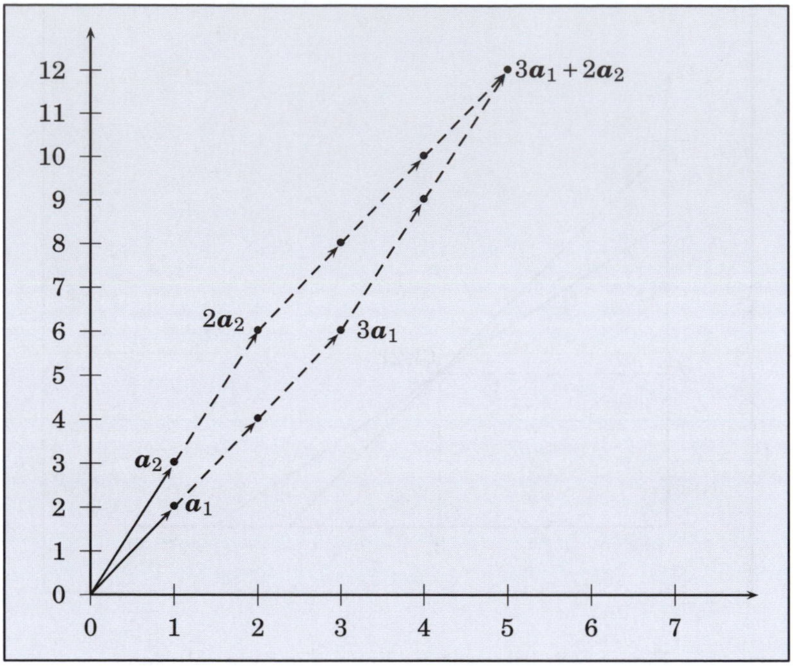

Figure 2.3: Combinations of basis vectors

In case (a), there are an infinite number of solutions. If $x+y=2$, then multiplying both sides by 2, $2\times(x+y)=2\times 2$ or $2x+2y=4$. So whenever the first equation is satisfied, the second equation is also satisfied. For example, $x=0$, $y=2$ or $x=1$, $y=1$ are solutions to the equations in (a). Figure 2.4 depicts the equations. In case (a) both equations give the same line in Figure 2.4(i) – connecting (0,2) to (2,0) so any point in the line satisfies both equations.

In case (b), there is no solution. If $x+y=2$, then multiplying both sides by 2, $2\times(x+y)=2\times 2$ or $2x+2y=4$. Since the second equation is $2x+2y=3$, whenever the first equation is satisfied, the second equation cannot be satisfied. In case (b) the first equation is unchanged and the second gives the line connecting $(\frac{3}{2},0)$ and $(0,\frac{3}{2})$ in Figure 2.4(i). Figure 2.4(ii) illustrates the situation in terms of combining columns of the coefficient matrix.

Finally, consider a slightly different perspective. Take the coefficient matrix and in both cases add an extra column, the vector $\boldsymbol{b}$ or $\boldsymbol{b}^*$. This gives:

$$(a)\begin{pmatrix}1 & 1 & 2\\ 2 & 2 & 4\end{pmatrix} \quad (b)\begin{pmatrix}1 & 1 & 2\\ 2 & 2 & 3\end{pmatrix}.$$

Notice that in case (a) the second row is twice the first, so the rank of the matrix is unchanged by the addition of the extra column. However, in case (b), the matrix

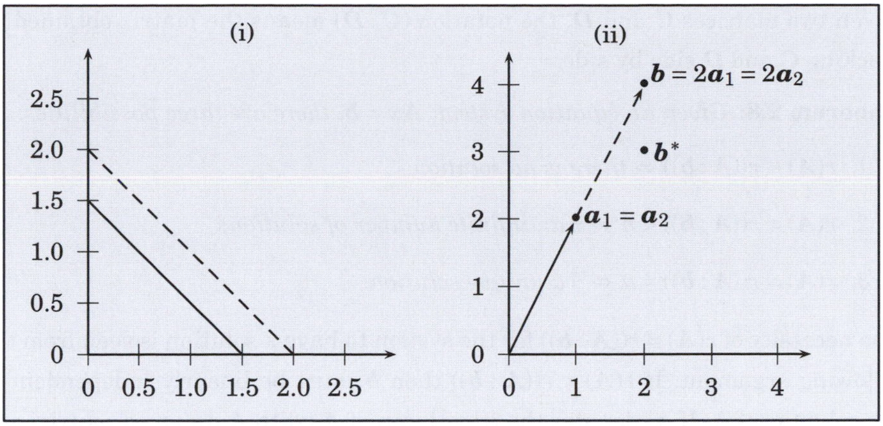

Figure 2.4: Overlapping or parallel lines: many or no solutions

rank is increased, from 1 to 2. This latter fact – an increase in rank – implies there is no solution in case (b), an issue taken up in the next section.

Finally note that in both cases (a) and (b), the coefficient matrix is singular and has no inverse. When the coefficient matrix of a square system is singular, there may be no solution, or there may be an infinite number of solutions. This issue is discussed next in Section 2.5.2.

2.5.2 Solutions of equation systems: rank and linear independence

Consider a system of equations: $\sum_{j=1}^{n} a_{ij}x_j = b_i$ for $i = 1,\dots,m$. This is a system of m equations in n unknowns, the x_j. In matrix terms, $\boldsymbol{Ax} = \boldsymbol{b}$, or written out more fully,

$$\begin{pmatrix} a_{11} & a_{12} & \cdots & a_{1n} \\ a_{21} & a_{22} & \cdots & a_{2n} \\ \vdots & \vdots & \ddots & \vdots \\ a_{m1} & a_{m2} & \cdots & a_{mn} \end{pmatrix} \begin{pmatrix} x_1 \\ x_2 \\ \vdots \\ x_n \end{pmatrix} = \begin{pmatrix} b_1 \\ b_2 \\ \vdots \\ b_m \end{pmatrix}.$$

Let $\boldsymbol{Ax} = \boldsymbol{b}$ be the system of equations, where the dimensions of the matrices and vectors are $\boldsymbol{A} - m \times n$, $\boldsymbol{x} - n \times 1$, $\boldsymbol{b} - m \times 1$, and where m may be greater, equal to or less than n. Does this system have a solution (some $\boldsymbol{x}^*$ such that $\boldsymbol{Ax}^* = \boldsymbol{b}$) and, if so, is the solution unique? A solution is unique if $\boldsymbol{Ax}^* = \boldsymbol{b}$ and $\boldsymbol{A\tilde{x}} = \boldsymbol{b}$ imply $\boldsymbol{x}^* = \boldsymbol{\tilde{x}}$. The following theorem answers this question in terms of ranks of matrices.

Given two matrices $\boldsymbol{C}$ and $\boldsymbol{D}$, the notation $(\boldsymbol{C}:\boldsymbol{D})$ means the matrix obtained by stacking $\boldsymbol{C}$ and $\boldsymbol{D}$ side by side.

Theorem 2.8: *Given an equation system, $\boldsymbol{Ax}=\boldsymbol{b}$, there are three possibilities:*

1. $r(\boldsymbol{A}) < r((\boldsymbol{A}:\boldsymbol{b})) \Leftrightarrow$ *there is no solution.*
2. $r(\boldsymbol{A}) = r((\boldsymbol{A}:\boldsymbol{b})) < n \Leftrightarrow \exists$ *an infinite number of solutions.*
3. $r(\boldsymbol{A}) = r((\boldsymbol{A}:\boldsymbol{b})) = n \Leftrightarrow \exists$ *a unique solution.*

The necessity of $r(\boldsymbol{A}) = r((\boldsymbol{A}:\boldsymbol{b}))$ for the system to have a solution is seen from the following argument. If $r(\boldsymbol{A}) < r((\boldsymbol{A}:\boldsymbol{b}))$ then $\boldsymbol{b}$ must be linearly independent of the columns of $\boldsymbol{A}$. If $\boldsymbol{a}_i$ denotes the ith column of $\boldsymbol{A}$ (with $\boldsymbol{A}$ an $m\times n$ matrix) and $\boldsymbol{x}=(x_1,x_2,\dots,x_n)'$, then the system $\boldsymbol{Ax}=\boldsymbol{b}$ can be written: $\sum_{i=1}^n \boldsymbol{a}_i x_i = \boldsymbol{b}$. For there to be a vector $\boldsymbol{x}$ satisfying $\boldsymbol{Ax}=\boldsymbol{b}$, $\boldsymbol{x}$ must satisfy $\sum_{i=1}^n \boldsymbol{a}_i x_i = \boldsymbol{b}$, which requires writing $\boldsymbol{b}$ as a linear combination of the columns of $\boldsymbol{A}$ and this is impossible when $\boldsymbol{b}$ is linearly independent of $\{\boldsymbol{a}_1,\boldsymbol{a}_2,\dots,\boldsymbol{a}_n\}$.

Next, consider the case where $r(\boldsymbol{A}) = r((\boldsymbol{A}:\boldsymbol{b}))$. Since $\boldsymbol{A} - m\times n$, the rank of $\boldsymbol{A}$ is no larger than m. If the rank of $\boldsymbol{A}$ (and hence $(\boldsymbol{A}:\boldsymbol{b})$) is less than m, then writing $(\boldsymbol{A}:\boldsymbol{b})_j$ for the jth row of $(\boldsymbol{A}:\boldsymbol{b})$, some row of $(\boldsymbol{A}:\boldsymbol{b})$ may be written as a linear combination of the other rows: $(\boldsymbol{A}:\boldsymbol{b})_k = \sum_{j\neq k}\theta_j\cdot(\boldsymbol{A}:\boldsymbol{b})_j$. With $\boldsymbol{z}'=(\boldsymbol{x},-1)$, x satisfies the jth equation if and only if $(\boldsymbol{A}:\boldsymbol{b})_j\cdot\boldsymbol{z}=0$. Thus, when this holds for $j\neq k$, then it also holds for k since $(\boldsymbol{A}:\boldsymbol{b})_k\cdot\boldsymbol{z} = \sum_{j\neq k}\theta_j\cdot(\boldsymbol{A}:\boldsymbol{b})_j\cdot\boldsymbol{z}=0$. Therefore, assume that $\boldsymbol{A}$ has rank m. Since $r(\boldsymbol{A})\le\min\{m,n\}$, it must be that $n\ge m$.

Consider the two possible cases. If $m=n$, then $\boldsymbol{A}$ is a square matrix of full rank and x is the unique solution given by $x=\boldsymbol{A}^{-1}\boldsymbol{b}$. If $m<n$, then write $\boldsymbol{A}=(\boldsymbol{A}_1:\boldsymbol{A}_2)$ where $\boldsymbol{A}_1$ is a square matrix of full rank m (if necessary, rearrange the equation system $\boldsymbol{A}x=\boldsymbol{b}$ so the first m columns of $\boldsymbol{A}$ are linearly independent) and the equation system may be written $\boldsymbol{A}x=\boldsymbol{b}=(\boldsymbol{A}_1:\boldsymbol{A}_2)\begin{pmatrix}x_1\\x_2\end{pmatrix}=\boldsymbol{b}$. Let x_2 be arbitrary and let $x_1=\boldsymbol{A}_1^{-1}(\boldsymbol{b}-\boldsymbol{A}_2x_2)$. Since x_2 is arbitrary, there are an infinite number of solutions.

EXAMPLE 2.23: The following examples illustrate the three cases.

1. $r(\boldsymbol{A}) < r((\boldsymbol{A}:\boldsymbol{b}))$.

$$\boldsymbol{A}=\begin{pmatrix}1&1\\2&2\end{pmatrix},\ \boldsymbol{b}=\begin{pmatrix}5\\12\end{pmatrix},\ (\boldsymbol{A}:\boldsymbol{b})=\begin{pmatrix}1&1&5\\2&2&12\end{pmatrix}.$$

Since $(\boldsymbol{A}:\boldsymbol{b})$ has rank 2 (the second and third column are linearly independent), this implies that $r(\boldsymbol{A}) < r((\boldsymbol{A}:\boldsymbol{b}))$ since $r(\boldsymbol{A})=1$. Note that any $\boldsymbol{x} = (y^*, z^*)'$ satisfying the first equation satisfies $y^* + z^* = 5$. To satisfy the second equation $2(y^* + z^*)$ must equal 12 and this is impossible. Hence, there is no solution.

2. $r(\boldsymbol{A}) = r((\boldsymbol{A} : \boldsymbol{b})) < n.$

$$\boldsymbol{A} = \begin{pmatrix} 1 & 1 \\ 2 & 2 \end{pmatrix}, \boldsymbol{b} = \begin{pmatrix} 5 \\ 10 \end{pmatrix}, (\boldsymbol{A} : \boldsymbol{b}) = \begin{pmatrix} 1 & 1 & 5 \\ 2 & 2 & 10 \end{pmatrix}.$$

Thus $r(\boldsymbol{A}) = r((\boldsymbol{A} : \boldsymbol{b})) = 1 < 2 = n$. Note that any $\boldsymbol{x} = (y, z)'$ such that $y + z = 5$ satisfies both equations and so there are an infinite number of solutions.

3. $r(\boldsymbol{A}) = r((\boldsymbol{A} : \boldsymbol{b})) = n.$

$$\boldsymbol{A} = \begin{pmatrix} 1 & 1 \\ 2 & 3 \end{pmatrix}, \boldsymbol{b} = \begin{pmatrix} 5 \\ 12 \end{pmatrix} \Rightarrow x^* = \begin{pmatrix} x_1^* \\ x_2^* \end{pmatrix} = \begin{pmatrix} 3 \\ 2 \end{pmatrix}.$$ ◊

Some more examples illustrate the theorem.

EXAMPLE 2.24: $3x_1 + 4x_2 = 7$ and $2.25x_1 + 3x_2 = 5.25$. Thus:

$$\boldsymbol{A} = \begin{pmatrix} 3 & 4 \\ 2.25 & 3 \end{pmatrix}, (\boldsymbol{A} : \boldsymbol{b}) = \begin{pmatrix} 3 & 4 & 7 \\ 2.25 & 3 & 5.25 \end{pmatrix}.$$

In this case, $r(\boldsymbol{A}) = r(\boldsymbol{A} : \boldsymbol{b}) = 1 < 2$ so there is an infinite number of solutions.

Suppose the coefficient matrix remains unchanged but the right-hand side is modified, giving $3x_1 + 4x_2 = 7$ and $2.25x_1 + 3x_2 = 1$. Thus

$$\boldsymbol{A} = \begin{pmatrix} 3 & 4 \\ 2.25 & 3 \end{pmatrix}, (\boldsymbol{A} : \boldsymbol{b}) = \begin{pmatrix} 3 & 4 & 7 \\ 2.25 & 3 & 1 \end{pmatrix}.$$

Now, $r(\boldsymbol{A}) = 1, r((\boldsymbol{A} : \boldsymbol{b})) = 2 \Rightarrow$ so there is no solution. ◊

The next example considers a three-equation system.

EXAMPLE 2.25: $3x_1 + 2x_2 + x_3 = 7$, $x_1 + \frac{1}{2}x_2 - x_3 = 4$ and $x_1 + \frac{3}{4}x_2 + x_3 = 5$.

$$\boldsymbol{A} = \begin{pmatrix} 3 & 2 & 1 \\ 1 & \frac{1}{2} & -1 \\ 1 & \frac{3}{4} & 1 \end{pmatrix}, (\boldsymbol{A} : \boldsymbol{b}) = \begin{pmatrix} 3 & 2 & 1 & 7 \\ 1 & \frac{1}{2} & -1 & 4 \\ 1 & \frac{3}{4} & 1 & 5 \end{pmatrix}.$$

Here $r(\mathbf{A}) = 2$ and $r((\mathbf{A} : \boldsymbol{b})) = 3$ (check that $|\mathbf{A}| = 0$). Thus, there is no solution. Note: (equation # 2) + 2(equation # 3) $\Rightarrow 3x_1 + 2x_2 + x_3 = 14$ – inconsistent with (equation # 1). ◊

2.6 Special matrices

The following discussion introduces symmetric, triangular, idempotent and diagonal matrices, along with the trace operation on square matrices.

Definition 2.13: *A square matrix* $\mathbf{A}$ *is symmetric if it equals its transpose:* $\mathbf{A} = \mathbf{A}'$. *For example:*

$$\mathbf{A} = \begin{pmatrix} 1 & 7 & 5 \\ 7 & 2 & 6 \\ 5 & 6 & 8 \end{pmatrix}.$$

Definition 2.14: *A square matrix is a triangular matrix if either all of the entries above the diagonal are 0, or all the entries below the diagonal are 0.*

A triangular matrix may be an upper or lower triangular matrix.

Definition 2.15: *A square matrix* $\mathbf{A}$ *is a lower triangular matrix if* $a_{ij} = 0$ *for all* (i, j) *pairs with* $i < j$. *Thus, a lower triangular matrix has the form:*

$$\mathbf{A} = \begin{pmatrix} a_{11} & 0 & 0 & \cdots & 0 \\ a_{21} & a_{22} & 0 & \cdots & 0 \\ \vdots & \vdots & \ddots & \vdots & 0 \\ a_{n1} & a_{n2} & & & a_{nn} \end{pmatrix}.$$

Definition 2.16: *A square matrix* $\mathbf{A}$ *is an upper triangular matrix if* $a_{ij} = 0$ *for all* (i, j) *pairs with* $i > j$. *Thus, an upper triangular matrix has the form:*

$$\mathbf{A} = \begin{pmatrix} a_{11} & a_{21} & \cdots & \cdots & a_{n1} \\ 0 & a_{22} & \cdots & \cdots & a_{n2} \\ \vdots & \vdots & \ddots & \vdots & \vdots \\ 0 & 0 & & 0 & a_{nn} \end{pmatrix}.$$

From the definition, the determinant of an upper or lower triangular matrix is given by $|\mathbf{A}| = \prod_{i=1}^{n} a_{ii} = a_{11} \times \cdots \times a_{nn}$. Therefore, a triangular matrix is non-singular if and only if all diagonal entries are non-zero.

Theorem 2.9: *If a lower (upper) triangular matrix is invertible, its inverse is lower (upper) triangular.*

PROOF: Consider a lower triangular matrix $\boldsymbol{L}$ and an entry l_{ij} with $i > j$. The minor matrix associated with this entry is obtained by deleting the ith row and jth column of $\boldsymbol{L}$. Deleting the jth column deletes l_{jj} on the diagonal so the minor matrix has at entry (j,j) the element of the $\boldsymbol{L}$ matrix in position $(j,j+1)$, which is 0. Therefore, the minor matrix is a lower triangular matrix with a 0 on the diagonal and hence has determinant $m_{ij} = 0$. Thus the corresponding adjoint matrix A^d has $a^d_{ji} = 0$ with $i < j$. The entries above the main diagonal in the adjoint matrix are 0 and so the inverse is lower triangular. A similar calculation applies for an upper triangular matrix. ∎

Definition 2.17: *A square matrix* $\boldsymbol{A}$ *admits a* $\boldsymbol{LU}$ *(lower-upper) decomposition if* $\boldsymbol{A}$ *may be written:* $\boldsymbol{A} = \boldsymbol{LU}$*, where* $\boldsymbol{L}$ *is a lower triangular and* $\boldsymbol{U}$ *an upper triangular matrix.*

Not all square matrices $\boldsymbol{A}$ admit a $\boldsymbol{LU}$ decomposition.

EXAMPLE 2.26: Let $\boldsymbol{A}$ be a 2×2 matrix that admits a $\boldsymbol{LU}$ decomposition, so that $\boldsymbol{A} = \boldsymbol{LU}$:

$$\boldsymbol{A} = \begin{pmatrix} a_{11} & a_{12} \\ a_{21} & a_{22} \end{pmatrix} = \begin{pmatrix} l_{11} & 0 \\ l_{21} & l_{22} \end{pmatrix} \begin{pmatrix} u_{11} & u_{12} \\ 0 & u_{22} \end{pmatrix}$$
$$= \begin{pmatrix} l_{11}u_{11} & l_{11}u_{12} \\ l_{21}u_{11} & l_{21}u_{12} + l_{22}u_{22} \end{pmatrix}.$$

If $\boldsymbol{A}$ is non-singular, then so are $\boldsymbol{L}$ and $\boldsymbol{U}$, so that both have all entries on the diagonal not equal to 0. Therefore, $a_{11} = l_{11}u_{11} \neq 0$. However, consider

$$\boldsymbol{A} = \begin{pmatrix} 0 & 1 \\ 1 & 0 \end{pmatrix} \quad \text{with} \quad \boldsymbol{A}^{-1} = \begin{pmatrix} 0 & 1 \\ 1 & 0 \end{pmatrix},$$

so $\boldsymbol{A}$ is non-singular with $a_{11} = 0$ and any $\boldsymbol{LU}$ decomposition of a non-singular matrix requires $a_{11} > 0$. There is no $\boldsymbol{LU}$ decomposition of $\boldsymbol{A}$. ◊

EXAMPLE 2.27: Let

$$\boldsymbol{A} = \begin{pmatrix} a_{11} & a_{12} \\ a_{21} & a_{22} \end{pmatrix}.$$

Observe that, provided $a_{11} \neq 0$, subtracting $\frac{a_{21}}{a_{11}} \times$ row 1 from row 2 creates a matrix with a 0 in the lower left corner and the second row becomes:

$$(a_{21}, a_{22}) - \frac{a_{21}}{a_{11}}(a_{11}, a_{12}) = \left(a_{21} - \frac{a_{21}}{a_{11}}a_{11}, a_{22} - \frac{a_{21}}{a_{11}}a_{12}\right)$$
$$= \left(0, a_{22} - \frac{a_{21}}{a_{11}}a_{12}\right).$$

A matrix transformation that achieves this while leaving row 1 unchanged is as follows:

$$\begin{pmatrix} 1 & 0 \\ -\frac{a_{21}}{a_{11}} & 1 \end{pmatrix} \begin{pmatrix} a_{11} & a_{12} \\ a_{21} & a_{22} \end{pmatrix} = \begin{pmatrix} a_{11} & a_{12} \\ 0 & a_{22} - \frac{a_{21}}{a_{11}}a_{12} \end{pmatrix}.$$

Observe that

$$\begin{pmatrix} 1 & 0 \\ -\frac{a_{21}}{a_{11}} & 1 \end{pmatrix}^{-1} = \begin{pmatrix} 1 & 0 \\ \frac{a_{21}}{a_{11}} & 1 \end{pmatrix}.$$

So,

$$\begin{pmatrix} a_{11} & a_{12} \\ a_{21} & a_{22} \end{pmatrix} = \begin{pmatrix} 1 & 0 \\ \frac{a_{21}}{a_{11}} & 1 \end{pmatrix} \begin{pmatrix} a_{11} & a_{12} \\ 0 & a_{22} - \frac{a_{21}}{a_{11}}a_{12} \end{pmatrix},$$

which expresses $\boldsymbol{A}$ in the form $\boldsymbol{A} = \boldsymbol{LU}$. ◊

Definition 2.18: *Given an $n \times n$ matrix A, for $j = 1, \ldots, n$, a principal minor is obtained from $\mathbf{A}$ by removing $0 \le k < n$ rows and the corresponding columns. If A_{jj} is the matrix obtained from $\mathbf{A}$ by deleting the last $n - j$ rows and columns, this matrix is called the leading principal minor of order j.*

Theorem 2.10: *An invertible matrix admits an $\boldsymbol{LU}$ decomposition if and only if the determinants of all leading principal minors are non-zero.*

A non-invertible matrix may have an $\boldsymbol{LU}$ decomposition (such as a triangular matrix with some 0s on the diagonal.)

Definition 2.19: *A square matrix $\mathbf{A}$ is idempotent if $\mathbf{A} = \mathbf{AA}$.*

The identity matrix is an idempotent matrix. Other examples: with a and b any numbers,

$$\boldsymbol{A} = \begin{pmatrix} 1 & a \\ 0 & 0 \end{pmatrix} \quad \text{or} \quad \boldsymbol{B} = \begin{pmatrix} 0 & a \\ 0 & 1 \end{pmatrix}$$

are idempotent matrices, as is

$$A = \begin{pmatrix} 1-a & \frac{a(1-a)}{b} \\ b & a \end{pmatrix}, \text{ since } AA = \begin{pmatrix} (1-a)^2 + a(1-a) & \frac{a(1-a)^2}{b} + a^2\frac{(1-a)}{b} \\ b(1-a)+ab & a(1-a)+a^2 \end{pmatrix},$$

with $\boldsymbol{AA}$ simplifying to equal $\boldsymbol{A}$. In the 3×3 case the matrix $\boldsymbol{A}$ is idempotent:

$$A = \begin{pmatrix} \frac{1}{3} & \frac{1}{3} & \frac{1}{3} \\ \frac{1}{3} & \frac{5}{6} & -\frac{1}{6} \\ \frac{1}{3} & -\frac{1}{6} & \frac{5}{6} \end{pmatrix}.$$

If $\boldsymbol{A}$ is idempotent, its determinant is either 0 or 1. Since: $\boldsymbol{A} = \boldsymbol{AA}$,

$$|\boldsymbol{A}| = |\boldsymbol{AA}| = |\boldsymbol{A}| \cdot |\boldsymbol{A}| = |\boldsymbol{A}|^2,$$

using the rule that the determinant of the product of two square matrices is the product of the determinants. And, since $|\boldsymbol{A}| = |\boldsymbol{A}|^2$, it must be that $|\boldsymbol{A}|$ is equal to 0 or 1. If $\boldsymbol{A}$ is idempotent, so is $\boldsymbol{I}-\boldsymbol{A}$, since $(\boldsymbol{I}-\boldsymbol{A})(\boldsymbol{I}-\boldsymbol{A}) = \boldsymbol{II} - \boldsymbol{IA} - \boldsymbol{AI} + \boldsymbol{AA} = \boldsymbol{I} - 2\boldsymbol{A} + \boldsymbol{A} = \boldsymbol{I} - \boldsymbol{A}$.

Idempotent matrices arise naturally in statistics. The following discussion introduces a matrix that is used to compute deviations around the mean and the sum of squared deviations – the main calculation in computing the variance of sample data. Let $\boldsymbol{\iota}$ be an $n \times 1$ vector of 1s, so that $\boldsymbol{\iota}' = (1,1,\dots,1)$. Then $\boldsymbol{\iota}'\boldsymbol{\iota} = n$, so that $(\boldsymbol{\iota}'\boldsymbol{\iota})^{-1} = \frac{1}{n}$. Define $\boldsymbol{Q} \stackrel{\text{def}}{=} \boldsymbol{\iota}(\boldsymbol{\iota}'\boldsymbol{\iota})^{-1}\boldsymbol{\iota}'$:

$$\boldsymbol{Q} = \boldsymbol{\iota}(\boldsymbol{\iota}'\boldsymbol{\iota})^{-1}\boldsymbol{\iota}' = \boldsymbol{\iota}\left(\frac{1}{n}\right)\boldsymbol{\iota}' = \left(\frac{1}{n}\right)\boldsymbol{\iota}\boldsymbol{\iota}' = \left(\frac{1}{n}\right)\begin{pmatrix} 1 \\ 1 \\ \vdots \\ 1 \end{pmatrix}\begin{pmatrix} 1 & 1 & \dots & 1 \end{pmatrix} = \left(\frac{1}{n}\right)\begin{pmatrix} 1 & 1 & \dots & 1 \\ 1 & 1 & \dots & 1 \\ \vdots & \vdots & \ddots & \vdots \\ 1 & 1 & \dots & 1 \end{pmatrix}.$$

Notice that $\boldsymbol{Q}$ is an idempotent matrix. It is symmetric and

$$\boldsymbol{QQ} = \boldsymbol{\iota}(\boldsymbol{\iota}'\boldsymbol{\iota})^{-1}\boldsymbol{\iota}'\,\boldsymbol{\iota}(\boldsymbol{\iota}'\boldsymbol{\iota})^{-1}\boldsymbol{\iota}' = \boldsymbol{\iota}(\boldsymbol{\iota}'\boldsymbol{\iota})^{-1}\overbrace{\boldsymbol{\iota}'\boldsymbol{\iota}(\boldsymbol{\iota}'\boldsymbol{\iota})^{-1}}^{nn^{-1}=1}\boldsymbol{\iota}' = \boldsymbol{\iota}(\boldsymbol{\iota}'\boldsymbol{\iota})^{-1}\boldsymbol{\iota}' = \boldsymbol{Q}.$$

Given an $n \times 1$ vector $\boldsymbol{x}$ so that $\boldsymbol{x}' = (x_1, x_2, \dots, x_n)$:

$$\boldsymbol{Qx} = \left(\frac{1}{n}\right)\begin{pmatrix} 1 & 1 & \dots & 1 \\ 1 & 1 & \dots & 1 \\ \vdots & \vdots & \ddots & \vdots \\ 1 & 1 & \dots & 1 \end{pmatrix}\begin{pmatrix} x_1 \\ x_2 \\ \vdots \\ x_n \end{pmatrix} = \left(\frac{1}{n}\right)\begin{pmatrix} x_1 + x_2 + \cdots + x_n \\ x_1 + x_2 + \cdots + x_n \\ \vdots \\ x_1 + x_2 + \cdots + x_n \end{pmatrix} = \begin{pmatrix} \frac{1}{n}\sum_{i=1}^n x_i \\ \frac{1}{n}\sum_{i=1}^n x_i \\ \vdots \\ \frac{1}{n}\sum_{i=1}^n x_i \end{pmatrix} = \begin{pmatrix} \bar{x} \\ \bar{x} \\ \vdots \\ \bar{x} \end{pmatrix}.$$

Thus, $\boldsymbol{Qx}$ is an $n \times 1$ vector with the mean of the data $(x_1, x_2, \ldots, x_n)$ in every position. Deviations from the mean may be computed by the matrix $\boldsymbol{I} - \boldsymbol{Q}$ where $\boldsymbol{I}$ is the $n \times n$ identity matrix:

$$(\boldsymbol{I} - \boldsymbol{Q})\boldsymbol{x} = \boldsymbol{I}\boldsymbol{x} - \boldsymbol{Q}\boldsymbol{x} = \begin{pmatrix} x_1 \\ x_2 \\ \vdots \\ x_n \end{pmatrix} - \begin{pmatrix} \bar{x} \\ \bar{x} \\ \vdots \\ \bar{x} \end{pmatrix} = \begin{pmatrix} x_1 - \bar{x} \\ x_2 - \bar{x} \\ \vdots \\ x_n - \bar{x} \end{pmatrix}.$$

Let $\boldsymbol{D} = \boldsymbol{I} - \boldsymbol{Q}$. From previous discussion, $\boldsymbol{D}$ is symmetric and idempotent so that:

$$(\boldsymbol{Dx})'(\boldsymbol{Dx}) = \boldsymbol{x}\boldsymbol{D}'\boldsymbol{D}\boldsymbol{x} = \boldsymbol{x}\boldsymbol{D}\boldsymbol{x} = \sum_{i=1}^{n} (x_i - \bar{x})^2.$$

More generally, suppose that $\boldsymbol{X}$ is an $n \times k$ matrix with $n \geq k$ and $\boldsymbol{X}$ of rank k. Then $\boldsymbol{X}(\boldsymbol{X}'\boldsymbol{X})^{-1}\boldsymbol{X}'$ is a symmetric idempotent matrix, as is $\boldsymbol{I} - \boldsymbol{X}(\boldsymbol{X}'\boldsymbol{X})^{-1}\boldsymbol{X}'$.

Given a square $n \times n$ matrix $\boldsymbol{A}$, the terms $a_{11}, a_{22}, \ldots, a_{nn}$ form the main diagonal – the entries along the diagonal from the top left corner to the bottom right corner. If a matrix has all entries that are not on the main diagonal equal to 0, the matrix is called a *diagonal matrix*. Given a collection of numbers, $d_1, d_2, \ldots, d_n$, $\text{diag}(d_1, d_2, \ldots, d_n)$ is the $n \times n$ matrix formed by arranging $d_1, d_2, \ldots, d_n$ in the main diagonal and placing 0s elsewhere.

$$\text{diag}(d_1, d_2, \ldots, d_n) = \begin{pmatrix} d_1 & 0 & \cdots & 0 \\ 0 & d_2 & \cdots & 0 \\ \vdots & \vdots & \ddots & \vdots \\ 0 & 0 & \cdots & d_n \end{pmatrix}.$$

Definition 2.20: *The trace of a square $n \times n$ matrix $\boldsymbol{A}$ is defined as the sum of the terms on the main diagonal and is denoted* $\text{tr}(\boldsymbol{A})$. *Thus:*

$$\text{tr}(\boldsymbol{A}) = \sum_{i=1}^{n} a_{ii}.$$

The trace operation has a number of useful properties.

- If $\boldsymbol{A}$ and $\boldsymbol{B}$ are two square $n \times n$ matrices then $\text{tr}(\boldsymbol{A} + \boldsymbol{B}) = \text{tr}(\boldsymbol{A}) + \text{tr}(\boldsymbol{B})$.
- $\text{tr}(\boldsymbol{A}) = \text{tr}(\boldsymbol{A}')$.
- $\text{tr}(c\boldsymbol{A}) = c \cdot \text{tr}(\boldsymbol{A})$, c a scalar.
- If $\boldsymbol{A}$ is idempotent, then the trace of $\boldsymbol{A}$ equals its rank: $\text{tr}(\boldsymbol{A}) = r(\boldsymbol{A})$.

- If $\boldsymbol{A}$ is $m \times n$ and $\boldsymbol{B}$ is $n \times m$, then $\mathrm{tr}(\boldsymbol{AB}) = \mathrm{tr}(\boldsymbol{BA})$. This follows because:

$$\mathrm{tr}(\boldsymbol{AB}) = \sum_{i=1}^{m}\sum_{j=1}^{n} a_{ij}b_{ji} = \sum_{j=1}^{n}\sum_{i=1}^{m} b_{ji}a_{ij} = \mathrm{tr}(\boldsymbol{BA}).$$

To see this more explicitly, let $\boldsymbol{C} = \boldsymbol{AB}$ and $\boldsymbol{D} = \boldsymbol{BA}$. Then:

$$\begin{aligned}
c_{11} &= a_{11}b_{11} + a_{12}b_{21} + \cdots + a_{1n}b_{n1} \\
c_{22} &= a_{21}b_{12} + a_{22}b_{22} + \cdots + a_{2n}b_{n2} \\
\vdots &= \qquad \vdots \\
c_{mm} &= a_{m1}b_{1m} + a_{m2}b_{2m} + \cdots + a_{mn}b_{nm} \\
d_{11} &= b_{11}a_{11} + b_{12}a_{21} + \cdots + b_{1m}a_{m1} \\
d_{22} &= b_{21}a_{12} + b_{22}a_{22} + \cdots + b_{2m}a_{m2} \\
\vdots &= \qquad \vdots \\
d_{nn} &= b_{n1}a_{1n} + b_{n2}a_{2n} + \cdots + b_{nm}a_{mn}
\end{aligned}$$

The diagonal sum of $\boldsymbol{C} = \boldsymbol{AB}$ is $\sum_{i=1}^{m} c_{ii}$ and the diagonal sum of $\boldsymbol{D} = \boldsymbol{BA}$ is $\sum_{i=1}^{n} d_{ii}$. From the expressions above, $\sum_{i=1}^{m} c_{ii} = \sum_{i=1}^{n} d_{ii}$. The following example illustrates:

EXAMPLE 2.28:

$$\boldsymbol{A} = \begin{pmatrix} 1 & 3 & 5 \\ 2 & 4 & 6 \end{pmatrix}, \ \boldsymbol{B} = \begin{pmatrix} 2 & 1 \\ 5 & 7 \\ 3 & 1 \end{pmatrix}, \text{ so } \boldsymbol{AB} = \begin{pmatrix} 32 & 27 \\ 42 & 36 \end{pmatrix}, \text{ and } \boldsymbol{BA} = \begin{pmatrix} 4 & 10 & 16 \\ 19 & 43 & 67 \\ 5 & 13 & 21 \end{pmatrix},$$

and observe that

$$\mathrm{tr}(\boldsymbol{AB}) = 32 + 36 = 4 + 43 + 21 = \mathrm{tr}(\boldsymbol{BA}). \qquad \diamond$$

Exercises for Chapter 2

Exercise 2.1 Consider the following three matrices:

$$\boldsymbol{A}=\begin{pmatrix}1 & 2\\ 3 & 4\end{pmatrix},\ \boldsymbol{B}=\begin{pmatrix}1 & 3 & 5\\ 2 & 4 & 6\\ 8 & 7 & 9\end{pmatrix},\ \boldsymbol{C}=\begin{pmatrix}1 & 2 & 5 & 4\\ 2 & 1 & 3 & 2\\ 6 & 4 & 3 & 1\\ 7 & 4 & 2 & 3\end{pmatrix}.$$

For each matrix, find (a) the determinant, (b) the matrix of minors, (c) the matrix of cofactors, (d) the adjoint matrix and (e) the inverse matrix.

Exercise 2.2 Consider the following matrix:

$$\boldsymbol{D}=\begin{pmatrix}1 & 2\\ 4 & 3\\ 5 & 1\end{pmatrix}.$$

What is the rank of $\boldsymbol{D}$, $\boldsymbol{D}'\boldsymbol{D}$ and $\boldsymbol{D}\boldsymbol{D}'$? Find (a) $\boldsymbol{D}'$ ($\boldsymbol{D}$ transpose), (b) $\boldsymbol{D}'\boldsymbol{D}$, (c) $(\boldsymbol{D}'\boldsymbol{D})^{-1}$ (d) $\boldsymbol{D}(\boldsymbol{D}'\boldsymbol{D})^{-1}\boldsymbol{D}'$. Call the last matrix you have calculated $\boldsymbol{E}$ (i.e., $\boldsymbol{E} = \boldsymbol{D}(\boldsymbol{D}'\boldsymbol{D})^{-1}\boldsymbol{D}'$). Find $\boldsymbol{ED}$ and $\boldsymbol{DE}$. Check that $\boldsymbol{ED} = \boldsymbol{D}$ and $\boldsymbol{D}'\boldsymbol{E} = \boldsymbol{D}'$ and explain why this is so.

Exercise 2.3 Given $\boldsymbol{A}=\begin{pmatrix}1 & 2\\ 0 & 3\end{pmatrix}$, find

(a) $\boldsymbol{A}^2 = \boldsymbol{AA}$, (b) $\boldsymbol{A}^3 = \boldsymbol{AAA}$, (c) $\boldsymbol{A}^{-1}$, (d) $2\boldsymbol{A}^3 + 3\boldsymbol{A}^2 + 5\boldsymbol{I}$.

Verify that (e) $(\boldsymbol{A} + \boldsymbol{A}')' = \boldsymbol{A}' + \boldsymbol{A}$ and (f) $(\boldsymbol{AA}')' = \boldsymbol{AA}'$.

Exercise 2.4 (a) For each of the following matrices, find the determinant and the inverse if it exists.

$$\boldsymbol{A}=\begin{pmatrix}1 & -1 & 2\\ 3 & 1 & -1\\ -1 & 2 & -4\end{pmatrix},\ \boldsymbol{B}=\begin{pmatrix}1 & 2 & 2\\ 2 & 1 & 4\\ 3 & 0 & 6\end{pmatrix},$$

$$\boldsymbol{C}=\begin{pmatrix}2 & 1 & 4\\ 3 & 2 & 5\\ 0 & -1 & 1\end{pmatrix},\ \boldsymbol{D}=\begin{pmatrix}1 & 0 & 1 & 4\\ -1 & 2 & 0 & 1\\ 0 & 4 & 2 & 5\\ 3 & 0 & 3 & 0\end{pmatrix},\quad \boldsymbol{E}=\begin{pmatrix}t-2 & -3\\ -4 & t-1\end{pmatrix}.$$

(b) Find the value of t, such that the determinant of matrix $\boldsymbol{E}$ is zero.

Exercise 2.5 Solve the following systems of equations by matrix inversion or Cramer's rule:

$$\text{(a)}\ \begin{aligned}3x+2y &= 16\\ x+2y &= 4\end{aligned}\qquad \text{(b)}\ \begin{aligned}2x-3y &= 18\\ x+3y &= 9\end{aligned}\qquad \text{(c)}\ \begin{aligned}x+2y &= 2\\ 3x+4y &= 12\end{aligned}$$

$$\text{(d)}\ \begin{aligned}-2x+3y &= 3\\ -5x+7y &= 6\end{aligned}\qquad \text{(e)}\ \begin{aligned}Q_s &= -c+dP\\ Q_d &= a-bP\\ Q_d &= Q_s\end{aligned}\qquad \text{(f)}\ \begin{aligned}-x+2y-3z &= 2\\ 2x+y\qquad &= 1\\ 4x-2y+5z &= 3\end{aligned}$$

$$\text{(g)}\ \begin{aligned}x-y+z &= 4\\ -x+2y-z &= 5\,.\\ 2x-y+z &= 2\end{aligned}$$

Exercise 2.6 Solve the following systems of equations by Cramer's rule:

$$\text{(a)}\ \begin{aligned}2y+4z\qquad &= 8\\ 2x+4y+2z &= 8\\ 3x+3y+z &= 4\end{aligned}\qquad \text{(b)}\ \begin{aligned}x+2y+z &= 6\\ -x+y+2z &= 3\\ x+z\qquad &= 2\end{aligned}\qquad \text{(c)}\ \begin{aligned}x+y-z\quad &= 1\\ 2x+4y-5z &= 2\,.\\ -2y+4z &= 1\end{aligned}$$

Exercise 2.7 Consider the following:

$\boldsymbol{X}$ is an $n\times k$ matrix

$\boldsymbol{e}$ is an $n\times 1$ vector of 1s (i.e., a column vector with n ones)

$\boldsymbol{A}$ is an $n\times n$ symmetric matrix

$$\boldsymbol{P}=\boldsymbol{X}(\boldsymbol{X}'\boldsymbol{X})^{-1}\boldsymbol{X}'$$

$$\boldsymbol{M}=\boldsymbol{I}_n-\boldsymbol{P}$$

$$Q = I_n - \left(\frac{1}{n}\right) ee'$$
$$F = X(X'AX)^{-1}X'A.$$

(a) Determine the order of $\boldsymbol{P}, \boldsymbol{M}, \boldsymbol{Q}, \boldsymbol{F}$ (i.e., the number of rows and columns).

(b) Determine which of these are idempotent ($\boldsymbol{B}$ is idempotent if $\boldsymbol{B}^2 = \boldsymbol{BB} = \boldsymbol{B}$).

(c) Determine which are symmetric ($\boldsymbol{B} = \boldsymbol{B}'$).

(d) Show that for an n vector $\boldsymbol{x}$

$$\boldsymbol{Qx} = \begin{pmatrix} x_1 - \bar{x} \\ x_2 - \bar{x} \\ \vdots \\ x_n - \bar{x} \end{pmatrix}, \text{ where } \bar{x} = \sum_{i=1}^{n} x_i.$$

Exercise 2.8 The holdings of stock of individuals I and II in companies A, B, C, D are given by the matrix:

$$\begin{matrix} & A & B & C & D \\ \text{I} & 2000 & 1000 & 500 & 5000 \\ \text{II} & 1000 & 500 & 2000 & 0 \end{matrix}.$$

Denote the stock price of stock A by p_A(and similarly for B, C and D). If the vector of stock prices (per share) is $(p_A, p_B, p_C, p_D) = (24, 47, 150, 14)$, calculate the value of the stock holdings of both individuals.

Exercise 2.9 (a) Plot the following pairs of equations:

$$\begin{aligned} x + 3y &= 9 \\ 4x + 2y &= 8 \end{aligned} \tag{1}$$

$$\begin{aligned} 3x + 3y &= 12 \\ 4x + y &= 8 \end{aligned} \tag{2}$$

$$\begin{aligned} 4x + 3y &= 4 \\ 8x + 6y &= 8 \end{aligned} \tag{3}$$

$$\begin{aligned} 2x + 5y &= 9 \\ 4x + 10y &= 10. \end{aligned} \tag{4}$$

If possible, determine the solution of each equation pair.

(b) Write the four pairs of equations in matrix form. Thus, for example,

$$\begin{aligned} ax + by &= \alpha \\ cx + dy &= \gamma \end{aligned}$$

is written:

$$\begin{pmatrix} a & b \\ c & d \end{pmatrix} \begin{pmatrix} x \\ y \end{pmatrix} = \begin{pmatrix} \alpha \\ \gamma \end{pmatrix}.$$

(c) Find the determinant of each of the four coefficient matrices in part (b).

Exercise 2.10 For the following matrices:

$$\boldsymbol{A} = \begin{pmatrix} 1 & 4 \\ 2 & 1 \end{pmatrix}, \quad \boldsymbol{B} = \begin{pmatrix} 3 & 2 \\ 1 & 6 \end{pmatrix}, \quad \boldsymbol{C} = \begin{pmatrix} 1 & 3 & 1 \\ 3 & 2 & 1 \end{pmatrix}, \quad \boldsymbol{D} = \begin{pmatrix} 2 & 1 \\ 4 & 4 \\ 6 & 2 \end{pmatrix},$$

(a) Find $\boldsymbol{AB}, \boldsymbol{BA}, |\boldsymbol{A}|, |\boldsymbol{B}|, |\boldsymbol{AB}|$ and compare $|\boldsymbol{AB}|$ with $|\boldsymbol{A}| \times |\boldsymbol{B}|$.

(b) Find $\boldsymbol{DC}, \boldsymbol{CD}, |\boldsymbol{CD}|$ and $|\boldsymbol{DC}|$.

(c) Find $\boldsymbol{A}'$ and $\boldsymbol{C}'$.

Exercise 2.11 Use Cramer's rule to solve the following two equation systems:

$$\begin{pmatrix} 1 & 2 \\ 3 & 4 \end{pmatrix} \begin{pmatrix} x_1 \\ x_2 \end{pmatrix} = \begin{pmatrix} 5 \\ 6 \end{pmatrix} \quad \text{and} \quad \begin{pmatrix} 1 & 2 & 3 \\ 3 & 2 & 1 \\ 2 & 3 & 1 \end{pmatrix} \begin{pmatrix} x_1 \\ x_2 \\ x_3 \end{pmatrix} = \begin{pmatrix} 2 \\ 4 \\ 6 \end{pmatrix}.$$

Exercise 2.12 Given:

$$\boldsymbol{A} = \begin{pmatrix} 3 & 5 \\ 2 & 3 \end{pmatrix}, \boldsymbol{B} = \begin{pmatrix} 2 & 4 \\ 4 & 2 \end{pmatrix}, \boldsymbol{C} = \begin{pmatrix} 8 & -1 & -3 \\ -5 & 1 & 2 \\ 10 & -1 & -4 \end{pmatrix}, \boldsymbol{D} = \begin{pmatrix} 3 & 0 & 3 \\ 0 & -1 & 2 \\ 6 & 3 & 0 \end{pmatrix}$$

$$\boldsymbol{y} = \begin{pmatrix} 2 \\ -7 \\ 4 \end{pmatrix}, \boldsymbol{x} = \begin{pmatrix} 1 & 3 & 7 \end{pmatrix},$$

find (a) $\boldsymbol{A} + \boldsymbol{B}$, (b) $\boldsymbol{C} - \boldsymbol{D}$, (c) $\boldsymbol{D}'$, (d) $5\boldsymbol{A} - 6\boldsymbol{B}$, (e) $\boldsymbol{A}^{-1}$, (f) $\boldsymbol{C}^{-1}$, (g) $\boldsymbol{CD}$, (h) $\boldsymbol{DC}$, (i) $(\boldsymbol{x}'\boldsymbol{x})$, (j) $(\boldsymbol{x}'\boldsymbol{x})\boldsymbol{y}$, (k) $\boldsymbol{y}'\boldsymbol{C}$, (l) the inverse of $\boldsymbol{D}$.

Exercise 2.13 In matrix notation, a linear regression equation can be written

$$\boldsymbol{y} = \boldsymbol{X\beta} + \boldsymbol{u},$$

where $\boldsymbol{y}$ is an $n \times 1$ vector of observations on the dependent variable.

$\boldsymbol{X}$ is a $n \times k$ matrix of observations on the exogenous variables.

$\boldsymbol{\beta}$ is a $k \times 1$ vector of coefficients.

$\boldsymbol{u}$ is an $n \times 1$ vector of error terms.

The (ordinary least squares OLS) estimator $\boldsymbol{b}$ is defined: $\boldsymbol{b} = (\boldsymbol{X}'\boldsymbol{X})^{-1}\boldsymbol{X}'\boldsymbol{y}$.

Suppose that

$$\boldsymbol{X} = \begin{pmatrix} 1 & 1 \\ 1 & 3 \\ 1 & 5 \end{pmatrix} \text{ and } \boldsymbol{y} = \begin{pmatrix} 2 \\ 4 \\ 3 \end{pmatrix}.$$

(a) Calculate $\boldsymbol{b}$.

(b) Calculate $\boldsymbol{P} = \boldsymbol{X}(\boldsymbol{X}'\boldsymbol{X})^{-1}\boldsymbol{X}'$, $\boldsymbol{M} = \boldsymbol{I}_n - \boldsymbol{P}$ (where $\boldsymbol{I}_n$ is a $n \times n$ identity matrix).

(c) Show or prove that $\boldsymbol{X}'\boldsymbol{M} = \boldsymbol{0}$ and explain why this is so.

Exercise 2.14 For the matrix

$$\boldsymbol{D} = \begin{pmatrix} 1 & 2 & 3 \\ 2 & 1 & 4 \\ 1 & 2 & 1 \end{pmatrix},$$

find the determinant and the matrices of minors and cofactors. Give the adjoint matrix and then the inverse matrix, $\boldsymbol{D}^{-1}$.

Exercise 2.15 Given the matrices

$$\boldsymbol{A} = \begin{pmatrix} 2 & 5 \\ 4 & 3 \end{pmatrix}, \quad \boldsymbol{B} = \begin{pmatrix} 1 & 4 \\ 3 & 2 \end{pmatrix}, \quad \boldsymbol{C} = \begin{pmatrix} 1 & 3 & 5 \\ 4 & 2 & 1 \end{pmatrix}, \quad \boldsymbol{D} = \begin{pmatrix} 2 & 1 \\ 3 & 4 \\ 6 & 2 \end{pmatrix},$$

$$\boldsymbol{E} = \begin{pmatrix} 4 & 3 & 2 \\ 6 & 6 & 5 \\ 4 & 9 & 5 \end{pmatrix}, \quad \boldsymbol{F} = \begin{pmatrix} 2 & 3 & 2 \\ 6 & 6 & 5 \\ 4 & 9 & 5 \end{pmatrix}:$$

(a) Find $\boldsymbol{A}'$, $\boldsymbol{B}'$, $\boldsymbol{B}'\boldsymbol{A}'$ and $(\boldsymbol{AB})'$. Confirm that $\boldsymbol{B}'\boldsymbol{A}'$ equals $(\boldsymbol{AB})'$. Is it also true that $\boldsymbol{D}'\boldsymbol{C}' = (\boldsymbol{CD})'$?

(b) Calculate the matrix $\boldsymbol{H}$, defined $\boldsymbol{H} = \boldsymbol{I} - \boldsymbol{D}(\boldsymbol{D}'\boldsymbol{D})^{-1}\boldsymbol{D}'$. Confirm that $\boldsymbol{H}' = \boldsymbol{H}$, $\boldsymbol{HD} = \boldsymbol{0}$ and that $\boldsymbol{HH} = \boldsymbol{H}$. Let the matrix $\boldsymbol{J}$ be defined: $\boldsymbol{J} = \boldsymbol{I} - \boldsymbol{C}'(\boldsymbol{CC}')^{-1}\boldsymbol{C}$. Show that $\boldsymbol{JC}' = \boldsymbol{0}$.

(c) Find the rank of $\boldsymbol{A}$, $\boldsymbol{E}$, $\boldsymbol{F}$, $\boldsymbol{CC}'$ and $\boldsymbol{DD}'$. Also find the determinants of $\boldsymbol{A}$, $\boldsymbol{B}$, $\boldsymbol{E}$ and $\boldsymbol{F}$.

(d) Using Cramer's rule, solve:

$$\begin{pmatrix} 2 & 5 \\ 4 & 3 \end{pmatrix}\begin{pmatrix} x_1 \\ x_2 \end{pmatrix} = \begin{pmatrix} 4 \\ 7 \end{pmatrix}, \quad \begin{pmatrix} 4 & 3 & 2 \\ 6 & 6 & 5 \\ 4 & 9 & 5 \end{pmatrix}\begin{pmatrix} x_1 \\ x_2 \\ x_3 \end{pmatrix} = \begin{pmatrix} 2 \\ 4 \\ 6 \end{pmatrix}.$$

Note that the coefficient matrices are $\boldsymbol{A}$ and $\boldsymbol{E}$.

Exercise 2.16 Let

$$\boldsymbol{A} = \begin{pmatrix} a_{11} & a_{12} & \cdots & a_{1n} \\ a_{21} & a_{22} & \cdots & a_{2n} \\ \vdots & \vdots & \ddots & \vdots \\ a_{m1} & a_{m2} & \cdots & a_{mn} \end{pmatrix}, \quad \boldsymbol{B} = \begin{pmatrix} b_{11} & b_{12} & \cdots & b_{1k} \\ b_{21} & b_{22} & \cdots & b_{2k} \\ \vdots & \vdots & \ddots & \vdots \\ b_{n1} & b_{n2} & \cdots & b_{nk} \end{pmatrix}.$$

Thus, $\boldsymbol{A}$ is an $m \times n$ matrix and $\boldsymbol{B}$ an $n \times k$ matrix. Show that $(\boldsymbol{A}')' = \boldsymbol{A}$ (the transpose of $\boldsymbol{A}'$ is $\boldsymbol{A}$), and $(\boldsymbol{AB})' = \boldsymbol{B}'\boldsymbol{A}'$. For the second part, the (j,i)th element of $(\boldsymbol{AB})'$ is the (i,j)th element of $\boldsymbol{AB}$. Thus, show that the (i,j)th element of $\boldsymbol{AB}$ is equal to the (j,i)th element of $\boldsymbol{B}'\boldsymbol{A}'$.

3

Linear algebra: applications

3.1 Introduction

This chapter discusses applications of matrix algebra to illustrate techniques developed earlier. Section 3.2 reviews the basic supply and demand model with two goods. Sections 3.4 to 3.7 consider various applications, illustrating techniques such as the use of matrix polynomials and convergence of matrix expansions. These techniques have widespread use (for example in the study of dynamic systems) and are developed further in later chapters.

3.2 The market model

The two-market model has demand and supply in each market given by:

$$\begin{aligned} &\text{Market 1:} \quad D_1 = a_0 + a_1 p_1 + a_2 p_2, \quad S_1 = b_0 + b_1 p_1 + b_2 p_2 \\ &\text{Market 2:} \quad D_2 = \alpha_0 + \alpha_1 p_1 + \alpha_2 p_2, \quad S_2 = \beta_0 + \beta_1 p_1 + \beta_2 p_2. \end{aligned}$$

Equating supply and demand in each market gives the system of equations determining equilibrium prices:

$$\begin{pmatrix} (a_1 - b_1) & (a_2 - b_2) \\ (\alpha_1 - \beta_1) & (\alpha_2 - \beta_2) \end{pmatrix} \begin{pmatrix} p_1 \\ p_2 \end{pmatrix} = \begin{pmatrix} (b_0 - a_0) \\ (\beta_0 - \alpha_0) \end{pmatrix}.$$

Thus,

$$\begin{pmatrix} p_1 \\ p_2 \end{pmatrix} = \begin{pmatrix} (a_1 - b_1) & (a_2 - b_2) \\ (\alpha_1 - \beta_1) & (\alpha_2 - \beta_2) \end{pmatrix}^{-1} \begin{pmatrix} (b_0 - a_0) \\ (\beta_0 - \alpha_0) \end{pmatrix}.$$

This is of the form discussed in the previous chapter with

$$\boldsymbol{A} = \begin{pmatrix} (a_1 - b_1) & (a_2 - b_2) \\ (\alpha_1 - \beta_1) & (\alpha_2 - \beta_2) \end{pmatrix}, \boldsymbol{x} = \begin{pmatrix} p_1 \\ p_2 \end{pmatrix}, \boldsymbol{b} = \begin{pmatrix} (b_0 - a_0) \\ (\beta_0 - \alpha_0) \end{pmatrix}.$$

EXAMPLE 3.1: Consider the following two-market model:

Market 1	*Market 2*
$D_1 = 6 + 5p_1 + 4p_2$	$D_2 = 5 + 4p_1 + 7p_2$
$S_1 = 12 + 2p_1 + 8p_2$	$S_2 = 9 + 6p_1 + 4p_2$

To find the equilibrium prices equate supply and demand. In the first market this gives:

$$6 + 5p_1 + 4p_2 = 12 + 2p_1 + 8p_2 \Leftrightarrow 3p_1 - 4p_2 = 6.$$

Similarly, for the second market:

$$5 + 4p_1 + 7p_2 = 9 + 6p_1 + 4p_2 \Leftrightarrow -2p_1 + 3p_2 = 4.$$

This gives the equation system (in matrix form):

$$\begin{pmatrix} 3 & -4 \\ -2 & 3 \end{pmatrix} \begin{pmatrix} p_1 \\ p_2 \end{pmatrix} = \begin{pmatrix} 6 \\ 4 \end{pmatrix},$$

so that

$$\begin{pmatrix} p_1 \\ p_2 \end{pmatrix} = \begin{pmatrix} 3 & -4 \\ -2 & 3 \end{pmatrix}^{-1} \begin{pmatrix} 6 \\ 4 \end{pmatrix} = \begin{pmatrix} 3 & 4 \\ 2 & 3 \end{pmatrix} \begin{pmatrix} 6 \\ 4 \end{pmatrix} = \begin{pmatrix} 34 \\ 24 \end{pmatrix}.$$

The example generalizes directly to the case of many markets. ◇

3.3 Adjacency matrices

A network of connections can be expressed as an adjacency matrix that summarizes connections between different nodes in a network. If there is a direct link from node i to node j, write $m_{ij} = 1$. If, in addition, there is a direct link back from j to i, then also, $m_{ji} = 1$. The network of links is directed in the sense that a link from i to j does not necessarily imply a link from j to i. The adjacency matrix is:

$$\boldsymbol{M} = \begin{pmatrix} m_{11} & m_{12} & \cdots & m_{1n} \\ m_{21} & m_{22} & \cdots & m_{2n} \\ \vdots & \vdots & \ddots & \vdots \\ m_{n1} & m_{n2} & \cdots & m_{nn} \end{pmatrix}.$$

Just as $\boldsymbol{M}$ identifies those pairs (i, j) such that j can be reached from i in one step ($m_{ij} = 1$), it is easy to identify those nodes connected by exactly two steps.

Consider $\boldsymbol{M}^2 = \boldsymbol{M}\boldsymbol{M}$. The entry in the (i,j)th position of $\boldsymbol{M}^2$ is

$$m^2_{ij} = \sum_{k=1}^{n} m_{ik} m_{kj}.$$

Note that $m_{ik}m_{kj} = 1$ if and only if there is a link from i to k and from k to j. Thus, m^2_{ij} is the number of distinct ways to go from i to j in two steps (using exactly two links).

This insight carries over to steps of greater length. Consider $\boldsymbol{M}^3 = \boldsymbol{M}\boldsymbol{M}^2$. The element in the (i,j)th position is $m^3_{ij} = \sum_{r=1}^{n} m_{ir} m^2_{rj}$. If there is a path from i to r, then $m_{ir} = 1$ and there are m^2_{rj} two-step paths (or $m_{ir}m^2_{rj}$ paths since $m_{ir} = 1$) from i through r to j. If there is no path from i to r, $m_{ir} = 0$ and there are $m_{ir}m^2_{rj} = 0$ paths from i through r to j. In total, there are $m_{ir}m^2_{rj}$ paths of length 3 from i to j that pass through r. Summing over all r, there are $\sum_{r=1}^{n} m_{ir}m^2_{rj}$ paths of length 3 that go from i to j. Thus, the entry m^3_{ij} in $\boldsymbol{M}^3$ gives the number of paths of length 3 that go from i to j. If the matrix $\boldsymbol{M}$ is multiplied k times, the same reasoning implies that the (i,j)th entry m^k_{ij} gives the number of paths of length k going from i to j.

Define the sum:

$$\boldsymbol{S}_k = \boldsymbol{M} + \boldsymbol{M}^2 + \boldsymbol{M}^3 + \cdots + \boldsymbol{M}^k = \sum_{j=1}^{k} \boldsymbol{M}^j.$$

The element s^k_{ij} in the (i,j)th position gives the number of paths of length k or less, from i to j.

EXAMPLE 3.2: Consider the network:

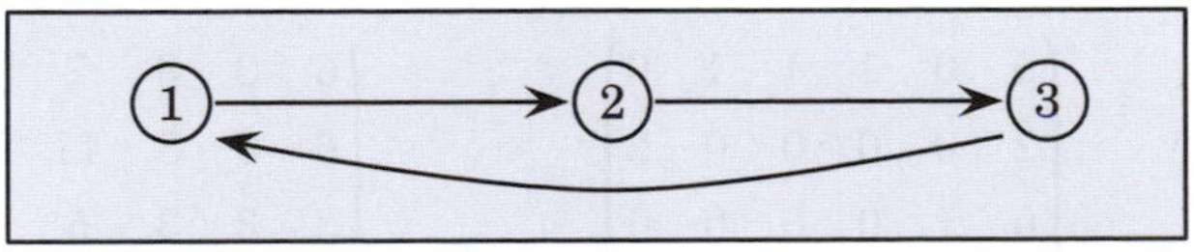

$$\boldsymbol{M} = \begin{pmatrix} 0 & 1 & 0 \\ 0 & 0 & 1 \\ 1 & 0 & 0 \end{pmatrix}, \quad \boldsymbol{M}^2 = \begin{pmatrix} 0 & 0 & 1 \\ 1 & 0 & 0 \\ 0 & 1 & 0 \end{pmatrix}, \quad \boldsymbol{M}^3 = \begin{pmatrix} 1 & 0 & 0 \\ 0 & 1 & 0 \\ 0 & 0 & 1 \end{pmatrix}.$$

The matrices $\boldsymbol{M}$, $\boldsymbol{M}^2$, $\boldsymbol{M}^3$ give the nodes reachable in one, two and three steps from any initial node.

$$\boldsymbol{S}_1 = \boldsymbol{M} = \begin{pmatrix} 0 & 1 & 0 \\ 0 & 0 & 1 \\ 1 & 0 & 0 \end{pmatrix}, \quad \boldsymbol{S}_2 = \boldsymbol{M} + \boldsymbol{M}^2 = \begin{pmatrix} 0 & 1 & 1 \\ 1 & 0 & 1 \\ 1 & 1 & 0 \end{pmatrix},$$

$$\boldsymbol{S}_3 = \boldsymbol{M} + \boldsymbol{M}^2 + \boldsymbol{M}^3 = \begin{pmatrix} 1 & 1 & 1 \\ 1 & 1 & 1 \\ 1 & 1 & 1 \end{pmatrix},$$

so, for example, considering row 1 of $\boldsymbol{S}_2$, there is one way to go from node 1 to node 2 using one or two steps. But there is no way to go from node 1 (back) to node 1 in one or two steps. From $\boldsymbol{S}_3$, there is one way to go from any node to any other in three or fewer steps. ◊

The next example illustrates similar points with a larger set of nodes.

EXAMPLE 3.3: In this example, the network consists of six nodes. The corresponding adjacency matrix is on the right.

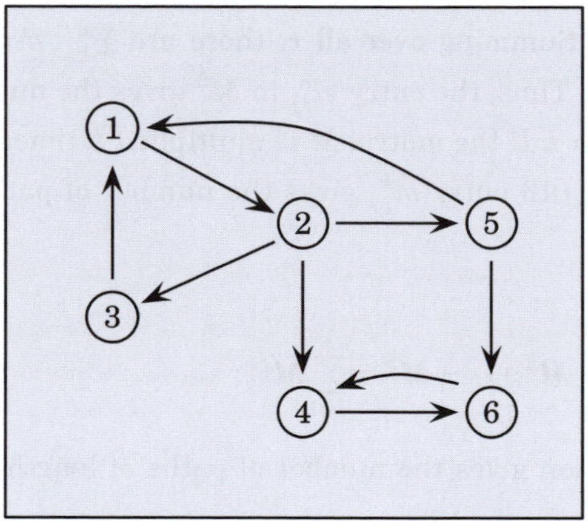

	1	2	3	4	5	6
1	0	1	0	0	0	0
2	0	0	1	1	1	0
3	1	0	0	0	0	0
4	0	0	0	0	0	1
5	1	0	0	0	0	1
6	0	0	0	1	0	0

Then $\boldsymbol{M}^2$ and the sum $\boldsymbol{S}_6 = \sum_{j=1}^{6} \boldsymbol{M}^j$ are:

$$\boldsymbol{M}^2 = \begin{pmatrix} 0 & 0 & 1 & 1 & 1 & 0 \\ 2 & 0 & 0 & 0 & 0 & 2 \\ 0 & 1 & 0 & 0 & 0 & 0 \\ 0 & 0 & 0 & 1 & 0 & 0 \\ 0 & 1 & 0 & 1 & 0 & 0 \\ 0 & 0 & 0 & 0 & 0 & 1 \end{pmatrix}, \qquad \boldsymbol{S}_6 = \begin{pmatrix} 6 & 3 & 3 & 7 & 3 & 8 \\ 6 & 6 & 3 & 11 & 3 & 10 \\ 3 & 3 & 3 & 5 & 3 & 4 \\ 0 & 0 & 0 & 3 & 0 & 3 \\ 3 & 3 & 3 & 8 & 3 & 7 \\ 0 & 0 & 0 & 3 & 0 & 3 \end{pmatrix}.$$

Thus, for example, from $\boldsymbol{M}^2$ there is exactly one path of length 2 from node 1 to node 5: $(1 \to 2 \to 5)$, indicated by $m_{15} = 1$. Similarly, there are exactly two paths of length 2 from node 2 to node 1 ($(2 \to 3 \to 1)$ and $(2 \to 5 \to 1)$), indicated by $m_{21} = 2$.

Considering S_6, s_{ij}^6, the (i,j)th entry gives the number of paths with 6 or fewer steps from i to j. So, for example, $s_{24} = 11$ indicates that there are 11 paths from 2 to 4 with 6 steps or less. ◊

REMARK 3.1: One issue concerns the possibility of moving from one node to another. The main observation in this regard is that with n nodes, if it is not possible to reach node j from node i in n steps, then it is impossible to reach j from i in any number of steps. This is easy to see.

Let $W_k(i) \subseteq \{1,2,\ldots,n\}$ be the set of nodes reachable from i in k or fewer steps. Write $\#W_k(i)$ to denote the number of nodes in $W_k(i)$. Then $W_k(i) \subseteq W_{k+1}(i)$: $W_k(i)$ is a subset of $W_{k+1}(i)$, so that if a node l is in $W_k(i)$ it is also in $W_{k+1}(i)$. This follows from the observation that if it is possible to reach j from i in k or fewer steps, it is certainly possible to reach j in no more than $k+1$ steps. Next, if $W_k(i) = W_{k+1}(i)$ then the same set of nodes is reached in k steps as in $k+1$ steps – additional steps reach no extra nodes. If $W_{k+1}(i) = W_k(i)$ then $W_{k+j}(i) = W_k(i)$ for $j = 1,2,\ldots$ so, at each iteration in k, either extra nodes are reached or the process ceases to add nodes. Suppose that $W_n(i) \neq \{1,2,\ldots,n\}$. Then for some $k < n$ it must be that $W_k(i) = W_{k+1}(i)$, otherwise the set $W_k(i)$ would have increased at each round k and, with n steps, would have n nodes. But, $W_k(i) = W_{k+1}(i)$ implies that $W_k(i) = W_{k+j}(i)$ for all $j \geq 1$. At $j = n-k$, $W_k(i) = W_n(i)$, and for any $r \geq 0$, with $j = n-k+r$, $W_k(i) = W_{n+r}(i)$. For any r, $W_k(i) = W_n(i) = W_{n+r}(i)$. If node l is not in $W_n(i)$ it can never be reached from i. Note that this includes i: if i is not in $W_n(i)$, there is no way to return to i. □

The possibility of moving from one node to any other is closely related to the concept of irreducibility, which arises in the study of dynamic processes.

Definition 3.1: *A matrix* $\mathbf{A}$ *is irreducible if and only if, for any* i, j, *there is a sequence* $i_1, i_2, \ldots, i_k$ *with* $i_1 = i$, $i_k = j$ *and* $a_{i_l i_{l+1}} \neq 0$ *for* $l = 1,\ldots,k-1$.

3.4 Input–output analysis

In input–output analysis, the economy is viewed as a number of interlinked industries (n industries). The output of an industry is used for final consumption, and in the production of other goods. For example, steel is used to produce trucks and transportation equipment, roads and bridges, and so on. Moreover, industries may be interlinked in complex ways: trucks are produced using steel but also transport iron ore used to make steel, giving rise to "circular" relationships in the production process – achieving a certain level of steel output implies an appropriate level of transportation capacity. Apart from output going from one industry to others, some steel is "consumed" – for example, people buy garden tools, kitchen utensils and so on. So, the total output of steel is allocated between that used in production in other industries (such as transportation and housing industries), an intermediate or "interindustry" demand; and consumption or final demand. Planning to achieve some (target) level of consumption or final demand indirectly implies production levels in each industry. Input–output analysis is concerned with determining the industry production levels required to sustain or achieve target levels across the entire range of final demands. A simple model is developed next to illustrate the procedure.

Let a_{ij} = amount of i required to produce 1 unit of j, x_i = output of i and f_i = final demand for i. Thus if x_j is the amount of good j produced, $a_{ij}x_j$ is the amount of i required in the production of j. (If one kilo of good j is produced, a_{ij} is the fraction of one kilo of i used in the production of the kilo of j.) Let all measurements be in dollars. (So, for example, a_{ij} is the number of dollars of good i required to produce one dollar in value of good j.) The output of industry i is allocated to production in other industries (the $a_{ij}x_j$) and final demand (f_i). Thus,

$$x_i = a_{i1}x_1 + a_{i2}x_2 + \cdots + a_{in}x_n + f_i, i = 1, \ldots, n.$$

This can be rearranged:

$$-a_{i1}x_1 - a_{i2}x_2 \cdots + (1 - a_{ii})x_i \cdots - a_{in}x_n = f_i, i = 1, \ldots, n,$$

or, in matrix notation:

$$\begin{pmatrix} (1-a_{11}) & -a_{12} & \cdots & -a_{1n} \\ -a_{21} & (1-a_{22}) & \cdots & -a_{2n} \\ \vdots & \vdots & \ddots & \vdots \\ -a_{n1} & -a_{n2} & \cdots & (1-a_{nn}) \end{pmatrix} \begin{pmatrix} x_1 \\ x_2 \\ \vdots \\ x_n \end{pmatrix} = \begin{pmatrix} f_1 \\ f_2 \\ \vdots \\ f_n \end{pmatrix}.$$

This expression can be written succinctly: $(\boldsymbol{I} - \boldsymbol{A})\boldsymbol{x} = \boldsymbol{f}$, where the matrices are:

$$\boldsymbol{I} = \begin{pmatrix} 1 & 0 & \cdots & 0 \\ 0 & 1 & \cdots & 0 \\ \vdots & \vdots & \ddots & \vdots \\ 0 & 0 & \cdots & 1 \end{pmatrix}, \quad \boldsymbol{A} = \begin{pmatrix} a_{11} & a_{12} & \cdots & a_{1n} \\ a_{21} & a_{22} & \cdots & a_{2n} \\ \vdots & \vdots & \ddots & \vdots \\ a_{n1} & a_{n2} & \cdots & a_{nn} \end{pmatrix},$$

$$\boldsymbol{I} - \boldsymbol{A} = \begin{pmatrix} 1-a_{11} & -a_{12} & \cdots & -a_{1n} \\ -a_{21} & 1-a_{22} & \cdots & -a_{2n} \\ \vdots & \vdots & \ddots & \vdots \\ -a_{n1} & -a_{n2} & \cdots & 1-a_{nn} \end{pmatrix}.$$

EXAMPLE 3.4:

$$A = \begin{pmatrix} 0.7 & 0.4 & 0.3 \\ 0.1 & 0.2 & 0.3 \\ 0.1 & 0.3 & 0.2 \end{pmatrix}$$

The matrix $\boldsymbol{A}$ corresponds to the system:

$$\begin{aligned} x_1 &= 0.7x_1 + 0.4x_2 + 0.3x_3 + f_1 \\ x_2 &= 0.1x_1 + 0.2x_2 + 0.3x_3 + f_2 \\ x_3 &= 0.1x_1 + 0.3x_2 + 0.2x_3 + f_3. \end{aligned}$$

So, to produce one dollar of x_1 requires 70 cents of x_1 as input, 10 cents of x_2 and 10 cents of x_3. Thus $(\boldsymbol{I} - \boldsymbol{A})\boldsymbol{x} = \boldsymbol{f}$ is:

$$\begin{pmatrix} 0.3 & -0.4 & -0.3 \\ -0.1 & 0.8 & -0.3 \\ -0.1 & -0.3 & 0.8 \end{pmatrix} \begin{pmatrix} x_1 \\ x_2 \\ x_3 \end{pmatrix} = \begin{pmatrix} f_1 \\ f_2 \\ f_3 \end{pmatrix}.$$

For a fixed value of $\boldsymbol{f}$ the unique solution to these equations is obtained: $\boldsymbol{x} = (\boldsymbol{I} - \boldsymbol{A})^{-1}\boldsymbol{f}$. For example,

$$\begin{aligned} \boldsymbol{f} = \begin{pmatrix} f_1 \\ f_2 \\ f_3 \end{pmatrix} = \begin{pmatrix} 90 \\ 75 \\ 80 \end{pmatrix} \Rightarrow \begin{pmatrix} x_1 \\ x_2 \\ x_3 \end{pmatrix} &= \begin{pmatrix} 0.3 & -0.4 & -0.3 \\ -0.1 & 0.8 & -0.3 \\ -0.1 & -0.3 & 0.8 \end{pmatrix}^{-1} \begin{pmatrix} 90 \\ 75 \\ 80 \end{pmatrix} \\ &= \begin{pmatrix} 1239.20 \\ 400.57 \\ 405.17 \end{pmatrix}. \end{aligned}$$

Thus, if the economy must meet the level of final demand given by the $\boldsymbol{f}$ vector, the required level of output in the industries is given by the $\boldsymbol{x}$ vector, $\boldsymbol{x}' = (1239.20, 400.57, 405.17)$. ◊

Provided $(\boldsymbol{I} - \boldsymbol{A})$ is non-singular, the industry output vector is $\boldsymbol{x} = (\boldsymbol{I} - \boldsymbol{A})^{-1}\boldsymbol{f}$. However, not only must $\boldsymbol{x}$ satisfy $\boldsymbol{x} = (\boldsymbol{I} - \boldsymbol{A})^{-1}\boldsymbol{f}$, it must also be the case that $\boldsymbol{x}$ is non-negative for it to be a feasible level of production; negative output cannot be produced. So, it must be that $x_i \geq 0$ for all i. Whether $\boldsymbol{x}$ is non-negative or not depends on the $\boldsymbol{A}$ matrix.

The following discussion shows that a sufficient condition for a non-negative solution to $\boldsymbol{x} = (\boldsymbol{I} - \boldsymbol{A})^{-1}\boldsymbol{f}$ is that the sum of elements in each column of $\boldsymbol{A}$ is less than 1. In this case, it turns out that all entries in the matrix $(\boldsymbol{I}-\boldsymbol{A})^{-1}$ are non-negative, so that $(\boldsymbol{I}-\boldsymbol{A})^{-1}\boldsymbol{f}$ is non-negative, since $\boldsymbol{f}$ is non-negative. When all column sums of $\boldsymbol{A}$ are less than 1, $\boldsymbol{A}^k$, the kth power of $\boldsymbol{A}$, approaches a matrix of 0s as k becomes large, and this fact is sufficient to ensure that $(\boldsymbol{I}-\boldsymbol{A})^{-1}$ has all non-negative entries. The next example illustrates this property; the key result is then developed.

EXAMPLE 3.5: To illustrate, consider the following input–output coefficient matrix $\boldsymbol{A}$ and final demand vector $\boldsymbol{f}$ given by:

$$\boldsymbol{A}=\begin{pmatrix}a_{11} & a_{12}\\ a_{21} & a_{22}\end{pmatrix}=\begin{pmatrix}0.2 & 0.5\\ 0.3 & 0.1\end{pmatrix},\qquad \boldsymbol{f}=\begin{pmatrix}f_1\\ f_2\end{pmatrix}=\begin{pmatrix}30\\ 60\end{pmatrix}.$$

Then,

$$\boldsymbol{I}-\boldsymbol{A}=\begin{pmatrix}1-a_{11} & -a_{12}\\ -a_{21} & 1-a_{22}\end{pmatrix}=\begin{pmatrix}0.8 & -0.5\\ -0.3 & 0.9\end{pmatrix}\quad\text{implies}\quad (\boldsymbol{I}-\boldsymbol{A})^{-1}=\begin{pmatrix}\frac{30}{19} & \frac{50}{57}\\ \frac{10}{19} & \frac{80}{57}\end{pmatrix}$$

so that

$$\begin{pmatrix}x_1\\ x_2\end{pmatrix}=\begin{pmatrix}\frac{30}{19} & \frac{50}{57}\\ \frac{10}{19} & \frac{80}{57}\end{pmatrix}\begin{pmatrix}30\\ 60\end{pmatrix}=\begin{pmatrix}100\\ 100\end{pmatrix}.$$

Notice that the solution is positive – both x_1 and x_2 are positive. This is not accidental: it is guaranteed by the fact that the column sums of the $\boldsymbol{A}$ matrix are both less than 1. With a view to motivating future discussion, consider multiplying $\boldsymbol{A}$ by itself multiple times: $\boldsymbol{A}^2=\boldsymbol{AA}$, $\boldsymbol{A}^3=\boldsymbol{AAA}$ and $\boldsymbol{A}^4=\boldsymbol{AAAA}$. Note that the column sums of the $\boldsymbol{A}$ matrix are $0.5=0.2+0.3$ (first column) and $0.6=0.5+0.1$ (second column), and that the largest of these two numbers is 0.6. Observe now that

$$\boldsymbol{A}^2=\begin{pmatrix}\frac{19}{100} & \frac{3}{20}\\ \frac{9}{100} & \frac{4}{25}\end{pmatrix},\quad \boldsymbol{A}^3=\begin{pmatrix}\frac{83}{1000} & \frac{11}{100}\\ \frac{33}{500} & \frac{61}{1000}\end{pmatrix},\quad \boldsymbol{A}^4=\begin{pmatrix}\frac{31}{625} & \frac{21}{400}\\ \frac{63}{2000} & \frac{391}{10000}\end{pmatrix},$$

and these matrices have column sums as follows. The column sums of $\boldsymbol{A}^2=\left(\frac{28}{100},\frac{31}{100}\right)=(0.28,0.31)$; column sums of $\boldsymbol{A}^3=\left(\frac{149}{1000},\frac{171}{1000}\right)=(0.149,0.171)$; and column sums of $\boldsymbol{A}^4=\left(\frac{811}{10000},\frac{916}{10000}\right)=(0.0811,0.0916)$. Finally, note that (a) the largest column sum in $\boldsymbol{A}^2$ is no larger than $(0.6)^2=0.36$, (b) the largest column sum in $\boldsymbol{A}^3$ is no larger than $(0.6)^3=0.216$ and (c) the largest column sum in $\boldsymbol{A}^4$ is no larger than $(0.6)^4=0.1296$. There is a good reason for this pattern and it has one particularly useful implication – that the solution, $\boldsymbol{x}$, is positive. These matters are discussed next and relate to the non-negativity of the solution to the input–output problem. ◇

Before proving the result (that $\boldsymbol{A}^k \to 0$ when all column sums of $\boldsymbol{A}$ are less than 1), a simple formula for calculating certain sums of scalars will provide useful insight. This formula is then generalized to matrices.

Let α be a real number ($\alpha \in \mathscr{R}$) with $|\alpha| < 1$. Then:

$$(1-\alpha)(1+\alpha+\alpha^2+\alpha^3+\alpha^4+\cdots+\alpha^n) = (1+\alpha+\alpha^2+\cdots+\alpha^n)-(\alpha+\alpha^2+\cdots+\alpha^{n+1}) = 1-\alpha^{n+1}.$$

Thus,

$$(1+\alpha+\alpha^2+\cdots+\alpha^n) = \frac{1-\alpha^{n+1}}{1-\alpha} = (1-\alpha)^{-1}(1-\alpha^{n+1}).$$

Since $|\alpha| < 1$, as n becomes large, $\alpha^{n+1} \to 0$. Consequently,

$$\lim_{n\to\infty}(1+\alpha+\alpha^2+\cdots+\alpha^n) = \lim_{n\to\infty}(1-\alpha)^{-1}(1-\alpha^{n+1}) = (1-\alpha)^{-1}.$$

Because $\alpha^n \to 0$ as $n \to \infty$, $(1+\alpha+\alpha^2+\cdots+\alpha^n+\cdots) = \frac{1}{1-\alpha}$.

This discussion applies to real numbers, but there is an analogous discussion for matrices. Let $\boldsymbol{A}^2 = \boldsymbol{A}\boldsymbol{A}$, $\boldsymbol{A}^3 = \boldsymbol{A}\boldsymbol{A}\boldsymbol{A}$ and so on. Then,

$$\begin{aligned}(\boldsymbol{I}-\boldsymbol{A})(\boldsymbol{I}+\boldsymbol{A}+\boldsymbol{A}^2+\boldsymbol{A}^3+\cdots+\boldsymbol{A}^n) &= (\boldsymbol{I}+\boldsymbol{A}+\boldsymbol{A}^2+\cdots+\boldsymbol{A}^n)\\ &\quad -(\boldsymbol{A}+\boldsymbol{A}^2+\cdots+\boldsymbol{A}^{n+1})\\ &= \boldsymbol{I}-\boldsymbol{A}^{n+1}.\end{aligned}$$

Thus

$$\boldsymbol{I}+\boldsymbol{A}+\boldsymbol{A}^2+\cdots+\boldsymbol{A}^n = (\boldsymbol{I}-\boldsymbol{A})^{-1}(\boldsymbol{I}-\boldsymbol{A}^{n+1}).$$

These calculations show that provided $\boldsymbol{A}^n \to \boldsymbol{0}$ as $n \to \infty$, $(\boldsymbol{1}+\boldsymbol{A}+\boldsymbol{A}^2+\boldsymbol{A}^3+\cdots+\boldsymbol{A}^n) \to (\boldsymbol{I}-\boldsymbol{A})^{-1}$. Thus, $(\boldsymbol{I}+\boldsymbol{A}+\boldsymbol{A}^2+\cdots+\boldsymbol{A}^n\cdots) = (\boldsymbol{I}-\boldsymbol{A})^{-1}$, if $\boldsymbol{A}^n \to \boldsymbol{0}$ as n becomes large.

Theorem 3.1: *A sufficient condition for $\boldsymbol{A}^n \to \boldsymbol{0}$ is that the sum of the elements in any column of $\boldsymbol{A}$ is less than one: $\sum_i a_{ij} < 1$ for all j.*

PROOF: If this condition holds then there is some number less than 1, and greater than any column sum: $\exists\, r < 1$ with $\sum_i a_{ij} \le r < 1, \forall j$. Denote the (i,j)th element of $\boldsymbol{A}^2$ by $a_{ij}^{(2)}$, the ith row of $\boldsymbol{A}$ by $\boldsymbol{a}_i$, and the jth column of $\boldsymbol{A}$ by $\boldsymbol{a}_j$. Then $a_{ij}^{(2)} = \boldsymbol{a}_i \cdot \boldsymbol{a}_j = \sum_k a_{ik}a_{kj}$. Thus $\sum_i a_{ij}^{(2)} = \sum_i \sum_k a_{ik}a_{kj} = \sum_k(\sum_i a_{ik}a_{kj})$

$= \sum_k(\sum_i a_{ik})a_{kj} \le r\sum_k a_{kj} \le r^2$, using the fact that $(\sum_i a_{ik}) \le r$. This implies that $a_{ij}^{(2)} \le r^2$. Proceeding in this way gives $a_{ij}^{(n)} \le r^n$ and since $r < 1$, $a_{ij}^{(n)} \to 0$ for all (i,j).

To see these calculations more clearly:

$$A^2 = \begin{pmatrix} \sum_k a_{1k}a_{k1} & \sum_k a_{1k}a_{k2} & \cdots & \sum_k a_{1k}a_{kn} \\ \sum_k a_{2k}a_{k1} & \sum_k a_{2k}a_{k2} & \cdots & \sum_k a_{2k}a_{kn} \\ \vdots & \vdots & \ddots & \vdots \\ \sum_k a_{nk}a_{k1} & \sum_k a_{nk}a_{k2} & \cdots & \sum_k a_{nk}a_{kn} \end{pmatrix}.$$

Consider the first column of A^2. Writing out all the terms:

$$\begin{array}{c} a_{11}a_{11} + a_{12}a_{21} + \cdots + a_{1n}a_{n1} \\ a_{21}a_{11} + a_{22}a_{21} + \cdots + a_{2n}a_{n1} \\ \vdots \qquad \vdots \qquad \vdots \\ a_{n1}a_{11} + a_{n2}a_{21} + \cdots + a_{nn}a_{n1} \end{array}.$$

Adding all these together:

$$a_{11}\overbrace{(a_{11} + a_{21} + \cdots + a_{n1})}^{\text{column 1 sum}} + a_{21}\overbrace{(a_{12} + a_{22} + \cdots + a_{n2})}^{\text{column 2 sum}}$$
$$+ \cdots + a_{n1}\overbrace{(a_{1n} + a_{2n} + \cdots + a_{nn})}^{\text{column n sum}}.$$

Since each column sum is no larger than r, this sum is no larger than $a_{11}r + a_{21}r + \cdots + a_{n1}r = (a_{11} + a_{21} + \cdots + a_{n1})r$, which for the same reason is no larger than r^2. The same reasoning applies to the other columns of A^2. ■

Returning to the discussion, $\lim_{n\to\infty} A^{n+1} = \mathbf{0}$, when the column sums of A are all less than 1. Thus $(\mathbf{1} + A + A^2 + \cdots) = (I - A)^{-1}$ and so $x = (I - A)^{-1}f = (I + A + A^2 + \cdots)f$. Since all the terms in the sum $f + Af + A^2f + A^3f + \cdots = (I + A + A^2 + \cdots)f$ are non-negative, this implies that x is non-negative.

EXAMPLE 3.6: This example illustrates how the polynomial expansion approximates the inverse.

$$A = \begin{pmatrix} 0.2 & 0.35 \\ 0.25 & 0.3 \end{pmatrix}, \quad A^2 = \begin{pmatrix} 0.1275 & 0.175 \\ 0.125 & 0.1775 \end{pmatrix},$$

$$A^3 = \begin{pmatrix} 0.0693 & 0.0971 \\ 0.0694 & 0.0970 \end{pmatrix}, \quad A^4 = \begin{pmatrix} 0.0381 & 0.0534 \\ 0.0381 & 0.0534 \end{pmatrix},$$

$$A^5 = \begin{pmatrix} 0.0210 & 0.0294 \\ 0.0210 & 0.0294 \end{pmatrix}, \; A^6 = \begin{pmatrix} 0.0115 & 0.0161 \\ 0.0115 & 0.0161 \end{pmatrix},$$

$$A^7 = \begin{pmatrix} 0.0063 & 0.0089 \\ 0.0063 & 0.0089 \end{pmatrix}, \; A^8 = \begin{pmatrix} 0.0035 & 0.0049 \\ 0.0035 & 0.0049 \end{pmatrix}.$$

So,

$$I + A + A^2 + A^3 + A^4 + A^5 + A^6 + A^7 + A^8 = \begin{pmatrix} 1.4772 & 0.7348 \\ 0.5248 & 1.6872 \end{pmatrix},$$

and

$$(I - A)^{-1} = \begin{pmatrix} 1.4815 & 0.7407 \\ 0.5291 & 1.6931 \end{pmatrix}.$$

Thus,

$$(I - A)^{-1} - (I + A + A^2 + A^3 + A^4 + A^5 + A^6 + A^7 + A^8) = \begin{pmatrix} 0.0043 & 0.0060 \\ 0.0043 & 0.0060 \end{pmatrix},$$

which is very small. So $(I+A+A^2+A^3+A^4+A^5+A^6+A^7+A^8)$ approximates $(I - A)^{-1}$ accurately to the first two decimal places. ◇

REMARK 3.2: Given $f \geq 0$, the key requirement for $x \geq 0$ is that $(I - A)^{-1}$ has all non-negative entries. This is a necessary condition: if the (i,j)th element of $(I - A)^{-1}$ is negative, a final consumption vector f with 1 in the ith row produces an x vector with a negative entry in the ith position. Consider the condition: $\exists\, x^* \geq 0$ with $x^* \gg Ax^*$, where $a \gg b$ means that $a_i > b_i$ for all i. Note that this implies that $x^* \gg 0$. Since $x^* \gg A^k x^*$ implies the elements of $A^k x^*$ are bounded, so $A^k x^*$ or some subsequence converges to a vector z giving $x^* \geq z \geq 0$. Since $x^* \gg Ax^*$, there is a $\lambda < 1$ with $\lambda x^* \gg Ax^*$. Therefore, since all entries in x^* and Ax^* are non-negative, $A(\lambda x^*) \geq AAx^*$ or $\lambda Ax^* \geq A^2 x^*$ ($\lambda Ax^* \gg A^2 x^*$ if each row of A is strictly positive). And, since $\lambda x^* \gg Ax^*$, $\lambda^2 x^* \gg \lambda Ax^*$. Therefore $\lambda^2 x^* \gg A^2 x^*$. Multiplying this expression by A with the same observations gives $\lambda^3 x^* \gg A^3 x^*$. With k iterations, $\lambda^k x^* \gg A^k x^*$ and since $\lambda^k x^* \to 0$, $A^k x^* \to 0$. And since $x^* \gg 0$, $A^k \to 0$.

Finally, suppose that the row sums of A are all less than 1. Let x^* be a vector of 1s, so that Ax^* is a vector with the row sums of A. Since the row sums are all less than 1, $x^* \gg Ax^*$, and the previous discussion implies that $A^k \to 0$. This may be seen directly. Recall that in A^2, the first row consisted of the terms:

$\{\sum_k a_{1k}a_{kj}\}_{j=1}^n$. Expanding these:

$$\begin{aligned}
\sum_k a_{1k}a_{k1} &= a_{11}a_{11} + a_{12}a_{21} + a_{13}a_{31} + \cdots + a_{1n}a_{n1} \\
\sum_k a_{1k}a_{k2} &= a_{11}a_{12} + a_{12}a_{22} + a_{13}a_{32} + \cdots + a_{1n}a_{n2} \\
\vdots &= \qquad\qquad \vdots \\
\sum_k a_{1k}a_{kn} &= a_{11}a_{1n} + a_{12}a_{2n} + a_{13}a_{3n} + \cdots + a_{1n}a_{nn}.
\end{aligned}$$

Adding these terms to obtain the first row sum in $\boldsymbol{A}^2$:

$$\begin{aligned}
&a_{11}(a_{11} + a_{12} + \cdots + a_{1n}) + a_{12}(a_{21} + a_{22} + \cdots + a_{2n}) \\
&\quad + \cdots + a_{1n}(a_{n1} + a_{n2} + \cdots + a_{nn}).
\end{aligned}$$

Letting d_i be the row sum of i, this is equal to $a_{11}d_1 + a_{12}d_2 + \cdots + a_{1n}d_n$. And letting $d = \max_i d_i = d < 1$, the sum is no larger than $(a_{11}+a_{12}+\cdots+a_{1n})d$, which in turn is no larger than d^2. Thus, the row sum of $\boldsymbol{A}^k$ is no larger than d^k, which converges to 0, so that $\boldsymbol{A}^k \to 0$. □

3.5 Inflation and unemployment dynamics

Earlier calculations using a model of the Phillips curve and Okun's law led to the equation system:

$$\begin{aligned}
\pi_t &= \frac{4}{5}\pi_{t-1} - \frac{2}{5}u_{t-1} + \left[\frac{1}{5}g + \frac{2}{5}\bar{u}\right] \\
u_t &= \frac{2}{5}\pi_{t-1} + \frac{4}{5}u_{t-1} + \left[-\frac{2}{5}g + \frac{1}{5}\bar{u}\right].
\end{aligned}$$

Rewriting in matrix form,

$$\begin{pmatrix}\pi_t \\ u_t\end{pmatrix} = \begin{pmatrix}\frac{4}{5} & -\frac{2}{5} \\ \frac{2}{5} & \frac{4}{5}\end{pmatrix}\begin{pmatrix}\pi_{t-1} \\ u_{t-1}\end{pmatrix} + \begin{pmatrix}\frac{1}{5}g + \frac{2}{5}\bar{u} \\ -\frac{2}{5}g + \frac{1}{5}\bar{u}\end{pmatrix}.$$

We can write this system succinctly as: $v_t = \boldsymbol{A}v_{t-1} + \delta$. At a steady state, $v_t = v_{t-1} = v^*$ and $v^* = \boldsymbol{A}v^* + \delta$ so that $v^* = (\boldsymbol{I} - \boldsymbol{A})^{-1}\delta$ or $(\boldsymbol{I} - \boldsymbol{A})v^* = \delta$. Here,

$$(\boldsymbol{I} - \boldsymbol{A})v^* = \delta \quad \Longleftrightarrow \quad \begin{pmatrix}\frac{1}{5} & \frac{2}{5} \\ -\frac{2}{5} & \frac{1}{5}\end{pmatrix}\begin{pmatrix}\pi^* \\ u^*\end{pmatrix} = \begin{pmatrix}\frac{1}{5}g + \frac{2}{5}\bar{u} \\ -\frac{2}{5}g + \frac{1}{5}\bar{u}\end{pmatrix}$$

or

$$\begin{aligned} \frac{1}{5}\pi^* + \frac{2}{5}u^* &= \frac{1}{5}g + \frac{2}{5}\bar{u} \\ -\frac{2}{5}\pi^* + \frac{1}{5}u^* &= -\frac{2}{5}g + \frac{1}{5}\bar{u}, \end{aligned}$$

with solution $\pi^* = g$ and $u^* = \bar{u}$.

Returning to the equation $v_t = \boldsymbol{A}v_{t-1} + \delta$, whether v_t converges or not depends on the matrix $\boldsymbol{A}$. Substituting for v_{t-1},

$$v_t = \boldsymbol{A}[\boldsymbol{A}v_{t-2} + \delta] + \delta = \boldsymbol{A}^2 v_{t-2} + \boldsymbol{A}\delta + \delta.$$

And continuing, $v_t = \boldsymbol{A}^t v_0 + [\boldsymbol{I} + \boldsymbol{A} + \boldsymbol{A}^2 + \cdots + \boldsymbol{A}^{t-1}]\delta$. If $\boldsymbol{A}^t \to 0$, then $[\boldsymbol{I} + \boldsymbol{A} + \boldsymbol{A}^2 + \cdots + \boldsymbol{A}^{t-1}] \to (\boldsymbol{I} - \boldsymbol{A})^{-1}$ and $v_t \to v^* = (\boldsymbol{I} - \boldsymbol{A})^{-1}\delta$. (The convergence of $\boldsymbol{A}^t$ is discussed in a Chapter 15, which deals with eigenvalues.) In this case, starting from any point (π_0, u_0), the system converges over time to $(\pi^*, u^*) = (g, \bar{u})$.

3.6 Stationary (invariant) distributions

The notion of a stationary distribution is most easily motivated by an example. Suppose that a worker currently employed has a 95% probability of being employed next period and a worker currently unemployed has a 25% probability of being employed next period. Letting E denote "employed" and U denote "unemployed"; write $P(U \mid E)$ to denote the probability of being unemployed next period given that one is currently employed. With this notation, $P(E \mid E) = 0.95$, $P(U \mid E) = 0.05$, $P(E \mid U) = 0.25$ and $P(U \mid U) = 0.75$. This defines a *transition matrix* or *Markov matrix*, $\boldsymbol{P}$:

$$\boldsymbol{P} = \begin{pmatrix} P(E \mid E) & P(U \mid E) \\ P(E \mid U) & P(U \mid U) \end{pmatrix} = \begin{pmatrix} 0.95 & 0.05 \\ 0.25 & 0.75 \end{pmatrix}$$

We are interested in the evolution of employment and unemployment in the workforce. Suppose that initially (currently), the fraction π_E of the workforce is employed, and the fraction π_U unemployed. Of workers currently employed, the fraction $P(E \mid E)$ will continue to be employed next period – so the probability of being currently employed and continuing to be employed next period is $\pi_E P(E \mid E)$. Of workers currently unemployed, the fraction $P(E \mid U)$ will find employment next period – so the probability of being currently unemployed and finding employment next period is $\pi_U P(E \mid U)$. The fraction of the workforce employed next period, $\hat{\pi}_E$, is equal to the fraction of those currently employed who retain employment, $\pi_E P(E \mid E)$, plus the fraction of those currently unemployed who gain employment, $\pi_U P(E \mid U)$: $\hat{\pi}_E = \pi_E P(E \mid E) + \pi_U P(E \mid U)$. Similarly, the fraction of the workforce unemployed next period is $\hat{\pi}_U = \pi_E P(U \mid E) + \pi_U P(U \mid U)$. For example, suppose that $\pi_E = 0.7$ and $\pi_U = 0.3$. Then $\hat{\pi}_E = (0.7)(0.95) + (0.3)(0.25) = 0.74$, and $\hat{\pi}_U = (0.7)(0.05) + (0.3)(0.75) = 0.26$. Notice that:

$$(\hat{\pi}_E, \hat{\pi}_U) = (\pi_E, \pi_U)\begin{pmatrix} P(E \mid E) & P(U \mid E) \\ P(E \mid U) & P(U \mid U) \end{pmatrix}.$$

Writing $\boldsymbol{\pi} = (\pi_E, \pi_U)$, this gives $\hat{\boldsymbol{\pi}} = \boldsymbol{\pi}\boldsymbol{P}$. Because this process proceeds over time, let $\boldsymbol{\pi}(t) = (\pi_E(t), \pi_U(t))$ denote the distribution of states (employment and unemployment) in period t, so the evolution of the process proceeds according to the formula $\boldsymbol{\pi}(t+1) = \boldsymbol{\pi}(t)\boldsymbol{P}$.

Proceeding, step by step, if $\boldsymbol{\pi}(0)$ denotes the current distribution of employed and unemployed, next period the distribution is determined by $\boldsymbol{\pi}(1) = \boldsymbol{\pi}(0)\boldsymbol{P}$. Repeating this reasoning, in the third period, the distribution is determined by $\boldsymbol{\pi}(2) = \boldsymbol{\pi}(1)\boldsymbol{P} = \boldsymbol{\pi}(0)\boldsymbol{P}\boldsymbol{P} = \boldsymbol{\pi}(0)\boldsymbol{P}^2$. In the $(t+1)$th period, this becomes $\boldsymbol{\pi}(t) = \boldsymbol{\pi}(0)\boldsymbol{P}^t$. In the case where the matrix $\boldsymbol{P}$ has all positive entries, $\boldsymbol{P}^t$ converges to a matrix, $\boldsymbol{P}^*$, where in any given column all entries are the same, or equivalently, all rows are identical (see Section 3.6.1). In the two state case,

$$\boldsymbol{P}^* = \begin{pmatrix} \alpha & \beta \\ \alpha & \beta \end{pmatrix},$$

where α and β are positive and sum to 1, $(\alpha + \beta = 1)$. Notice that

$$\boldsymbol{\pi}(0)\boldsymbol{P}^* = (\pi_E(0), \pi_U(0))\begin{pmatrix} \alpha & \beta \\ \alpha & \beta \end{pmatrix} = (\alpha, \beta).$$

The important point to observe is that $\boldsymbol{\pi}(0)\boldsymbol{P}^* = (\alpha, \beta)$, *regardless of the particular value of* $\boldsymbol{\pi}(0)$. Write $\boldsymbol{\pi}^* = \boldsymbol{\pi}(0)\boldsymbol{P}^*$ to denote this distribution, the long run *stationary* or *invariant* distribution. Another way of viewing this is to observe that $\boldsymbol{\pi}^*$ is the *only* distribution that satisfies the equation $\boldsymbol{\pi} = \boldsymbol{\pi}\boldsymbol{P}^*$: $\boldsymbol{\pi}^* = \boldsymbol{\pi}^*\boldsymbol{P}^*$ and $\hat{\boldsymbol{\pi}} \neq \boldsymbol{\pi}^*$ implies that $\hat{\boldsymbol{\pi}} \neq \hat{\boldsymbol{\pi}}\boldsymbol{P}^*$, since $\hat{\boldsymbol{\pi}}\boldsymbol{P}^* = \boldsymbol{\pi}^*$.

The following calculations illustrate these observations using the example.

$$\boldsymbol{P} = \begin{pmatrix} 0.95 & 0.05 \\ 0.25 & 0.75 \end{pmatrix}, \boldsymbol{P}^2 = \begin{pmatrix} 0.915 & 0.085 \\ 0.425 & 0.575 \end{pmatrix},$$

$$\boldsymbol{P}^3 = \begin{pmatrix} 0.8905 & 0.1095 \\ 0.5475 & 0.4525 \end{pmatrix}, \boldsymbol{P}^{10} = \begin{pmatrix} 0.8380 & 0.1620 \\ 0.8098 & 0.1902 \end{pmatrix},$$

$$\boldsymbol{P}^{20} = \begin{pmatrix} 0.8335 & 0.1665 \\ 0.8327 & 0.1673 \end{pmatrix} \boldsymbol{P}^{50} = \begin{pmatrix} 0.8333 & 0.1667 \\ 0.8333 & 0.1667 \end{pmatrix}, \boldsymbol{P}^{100} = \begin{pmatrix} 0.8333 & 0.1667 \\ 0.8333 & 0.1667 \end{pmatrix}.$$

If the initial distribution is $\pi(0) = (0.7, 0.3)$, then

$$\begin{aligned}
\boldsymbol{\pi}(1) &= \boldsymbol{\pi}(0)\boldsymbol{P} = \boldsymbol{\pi}(0)\boldsymbol{P} &&= (0.740, 0.260),\\
\boldsymbol{\pi}(2) &= \boldsymbol{\pi}(1)\boldsymbol{P} = \boldsymbol{\pi}(0)\boldsymbol{P}^2 &&= (0.7680, 0.2320)\\
\boldsymbol{\pi}(3) &= \boldsymbol{\pi}(2)\boldsymbol{P} = \boldsymbol{\pi}(0)\boldsymbol{P}^2 &&= (0.7876, 0.2124),\\
\boldsymbol{\pi}(10) &= \boldsymbol{\pi}(9)\boldsymbol{P} = \boldsymbol{\pi}(0)\boldsymbol{P}^9 &&= (0.8296, 0.1704)\\
\boldsymbol{\pi}(50) &= \boldsymbol{\pi}(49)\boldsymbol{P} = \boldsymbol{\pi}(0)\boldsymbol{P}^{50} &&= (0.8333, 0.1667),\\
\boldsymbol{\pi}(100) &= \boldsymbol{\pi}(99)\boldsymbol{P} = \boldsymbol{\pi}(0)\boldsymbol{P}^{100} &&= (0.8333, 0.1667).
\end{aligned}$$

Therefore, in the long run, regardless of the initial employment–unemployment distribution, the long-run distribution is $\boldsymbol{\pi}^* = (0.8333, 0.1667) = \left(\frac{5}{6}, \frac{1}{6}\right)$. The fraction of the workforce employed in the long run is $\frac{5}{6}$.

Although this discussion has centered around the two-state case (the states being employed and unemployed), the principles apply generally. If there are n states, labeled $s_1, \ldots, s_n$ with probability p_{ij} of going from state s_i to s_j, then the corresponding transition matrix is:

$$\boldsymbol{P} = \begin{pmatrix} p_{11} & p_{12} & \cdots & p_{1n} \\ p_{21} & p_{22} & \cdots & p_{2n} \\ \vdots & \vdots & \ddots & \vdots \\ p_{n1} & p_{n2} & \cdots & p_{nn} \end{pmatrix},$$

where each row sums to 1: for each i, $\sum_{j=1}^{n} p_{ij} = 1$. If $p_{ij} > 0$ for all (i,j) then $\boldsymbol{P}^t$ converges to a matrix $\boldsymbol{P}^*$ as t becomes large, where the matrix $\boldsymbol{P}^*$ has all rows equal or, equivalently, the numbers in each column are the same. If $\boldsymbol{\pi}^*$ is the invariant distribution of $\boldsymbol{P}$ so $\boldsymbol{\pi}^* = \boldsymbol{\pi}^*\boldsymbol{P}$, then

$$\boldsymbol{P}^* = \begin{pmatrix} \boldsymbol{\pi}^* \\ \vdots \\ \boldsymbol{\pi}^* \end{pmatrix},$$

a matrix with n rows, each one equal to the invariant distribution. This raises the question: Why does this convergence take place?

3.6.1 Convergence to a stationary distribution

Let $\boldsymbol{P}$ be a transition matrix with representative entry p_{ij}: p_{ij} is the probability that state j will be reached next time, given that the current state is i. Let M_j be the largest number in column of j, and m_j be the smallest number in column j.

The matrix $\boldsymbol{P}^2 = \boldsymbol{P}\boldsymbol{P}$ has entry $p_{ij}^{(2)}$ in the (i,j)th position, where $p_{ij}^{(2)} = \sum_k p_{ik} p_{kj}$. Say that $\boldsymbol{P}$ is strictly positive if there is some (small) strictly positive number ϵ such that $p_{ij} \geq \epsilon$ for every i, j.

Theorem 3.2: *Suppose that $\boldsymbol{P}$ is strictly positive. Then $\boldsymbol{P}^t$ converges to a matrix $\boldsymbol{P}^*$ where all rows of $\boldsymbol{P}^*$ are identical.*

PROOF: Considering column j of $\boldsymbol{P}$, since $p_{kj} \geq m_j$ for all k and for some k, say k^*, $p_{k^*j} = M_j$, we see that

$$p_{ij}^{(2)} = \sum_k p_{ik} p_{kj} = p_{ik^*} p_{k^*j} + \sum_{k \neq k^*} p_{ik} p_{kj} = p_{ik^*} M_j + \sum_{k \neq k^*} p_{ik} p_{kj}.$$

Every element in column j is at least as large as m_j, $p_{kj} \geq m_j$, so that

$$\sum_{k \neq k^*} p_{ik} p_{kj} \geq \sum_{k \neq k^*} p_{ik} m_j = m_j \sum_{k \neq k^*} p_{ik} = m_j(1 - p_{ik^*}).$$

(In this calculation, because each row sums to 1, $\sum_{k \neq k^*} p_{ik} + p_{ik^*} = 1$, $\sum_{k \neq k^*} p_{ik} = 1 - p_{ik^*}$.) Therefore,

$$p_{ij}^{(2)} = \sum_k p_{ik} p_{kj} \geq p_{ik^*} M_j + (1 - p_{ik^*}) m_j \geq \epsilon M_j + (1 - \epsilon) m_j = m_j + \epsilon(M_j - m_j),$$

where the inequality follows because $p_{ik^*} \geq \epsilon$. Exactly the same reasoning shows that:

$$p_{ij}^{(2)} = \sum_k p_{ik} p_{kj} \leq (1 - \epsilon) M_j + \epsilon m_j = M_j - \epsilon(M_j - m_j).$$

Therefore, every entry of the jth column of $\boldsymbol{P}^2$ is at least as large as $m_j + \epsilon(M_j - m_j)$, and no entry of the jth column of $\boldsymbol{P}^2$ is larger than $M_j - \epsilon(M_j - m_j)$. This shows that elements in the columns of $\boldsymbol{P}^2$ are "closer" together than elements in the columns of $\boldsymbol{P}$. If M_j^2 and m_j^2 are the largest and smallest entries in column j of $\boldsymbol{P}^2$ then

$$\{m_j + \epsilon(M_j - m_j)\} \leq m_j^2 \leq M_j^2 \leq \{M_j - \epsilon(M_j - m_j)\},$$

so

$$M_j^2 - m_j^2 \leq \{M_j - \epsilon(M_j - m_j)\} - \{m_j + \epsilon(M_j - m_j)\} = (1 - 2\epsilon)(M_j - m_j).$$

The same reasoning shows that with M_j^t and m_j^t the largest and smallest entries in column j of $\boldsymbol{P}^t$, $M_j^{t+1} - m_j^{t+1} \leq (1 - 2\epsilon)^t (M_j - m_j) \to 0$ as t becomes large. ■

The intuition is simple. In the jth column of $\boldsymbol{P}^2$, entry (i,j) is $P_{ij}^{(2)} = \sum_k p_{ik}p_{kj}$. Therefore, $P_{ij}^{(2)}$ is a weighted average of the elements of column j of $\boldsymbol{P}$. Since all weights are strictly positive, this weighted average must be smaller than the largest entry in column j and larger than the smallest entry in this column. (Assuming not all entries of column j are equal.)

It is worth remarking that the assumption that every entry in $\boldsymbol{P}$ is positive gives a *sufficient* condition for $\boldsymbol{P}^t$ to converge to a matrix $\boldsymbol{P}^*$ with all rows equal. But it is not necessary – weaker conditions on $\boldsymbol{P}$ are sufficient to ensure convergence, but are more complex to describe.

3.6.2 Computation of the invariant distribution

Finding the long-run stationary distribution would be tedious if computed by calculating $\boldsymbol{P}^t$ for large values of t. There is an easy way to find the stationary distribution. Let $\boldsymbol{I}$ be the identity matrix, and define $\boldsymbol{Q} = \boldsymbol{I} - \boldsymbol{P}$. Let c_{ii} be the cofactor in the ith row and column of the cofactor matrix of $\boldsymbol{Q}$, and let $c = \sum_{i=1}^n c_{ii}$. Then $\pi^* = (\pi_1^*, \pi_2^*, \ldots, \pi_n^*)$ is given by

$$\pi_i^* = \frac{c_{ii}}{c}, \; i = 1, \ldots, n.$$

Applying this to the example discussed earlier,

$$\boldsymbol{Q} = \boldsymbol{I} - \boldsymbol{P} = \begin{pmatrix} 1-0.95 & -0.05 \\ -0.25 & 1-0.75 \end{pmatrix} = \begin{pmatrix} 0.05 & -0.05 \\ -0.25 & 0.25 \end{pmatrix}.$$

Thus, $c_{11} = 0.25$, $c_{22} = 0.05$, $c = 0.25 + 0.05 = 0.3$, so that $\pi_1^* = \frac{0.25}{0.3} = \frac{5}{6}$ and $\pi_2^* = \frac{0.05}{0.3} = \frac{1}{6}$.

EXAMPLE 3.7: Suppose that the following statistics for a baseball player are known. When batting, there is a $\frac{3}{5}$ chance that he will hit the ball (on a swing), given that he hit the ball on his most recent swing. Write this as $P(H \mid H) = \frac{3}{5}$. Thus, the probability that he will miss on a swing, given that he hit the ball on his most recent swing, is $P(M \mid H) = \frac{2}{5}$. The probabilities of hit and miss given that he missed on the most recent swing are $P(H \mid M) = \frac{1}{5}$, $P(M \mid M) = \frac{4}{5}$. The transition matrix is then:

$$\boldsymbol{T} = \begin{pmatrix} P(H \mid H) & P(M \mid H) \\ P(H \mid M) & P(M \mid M) \end{pmatrix} = \begin{pmatrix} \frac{3}{5} & \frac{2}{5} \\ \frac{1}{5} & \frac{4}{5} \end{pmatrix}$$

What is the (unique) stationary or invariant distribution of this transition matrix? On average (over the season), how often does the batter miss the ball on a swing?

First:

$$\boldsymbol{I}-\boldsymbol{T}=\begin{pmatrix}1-\frac{3}{5} & -\frac{2}{5}\\ -\frac{1}{5} & 1-\frac{4}{5}\end{pmatrix}=\begin{pmatrix}\frac{2}{5} & -\frac{2}{5}\\ -\frac{1}{5} & \frac{1}{5}\end{pmatrix}$$

The cofactors of $\boldsymbol{I}-\boldsymbol{T}$ are $c_{11}=\frac{1}{5}$, $c_{22}=\frac{2}{5}$ so that $c=c_{11}+c_{22}=\frac{3}{5}$. The invariant distribution is: $\left(\frac{1}{3},\frac{2}{3}\right)$. ◊

EXAMPLE 3.8: Students never sleep – they spend all their time in the restaurant (R) or library (L). At any given time, a student in the library has a 75% chance of being in the library an hour later (and a 25% chance of being in the restaurant). A student in the restaurant has an 80% chance of being in the restaurant an hour later (and a 20% chance of being in the library). These probabilities determine the following transition matrix:

$$\boldsymbol{P}=\begin{pmatrix}P(L\mid L) & P(R\mid L)\\ P(L\mid R) & P(R\mid R)\end{pmatrix}=\begin{pmatrix}\frac{3}{4} & \frac{1}{4}\\ \frac{1}{5} & \frac{4}{5}\end{pmatrix}$$

On "average", what percentage of the student population is in the library (i.e., find the stationary distribution of the transition matrix)? In this case,

$$\boldsymbol{Q}=\boldsymbol{I}-\boldsymbol{P}=\begin{pmatrix}\frac{1}{4} & -\frac{1}{4}\\ -\frac{1}{5} & \frac{1}{5}\end{pmatrix}$$

Thus, the cofactors are $c_{11}=\frac{1}{5}$ and $c_{22}=\frac{1}{4}$, so that $c=c_{11}+c_{22}=\frac{9}{20}$. Therefore $\pi_1=\frac{4}{9}$ and $\pi_2=\frac{5}{9}$. So, on average $\frac{4}{9}$ or 44.4% of students are in the library. ◊

3.7 Econometrics

A common problem in statistics and economics is the measurement of the relation between different variables. Suppose there is a collection of observations on a variable, y_t: $(y_1,y_2,\ldots,y_T)$ and on a second variable x_t: $(x_1,x_2,\ldots,x_T)$. Suppose also that these variables are linearly related. The discussion here describes a statistical process for measuring this relationship.

When the data sets consist of just two points, (y_1,y_2) and (x_1,x_2), then these may be arranged in pairs, (x_1,y_1) and (x_2,y_2), and placed on a graph as in Figure 3.1(A). Because there are only two points, one can draw a straight line

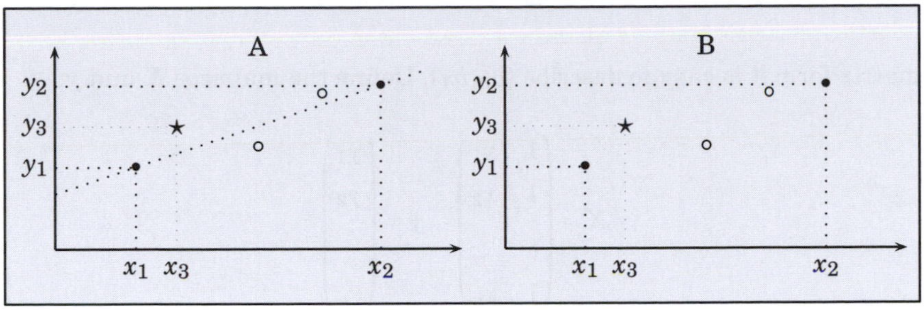

Figure 3.1: Fitting a line

through both points as indicated. The slope is given by $b = \frac{y_2-y_1}{x_2-x_1}$. Since the equation satisfies $y_i = a+bx_i$, the intercept is given by $a = y_1-bx_1 = y_2-bx_2$. Thus, the relationship between y and x is fully described by the intercept and slope, (a,b).

Suppose now there is a third point (x_3,y_3), located at $\star$ on the graph. This point does not lie on the line, and it is not possible to draw a straight line through these three points. And one could have still more points – such as those indicated by $\circ$. In the graph, it seems reasonable to say that x and y are positively related – when x increases, so does y. Also, the relationship appears "linear". From a different perspective, one can imagine starting with the data as in Figure 3.1(B), and trying to draw a straight line as close to all the points as possible. This problem arises in econometric theory, where there is a presumed relationship between y and x variables of the form:

$$y_i = \beta_0 + \beta_1 x_i + \varepsilon_i$$

If the x variable takes on the value x_i, the corresponding y variable is determined by multiplying x_i by a constant β_1, and adding an intercept β_0 and a random error term ε_i. The theory is that y is determined linearly from x, plus a small random shock. Given a finite collection of observations, $(y_1,x_1),\cdots,(y_n,x_n)$, how should one estimate β_0 and β_1? Suppose b_0 and b_1 are the estimates for β_0 and β_1. Then, the ith observation has a gap between y_i and its estimated value (according to the estimated line given the intercept b_0 and slope b_1) of $b_0 + b_1 x_i$ equal to

$$e_i = y_i - b_0 - b_1 x_i.$$

Ordinary least squares requires minimizing the sum of squares of these errors by choice of b_0 and b_1.

$$S = \sum_{i=1}^{n} (y_i - b_0 - b_1 x_i)^2. \tag{3.1}$$

In matrix form it is easy to describe (b_0, b_1). Define the matrices $\boldsymbol{X}$ and $\boldsymbol{y}$:

$$\boldsymbol{X} = \begin{pmatrix} 1 & x_1 \\ 1 & x_2 \\ \vdots & \vdots \\ 1 & x_n \end{pmatrix} \quad \boldsymbol{y} = \begin{pmatrix} y_1 \\ y_2 \\ \vdots \\ y_n \end{pmatrix}$$

It turns out that the choice of b_0 and b_1 that minimizes the sum of squared errors in Equation 3.1 is given by the following expression:

$$\begin{pmatrix} b_0 \\ b_1 \end{pmatrix} = \boldsymbol{b} = (\boldsymbol{X}'\boldsymbol{X})^{-1}\boldsymbol{X}'\boldsymbol{y}.$$

EXAMPLE 3.9: Suppose that there are five observations on (x, y) pairs: $x = (3,4,6,2,5)$ and $y = (7,9,12,5,11)$. Thus:

$$\boldsymbol{X} = \begin{pmatrix} 1 & 3 \\ 1 & 4 \\ 1 & 6 \\ 1 & 2 \\ 1 & 5 \end{pmatrix}, \quad \boldsymbol{y} = \begin{pmatrix} 7 \\ 9 \\ 12 \\ 5 \\ 11 \end{pmatrix}$$

and

$$\boldsymbol{X}'\boldsymbol{X} = \begin{pmatrix} 5 & 20 \\ 20 & 90 \end{pmatrix}, \quad (\boldsymbol{X}'\boldsymbol{X})^{-1} = \begin{pmatrix} \frac{9}{5} & -\frac{2}{5} \\ -\frac{2}{5} & \frac{1}{10} \end{pmatrix}, \quad \boldsymbol{X}'\boldsymbol{y} = \begin{pmatrix} 44 \\ 194 \end{pmatrix}.$$

Therefore,

$$\boldsymbol{b} = \begin{pmatrix} b_0 \\ b_1 \end{pmatrix} = (\boldsymbol{X}'\boldsymbol{X})^{-1}\boldsymbol{X}'\boldsymbol{y} = \begin{pmatrix} \frac{9}{5} & -\frac{2}{5} \\ -\frac{2}{5} & \frac{1}{10} \end{pmatrix} \begin{pmatrix} 44 \\ 194 \end{pmatrix} = \begin{pmatrix} \frac{8}{5} \\ \frac{9}{5} \end{pmatrix}.$$

◇

REMARK 3.3: This calculation assumes that $(\boldsymbol{X}'\boldsymbol{X})^{-1}$ exists – that $\boldsymbol{X}'\boldsymbol{X}$ has an inverse. What if $(\boldsymbol{X}'\boldsymbol{X})^{-1}$ does not exist? If $r(\boldsymbol{X}'\boldsymbol{X}) < 2$ then since

$r(\boldsymbol{X}'\boldsymbol{X}) = \min\{r(\boldsymbol{X}'), r(\boldsymbol{X})\}$ and $r(\boldsymbol{X}') = r(\boldsymbol{X})$, $r(\boldsymbol{X}) < 2$. This implies that the columns of $\boldsymbol{X}$ are linearly dependent so that $(x_1,\ldots,x_n) = \alpha(1,\ldots,1)$, with $\alpha \neq 0$. Thus, $x_i = \alpha, \forall i$ and so $\sum x_i = n\alpha$, $\bar{x} = \frac{1}{n}\sum x_i = \alpha$, so $\bar{x} = x_i, \forall i$ and $\sum(x_i - \bar{x})^2 = 0$. In this case, b_1 cannot be computed. Thus, a sufficient condition for the estimator to exist is that $x \neq \alpha(1,\ldots,1)$ for any α.

The general case with many explanatory variables is similar.

$$y_i = \beta_0 + \beta_1 x_{1i} + \beta_2 x_{2i} + \cdots + \beta_k x_{ki} + \epsilon_i.$$

In vector notation:

$$\boldsymbol{y} = \boldsymbol{X}\beta + \epsilon,$$

where $\boldsymbol{y} - n \times 1$, $\boldsymbol{x} - n \times (k+1)$, $\epsilon - n \times 1$. The least squares estimator is (as before):

$$\boldsymbol{b} = (\boldsymbol{X}'\boldsymbol{X})^{-1}\boldsymbol{X}'\boldsymbol{y}, \text{where } \boldsymbol{X}'\boldsymbol{X} - (k+1) \times (k+1).$$

For $(\boldsymbol{X}'\boldsymbol{X})^{-1}$ to exist, we require that $r(\boldsymbol{X}'\boldsymbol{X}) = k+1$. Since $r(\boldsymbol{X}'\boldsymbol{X}) = r(\boldsymbol{X}) \leq \min(k+1, n)$, in particular, this requires $n \geq k+1$ (the number of variables to be estimated, k, is no greater than the number of observations). □

The following discussion clarifies some details on the computation of the least squares estimator. This requires knowledge calculus.

3.7.1 Derivation of the least squares estimator

The sum of squared errors is given by:

$$\begin{aligned} S = \sum_{i=1}^{n} e_i^2 &= \sum_{i=1}^{n} (y_i - b_0 - b_1 x_i)^2 \\ &= \sum_{i=1}^{n} (y_i^2 - 2b_0 y_i - 2b_1 x_i y_i + b_0^2 + 2b_0 b_1 x_i + b_1^2 x_1^2). \end{aligned}$$

Minimization of this expression by choice of β_0 and β_1 requires calculus – discussed in a later chapter. To minimize this expression, differentiate S with respect to b_0 and b_1 in turn and set the partial derivatives to 0. This gives:

$$\begin{aligned} \frac{\partial S}{\partial b_0} &= 2\sum_{i=1}^{n} y_i - 2\sum_{i=1}^{n} b_0 - 2b_1 \sum_{i=1}^{n} x_i = 0 \\ \frac{\partial S}{\partial b_1} &= -2\sum_{i=1}^{n} x_i y_i + 2b_0 \sum_{i=1}^{n} x_i + 2b_1 \sum_{i=1}^{n} x_i^2 = 0 \end{aligned}$$

Dividing both equations by 2 and rearranging:

$$\sum_{i=1}^{n} y_i = \sum_{i=1}^{n} b_0 + b_1 \sum_{i=1}^{n} x_i = nb_0 + b_1 \sum_{i=1}^{n} x_i$$
$$\sum_{i=1}^{n} x_i y_i = b_0 \sum_{i=1}^{n} x_i + b_1 \sum_{i=1}^{n} x_i^2.$$

So, the least squares estimators b_0 and b_1 satisfy the pair of equations:

$$\sum_{i=1}^{n} y_i = \sum_{i=1}^{n} b_0 + b_1 \sum_{i=1}^{n} x_i$$
$$\sum_{i=1}^{n} x_i y_i = b_0 \sum_{i=1}^{n} x_i + b_1 \sum_{i=1}^{n} x_i^2,$$

or, in matrix form:

$$\begin{pmatrix} \sum_{i=1}^{n} y_i \\ \sum_{i=1}^{n} x_i y_i \end{pmatrix} = \begin{pmatrix} n & \sum_{i=1}^{n} x_i \\ \sum_{i=1}^{n} x_i & \sum_{i=1}^{n} x_i^2 \end{pmatrix} \begin{pmatrix} b_0 \\ b_1 \end{pmatrix}.$$

Then

$$\boldsymbol{X}'\boldsymbol{X} = \begin{pmatrix} 1 & 1 & \cdots & 1 \\ x_1 & x_2 & \cdots & x_n \end{pmatrix} \begin{pmatrix} 1 & x_1 \\ 1 & x_2 \\ \vdots & \vdots \\ 1 & x_n \end{pmatrix} = \begin{pmatrix} n & \sum_{i=1}^{n} x_i \\ \sum_{i=1}^{n} x_i & \sum_{i=1}^{n} x_i^2 \end{pmatrix}$$

and

$$\boldsymbol{X}'\boldsymbol{y} = \begin{pmatrix} 1 & 1 & \cdots & 1 \\ x_1 & x_2 & \cdots & x_n \end{pmatrix} \begin{pmatrix} y_1 \\ y_2 \\ \vdots \\ y_n \end{pmatrix} = \begin{pmatrix} \sum_{i=1}^{n} y_i \\ \sum_{i=1}^{n} x_i y_i \end{pmatrix}.$$

So, the least squares estimators $\boldsymbol{b}' = (b_0, b_1)$ satisfy the equations:

$$\boldsymbol{X}'\boldsymbol{X}\boldsymbol{b} = \boldsymbol{X}'\boldsymbol{y}.$$

Rearranging, on the assumption that $(\boldsymbol{X}'\boldsymbol{X})$ is non-singular,

$$\boldsymbol{b} = (\boldsymbol{X}'\boldsymbol{X})^{-1}\boldsymbol{X}\boldsymbol{y}.$$

To compute $(\boldsymbol{X}'\boldsymbol{X})^{-1}$ we need the determinant, minors, cofactors and adjoint. The determinant is $|\,\boldsymbol{X}'\boldsymbol{X}\,| = n\Sigma x_i^2 - (\Sigma x_i)^2 = n\Sigma x_i^2 - n^2\bar{x}\ (= n(\Sigma x_i^2 - n\bar{x}^2))$.

Minors	Cofactors	Adjoint
$\begin{pmatrix} \sum x_i^2 & \sum x_i \\ \sum x_i & n \end{pmatrix}$	$\begin{pmatrix} \sum x_i^2 & -\sum x_i \\ -\sum x_i & n \end{pmatrix}$	$\begin{pmatrix} \sum x_i^2 & -\sum x_i \\ -\sum x_i & n \end{pmatrix}$

Thus,

$$(\boldsymbol{X}'\boldsymbol{X})^{-1} = \frac{1}{n(\sum x_i^2 - n\bar{x}^2)} \begin{pmatrix} \sum x_i^2 & -\sum x_i \\ -\sum x_i & n \end{pmatrix}, \ \boldsymbol{X}'\boldsymbol{y} = \begin{pmatrix} 1 \cdots 1 \\ x_1 \cdots x_n \end{pmatrix} \begin{pmatrix} y_1 \\ \vdots \\ y_n \end{pmatrix} = \begin{pmatrix} \sum y_i \\ \sum x_i y_i \end{pmatrix}.$$

Thus,

$$\begin{aligned} (\boldsymbol{X}'\boldsymbol{X})^{-1}\boldsymbol{X}'\boldsymbol{y} &= \frac{1}{n(\sum x_i^2 - n\bar{x}^2)} \begin{pmatrix} \sum x_i^2 & -\sum x_i \\ -\sum x_i & n \end{pmatrix} \begin{pmatrix} \sum y_i \\ \sum x_i y_i \end{pmatrix} \\ &= \frac{1}{n(\sum x_i^2 - n\bar{x}^2)} \begin{pmatrix} (\sum x_i^2)(\sum y_i) - (\sum x_i)(\sum x_i y_i) \\ -(\sum x_i)(\sum y_i) + n(\sum x_i y_i) \end{pmatrix}. \end{aligned}$$

So

$$\begin{aligned} b_0 &= \frac{(\sum x_i^2)(n\bar{y}) - (n\bar{x})(\sum x_i y_i)}{n(\sum x_i^2 - n\bar{x}^2)} = \frac{[(\sum x_i^2)(\bar{y}) - (\bar{x})(\sum x_i y_i)]}{(\sum x_i^2 - n\bar{x}^2)} \\ b_1 &= \frac{[n\sum x_i y_i - n^2\bar{x}\bar{y}]}{n(\sum x_i^2 - n\bar{x}^2)} = \frac{[\sum x_i y_i - n\bar{x}\bar{y}]}{(\sum x_i^2 - n\bar{x}^2)} = \frac{\sum(x_i - \bar{x})(y_i - \bar{y})}{\sum(x_i - \bar{x})^2}. \end{aligned}$$

Now,

$$\begin{aligned} [(\Sigma x_i^2)\bar{y} - \bar{x}\Sigma x_i y_i] &= [(\Sigma x_i^2)\bar{y} - n\bar{x}^2\bar{y} - \bar{x}(\Sigma x_i y_i - n\bar{x}\bar{y})] \\ &= \bar{y}(\Sigma x_i^2 - n\bar{x}^2) - \bar{x}(\Sigma x_i y_i - n\bar{x}\bar{y}) \end{aligned}$$

$$b_0 = \frac{[(\sum x_i^2)(\bar{y}) - (\bar{x})(\sum x_i y_i)]}{(\sum x_i^2 - n\bar{x}^2)} = \frac{(\sum x_i^2)\bar{y} - \bar{x}\sum x_i y_i}{(\sum x_i^2 - n\bar{x}^2)} = \bar{y} - \bar{x}b_1.$$

Thus, without the use of matrices, we can summarize:

$$b_0 = \bar{y} - \bar{x}b_1, \text{ and } b_1 = \frac{\sum(x_i - \bar{x})(y_i - \bar{y})}{\sum(x_i - \bar{x})^2}.$$

Exercises for Chapter 3

Exercise 3.1 For the (supply-and-demand) system of equations given below, find the equilibrium prices:

$$\begin{aligned} 2p_1 + p_2 + 6p_3 &= 26 \\ p_1 + p_2 + p_3 &= 12 \\ 3p_1 + 2p_2 - 2p_3 &= 2. \end{aligned}$$

Exercise 3.2

$$\boldsymbol{A} = \begin{pmatrix} 0.7 & 0.4 & 0.3 \\ 0.1 & 0.2 & 0.3 \\ 0.1 & 0.3 & 0.2 \end{pmatrix}, \qquad \boldsymbol{f} = \begin{pmatrix} f_1 \\ f_2 \\ f_3 \end{pmatrix} = \begin{pmatrix} 90 \\ 75 \\ 80 \end{pmatrix}:$$

find the vector of interindustry outputs.

Exercise 3.3 For the input–output model with technology matrix $\boldsymbol{A} = \begin{pmatrix} 0.2 & 0.3 \\ 0.2 & 0.25 \end{pmatrix}$, suppose that the final demand vector is $\boldsymbol{f} = \begin{pmatrix} 350 \\ 250 \end{pmatrix}$.

(a) Find the appropriate industry output levels (the $\boldsymbol{x}$ vector).

(b) Find $\boldsymbol{A}^2 (= \boldsymbol{AA})$, $\boldsymbol{A}^3 (= \boldsymbol{AAA})$, $\boldsymbol{A}^4 (= \boldsymbol{AAAA})$ and $\boldsymbol{A}^5 (= \boldsymbol{AAAAA})$. Then calculate the matrix $\boldsymbol{G} = (\boldsymbol{I} + \boldsymbol{A} + \boldsymbol{A}^2 + \boldsymbol{A}^3 + \boldsymbol{A}^4 + \boldsymbol{A}^5)$ and compute $\boldsymbol{Gf}$.

(c) Observe that $\boldsymbol{G}$ is close to $(\boldsymbol{I} - \boldsymbol{A})^{-1}$ and your answer in (a) is close to $\boldsymbol{Gf}$.

Exercise 3.4 Given the input–output model with technology matrix $\boldsymbol{A} = \begin{pmatrix} 0.4 & 0.0 & 0.2 \\ 0.2 & 0.1 & 0.0 \\ 0.2 & 0.2 & 0.3 \end{pmatrix}$ and final demand $\boldsymbol{f} = \begin{pmatrix} 30 \\ 40 \\ 50 \end{pmatrix}$, find the equilibrium industry outputs.

Exercise 3.5 A simple economy consists of two sectors: agriculture (A) and transportation (T). The input–output matrix for this economy is given by

$$\begin{array}{cc} & \begin{array}{cc} A & T \end{array} \\ \begin{array}{c} A \\ B \end{array} & \begin{pmatrix} 0.4 & 0.1 \\ 0.2 & 0.2 \end{pmatrix}. \end{array}$$

(a) Find the gross output of agricultural products needed to satisfy a consumer demand for \$50 million worth of agriculture products and \$10 million worth of transportation.

(b) Find the value of agriculture products and transportation consumed in the internal process of production in order to meet the gross output.

Exercise 3.6 Given the input–output matrix $\begin{pmatrix} 0.2 & 0.2 & 0.1 \\ 0.2 & 0.4 & 0.1 \\ 0.1 & 0.2 & 0.3 \end{pmatrix}$ and final demand $\boldsymbol{f} = \begin{pmatrix} 100 \\ 80 \\ 50 \end{pmatrix}$, compute the interindustry flows necessary to sustain this final demand.

Exercise 3.7 Given an input–output matrix $\boldsymbol{A}$, if $\boldsymbol{x}$ is the vector of industry outputs and $\boldsymbol{f}$ the vector of final demands, then $\boldsymbol{x}$ and $\boldsymbol{f}$ satisfy $\boldsymbol{x} = \boldsymbol{A}\boldsymbol{x} + \boldsymbol{f}$. In the case where

$$\boldsymbol{A} = \begin{pmatrix} \frac{3}{12} & \frac{4}{12} & 0 \\ \frac{1}{12} & 0 & \frac{6}{12} \\ 0 & \frac{4}{12} & \frac{2}{12} \end{pmatrix}, \qquad \boldsymbol{f} = \begin{pmatrix} 20 \\ 30 \\ 40 \end{pmatrix},$$

find the solution value of $\boldsymbol{x}$.

Suppose that each column sum of $\boldsymbol{A}$ is less than 1: for each j, $\sum_{i=1}^{n} a_{ij} < 1$. Show that this implies that $(\boldsymbol{I} - \boldsymbol{A})^{-1}$ exists and equals $(\boldsymbol{I} + \boldsymbol{A} + \boldsymbol{A}^2 + \boldsymbol{A}^3 + \cdots) = \boldsymbol{I} + \sum_{t=1}^{\infty} \boldsymbol{A}^t$. Explain why this implies that for a given final demand $\boldsymbol{f}$ (a vector with non-negative entries), the solution value of $\boldsymbol{x}$ is non-negative.

Exercise 3.8 Consider the transition matrix:

$$\boldsymbol{T} = \begin{pmatrix} \frac{1}{2} & \frac{1}{4} & \frac{1}{4} \\ \frac{1}{3} & \frac{1}{3} & \frac{1}{3} \\ \frac{1}{4} & \frac{1}{2} & \frac{1}{4} \end{pmatrix}.$$

(a) Find the unique stationary distribution (using the cofactors) of $\boldsymbol{T}$ i.e., find the 1×3 vector $\boldsymbol{q}$ satisfying $\boldsymbol{q}\boldsymbol{T} = \boldsymbol{q}$.

(b) Find $\boldsymbol{T}^2$ and compare with $\boldsymbol{T}$.

(c) Find the determinant of $\boldsymbol{T}$. Is $\boldsymbol{T}$ invertible? Why?

Exercise 3.9 Let

$$\boldsymbol{x} = \begin{pmatrix} 2 \\ 3 \\ 9 \\ 3 \\ 6 \\ 11 \\ 4 \end{pmatrix}, \quad \boldsymbol{y} = \begin{pmatrix} 7 \\ 12 \\ 26 \\ 13 \\ 18 \\ 36 \\ 13 \end{pmatrix}.$$

Find the least squares estimator.

Exercise 2.8 Consider the transition matrix

$$T = [illegible]$$

(a) Find the unique stationary distribution using the [illegible] of T [illegible] and the 1×3 vector q satisfying $qT = q$.

(b) Find T^{n} and compare with [illegible]

(c) Find the [illegible] of T. Is T [illegible]? Why?

Exercise 2.9 Let [illegible]

[illegible]

Find the [illegible]

4

Linear programming

4.1 Introduction

Linear programming is an optimization technique for finding optimal solutions to problems where the objective and constraints are linear. In what follows, the general linear programming problem and methods of solution are described, along with various features of this class of problem. A linear program is characterized by a linear objective, linear constraints and restrictions on the signs of choice variables. Linearity of the constraint set leads to a particular geometric structure on the set possible choice variables. For low-dimensional programs this permits a graphical treatment of the problem and, more generally, the geometric structure leads naturally to an algorithmic method for finding solutions. Section 4.1 develops the basic features of a linear program. This consists of formulation of the programming problem, description of key issues in solving these programs and the valuation of resource constraints through the dual prices. A broader understanding of linear programming requires knowledge of basic solutions, which are discussed in Section 4.2. Here the geometric view described in Section 4.1 is reinterpreted in terms of equation systems. Section 4.3 describes classic results in linear programming where the original program (the primal) is related to a second program (the dual). Before developing the discussion, a simple example will illustrate how these programs arise.

EXAMPLE 4.1: A firm produces two goods, 1 and 2. Production requires the application of two techniques: processing and finishing. A unit of good 1 requires two hours of processing time and four hours of finishing time; a unit of good 2 requires three hours of processing and three hours of finishing time. The total amount of time that the processing machine can run for each day is 13 hours, while the finishing machine can run for 17 hours. Each unit of good 1 sells for \$3 and each unit of good 2 sells for \$4. The problem is to decide how much of each good should be produced to maximize revenue – a problem that may be formulated as a linear program.

Let x denote the number of units of good 1 produced, and let y denote the number of units of good 2 produced. Then the revenue is equal to $3x+4y$, since each unit of 1 generates \$3 of revenue and each unit of 2 generates \$4 of revenue. This defines the linear objective. Next, x units of good 1 require two hours of processing time, while y units of good 2 require three hours of processing time: altogether, producing x units of 1 and y units of 2 requires $2x+3y$ hours of processing time. Similarly, producing x units of 1 and y units of 2 requires $4x+3y$ hours of finishing time. The total amount of processing time is 13 hours, so the production plan must satisfy $2x+3y \leq 13$, the first linear constraint, and likewise $4x+3y \leq 17$ defines the finishing constraint. Finally, negative amounts cannot be produced: $x, y \geq 0$. Putting all these conditions together, the full problem may be summarized:

$$\begin{aligned} \max \quad & 3x+4y \\ \text{subject to:} \quad & \\ & 2x+3y \leq 13 \quad \text{(processing)} \\ & 4x+3y \leq 17 \quad \text{(finishing)} \\ & x, y \geq 0. \end{aligned}$$

◊

4.1.1 Formulation

In a linear-programming problem a collection of variables $(x_1, x_2, \cdots, x_n)$ must be chosen to maximize (or minimize) a linear function,

$$f(x_1, x_2, \cdots, x_n) = c_1x_1 + c_2x_2 + \cdots + c_nx_n,$$

subject to a finite number of linear constraints of the form:

$$a_1x_1 + a_2x_2 + \cdots + a_nx_n \leq b, \quad a_1x_1 + a_2x_2 + \cdots + a_nx_n = b, \quad \text{or}$$
$$a_1x_1 + a_2x_2 + \cdots + a_nx_n \geq b.$$

In addition, the x_i variables are required to be chosen non-negative. A typical linear program may be written:

$$\begin{array}{ll} \max & c_1x_1 + c_2x_2 + \cdots + c_nx_n \\ \text{subject to:} & \\ & \begin{array}{rcl} a_{11}x_1 + a_{12}x_2 + \cdots + a_{1n}x_n & \leq & b_1 \\ a_{21}x_1 + a_{22}x_2 + \cdots + a_{2n}x_n & \leq & b_2 \\ \vdots & \vdots & \vdots \\ a_{m1}x_1 + a_{m2}x_2 + \cdots + a_{mn}x_n & \leq & b_m \\ x_1, x_2, \cdots, x_n & \geq & 0. \end{array} \end{array}$$

This is a linear program with n variables and m constraints. If we write $\boldsymbol{c} = (c_1, c_2, \cdots, c_n)$,

$$\boldsymbol{A} = \begin{pmatrix} a_{11} & a_{12} & \cdots & a_{1n} \\ a_{21} & a_{22} & \cdots & a_{2n} \\ \vdots & \vdots & \ddots & \vdots \\ a_{m1} & a_{m2} & \cdots & a_{mn} \end{pmatrix}, \quad \boldsymbol{b} = \begin{pmatrix} b_1 \\ b_2 \\ \vdots \\ b_m \end{pmatrix}, \quad \boldsymbol{x} = \begin{pmatrix} x_1 \\ x_2 \\ \vdots \\ x_n \end{pmatrix},$$

then the program may be written in matrix form:

$$\begin{array}{ll} \max & \boldsymbol{c} \cdot \boldsymbol{x} \\ \text{subject to:} & \\ & \boldsymbol{A}\boldsymbol{x} \leq \boldsymbol{b} \\ & \boldsymbol{x} \geq \boldsymbol{0}. \end{array} \tag{4.1}$$

In parallel with the maximization problem, there is minimization formulation:

$$\begin{array}{ll} \min & c_1x_1 + c_2x_2 + \cdots + c_nx_n \\ \text{subject to:} & \\ & \begin{array}{rcl} a_{11}x_1 + a_{12}x_2 + \cdots + a_{1n}x_n & \geq & b_1 \\ a_{21}x_1 + a_{22}x_2 + \cdots + a_{2n}x_n & \geq & b_2 \\ \vdots & \vdots & \vdots \\ a_{m1}x_1 + a_{m2}x_2 + \cdots + a_{mn}x_n & \geq & b_m \\ x_1, x_2, \cdots, x_n & \geq & 0. \end{array} \end{array}$$

In matrix form:

$$\begin{array}{ll} \min & \boldsymbol{c} \cdot \boldsymbol{x} \\ \text{subject to:} & \\ & \boldsymbol{A}\boldsymbol{x} \geq \boldsymbol{b} \\ & \boldsymbol{x} \geq \boldsymbol{0}. \end{array} \tag{4.2}$$

4.1.1.1 Mixed constraints and unrestricted variable signs

The formulations for maximization and minimization given in Equations 4.1 and 4.2 have a standardized form with non-negativity constraints. In general, maximization and minimization problems can be cast in this way. The following discussion illustrates. Consider the program:

$$\begin{aligned} \max \quad & c_1x_1 + c_2x_2 \\ \text{subject to:} \quad & \\ & a_{11}x_1 + a_{12}x_2 \leq b_1 \\ & a_{21}x_1 + a_{22}x_2 \geq b_2 \\ & x_1 \geq 0. \end{aligned}$$

In this program, both types of inequality appear, and only x_1 is required to be non-negative. Replacing x_2 with $y - z$ and requiring y and z to be non-negative overcomes the second issue – picking $y = 0$ and z any desired positive number gives x_2 equal to the negative of that number.

$$\begin{aligned} \max \quad & c_1x_1 + c_2y - c_2z \\ \text{subject to:} \quad & \\ & a_{11}x_1 + a_{12}y - a_{12}z \leq b_1 \\ & a_{21}x_1 + a_{22}y - a_{22}z \geq -b_2 \\ & x_1, y, z \geq 0. \end{aligned}$$

The first issue is solved by multiplying the relevant constraint by -1 to reverse the inequality: for example $5 \geq 4$ is equivalent to $-5 \leq -4$ and $ax \geq b$ is equivalent to $-ax \leq -b$. In this example, multiplying $a_{21}x_1 + a_{22}y - a_{22}z \geq b_2$ by -1 gives:

$$\begin{aligned} \max \quad & c_1x_1 + c_2y - c_2z \\ \text{subject to:} \quad & \\ & a_{11}x_1 + a_{12}y - a_{12}z \leq b_1 \\ & -a_{21}x_1 - a_{22}y + a_{22}z \leq b_2 \\ & x_1, y, z \geq 0. \end{aligned}$$

Similarly, given an equality constraint $\sum a_ix_i = b_i$, this is equivalent to two constraints, $\sum a_ix_i \leq b_i$ and $\sum a_ix_i \geq b_i$. Note that the convention is to formulate the minimization program with the inequalities having the "≥" form: in matrix terms, $\boldsymbol{Ax} \geq \boldsymbol{b}$. In the linear program, constraints of the form $\boldsymbol{Ax} \leq \boldsymbol{b}$ or $\boldsymbol{Ax} \geq \boldsymbol{b}$ are called feasibility constraints; the constraints that $x_i \geq 0$ for all i are called non-negativity constraints. Taking both types of constraint together, refer to them as the program constraints or constraints.

The next sections consider a number of important concepts in linear programming. Section 4.1.2 discusses the feasible region – those choices that satisfy all the constraints of the problem. Following this, Section 4.1.3 considers the problem of finding an optimal solution. In Section 4.1.4, *dual prices* and *slack variables* are introduced. Dual prices attach a price or valuation to each constraint, while slack variables indicate whether constraints are binding or not. Finally, the *dual program* associated with a linear program (called "the primal") is introduced (Section 4.1.6). The dual program is derived from the primal program, is closely related and provides additional perspective on the primal program.

4.1.2 The feasible region

A vector $\boldsymbol{x} = (x_1,\dots,x_n)$ is called feasible if it satisfies the linear constraint inequalities and the non-negativity constraints.

Definition 4.1: *The set of $\boldsymbol{x}$ vectors that satisfy the constraints $\mathbf{A}\boldsymbol{x} \le \boldsymbol{b}$ and $\boldsymbol{x} \ge \mathbf{0}$ is called the feasible set or feasible region.*

Geometrically, it has a special shape – it is a convex polyhedron. The following examples give the feasible regions defined by inequality constraints (on the left side), and the corresponding graphs show the feasible regions as the shaded areas.

EXAMPLE 4.2: In this example, the feasible region is defined by two inequality constraints and the non-negativity constraints.

$$\begin{aligned} 2x + y &\le 6 \\ 3x + 4y &\le 12 \\ x, y &\ge 0 \end{aligned}$$

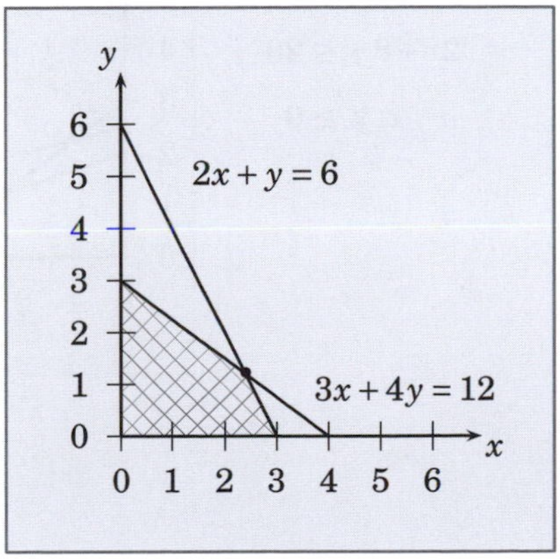

◊

The next example has three constraints. The additional constraint adds corners to the feasible region, and indicates the impact of additional constraints.

EXAMPLE 4.3:

$$2x + 2y \leq 10$$
$$3x + 6y \leq 18$$
$$2x + 12y \leq 24$$
$$x, y \geq 0$$

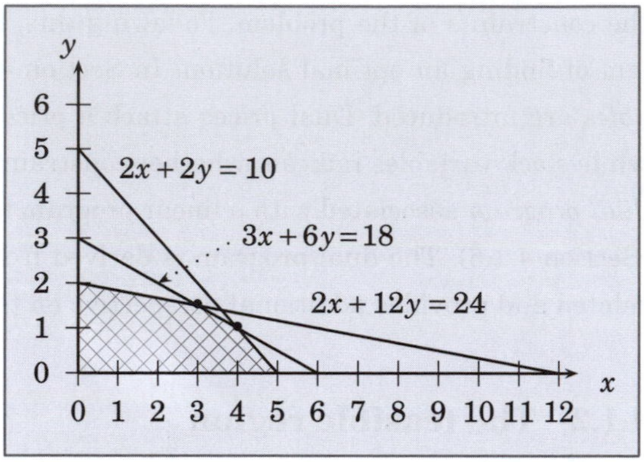

◊

The feasible region is defined by linear equations, but this region may be quite complex. The next example illustrates increasing complexity. In this example, the second constraint is written as $2x + 6y \geq 12$, which is equivalent to $-2x - 6y \leq 12$.

EXAMPLE 4.4:

$$3x + 4y \leq 12$$
$$2x + 6y \geq 12$$
$$2x + 8y \leq 20$$
$$x, y \geq 0$$

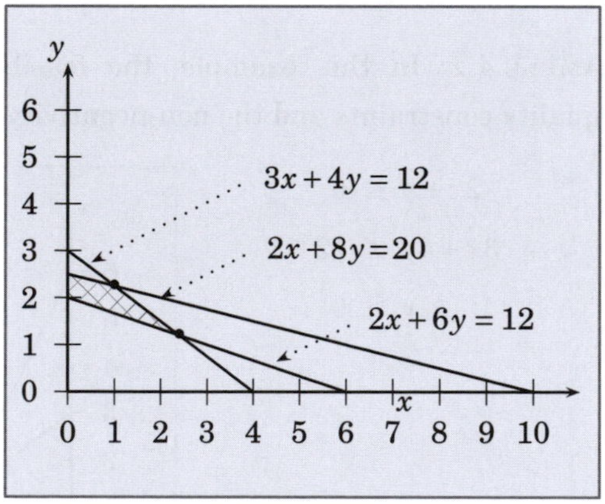

◊

4.1.3 Finding an optimal solution

Once the feasible region is identified, one can begin the search for an optimal solution: a point in the feasible region such that there is no other point in the feasible region that improves the objective value.

Definition 4.2: *Consider the linear program:* max $\mathbf{c}\cdot\boldsymbol{x}$ *subject to* $\mathbf{A}\boldsymbol{x} \le \boldsymbol{b}$ *and* $\boldsymbol{x} \ge \mathbf{0}$. *If* $\boldsymbol{x}^*$ *is feasible and* $\mathbf{c}\cdot\boldsymbol{x}^* \ge \mathbf{c}\cdot\boldsymbol{x}$ *for all feasible* $\boldsymbol{x}$*, then* $\boldsymbol{x}^*$ *is called an optimal solution. Similarly, in the program:* min $\mathbf{c}\cdot\boldsymbol{x}$ *subject to* $\mathbf{A}\boldsymbol{x} \ge \boldsymbol{b}$ *and* $\boldsymbol{x} \ge \mathbf{0}$. *If* $\boldsymbol{x}^*$ *is feasible and* $\mathbf{c}\cdot\boldsymbol{x}^* \le \mathbf{c}\cdot\boldsymbol{x}$ *for all feasible* $\boldsymbol{x}$*, then* $\boldsymbol{x}^*$ *is called an optimal solution.*

When it is possible to graph the feasible region (with two choice variables), and there are not too many constraints, the problem can be solved by inspection. To illustrate, consider Example 4.1 again.

EXAMPLE 4.5: Recall the program from Example 4.1:

$$\begin{aligned} \max \quad & 3x + 4y \\ \text{subject to:} \quad & \\ & 2x + 3y \le 13 \\ & 4x + 3y \le 17 \\ & x, y \ge 0. \end{aligned}$$

Consider the objective function, $\pi(x,y) = 3x + 4y$. An equation such as $\pi(x,y) = 2$ defines a line giving those combinations of (x,y) such that $\pi(x,y) = 2$. For example, because $3\cdot 0 + 4\cdot\frac{1}{2} = 2$ and $3\cdot\frac{2}{3} + 4\cdot 0 = 2$, the points $(x,y) = \left(0,\frac{1}{2}\right)$ and $(x,y) = \left(\frac{2}{3},0\right)$ are on this line. Referring to π as profit, the set $\{(x,y) \mid \pi(x,y) = k\}$ is called an "iso-profit" line with profit k. Iso-profit lines for the objective function, along with the feasible region, may easily be graphed. In the figure, the iso-profit lines are drawn for $k = 8, 12$ and 18. Plotting these and the constraints gives:

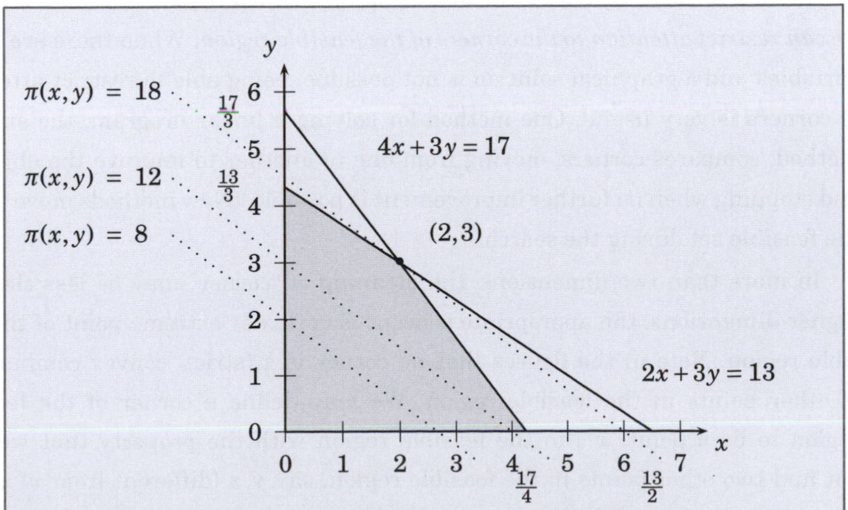

From the figure, the highest value for the objective, 18, is achieved at (2,3), the corner of the feasible region determined by equations $2x+3y=13$ and $4x+3y=17$. Using Cramer's rule:

$$\begin{vmatrix} 2 & 3 \\ 4 & 3 \end{vmatrix} = 6-12=-6, \quad \begin{vmatrix} 13 & 3 \\ 17 & 3 \end{vmatrix} = 39-51=-12, \quad \begin{vmatrix} 2 & 13 \\ 4 & 17 \end{vmatrix} = 34-52=-18.$$

Thus, the solution is $\boldsymbol{x}^* = (x^*, y^*) = \left(\frac{-12}{-6}, \frac{-18}{-6}\right) = \left(\frac{12}{6}, \frac{18}{6}\right) = (2,3)$. Note that the feasible region is defined by its corners. Determining the objective value at each of the corners gives the table:

Corner : (x,y)	*Objective value*
$(0,0)$	$0 = 3(0)+4(0)$
$\left(0,\frac{13}{3}\right)$	$17\frac{1}{3} = 3(0)+4\left(\frac{13}{3}\right)$
$(2,3)^*$	$18 = 3(2)+4(3)$
$\left(\frac{17}{4},0\right)$	$12\frac{3}{4} = 3\left(\frac{17}{4}\right)+4(0)$

By comparing the objective value at each corner, observe that the highest value occurs at corner (2,3). This is not accidental. An optimal solution can always be found at a corner. ◊

An important theorem in linear programming (see Theorem 4.2, later) says that if a linear program has an optimal solution, then it has an optimal solution at a corner. This does not say that every optimal solution is at a corner – just that there is an optimal solution at a corner (assuming the problem has an optimal solution). *The key benefit of the theorem is that in searching for an optimal solution we can restrict attention to the corners of the feasible region.* When there are many variables and a graphical solution is not possible, being able to restrict attention to corners is very useful. One method for solving a linear program, the simplex method, compares corners, moving from one to another to improve the objective and stopping when no further improvement is possible. (New methods move inside the feasible set during the search.)

In more than two dimensions, the meaning of "corner" may be less clear. In higher dimensions, the appropriate concept is vertex or extreme point of the feasible region. Note in the figures that no corner is a (strict) convex combination of other points in the feasible region. We may define a corner of the feasible region to be a point, $\boldsymbol{x}^*$, in the feasible region with the property that we cannot find two other points in the feasible region, say $\boldsymbol{y}, \boldsymbol{z}$ (different from $\boldsymbol{x}^*$), such that $\boldsymbol{x}^*$ lies on a straight line connecting these two points. The only points in the figures satisfying this criterion are the corners. This criterion is independent of the dimension of the feasible region (the number of elements in the x vector) and defines a vertex or extreme point of the feasible set in general. This is discussed further in Section 4.2.

4.1.4 Dual prices and slack variables

In a linear program, the inequality constraints identify resource constraints: usage of a particular resource may not exceed a particular level. Subject to these constraints, the objective is optimized. Therefore, when a resource constraint is varied (some right-hand side parameter b_i is altered), this generally has an impact on the solution and corresponding objective value. This change in the objective value is measured by the dual price.

4.1.4.1 Dual prices

The *dual* or *shadow price* of the ith constraint is defined as the change in the value of the optimized objective due to a unit change in the right-hand side of the ith constraint. For example, suppose that the program has been solved and the solution is at a corner $\boldsymbol{x}^*$ with associated objective value $\boldsymbol{c} \cdot \boldsymbol{x}^*$. Then, $\boldsymbol{x}^*$ is located at the intersection of some set of constraints. Changing the right-hand side (RHS) of one of those constraints slightly will change the intersection point slightly, to say $\boldsymbol{x}'$. Suppose that the RHS of the ith constraint is changed from b_i to b_i' and suppose that the corner at which the solution is located is defined by the same set of constraints (the solution occurs at the $\boldsymbol{x}'$). In Example 4.6, figure A, suppose the original solution is at (2,3). If the RHS of constraint 1 is changed from 13 to 14 and the solution will move to $\left(\frac{9}{6}, \frac{22}{6}\right)$ – still defined by the intersection of constraints 1 and 2.

(This is the typical situation. Formally, the solution basis does not change. See Example 4.6, Remark 4.1 and the discussion of basic solutions in Section 4.2.) Let the new solution be $\boldsymbol{x}'$. Then the dual price of the ith constraint is defined:

$$DP_i = \frac{\text{Change in objective}}{\text{Change in RHS of constraint } i} = \frac{\boldsymbol{c} \cdot \boldsymbol{x}' - \boldsymbol{c} \cdot \boldsymbol{x}^*}{b_i' - b_i}$$

EXAMPLE 4.6: To illustrate, continuing with the previous example, consider the impact of changing the right-hand side (RHS_1) of the first constraint from 13 to 14. In the notation above, $b_1 = 13$, $b_1' = 14$. This defines a new program:

$$\begin{array}{l} \max\ 3x+4y \\ \text{subject to:} \\ \quad 2x+3y \le 13\ (1) \\ \quad 4x+3y \le 17\ (2) \\ \quad x,y \ \ge 0 \end{array}$$

$$\Longrightarrow \begin{array}{l} \max\ 3x+4y \\ \text{subject to:} \\ \quad 2x+3y \le 14\ (1') \\ \quad 4x+3y \le 17\ (2) \\ \quad x,y \ \ge 0 \end{array}$$

When the RHS of constraint 1 changes to 13 from 14, the new solution using Cramer's rule is:

$$\begin{vmatrix} 2 & 3 \\ 4 & 3 \end{vmatrix} = 6-12=-6, \quad \begin{vmatrix} 14 & 3 \\ 17 & 3 \end{vmatrix} = 42-51=-9, \quad \begin{vmatrix} 2 & 14 \\ 4 & 17 \end{vmatrix} = 34-56=-22.$$

Then the new solution is at $\boldsymbol{x} = (x', y') = \left(\frac{9}{6}, \frac{22}{6}\right)$ (see figure A).

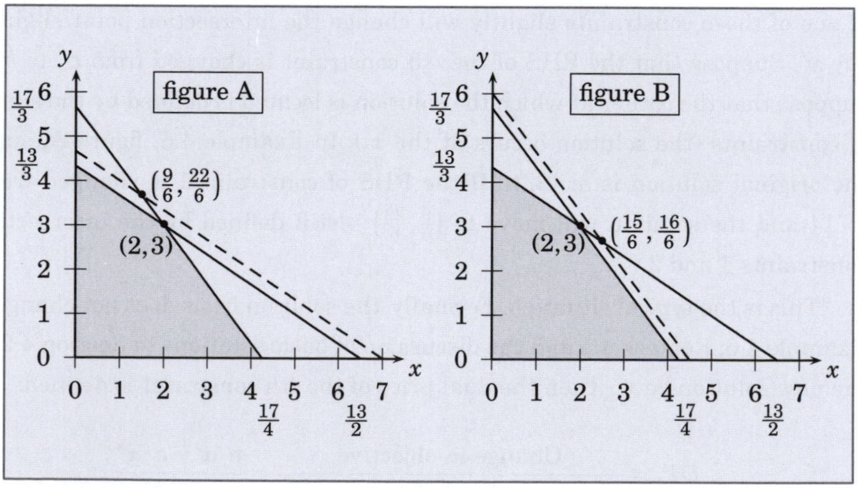

The objective value at the new solution is:

$$\boldsymbol{c} \cdot \boldsymbol{x}' = (3,4) \cdot \left(\frac{9}{6}, \frac{22}{6}\right) = \left(\frac{1}{6}\right)[3(9)+4(22)] = \left(\frac{1}{6}\right)115 = 19.1667 = 19\frac{1}{6}.$$

Therefore,

$$DP_1 = \frac{\Delta \text{objective}}{\Delta RHS_1} = \frac{\boldsymbol{c}\cdot\boldsymbol{x}' - \boldsymbol{c}\cdot\boldsymbol{x}^*}{b_1' - b_1} = \frac{19.1667 - 18}{14 - 13} = 1.1667 = 1\tfrac{1}{6}.$$

The same reasoning applies to constraint 2. Consider raising the RHS (RHS_2) of constraint 2 from 17 to 18. This is depicted in figure B.

$$\boxed{\begin{array}{l} \quad\quad \max\ 3x+4y \\ \text{subject to:} \\ \quad\quad\quad\quad 2x+3y \le 13\ (1) \\ \quad\quad\quad\quad 4x+3y \le 17\ (2) \\ \quad\quad\quad\quad\quad x,y \ \ \ge 0 \end{array}}$$

$$\Longrightarrow \quad \boxed{\begin{array}{l} \quad\quad \max\ 3x+4y \\ \text{subject to:} \\ \quad\quad\quad\quad 2x+3y \le 13\ (1) \\ \quad\quad\quad\quad 4x+3y \le 18\ (2') \\ \quad\quad\quad\quad\quad x,y \ \ \ge 0 \end{array}}$$

The new solution is at $\boldsymbol{x}^\dagger = (x^\dagger, y^\dagger) = \left(\frac{15}{6}, \frac{16}{6}\right)$ with objective value:

$$\boldsymbol{c}\cdot\boldsymbol{x}^\dagger = (3,4)\cdot\left(\tfrac{15}{6}, \tfrac{16}{6}\right) = \left(\tfrac{1}{6}\right)[3(15)+4(16)] = \left(\tfrac{1}{6}\right)109 = 18.1667$$

Using Cramer's rule, the calculations for determining the solution are:

$$\begin{vmatrix} 2 & 3 \\ 4 & 3 \end{vmatrix} = 6 - 12 = -6, \quad \begin{vmatrix} 13 & 3 \\ 18 & 3 \end{vmatrix} = 39 - 54 = -15, \quad \begin{vmatrix} 2 & 13 \\ 4 & 18 \end{vmatrix} = 36 - 52 = -16.$$

giving $x' = \frac{-15}{-6} = \frac{15}{6}$ and $y' = \frac{-16}{-6} = \frac{16}{6}$. Therefore,

$$DP_2 = \frac{\Delta \text{objective}}{\Delta RHS_2} = \frac{\boldsymbol{c}\cdot\boldsymbol{x}^\dagger - \boldsymbol{c}\cdot\boldsymbol{x}^*}{b_2' - b_2} = \frac{18.1667 - 18}{18 - 17} = 0.1667 = \tfrac{1}{6}$$

Note: One can also consider the impact of changes in other parameters of the problem. For example, suppose that the objective were $8x+6y$ instead of $3x+4y$. In this case, the objective would run parallel to the second constraint and any point on the line between $(x,y) = (2,3)$ and $(x,y) = \left(\frac{17}{4},0\right)$ would be optimal, including the two optimal corners. ◊

4.1.4.2 Slack variables

Given the constraints $\boldsymbol{Ax} \le \boldsymbol{b}$ and $\boldsymbol{x} \ge \boldsymbol{0}$, suppose that the optimal solution has been found and call it $\boldsymbol{x}^*$. The *slack variables* are defined as the vector $\boldsymbol{s}^*$ that satisfies $\boldsymbol{Ax}^* + \boldsymbol{s}^* = \boldsymbol{b}$ or $\boldsymbol{s}^* = \boldsymbol{b} - \boldsymbol{Ax}^*$. (In the case where the constraints have the form $\boldsymbol{Ax} \ge \boldsymbol{b}$ and $\boldsymbol{x} \ge \boldsymbol{0}$, the slack variables are defined: $\boldsymbol{Ax}^* - \boldsymbol{s}^* = \boldsymbol{b}$ or $\boldsymbol{s}^* = \boldsymbol{Ax}^* - \boldsymbol{b}$. This convention ensures that the slack variables are non-negative.)

Continuing with the example, $\boldsymbol{s}^* = (s_1^*, s_2^*)$ is defined by the equations:

$$\begin{aligned} 2x^* + 3y^* + s_1^* &= 13 \\ 4x^* + 3y^* + s_2^* &= 17 \end{aligned} \quad \text{or} \quad \begin{aligned} 2(2) + 3(3) + s_1^* &= 13 \\ 4(2) + 3(3) + s_2^* &= 17 \end{aligned} \quad \text{or} \quad \begin{aligned} 13 + s_1^* &= 13 \\ 17 + s_2^* &= 17 \end{aligned}$$

so that $\boldsymbol{s}^* = (s_1^*, s_2^*) = (0,0)$. When a slack variable is 0, the corresponding constraint is binding. Here both slack variables are 0, and the constraints bind. In general, changing the RHS of a binding constraint changes the solution and the value of the objective. Therefore, in general, if the slack variable associated with a constraint is 0, the corresponding dual price is non-zero. Conversely, if the slack variable associated with a constraint is positive, then the constraint is not binding: changing it has no effect on the solution and so the dual price is zero.

4.1.5 Some examples

In this section, two examples are considered in detail to summarize the discussion so far.

EXAMPLE 4.7: Consider the following linear program:

$$\begin{aligned} \max \quad & 4x + 7y \\ \text{subject to:} \quad & \\ & 2x + 6y \le 28 \quad (1) \\ & 3x + 4y \le 22 \quad (2) \\ & 5x + 4y \le 34 \quad (3) \\ & x, y \ge 0. \end{aligned}$$

- Plot: A plot of the constraints gives:

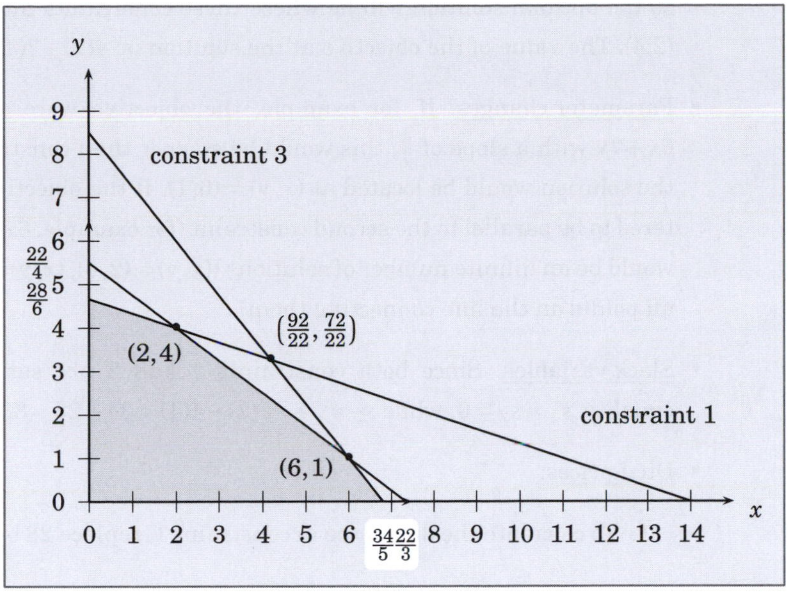

- Solution: Constraints 1 and 2 intersect at $(x,y) = (2,4)$; constraints 1 and 3 at $(x,y) = \left(\frac{92}{22}, \frac{72}{22}\right)$; and constraints 2 and 3 at $(x,y) = (6,1)$. These are found by solving each pair of equations in turn. For example, $2x+6y = 28$ and $3x+4y = 22$ solve to give $(x,y) = (2,4)$, which turns out to be the optimal solution:

$$\begin{pmatrix} 2 & 6 \\ 3 & 4 \end{pmatrix}\begin{pmatrix} x \\ y \end{pmatrix} = \begin{pmatrix} 28 \\ 22 \end{pmatrix}.$$

Thus

$$\begin{vmatrix} 2 & 6 \\ 3 & 4 \end{vmatrix} = -10, \quad \begin{vmatrix} 28 & 6 \\ 22 & 4 \end{vmatrix} = -20, \quad \begin{vmatrix} 2 & 28 \\ 3 & 22 \end{vmatrix} = -40.$$

This gives $(x,y) = \left(\frac{-20}{-10}, \frac{-40}{-10}\right) = (2,4)$. To see that it is optimal, recall from earlier discussion that comparing the objective value at each feasible corner is one way to solve the program. Another way is to compare the slopes of the constraints and the objective.

Constraint	*Slope*	*Objective*
Constraint 1	$-\frac{1}{3}$	
		$-\frac{4}{7}$
Constraint 2	$-\frac{3}{4}$	
Constraint 3	$-\frac{5}{4}$	

The slope of the objective is between the slope of constraints 1 and 2, so the optimal solution will be where these constraints intersect – at (2,4). The value of the objective at the solution is: $4(2)+7(4)=36$.

- Parameter changes: If, for example, the objective were changed to $6x+7y$ with a slope of $\frac{6}{7}$, this would be steeper than constraint 2 and the solution would be located at $(x,y)=(6,1)$. If the objective were altered to be parallel to the second constraint (for example, $6x+8y$) there would be an infinite number of solutions ($(x,y)=(2,4)$, $(x,y)=(6,1)$ and all points on the line connecting them).
- Slack variables: Since both constraints 1 and 2 are satisfied with equality, $s_1=s_2=0$, while $s_3=34-5(2)-4(4)=34-26=8$.
- Dual prices:
  1. To calculate the dual price in constraint 1, replace 28 by 29. Then:
  $$\begin{vmatrix} 29 & 6 \\ 22 & 4 \end{vmatrix} = -16, \quad \begin{vmatrix} 2 & 29 \\ 3 & 22 \end{vmatrix} = -43$$
  and the new solution is $(x,y)=\left(\frac{16}{10},\frac{43}{10}\right)$. The value of the objective is $4\left(\frac{16}{10}\right)+7\left(\frac{43}{10}\right)=\frac{365}{10}=36\frac{1}{2}$. Consequently, $DP_1=\frac{1}{2}$.
  2. To calculate the dual price in constraint 2, replace 22 by 23. Then:
  $$\begin{vmatrix} 28 & 6 \\ 23 & 4 \end{vmatrix} = -26, \quad \begin{vmatrix} 2 & 28 \\ 3 & 23 \end{vmatrix} = -38$$
  and the new solution is $(x,y)=\left(\frac{26}{10},\frac{38}{10}\right)$. The value of the objective is $4\left(\frac{26}{10}\right)+7\left(\frac{38}{10}\right)=\frac{370}{10}=37$. Consequently, $DP_2=1$.
  3. Since changing the right-hand side of constraint 3 does not affect the solution, the associated dual price is 0. ◊

EXAMPLE 4.8: Consider the following linear program:

$$\begin{aligned} \min \quad & 4x+2y \\ \text{subject to:} \quad & \\ & 4x+8y \geq 40 \quad (1) \\ & 3x+2y \geq 24 \quad (2) \\ & 6x+2y \geq 30 \quad (3) \\ & x,y \geq 0. \end{aligned}$$

- Plot:

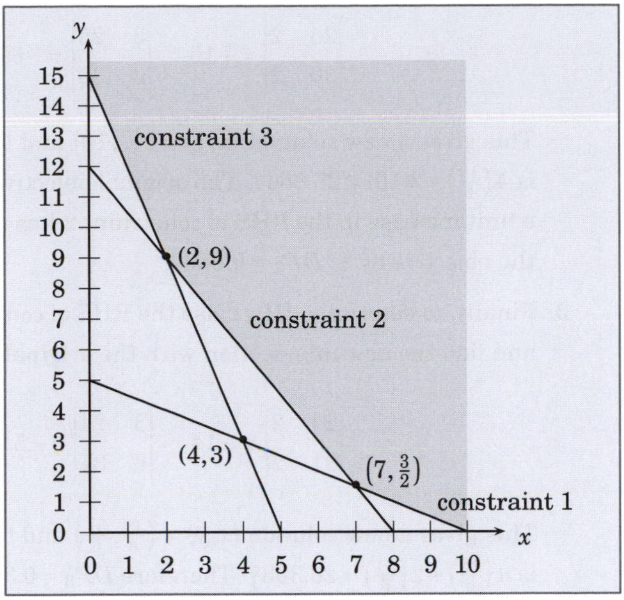

- Solution: Comparing the slopes of the objective and constraints gives the solution at the intersection of constraints 2 and 3.

$$\begin{vmatrix} 3 & 2 \\ 6 & 2 \end{vmatrix} = -6, \quad \begin{vmatrix} 24 & 2 \\ 30 & 2 \end{vmatrix} = -12, \quad \begin{vmatrix} 3 & 24 \\ 6 & 30 \end{vmatrix} = -54.$$

Thus, $(x,y) = (2,9)$. The value of the objective at the solution is: $4(2) + 2(9) = 26$.

- Slack variables: The values of the slack variables at the optimal solution, (x^*, y^*) are defined from the equations:

$$4x^* + 8y^* - s_1^* = 40$$
$$3x^* + 2y^* - s_2^* = 24$$
$$6x^* + 2y^* - s_3^* = 30,$$

giving $s_1^* = 4x^* + 8x^* - 40 = 4(2) + 8(9) - 40 = 80 - 40 = 40$, and $s_2^* = 0$, $s_3^* = 0$.

- Dual prices:

  1. Because constraint 1 is not binding, its shadow price is 0: $DP_1 = 0$.
  2. To determine the dual price of the second constraint, increase RHS_2 by 1 from 24 to 25. Solving for the intersection of this new

constraint 2 and the original constraint 3:

$$\begin{vmatrix} 25 & 2 \\ 30 & 2 \end{vmatrix} = -10, \quad \begin{vmatrix} 3 & 25 \\ 6 & 30 \end{vmatrix} = -60.$$

This gives a new solution $(x,y) = \left(\frac{10}{6}, 10\right)$ and the objective value is $4\left(\frac{10}{6}\right) + 2(10) = 26.6667$. The original objective value was 26, so a unit increase in the RHS of constraint 2 has raised the value of the objective by $\frac{2}{3}$: $DP_2 = 0.6667$.

3. Finally, to determine DP_3 raise the RHS of constraint 3 by 1 unit and find the new intersection with the original constraint 2:

$$\begin{vmatrix} 24 & 2 \\ 31 & 2 \end{vmatrix} = -14, \quad \begin{vmatrix} 3 & 24 \\ 6 & 31 \end{vmatrix} = -51.$$

This gives a new solution $(x,y) = \left(\frac{14}{6}, \frac{51}{6}\right)$ and the objective value is $4\left(\frac{14}{6}\right) + 2\left(\frac{51}{6}\right) = 26.3337$. Therefore $DP_3 = 0.3337$. ◊

4.1.6 Duality reconsidered

The following discussion provides a different perspective on dual prices and their computation. An example will motivate the discussion. The next example takes an earlier example and rearranges the parameters and objective to give a program called the *dual program*. Every program has a dual program, giving the terminology *primal* and *dual* to refer to the initial and transformed programs. It turns out that the optimal solution payoff values are the same in both programs, the dual prices in the primal are the solution values in the dual, and the solution values in the primal are the dual prices in the dual program.

EXAMPLE 4.9: Consider the following linear program:

Program A

$$\begin{aligned} \min \quad & 13w + 17z \\ \text{subject to:} \quad & \\ & 2w + 4z \geq 3 \quad (1) \\ & 3w + 3z \geq 4 \quad (2) \\ & w, z \geq 0 \quad . \end{aligned}$$

Observe that this program bears a close relation to one discussed earlier (see Examples 4.5 and 4.6):

Program B

$$\begin{aligned} \max \quad & 3x+4y \\ \text{subject to:} \quad & \\ & 2x+3y \le 13 \quad (1) \\ & 4x+3y \le 17 \quad (2) \\ & x,y \ge 0. \end{aligned}$$

Program A is called the dual of program B. Notice that A is obtained from B by exchanging the objective coefficients and RHS constraint parameters, and taking the transpose of the coefficient matrix. The remarkable fact relating these programs is the following. Let (x^*, y^*) solve program B, with objective value π_B and dual prices (DP_1^B, DP_2^B). Similarly, let (w^*, z^*) solve program A, with objective value π_A and dual prices (DP_1^A, DP_2^A). Then

$$(1) \quad \pi_B = \pi_A, \quad (2) \quad (x^*, y^*) = (DP_1^A, DP_2^A), \quad (3) \quad (w^*, z^*) = (DP_1^B, DP_2^B).$$

Recall from Examples 4.5 and 4.6, that the solution to program B has: $\pi_B = 18$, $(x^*, y^*) = (2,3)$ and $(DP_1, DP_2) = \left(1\frac{1}{6}, \frac{1}{6}\right)$. To solve program A, plot the constraints:

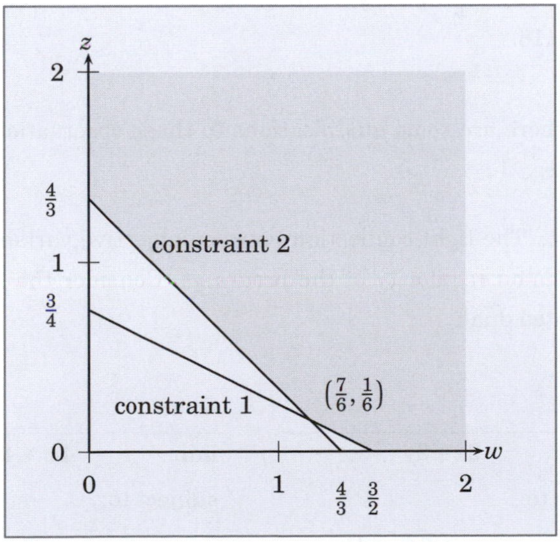

The solution is at the intersection of the two constraints, and using Cramer's rule:

$$\begin{vmatrix} 2 & 4 \\ 3 & 3 \end{vmatrix} = -6, \quad \begin{vmatrix} 3 & 4 \\ 4 & 3 \end{vmatrix} = -7, \quad \begin{vmatrix} 2 & 3 \\ 3 & 4 \end{vmatrix} = -1$$

gives solution $(w^*,z^*) = \left(\frac{7}{6},\frac{1}{6}\right) = \left(1\frac{1}{6},\frac{1}{6}\right)$. The objective value is $\pi_A = 13\left(\frac{7}{6}\right) + 17\left(\frac{1}{6}\right) = \frac{108}{6} = 18$.

To calculate the dual prices consider first an increase of $\frac{1}{4}$ in the RHS of constraint 1 and solve for the new intersection:

$$\begin{vmatrix} 2 & 4 \\ 3 & 3 \end{vmatrix} = -6, \quad \begin{vmatrix} 3\frac{1}{4} & 4 \\ 4 & 3 \end{vmatrix} = -6\tfrac{1}{4}, \quad \begin{vmatrix} 2 & 3\frac{1}{4} \\ 3 & 4 \end{vmatrix} = -1\tfrac{3}{4}.$$

Thus, the new solution is: $(w',z') = \left(\frac{6\frac{1}{4}}{6},\frac{1\frac{3}{4}}{6}\right) = \left(\frac{25}{24},\frac{7}{24}\right)$. The new objective value is then $13\left(\frac{25}{24}\right) + 17\left(\frac{7}{24}\right) = \frac{444}{24} = 18\frac{1}{2}$. Thus, $DP_1^A = \frac{\frac{1}{2}}{\frac{1}{4}} = 2$.

For constraint 2, to determine DP_2^A, change the RHS to $4\frac{1}{4}$:

$$\begin{vmatrix} 2 & 4 \\ 3 & 3 \end{vmatrix} = -6, \quad \begin{vmatrix} 3 & 4 \\ 4\frac{1}{4} & 3 \end{vmatrix} = -8, \quad \begin{vmatrix} 2 & 3 \\ 3 & 4\frac{1}{4} \end{vmatrix} = -\tfrac{1}{2}.$$

The new solution is now $(w^\dagger,z^\dagger) = \left(\frac{8}{6},\frac{1}{12}\right)$. The new objective value is then $\left[13\left(\frac{16}{12}\right) + 17\left(\frac{1}{12}\right)\right] = \frac{225}{12} = 18\frac{3}{4}$. Thus, $DP_2^A = \frac{\frac{3}{4}}{\frac{1}{4}} = 3$. Summarizing: $DP_1^A = 2$, $DP_2^A = 3$. Thus, $(x^*,y^*) = (DP_1^A, DP_2^A) = (2,3)$.

Recall that the dual prices for program B were $DP_1^B = 1.1667 = \frac{7}{6}$ and $DP_2^B = 0.1667 = \frac{1}{6}$ and $\pi_B = 18$. Thus, $(w^*,z^*) = (DP_1^B, DP_2^B) = \left(\frac{7}{6},\frac{1}{6}\right)$ and $\pi_A = \pi_B = 18$. ◇

However, there are some qualifications to these observations, as the next remark illustrates.

REMARK 4.1: The tight connection between objective variables and dual variables depends on regularity of the programs. Consider the following program with associated dual:

Primal			*Dual*		
max	$2x+2y$		min	$4w+4v+6z$	
subject to:			subject to:		
	$x \le 4$	(1)		$w+z \ge 2$	(1)
	$y \le 4$	(2)		$v+z \ge 2$	(2)
	$x+y \le 6$	(3)		$w,v,z \ge 0$	
	$x,y \ge 0$				

.

The primal has an objective parallel to the third constraint. Any point on the line connecting (2,4) and (4,2), such as (3,3), is an optimal solution to the linear program. The value of the objective is 12. Furthermore, for either constraints 1 or 2, raising the RHS by 1 or reducing it by 1, leaves (3,3) both feasible and optimal, so that there is no impact on the objective. The dual prices of constraints 1 and 2 are 0. The third constraint in the primal has a dual price of 2.

In the dual program, $w = v = 0$ and $z = 2$ is the unique optimal solution. To see this, consider changing these values. If z is reduced by Δ, then both x and y must increase by Δ to satisfy the constraints. The reduction in the objective from the change in z, -6Δ, is outweighed by the gain from the increase in x and y: $8\Delta = 4\Delta + 4\Delta$. Turning to shadow prices, if the RHS of the first constraint in the dual is raised to $2+\gamma$, γ small, then it is optimal to raise w from 0 to γ. Raising w raises the objective by 4γ, whereas increasing z would raise the cost by 6γ. The change in the objective divided by the change in the RHS of 1 is 4, suggesting $DP_1 = 4$ in the dual. The same reasoning suggests that $DP_2 = 4$ in the dual, but the pair (4,4) correspond to no solution in the primal. □

4.2 Basic solutions

The previous discussion dealt with programs having constraints of the form $\boldsymbol{Ax} \leq \boldsymbol{b}$ or $\boldsymbol{Ax} \geq \boldsymbol{b}$. In what follows, it is shown that such inequality-constrained problems may be reformulated to equality-constrained programs. The equality-constrained formulation is then used to discuss basic solutions.

4.2.1 Equality-constrained programs

Consider the linear program: $\max \boldsymbol{c}\cdot\boldsymbol{x}$ subject to $\boldsymbol{Ax} \leq \boldsymbol{b}$ and $\boldsymbol{x} \geq \boldsymbol{0}$. Here $\boldsymbol{c} - n \times 1$, $\boldsymbol{x} - n \times 1$, $\boldsymbol{A} - m \times n$, $\boldsymbol{s} - m \times 1$ and $\boldsymbol{I} - m \times m$, the identity matrix. The program may be written in a variety of equivalent ways:

I		*II*	
max	$\boldsymbol{c}\cdot\boldsymbol{x}$	max	$\boldsymbol{c}\cdot\boldsymbol{x} + \boldsymbol{0}\cdot\boldsymbol{s}$
subject to:		subject to:	
	$\boldsymbol{Ax} \leq \boldsymbol{b}$		$\boldsymbol{Ax} + \boldsymbol{Is} = \boldsymbol{b}$
	$\boldsymbol{x} \geq \boldsymbol{0}$		$\boldsymbol{x}, \boldsymbol{s} \geq \boldsymbol{0}$

III		*IV*	
max	$(\boldsymbol{c},\boldsymbol{0})\cdot(\boldsymbol{x},\boldsymbol{s})$	max	$\bar{\boldsymbol{c}}\cdot\boldsymbol{z}$
subject to:		subject to:	
	$(\boldsymbol{A}:\boldsymbol{I})(\boldsymbol{x},\boldsymbol{s})=\boldsymbol{b}$		$\bar{\boldsymbol{A}}\boldsymbol{z}=\boldsymbol{b}$
	$(\boldsymbol{x},\boldsymbol{s})\geq\boldsymbol{0}$		$\boldsymbol{z}\geq\boldsymbol{0}.$

Formulation *II* can be expressed succinctly as follows. Define new matrices and vectors: $\bar{\boldsymbol{c}}=(\boldsymbol{c},\boldsymbol{0})$, $\bar{\boldsymbol{A}}=(\boldsymbol{A}:\boldsymbol{I})$, $\boldsymbol{z}=(\boldsymbol{x},\boldsymbol{s})$. In the initial problem, the dimensions of the matrices and vectors were $\boldsymbol{c}-n\times 1$, $\boldsymbol{x}-n\times 1$, $\boldsymbol{A}-m\times n$. Since $\boldsymbol{A}$ is an $m\times n$ matrix, this means that there are m constraints (leaving aside the non-negativity constraints on the $\boldsymbol{x}$). The equality-constrained model is obtained by adding "slack" variables – $\boldsymbol{s}$ – one for each constraint. So there are m slack variables added to give formulation *II*. Thus $\bar{\boldsymbol{c}}$ is a lengthened vector with zeros added to $\boldsymbol{c}$; likewise, $\boldsymbol{z}$ involves lengthening the $\boldsymbol{x}$ vector with $\boldsymbol{s}$, and $\bar{\boldsymbol{A}}$ is a matrix formed by placing $\boldsymbol{A}$ and $\boldsymbol{I}$, the $m\times m$ identity matrix, side by side. Therefore, $\boldsymbol{z}-(n+m)\times 1$, $\bar{\boldsymbol{c}}-(n+m)\times 1$ and $\bar{\boldsymbol{A}}-m\times(n+m)$. Programs *II*, *III* and *IV* are equivalent problems – just involving relabeling. Furthermore, *I* and *II* are equivalent problems, so that the inequality-constrained program in *I* may be replaced by the equality-constrained program, *IV*.

Theorem 4.1: *The linear programs I and II have the same solutions: if* $\boldsymbol{x}^*$ *solves I then for some* $\boldsymbol{s}^*\geq 0,(\boldsymbol{x}^*,\boldsymbol{s}^*)$ *solves II, and if* $(\tilde{\boldsymbol{x}},\tilde{\boldsymbol{s}})$ *solves II then* $\tilde{\boldsymbol{x}}$ *solves I.*

PROOF: Let $\boldsymbol{x}^*$ solve *I* and suppose that there is no $\boldsymbol{s}^*\geq 0$ such that $(\boldsymbol{x}^*,\boldsymbol{s}^*)$ solves *II*. If this is the case, then for any $\boldsymbol{s}^*\geq\boldsymbol{0},\exists(\tilde{\boldsymbol{x}},\tilde{\boldsymbol{s}})$ with:

$$\boldsymbol{c}\cdot\tilde{\boldsymbol{x}}=\boldsymbol{c}\cdot\tilde{\boldsymbol{x}}+\boldsymbol{0}\cdot\tilde{\boldsymbol{s}}>\boldsymbol{c}\cdot\boldsymbol{x}^*+\boldsymbol{0}\cdot\boldsymbol{s}^*=\boldsymbol{c}\cdot\boldsymbol{x}^* \text{ and } \boldsymbol{A}\tilde{\boldsymbol{x}}+\boldsymbol{I}\tilde{\boldsymbol{s}}=\boldsymbol{b},(\tilde{\boldsymbol{x}},\tilde{\boldsymbol{s}})\geq\boldsymbol{0}.$$

However, from the previous equation,

$$\boldsymbol{0}\leq\tilde{\boldsymbol{s}}=\tilde{\boldsymbol{I}\boldsymbol{s}}=\boldsymbol{b}-\boldsymbol{A}\tilde{\boldsymbol{x}} \text{ implies } \boldsymbol{A}\tilde{\boldsymbol{x}}\leq\boldsymbol{b}.$$

Therefore, $\tilde{\boldsymbol{x}}$ is feasible for the first problem and $\boldsymbol{c}\cdot\tilde{\boldsymbol{x}}>\boldsymbol{c}\cdot\boldsymbol{x}^*$, contradicting the assumption that $\boldsymbol{x}^*$ solves problem *I*.

Conversely, suppose that $(\tilde{\boldsymbol{x}},\tilde{\boldsymbol{s}})$ solves *II* but $\tilde{\boldsymbol{x}}$ does not solve *I*. Then $\exists\boldsymbol{x}^*\geq\boldsymbol{0}$, such that $\boldsymbol{c}\cdot\boldsymbol{x}^*>\boldsymbol{c}\cdot\tilde{\boldsymbol{x}}$ and $\boldsymbol{A}\boldsymbol{x}^*\leq\boldsymbol{b}$. Set $\boldsymbol{s}^*=\boldsymbol{b}-\boldsymbol{A}\boldsymbol{x}^*$ so that $\boldsymbol{s}^*\geq\boldsymbol{0}$. Then $(\boldsymbol{x}^*,\boldsymbol{s}^*)$ is feasible and improving over $(\tilde{\boldsymbol{x}},\tilde{\boldsymbol{s}})$ in *II*: $\boldsymbol{c}\cdot\boldsymbol{x}^*+\boldsymbol{0}\cdot\boldsymbol{s}^*>\boldsymbol{c}\cdot\tilde{\boldsymbol{x}}+\boldsymbol{0}\cdot\tilde{\boldsymbol{s}}$. This contradicts the assumption that $(\tilde{\boldsymbol{x}},\tilde{\boldsymbol{s}})$ solves *II*. Therefore, solving *I* and *II* are equivalent. This completes the proof. ■

REMARK 4.2: The minimization problem is analogous.

I		*II*	
min	$\boldsymbol{c}\cdot\boldsymbol{x}$	min	$\boldsymbol{c}\cdot\boldsymbol{x}+\boldsymbol{0}\cdot\boldsymbol{s}$
subject to:		subject to:	
	$\boldsymbol{A}\boldsymbol{x}\geq\boldsymbol{b}$		$\boldsymbol{A}\boldsymbol{x}-\boldsymbol{I}\boldsymbol{s}=\boldsymbol{b}$
	$\boldsymbol{x}\geq\boldsymbol{0}$		$\boldsymbol{x},\boldsymbol{s}\geq\boldsymbol{0}.$

This can be expressed:

$$\begin{aligned} &\min && \bar{\boldsymbol{c}}\cdot\boldsymbol{z} \\ &\text{subject to:} && \\ & && \bar{\boldsymbol{A}}\boldsymbol{z}=\boldsymbol{b} \\ & && \boldsymbol{z}\geq\boldsymbol{0}, \end{aligned}$$

where $\bar{\boldsymbol{A}}=(\boldsymbol{A}:-\boldsymbol{I})$, $\bar{\boldsymbol{c}}=(\boldsymbol{c},\boldsymbol{0})$, and $\boldsymbol{z}=(\boldsymbol{x},\boldsymbol{s})$. Again, solving *I* and *II* yields the same solution. □

4.2.2 Definition and identification of basic solutions

In this section, basic solutions are introduced and related to the vertices of the feasible region. In view of Theorem 4.1, we can focus on the equality-constrained formulation for the following discussion (and discard the "bar" notation). Thus, the linear program will be written: max $\boldsymbol{c}\cdot\boldsymbol{x}$, subject to : $\boldsymbol{A}\boldsymbol{x}=\boldsymbol{b}$ and $\boldsymbol{x}\geq\boldsymbol{0}$. That is,

$$\begin{aligned} &\max && \boldsymbol{c}\cdot\boldsymbol{x} \\ &\text{subject to:} && \boldsymbol{A}\boldsymbol{x}=\boldsymbol{b} \\ & && \boldsymbol{x}\geq\boldsymbol{0}. \end{aligned}$$

Recall that a *feasible solution* in the inequality-constrained program is any vector $\boldsymbol{x}$ satisfying the constraints: $\boldsymbol{A}\boldsymbol{x}\leq\boldsymbol{b}$ and $\boldsymbol{x}\geq\boldsymbol{0}$. A feasible solution $\hat{\boldsymbol{x}}$ is an *optimal feasible solution* if there is no other feasible $\tilde{\boldsymbol{x}}$, such that $\boldsymbol{c}\cdot\tilde{\boldsymbol{x}}>\boldsymbol{c}\cdot\hat{\boldsymbol{x}}$ in the case of a maximization problem and $\boldsymbol{c}\cdot\tilde{\boldsymbol{x}}<\boldsymbol{c}\cdot\hat{\boldsymbol{x}}$ in the case of a minimization problem. (Alternatively, the definitions of feasible solution and optimal feasible solution can be given in terms of the equality-constrained program – it makes no difference.) With the equality-constrained program, a vector $\boldsymbol{x}$ satisfying $\boldsymbol{A}\boldsymbol{x}=\boldsymbol{b}$ and $\boldsymbol{x}\geq\boldsymbol{0}$ is said to be *feasible*. Let m be the number of rows in $\boldsymbol{A}$ and recall that $\boldsymbol{A}$ has more columns than rows.

We assume that $\boldsymbol{A}$ has full rank and since $\boldsymbol{A}$ has more columns than rows this means that the rank of $\boldsymbol{A}$ equals the number of rows: m. This is called the *full rank assumption*: the m rows of $\boldsymbol{A}$ are linearly independent. This assumption involves no loss of generality. If $\boldsymbol{A}$ does not have full rank then there are two possibilities. Either some of the constraints in $\boldsymbol{A}\boldsymbol{x}=\boldsymbol{b}$ are redundant (when some subset of constraints are satisfied, so are the remainder), or the constraints are inconsistent – there is no $\boldsymbol{x}$ such that $\boldsymbol{A}\boldsymbol{x}=\boldsymbol{b}$. To see this more clearly, let $\boldsymbol{a}_i$ denote the ith row of $\boldsymbol{A}$. Suppose that $\boldsymbol{A}$ does not have full rank. Then some row of $\boldsymbol{A}$, say row i, can be written as a linear combination of the other rows: $\boldsymbol{a}_i=\sum_{j\neq i}\theta_j\boldsymbol{a}_j$. Now, suppose $\boldsymbol{a}_j\cdot\boldsymbol{x}=b_j, j\neq i$, so that $\sum_{j\neq i}\theta_j(\boldsymbol{a}_j\cdot\boldsymbol{x})=\sum_{j\neq i}\theta_j b_j$. Since $\sum_{j\neq i}(\theta_j\boldsymbol{a}_j)\cdot\boldsymbol{x}=\boldsymbol{a}_j\cdot\boldsymbol{x}$, this implies that $\boldsymbol{a}_i\cdot\boldsymbol{x}=\sum_{j\neq i}\theta_j b_j$. However, the ith equation specifies that $\boldsymbol{x}$ must satisfy $\boldsymbol{a}_i\cdot\boldsymbol{x}=b_i$. Thus, there are two possibilities: (a) $b_i=\sum_{j\neq i}\theta_j b_j$, in which case any $\boldsymbol{x}$ satisfying all but the ith constraint will also satisfy the ith constraint or (b) $b_i\neq\sum_{j\neq i}\theta_j b_j$, so that any $\boldsymbol{x}$ satisfying all but the ith constraint cannot satisfy the ith constraint. In case (a), the ith constraint is redundant, while in case (b) the ith constraint is inconsistent with the other constraints. Therefore, in case (b) the problem has no solution while in case (a) the ith constraint may be dropped from the problem and the problem remains unchanged. From now on we assume that $\boldsymbol{A}$ has full rank – equal to the number of rows (m) in $\boldsymbol{A}$.

Suppose we select m linearly independent columns from the matrix $\boldsymbol{A}$: say these columns together form a matrix $\boldsymbol{B}$. Such a set of columns is called a *basis.* It is possible that not all sets of m columns of $\boldsymbol{A}$ are linearly independent, but since $\boldsymbol{A}$ has full rank there is at least one set of m linearly independent columns.

Since $\boldsymbol{B}$ is non-singular, the equation system $\boldsymbol{B}\boldsymbol{x}_B=\boldsymbol{b}$ has a unique solution. For example, suppose that the first m columns of $\boldsymbol{A}$ are linearly independent and let $\boldsymbol{B}$ be the matrix formed by these first m columns. Let the remaining columns of $\boldsymbol{A}$ form some matrix $\boldsymbol{C}$, so that $\boldsymbol{A}=(\boldsymbol{B}:\boldsymbol{C})$. Then if $\boldsymbol{x}_B$ satisfies $\boldsymbol{B}\boldsymbol{x}_B=\boldsymbol{b}$, let $\boldsymbol{x}=(\boldsymbol{x}_B,\boldsymbol{0})$ and observe that $\boldsymbol{A}\boldsymbol{x}=\boldsymbol{B}\boldsymbol{x}_B+\boldsymbol{c}\cdot\boldsymbol{0}=\boldsymbol{B}\boldsymbol{x}_B=\boldsymbol{b}$. Thus $\boldsymbol{x}=(\boldsymbol{x}_B:\boldsymbol{0})$ satisfies $\boldsymbol{A}\boldsymbol{x}=\boldsymbol{b}$. However, this $\boldsymbol{x}$ may not satisfy $\boldsymbol{x}\geq 0$ – some elements of the vector $\boldsymbol{x}$ may be negative. This problem is taken up later. Note that when the $\boldsymbol{B}$ matrix is selected, a corresponding set of $\boldsymbol{x}$ is selected. In the example above, selection of the first m columns of $\boldsymbol{A}$ implicitly led to the selection of the first m elements of $\boldsymbol{x}$, with the remainder set to 0.

Given $\boldsymbol{A}\boldsymbol{x}=\boldsymbol{b}$, let $\boldsymbol{B}$ be *any* non-singular $m\times m$ submatrix of columns of $\boldsymbol{A}$. If all elements of $\boldsymbol{x}$ not associated with columns of $\boldsymbol{B}$ are set to $\boldsymbol{0}$, and the resulting set of equations solved, the unique $\boldsymbol{x}$ determined in this way is called a *basic solution* of the system $\boldsymbol{A}\boldsymbol{x}=\boldsymbol{b}$ with respect to the basis $\boldsymbol{B}$. The components of $\boldsymbol{x}$ associated with columns of $\boldsymbol{B}$ are called *basic variables*. Thus, in the discussion above where the first m columns of $\boldsymbol{A}$ are chosen (and are linearly independent), $\boldsymbol{x} = (\boldsymbol{x}_B, \mathbf{0})$ is a basic solution (with respect to the basis $\boldsymbol{B}$) and $\boldsymbol{x}_B$ the *basic variables*.

If one or more of the basic variables in a basic solution has a value of $\mathbf{0}$, that solution is said to be a *degenerate basic solution*. Therefore, in a non-degenerate basic solution there are m non-zero components (and the remaining components are $\mathbf{0}$). In a degenerate basic solution, there are fewer than m components non-zero.

Recall that if $\boldsymbol{x}$ satisfies both $\boldsymbol{A}\boldsymbol{x} = \boldsymbol{b}$ and $\boldsymbol{x} \geq 0$, $\boldsymbol{x}$ is said to be *feasible* (for these constraints). If $\boldsymbol{x}$ is a basic solution satisfying constraint 3 it is called a *basic feasible solution*. If this solution is also degenerate, it is called a *degenerate basic feasible solution*. The following result, called the *fundamental theorem of linear programming*, is central in linear programming.

Theorem 4.2: *Given a linear program of the form I or II,*

I		*II*	
max	$\boldsymbol{c} \cdot \boldsymbol{x}$	max	$\boldsymbol{c} \cdot \boldsymbol{x}$
subject to:		subject to:	
	$\boldsymbol{A}\boldsymbol{x} \leq \boldsymbol{b}$		$\boldsymbol{A}\boldsymbol{x} = \boldsymbol{b}$
	$\boldsymbol{x} \geq \mathbf{0}$		$\boldsymbol{x} \geq \mathbf{0}$.

1. *If there is a feasible solution there is a basic feasible solution.*

2. *If there is an optimal feasible solution, there is an optimal basic feasible solution.*

The importance of this theorem (from the computational point of view) is that the search for an optimal solution can be restricted to the basic solutions. Since there are only a finite number of these, there are only a finite number of $\boldsymbol{x}$ to be considered in the search for an optimal solution. The following example illustrates the use of basic solutions in locating an optimal solution.

EXAMPLE 4.10:

$$
\begin{array}{lrl}
\text{max} & 2x + 5y & \\
\text{subject to:} & 4x + y & \leq 24 \\
 & x + 3y & \leq 21 \\
 & x + y & \leq 9 \\
 & x, y & \geq 0
\end{array}
$$

This is rearranged in equality constraint form as:

$$\begin{aligned}
\max \quad & 2x+5y+0s_1+0s_2+0s_3 \\
\text{subject to:} \quad & 4x+y+s_1 = 24 \\
& x+3y+s_2 = 21 \\
& x+y+s_3 = 9 \\
& x,y,s_1,s_2,s_3 = 0.
\end{aligned}$$

In matrix form, let:

$$\bar{\boldsymbol{A}} = \begin{pmatrix} 4 & 1 & 1 & 0 & 0 \\ 1 & 3 & 0 & 1 & 0 \\ 1 & 1 & 0 & 0 & 1 \end{pmatrix}, \ \boldsymbol{b} = \begin{pmatrix} 24 \\ 21 \\ 9 \end{pmatrix}, \ \bar{\boldsymbol{c}} = (2,5,0,0,0), \text{ and } \boldsymbol{z} = (x,y,s_1,s_2,s_3).$$

The problem may then be written:

$$\begin{aligned}
\max \quad & \bar{\boldsymbol{c}} \cdot \boldsymbol{z} \\
\text{subject to:} \quad & \bar{\boldsymbol{A}}\boldsymbol{z} = \boldsymbol{b} \\
& \boldsymbol{z} \geq \boldsymbol{0}.
\end{aligned}$$

◊

In general, given a matrix with m rows and p columns ($p > m$), the number of $m \times m$ submatrices that can be formed is $\frac{p!}{m!(p-m)!}$ (the binomial coefficient). In the present case, the coefficient matrix has $m = 3$ and $p = 5$, and so the maximum number of distinct 3×3 matrices (giving the maximum number of bases) is $\frac{5!}{3!(5-3)!} = 10$. Consider the $\bar{\boldsymbol{A}}$ matrix. The constraints are plotted in Figure 4.1 (the numbering identifies the basic solutions). To find the first basic solution, begin with the first three columns of $\bar{\boldsymbol{A}}$. Provided these three columns are linearly independent (which they are), set the coefficients on the last two columns to 0 (s_2 and s_3) and solve for the coefficients of the first three columns (x,y,s_1).

$$\bar{\boldsymbol{A}} = \begin{matrix} & x & y & s_1 & s_2 & s_3 \\ & \end{matrix}$$
$$\bar{\boldsymbol{A}} = \begin{pmatrix} 4 & 1 & 1 & 0 & 0 \\ 1 & 3 & 0 & 1 & 0 \\ 1 & 1 & 0 & 0 & 1 \end{pmatrix}.$$

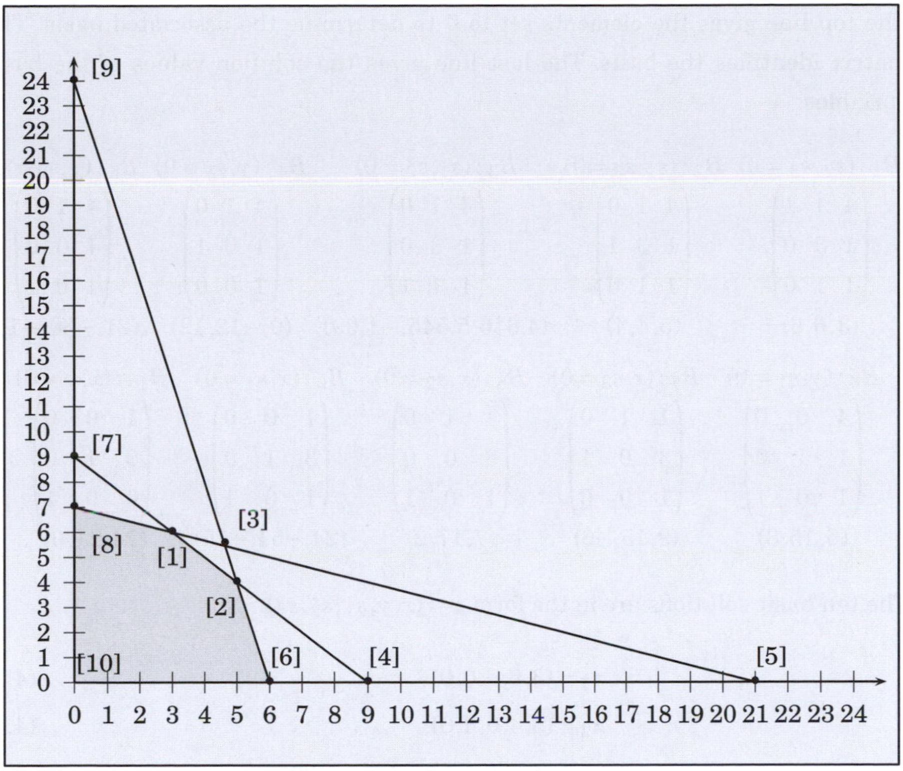

Figure 4.1: Basic solutions

So, suppose $s_2 = s_3 = 0$, and solve for the remaining variables:

$$B_1 = \begin{pmatrix} 4 & 1 & 1 \\ 1 & 3 & 0 \\ 1 & 1 & 0 \end{pmatrix} \Rightarrow \begin{pmatrix} 4 & 1 & 1 \\ 1 & 3 & 0 \\ 1 & 1 & 0 \end{pmatrix} \begin{pmatrix} x \\ y \\ s_1 \end{pmatrix} = \begin{pmatrix} 24 \\ 21 \\ 9 \end{pmatrix}.$$

The solution is $(x,y,s_1) = (3,6,6)$. In the terminology introduced earlier, the set (x,y,s_1) are called basic variables and (3,6,6,0,0) is a basic solution. The basis B_1 was obtained by setting s_2 and s_3 to 0. To identify how the basis was obtained, denote this situation by $B_1 : (s_2, s_3 = 0)$. The notation is:

$$\begin{gathered} B_1 : (s_2, s_3 = 0) \\ \begin{pmatrix} 4 & 1 & 1 \\ 1 & 3 & 0 \\ 1 & 1 & 0 \end{pmatrix} \\ (3,6,6) \end{gathered}$$

The top line gives the elements set to 0 to determine the associated basis. The matrix identifies the basis. The last line gives the solution values of the basic variables:

$B_1:(s_2,s_3=0)$ $\begin{pmatrix}4&1&1\\1&3&0\\1&1&0\end{pmatrix}$ $(3,6,6)$

$B_2:(s_1,s_3=0)$ $\begin{pmatrix}4&1&0\\1&3&1\\1&1&0\end{pmatrix}$ $(5,4,4)$

$B_3:(s_1,s_2=0)$ $\begin{pmatrix}4&1&0\\1&3&0\\1&1&1\end{pmatrix}$ $(4.636,5.545,-1.09)$

$B_4:(y,s_3=0)$ $\begin{pmatrix}4&1&0\\1&0&1\\1&0&0\end{pmatrix}$ $(9,-12,12)$

$B_5:(y,s_2=0)$ $\begin{pmatrix}4&1&0\\1&0&0\\1&0&1\end{pmatrix}$ $(21,-60,-12)$

$B_6:(y,s_1=0)$ $\begin{pmatrix}4&0&0\\1&1&0\\1&0&1\end{pmatrix}$ $(6,15,3)$

$B_7:(x,s_3=0)$ $\begin{pmatrix}1&1&0\\3&0&1\\1&0&0\end{pmatrix}$ $(9,15,-6)$

$B_8:(x,s_2=0)$ $\begin{pmatrix}1&1&0\\3&0&0\\1&0&1\end{pmatrix}$ $(7,17,2)$

$B_9:(x,s_1=0)$ $\begin{pmatrix}1&0&0\\3&1&0\\1&0&1\end{pmatrix}$ $(24,-51,-15)$

$B_{10}:(x,y=0)$ $\begin{pmatrix}1&0&0\\0&1&0\\0&0&1\end{pmatrix}$ $(24,21,9)$.

The ten basic solutions are in the form $\boldsymbol{z}=(x,y,s_1,s_2,s_3)$:

$$\boldsymbol{z}_1=(3,6,6,0,0) \tag{4.3}$$

$$\boldsymbol{z}_2=(5,4,0,4,0) \tag{4.4}$$

$$\boldsymbol{z}_3=(4.636,5.545,0,0,-1.09) \tag{4.5}$$

$$\boldsymbol{z}_4=(9,0,-12,12,0) \tag{4.6}$$

$$\boldsymbol{z}_5=(21,0,-60,0,-12) \tag{4.7}$$

$$\boldsymbol{z}_6=(6,0,0,15,3) \tag{4.8}$$

$$\boldsymbol{z}_7=(0,9,15,-6,0) \tag{4.9}$$

$$\boldsymbol{z}_8=(0,7,17,0,2) \tag{4.10}$$

$$\boldsymbol{z}_9=(0,24,0,-51,-15) \tag{4.11}$$

$$\boldsymbol{z}_{10}=(0,0,24,21,9). \tag{4.12}$$

This list of $\boldsymbol{z}$ is the list of basic solutions. Note now that there are some basic solutions that have negative entries. Once we eliminate those with negative entries, we are left with the basic *feasible* solutions. Recall that all basic solutions are chosen to satisfy $\boldsymbol{Az}=\boldsymbol{b}$ and the constraint $\boldsymbol{z}\geq\boldsymbol{0}$ is ignored. Eliminating those basic solutions with negative entries amounts to eliminating those basic solutions, $\boldsymbol{z}$, that fail to satisfy $\boldsymbol{z}\geq\boldsymbol{0}$. The non-feasible basic solutions are: $\boldsymbol{z}_3,\boldsymbol{z}_4,\boldsymbol{z}_5,\boldsymbol{z}_7$ and $\boldsymbol{z}_9$. Observe that those points on the graph are those at the intersection of constraints at points outside the feasible region. The basic feasible solutions are $\boldsymbol{z}_1,\boldsymbol{z}_2,\boldsymbol{z}_6,\boldsymbol{z}_8,\boldsymbol{z}_{10}$. Denoting the objective value at $\boldsymbol{z}$ by $\pi(\boldsymbol{z})$, $\pi(\boldsymbol{z}_1)=2(3)+5(6)=36$,

$\pi(\boldsymbol{z}_2) = 2(5)+5(4) = 30$, $\pi(\boldsymbol{z}_6) = 2(6)+5(0) = 12$, $\pi(\boldsymbol{z}_8) = 2(0)+5(7) = 35$, $\pi(\boldsymbol{z}_{10}) = 2(0)+5(0) = 0$. So, the solution is at $\boldsymbol{z}$, with $\boldsymbol{x} = 3$ and $\boldsymbol{y} = 6$.

4.3 Duality principles

There is a general principle behind the close connection between the primal and dual programs. It turns out that both have the same objective values. The following discussion formalizes these observations. Consider the primal (P) and dual (D):

P		D	
max	$\boldsymbol{c}'\boldsymbol{x}$	min	$\boldsymbol{b}'\boldsymbol{y}$
subject to:		subject to:	
	$\boldsymbol{A}\boldsymbol{x} \le \boldsymbol{b}$		$\boldsymbol{A}'\boldsymbol{y} \ge \boldsymbol{c}$
	$\boldsymbol{x} \ge \boldsymbol{0}$		$\boldsymbol{y} \ge \boldsymbol{0}$.

The dual is formed from the primal by exchanging objective coefficients with the RHS vector; reversing the inequalities in the linear constraints; and transposing the constraint matrix. Taking the dual program associated with the dual gives the primal.

Theorem 4.3: *For any feasible $\boldsymbol{x}$ and any feasible $\boldsymbol{y}$, $\boldsymbol{b}'\boldsymbol{y} \ge \boldsymbol{c}'\boldsymbol{x}$.*

PROOF: Taking any $\boldsymbol{y}$ feasible in the dual and pre-multiplying $\boldsymbol{A}\boldsymbol{x} \le \boldsymbol{b}$ (from the primal) gives $\boldsymbol{y}'\boldsymbol{A}\boldsymbol{x} \le \boldsymbol{y}'\boldsymbol{b} = \boldsymbol{b}'\boldsymbol{y}$. Similarly, taking any $\boldsymbol{x}$ feasible in the primal and pre-multiplying $\boldsymbol{A}'\boldsymbol{y} \ge \boldsymbol{c}$ from the dual gives $\boldsymbol{x}'\boldsymbol{A}'\boldsymbol{y} \ge \boldsymbol{x}'\boldsymbol{c} = \boldsymbol{c}'\boldsymbol{x}$. So, for any feasible $\boldsymbol{x}$ and $\boldsymbol{y}$,

$$\boldsymbol{b}'\boldsymbol{y} \ge \boldsymbol{y}'\boldsymbol{A}\boldsymbol{x} = \boldsymbol{x}'\boldsymbol{A}'\boldsymbol{y} \ge \boldsymbol{c}'\boldsymbol{x}. \tag{4.13}$$

Therefore, for any feasible solution, the value of the primal is no larger than the value of the dual: $\boldsymbol{b}'\boldsymbol{y} \ge \boldsymbol{c}'\boldsymbol{x}$. ■

Perhaps surprisingly, when there is equality, the corresponding $\boldsymbol{x}$ and $\boldsymbol{y}$ are optimal in their respective programs.

Theorem 4.4: *If $\boldsymbol{x}^*$ and $\boldsymbol{y}^*$ are feasible and satisfy $\boldsymbol{c}'\boldsymbol{x}^* = \boldsymbol{b}'\boldsymbol{y}^*$, then $\boldsymbol{x}^*$ is optimal in the primal and $\boldsymbol{y}^*$ is optimal in the dual.*

PROOF: To see this, observe that from the inequalities above in Equation 4.13, for all feasible $\boldsymbol{x}$, $\boldsymbol{c}'\boldsymbol{x} \le \boldsymbol{b}'\boldsymbol{y}^* = \boldsymbol{c}'\boldsymbol{x}^*$, so that $\boldsymbol{x}^*$ gives at least as high a payoff as any feasible $\boldsymbol{x}$. Similarly, for all feasible $\boldsymbol{y}$, $\boldsymbol{b}'\boldsymbol{y} \geq \boldsymbol{c}'\boldsymbol{x}^* = \boldsymbol{b}'\boldsymbol{y}^*$, so that $\boldsymbol{y}^*$ is optimal for the dual. ■

For programs that have a solution, the equality $\boldsymbol{c}'\boldsymbol{x}^* = \boldsymbol{b}'\boldsymbol{y}^*$ always holds, as stated in the next theorem, Theorem 4.5. The proof makes use of Farkas' lemma.

Farkas' lemma asserts that if $\boldsymbol{Q}$ is an $m \times n$ matrix and $\boldsymbol{\alpha}$ an $m \times 1$ vector then:

1. $[\boldsymbol{Q}\boldsymbol{v} \leq \boldsymbol{\alpha}$ and $\boldsymbol{v} \geq \boldsymbol{0}]$ has a solution, or

2. $[\boldsymbol{Q}'\boldsymbol{z} \geq \boldsymbol{0}$, $\boldsymbol{\alpha}'\boldsymbol{z} < 0$, and $\boldsymbol{z} \geq \boldsymbol{0}]$ has a solution, but not both.

Theorem 4.5: *If $\boldsymbol{x}^*$ and $\boldsymbol{y}^*$ are solutions to the primal and dual respectively, then $\boldsymbol{c}'\boldsymbol{x}^* = \boldsymbol{b}'\boldsymbol{y}^*$.*

PROOF: From Theorem 4.3, $\boldsymbol{b}'\boldsymbol{y}^* \geq \boldsymbol{c}'\boldsymbol{x}^*$. Let $\delta = \boldsymbol{b}'\boldsymbol{y}^*$, so it is necessary to confirm that $\boldsymbol{c}'\boldsymbol{x}^* = \delta$. For $\epsilon > 0$, there is no solution to the system:

$$\begin{aligned} \boldsymbol{b}'\boldsymbol{y} &\leq \delta - \epsilon \\ \boldsymbol{A}'\boldsymbol{y} &\geq \boldsymbol{c} \\ \boldsymbol{y} &\geq 0, \end{aligned}$$

otherwise, $\boldsymbol{y}^*$ would not be optimal in the dual program. Rearranging, there is no $\boldsymbol{y} \geq 0$ with:

$$\begin{pmatrix} \boldsymbol{b}' \\ -\boldsymbol{A}' \end{pmatrix} \boldsymbol{y} \leq \begin{pmatrix} \delta - \epsilon \\ -\boldsymbol{c} \end{pmatrix}.$$

In the statement of Farkas' lemma (point 1), let

$$\boldsymbol{Q} = \begin{pmatrix} \boldsymbol{b}' \\ -\boldsymbol{A}' \end{pmatrix}, \quad \boldsymbol{\alpha} = \begin{pmatrix} \delta - \epsilon \\ -\boldsymbol{c} \end{pmatrix},$$

so that from Farkas' lemma (point 2), this implies that there is $\hat{\boldsymbol{z}} \geq 0$ with

$$(\boldsymbol{b} : -\boldsymbol{A})\hat{\boldsymbol{z}} \geq 0, \ (\delta - \epsilon, -\boldsymbol{c}')\hat{\boldsymbol{z}} < 0, \ \hat{\boldsymbol{z}} \geq 0.$$

Let $\hat{\boldsymbol{z}} = (z_0, \boldsymbol{z})$ so that these expressions say that there is a $(z_0, \boldsymbol{z})$ with:

$$\boldsymbol{b}z_0 - \boldsymbol{A}\boldsymbol{z} \geq 0, (\delta - \epsilon)z_0 - \boldsymbol{c}'\boldsymbol{z} < 0, (z_0, \boldsymbol{z}) \geq 0.$$

Next, observe that $z_0 > 0$. For if not ($z_0 = 0$), then $-\boldsymbol{A}\boldsymbol{z} \geq 0$ and $-\boldsymbol{c}'\boldsymbol{z} < 0$, or equivalently, $\boldsymbol{A}\boldsymbol{z} \leq 0$ and $\boldsymbol{c}'\boldsymbol{z} > 0$. And if this were so, $\boldsymbol{x}^*$ would not be optimal for the primal (since $(\boldsymbol{x}^* + \boldsymbol{z})$ would be feasible $\boldsymbol{A}(\boldsymbol{x}^* + \boldsymbol{z}) \le \boldsymbol{b}$, $(\boldsymbol{x}^* + \boldsymbol{z}) \ge 0$ and yield a higher payoff than $\boldsymbol{x}^*$, $\boldsymbol{c}'(\boldsymbol{x}^* + \boldsymbol{z}) > \boldsymbol{c}'\boldsymbol{x}^*$). Therefore, since $z_0 > 0$, in the expressions, $\boldsymbol{b}z_0 - \boldsymbol{A}\boldsymbol{z} \ge 0$ and $z_0(\delta - \epsilon) - \boldsymbol{c}\boldsymbol{z} < 0$ divide by z_0 to get:

$$\boldsymbol{b} - \boldsymbol{A}\tilde{\boldsymbol{z}} \ge 0 \quad \text{and} \quad (\delta - \epsilon) - \boldsymbol{c}\tilde{\boldsymbol{z}} < 0,$$

where $\tilde{\boldsymbol{z}} = \boldsymbol{z}\frac{1}{z_0}$. Rearranging, there is $\tilde{\boldsymbol{z}} \ge 0$ with

$$\boldsymbol{A}\tilde{\boldsymbol{z}} \le \boldsymbol{b}, \quad \text{and} \quad \boldsymbol{c}\tilde{\boldsymbol{z}} > (\delta - \epsilon).$$

Hence, a solution to the dual program $\boldsymbol{x}^*$ satisfies $\boldsymbol{c}'\boldsymbol{x}^* > \delta - \epsilon$ for all $\epsilon > 0$, so that $\boldsymbol{c}'\boldsymbol{x}^* \ge \delta$. Therefore, $\delta = \boldsymbol{b}'\boldsymbol{y}^* \ge \boldsymbol{c}'\boldsymbol{x}^* \ge \delta$, so that $\boldsymbol{b}'\boldsymbol{y}^* = \boldsymbol{c}'\boldsymbol{x}^*$. ■

For the next theorem, let $F_P = \{\boldsymbol{x} \mid \boldsymbol{A}\boldsymbol{x} \le \boldsymbol{b},\ \boldsymbol{x} \ge 0\}$ and $F_D = \{\boldsymbol{y} \mid \boldsymbol{A}'\boldsymbol{y} \ge \boldsymbol{c},\ \boldsymbol{y} \ge 0\}$, the sets of feasible points for the primal and dual. Theorem 4.6 asserts that if there is no feasible solution for the primal ($F_P = \emptyset$, the empty set), then there is no optimal solution for the dual: if $F_D \ne \emptyset$, then the dual program is unbounded.

Theorem 4.6: *Suppose that $F_P = \emptyset$, so that there is no $\boldsymbol{x}$ feasible for the primal. Then there is no $\hat{\boldsymbol{y}} \in F_D$ satisfying $\boldsymbol{b}'\hat{\boldsymbol{y}} \le \boldsymbol{b}\boldsymbol{y}$ for all $\boldsymbol{y} \in F_D$.*

PROOF: Suppose that $F_P = \emptyset$, so $\boldsymbol{A}\boldsymbol{x} \le \boldsymbol{b}$ for some $\boldsymbol{x} \ge \boldsymbol{0}$ has no solution. This implies, by Farkas' theorem, that there is a $\tilde{\boldsymbol{y}}$ with $\boldsymbol{A}'\tilde{\boldsymbol{y}} \ge \boldsymbol{0}$, $\tilde{\boldsymbol{y}} \ge \boldsymbol{0}$ and $\beta = \boldsymbol{b}'\tilde{\boldsymbol{y}} < 0$.

If $F_D = \emptyset$, there is nothing to prove. If $F_D \ne \emptyset$, $\boldsymbol{A}'\boldsymbol{y} \ge \boldsymbol{c}$ and $\boldsymbol{y} \ge \boldsymbol{0}$ has a solution, $\hat{\boldsymbol{y}}$: $\boldsymbol{A}'\hat{\boldsymbol{y}} \ge \boldsymbol{c}$ and $\hat{\boldsymbol{y}} \ge \boldsymbol{0}$. Then, for any $\delta > 0$, $\hat{\boldsymbol{y}} + \delta\tilde{\boldsymbol{y}}$ is feasible: $\boldsymbol{A}'(\hat{\boldsymbol{y}} + \delta\tilde{\boldsymbol{y}}) \ge \boldsymbol{c}$, $\hat{\boldsymbol{y}} + \delta\tilde{\boldsymbol{y}} \ge \boldsymbol{0}$. Furthermore, $\boldsymbol{b}'(\hat{\boldsymbol{y}} + \delta\tilde{\boldsymbol{y}}) = \boldsymbol{b}'\hat{\boldsymbol{y}} + \delta\boldsymbol{b}'\tilde{\boldsymbol{y}} = \boldsymbol{b}'\hat{\boldsymbol{y}} + \delta\beta$. This can be made arbitrarily small (large negative value) by choosing δ sufficiently large. ■

The argument is symmetric if $F_D = \emptyset$. Then, either $F_P = \emptyset$ or $F_P \ne \emptyset$, and the primal program, P, has no upper bound on the objective. It is useful to summarize the possible cases in terms of feasibility:

1. $F_P = \emptyset$ and $F_D = \emptyset$.
2. $F_P = \emptyset$ and $F_D \ne \emptyset$.
3. $F_P \ne \emptyset$ and $F_D = \emptyset$.
4. $F_P \ne \emptyset$ and $F_D \ne \emptyset$.

In case 1, there is nothing to add. In case 2, according to Theorem 4.6, the dual program is unbounded. Case 3 is the same as case 2, except now the primal program is unbounded. Finally, considering case 4, since $\boldsymbol{b}'\boldsymbol{y} \geq \boldsymbol{c}'\boldsymbol{x}$ for all $\boldsymbol{y} \in F_P$ both programs are bounded.

Thus, if either program has an optimal solution, then so does the other, and the objective values are equal: if $\boldsymbol{x}^*$ solves the primal and $\boldsymbol{y}^*$ solves the dual, then $\boldsymbol{c} \cdot \boldsymbol{x}^* = \boldsymbol{b} \cdot \boldsymbol{y}^*$. If one program is not feasible, the other has no optimal solution.

EXAMPLE 4.11: Let

$$\boldsymbol{A} = \begin{pmatrix} 1 & 0 \\ 0 & -1 \end{pmatrix}, \quad \boldsymbol{b} = \begin{pmatrix} b_1 \\ b_2 \end{pmatrix}, \quad \boldsymbol{c} = \begin{pmatrix} c_1 \\ c_2 \end{pmatrix}$$

Then $\boldsymbol{Ax} \leq \boldsymbol{b}$ has no non-negative solution $\boldsymbol{x}$, if $b_1 < 0$; and $\boldsymbol{A}'\boldsymbol{y} \geq \boldsymbol{c}$ has no non-negative solution y if $c_2 > 0$. In this case, $F_P = F_D = \emptyset$. If b_1, b_2 and c_2 are all positive, then $F_P \neq \emptyset$, while $F_D = \emptyset$. ◇

REMARK 4.3: Complementary slackness follows from the following considerations. Recall that $\boldsymbol{y}$ feasible for the dual and $\boldsymbol{Ax} \leq \boldsymbol{b}$ give $\boldsymbol{y}'\boldsymbol{Ax} \leq \boldsymbol{y}'\boldsymbol{b}$, and $\boldsymbol{x}$ feasible for the primal and $\boldsymbol{A}'\boldsymbol{y} \geq \boldsymbol{c}$ give $\boldsymbol{x}'\boldsymbol{A}'\boldsymbol{y} \geq \boldsymbol{x}'\boldsymbol{c}$, so $\boldsymbol{b}'\boldsymbol{y} \geq \boldsymbol{y}'\boldsymbol{Ax} = \boldsymbol{x}'\boldsymbol{A}'\boldsymbol{y} \geq \boldsymbol{c}'\boldsymbol{x}$. Since, at a solution pair $(\hat{\boldsymbol{x}}, \hat{\boldsymbol{y}})$, $\boldsymbol{c}'\hat{\boldsymbol{x}} = \boldsymbol{b}'\hat{\boldsymbol{y}}$, the previous inequalities imply $\boldsymbol{b}'\hat{\boldsymbol{y}} = \hat{\boldsymbol{y}}'\boldsymbol{A}\hat{\boldsymbol{x}} = \hat{\boldsymbol{x}}'\boldsymbol{A}'\hat{\boldsymbol{y}} = \boldsymbol{c}'\hat{\boldsymbol{x}}$. From this, $\hat{\boldsymbol{y}}'(\boldsymbol{b} - \boldsymbol{A}\hat{\boldsymbol{x}}) = \boldsymbol{0}$ and $\hat{\boldsymbol{x}}'(\boldsymbol{c} - \boldsymbol{A}'\hat{\boldsymbol{y}}) = \boldsymbol{0}$. □

4.3.1 Duality and dual prices

Theorem 4.4 may be used to provide insight on dual prices. Note that in examples such as Example 4.9 in Section 4.1.6, changing the right-hand sides a "small" amount leaves the solution at the intersection of the same constraints – with the solution point moved to the new intersection. For example, changing the RHS of constraint 2 in program B from $b_2 = 17$ to $b_2 + \Delta b_2 = 18$, moved the solution from $(2,3)$ (defined by the intersection of $2x + 3y = 13$ and $3x + 4y = 17$) to $\left(\frac{15}{6}, \frac{16}{6}\right)$ (defined by the intersection of $2x+3y = 13$ and $3x+4y = 18$). Furthermore, considering the dual program, program A, the slopes of the constraints are $-\frac{4}{2} = -2$ and $-\frac{3}{3} = -1$, while the slope of the objective is $-\frac{17}{13}$. So changing the coefficient on the objective (from $b_2 = 17$ to $b_2 = 18$) leaves the objective slope between -1 and -2 and the solution unchanged at $\left(\frac{7}{6}, \frac{1}{6}\right)$.

More generally, if the parameters $\boldsymbol{b}$ in the dual objective do not define objective iso-cost lines that are parallel to any of the constraints, then small changes in $\boldsymbol{b}$

lead to *no* change in the optimal solution of the dual. So, given $\boldsymbol{b}$ with solution $\boldsymbol{y}^*$ to the dual program, if some b_i is changed to $b_i + \Delta b_i$, then the solution to the dual remains unchanged at $\boldsymbol{y}^*$ and the value of the dual objective at the optimal solution to the new program is $\sum_{j\neq i} b_j y_j^* + (b_i + \Delta b_i) y_i^*$. Turning to the primal, with initial RHS, $\boldsymbol{b}$, the optimal solution is $\boldsymbol{x}^*$. Changing the RHS of the ith equation to $b_i + \Delta b_i$ determines a new solution with each x_i^* shifted by a small amount – so that x_i^* moves to $x_i^* + \Delta x_i$, say. From Theorem 4.4,

$$c_1 x_1^* + c_2 x_2^* + \ldots + c_n x_n^*$$
$$= b_1 y_1^* + b_2 y_2^* + \ldots + b_m y_m^*$$

and after the RHS change:

$$c_1(x_1^* + \Delta x_1) + c_2(x_2^* + \Delta x_2) + \ldots + c_n(x_n^* + \Delta x_n) = b_1 y_1^* + b_2 y_2^* + \ldots + (b_i + \Delta b_i) y_i + \ldots + b_m y_m^*.$$

Using the pre-change equality:

$$c_1 \Delta x_1 + c_2 \Delta x_2 + \ldots + c_n \Delta x_n = \Delta b_j y_j.$$

Dividing:

$$c_1 \frac{\Delta x_1}{\Delta b_j} + c_2 \frac{\Delta x_2}{\Delta b_j} + \ldots + c_n \frac{\Delta x_n}{\Delta b_j} = y_j.$$

Writing the left side as $\frac{\Delta \pi}{\Delta b_j}$, this gives

$$\frac{\Delta \pi}{\Delta b_j} = y_j.$$

So, for the "regular" case, the change in the objective resulting from a change in the RHS is given by the corresponding dual price.

Exercises for Chapter 4

Exercise 4.1 You have a choice of two foods to eat per day. Food A contains 3 grams of vitamin A per ounce, 4 grams of vitamin C and 1 gram of vitamin E. Food B contains 1 gram of vitamin A, 5 grams of vitamin C and 3 grams of vitamin E. The minimum daily requirements of vitamins A, C and E are 22, 66 and 20 grams respectively. You are also on a diet, so you want to minimize your calorie intake. An ounce of food A has 30 calories and an ounce of food B has 20.

(a) Formulate the linear programming problem.

(b) What are the vertices of the convex set bounded by the constraints?

(c) At what point do you minimize your calorie intake subject to the nutritional requirements?

(d) Assume that each calorie you consume costs you $2. What would you pay for a vitamin pill that would reduce the amount of vitamin A you need from food to 20 grams?

Exercise 4.2 For the following systems of constraints, determine and plot the feasible regions.

(a)
$$\begin{aligned} 4x+3y &\geq 12 \\ x-y &\leq 0 \end{aligned}$$

(b)
$$\begin{aligned} x &\geq 0 \\ y &\geq 0 \\ x+y-6 &\leq 0 \\ 2x+y-8 &\leq 0 \end{aligned}$$

(c)
$$\begin{aligned} 2x+y &\geq 50 \\ x+2y &\geq 40 \\ x &\geq 0 \\ y &\geq 0 \end{aligned}$$

$$\text{(d)}\quad \begin{aligned} x+2y &\le 12 \\ 6x+4y &\le 48 \\ -x+2y &\le 8 \\ x &\ge 0 \\ y &\ge 0 \end{aligned} \qquad \text{(e)}\quad \begin{aligned} 3x+y &\ge 12 \\ 6x+5y &\ge 30 \\ x+2y &\ge 14 \\ x &\ge 0 \\ y &\ge 0 \end{aligned}$$

$$\text{(f)}\quad \begin{aligned} 20x+15y &\le 60,000 \\ 4x+9y &\le 25,200 \\ 4x+6y &\le 18,000 \\ 30x+20y &\le 87,000 \end{aligned}$$

Exercise 4.3 Solve the following linear programming problems:

$$\begin{aligned} \max \quad & 5x+4y \\ \text{such that:} \quad & x+2y \le 12 \\ & 6x+4y \le 48 \\ & -x+2y \le 8 \\ & x,y \ge 0, \end{aligned}$$

$$\begin{aligned} \min \quad & x+3y \\ \text{such that:} \quad & 3x+y \ge 12 \\ & 6x+5y \ge 30 \\ & x+2y \ge 14 \\ & x,y \ge 0. \end{aligned}$$

Part I.

In the maximization problem, calculate the change in the value of the objective as a result of:

(a) Changing the RHS (right-hand side) from 12 to 10, and from this deduce the dual price associated with the first constraint.

(b) Changing the RHS from 48 to 40, and from this deduce the dual price associated with the second constraint.

(c) Changing the RHS from 8 to 7, and from this deduce the dual price associated with the third constraint.

Part II.

In the minimization problem, calculate the change in the value of the objective as a result of:

(d) Changing the RHS from 12 to 11, and from this deduce the dual price associated with the first constraint.

(e) Changing the RHS from 30 to 32, and from this deduce the dual price associated with the second constraint.

(f) Changing the RHS from 14 to 15, and from this deduce the dual price associated with the third constraint.

Part III.

For both problems, introduce slack variables and write the programs as equality-constrained problems. In both cases:

(g) Identify all the basic solutions.

(h) Identify all the basic feasible solutions.

Exercise 4.4 Consider the following problem

$$\begin{aligned} \max \quad & x+y \\ \text{subject to:} \quad & 3x+5y+8z \le 30 \\ & 4x+5y-10z \le 20 \\ & x,y,z \ge 0. \end{aligned}$$

(a) Reformulate the problem using equalities.

(b) What are all the basic solutions?

(c) What are all the basic feasible solutions?

(d) What is the optimal basic feasible solution?

Exercise 4.5 Consider the following linear program:

$$\begin{aligned} \max \quad & x+y \\ \text{subject to:} \quad & 2x+4y \le 12 && (4.1) \\ & 4x+2y \le 12 && (4.2) \\ & x,y \ge 0. \end{aligned}$$

(a) Plot the constraints and solve the program.

(b) Write the program in equality constraint form ($\boldsymbol{Az} = \boldsymbol{b}$) by introducing slack variables, s_1 and s_2.

(c) The "$\boldsymbol{A}$" matrix in (b) has two rows and four columns. Write down all the submatrices having two rows and two columns. There are six such matrices: $\boldsymbol{A}_1, \boldsymbol{A}_2, \dots, \boldsymbol{A}_6$. Find the six basic solutions by setting two of the four variables (x, y, s_1, s_2) to zero. (e.g. $\boldsymbol{A}_1 \begin{bmatrix} x \\ y \end{bmatrix} = \begin{bmatrix} 12 \\ 12 \end{bmatrix}$, $\boldsymbol{A}_1 = \begin{bmatrix} 2 & 4 \\ 4 & 2 \end{bmatrix}$, when $s_1 = s_2 = 0$).

(d) Locate each of the six solutions found in (c) on a graph. Each solution has the form (x^*, y^*, s_1^*, s_2^*), but in all six cases the (x^*, y^*) differ – so you can see where each solution is on your x–y graph. Now eliminate those solutions that contain a negative term and relate the remaining solutions to the feasible region.

Exercise 4.6 Consider the following linear program:

$$\begin{aligned} \max \quad & \Pi = 2x + 2y \\ \text{subject to:} \quad & 2x + 4y \le 16 \\ & 3x + y \le 9 \\ & x \ge 0, y \ge 0. \end{aligned}$$

(a) Find the optimal values of x and y.

(b) What is the value of the objective at the (optimal) solution?

(c) What are the values of the slack (surplus) variables at the solution?

(d) Suppose that in the second constraint, 9 is replaced by 10. How are your answers in (a) and (b) affected?

Exercise 4.7 For the following linear program:

$$\begin{aligned} \min \quad & C = 4x + 3y \\ \text{subject to:} \quad & 3x + 2y \ge 158 \\ & 5x + 2y \ge 200 \\ & x + 3y \ge 90 \\ & x \ge 0, y \ge 0. \end{aligned}$$

(a) Find the optimal values of x and y.

(b) What is the value of the objective at the (optimal) solution?

(c) What are the values of the slack (surplus) variables at the solution?

(d) Calculate the dual prices associated with each constraint.

Exercise 4.8 For the following linear program:

$$\begin{aligned} \max \quad & \Pi = 10x + 24y \\ \text{subject to:} \quad & x + 2y \le 120 \\ & x + 4y \le 180 \\ & x \ge 0, y \ge 0. \end{aligned}$$

(a) Find the optimal values of x and y.

(b) What is the value of the objective at the (optimal) solution?

(c) What are the values of the slack (surplus) variables at the solution?

(d) Calculate the dual prices associated with each constraint.

Exercise 4.9 Consider the following linear program:

$$\begin{aligned} \max \quad & 2x+5y \\ \text{subject to:} \quad & 4x+1y \le 24 && (4.1) \\ & 1x+3y \le 21 && (4.2) \\ & 1x+1y \le 9 && (4.3) \\ & x,y \ge 0. \end{aligned}$$

(a) Solve the linear program – find the (x,y) pair that solves the problem.

(b) Find the slack (surplus) values.

(c) What is the effect on the value of the objective if the third constraint is replaced by $1x+1y \le 8$? From this, give the dual price associated with the third constraint.

(d) Calculate the dual prices for the other two constraints, (1) and (2).

(e) Would the optimal values of the variables (x,y) be different if the objective were $1x+5y$?

Exercise 4.10 Consider the linear program:

$$\begin{aligned} \max \quad & 3x+2y \\ \text{subject to:} \quad & 4x+2y \le 14 \\ & 4x+3y \le 17 \\ & 6x+y \le 19 \\ & x \ge 0, y \ge 0. \end{aligned}$$

(a) Plot the constraints.

(b) Find the optimal solution, (x,y).

(c) Calculate the dual prices associated with each constraint.

(d) Find all the basic solutions.

(e) Identify all the basic *feasible* solutions.

Exercise 4.11 Consider the linear program:

$$\begin{aligned} \max \quad & 3x+4y \\ \text{subject to:} \quad & 3x+y \le 16 \\ & x+3y \le 16 \\ & x+y \le 6 \\ & x \ge 0, y \ge 0. \end{aligned}$$

(a) Plot the constraints.

(b) Find the optimal solution, (x,y).

(c) Calculate the dual prices associated with each constraint.

(d) Find all the basic solutions.

(e) Identify all the basic *feasible* solutions.

Exercise 4.12 Consider the following linear program:

$$\begin{aligned} \max \quad & 3x+4y \\ \text{subject to:} \quad & 2x+3y \le 13 && (4.1) \\ & 3x+4y \le 17 && (4.2) \\ & x,y \ge 0. \end{aligned}$$

Plot the program and solve, giving the slack variables, objective value and dual prices.

Exercise 4.13 Consider the following linear programming problem.

$$\begin{aligned} \max \quad & 3x+2y \\ \text{subject to:} \quad & x+y \le 12 \\ & 2x+y \le 20 \\ & x \ge 0, y \ge 0. \end{aligned}$$

(a) Plot the constraints and the feasible region for this linear programming problem.

(b) Solve for the optimal choice of x and y.

Exercise 4.14 Consider the following linear program:

$$\begin{aligned} \max \quad & x+y \\ \text{subject to:} \quad & 7x+2y \le 28 && (4.1) \\ & 4x+3y \le 24 && (4.2) \\ & 2x+6y \le 42 && (4.3) \\ & x,y \ge 0. \end{aligned}$$

Plot the program and solve, giving the slack variables, objective value and dual prices.

Exercise 4.15 Consider the problem of maximizing xy subject to the constraint $x^2+y^2 \le 2$ and $x \ge 0, y \ge 0$. Let the feasible region be the (x,y) pairs satisfying $x^2+y^2 \le 2$ and $x \ge 0, y \ge 0$.

(a) Plot the feasible region.

(b) Plot the set of (x,y) pairs such that (i) $xy = 1$, (ii) $xy = 2$ and (iii) $xy = 3$.

(c) Indicate the solution on a graph and find the solution.

(d) How does your answer change if the constraint changes to $x^2 + y^2 \leq 3$?

Exercise 4.16 (a) Consider the linear program:

$$
\begin{aligned}
\max \quad & 3x + 5y \\
\text{subject to:} \quad & x + 2y \leq 10 && (4.1) \\
& 2x + 3y \leq 17 && (4.2) \\
& 3x + 2y \leq 24 && (4.3) \\
& x \geq 0, y \geq 0.
\end{aligned}
$$

(i) Plot the constraints.

(ii) Find the optimal solution, (x, y).

(iii) Calculate the dual prices associated with each constraint.

(b) Consider the linear program:

$$
\begin{aligned}
\max \quad & x + y \\
\text{subject to:} \quad & 2x + y \leq 6 \\
& x + 2y \leq 6 \\
& x \geq 0, y \geq 0.
\end{aligned}
$$

(i) Find all the basic solutions.

(ii) Identify all the basic *feasible* solutions.

5

Functions of one variable

5.1 Introduction

Functions arise naturally in the study of economic problems. For example, a demand function relates quantity demanded to price. Properties of such functions are routinely studied – the slope and elasticity measure responsiveness of demand to price variations. The techniques presented here introduce the tools required for such calculations. Section 5.2 describes a variety of functions arising in different areas (taxation, finance and present value calculations). Section 5.3 introduces the basic notions of continuity and differentiability of a function. Certain families of functions, such as logarithmic, polynomial and exponential functions, appear frequently in economic applications; some of these are introduced in Section 5.4 and their shapes described. Growth rates and the algebra of growth calculations are developed in Section 5.5. Finally, Section 5.6 provides rules for the differentiation where the function is itself formed from other functions (such as the ratio or product of functions).

A *function*, f, from a set X to a set Y is a rule associating with each element of a set X some element in the set Y. This is written $f: X \to Y$. In words, to each point in X, f associates a point, $y = f(x)$ in Y. The set X is called the domain and the set Y is called the range. The inverse demand function, relating price to quantity demanded, $p(Q)$, is a familiar example. The domain is the set of non-negative numbers, $\mathscr{R}_+$ (quantity must be positive), and the range is also the set of non-negative numbers (price is non-negative). Thus, $p: \mathscr{R}_+ \to \mathscr{R}_+$.

5.2 Some examples

The following examples show a variety of ways in which functions arise in economics.

5.2.1 Demand functions

Consumer demand theory describes an individual in terms of utility derived from consumption of a vector of goods $(x_1,x_2,\dots,x_n)$. With a fixed income y and price p_i for good i, utility maximization leads to choices for the goods that depend on prices and income: $x_i(p_1,p_2,\dots,p_n,y)$. See Figure 5.1 for the two-good case.

In the two-good case, utility maximization leads to the selection of (x_1^*,x_2^*). The values of x_1^* and x_2^* depend on the prices (p_1,p_2) and income, y. These are called the demand functions and are written $x_1(p_1,p_2,y)$ and $x_2(p_1,p_2,y)$.

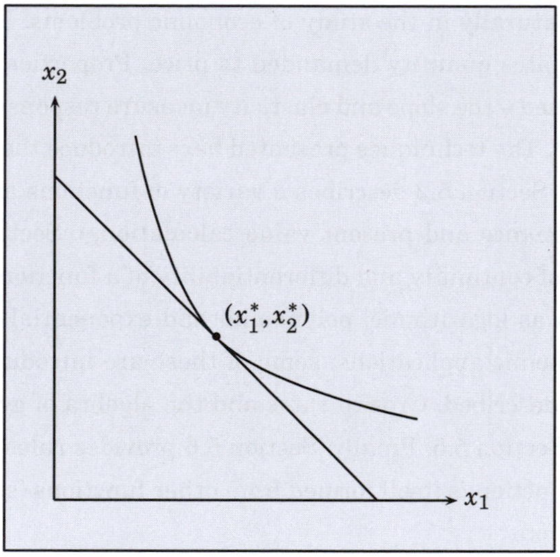

Figure 5.1: Choices determine demand

With J individuals, write the demand for good i by individual j as $x_i^j(p_1,p_2,\dots,p_n,y^j)$, where y^j is income of individual j. Then, the aggregate demand for good i is:

$$X_i(p_1,p_2,\dots,p_n,y^1,y^2,\dots,y^J)=\sum_{j=1}^{J}x_i^j(p_1,p_2,\dots,p_n,y^j).$$

The function X_i depends on the prices and incomes of each individual. If we consider the case where all prices other than that of good j, $p_j=\bar{p}_j$, $j\neq i$, and all incomes, $y^j = \bar{y}^j$, are fixed then the aggregate demand is:

$$D_i(p_i) = X_i(\bar{p}_1, \bar{p}_2, \ldots, \bar{p}_{i-1}, p_i, \bar{p}_{i+1}, \ldots, p_n, \bar{y}^1, \bar{y}^2, \ldots, \bar{y}^J).$$

For simplicity, ignore the subscripts and write, $D(p)$, for the aggregate demand as a function of own price. With this, one may study the behavior of aggregate demand as own price varies. Writing Q for the quantity demanded, $Q = D(p)$. This gives aggregate demand for the good as a function of the good's price. With this, the responsiveness of demand to price variation may be considered. Consider a price $p' \neq p$ with corresponding demand Q': the variation in quantity resulting from the price variation is $Q' - Q = D(p') - D(p)$. Thus, for example, the change in aggregate quantity demanded per unit change in price is given by

$$\frac{Q' - Q}{p' - p} = \frac{D(p') - D(p)}{p' - p}$$

and gives the slope of the demand function. Similarly, the percentage change in quantity resulting from a percentage change in price is given by:

$$\frac{\frac{Q'-Q}{Q}}{\frac{p'-p}{p}} = \frac{\frac{D(p')-D(p)}{D(p)}}{\frac{p'-p}{p}} = \frac{p}{D(p)} \frac{D(p') - D(p)}{p' - p}.$$

The negative of this term gives the elasticity of demand. Note that the elasticity of demand is computed at a point on the demand curve: at different points, the value of the elasticity will vary. Thus, we may write:

$$\epsilon(p) = -\left[\frac{p}{D(p)}\right]\left[\frac{D(p') - D(p)}{p' - p}\right].$$

5.2.2 Present value and the interest rate

The present value of a cash flow may be viewed as a function of the interest rate. A bond is characterized by a face value P, an interest rate ρ and a maturity date T, and it pays the holder a fixed payment, $f = \rho P$, each time period (such as every six months or every year), and the face value P at the end of the Tth period. Thus, the cash flow generated by the bond is: $(f, f, f, \ldots, f, P)$, where f is paid at the end of periods 1, 2, up to $T - 1$. What is the value of this cash flow? If the market interest rate is r, x dollars invested today for one period is worth $y = x(1 + r)$ at the end of the period. Put differently, to have y dollars at the end of the period, $x = \frac{y}{(1+r)}$ dollars should be invested now. How much should one pay for a payoff of y dollars in one period from now? If the cost is $c < x$, borrowing c will lead to a payoff y next period and a debt of $c(1+r)$. Since $c(1+r) < x(1+r) = y$, this generates a positive net payoff. Similarly, if $c > x$ the loan repayment in one period will exceed the payoff: $c(1+r) > y$. So, $c < x$ undervalues the return y in one period; and $c > x$ overvalues the return y in one period. For market equilibrium, $c = x = \frac{y}{1+r}$. This is the current value of y dollars, one period from now. The market value of the return y in one period is $\frac{y}{1+r}$. The same reasoning shows that the current value of a payoff y two periods from now is $\frac{y}{(1+r)^2}$, and t periods from now it is $\frac{y}{(1+r)^t}$. Therefore, the flow $(f,f,f,\ldots,f,P)$ has current market value of:

$$\begin{aligned}V &= \frac{f}{1+r} + \frac{f}{(1+r)^2} \cdots \frac{f}{(1+r)^t} + \cdots + \frac{f}{(1+r)^{T-1}} + \frac{P}{(1+r)^T} \\ &= f\left[\frac{1}{1+r} + \frac{1}{(1+r)^2} \cdots \frac{1}{(1+r)^t} + \cdots + \frac{1}{(1+r)^{T-1}}\right] + \frac{P}{(1+r)^T}.\end{aligned}$$

One simple case occurs when the payments continue indefinitely (and there is no end period at which P would be paid). In this case,

$$V = f\left[\frac{1}{1+r}\right]\left[1 + \frac{1}{1+r} + \frac{1}{(1+r)^2} \cdots \frac{1}{(1+r)^t} + \cdots\right].$$

Using the fact that for $0 < \alpha < 1$, $1 + \alpha + \alpha^2 + \cdots = \frac{1}{1-\alpha}$, when $\alpha = \frac{1}{1+r}$, the infinite sum becomes $\frac{1}{1-\frac{1}{1+r}} = \frac{1+r}{r}$. Therefore, $V = f\frac{1}{1+r}\frac{1+r}{r} = \frac{f}{r}$ (Figure 5.2). At the time the bond is issued, the rate on the bond, ρ, will typically equal the market rate r. Afterwards, while the bond rate is fixed, the market rate will vary, causing the value of the cash flow provided by the bond to vary. Write $V(r)$ to denote this dependence, the valuing of the bond as a function of the current market interest rate:

$$V(r) = \frac{f}{r}.$$

From this, one may, for example, calculate the impact on the value of the bond of a change in the market interest rate as: $\Delta V = V(r') - V(r) = \frac{f}{r'} - \frac{f}{r} = f\left[\frac{1}{r'} - \frac{1}{r}\right]$ $f\left[\frac{r}{rr'} - \frac{r'}{rr'}\right] = f\left[\frac{r-r'}{rr'}\right]$.

5.2.3 Taxation

In a market with supply and demand curves $S(p)$ and $D(p)$, the introduction of a sales tax creates a wedge between the price paid and the price received. How do these changes relate to the pre-tax price and quantity? With $D(p)$ and supply $S(p)$, the market clearing price is p_0, determined as the solution to: $D(p_0) = S(p_0)$.

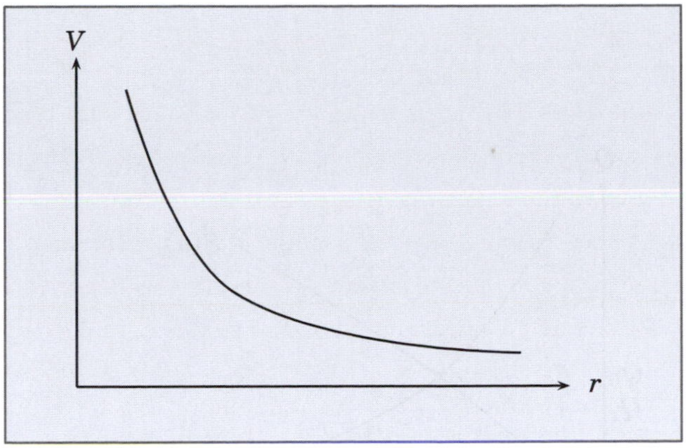

Figure 5.2: The present value function

Suppose now that an ad-valorem tax is introduced, so that if the seller receives p_t, the buyer pays $(1+t)p_t$.

Let $t_0 = 0$ be the tax initially (no tax) and t the tax imposed, so the tax increment is $\Delta t = t - t_0 = t$. Now, market clearing requires that the amount supplied at price p_t equals the amount demanded at price $(1+t)p_t$: $D((1+\Delta t)p_t) = S(p_t)$. With $Q(0) = D(p_0) = S(p_0)$ and $Q(t) = D((1+\Delta t)p_t) = S(p_t)$, the variation on quantity traded as a result of the tax increment is $Q(t) - Q(0) < 0$, and the rate of change of quantity traded relative to the tax variation is given by $\frac{Q(t)-Q(0)}{\Delta t}$.

From Figure 5.3, the magnitude of this depends on the slopes of the supply and demand curves. Similarly, the extent to which p_t lies below p_0 and $(1+t)p_t$ lies above p_0 depends on the slopes of the curves and determines the incidence of the tax (which party, supplier or demander, pays most of the tax).

5.2.4 Options

At any point in time, a stock trades at some price p, which varies over time. Derivative assets can be created as a function of the stock price p, $f(p)$. For example, consider a contract offering $f(p)$ dollars if the stock price is p dollars or more at the end of trading, and 0 otherwise. For example, if $f(p) = 5$, then the contract pays \$5 if the underlying stock has a price of at least p at the end of trading. Such derivative assets are called "options".

In finance, an option gives the owner the right to buy or sell an asset. An option giving the right to purchase an asset at some specified price, p^*, called the

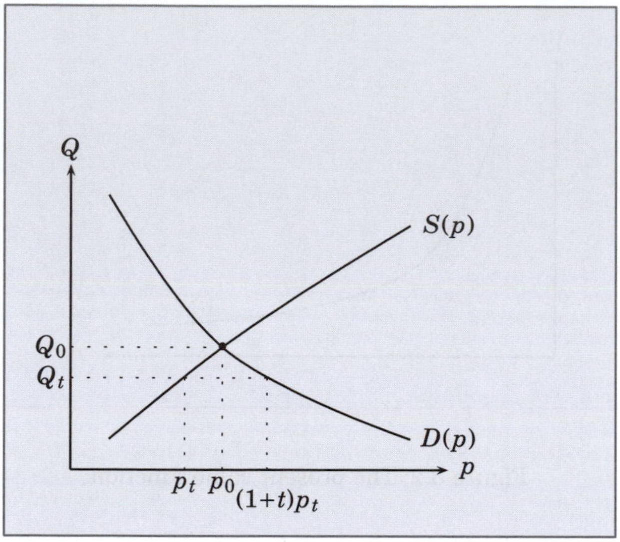

Figure 5.3: The impact of taxation in a market

exercise or *strike price*, at (or before) a specified date, T, is a *call option*. An option giving the right to sell is called a *put option*. When the right can be exercised only at the date T, the option is a *European option*. When the right can be exercised on or before the date T, the option is an *American option*. Such options define assets that are functions of the underlying assets and so are called derivative assets. Suppose that the strike price of a European call option is $\bar{p}$ and that $\tilde{p}$ is the (as yet unknown) stock price at period T (in the future). Then, if the price $\tilde{p}$ turns out to be above $\bar{p}$, the owner of the call option can exercise the right to purchase one unit of the stock at price $\bar{p}$. That unit of stock can be sold on the market at price $\tilde{p}$, so the net gain is $\tilde{p}-\bar{p}$. If the price $\tilde{p}$ turns out to be below $\bar{p}$, the owner of the call option will choose not to exercise the purchase right, and the net benefit is 0. So, the owner of the option will get a return of $\tilde{p}-\bar{p}$ if $\tilde{p}>\bar{p}$, and 0 otherwise. Therefore, the return per call option owned is $\varphi(\tilde{p}) = \max\{\tilde{p}-\bar{p},0\}$, the larger of $\tilde{p}-\bar{p}$ and 0 ($\max\{x,y\}$ equals the larger of the two numbers, x and y). If the individual owns k such options, then k units of the stock may be purchased at price $\bar{p}$ and so the return is then $k\varphi(\tilde{p})$. Figure 5.4 depicts the return of a call and put option where the striking price is $\bar{p}$.

In contrast, a put option gives the right to *sell* a unit of the asset at a (strike) price, $\hat{p}$, say. In this case, net benefit arises when the price turns out to be low, so that the asset may be sold at the (higher) strike price $\hat{p}$. This has a return of $\theta(\tilde{p}) = \max\{\hat{p}-\tilde{p},0\}$. Such options may be combined to provide more complex return streams.

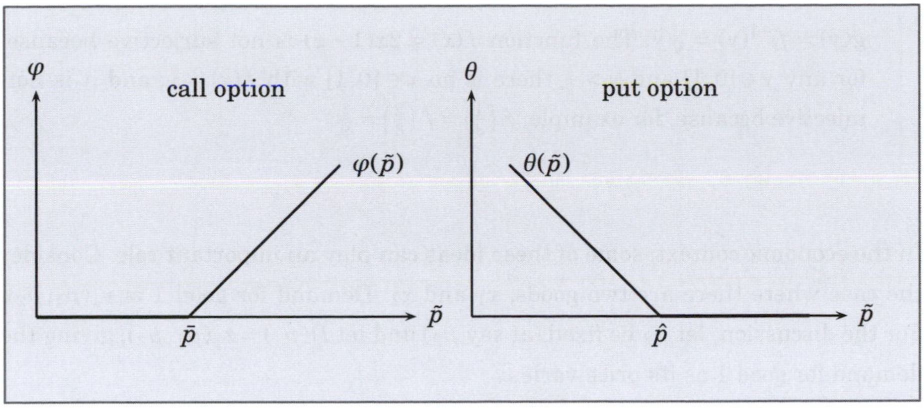

Figure 5.4: Options

5.3 Continuity and differentiability

5.3.1 Preliminaries

Before discussing continuity and differentiability, it is necessary to give some background definitions, describe types of function and introduce the notion of distance between numbers.

Definition 5.1: *Functions may be categorized in terms of surjectivity, injectivity and bijectivity.*

- *If* $f(X) = Y$*, where* $f(X) = \{y \mid \text{there is an } x \in X, f(x) = y\}$*, we say* f *is* onto Y *or* surjective. *In words, for any point* $y \in Y$*, there is some* $x \in X$ *with* $f(x) = y$*.*
- *The function* f *is called 1-1 (one-to-one) or* injective *if* $f(x_1) = f(x_2)$ *implies that* $x_1 = x_2$*; a 1-1 function cannot have the same value at two different points in the domain.*
- *If* f *is both 1-1 and onto, it is called* bijective. *If* f *is a bijective mapping,* $\exists$ *a function* $g : Y \to X$*, called the inverse of* f *with* $g(f(x)) = x$ *and* $f(g(y)) = y$*. The function* g *associates a unique point in* X *to each point in* Y*, and is written* f^{-1}*.*

EXAMPLE 5.1: For example, with $X = Y = [0,1]$, $f(x) = x^2$ is surjective since for $y \in Y$ there is an x with $f(x) = y$ ($x = \sqrt{y}$); and injective because $x \neq \bar{x}$ and $x, \bar{x} \in [0,1]$ imply $f(x) = x^2 \neq \bar{x}^2 = f(\bar{x})$. This function is bijective with inverse

$g(y) = f^{-1}(y) = \sqrt{y}$. The function $f(x) = 2x(1-x)$ is not surjective because for any $y \in [0,1]$ and $y > \frac{1}{2}$ there is no $x \in [0,1]$ with $f(x) = y$; and it is not injective because, for example, $f\left(\frac{1}{3}\right) = f\left(\frac{2}{3}\right) = \frac{4}{9}$. ◊

In the economic context, some of these ideas can play an important role. Consider the case where there are two goods, x_1 and x_2. Demand for good 1 is $x_1(p_1, p_2)$. For the discussion, let p_2 be fixed (at say $\bar{p}_2$) and let $D(p_1) = x_1(p_1, \bar{p}_2)$, giving the demand for good 1 as its price varies.

In Figure 5.5, an increase in the price of good 1 from p_1 to p_1' *raises* the demand for good 1 from x_1 to x_1'. Such a good is called a "Giffen" good, and has the odd feature that the demand curve slopes upward in some regions. The second figure illustrates such a demand curve. Here, there are multiple prices that lead to the same level of demand. In the figure, $x_1 = D(p_1^\circ) = D(p_1) = D(p_1'')$: the demand equals x_1 at three different prices, p_1°, p_1 and p_1''. The demand function

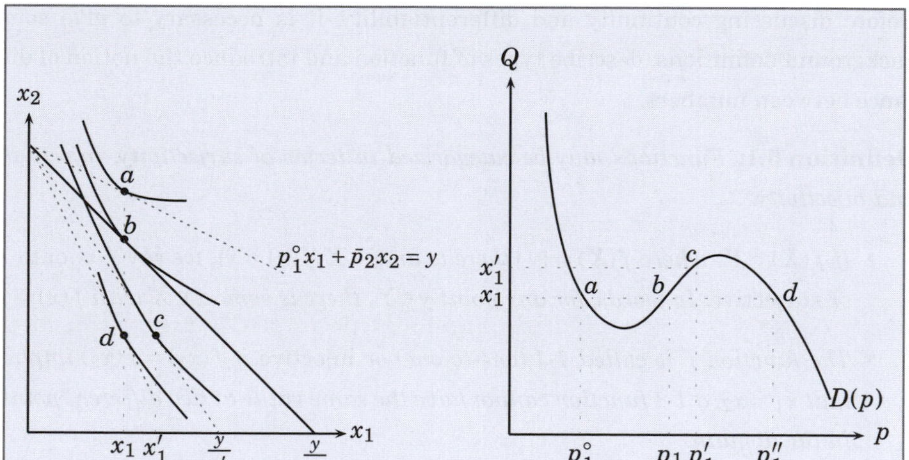

Figure 5.5: Giffen good; a non-injective demand function

is not injective and we cannot associate a unique price to a demand level such as x_1. Technically, the demand function does not have an inverse – there is no inverse demand function, $p(x_1)$. Such a function cannot be defined, since, at x_1 for example, there are multiple prices associated with this demand level (p_1°, p_1 and p_1'').

5.3.2 Inequalities and absolute value

Informally, a function f is continuous if when x is close to x', $f(x)$ is close to $f(x')$. Formally defining continuity requires a notion of closeness – the distance between different points or numbers. For example, considering the numbers 4, 6 and 9, four is closer to 6 than 9 is to 6. Since $6-4=2$ while $9-6=3$. Here, the subtraction is done so that the result is positive: if in the second calculation, we write $6-9=-3$; this is smaller than 2 ($-3<2$), but 9 is not closer than 4 is to 6. Clearly, here the magnitude is what is important, and this is measured by the *absolute value* of a number. Comparing $9-6=3$ and $6-9=-3$, the magnitude of the difference is 3; both 3 and -3 indicate that 9 and 6 are separated by three units.

When comparing two numbers, one can always say which one is larger or smaller, or else they are equal. The expressions "≥" and "≤" denote "greater than or equal to" and "less than or equal to", respectively. Thus, "a is at least as large as b" is written $a \geq b$, while "a is greater (strictly) than b" is written $a > b$. Note that if $a \geq b$, then $-a \leq -b$: multiplying both sides of an inequality reverses the inequality (consider $7>5$ and $-7<-5$).

For a real number a, the expression $|a|$ is called the absolute value of a, and is defined: $|a|=a$, if $a \geq 0$ and $|a|=-a$, if $a<0$:

$$|a| = \begin{cases} -a & \text{if } a<0, \\ a & \text{if } a \geq 0. \end{cases}$$

In words, $|a|$ equals a if a is non-negative; and equals $-a$ if a is negative. So, $|a|$ is always non-negative, and strictly positive unless $a=0$.

PROPOSITION 5.1: The absolute value operation satisfies a number of simple rules. For real numbers, a, b and c:

1. If $a \geq 0$, then $|x| \leq a$ if *and only* if $-a \leq x \leq a$;
2. $|a+b|=|b+a|$;
3. $|a \times b|=|a||b|$;
4. $|a+b| \leq |a|+|b|$;
5. $|a+b| \geq |a|-|b|$;
6. $|\frac{a}{b}|=\frac{|a|}{|b|}$, if $b \neq 0$. ◊

PROOF: (1) From the definition of absolute value, $-|x| \leq x \leq |x|$, since $x = |x|$ or $-|x|$. If we assume $|x| \leq a$, then $-a \leq -|x| \leq x \leq |x| \leq a$. Conversely, if we assume $-a \leq x \leq a$ then $x \geq 0$ implies $|x| = x \leq a$. On the other hand, if $x < 0$, $-a \leq x$ implies $-x \leq a$ and $|x| = x$ so $|x| \leq a$. Either way, $|x| \leq a$. Thus, $|x| \leq a$ is equivalent to $-a \leq x \leq a$. Points (2) and (3) follow from the definition. Concerning (4), to see that $|a+b| \leq |b| + |a|$, observe that $-|a| \leq a \leq |a|$ and $-|b| \leq b \leq |b|$: these inequalities imply $-(|a|+|b|) \leq a+b \leq (|a|+|b|)$ and $|a+b| \leq |a|+|b|$. To see that $|a+b| \geq |a|-|b|$, using rule (4), note that $|x|+|y| \geq |x+y|$ and, rearranging, $|x| \geq |x+y|-|y|$. Set $x = a+b$, and $y = -b$ in this expression to give the result $|a+b| \geq |a|-|b|$, for any a,b. Finally, (6) follows from the definition. ■

Rules 4 and 5 generalize directly. Consider $|\sum_{i=1}^n x_i| \leq \sum_{i=1}^n |x_i|$. To confirm this, put $a = x_1, b = \sum_{i=2}^n x_i$ so that $|\sum_{i=1}^n x_i| \leq |x_1| + |\sum_{i=2}^n x_i|$, using rule 4. Repeat with $|\sum_{i=2}^n x_i|$ to give $|\sum_{i=1}^n x_i| \leq |x_1| + |x_2| + |\sum_{i=3}^n x_i|$. Proceeding in this way gives $|\sum_{i=1}^n x_i| \leq |x_1| + |x_2| + \cdots + |x_n|$.

Similarly, $|\sum_{i=1}^n x_i| \geq |x_1| - |x_2| - |x_3| - \cdots - |x_n|$. Given $\sum_{i=1}^n x_i$, set $a = x_1$ and $b = \sum_{i=2}^n x_i$ to get $|\sum_{i=1}^n x_i| \geq |x_1| - |\sum_{i=2}^n x_i|$, using rule 5. The previous discussion gives $|\sum_{i=2}^n x_i| \leq |x_2| + |x_3| + \cdots + |x_n|$, so $-|\sum_{i=2}^n x_i| \geq -|x_2| - |x_3| - \cdots - |x_n|$. Therefore, $|\sum_{i=1}^n x_i| \geq |x_1| - |\sum_{i=2}^n x_i| \geq |x_1| - |x_2| - |x_3| - \cdots - |x_n|$.

5.3.3 Continuity

Continuity of a function is the requirement that when points are close in the domain, the function should take those points to the same neighborhood in the range. This is formalized next.

Definition 5.2: *Let $x_1, x_2, \ldots, x_n = \{x_n\}_{n\geq 1}$ be a sequence of real numbers. We say that x_n converges to a real number c if given any positive number, $\epsilon > 0$, however small, there is an integer $\bar{n}$, such that $|x_n - c| \leq \epsilon$, whenever $n \geq \bar{n}$.*

This is written $|x_n - c| \to 0$ as $n \to \infty$, or $x_n \to c$, or $\lim_{n\to\infty} x_n = c$. Intuitively, $|x_n - c| \to 0$ means that $|x_n - c|$ gets "closer and closer" to 0 as n gets "larger and larger" ($n \to \infty$).

EXAMPLE 5.2: The following sequences illustrate convergence:

1. $x_n = c + \frac{1}{n}$, so that $|x_n - c| = \frac{1}{n} \to 0$, as $n \to \infty$, so that x_n converges to c.

2. $x_n = c+(-1)^n$. Thus x_n is equal to $c+1$ when n is even and $c-1$ when n is odd. In this case $|x_n - c| = 1$ for all n and so x_n does not converge to c.

3. $x_n = c+n$, so that $|x_n - c| = n$ and so x_n does not converge to c. ◊

Definition 5.3: *A function f is continuous at c if given $\epsilon > 0$, $\exists\delta > 0$ such that*

$$|f(x) - f(c)| < \epsilon \text{ whenever } |x - c| < \delta.$$

In this case, write $\lim_{x\to c} f(x) = f(c)$. As x gets "closer and closer" to c, $f(x)$ gets "closer and closer" to $f(c)$. Call f *continuous*, if f is continuous at every $c \in X$.

EXAMPLE 5.3: The following functions illustrate continuity:

1. $X = (0,\infty), f(x) = \frac{1}{x}$ is continuous; but f is not continuous if the domain is $X = [0,\infty)$.

2. $X = (-\infty,\infty), f(x) = 2x^3 + x^2 + 5$ is continuous.

3. $X = [0,1]$, $f(x) = x$, if $x < \frac{1}{2}$ and $f(x) = \frac{1}{4} + x^2$, if $x \geq \frac{1}{2}$. In this example, two separate functions (x and $\frac{1}{4} + x^2$) are pieced together at the point $x = \frac{1}{2}$. ◊

Some simple rules follow from the definitions. If $\lim_{x\to c} f(x) = a$ and $\lim_{x\to c} g(x) = b$, then:

1. For any constant α, $\lim_{x\to c} \alpha f(x) = \alpha a$.

2. $\lim_{x\to c}[f(x) + g(x)] = a + b$,

3. $\lim_{x\to c} f(x)\cdot g(x) = ab$,

4. $\lim_{x\to c} \frac{f(x)}{g(x)} = \frac{a}{b}$, if $b \neq 0$.

5. If $f: X \to Y$, and $g: Y \to Z$ are continuous functions, then $h: X \to Z$, defined by $h(x) = g(f(x))$ is a continuous function: $\lim_{x\to c} h(x) = h(c)$.

Therefore, multiplying a continuous function by a constant gives a continuous function; the sum and product of continuous functions is a continuous function;

and the ratio of continuous functions is continuous, provided the denominator is non-zero. If, for example, an inverse demand function $p(q)$ and cost function $c(q)$ are continuous, then so is the resulting profit function $\pi(q) = p(q)q - c(q)$.

5.3.4 Differentiability

The derivative of a function relates to the "slope" of the function, measuring the rate of increase or decrease in the function resulting from changes in the argument of the function (x). Suppose the continuous function f is increasing: if $x' > x$, then $f(x') > f(x)$. If x' is slightly larger than x, then $f(x')$ is slightly larger than $f(x)$, and $f(x') - f(x)$ measures the increase in f resulting from the increase from x to x'. More generally, whether f is increasing or not, $f(x') - f(x)$ measures the change in f resulting from the change in x.

Definition 5.4: *The function f is differentiable at x if, as x' converges to x, $\frac{f(x')-f(x)}{(x'-x)}$ converges to a number. Formally, f is differentiable at x if the following limit exists:*

$$\lim_{x' \to x} \frac{f(x') - f(x)}{x' - x} = \lim_{\Delta x \to 0} \frac{f(x + \Delta x) - f(x)}{\Delta x} = \lim_{\epsilon \to 0} \frac{f(x + \epsilon) - f(x)}{\epsilon}.$$

If the limit exists it is denoted

$$\frac{df(x)}{dx} \text{ or } f'(x)$$

and is called the derivative of f.

In Figure 5.6, $f'(x)$ is equal to the slope of the tangent to the curve; $\frac{f(x+\Delta x)-f(x)}{\Delta x}$ is equal to the slope of the hypotenuse of the triangle. As Δx becomes smaller, the two slopes get closer.

When the derivative is equal to 0 at some point $\bar{x}$, the function is flat at that point. This is important in optimization, where one wishes to locate the value of x than makes $f(x)$ as large as possible. The function $f(x) = x(1-x)$ has a plot similar to Figure 5.6 and has $f'(x) = 0$ at $x = \frac{1}{2}$.

The following functions illustrate differentiability.

EXAMPLE 5.4: If $f(x) = x^2$, then $f(x + \Delta x) = (x + \Delta x)^2 = x^2 + 2x\Delta x + (\Delta x)^2$ so that $f(x + \Delta x) - f(x) = x^2 + 2x\Delta x + (\Delta x)^2 - x^2 = 2x\Delta x + (\Delta x)^2$. Therefore

$$\frac{f(x + \Delta x) - f(x)}{\Delta x} = \frac{2x\Delta x + (\Delta x)^2}{\Delta x} = 2x + (\Delta x).$$

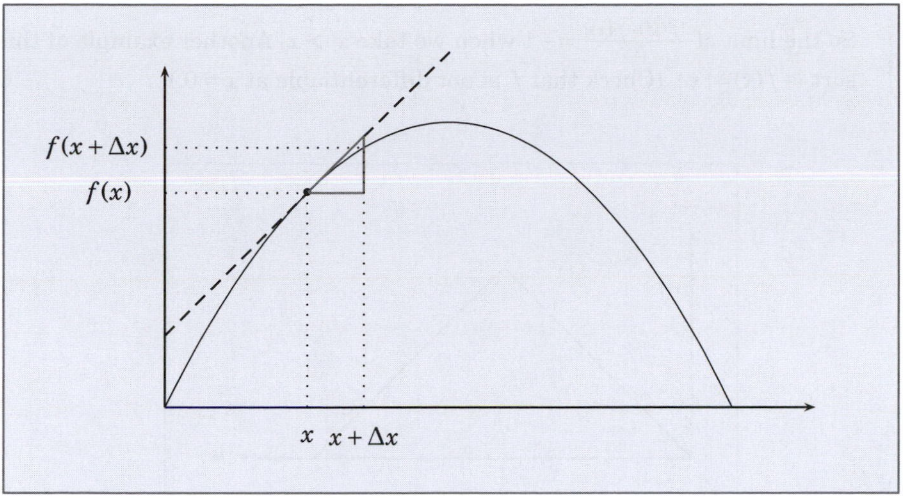

Figure 5.6: The derivative

Since the limit of this as Δx goes to 0 is $2x$, this shows that $f'(x) = 2x$. For example, take $x = \frac{1}{2} = 0.5$ and $\Delta x = 0.05$, so that $x + \Delta x = 0.55$. Then $f(x + \Delta x) = f(0.55) = (0.55)^2 = 0.3025$ and $f(x) = f(0.5) = 0.25$ so that

$$\frac{f(x+\Delta x)-f(x)}{\Delta x} = \frac{f(0.55)-f(0.5)}{0.05} = \frac{0.3025-0.25}{0.05} = \frac{0.0525}{0.05} = 1.05.$$

At $x = 0.5$, the derivative is $f'(x) = 2x = 2(0.5) = 1$, so the approximation is good at $\Delta x = 0.05$. ◇

The next example gives a continuous function that does not have a derivative at $x = \frac{1}{2}$ (see Figure 5.7).

EXAMPLE 5.5: Let $f(x) = \min\{x, 1-x\}$, so that $f(x)$ is the smaller of x and $1-x$. Take $x = \frac{1}{2}$ and let $x' < \frac{1}{2}$ so that $\Delta x = x' - x < 0$, $f(x) = f\left(\frac{1}{2}\right) = \frac{1}{2} = x$ and $f(x') = x'$ since $x' < \frac{1}{2}$. Now, observe that $f(x+\Delta x) - f(x) = (x+\Delta x) - x = \Delta x$, so that

$$\frac{f(x')-f(x)}{\Delta x} = \frac{\Delta x}{\Delta x} = 1.$$

Thus $\lim_{\Delta x \to 0} \frac{f(x+\Delta x)-f(x)}{\Delta x} = 1$, when we take $x' < x$ ($\Delta x < 0$).

Now, take $x' > x$, so that $x' = x + \Delta x$, $\Delta x > 0$. Then $f(x') = 1 - x' = 1 - (x + \Delta x) = (1-x) - \Delta x$. Since $f(x) = x = \frac{1}{2} = (1-x)$,

$$\frac{f(x')-f(x)}{\Delta x} = \frac{f(x+\Delta x)-f(x)}{\Delta x} = \frac{[(1-x)-\Delta x-(1-x)]}{\Delta x} = \frac{-\Delta x}{\Delta x} = -1$$

So the limit of $\frac{[f(x')-f(x)]}{\Delta x} = -1$ when we take $x' > x$. Another example of this sort is $f(x) = |x|$. (Check that f is not differentiable at $x = 0$.) ◇

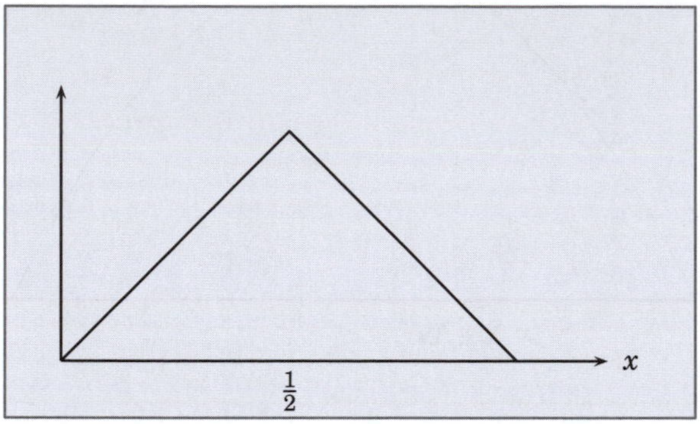

Figure 5.7: Plot of $f(x) = \min\{x, 1-x\}$

REMARK 5.1: It is always the case that differentiability of a function implies that the function is continuous. To see this, suppose that $\lim_{\Delta x \to 0} \frac{f(x_0+\Delta x)-f(x_0)}{\Delta x}$ exists. As the denominator (Δx) approaches 0, for the ratio $\frac{f(x_0+\Delta x)-f(x_0)}{\Delta x}$ to have a limit requires that $(f(x_0+\Delta x)-f(x_0))$ also approaches 0. Therefore, $\lim_{\Delta x \to 0}(f(x_0+\Delta x)-f(x_0)) = 0$. The converse is not true. For example, $|x|$ is continuous but not differentiable. □

REMARK 5.2: Differentiability plays an important role in locating the maximum or minimum of a function. Consider $f(x) = x(1-x) = x - x^2$, so $f'(x) = 1-2x$. Thus $f'(x)$ is positive when $x < \frac{1}{2}$, negative when $x > \frac{1}{2}$ and equal to 0 at $x = \frac{1}{2}$. The derivative of f measures the slope of f: the slope is positive to the left of $\frac{1}{2}$, 0 at $\frac{1}{2}$ and negative to the right of $\frac{1}{2}$. For certain functions (such as this one), there is a unique point at which the derivative is 0 $\left(\frac{1}{2}\right.$ for the function $\left.f(x) = x(1-x)\right)$. Note that this occurs at the highest point of f so by solving $f'(x) = 0 = 1-2x$, we find the value of x that maximizes $f: 1-2x = 0 \Rightarrow x = \frac{1}{2}$. This suggests that the use of derivatives will be helpful in finding maxima (and minima) of functions.□

5.4 Types of function

In principle, a function f is a rule that associates to each point x in the domain, a point y in the range. In practice, it is useful to use specific functional forms that reflect specific patterns in relationship between x and y.

EXPONENTIAL FUNCTIONS: The exponential function is defined $f(x) = e^x$, where e is a constant (known as Euler's number, $e \approx 2.718$), discussed later. More generally, exponential functions have the functional form $f(x) = a^x$ or $f(x) = ba^{cx}$ where a, b and c are constants.

POLYNOMIAL FUNCTIONS: A polynomial function is a function of the form: $f(x) = a_n x^n + a_{n-1}x^{n-1} + \cdots + a_1 x + a_0$, where $\{a_k\}_{k=0}^n$ are constants.

TRIGONOMETRIC FUNCTIONS: The three primary trigonometric functions are the sine, cosine and tangent functions, $f(x) = \sin(x)$, $f(x) = \cos(x)$, $f(x) = \tan(x)$. Trigonometric functions are functions of an angle x, which may be measured in degrees or radians. When measured in degrees, a $360°$ angle represents one full rotation in a circle. When measured in radians, a radian is that angle, such that 2π such angles is a full rotation in the circle. It is common to plot these functions from $-\pi$ to π, after which they repeat, so in this case the unit of measure is radians.

LOGARITHMIC FUNCTIONS: The logarithm of a number is defined relative to some given base, and is said to be the logarithm of the number to that base. The logarithm of x to the base b is that number y such that $x = b^y$, that is the power to which the base must be raised to equal that number. So, for example, the logarithm of 100 to the base 10 is 2, because $100 = 10^2$. The log of x to the base b is written $\log_b x$. One special case is where $b = e$, Euler's number. Routinely, $\log_e x$ is written $\ln x$.

5.4.1 Plots of various functions

The following table gives some well known functions and their derivatives.

$f(x)$	x^a	a^x	$\log_e x$	e^{ax}	$\sin(x)$	$\cos(x)$	$\tan(x)$	$e^{g(x)}$
$f'(x)$	ax^{a-1}	$a^x \cdot \ln(a)$	$\frac{1}{x}$	ae^{ax}	$\cos(x)$	$-\sin(x)$	$\frac{1}{\cos^2(x)}$	$g'(x) \cdot e^{g(x)}$

For some functions, parameter variation may change the plot of the function substantially. In Figure 5.8, different values of a vary the appearance of the function significantly.

Note that the function $f(x) = x^a$ may only be defined for some values of x. For example, when $a = \frac{1}{2}$, then $f(x) = \sqrt{x}$ is defined only for real $x \geq 0$ and when $a = -1$ it is defined for all $x \neq 0$.

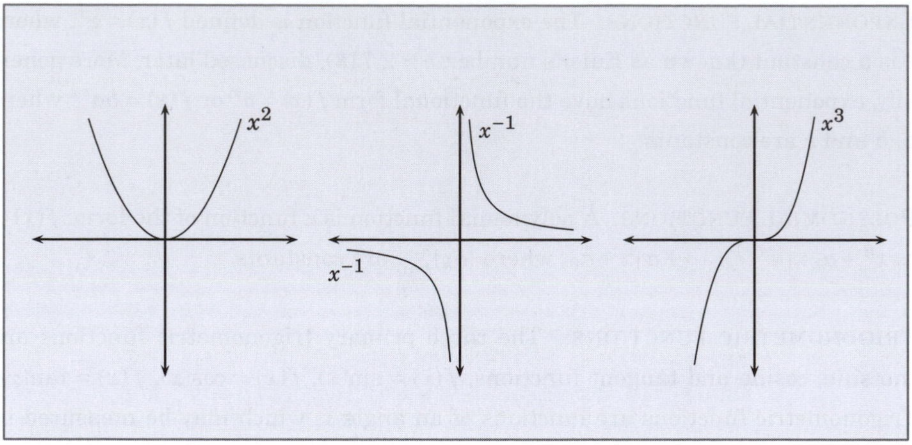

Figure 5.8: $f(x) = x^a$, for various values of a

The function $f(x) = a^x$ (Figure 5.9) is defined for all real x when $a > 0$. When $a < 0$, a^x may or may not be defined. For example, $(-8)^{-\frac{1}{3}} = -2$, but $(-1)^{-\frac{1}{2}}$ is not a real number.

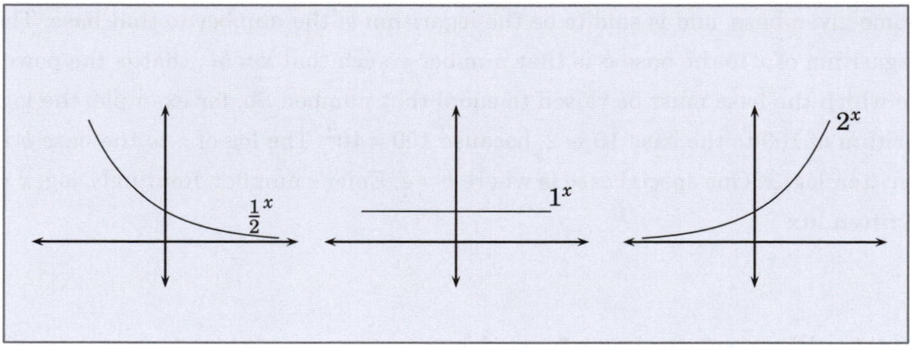

Figure 5.9: Plot of $f(x) = a^x$, for various values of a

The trigonometric functions are typified by repeating patterns. For any angle x, measured in degrees, this angle is the same angle as $x + 360°$ so that any function of the angle will have the same value at x and $x + 360°$. Alternatively, angles may be measured in radians. A radian is that angle such that 2π radians equals one rotation in the circle (or 360°). In Figure 5.10 the angles are measured in radians.

These functions may be modified (Figure 5.11). Consider the function $f(x) = A\sin(bx - c) + d = A\sin\left(b\left[x - \frac{c}{b}\right]\right) + d$. The parameter A is called the amplitude: if $A = 2$, the height of the fluctuation is doubled. The parameter $\frac{c}{b}$ measures

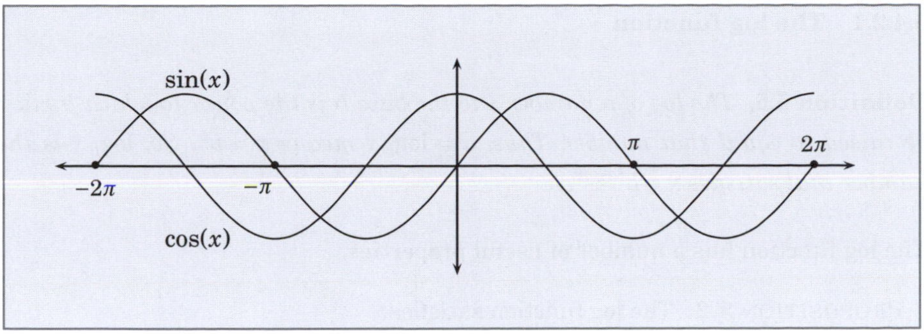

Figure 5.10: Plot of sin(x) and cos(x) for $-2\pi \le x \le 2\pi$

horizontal shift. Comparing $f(x)$ with $g(x) = A\sin(bx) + d$, $f\left(\frac{c}{b}\right) = g(0)$. The ratio $\frac{c}{b}$ is called the phase shift (the horizontal displacement of the function from its initial position). Finally, the period of a function is the distance between repeating points and is given by $\frac{2\pi}{b}$.

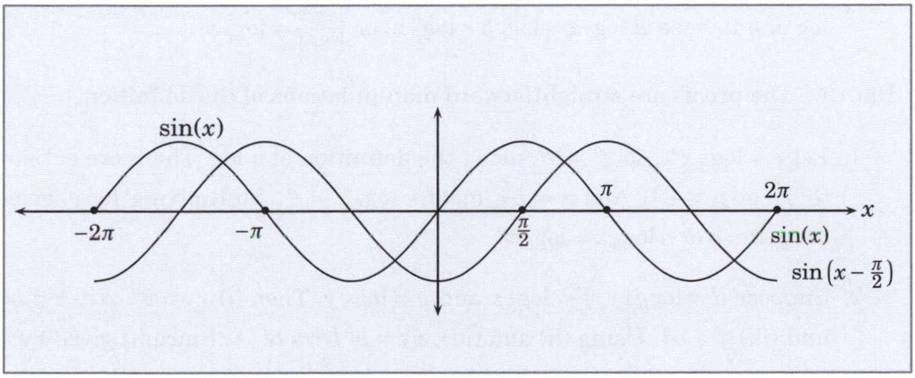

Figure 5.11: Plot of sin(x) and $\sin\left(x - \frac{\pi}{2}\right)$ for $-2\pi \le x \le 2\pi$

5.4.2 The log and exponential functions

Both log and exponential functions are widely used in economics because, for example, growth formula and growth calculations utilize these functions. The log of x to the base b is denoted $\log_b x$ giving that number a such that $x = b^a$. The log function is defined only for positive numbers (for $x > 0$). Two particularly important bases are $b = 10$ and $b = e$, where e is Euler's number with $e \approx 2.718$. Because the function $\log_e x$ is widely used, for notational convenience this is often written as $\ln x$: $\ln x \overset{\text{def}}{=} \log_e x$.

5.4.2.1 The log function

Definition 5.5: *The log of a number x to the base b is the power to which b must be raised to equal that number. Thus, $a = \log_b x$ means $x = b^a$. So, $\log_b x$ is the number that satisfies $x = b^{\log_b x}$.*

The log function has a number of useful properties.

PROPOSITION 5.2: The log function satisfies:

1. The log of x^a to base b is equal to a times the log of x to the base b: $\log_b x^a = a \log_b x$.
2. The log of a product is equal to the sum of the logs: $\log_b xy = \log_b x + \log_b y$.
3. The log of a ratio is equal to the difference of the logs: $\log_b \frac{x}{y} = \log_b x - \log_b y$.
4. Change of base. The log of x to base c equals the log of x to base b times log of b to base c: $\log_c x = \log_c b \times \log_b x$, or $\frac{\log_c x}{\log_b x} = \log_c b$. ◇

PROOF: The proofs are straightforward manipulations of the definition.

1. Let $c = \log_b x^a$, so $x^a = b^c$, using the definition of a log. Therefore $(x^a)^{\frac{1}{a}} = (b^c)^{\frac{1}{a}}$ or $x = c^{\frac{c}{a}}$. And $x = b^{\frac{c}{a}}$ means $\log_b x = \frac{c}{a}$. Multiplying by a gives $a \log_b x = c$ or $a \log_b x = \log_b x^a$.
2. Suppose $d = \log_b xy$, $f = \log_b x$ and $g = \log_b y$. Then (i) $xy = b^d$, (ii) $x = b^f$ and (iii) $y = b^g$. Using (ii) and (iii), $xy = b^f b^g = b^{f+g}$. Since (i) gives $xy = b^d$, it must be that $d = f + g$ or $\log_b xy = \log_b x + \log_b y$.
3. Suppose $d = \log_b \frac{x}{y}$, $f = \log_b x$ and $g = \log_b y$. Then (i) $\frac{x}{y} = b^d$, (ii) $x = b^f$ and (iii) $y = b^g$. Using (ii) and (iii), $\frac{x}{y} = \frac{b^f}{b^g} = b^{f-g}$. Since (i) gives $\frac{x}{y} = b^d$, it must be that $d = f - g$ or $\log_b \frac{x}{y} = \log_b x - \log_b y$.
4. Let $d = \log_b x$ so that $x = b^d$. Then $\log_c x = \log_c b^d$. From rule 1, $\log_c b^d = d \log_c b$. So, $\log_c x = d \log_c b = \log_b x \cdot \log_c b$, and thus $\frac{\log_c x}{\log_b x} = \log_c b$. ■

5.4.2.2 Choice of base for the log function

There are two commonly used bases: 10 and e. For example, if $\alpha = \log_{10} x$ then $10^\alpha = x$. Thus, $\log_{10} 10 = 1$, $\log_{10} 100 = 2$, $\log_{10} 1000 = 3$, and so on. The number e is defined as $e = \lim_{x\to\infty} \left(1 + \frac{1}{x}\right)^x \approx 2.71828$. It arises naturally in differentiation and integration. The functions are related by the following equations: $\log_e x = \frac{\log_{10} x}{\log_{10} e}$ and $\log_{10} x = \frac{\log_e x}{\log_e 10}$. For example, if $\alpha = \log_{10} x$, then $10^\alpha = x$ and $\log_e(10^\alpha) = \log_e x$, and $\log_e(10^\alpha) = \alpha \cdot \log_e 10 = \log_{10} x \cdot \log_e 10$, giving the second relation.

It is common practice to write $\ln x$ for $\log_e x$. One useful feature of the ln function is that the derivative of $\ln x$ is $\frac{1}{x}$. If $y = \log_e x$, then $e^y = e^{\log_e x} = x$ so that $e^y \frac{dy}{dx} = 1$ or $x\frac{dy}{dx} = 1$ or $\frac{dy}{dx} = \frac{1}{x}$. For the formulae above: $\log_{10} e \cdot \ln x = \log_{10} x$ and $\log_{10} x = \frac{\log_e x}{\log_e 10}$ so the derivative of $\log_{10} x$ can be written as $\log_{10} e \cdot \frac{1}{x}$ and as $\frac{1}{\log_e 10} \cdot \frac{1}{x}$.

REMARK 5.3: The mathematical constant $e \approx 2.71828$, Euler's number, may be defined in a number of ways. Define e to be that number that makes the area under the curve $\frac{1}{x}$ between 1 and e equal to 1. The number e may also be defined as:

$$e = 1 + \frac{x}{1!} + \frac{x^2}{2!} + \frac{x^3}{3!} \cdots = \sum_{j=0}^{\infty} \frac{x^j}{j!}.$$

Yet another equivalent definition gives e as the limit of a sequence of numbers:

$$e = \lim_{x\to\infty} \left(1 + \frac{1}{x}\right)^x.$$

The latter formula makes the connection with continuous-time discounting. □

5.5 Growth rates

Suppose an investor invests one dollar for a year at a 100% interest rate, $r = 100$. If the interest is calculated once at the end of the year, the loan is worth $1\times(1+r) = 1\times 2 = 2$. Suppose instead interest is compounded twice a year, with the half yearly rate set at $\frac{r}{2} = \frac{1}{2}$. Then after six months the loan is worth $1 \times \left(1+\frac{1}{2}\right)$ and at the end of twelve months it is worth $1 \times \left(1+\frac{1}{2}\right)\left(1+\frac{1}{2}\right) = \left(1+\frac{1}{2}\right)^2 = 2.25$. Similarly, if the interest is calculated every four months at the four-monthly rate $\frac{r}{3} = \frac{1}{3}$, then at the end of the year the value of the loan is $\left(1+\frac{1}{3}\right)^3 = 2.36$. Calculating interest on a monthly basis, the value of the loan would be $\left(1+\frac{1}{12}\right)^{12} = (1.08333)^{12} = 2.613$. In general, with n rounds of interest calculation, the value of the loan at the end of the year is $\left(1+\frac{1}{n}\right)^n$. As n becomes large, this approaches e.

$$\left(1+\frac{1}{n}\right)^n \longrightarrow e.$$

With continuous interest accumulation (from one instant to the next, a tiny amount of interest is added to the principal), the value of the loan at the end of the year is e.

Suppose now that the interest rate is r instead of 1 and note that:

$$\left(1+\frac{r}{n}\right)^n = \left(1+\frac{1}{\left(\frac{n}{r}\right)}\right)^n = \left(1+\frac{1}{\left(\frac{n}{r}\right)}\right)^{\left(\frac{n}{r}\right)r} = \left\{\left(1+\frac{1}{\left(\frac{n}{r}\right)}\right)^{\left(\frac{n}{r}\right)}\right\}^r \to e^r.$$

(Recall some rules of manipulation: $x^a x^b = x^{a+b}$, $(x^a)^b = x^{a \cdot b}$.) With t periods, and an interest rate r each period, the value after t periods is $(1+r)^t$. With interest calculated twice each period, the final value is $\left([(1+\frac{r}{2})]^2\right)^t$ and with interest calculated n times each period the final value is $\left([(1+\frac{r}{n})]^n\right)^t$. From the previous calculations:

$$\left(\left[\left(1+\frac{r}{n}\right)\right]^n\right)^t \to (e^r)^t = e^{rt}.$$

Thus, the value of P dollars invested now with continuous interest accumulation is $V = Pe^{rt}$. Conversely, V dollars in t periods with continuous discounting is worth $Ve^{-rt}(=P)$ now.

5.5.1 Additivity of continuous growth rates

One advantage of using continuous compound rates is that growth rates are additive. Consider three time periods, $s < t < u$, with some variable (such as income) observed at those three dates: y_s, y_t, y_u. In the discrete calculation, the growth rate from s to t is the number r_{st} determined by $\frac{y_t}{y_s} = (1+r_{st})$, and the growth rate from t to u is the number r_{tu} determined by $\frac{y_u}{y_t} = (1+r_{tu})$. Over the period from s to u the growth rate is r_{su}, defined from $\frac{y_u}{y_s} = (1+r_{su})$. However,

$$(1+r_{st})(1+r_{tu}) = (1+r_{st}+r_{tu}+r_{st}r_{tu}) = (1+r_{su}),$$

so the overall growth is not the sum of the growths in the two periods: $r_{st}+r_{tu} \neq r_{su}$. By contrast, using the continuous growth formula:

$$\frac{y_t}{y_s} = e^{\rho_{st}}, \quad \frac{y_u}{y_t} = e^{\rho_{tu}}, \quad \frac{y_u}{y_s} = e^{\rho_{su}},$$

where ρ_{ij} is the continuous growth rate from time i to time j. Then, multiplying,

$$e^{\rho_{su}} = \frac{y_u}{y_s} = \frac{y_t}{y_s}\frac{y_u}{y_t} = e^{\rho_{st}}e^{\rho_{tu}} = e^{\rho_{st}+\rho_{tu}},$$

so that $\rho_{su} = \rho_{st} + \rho_{tu}$. Thus, the growth rate over the full period from t to u is equal to the sum of the growth rates in the two subperiods.

5.5.2 The functions $\ln x$ ($\log_e x$) and e^{kx}

Figure 5.12 shows the functions e^x and $\log_e x$. Both are increasing in x, but e^x increases rapidly (at an increasing rate), whereas $\log_e x$ increases at a decreasing rate. The following section provides additional discussion on these important functions.

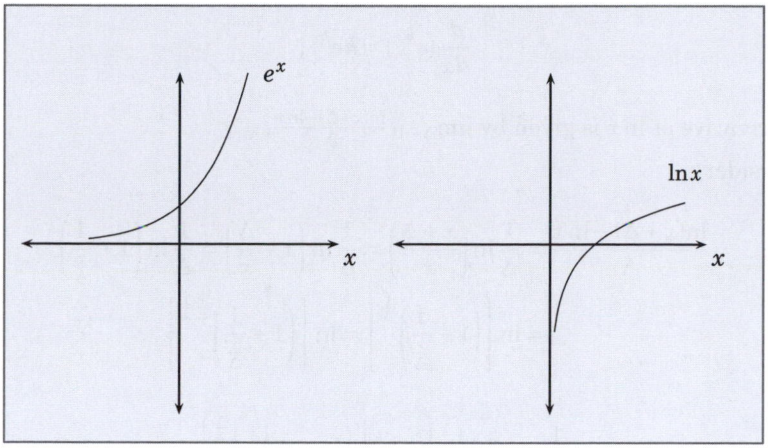

Figure 5.12: The functions e^x and $\ln x$ ($\equiv \log_e(x)$)

5.5.2.1 The derivatives of e^x and $\ln x$

Given a constant k, the derivative of e^{kx} is given by $\lim_{\Delta\to 0}\frac{e^{k(x+\Delta)}-e^{kx}}{\Delta}$. Consider:

$$\frac{e^{k(x+\Delta)}-e^{kx}}{\Delta} = \frac{e^{(kx+k\Delta)}-e^{kx}}{\Delta} = \frac{e^{kx}e^{k\Delta}-e^{kx}}{\Delta}$$
$$= \frac{e^{kx}[e^{k\Delta}-1]}{\Delta} = e^{kx}\frac{[e^{k\Delta}-1]}{\Delta}.$$

Recalling that one definition of e^z is $e^z = 1+\frac{z}{1!}+\frac{z^2}{2!}+\cdots$

$$e^{k\Delta} = 1+\frac{k\Delta}{1!}+\frac{k^2\Delta^2}{2!}+\frac{k^3\Delta^3}{3!}\cdots$$
$$= 1+k\Delta+\Delta^2k^2\left[\frac{1}{2!}+\frac{k\Delta}{3!}+\frac{k^2\Delta^2}{4!}\cdots\right].$$

Therefore,

$$\frac{[e^{k\Delta}-1]}{\Delta} = k+\Delta k^2\left[\frac{1}{2!}+\frac{k\Delta}{3!}+\frac{k^2\Delta^2}{4!}\cdots\right].$$

Since the term $[\cdots]$ is bounded for $|k\Delta| < 1$:

$$\lim_{\Delta \to 0} \frac{[e^{k\Delta} - 1]}{\Delta} = k.$$

(The term $[\cdots]$ is bounded because $\left|\left[\frac{1}{2!} + \frac{k\Delta}{3!} + \frac{k^2\Delta^2}{4!} \cdots\right]\right| \le \frac{1}{2!} + \frac{|k\Delta|}{3!} + \frac{|k\Delta|^2}{4!} \cdots < 1 + |k\Delta| + |k\Delta|^2 + \cdots = \frac{1}{1-|k\Delta|}$.) Therefore:

$$\frac{d}{dx}\{e^{kx}\} = ke^{kx}.$$

The derivative of $\ln x$ is given by $\lim_{\Delta\to 0} \frac{\ln(x+\Delta)-\ln x}{\Delta}$.

Consider:

$$\begin{aligned}\frac{\ln(x+\Delta) - \ln x}{\Delta} &= \frac{1}{\Delta}\ln\left(\frac{x+\Delta}{x}\right) = \frac{1}{\Delta}\ln\left(1 + \frac{\Delta}{x}\right) = \frac{1}{\Delta}\ln\left(1 + \frac{1}{\frac{x}{\Delta}}\right) \\ &= \ln\left[\left(1 + \frac{1}{\frac{x}{\Delta}}\right)^{\frac{1}{\Delta}}\right] = \ln\left[\left(1 + \frac{1}{\frac{x}{\Delta}}\right)^{\frac{x}{\Delta}\frac{1}{x}}\right].\end{aligned}$$

Rearranging brackets, $\ln\left[\left(1+\frac{1}{\frac{x}{\Delta}}\right)^{\frac{x}{\Delta}\frac{1}{x}}\right] = \ln\left[\left\{\left(1+\frac{1}{\frac{x}{\Delta}}\right)^{\frac{x}{\Delta}}\right\}^{\frac{1}{x}}\right]$. The $\ln x$ function is continuous for $x > 0$ and since $\left(1+\frac{1}{\frac{x}{\Delta}}\right)^{\frac{x}{\Delta}} > 1$, therefore $\left\{\left(1+\frac{1}{\frac{x}{\Delta}}\right)^{\frac{x}{\Delta}}\right\}^{\frac{1}{x}} > 1$. (Recall, if f is continuous at x_0 then $\lim_{x\to x_0} f(x) = f(x_0)$.) Therefore,

$$\lim_{\Delta\to 0} \frac{\ln(x+\Delta) - \ln x}{\Delta} = \lim_{\Delta\to 0} \ln\left[\left\{\left(1+\frac{1}{\frac{x}{\Delta}}\right)^{\frac{x}{\Delta}}\right\}^{\frac{1}{x}}\right] = \ln\left[\lim_{\Delta\to 0}\left\{\left(1+\frac{1}{\frac{x}{\Delta}}\right)^{\frac{x}{\Delta}}\right\}^{\frac{1}{x}}\right] = \ln\left[e^{\frac{1}{x}}\right] = \frac{1}{x},$$

and so,

$$\frac{d}{dx}\{\ln x\} = \frac{1}{x}.$$

5.6 Some rules of calculus

Some basic rules of calculus apply for the differentiation of the product, ratio and composition of continuous functions.

1. The derivative of a product of functions:

$$\frac{d}{dx}\{f(x)g(x)\} = f(x)g'(x) + g(x)f'(x).$$

For example, if $f(x) = x^\alpha$ and $g(x) = \cos x$. Then $\frac{d}{dx}\{x^\alpha \cos(x)\} = x^\alpha \sin(x) + \alpha x^{\alpha-1} \cos x$.

2. The derivative of a ratio of functions ($g(x) \neq 0$):

$$\frac{d}{dx}\left\{\frac{f(x)}{g(x)}\right\} = \frac{g(x)f'(x) - f(x)g'(x)}{[g(x)]^2}.$$

3. The chain rule (differentiating the composition of functions).

$$\frac{d}{dx}\{f(g(x))\} = f'(g(x))g'(x).$$

EXAMPLE 5.6: Consider: $f(x) = \ln x$ and $g(x) = x^2$. (Note that the ln function is defined only for $x > 0$.) Then

$$f'(x) = \frac{d}{dx}\{\ln(x)\} = \frac{1}{x}$$
$$g'(x) = \frac{d}{dx}\{x^2\} = 2x.$$

1. Product rule: If $h(x) = f(x)g(x)$, then $h'(x) = (\ln x)\cdot(2x) + \frac{1}{x}\cdot x^2$.
2. Ratio rule: If $h(x) = \frac{f(x)}{g(x)}$, then $h'(x) = \frac{x^2 \cdot \frac{1}{x} - (\ln x)\cdot(2x)}{(x^2)^2}$
3. Chain rule: Put $h(x) = g(f(x)) = (\ln x)^2$. Differentiating directly, $h'(x) = (2\ln x)\left(\frac{1}{x}\right)$. Or, from the chain rule,

$$h'(x) = g'(f(x))f'(x) = (2\ln x)\left(\frac{1}{x}\right) = (2\ln x)\left(\frac{1}{x}\right) \qquad \diamond$$

EXAMPLE 5.7 (RATIO RULE): The function $\tan x$ is defined $\tan x = \frac{\sin x}{\cos x}$, and using the ratio formula:

$$\frac{d}{dx}\tan(x) = \frac{\cos(x)\cos(x) + \sin(x)\sin(x)}{(\cos(x))^2} = \frac{\cos^2(x) + \sin^2(x)}{\cos^2(x)} = \frac{1}{\cos^2(x)},$$

utilizing the fact that the sine of an angle is the opposite (a) over the hypotenuse (h) and the cosine is the base (b) divided by the hypotenuse (in a right-angled triangle). Thus, $\cos x = \frac{b}{h}$, $\sin x = \frac{a}{h}$ and $\tan x = \frac{\sin x}{\cos x} = \frac{a}{b}$.

The figure illustrates:

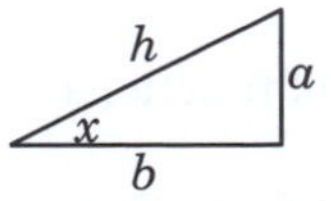

. ◊

EXAMPLE 5.8 (CHAIN RULE): Let $g(y) = \sin y$ and $f(x) = \sqrt{x}$. Then

$$g'(y) = \cos y$$
$$f'(x) = \frac{1}{2}x^{-\frac{1}{2}}.$$

If $h(x) = g(f(x)) = \sin(\sqrt{x})$ then $h'(x) = \cos(\sqrt{x})\frac{1}{2}x^{-\frac{1}{2}} = \frac{\cos(\sqrt{x})}{2\sqrt{x}}$.

Let $g(y) = \alpha^y$ and $f(x) = x^2$, so that $g(f(x)) = \alpha^{x^2}$. This may be written $e^{\ln\alpha \cdot x^2}$, so that the derivative is $2x\ln\alpha e^{\ln\alpha x^2} = 2x\ln\alpha \cdot \alpha^{x^2}$. ◊

In the context of the exponential and log functions, the chain rule gives:

$$\frac{d}{dx}e^{f(x)} = f'(x)e^{f(x)}, \quad \frac{d}{dx}\ln f(x) = \frac{1}{f(x)}f'(x) \text{ given } f(x) > 0.$$

The next example illustrates a more abstract calculation using the chain rule.

EXAMPLE 5.9: Suppose that $f(\alpha) = g(\alpha)^{h(\alpha)}$, where $g(\alpha) > 0$ for all α. To find the derivative of f, take the ln of f and differentiate it. $\ln f(\alpha) = h(\alpha)\ln g(\alpha)$. Thus, differentiating both sides:

$$\frac{d\ln f(\alpha)}{d\alpha} = \left(\frac{1}{f(\alpha)}\right)f'(\alpha) = \left[h'(\alpha)\ln g(\alpha) + h(\alpha)\left(\frac{1}{g(\alpha)}\right)g'(\alpha)\right].$$

So,

$$f'(\alpha) = f(\alpha)h'(\alpha)\ln g(\alpha) + f(\alpha)h(\alpha)\left(\frac{1}{g(\alpha)}\right)g'(\alpha)$$
$$f'(\alpha) = g(\alpha)^{h(\alpha)}h'(\alpha)\ln g(\alpha) + g(\alpha)^{h(\alpha)}h(\alpha)\left(\frac{1}{g(\alpha)}\right)g'(\alpha).$$

Alternatively, observe that $f(\alpha) = e^{h(\alpha)\ln g(\alpha)}$, so that (using the rule $\frac{d}{dx}e^{r(x)} = e^{r(x)}r'(x)$)

$$f'(x) = e^{h(\alpha)\ln g(\alpha)}\left[h'(\alpha)\ln g(\alpha) + h(\alpha)\left(\frac{1}{g(\alpha)}\right)g'(\alpha)\right].$$ ◊

5.6.1 l'Hospital's rule

Sometimes it is necessary to consider the behavior of the ratio of functions, $\frac{f(x)}{g(x)}$, in the neighborhood of a point where the ratio is not defined. Consider the behavior of the ratio $\frac{f(x)}{g(x)}$ in the neighborhood of a point a. Suppose that $x \to a$; how does $\frac{f(x)}{g(x)}$ behave?

There are a few cases where the answer is obvious. If $f(x) \to c$ and $g(x) \to d$ with $d \neq 0$, then $\frac{f(x)}{g(x)} \to \frac{c}{d}$. Similarly, if $f(x) \to c$ and $g(x) \to d$ with $d = 0$ and $c \neq 0$, then the ratio tends to $+\infty$ or $-\infty$.

However, there are two specific cases where the answer is not obvious: When both $f(x)$ and $g(x)$ go to 0, and when both $f(x)$ and $g(x)$ go to $\pm\infty$. In such cases, determining the limit may be tedious. However, l'Hospital's rule provides a means of evaluating the limit.

Theorem 5.1: *Suppose that (i) f and g are differentiable on $(a-\delta, a+\delta)\setminus\{a\}$ for some $\delta > 0$ (where the notation $x\setminus\{y\}$ means the set x with the point y removed), (ii) $\lim_{x\to a} f(x) = 0$ and $\lim_{x\to a} g(x) = 0$, (iii) $g'(x) \neq 0$ on $(a-\delta, a+\delta)\setminus\{a\}$. Then,*

$$\lim_{x\to a} \frac{f'(x)}{g'(x)} = k \quad \textit{implies} \quad \lim_{x\to a} \frac{f(x)}{g(x)} = k.$$

Similarly, suppose that (i) f and g are differentiable on $(a-\delta, a+\delta)\setminus\{a\}$ for some $\delta > 0$, (ii) $\lim_{x\to a} f(x) = \infty$ and $\lim_{x\to a} g(x) = \infty$, (iii) $g'(x) \neq 0$ on $(a-\delta, a+\delta)\setminus\{a\}$. Then,

$$\lim_{x\to a} \frac{f'(x)}{g'(x)} = k \quad \textit{implies} \quad \lim_{x\to a} \frac{f(x)}{g(x)} = k.$$

Thus, under general conditions, the limit of the ratio is equal to the limit of the ratio of the derivatives.

EXAMPLE 5.10 (L'HOSPITAL'S RULE): Let $f(x) = x^2$ and $g(x) = e^x - 1$. As x approaches 0, the values of the functions approach 0: $\lim_{x\to a} f(x) = \lim_{x\to a} g(x) = 0$. How does $\frac{f(x)}{g(x)}$ behave as $x \to 0$? Considering the derivatives of both functions, $f'(x) = 2x$ and $g'(x) = e^x$. Since $2x$ goes to 0 as x approaches 0 and e^x goes to 1 as x goes to 0,

$$\lim_{x\to 0} \frac{x^2}{e^x - 1} = \lim_{x\to 0} \frac{2x}{e^x} = 0. \qquad \diamond$$

5.6.2 Higher derivatives

For notational convenience, the second derivative is denoted $f''(x)$. Technically, the second derivative is defined in exactly the same way as the first derivative:

$$f''(x) = \lim_{\Delta \to 0} \frac{f'(x+\Delta) - f'(x)}{\Delta}.$$

Thus, the second derivative is the derivative of the first derivative. Likewise, higher-order derivatives may be determined step by step. As noted earlier, a continuous function may not be differentiable. Similarly, a function may have a first derivative, but not a second derivative; and so on. Properties of the second derivative are central in the discussion of optimization. If the second derivative is positive, the *slope* of the function is increasing at that point; if the second derivative is negative, the *slope* of the function is decreasing. Third-order derivatives are calculated similarly:

$$f'''(x) = \lim_{\Delta \to 0} \frac{f''(x+\Delta) - f''(x)}{\Delta}.$$

Higher-order derivatives are calculated by repeated differentiation.

Exercises for Chapter 5

Exercise 5.1 Consider the function F, defined:

$$f(x) = \begin{cases} 2x^2, & x < \frac{1}{2} \\ x, & x \geq \frac{1}{2}. \end{cases}$$

Show that f is continuous but not differentiable at $x = \frac{1}{2}$.

Exercise 5.2 Let

$$f(x) = \begin{cases} ax^3 - 2x + b, & x < 1 \\ x^2 - bx, & x \geq 1. \end{cases}$$

Find values of a and b so that f is differentiable.

Exercise 5.3 Consider the ratio $\frac{f(x)}{g(x)}$. Suppose that $g(x) > 0$ for all x and $\lim_{x \to a} \frac{f(x)}{g(x)} = 1$. Suppose that $\lim_{x \to a} g(x) = \bar{g} \neq 0$, a finite number, show that $\lim_{x \to a}[f(x) - g(x)] = 0$.

Exercise 5.4 Use l'Hospital's rule to find:

(a) $\lim_{x \to 1} \frac{\sqrt{x}-1}{x^2-1}$.

(b) $\lim_{x \to 4} \frac{x-4}{\sqrt{x}-2}$.

(c) $\lim_{x \to 0} \frac{1-\cos(x)}{\sin(x)}$.

Exercise 5.5 Let $y_{2000} = \$150$ billion, $y_{2001} = \$165$ billion, and $y_{2010} = \$225$ billion. Let ρ_1 denote the continuous compound growth over the 1 year period,

2000–2001, ρ_2 denote the continuous compound growth over the 9 year period, 2001–2010 and ρ_3 denote the continuous compound growth over the 10 year period, 2000–2010. Thus, for example, $y_{2001} \cdot e^{\rho_2 \cdot 9} = y_{2010}$. Confirm algebraically and by numerical calculation that $\rho_1 \cdot 1 + \rho_2 \cdot 9 = \rho_3 \cdot 10$. Define r_1, r_2 and r_3 as the corresponding yearly compounded growth rates – so for example, $y_{2001} \cdot (1+r_2)^9 = y_{2010}$. Confirm that $r_1 \cdot 1 + r_2 \cdot 9 \neq r_3 \cdot 10$.

Exercise 5.6 Let demand be given by $P(Q) = a - bQ$. Suppose the market price is p. Consumer surplus is defined as the area under the demand curve above price. Denote the consumer surplus as C; this depends on the parameters of the demand curve and the market price p. Show that the impact of a small price increase on C is the negative of the impact of an increase in a.

Exercise 5.7 Let

$$f(x) = \frac{g(x)}{h(x)} = \frac{x^2 - 2x + 5}{4x}, \; x > 0.$$

Find the derivative of f with respect to x, and show that the derivative is negative for $x < \sqrt{5}$ and positive for $x > \sqrt{5}$.

Exercise 5.8 Consider the ratio $\frac{\ln(t)}{t}$: show that the limit as $t \to \infty$ is 0: $\lim_{t\to\infty} \frac{\ln(t)}{t} = 0$. Next, consider the function $f(t) = t^k e^{-at}$, $t > 0$, where $a > 0$. Show that for any positive number k, $\lim_{t\to\infty} f(t) = 0$.

Exercise 5.9 Given a function $f(x)$, the first derivative is a function $f'(x)$. Differentiating this function $f'(x)$ gives a function denoted $f''(x)$, the second derivative of f. Plot the function $f(x) = 8[x^3 - 2x^2 + x]$ and its first and second derivatives.

Exercise 5.10 Suppose that revenue as a function of output is given by $R(Q) = \ln(Q) - \frac{1}{4}Q^2$. Output, in turn, depends on input K according to the equation $Q(K) = 3 - e^{-aK}$. Thus, $r(K) = R(Q(K))$ gives revenue in terms of K. Find $\frac{dr}{dK}$.

Exercise 5.11 Suppose the inverse demand for a good is given by the function $P(Q)$ where $P'(Q) < 0$. This identifies the price that corresponds to a given level of demand. Conversely, the demand function $Q(P)$ associates a level of demand to each price. So, given a quantity Q, $P(Q)$ is the price that would lead to that level of demand. Mathematically, $Q = Q(P(Q))$ for each Q. The elasticity of demand is defined at $Q = P(Q)$ as

$$\epsilon(Q) = -\frac{P}{Q}\frac{dQ}{dP} = -\frac{P}{Q(P)}\frac{dQ(P)}{dP}.$$

Use the chain rule to show that

$$\frac{1}{\epsilon(Q)} = -\frac{Q}{P(Q)}\frac{dP(Q)}{dQ}.$$

Exercise 5.12 Suppose that output Q is produced at cost $C(Q) = \ln(Q+1) + e^{-Q}$. Therefore, the average cost function is:

$$A(Q) = \frac{C(Q)}{Q} = \frac{\ln(Q+1) + e^{-Q}}{Q}.$$

Calculate $A'(C)$.

Exercise 5.13 Suppose that output Q is produced at cost $C(Q)$. The average cost of the output is $A(C) = \frac{C(Q)}{Q}$. Show that average cost is declining if marginal cost is less than average cost.

Exercise 5.14 A firm pays tax on gross revenue $R(Q)$ at a rate that depends on output volume $t(Q)$. Suppose that $R(Q) = k\ln(Q+1) - aQ^{\alpha}$ and $t(Q) = 1 - e^{-bQ}$, then total tax paid is $T(Q) = R(Q)t(Q) = [k\ln(Q+1) - aQ^{\alpha}](1 - e^{-bQ})$. Use the product rule to find $T'(Q)$.

6

Functions of one variable: applications

6.1 Introduction

In this chapter, calculus techniques are applied to the study of a variety of problems, beginning with optimization in Section 6.2. In an optimization problem, given a function f of one variable x, the problem is to find the value of x, x^* that maximizes f. Such an x^* is characterized by the condition that $f(x^*) \geq f(x)$ for all x. If the choice of x is restricted to some set X, then the requirement is that $f(x^*) \geq f(x)$ for all $x \in X$. Calculus techniques are used to find x^* by considering how f varies with x. These techniques are local in nature and so locate "local" maxima (or minima), where there is no alternative choice close to the candidate optimum that yields a higher value of the function f (or lower value if the task is to minimize the value of f). In contrast, x^* is a global maximizer if $f(x^*)$ is as large as $f(x)$ for any possible x. Identifying whether a value x^* is a local or global maximizer is considered in Section 6.3. This discussion focuses on the shape of the function f, and the use of calculus techniques to identify the shape.

Following this discussion of optimization, a number of applications are examined. The discussion of externalities in Section 6.4 illustrates the use of these techniques to study the welfare implications of traffic congestion. Comparative statics, introduced in Section 6.5, concerns the change in a variable resulting from an underlying parameter change. For example, a change in the intercept of a demand curve will change the intersection of the supply and demand curves and hence the equilibrium quantity and price. In the optimization context, if the function f depends on x and a fixed parameter α, then the optimal choice of x to maximize $f(x,\alpha)$ will depend on α: $x(\alpha)$. Therefore, any change in α will result in a change in the value of $x(\alpha)$. The study of how $x(\alpha)$ changes as α changes is considered in this section.

Subsequent sections consider comparative static calculations in a taxation context – in Sections 6.7 and 6.8. A key measurement in economics is the elasticity, measuring the response of a dependent variable to a change in another variable – for example, measuring how demand responds to a price change. In the context of demand, this is introduced in Section 6.6. Finally, Section 6.9 considers Ramsey pricing, using variational arguments to study efficient taxation.

6.2 Optimization

Making optimal choices involves maximization or minimization of some objective. In the present context, given a function f where $f: X \to \mathcal{R}$, suppose that we wish to find a value for x, that maximizes (or minimizes) f on X. For most of this discussion, X will be a subset of $\mathcal{R}$.

Definition 6.1: *Say that $\bar{x}$ is a global maximum if $f(\bar{x}) \geq f(x)$ for all $x \in X$, and a global minimum if $f(\bar{x}) \leq f(x)$ for all $x \in X$. If $f(\bar{x}) \geq f(x)$ for all $x \in X$, say that $\bar{x}$ maximizes f. If $f(\bar{x}) \leq f(x)$ for all $x \in X$, say that $\bar{x}$ minimizes f.*

The discussion here focuses on the problem of maximizing f (to minimize f, maximize $-f$). Furthermore, throughout the discussion only unconstrained optimization is considered: at any point x, one can move in any direction, to alternative choices. That fact allows one to use derivatives to characterize optimal choices. (An example of a constrained optimization problem is: $\max f(x) = x^2$ subject to the constraint that $0 \leq x \leq 1$. In that case, the maximum occurs at $x = 1$, where $f'(x) = 2$.)

To motivate some of the key ideas, consider Figure 6.1, where the function f is increasing to the left of b (at point a, for example), is flat at b and decreasing to the right of b (such as at c). Corresponding to these movements, the derivative is positive at a, 0 at b and negative at c. The function is highest at b (and so is maximized at b) and this corresponds to the point where the derivative is 0. When the derivative is not 0, such as at a or c, rightward or leftward movements raise the value of the function. So, a first derivative equal to 0 is a necessary condition for a point to be a candidate for maximizing the function. Note also that, in this example, the derivative is *decreasing*. If the derivative is decreasing, then the slope of the derivative is negative – the derivative of the first derivative is negative or, equivalently, the second derivative is negative. So, in this case, a point maximizes the function if the derivative of the function at that point is 0, and in addition, the second derivative is negative.

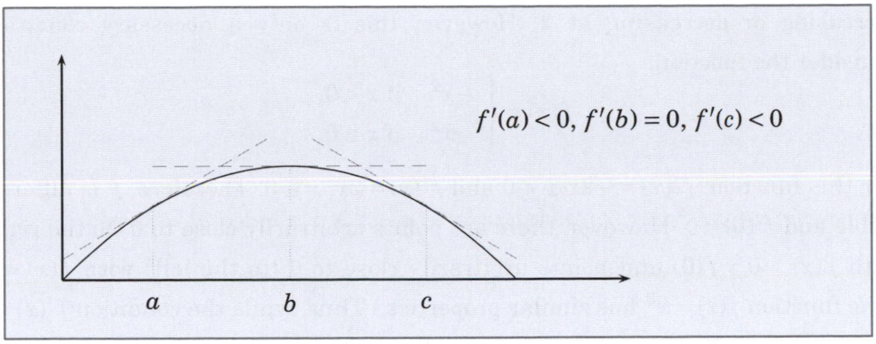

Figure 6.1: Changing slopes

A necessary condition for a point $\bar{x}$ to be an optimum is that $f'(\bar{x}) = 0$, so that the function is flat at $\bar{x}$: if $f'(\bar{x}) > 0$ increasing x raises the value of the function, and if $f'(\bar{x}) < 0$ reducing x raises the value of the function. This observation may be expressed mathematically.

Theorem 6.1: *Let f be a differentiable function. Then if $\bar{x}$ maximizes $f(x)$, $f'(\bar{x}) = 0$.*

PROOF: Suppose that $f'(\bar{x}) = k \neq 0$. The derivative of f at $\bar{x}$ is $\lim_{\Delta x \to 0} \frac{f(\bar{x}+\Delta x) - f(\bar{x})}{\Delta x} = k$. If $k > 0$, taking $\Delta x > 0$ and small implies that $f(\bar{x}+\Delta x) - f(\bar{x}) > 0$; and if $k < 0$, taking $\Delta x < 0$ and small implies that $f(\bar{x}+\Delta x) - f(\bar{x}) > 0$. In either case, $\bar{x}$ is not a maximizer of f. ■

If no point $\bar{x}$ exists with $f'(\bar{x}) = 0$, then the function has no maximum. Of course, if the function is not differentiable, then it may have a maximum at a point where the derivative is not defined. For example $f(x) = \min\{x, 1-x\}$ is a tent shaped function with a maximum at $\frac{1}{2}$, but the function is not differentiable at $\frac{1}{2}$. While $f'(\bar{x}) = 0$ is a necessary condition for a maximum of a differentiable function, it is not sufficient. Figure 6.1 gives a case where the function has a unique point, $\bar{x}$, where the first derivative is 0 and this point is the point that maximizes f. The condition, $f'(\bar{x}) = 0$, that the slope is zero, is called the *first-order condition* for an optimum.

6.2.1 The first-order condition: issues

The first-order condition is a necessary condition for a point $\bar{x}$ to be either a maximum or a minimum, for if $f'(\bar{x}) \neq 0$ it means that the function is either increasing or decreasing at $\bar{x}$. However, this is only a necessary condition. Consider the function:

$$f(x) = \begin{cases} -x^2 & \text{if } x < 0, \\ x^2 & \text{if } x \geq 0. \end{cases}$$

For this function, $f'(x) = -2x, x < 0$ and $f'(x) = 2x$, $x \geq 0$. Therefore, f is differentiable and $f'(0) = 0$. However, there are points arbitrarily close to 0 (to the right) with $f(x) > 0 = f(0)$ and points arbitrarily close to 0 (to the left) with $f(x) < 0$. (The function $f(x) = x^3$ has similar properties.) Thus, while the condition $f'(x) = 0$ locates an x such that the function f is flat at that point, additional conditions must be checked to determine if x is optimal.

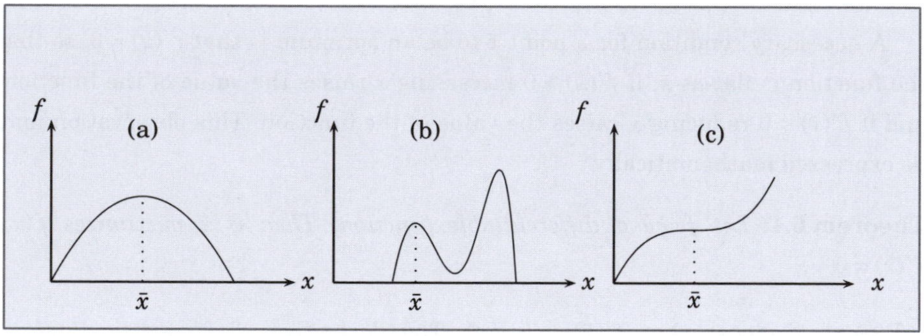

Figure 6.2: Optimization: some possibilities I

Figure 6.2 illustrates some of the many possibilities where a solution to $f'(x) = 0$ may or may not yield a maximum. In Figure 6.2, case (a) is "well-behaved", in that there is a unique point where $f'(x) = 0$, and this corresponds to a global maximum. In case (b), at $\bar{x}$, $f'(\bar{x}) = 0$, but while $\bar{x}$ is a local maximum, it is not a global maximum. In case (c), the point $\bar{x}$ is called a point of inflection, where the function is increasing, then flattens and then continues increasing.

The actual functions plotted in Figure 6.2 are:

(a) $f(x) = 3x - x^2$, (b) $f(x) = 1 - 4x(x-1)(x-2)^2(x-3)$, (c) $f(x) = 3x - 6x^2 + 4x^3$.

REMARK 6.1: It may be worth noting that $f'(x) = 0$ is consistent with the function being strictly increasing at x (f is strictly increasing if $y > x$ implies $f(y) > f(x)$). Considering case (c), $f(x) = 3x - 6x^2 + 4x^3$, $f'(\bar{x}) = 0$ but $\bar{x}$ is not a local maximum. In fact, this function is strictly increasing, despite the fact that $f'(\bar{x}) = 0$. To see this, observe that $f'(x) = 3 - 12x + 12x^2 = 3 - 12x(1-x)$, so that at $\bar{x} = \frac{1}{2}$, $f'\left(\frac{1}{2}\right) = 3 - 12\left(\frac{1}{4}\right) = 0$. Observe that $\bar{x} = \frac{1}{2}$, $f(\bar{x}) = \frac{1}{2}$, while for $x < \bar{x}$, $f(x) < \frac{1}{2}$ and for $x > \bar{x}$, $f(x) > \frac{1}{2}$, so f is a strictly increasing function. To see this more clearly,

$$\begin{aligned} f(x+z) - f(x) &= [3(x+z) - 6(x+z)^2 + 4(x+z)^3] - [3x - 6x^2 + 4x^3] \\ &= 3z - 12xz - 6z^2 + 12x^2z + 12xz^2 + 4z^3 \\ &= z[3 - 12x + 12x^2 - 6z + 12xz + 4z^2] \\ &= z[3 - 12x(1-x) + z(-6 + 12x + 4z)] \\ f\left(\tfrac{1}{2}+z\right) - f\left(\tfrac{1}{2}\right) &= 4z^3. \end{aligned}$$

Thus, $z > 0$ implies $f\left(\frac{1}{2}+z\right) - f\left(\frac{1}{2}\right) > 0$, and $z < 0$ implies $f\left(\frac{1}{2}+z\right) - f\left(\frac{1}{2}\right) < 0$, so the function f is strictly increasing. □

The next group of functions illustrate additional possibilities.

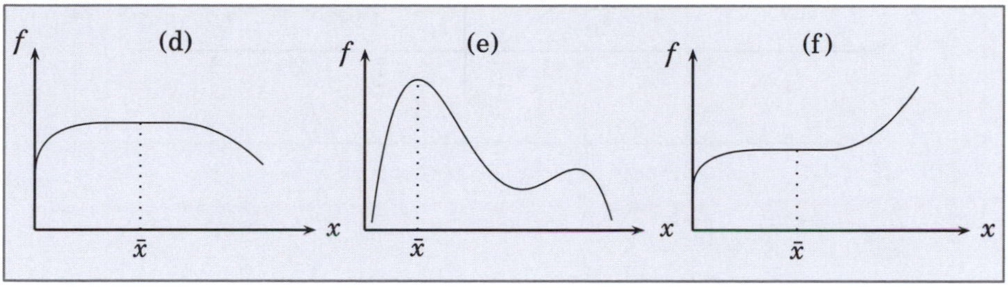

Figure 6.3: Optimization: some possibilities II

The functions plotted in Figure 6.3 are:

$$\text{(d)} \qquad f(x) = \begin{cases} 2\sqrt{x} - x + 1, & 0 \le x < 1 \\ 2, & 1 \le x < 2 \\ 2x\left(1 - \frac{1}{4}x\right), & x \ge 2, \end{cases}$$

$$\text{(e)} \qquad f(x) = (10x^3 - 40x^2 + 40x - 5)(2 - x) + 5,$$

$$\text{(f)} \qquad f(x) = \begin{cases} 2\sqrt{x} - x + 1, & 0 \le x < 1 \\ 2, & 1 \le x < 2 \\ (x-2)^2 + 2, & x \ge 2. \end{cases}$$

Observe that in all cases depicted in the figures $f'(\bar{x}) = 0$. However, in cases (b), (c) and (f), $\bar{x}$ does not give a global maximum of f. The following numerical examples further illustrate some of these functional forms.

Example 6.1 gives a function with three points having a derivative of 0.

EXAMPLE 6.1: Consider the function $f(x) = x^2(12 - x - x^2)$. The function has derivative $f'(x) = 24x - 3x^2 - 4x^3$ and $f'(x) = 0$ at $x = -2.853$, $x = 0$, and $x = 2.103$. The second derivative is $f''(x) = 24 - 6x - 12x^2$ and $f''(-2.853) = -56.559$, $f''(0) = 24$, and $f''(2.103) = -41.691$. The points $x = -2.853$, and $x = 2.103$ are local maxima; the point 0 is a local minimum. At the local maxima the second derivative is negative; at the local minimum the second derivative is positive. Finally, the point $x = -2.853$ is a global maximum.

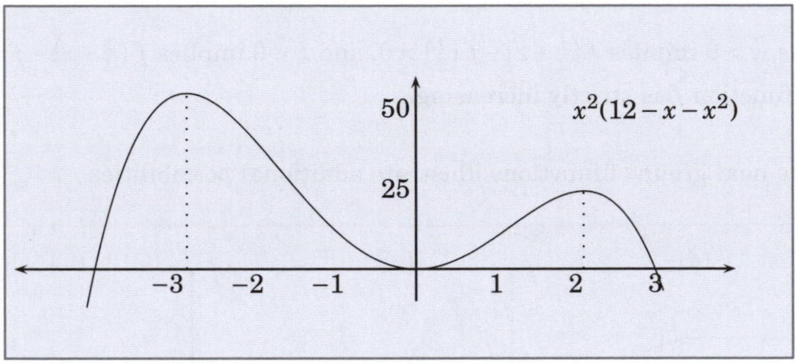

◇

Example 6.2 gives a function with just one point, $x = 1$, having a derivative of 0. The function decreases to the left and increases to the right so $x = 1$ is neither a local maximum or a local minimum. The point is an *inflection* point.

EXAMPLE 6.2: Consider the function $f(x) = 1 + (x - 1)^3$. The function has $f'(x) = 3(x - 1)^2$ and $f''(x) = 6(x - 1)$. At $x = 1$, $f'(x) = 0$, but this point is neither a minimum or maximum.

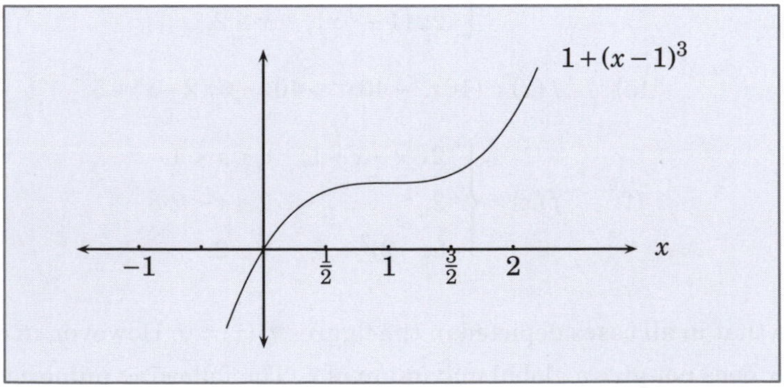

◇

In Figure 6.4, some figures for the minimization case are depicted, identifying unique minima, local minima and inflection points.

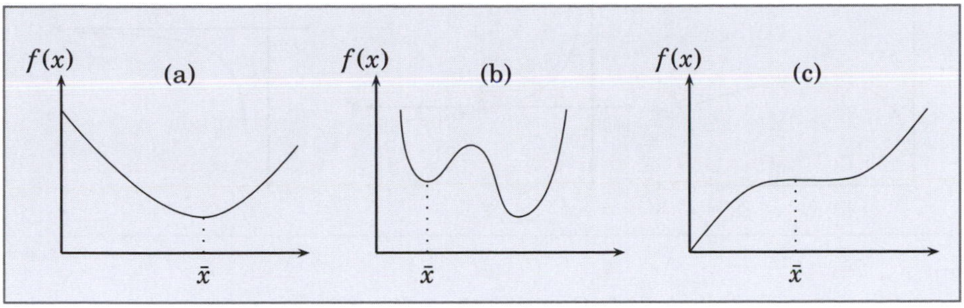

Figure 6.4: Minimization and inflection points

The main observation to be drawn from these examples is that for a local maximum or minimum, the first derivative must be 0 and the second derivative positive or negative, depending on whether the candidate point is a local minimum or maximum. Considering Figure 6.1, the function increases to the point where it is at a maximum, and then decreases. This property is satisfied at $\bar{x}$ in Figure 6.2 (a) and (b); and in Figure 6.3 (e). Around $\bar{x}$, f' is positive on the left and negative on the right. So, f' decreases as x increases: f' is a decreasing function around $\bar{x}$: $f'' < 0$. This is illustrated by the following function (defined for $x \geq 0$ and with first and second derivatives defined for $x > 0$):

$$f(x) = 3\sqrt{x} - x, \quad \text{with} \quad f'(x) = \frac{3}{2\sqrt{x}} - 1, \quad \text{and} \quad f''(x) = -\frac{3}{4x^{\frac{3}{2}}}.$$

The maximizing value of x is the solution to $f'(x) = 0$, giving:

$$\frac{3}{2\sqrt{x}} - 1 = 0 \quad \text{or} \quad 3 = 2\sqrt{x} \quad \text{or} \quad \sqrt{x} = \left(\tfrac{3}{2}\right) \quad \text{or} \quad x = \left(\tfrac{3}{2}\right)^2 = \tfrac{9}{4} = 2\tfrac{1}{4}.$$

These considerations motivate the second-order condition, giving a sufficient condition for a local maximum or minimum.

6.2.2 The second-order condition

Intuitively, if $f''(x) < 0$, then the slope of the function is decreasing as x increases, so that around a point where $f'(x) = 0$, with Δ a small positive number: $f'(x - \Delta) > f'(x) > f'(x + \Delta)$, so that the graph of f' appears as in Figure 6.5 (ii). These ideas can be formalized using Taylor series expansions.

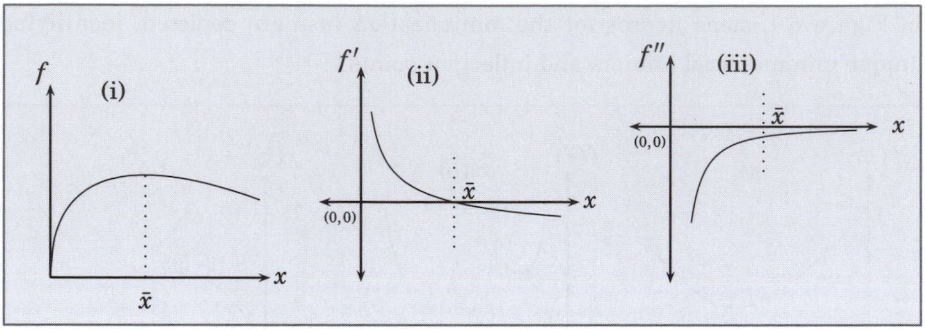

Figure 6.5: Optimization: $f(x) = 3\sqrt{x} - x$

Theorem 6.2: *Suppose that f is a continuous function with continuous first and second derivatives, $f'(x)$ and $f''(x)$ (these are continuous functions of x). Then*

$$f(\bar{x} + \Delta x) = f(\bar{x}) + f'(\bar{x})\Delta x + \tfrac{1}{2} f''(p)(\Delta x)^2 \qquad (6.1)$$

for some p between $\bar{x}$ and $\bar{x} + \Delta x$. (That is, p satisfies $\bar{x} - |\Delta x| \le p \le \bar{x} + |\Delta x|$.)

(This theorem is discussed in detail later.) Because Δx may be positive or negative, p between $\bar{x}$ and $\bar{x} + \Delta x$ means that $\bar{x} - |\Delta x| \le p \le \bar{x} + |\Delta x|$. With this result, necessary and sufficient conditions may be given for a local optimum.

Theorem 6.3: *Let f be a continuous function with continuous first and second derivatives. Suppose that $f'(\bar{x}) = 0$. Then:*

1. *$f''(\bar{x}) < 0$ implies that $\bar{x}$ is a local maximum.*

2. *$f''(\bar{x}) > 0$ implies that $\bar{x}$ is a local minimum.*

PROOF: Consider the first case and suppose that $f'(\bar{x}) = 0$ and $f''(\bar{x}) < 0$. Since f'' is continuous, there is an $\epsilon > 0$ such that $f''(z) < 0$ for all z with $\bar{x} - \epsilon < z < \bar{x} + \epsilon$. Pick Δx so that $|\Delta x| < \epsilon$. Then, from Theorem 6.2,

$$\begin{aligned} f(\bar{x} + \Delta x) &= f(\bar{x}) + f'(\bar{x})\Delta x + \tfrac{1}{2} f''(p)(\Delta x)^2 \\ &= f(\bar{x}) + \tfrac{1}{2} f''(p)(\Delta x)^2, \quad \text{since } f'(\bar{x}) = 0, \end{aligned}$$

with p satisfying:

$$\bar{x} - \epsilon < \bar{x} - |\Delta x| < p < \bar{x} + |\Delta x| < \bar{x} + \epsilon.$$

Because $\bar{x} - \epsilon < p < \bar{x} + \epsilon$, $f''(p) < 0$, and therefore $\frac{1}{2} f''(p)(\Delta x)^2 < 0$. Consequently, $f(\bar{x} + \Delta x) < f(\bar{x})$ when $|\Delta x| < \epsilon$. The second part of the theorem is proved in the same way. ■

The condition $f''(x) < 0$ is called a *second-order* condition.

REMARK 6.2: Suppose that $f'(\bar{x}) = 0$ but f'' (only) satisfies $f''(x) \leq 0$ in a *neighborhood* of $\bar{x}$ (rather than $f''(x) < 0$). That is, for some ϵ, if $\bar{x} - \epsilon < z < \bar{x} + \epsilon$, then $f''(z) \leq 0$. If $|\Delta x| < \epsilon$, then according to the Taylor series expansion, $f(x + \Delta x) = f(\bar{x}) + \frac{1}{2} f''(p)(\Delta x)^2$, with $f''(p) \leq 0$, so that $f(x + \Delta x) \leq f(\bar{x})$. But the inequality may not be strict. In this case $\bar{x}$ is a local maximum, but it may not be unique (in any neighborhood of $\bar{x}$ there may be $x' \neq \bar{x}$ and $f(x') = f(\bar{x})$).

Conversely, if $f'(\bar{x}) = 0$ and f'' satisfies $f''(x) \geq 0$ in a *neighborhood* of $\bar{x}$, then $\bar{x}$ is a local minimum. Again, if $f''(\bar{x}) > 0$, then continuity of f'' implies that $f''(p) > 0$ for p sufficiently close to $\bar{x}$, and then $f(x + \Delta x) > f(\bar{x})$, implying that $\bar{x}$ is a unique local minimum. If, in fact, $f''(x) \leq 0$ for all x, then $f'(\bar{x}) = 0$ implies that $\bar{x}$ is a global maximum. This observation is developed in Section 6.3. The property $f''(x) \leq 0$ for all x is closely related to concavity (see Section 6.3 and Theorem 6.6). □

6.2.3 Applications: profit maximization and related problems

A simple application of these results is the monopoly revenue optimization problem, where inverse demand is given by $p = p(Q)$ and the cost by $c(Q)$. Thus, total revenue is $R(Q) = Qp(Q)$ and profit is $\pi(Q) = R(Q) - c(Q)$. These functions are depicted in Figure 6.6.

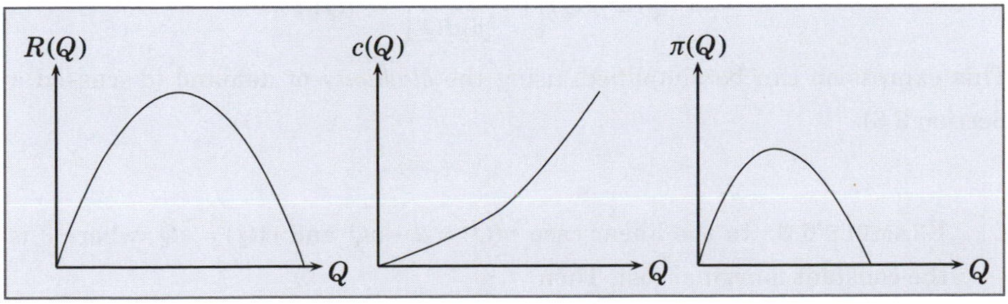

Figure 6.6: Revenue, cost and profit

The first-order condition for profit maximization is:

$$\pi'(Q) = R'(Q) - c'(Q) = 0.$$

If we assume that $R''(Q) \leq 0$ and $c''(Q) \geq 0$ for all Q, then the profit function has a non-positive second derivative: $\pi''(Q) = R''(Q) - c''(Q) \leq 0$. In this case, to find a

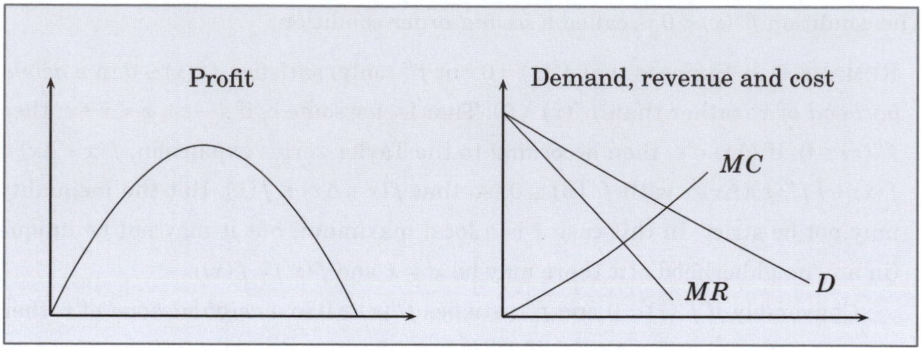

Figure 6.7: Marginal cost and marginal revenue

maximum of π it is sufficient to find a value $\bar{Q}$ such that $\pi'(\bar{Q}) = 0$. The intuition is simple: increase output to the point where the increase in revenue matches the increase in cost. (Beyond that point the increase in cost is greater.) This yields the profit-maximizing rule: set marginal revenue equal to marginal cost. Figure 6.7 depicts the profit-maximizing choice.

The calculations may be taken a step further. With $\pi(Q) = R(Q) - c(Q) = Qp(Q) - c(Q)$, differentiating gives

$$\pi'(Q) = [p(Q) + Qp'(Q)] - c'(Q)$$

and rearranging:

$$\pi'(Q) = p(Q)\left[1 + \frac{Q}{p}\frac{dp}{dQ}\right] - c'(Q).$$

This expression can be simplified, using the *elasticity* of demand (discussed in Section 6.6).

EXAMPLE 6.3: In the linear case $p(Q) = a - bQ$ and $c(Q) = cQ$, where c is the constant marginal cost. Then

$$R(Q) = Qp(Q) = Q(a - bQ) = aQ - bQ^2$$

and

$$\pi(Q) = aQ - bQ^2 - cQ = (a - c)Q - bQ^2.$$

Also, $\pi'(Q) = (a-c) - 2bQ = R'(Q) - c'(Q) = a - 2bQ - c$. Setting $\pi'(Q) = (a-c) - 2bQ = 0$, the solution for Q is $Q_m = \frac{(a-c)}{2b}$. The second-order condition is $\pi''(Q) \leq 0$ and in the linear case $\pi''(Q) = -2b < 0$. Therefore, $Q_m = \frac{(a-c)}{2b}$

maximizes π. This may be compared with the competitive model, where quantity (Q_c) is such that price equals marginal cost: $p(Q) = c'(Q)$. With linearity $a - bQ = c$ or $Q_c = \frac{(a-c)}{b}$ for the competitive output. $\diamondsuit$

EXAMPLE 6.4: Let $f(x) = \beta \ln(\alpha + x) - (\alpha + x)^2$, where $\alpha, \beta > 0$. Then

$$f'(x) = \frac{\beta}{\alpha + x} - 2(\alpha + x)$$

and

$$f''(x) = -\frac{\beta}{(\alpha + x)^2} - 2 < 0.$$

Solving for the first-order condition: $0 = f'(x) = \frac{\beta}{\alpha+x} - 2(\alpha + x)$ gives $\frac{\beta}{2} = (\alpha + x)^2$ and so $\sqrt{\frac{\beta}{2}} = (\alpha + x)$ and $x = \sqrt{\frac{\beta}{2}} - \alpha$ (assume $\sqrt{\frac{\beta}{\alpha}} - \alpha > 0$). $\diamondsuit$

The next example is a variation of the monopoly problem.

EXAMPLE 6.5: A monopolist has demand function $p(Q) = \frac{1}{Q^\alpha} = Q^{-\alpha}$, $0 < \alpha < 1$, and linear cost function $c(Q) = cQ$. The monopolist wishes to maximize profit, π, (revenue minus cost). Thus,

$$\pi(Q) = p(Q)Q - cQ = Q^{-\alpha}Q - cQ = Q^{1-\alpha} - cQ.$$

So,

$$\pi'(Q) = (1 - \alpha)Q^{-\alpha} - c.$$

Setting this to 0 gives $(1-\alpha)Q^{-\alpha} - c = 0$, or $(1-\alpha)Q^{-\alpha} = c$, or $Q^{-\alpha} = \frac{c}{(1-\alpha)}$, so that

$$Q^\alpha = \frac{(1-\alpha)}{c}.$$

Therefore, the value of Q that satisfies $\pi'(Q) = 0$ is $Q = \left(\frac{1-\alpha}{c}\right)^{\frac{1}{\alpha}}$. Is this an optimum? Consider the second derivative, $\pi''(Q) = -\alpha(1-\alpha)Q^{-\alpha-1} = -\alpha(1-\alpha)Q^{-(1+\alpha)}$. Since $0 < \alpha < 1$, $-\alpha(1-\alpha) < 0$ so that $-\alpha(1-\alpha)Q^{-(1+\alpha)} < 0$. Therefore, the second order condition for a local maximum is satisfied. In fact, in this case, the function is concave and the first-order solution gives a global maximum. $\diamondsuit$

EXAMPLE 6.6: Suppose that (inverse) demand for a product is given by $p(Q) = Q^{-\alpha}$, where α is a number between 0 and 1. There is a single firm supplying the market with cost function $c(Q) = cQ^{\beta}$ where $\beta > 1$. The firm maximizes profit (revenue minus cost) giving the objective:

$$\pi(Q) = p(Q)Q - c(Q) = QQ^{-\alpha} - cQ^{\beta} = Q^{1-\alpha} - cQ^{\beta}.$$

Then

$$\pi'(Q) = (1-\alpha)Q^{-\alpha} - \beta c Q^{\beta-1}$$
$$\pi''(Q) = -\alpha(1-\alpha)Q^{-\alpha-1} - \beta(\beta-1)cQ^{\beta-2}.$$

Since $0 < \alpha < 1$ and $\beta > 1$, $-\alpha(1-\alpha) < 0$ and $-\beta(\beta-2) < 0$, and for $Q > 0$, $Q^{-\alpha-1} > 0$ and $Q^{\beta-2} > 0$. Therefore, for all $Q > 0$, $\pi''(Q) < 0$. Consequently, the solution to $\pi(Q) = 0$ gives a local optimum (in fact, a global optimum). And,

$$\pi'(Q) = (1-\alpha)Q^{-\alpha} - \beta c Q^{\beta-1} = 0 \quad \text{implies} \quad (1-\alpha) - \beta c Q^{\alpha+\beta-1} = 0.$$

Rearranging:

$$Q^{\alpha+\beta-1} = \frac{(1-\alpha)}{\beta c}$$
$$Q = \left[\frac{(1-\alpha)}{\beta c}\right]^{\frac{1}{\alpha+\beta-1}}.$$

◊

EXAMPLE 6.7: Consider a tree harvesting problem. Suppose that the cost of planting a tree is c. If the tree is sold when harvested at age t, its value is $f(t)$. Therefore, the present discounted value (PDV) of a tree harvested at age t is $P(t) = e^{-rt}f(t) - c$, where r is the interest rate.

The first-order condition for present value maximization is

$$P'(t) = 0 = e^{-rt}f'(t) - re^{-rt}f(t) = 0$$

(using the product rule). This may be written $P'(t) = 0 = e^{-rt}[f'(t) - rf(t)] = 0$, and since $e^{-rt} \neq 0$ this gives $[f'(t) - rf(t)] = 0$ as the first-order condition. Since no functional form has been specified (yet) for $f(t)$, this cannot be solved.

The second-order condition, which ensures that the solution above gives a maximum, is $P''(t) < 0$, and here this requires that

$$P''(t) = -re^{-rt}f'(t) + e^{-rt}f''(t) + r^2e^{-rt}f(t) - re^{-rt}f'(t) < 0,$$

which simplifies to

$$e^{-rt}\{-rf'(t) + f''(t) + r^2 f(t) - rf'(t)\} < 0,$$

or, since $e^{-rt} \neq 0$,

$$\{r^2 f(t) - 2rf'(t) + f''(t)\} < 0.$$

To consider a specific example, suppose that $f(t) = \sqrt{t} = t^{\frac{1}{2}}$. In this case

$$0 = [f'(t) - rf(t)] = \frac{1}{2}t^{-\frac{1}{2}} - rt^{\frac{1}{2}}.$$

Multiply by $t^{\frac{1}{2}}$ to obtain $\frac{1}{2} - rt = 0$, or $t = \frac{1}{2r}$. Since $f(t) = \sqrt{t} > 0, f'(t) = \frac{1}{2}\frac{1}{\sqrt{t}} > 0$ and $f''(t) = -\frac{1}{4}t^{-(\frac{3}{2})} < 0$, the second-order condition is:

$$r^2 t^{\frac{1}{2}} - rt^{-\frac{1}{2}} - \frac{1}{4}t^{-\frac{3}{2}} < 0.$$

Multiplying this by $t^{\frac{1}{2}}$ leaves the sign unchanged, so the condition is equivalent to:

$$r^2 t - r - \frac{1}{4}t^{-1} < 0.$$

And, substituting $t = \frac{1}{2r}$ gives the condition:

$$\frac{r}{2} - r - \frac{1}{2}r = -r < 0.$$

The second-order condition is satisfied. ◊

The circumstances under which the second-order conditions hold are closely related to the shape of the function.

6.3 Concavity and optimization

One issue with the use of second-order conditions is that such conditions do not determine whether $\bar{x}$ is a local or a global maximum. In cases (a) and (b) of Figure 6.2, the conditions $f'(\bar{x}) = 0$ and $f''(\bar{x}) < 0$ are satisfied but case (a) corresponds to a global maximum while case (b) corresponds to a local maximum.

As the figures plotted earlier show, the overall shape of the function f is crucial in determining whether these properties of the derivatives of the function identify local or global maxima. In Figure 6.2, the function depicted in (a) has a shape that leads to the solution to $f'(x) = 0$ identifying a global maximum. This function is an example of a concave function.

Definition 6.2: *A function f is concave if for any x, y and any $\theta \in [0,1]$*

$$f(\theta x + (1-\theta)y) \geq \theta f(x) + (1-\theta)f(y).$$

A concave function is "dome-shaped", as in Figure 6.8.

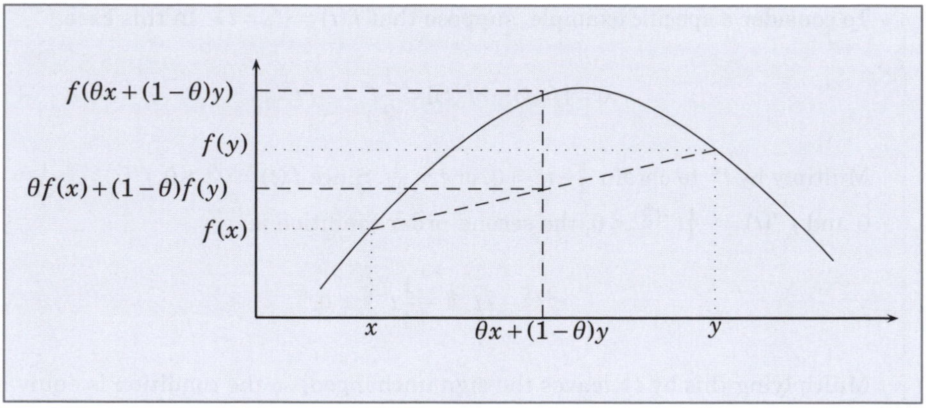

Figure 6.8: Concavity

When the shape is "reversed" the function is convex. Convexity is defined:

Definition 6.3: *A function g is convex if for any x, y and any $\theta \in [0,1]$*

$$g(\theta x + (1-\theta)y) \leq \theta g(x) + (1-\theta)g(y).$$

A convex function is "bowl-shaped" – the reverse of a concave function. Intuitively, a concave function is shaped like a hill; a convex function is shaped like a valley. Figure 6.9 depicts both types of function. A more extensive discussion of concavity and convexity is given in Chapter 8, when Taylor series expansions are considered in detail. (The functions plotted in Figure 6.9 are $f(x) = 5\ln(1+x) - x$ and $g(x) = -15x + 8e^x - 5$.) The properties of concavity and convexity are closely related to global optima.

Theorem 6.4: *If f is concave and $f'(\bar{x}) = 0$, then $\bar{x}$ is a global maximum. If f is convex and $f'(\bar{x}) = 0$, then $\bar{x}$ is a global minimum.*

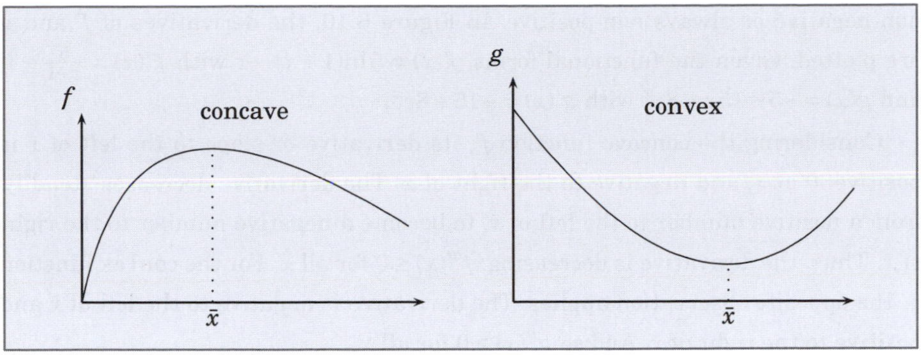

Figure 6.9: Concave and convex functions

PROOF: Considering the concave case, suppose that f is concave and $\bar{x}$ satisfies $f'(\bar{x}) = 0$ but $\bar{x}$ is not a global maximum. Then there is some y such that $f(y) > f(\bar{x})$. Using concavity of f:

$$f(\theta y + (1-\theta)\bar{x}) \geq \theta f(y) + (1-\theta) f(\bar{x}), \ \theta \in [0,1].$$

Rearrange this expression as

$$f(\theta y + (1-\theta)\bar{x}) - f(\bar{x}) \geq \theta[f(y) - f(\bar{x})], \theta \in [0,1].$$

Since $\theta y + (1-\theta)\bar{x} = \bar{x} + \theta(y - \bar{x})$, $f(\bar{x} + \theta(y - \bar{x})) - f(\bar{x}) \geq \theta[f(y) - f(\bar{x})]$, $\theta \in [0,1]$ or

$$\frac{f(\bar{x} + \theta(y - \bar{x})) - f(\bar{x})}{\theta} \geq [f(y) - f(\bar{x})],$$

and, multiplying and dividing the left side by $y - \bar{x}$, this may be written

$$(y - \bar{x}) \left[\frac{f(\bar{x} + \theta(y - \bar{x})) - f(\bar{x})}{\theta(y - \bar{x})} \right] \geq [f(y) - f(\bar{x})].$$

Finally, letting $\theta \to 0$, gives

$$(y - \bar{x}) f'(\bar{x}) \geq [f(y) - f(\bar{x})] > 0$$

(here $\Delta x = \theta(y - \bar{x})$). This contradicts the fact that $f'(\bar{x}) = 0$. Therefore, concavity of f and the condition $f'(x) = 0$ imply that $\bar{x}$ is a global maximum. A similar argument shows that with convexity, the condition $f'(x) = 0$ is sufficient for a global minimum. ■

In Figure 6.9, the concave function, f, has a maximum at $\bar{x}$; the convex function, g, has a minimum at $\bar{x}$. A key feature of either a concave function or a convex function is that the second derivative never changes sign: it is always non-negative or always non-positive. In Figure 6.10, the derivatives of f and g are plotted. Given the functional forms, $f(x) = 5\ln(1+x) - x$ with $f'(x) = \frac{5}{x+1} - 1$, and $g(x) = -5 - 15x + 8e^x$ with $g'(x) = -15 + 8e^x$.

Considering the concave function f, its derivative or slope to the left of $\bar{x}$ is positive, 0 at $\bar{x}$, and negative to the right of $\bar{x}$. The derivative decreases steadily, from a positive number to the left of $\bar{x}$, to become a negative number to the right of $\bar{x}$. Thus, the derivative is decreasing: $f''(x) \le 0$ for all x. For the convex function g, the opposite observation applies. The derivative is negative to the left of $\bar{x}$ and positive to the right of $\bar{x}$. And so $g''(x) \ge 0$ for all x.

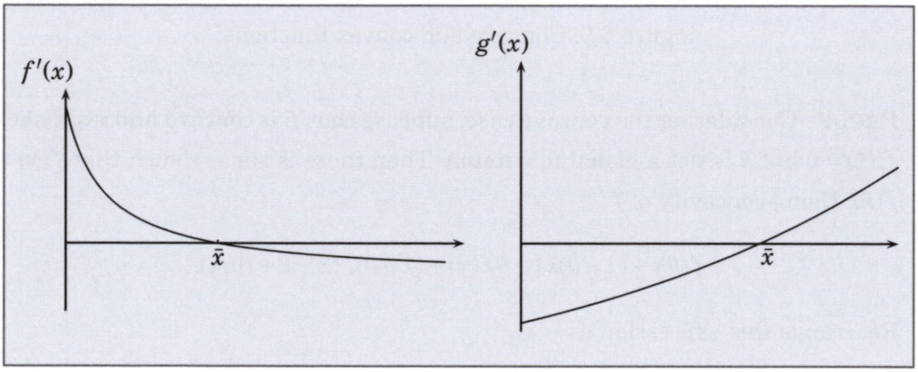

Figure 6.10: Plot of derivatives $f'(x)$, $g'(x)$

Notice that with concavity, $f'(\bar{x}) = 0$ is sufficient for a global maximum: whether $f''(\bar{x}) \le 0$ or $f''(\bar{x}) < 0$ is irrelevant. Recall in Figure 6.3(d), the function depicted there is concave but any number x in the interval $[1,2]$ is an optimum. While the function is concave, it is not strictly concave.

Definition 6.4: *A function f is* strictly concave *if for $x \neq y$*

$$f(\theta x + (1-\theta)y) > \theta f(x) + (1-\theta)f(y),\ \theta \in (0,1).$$

Strict concavity guarantees a unique optimum. (The definition of strict convexity reverses these strict inequalities.)

Theorem 6.5: *If f is a strictly concave function then f has a unique maximum.*

PROOF: Suppose that $\bar{x}_1$ and $\bar{x}_2$ are two maxima. Then

$$f(\theta\bar{x}_1 + (1-\theta)\bar{x}_2) > \theta f(\bar{x}_1) + (1-\theta)f(\bar{x}_2),\ \theta \in (0,1)$$

and this contradicts the assumption that $\bar{x}_1$ and $\bar{x}_2$ are maxima. ∎

REMARK 6.3: To summarize the discussion so far, if $f'(x)=0$ and $f'(x)<0$ or $f''(x)\le 0$ on a neighborhood of x, then x is a local maximum. And if f is concave, $f'(x)=0$ implies that x is a global maximum, while strict concavity implies that x is a unique global maximum. □

The following discussion clarifies the relation between concavity and the second derivative. The main conclusions are that concavity implies that $f''(x)\le 0$ for all x (proved in a later chapter); and $f''(x)\le 0$ for all x implies concavity. If $f''(x)<0$ for all x then f is strictly concave, but strict concavity of f does *not* imply that $f''(x)<0$ for all x. (Note that the proof of this theorem requires integration, a subject not developed until later.)

Theorem 6.6: *Given the function f, suppose that $f''(x)<0$ everywhere, then the function is strictly concave. If $f''(x)\le 0$ for all x, the function is concave.*

PROOF: With $f''(x)<0$ for all x, if $y>x$, $f'(y)-f'(x)=\int_x^y f''(\xi)d\xi<0$, so that $y>x$ implies $f'(y)<f'(x)$, and $f'(\cdot)$ is a strictly decreasing function. So, for $y>x$,

$$f(y)-f(x)=\int_x^y f'(\xi)d\xi<f'(x)[y-x] \tag{6.2}$$

$$f(y)-f(x)=\int_x^y f'(\xi)d\xi>f'(y)[y-x], \tag{6.3}$$

since $f'(x)>f'(z)>f'(y)$ for all z with $x<z<y$. Taking $x<y$, define $z(\theta)=\theta x+(1-\theta)y$, $0<\theta<1$. From Equation 6.3, since $x<z(\theta)<y$,

$$f(z(\theta))-f(x)>f'(z(\theta))[z(\theta)-x].$$

From Equation 6.2

$$f(y)-f(z(\theta))<f'(z(\theta))[y-z(\theta)].$$

Rearranging these expressions:

$$f(z(\theta))>f(x)+f'(z(\theta))[z(\theta)-x] \tag{6.4}$$

$$f(z(\theta))>f(y)+f'(z(\theta))[z(\theta)-y]. \tag{6.5}$$

Multiplying these expressions by θ and $1-\theta$, respectively, and adding gives:

$$[\theta+(1-\theta)]f(z(\theta))>\theta f(x)+(1-\theta)f(y)+f'(z(\theta))\{\theta[z(\theta)-x]+(1-\theta)[z(\theta)-y]\}. \tag{6.6}$$

Since $[\theta+(1-\theta)]=1$ and $\theta[z(\theta)-x]+(1-\theta)[z(\theta)-y]=z(\theta)-[\theta x+(1-\theta)y]=0$,

$$f(z(\theta))>\theta f(x)+(1-\theta)f(y). \tag{6.7}$$

Therefore, f is strictly concave if $f''(x)<0$ for all x. If f satisfies the weaker condition $f''(x)\le 0$ for all 0, the previous arguments go through with weak inequalities replacing strict inequalities everywhere to yield: $f(z(\theta))\ge\theta f(x)+(1-\theta)f(y)$. ■

REMARK 6.4: Because $f''(x)\le 0$ for all x implies concavity, the conditions $f'(\bar{x})=0$ and $f''(x)\le 0$ for all x are sufficient to ensure that $\bar{x}$ is a global maximum. □

However, strict concavity of f does not imply an everywhere strictly negative second derivative.

REMARK 6.5: Consideration of the strictly concave function $f(x)=-(x-1)^4$ (Figure 6.11) shows that strict concavity does not imply that $f''(x)<0$ everywhere, since $f'(x)=-4(x-1)^3$ and $f''(x)=-12(x-1)^2$, so that $f''(1)=0$. (That this function is strictly concave is shown in a later chapter.) □

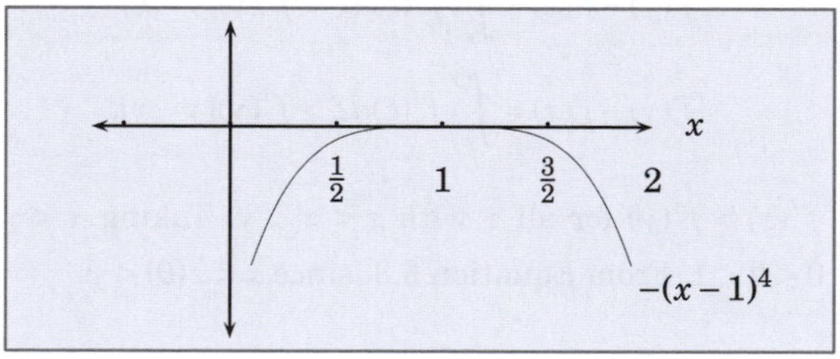

Figure 6.11: Strictly concave function, f, with $f''=0$ at $x=1$

The previous observations may be gathered to give:

Theorem 6.7: *1. A twice differentiable function f is concave if and only if $f''(x)\le 0,\forall x$.*

2. *The function f is convex if and only if $f''(x)\ge 0,\forall x$.*

3. *Strict concavity of f implies that $f''(x)\le 0$ for all x, and $f''(x)<0$ for all x implies strict concavity.*

4. *Strict convexity of f implies that $f''(x)\ge 0$ for all x, and $f''(x)>0$ for all x implies strict convexity.*

There is an analogous result in higher dimensions.

REMARK 6.6: For the multidimensional case, similar results apply. A function f on $\mathscr{R}^n$ is concave if and only if the matrix of second partial derivatives (the Hessian matrix) is negative semi-definite everywhere; strict concavity does not imply negative definiteness, but negative definiteness everywhere implies strict concavity. □

6.3.1 Summary

The following set of points summarizes the previous discussion on optimization.

- $f'(\bar{x}) = 0$ is necessary for a maximum or minimum.
- $f''(\bar{x}) < 0$ and $f'(\bar{x}) = 0$ is sufficient for $\bar{x}$ to be a unique local maximum.
- $f''(x) \leq 0$ for a neighborhood of $\bar{x}$ and $f'(\bar{x}) = 0$ is sufficient for $\bar{x}$ to be a local maximum.
- $f''(x) \leq 0$ for all x and $f'(\bar{x}) = 0$ implies that $\bar{x}$ is a global maximum.
- $f(x)$ concave implies that $f''(x) \leq 0$ for all x.
- $f''(x) < 0$ for all x implies f is strictly concave, which implies that f has a unique global maximum.

6.4 Externalities

In many circumstances, the actions of one individual affect the welfare of another, sometimes in a positive way and sometimes in a negative way – but the individual taking the action does not consider such matters. So the overall impact on welfare resulting from the action is not considered in making the choice and this results in an inefficient outcome. One example of this type of issue arises in the use of public goods. When individuals can use a public facility for free there is either over-use or under-provision of the public facility. This over-use is called the free-rider problem.

6.4.1 The free-rider problem

Consider a public bridge that drivers can use for free. For each individual driver, the benefit from using the bridge is b. With n drivers using the bridge, the cost (in terms of delay) to each driver is $c(n)$, so when there are n drivers in total crossing the bridge the net benefit or utility to *each* of those drivers is

$$u = b - c(n).$$

A driver who does not cross the bridge does not pay the cost or get any benefit and has a utility of 0. Assume that $c(0) = 0$, $c'(n) > 0$ and $c''(n) > 0$ for all $n \geq 0$. Thus, there is little delay with few drivers on the bridge, but delay increases with the number of drivers, getting progressively worse as the congestion increases. (Although in reality n must be an integer, assume for the calculations that n is a real number. If $c'(n) > 0$ is small until n is very large, rounding to the nearest integer will not be a significant issue.)

6.4.1.1 Equilibrium behavior

The marginal social benefit (MSB) from increasing the number of drivers crossing the bridge is b, the extra utility from an extra person making the journey. The social cost from bridge usage is the cost per user times the number of users,

$$s(c) = nc(n),$$

and the total benefit is nb, the benefit per user times the number of users. Therefore, the social welfare is:

$$w(n) = nb - s(c) = nb - nc(n).$$

Maximizing this:

$$w'(n) = b - s'(c) = b - [nc'(n) + c(n)].$$

The intuition for this is that an extra user gets benefit, b. There are two components to marginal or social cost. The extra user incurs cost $c(n)$ and, in addition, the marginal increase in congestion, measured by $c'(n)$, is incurred by all n people, for a total cost $nc'(n)$. This gives a total marginal social cost of $\text{MSC} = s'(c) = nc'(n) + c(n)$.

Socially efficient usage of the bridge occurs when the number of drivers crossing the bridge maximizes total welfare: the sum of utilities less costs of those crossing the bridge. This occurs when $w'(n) = 0$. Let the solution to this equation be n^*. The choice of n^* is the value of n satisfying:

$$[b - c(n)] - nc'(n) = 0$$

or

$$b = c(n^*) + n^* c'(n^*). \tag{6.8}$$

But drivers independently decide whether to use the bridge or not, and usage increases as long as $b > c(n)$, so the equilibrium usage level ($\hat{n}$) is determined as the solution $b - c(n) = 0$. Comparing these, the market equilibrium level of

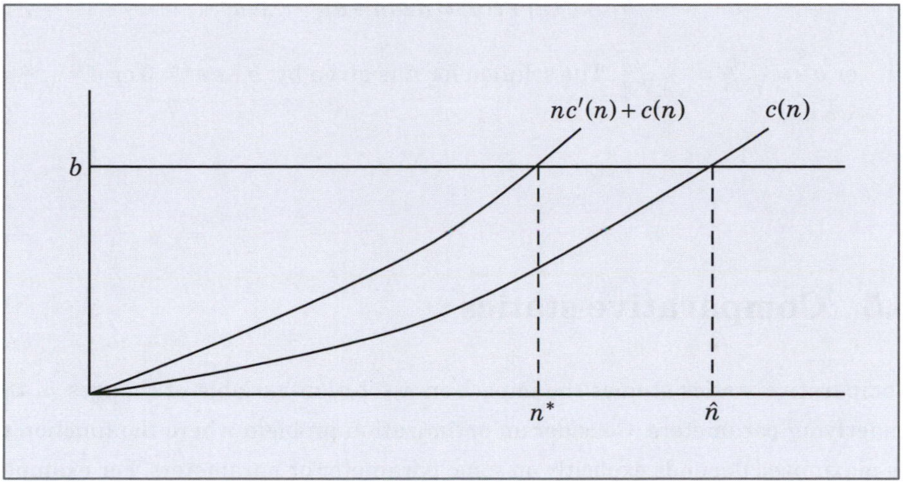

Figure 6.12: Congestion externalities

usage $\hat{n}$ is larger than the socially optimal level, n^* (see Figure 6.12). The welfare implications of the difference between n^* and $\hat{n}$ are clear – usage is excessive from a social point of view.

6.4.1.2 A policy solution

Suppose the government could charge a toll for bridge usage so a driver must pay t to cross the bridge. In particular, suppose the bridge manager solves the program $\max_n w(n)$ and determines that optimal usage is n^*. The manager then charges a toll of $t = n^* c'(n^*)$. So, if n people use the bridge, each individual compares b with $c(n) + n^* c'(n^*)$, the travel cost plus the toll charge. If

$$b > c(n) + n^* c'(n^*),$$

the benefit outweighs the cost and usage increases. By contrast, if

$$b < c(n) + n^* c'(n^*),$$

the opposite is true, and usage will decrease. So, with the tax, equilibrium bridge usage is determined by the value of n that satisfies $b = c(n)+n^* c'(n^*)$, which is n^*.

EXAMPLE 6.8: Consider the case where $c(n) = an^2$. Then $c'(n) = 2an$ and $c''(n) = 2a$. The solution for n^*, the socially optimal level of usage, is determined by

$$b = nc'(n) + c(n) = n2an + an^2 = 3an^2$$

or $n^* = \sqrt{\frac{b}{3a}} = \frac{1}{\sqrt{3}}\sqrt{\frac{b}{a}}$. The solution for $\hat{n}$ is given by: $b - an^2 = 0$ or $\hat{n} = \sqrt{\frac{b}{a}} = \sqrt{3}\,n^*$. ◊

6.5 Comparative statics

Comparative statics studies the impact on a solution variable of changes in the underlying parameters. Consider an optimization problem where the function to be maximized depends explicitly on some parameter or parameters. For example, consider the maximization of the function

$$f(x;a,c) = a \cdot \ln(x+1) - cx.$$

Writing the function this way makes explicit the dependence of the problem on both a and c. Maximizing gives:

$$a\frac{1}{x+1} - c = 0$$

and the second-order condition for a minimum, $-a\frac{1}{(x+1)^2} < 0$, is satisfied. From the first-order condition, the solution is $x(a,c) = \frac{a}{c} - 1$. So, for example, taking c to be a constant, $\frac{dx}{da} = \frac{1}{c}$, giving the impact on the solution of a change in the parameter, a.

In Figure 6.13, the impact of an increase in a to a' is depicted. At parameter value a, the optimal choice of x is at x^*; when a is increased to a', the optimal choice of x moves from x^* to x'. Thus, the change in x per unit change in a is given by:

$$\frac{\Delta x}{\Delta a} = \frac{x' - x^*}{a' - a},$$

measuring the change in the optimal choice resulting from the parameter change.

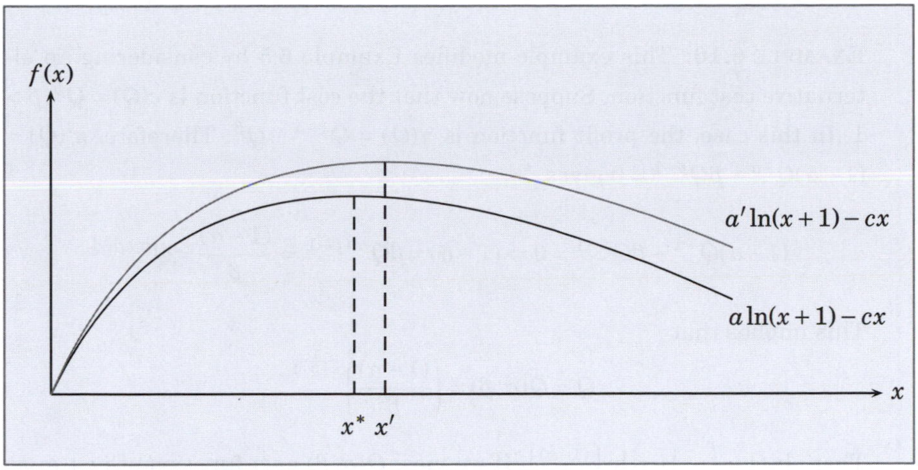

Figure 6.13: Impact of a parameter change on choice variable

The following examples illustrate these ideas.

EXAMPLE 6.9: Consider the monopoly profit maximization (Example 6.5). Note that the solution depends on the parameters c and α, $Q(\alpha,c) = \left(\frac{1-\alpha}{c}\right)^{\frac{1}{\alpha}}$, so that one may consider the impact of variations in α or c on the value of Q. To calculate the derivative $\frac{dQ}{d\alpha}$, we view Q as a function of α alone, taking marginal cost to be a constant. First, to simplify matters, take the ln of Q and use the fact that $\frac{d\ln Q}{d\alpha} = \frac{1}{Q}\frac{dQ}{d\alpha}$ so that $\frac{dQ}{d\alpha} = Q\frac{d\ln Q}{d\alpha}$. Since $\ln Q = \frac{1}{\alpha}\ln\frac{(1-\alpha)}{c}$,

$$\frac{d\ln Q}{d\alpha} = \left(\frac{-1}{\alpha^2}\right)\ln\frac{(1-\alpha)}{c} + \frac{1}{\alpha}\left(\frac{1}{\frac{1-\alpha}{c}}\right)\frac{-1}{c}$$
$$= \left(\frac{-1}{\alpha^2}\right)\ln\left(\frac{(1-\alpha)}{c}\right) - \left(\frac{1}{\alpha(1-\alpha)}\right).$$

Multiplying this expression by $Q = \left(\frac{1-\alpha}{c}\right)^{\frac{1}{\alpha}}$ gives $\frac{dQ}{d\alpha} = Q\frac{d\ln Q}{d\alpha}$:

$$\frac{dQ}{d\alpha} = \left(\frac{(1-\alpha)}{c}\right)^{\frac{1}{\alpha}}\left(\frac{-1}{\alpha^2}\right)\ln\left(\frac{(1-\alpha)}{c}\right) - \left(\frac{(1-\alpha)}{c}\right)^{\frac{1}{\alpha}}\left(\frac{1}{\alpha(1-\alpha)}\right),$$

giving the impact of a change in α on the optimal choice of Q. ◊

The next example generalizes the cost function:

EXAMPLE 6.10: This example modifies Example 6.5 by considering an alternative cost function. Suppose now that the cost function is $c(Q) = Q^{\beta}, \beta > 1$. In this case, the profit function is $\pi(Q) = Q^{1-\alpha} - Q^{\beta}$. Therefore, $\pi'(Q) = (1-\alpha)Q^{-\alpha} - \beta Q^{\beta-1} = 0$ gives

$$(1-\alpha)Q^{-\alpha} - \beta Q^{\beta-1} = 0 \Rightarrow (1-\alpha) = \beta Q^{\alpha+\beta-1} \Rightarrow \frac{(1-\alpha)}{\beta} = Q^{\alpha+\beta-1}.$$

This implies that

$$Q = Q(\alpha, \beta) = \left(\frac{(1-\alpha)}{\beta}\right)^{\frac{1}{\alpha+\beta-1}}.$$

Then, $\ln Q = \left(\frac{1}{\alpha+\beta-1}\right)\ln\left(\frac{(1-\alpha)}{\beta}\right)$. If we view $Q(\alpha,\beta)$ as a function of just α, so that β is held fixed for the entire discussion, then we can consider $\frac{dQ}{d\alpha}$:

$$\frac{dQ}{d\alpha} = -\left(\frac{1}{\alpha+\beta-1}\right)^2 \ln\left(\frac{(1-\alpha)}{\beta}\right) - \left(\frac{1}{\alpha+\beta-1}\right)\left(\frac{1}{\frac{1-\alpha}{\beta}}\right)\left(\frac{1}{\beta}\right),$$

calculated in the same way as in Example 6.9. ◊

Example 6.11 applies the same procedure to the tree harvesting problem (Example 6.7) to consider how the optimal time to harvest varies with the interest rate.

EXAMPLE 6.11: Continuing with the tree harvesting problem of Example 6.7, one may also consider, in the abstract, the effect of variations in the interest rate on the solution value of t. The condition $[f'(t) - rf(t)] = 0$ must hold for any r – so that as r varies, so must t. Therefore,

$$f''(t)dt - f(t)dr - rf'(t)dt = [f''(t) - rf'(t)]dt - f(t)dr = 0.$$

Thus

$$\frac{dt}{dr} = \frac{f(t)}{[f''(t) - rf'(t)]}, \tag{6.9}$$

and if we assume that $f > 0, f' > 0$, so that the value of the tree increases with age and at a diminishing rate so that $f'' < 0$, then this expression is negative.

To present a different perspective, this result may also be derived by approximation. The first-order condition is $f'(t) - rf(t) = 0$. If r varies to

$r+\Delta r$, then t must vary to, say, $t+\Delta t$ so that the first-order condition is satisfied at the new value of r:

$$f'(t+\Delta t)-(r+\Delta r)f(t+\Delta t)=0.$$

Thus,

$$[f'(t+\Delta t)-(r+\Delta r)f(t+\Delta t)]-[f'(t)-rf(t)]=0$$
$$\{f'(t+\Delta t)-f'(t)\}-r\{f(t+\Delta t)-f(t)\}-\Delta r f(t+\Delta t)=0.$$

Multiplying the last expression by $\frac{1}{\Delta t}\frac{\Delta t}{\Delta r}=\frac{1}{\Delta r}$ gives:

$$\frac{f'(t+\Delta t)-f'(t)}{\Delta t}\frac{\Delta t}{\Delta r}-r\frac{f(t+\Delta t)-f(t)}{\Delta t}\frac{\Delta t}{\Delta r}-\Delta r\frac{1}{\Delta r}f(t+\Delta t)=0$$
$$\left[\frac{f'(t+\Delta t)-f'(t)}{\Delta t}\right]\frac{\Delta t}{\Delta r}-r\left[\frac{f(t+\Delta t)-f(t)}{\Delta t}\right]\frac{\Delta t}{\Delta r}-f(t+\Delta t)=0$$

As $\Delta r\to 0$ with $t\to 0$, varying smoothly in r,

$$f''(t)\frac{dt}{dr}-rf'(t)\frac{dt}{dr}-f(t)=0.$$

For example, let $f(t)=kt^{\alpha}$, so that $P(t)=kt^{\alpha}e^{-rt}-c$ and

$$P'(t)=\alpha kt^{\alpha-1}e^{-rt}-rkt^{\alpha}e^{-rt}$$
$$=kt^{\alpha}e^{-rt}[\alpha t^{-1}-r].$$

Setting this to 0 gives $\alpha t^{-1}-r=0$ or $t=\frac{\alpha}{r}$. Then $\frac{dt}{dr}=-\frac{\alpha}{r^2}$.

To illustrate numerically, consider the case where $k=20$, $\alpha=0.9$, $c=70$ and $r=0.05$, so that $f(t)=20t^{0.9}$. The objective is then $P(t)=20t^{0.9}e^{-0.05t}-70$. Therefore, $f'(t)=(0.9)20t^{-0.1}$, $f''(t)=-(0.1)(0.9)20t^{-1.1}$. According to the formula given earlier,

$$\frac{dt}{dr}=\frac{f(t)}{[f''(t)-rf'(t)]}=\frac{20t^{0.9}}{-(0.1)(0.9)20t^{-1.1}-0.05(0.9)20t^{-0.1}}$$
$$=\frac{t^{0.9}}{-(0.1)(0.9)t^{-1.1}-0.05(0.9)t^{-0.1}}$$
$$=-\frac{t}{(0.09)t^{-1}+0.045}.$$

Using the fact that the solution is $t=\frac{\alpha}{r}=\frac{0.9}{0.05}=18$, this formula gives $\frac{dt}{dr}=360$. (Notice that this is the same result as would be obtained from $\frac{dt}{dr}=\frac{0.9}{0.05^2}=\frac{0.9}{0.0025}=360$.) ◊

6.6 Elasticity of demand

The elasticity of demand measures the responsiveness of demand to price changes and is defined as the percentage change in quantity demanded due to a percentage change in price and denoted ϵ:

$$\epsilon = \frac{\%\,\text{change in}\,Q}{\%\,\text{change in}\,P} = \frac{\frac{\Delta Q}{Q}}{\frac{\Delta P}{P}} = \frac{P\Delta Q}{Q\Delta P} \simeq \frac{P}{Q}\frac{dQ}{dP} = \frac{\frac{dQ}{dP}}{\frac{Q}{P}}.$$

REMARK 6.7: Often, because demand is downward sloping, the elasticity is defined as:

$$-\frac{\%\,\text{change in}\,Q}{\%\,\text{change in}\,P} \simeq -\frac{P}{Q}\frac{dQ}{dP},$$

so that the elasticity is expressed as a positive number for downward sloping demand. The elasticity definition consists of a slope term $\frac{dQ}{dP}$ and the relative magnitude of the variables, $\frac{P}{Q}$. For any given $\frac{dQ}{dP}$, the larger is Q relative to P, the smaller is the elasticity. For example, if demand is equal to 10 units and the slope is -1, a unit price increase causes a unit drop in demand, which is 10% of total demand. If, however, demand is 100 units, and the slope is unchanged, a unit price increase again will reduce demand by 1 unit, which is 1% of demand. The demand elasticity reflects the importance of the levels of the variables, whereas the slope does not. □

The calculations above use the fact that $\frac{\frac{a}{b}}{\frac{c}{d}} = \frac{ad}{bc}$. The elasticity approximation "$\simeq$" becomes exact as the change in price becomes small. The elasticity of demand has a graphical interpretation as the ratio of the slope of the demand curve to the angle of the line from the origin to the point on the demand curve (see Figure 6.14).

REMARK 6.8: A point worth noting is that the reciprocal of the elasticity satisfies:

$$\frac{1}{\epsilon} = \frac{Q}{P}\frac{dP}{dQ}.$$

This is the case if

$$\frac{dP}{dQ} = \frac{1}{\frac{dQ}{dP}},$$

since in that case, from the definition of ϵ,

$$\frac{1}{\epsilon} = \frac{1}{\frac{P}{Q}\frac{dQ}{dP}} = \frac{1}{\frac{P}{Q}}\frac{1}{\frac{dQ}{dP}} = \frac{Q}{P}\frac{1}{\frac{dQ}{dP}} = \frac{Q}{P}\frac{dP}{dQ}.$$

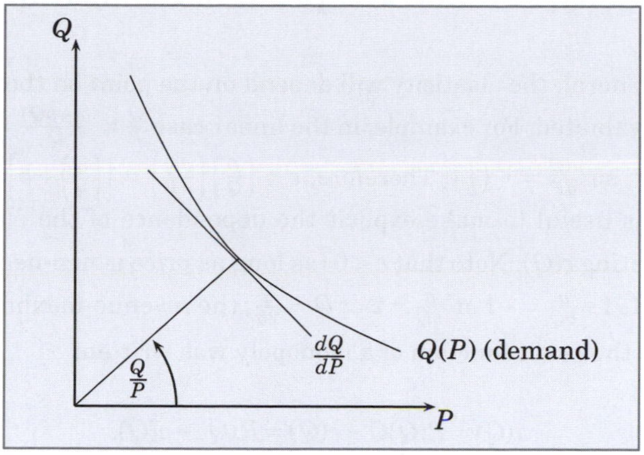

Figure 6.14: Demand elasticity

Because the price function is strictly decreasing (larger Q corresponds to lower P, a downward sloping demand), we can write the relation between quantity and price as $Q = Q(P)$. This gives two alternative representations of the demand function: $P = P(Q)$ and $Q = Q(P)$. If we substitute one into the other we have an identity: $Q \equiv Q(P(Q))$. In words, for any Q, the price at which this quantity will be purchased, $P(Q)$, is the price such that the quantity demanded at this price is Q. Because this equation holds for every Q, we can differentiate to get $\frac{dQ}{dQ} = \frac{dQ}{dP}\frac{dP}{dQ}$ and since $dQ/dQ = 1$, this implies that $\frac{dP}{dQ} = 1/\frac{dQ}{dP}$. Since $\epsilon = \frac{P}{Q}\frac{dQ}{dP}$,

$$\frac{1}{\epsilon} = \frac{1}{\left(\frac{P}{Q}\frac{dQ}{dP}\right)} = \left[\frac{1}{\frac{P}{Q}}\right]\left[\frac{1}{\frac{dQ}{dP}}\right] = \frac{Q}{P}\left[\frac{1}{\frac{dQ}{dP}}\right] = \frac{Q}{P}\frac{dP}{dQ}.$$ □

EXAMPLE 6.12: Let $P(Q) = \ln\left(1+\frac{1}{Q}\right)$. Then $\frac{dP}{dQ} = P'(Q)$ is:

$$\frac{dP}{dQ} = \frac{1}{\frac{1+Q}{Q}} \cdot \left(\frac{-1}{Q^2}\right) = \frac{-1}{(Q+1)Q}.$$

Since $P(Q) = \ln\left(1+\frac{1}{Q}\right)$, $\left(1+\frac{1}{Q}\right) = e^P$ or $\frac{1}{Q} = e^P - 1$ or $Q = \frac{1}{(e^P-1)}$. Thus $\frac{dQ}{dP} = Q'(P)$ is:

$$\frac{dQ}{dP} = \frac{-1}{(e^P-1)^2} \cdot e^P = \frac{-1}{\left(\frac{1}{Q}\right)^2} \cdot \left(1+\frac{1}{Q}\right) = -Q^2 \cdot \left(1+\frac{1}{Q}\right) = -Q(1+Q),$$

therefore $\frac{dQ}{dP} = \frac{1}{\frac{dP}{dQ}}$. ◊

6.6.1 Elasticity, revenue and profit maximization

Note that, in general, the elasticity will depend on the point on the demand curve at which it is evaluated. For example, in the linear case $\frac{Q}{P} = \frac{(a-bQ)}{Q} = \left[\left(\frac{a}{Q}\right) - b\right]$ and $Q = \left(\frac{a}{b}\right) - \left(\frac{1}{b}\right)P$, so $\frac{dQ}{dP} = -\left(\frac{1}{b}\right)$. Therefore, $\epsilon = \left(\frac{P}{Q}\right)\left(\frac{dQ}{dP}\right) = \left[\left(\frac{a}{Q}\right) - b\right]\left(-\frac{1}{b}\right) = 1 - \frac{a}{bQ}$. Sometimes it is useful to make explicit the dependence of the elasticity on the quantity by writing $\epsilon(Q)$. Note that $\epsilon < 0$ (as long as price is non-negative: $a - bQ \geq 0$). When $\epsilon = -1$, $1 - \frac{a}{bQ} = -1$ or $\frac{a}{bQ} = 2$ or $Q = \frac{a}{2b}$, the revenue-maximizing quantity.

Recall that the profit function of a monopoly was written:

$$\pi(Q) = P(Q)Q - c(Q) = R(Q) - c(Q).$$

From this expression:

$$\begin{aligned}\pi'(Q) &= R'(Q) - c'(Q)\\ &= P(Q)\left[1 + \frac{Q}{P}\frac{dP}{dQ}\right] - c'(Q)\\ &= P(Q)\left[1 + \frac{1}{\epsilon}\right] - c'(Q).\end{aligned}$$

Since a profit-maximizing monopolist sets $\pi'(Q) = 0$ and solves for Q, this implies that the solution satisfies:

$$P(Q)\left[1 + \frac{1}{\epsilon(Q)}\right] = c'(Q),$$

and since $\epsilon(Q) < 0$, this implies that $P(Q) > c'(Q)$ at the solution so that the profit-maximizing monopolist sets a price greater than marginal cost. Rearranging this expression,

$$\frac{P(Q) - c'(Q)}{P(Q)} = \frac{1}{\epsilon(Q)},$$

so that the markup of price over marginal cost is equal to the reciprocal of the elasticity. Finally, note that marginal revenue is given by

$$MR(Q) = R'(Q) = P(Q)\left[1 + \frac{1}{\epsilon}\right] = P(Q) + \frac{1}{\epsilon}P(Q) < P(Q)$$

and the marginal revenue curve lies below the demand curve $P(Q)$ by the amount $\frac{1}{\epsilon}P(Q)$.

6.7 Taxation and monopoly

Suppose that the government wishes to induce a monopolist to produce the quantity at which price equals marginal cost. To achieve this outcome, suppose that the government gives a per-unit subsidy of s and imposes a lump sum tax of T. The profit of the firm is then:

$$\pi_g(Q) = Q(P(Q)+s) - cQ - T.$$

Taking s and T as given, the firm maximizes profit – the first-order condition is

$$\pi_g'(Q) = 0 = P(Q) + QP'(Q) + s - c = 0.$$

By an earlier calculation, this is equivalent to: $P(Q)\left[1+\frac{1}{\epsilon(Q)}\right] = c - s$. Write the solution to this equation as $Q(s)$, to denote dependence on the level of the subsidy. For any given s, this equation has a unique solution, provided $R''(Q) < 0$, where $R(Q) = QP(Q)$.

In the particular case where $s = 0$, the solution is the monopoly level of output, Q_m: $Q(0) = Q_m$. Suppose that the subsidy level s^* is such that $Q(s^*)$ is the competitive quantity $Q(s^*) = Q_c$ so that price equals marginal cost, $P(Q(s^*)) = c$. Then the first-order condition at s^* is

$$\pi_g'(Q(s^*)) = 0 = P(Q(s^*)) + Q(s^*)P'(Q(s^*)) + s^* - c = 0,$$

and because $P(Q(s^*)) = c$, this reduces to

$$\pi_g'(Q(s^*)) = Q(s^*)P'(Q(s^*)) + s^* = 0.$$

Now, recall that $\frac{1}{\epsilon} = \frac{Q}{P}P'(Q)$, so this condition may be written:

$$s^* = -Q(s^*)P'(Q(s^*)) = -P(Q(s^*))\left(\frac{Q(s^*)}{P(Q(s^*))}P'(Q(s^*))\right) = -\frac{P(Q(s^*))}{\epsilon(Q(s^*))} = -\frac{P(Q_c)}{\epsilon(Q_c)}.$$

So, if the subsidy is chosen equal to the price–elasticity ratio at the competitive output, the monopolist's optimal choice is the competitive output. Unraveling the expression,

$$P(Q_c) = -s^*\epsilon(Q_c) = -s^*\left(\frac{P(Q_c)}{Q_c}\right)\left(\frac{dQ}{dP}\right).$$

Multiply both sides by $\frac{Q_c}{P(Q_c)}$ to get $Q_c = -s\left(\frac{dQ}{dP}\right)$. In the case of linear demand $\frac{dQ}{dP} = -\frac{1}{b}$, so that s^* should be chosen to satisfy $Q_c = \frac{s^*}{b}$ or $s^* = bQ_c$. Since

$Q_c = \frac{(a-c)}{b}$, the unit subsidy $s^* = a - c$ leads the monopolist to produce at the competitive output level.

In this case, the total subsidy is $s^* Q_c = (a-c)\left(\frac{(a-c)}{b}\right) = \frac{(a-c)^2}{b}$. Because the first-order condition for a maximum does not depend on the lump sum tax T, the output of the firm depends only on the policy parameter s. So, the subsidy may be recouped by a lump sum tax, T, chosen to equal $s^* Q_c = (a-c)\left(\frac{(a-c)}{b}\right) = \frac{(a-c)^2}{b}$.

Since the solution depends on government-set parameters, comparative statics calculations arise naturally in considering the impact of government policies. The next example discusses the impact of various tax changes on a monopolist's choice of quantity.

EXAMPLE 6.13: Suppose that a government imposes a revenue tax t and output subsidy s. Thus, if the firm receives revenue R and produces quantity Q, the tax paid is tR and the subsidy received sQ. With demand given by $P(Q) = \frac{1}{Q^\alpha} = Q^{-\alpha}, 0 < \alpha < 1$, and linear cost ($c(Q) = cQ$), the profit function is

$$\begin{aligned}\pi(Q) &= (1-t)R(Q) - cQ + sQ \\ &= (1-t)Q^{1-\alpha} - (c-s)Q.\end{aligned}$$

Thus, the first-order condition is $\pi'(Q) = (1-t)(1-\alpha)Q^{-\alpha} - (c-s) = 0$. Rearranging,

$$Q^{-\alpha} = \frac{(c-s)}{(1-t)(1-\alpha)} \Rightarrow Q^\alpha = \frac{(1-t)(1-\alpha)}{(c-s)} \Rightarrow Q = \left(\frac{(1-t)(1-\alpha)}{(c-s)}\right)^{\frac{1}{\alpha}}.$$

It is of interest to know the impact of changes in t and s on output. Some calculation gives:

$$\frac{dQ}{dt} = \frac{1}{\alpha}\left(\frac{(1-t)(1-\alpha)}{(c-s)}\right)^{(\frac{1}{\alpha}-1)}\left(\frac{-(1-\alpha)}{(c-s)}\right)$$

and

$$\begin{aligned}\frac{dQ}{ds} &= \frac{1}{\alpha}\left(\frac{(1-t)(1-\alpha)}{(c-s)}\right)^{(\frac{1}{\alpha}-1)}\left(\frac{(1-t)(1-\alpha)}{(c-s)^2}\right) \\ &= \frac{1}{\alpha}\left(\frac{(1-t)(1-\alpha)}{(c-s)}\right)^{\frac{1}{\alpha}}\left(\frac{1}{(c-s)}\right).\end{aligned}$$

◊

6.8 Taxation incidence in supply and demand

When a tax is imposed on a good, this generally impacts both the buyer and seller. The price paid by the buyer will typically rise, and the price received by the seller will typically fall – with the gap equal to the tax. Relative to the pre-tax price P_0, the extent to which the buyer price after tax, P_t^b, exceeds P_0 and the seller price received, P_t^s, falls below P_0 measures the tax incidence. With a unit tax, $P_t^b = P_t^s + t$. If, for example, $P_t^b - P_0$ is much larger than $P_0 - P_t^s$, then the tax incidence is borne mostly by the buyer. This is discussed next.

Suppose that in some market the government charges a tax t per unit sold. Thus the price paid by the purchaser exceeds the price received by the seller by the amount $t : P_d(Q) = P_s(Q) + t$. Here P_d denotes the price paid by the demander and P_s denotes the price paid by the seller. Prior to the imposition of the tax, the equilibrium quantity Q_e is determined by the equation $P_d(Q_e) = P_s(Q_e)$. Demand is downward sloping and supply upward sloping, so $P'_d(Q_e) < 0$, $P'_s(Q_e) > 0$ and there is a unique point of intersection. The situation before and after tax is depicted in Figure 6.15.

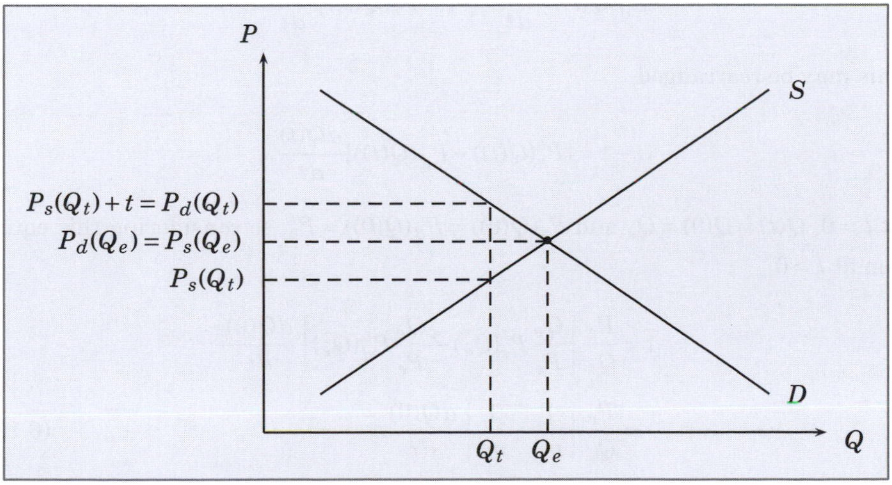

Figure 6.15: The impact of taxation on price and quantity

When a unit tax is imposed on a good, the amount of the tax is independent of the selling price, so that the gap between P_s and p_d is equal to the tax: $p_d(Q) = P_s(Q) + t$. In contrast, with an ad-valorem tax, the amount of the tax depends on the price. In this case, the selling price is scaled up by the addition of a tax that is a percentage of the price and then $p_d(Q) = (1+t)P_s(Q)$. In either case, the result is to reduce the equilibrium amount traded. Let $Q(t)$ (or Q_t) be the amount traded at tax level t. This is a function of t, whether the tax is a unit or ad-valorem tax. What is $\frac{dQ}{dt}$? This question may be answered directly by solving for Q as a function of t and then computing the derivative. It is also possible to derive a general formula for $\frac{dQ}{dt}$ in terms of elasticities, and that approach provides some insight into how elasticities affect the responsiveness of quantity to variation in the tax. These matters are discussed next. The analyses for the unit and ad-valorem taxes are provided separately, although the two cases are quite similar.

6.8.1 A unit tax

The impact of the imposition of a unit tax on quantity can be determined as follows. Given an initial value for t, the corresponding quantity, $Q(t)$, satisfies

$$P_d(Q(t)) = P_s(Q(t)) + t, \quad (Q(0) = Q_e).$$

Differentiating gives

$$P'_d(Q(t))\frac{dQ(t)}{dt} - 1 = P'_s(Q(t))\frac{dQ(t)}{dt}.$$

This may be rearranged,

$$-1 = [P'_s(Q(t)) - P'_d(Q(t))]\frac{dQ(t)}{dt}.$$

At $t = 0$, $Q(t) = Q(0) = Q_e$ and $P_s(Q(0)) = P_d(Q(0)) = P_e$, so considering this equation at $t = 0$,

$$\begin{aligned} -1 &= \frac{P_e}{Q_e}\left[\frac{Q_e}{P_e}P'_s(Q_e) - \frac{Q_e}{P_e}P'_d(Q_e)\right]\frac{dQ(0)}{dt} \\ &= \frac{P_e}{Q_e}\left[\frac{1}{\eta_s} - \frac{1}{\eta_d}\right]\frac{dQ(0)}{dt}. \end{aligned} \tag{6.10}$$

Let the terms η_s and η_d denote the elasticities of supply and demand respectively. Since $P'_d(Q) < 0$ and $P'_s(Q) > 0$, $\eta_d < 0$ and $\eta_s > 0$. Solving for $\frac{dQ}{dt}$ (evaluated at $t = 0$, $\frac{dQ(0)}{dt}$) gives

$$\left.\frac{dQ}{dt}\right|_{t=0} = \frac{dQ(0)}{dt} = \frac{dQ}{dt} = \frac{-1}{\frac{P}{Q}\left[\frac{1}{\eta_s} - \frac{1}{\eta_d}\right]}.$$

Since $\eta_s > 0$ and $\eta_d < 0$, $\frac{dQ}{dt} < 0$. Notice that when the elasticities are small, the impact on Q of a change in t is large. This calculation gives the impact on Q of a change in t starting from the initial point where $t = 0$.

EXAMPLE 6.14: Consider a one-market model for good Q, where demand is given by $P_d(Q) = Q^{-\alpha}$, $0 < \alpha < 1$ and supply is given $P_s(Q) = bQ$, $b > 0$. The market-clearing quantity is given by the solution to: $Q^{-\alpha} = bQ$, or $1 = bQ^{1+\alpha}$, or $Q^{1+\alpha} = \frac{1}{b}$, or $Q = \left(\frac{1}{b}\right)^{\frac{1}{1+\alpha}} = b^{-\frac{1}{1+\alpha}}$.

With a unit tax, market clearing becomes $Q^{-\alpha} = bQ + t$. This expression is difficult to work with. However, the elasticities formula above allows a direct computation of $\frac{dQ}{dt}$ at the market-clearing level. The elasticity of demand and supply may be computed: $P_d'(Q) = -\alpha Q^{-(1+\alpha)}$, so that

$$\frac{1}{\eta_d} = \left(\frac{Q}{P_d}\right) P_d'(Q) = \left(\frac{Q}{Q^{-\alpha}}\right)(-\alpha)Q^{-(1+\alpha)} = Q^{(1+\alpha)}Q^{-(1+\alpha)}(-\alpha) = (-\alpha).$$

Similarly $P_s(Q) = bQ$, so that $P_s'(Q) = b$, and

$$\frac{1}{\eta_s} = \frac{Q}{P_s} P_s'(Q) = \left(\frac{Q}{P_s}\right) b = \left(\frac{1}{b}\right) b = 1.$$

Summarizing, $\eta_d = \frac{P_d}{Q}\frac{dQ}{dP_d} = -\frac{1}{\alpha}$ and $\frac{1}{\eta_d} = -\alpha$. Similarly, $\eta_s = 1$ and $\frac{1}{\eta_s} = 1$. Finally, observe that $\frac{P}{Q} = b$ since $P_e = bQ_e$, so:

$$\frac{dQ}{dt} = -\frac{1}{b(1+\alpha)}.$$ ◊

As mentioned, a unit tax is independent of the selling price and is just added to the selling price so that market clears when $P_d(Q_t) = P_s(Q_t) + t$. In contrast, with the ad-valorem tax, the tax paid is a percentage of the sale price.

6.8.2 An ad-valorem tax

Suppose that the government imposes an ad-valorem tax t, so that in (the after-tax) equilibrium $(1+t)P_s(Q(t)) = P_d(Q(t))$, where $Q(t)$ is the after-tax quantity traded. Let $Q(t)$ denote the equilibrium quantity after tax. Note that $Q(0) = Q_e$. With an ad-valorem tax, the elasticity formula is slightly different. Differentiate $(1+t)P_s(Q(t)) = P_d(Q(t))$ with respect to t, to get

$$P_s(Q(t)) + (1+t)P_s'(Q(t))\frac{dQ(t)}{dt} = P_d'(Q(t))\frac{dQ(t)}{dt}$$

or

$$[(1+t)P_s'(Q(t)) - P_d'(Q(t))]\frac{dQ(t)}{dt} = -P_s(Q(t))$$

or

$$\left[(1+t)\frac{P_s(Q(t))}{Q(t)}\frac{Q(t)}{P_s(Q(t))}P_s'(Q(t)) - \frac{P_d(Q(t))}{Q(t)}\frac{Q(t)}{P_d(Q(t))}P_d'(Q(t))\right]\frac{dQ(t)}{dt} = -P_s(Q(t)).$$

Let $\frac{1}{\eta_s(Q(t))} = \frac{Q(t)}{P_s(Q(t))}P_s'(Q(t))$ and $\frac{1}{\eta_d(Q(t))} = \frac{Q(t)}{P_d(Q(t))}P_d'(Q(t))$ be the reciprocals of the elasticity of supply and demand at $Q(t)$. Then the expression may be written as:

$$\left[(1+t)\frac{P_s(Q(t))}{Q(t)}\frac{1}{\eta_s(Q(t))} - \frac{P_d(Q(t))}{Q(t)}\frac{1}{\eta_d(Q(t))}\right]\frac{dQ(t)}{dt} = -P_s(Q(t)).$$

This expression is valid at each value of t. However, it is simpler at $t = 0$, since $P_s(Q(0)) = P_d(Q(0)) = P_e$, so that $\frac{P_s(Q(t))}{Q(t)}$ equals $\frac{P_s(Q(0))}{Q(0)}$ or $\frac{P_e}{Q_e}$. Likewise, at $t = 0$, $\frac{P_d(Q(t))}{Q(t)}$ equals $\frac{P_d(Q(0))}{Q(0)}$ or $\frac{P_e}{Q_e}$. In this case, the expression becomes:

$$\left[\frac{P_e}{Q_e}\frac{1}{\eta_s(Q_e)} - \frac{P_e}{Q_e}\frac{1}{\eta_d(Q_e)}\right]\frac{dQ(0)}{dt} = -P_s(Q(0)) = -P_e.$$

So,

$$\frac{P_e}{Q_e}\left[\frac{1}{\eta_s} - \frac{1}{\eta_d}\right]\frac{dQ(0)}{dt} = -P_e,$$

where the elasticities are evaluated at Q_e. Rearranging:

$$\begin{aligned}\frac{dQ(0)}{dt} &= -\frac{P_e}{\frac{P_e}{Q_e}\left[\frac{1}{\eta_s} - \frac{1}{\eta_d}\right]}\\ &= -\frac{Q_e}{\left[\frac{1}{\eta_s} - \frac{1}{\eta_d}\right]};\end{aligned}$$

$$\frac{P_e}{Q_e(0)}\left[\frac{1}{\eta_s} - \frac{1}{\eta_d}\right]\frac{dQ}{dt} = -P_e,$$

$$\frac{dQ}{dt} = -\frac{P}{\frac{P_e}{Q_e}\left[\frac{1}{\eta_s} - \frac{1}{\eta_d}\right]} = -\frac{Q_e}{\left[\frac{1}{\eta_s} - \frac{1}{\eta_d}\right]},$$

where all functions are evaluated at $t = 0$. Applying this formula to the demand and supply functions of Example 6.14 gives:

$$\frac{dQ}{dt} = -\frac{1}{(1+\alpha)}\left(\frac{1}{b}\right)^{\frac{1}{1+\alpha}} = -\frac{1}{(1+\alpha)}b^{-\left(\frac{1}{1+\alpha}\right)}.$$

This calculation may be confirmed directly by solving for $Q(t)$ and differentiating. With an ad-valorem tax, the equation $Q(t)^{-\alpha} = (1+t)bQ(t)$ sets demand equal to supply. Rearranging,

$$Q(t)^{-\alpha} = (1+t)bQ(t) \text{ or } \frac{1}{(1+t)b} = Q(t)^{1+\alpha} \text{ or } Q(t) = \left(\frac{1}{(1+t)b}\right)^{\frac{1}{1+\alpha}} = ((1+t)b)^{-\frac{1}{1+\alpha}}.$$

Thus,

$$\frac{dQ(t)}{dt} = -\left(\frac{1}{1+\alpha}\right)[(1+t)b]^{-\frac{1}{1+\alpha}-1}\cdot b.$$

At $t = 0$,

$$\frac{dQ(t)}{dt} = -\frac{1}{(1+\alpha)} b^{-\frac{1}{1+\alpha}}.$$

So, the two calculations yield the same result.

EXAMPLE 6.15: As a final example, these calculations may be repeated, replacing $P_s(Q) = bQ, b > 0$ with $P_s(Q) = Q^\beta, \beta > 0$. In this case $\eta_d = -\frac{1}{\alpha}$ and $\eta_s = \frac{1}{\beta}$ so $\frac{1}{\eta_d} = -\alpha$ and $\frac{1}{\eta_s} = \beta$, and so $\frac{1}{\eta_s} - \frac{1}{\eta_d} = \alpha + \beta$. The equilibrium quantity satisfies $Q^{-\alpha} = Q^\beta$ or $1 = Q^{\alpha+\beta}$ or $Q = 1^{\frac{1}{\alpha+\beta}} = 1$. Likewise, $P = 1$ is the equilibrium price. Then, using the formula,

$$\frac{dQ}{dt} = -\frac{1}{\alpha+\beta}.$$

By contrast, solving $P_d(Q(t)) = Q(t)^{-\alpha} = (1+t)Q(t)^\beta$ gives $Q(t)^{\alpha+\beta} = \frac{1}{1+t} = (1+t)^{-1}$ and $Q(t) = (1+t)^{-\frac{1}{\alpha+\beta}}$. Differentiating:

$$\frac{dQ}{dt} = -\frac{1}{\alpha+\beta}(1+t)^{-\frac{1}{\alpha+\beta}-1}$$

and this equals $-\frac{1}{\alpha+\beta}$ at $t = 0$. ◊

6.8.3 Elasticities and taxation

Considering the formula associated with the unit tax,

$$\frac{dQ}{dt} = \frac{-1}{\frac{P}{Q}\left[\frac{1}{\eta_s} - \frac{1}{\eta_d}\right]},$$

the following observations highlight some aspects of elasticities and their relationship to the responsiveness of demand and price to changes in taxation. If η_s is very large, the quantity supplied is very responsive to price changes, so the supply curve is flat. When η_d is also very large, then the demand curve is also flat. In this case, both $\frac{1}{\eta_s}$ and $\frac{1}{\eta_d}$ are very small so that $\frac{dQ}{dt}$ is a large negative number (−1 divided by a small positive number). Therefore, large elasticities lead to a large change in quantity traded as a result of the tax. (See Figure 6.16.)

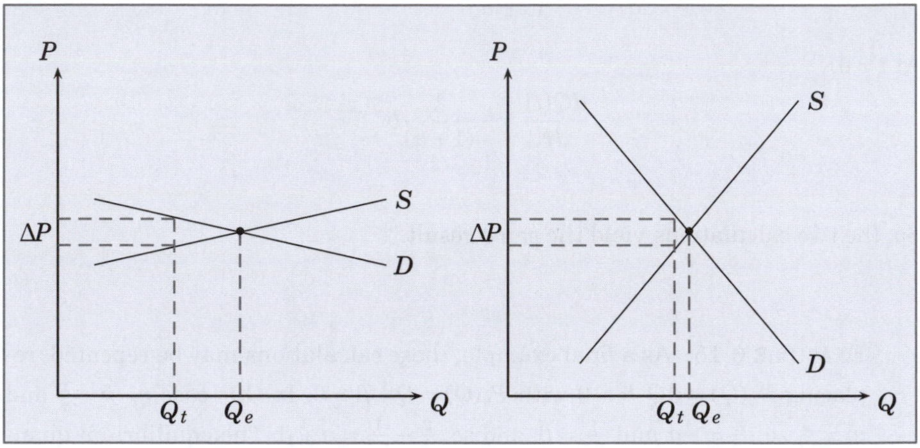

Figure 6.16: Large and small changes in quantity

When both η_s and η_d are small, this corresponds to the situation where neither supply nor demand is responsive to price changes – the curves are both steep. In this case, $\frac{1}{\eta_s}$ and $\frac{1}{\eta_d}$ are large so that $\frac{dQ}{dt}$ is a small negative number. (See Figure 6.16.) Other possibilities arise when only one of the elasticities is relatively small. If η_s is very large (so that $\frac{1}{\eta_s} \simeq 0$) and η_d is very small, then in this case the supply curve is close to horizontal and the demand curve steep. (See Figure 6.17.) When η_d is very large (so that $\frac{1}{\eta_d} \simeq 0$) and η_s is very small, the supply curve is steep and the demand curve flat. (See Figure 6.17.) While the price change is the same in both cases in Figure 6.16, the quantity change is much smaller when the elasticities are small.

In Figure 6.16, the demand and supply curves are similarly sloped; in one case they are relatively elastic so that large quantity changes occur as a result of the tax and in the other a less significant change occurs. Because they have similar slopes, the incidence of the tax is equally borne by demanders and suppliers – the rise in price of one is roughly equal to the drop in price for the other. In Figure 6.17, the demand and supply curves have dissimilar slopes, so the taxation incidence is unequally distributed.

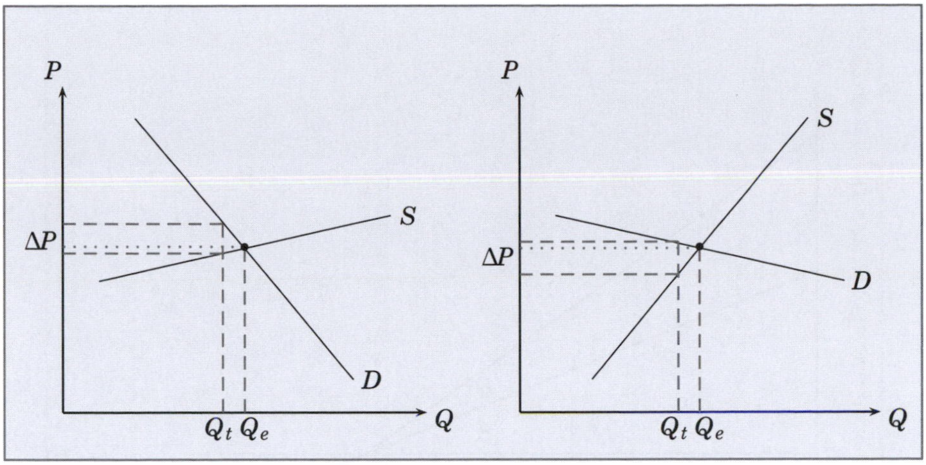

Figure 6.17: Taxation with elastic demand or supply

6.9 Ramsey pricing

Consider the efficiency problem that arises when there is a fixed cost in production, so that pricing at marginal cost generates a loss. In the absence of a subsidy to production, price must be set above marginal cost. Consider the task of recovering fixed cost by pricing above marginal cost in different markets where fixed cost is jointly shared in production, while minimizing loss of consumer and producer surplus. The issue then is to set prices in each of the markets to recover fixed cost, but in a way that minimizes overall dead-weight loss.

Consider two markets with demands $p_x(x) = a - bx$ and $p_y(y) = \alpha - \beta y$, where both items are produced by a firm with cost function $c(x, y) = F + c[x + y]$, so that there is common constant marginal cost c and a fixed cost F. Pricing at marginal cost is efficient but generates a loss for the firm. If prices are set above marginal cost so that the fixed cost F is recouped, what is the optimal pricing scheme (minimizing loss of surplus, subject to raising the fixed cost F)? In Figure 6.18, to simplify, the parameters are chosen so that the quantity, q, setting price equal to marginal cost is the same in both markets.

Consider reducing the quantity in each market by the same amount. In market X this leads to price d and revenue $R_x = (d - c)q'$. The dead-weight loss is given by the triangle $DWL_x = D + E = \frac{1}{2}(d - c)(q - q')$. In market Y, the price is e and the revenue raised is $R_y = E = (e - c)q'$. Also, the dead-weight loss is $DWL_y = E = \frac{1}{2}(e - c)(q - q')$.

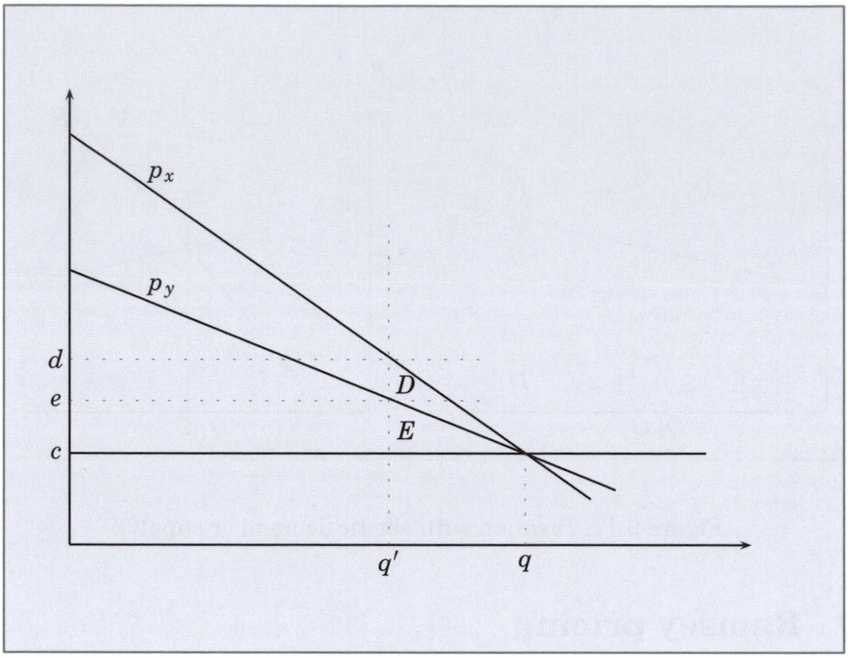

Figure 6.18: Ramsey pricing

Now, consider the revenue generated per unit of dead-weight loss in each market: ρ_x and ρ_y.

$$\rho_x = \frac{R_x}{DWL_x} = \frac{(d-c)q'}{\frac{1}{2}(d-c)(q-q')} = 2\frac{q'}{q-q'}, \quad \text{and}$$

$$\rho_y = \frac{R_y}{DWL_y} = \frac{(e-c)q'}{\frac{1}{2}(e-c)(q-q')} = 2\frac{q'}{q-q'}.$$

Let $\rho = \rho_x = \rho_y$. Viewed as a function of (q',q), $\rho(q',q)$ falls as q' falls, so the revenue gain per unit of dead-weight loss is greatest when q' is close to q. And the optimal quantity reduction is the same in both markets: if, for example, the reductions were $q'_x < q'_y$ then the revenue gain per unit of dead-weight loss is smaller in market X than Y.

In the case where marginal costs are constant but different and the price-equal-marginal-cost conditions give different quantities, the calculations are similar. When quantities are reduced from q_α to q'_α and corresponding prices following the quantity reductions are p'_α, $\alpha = x,y$, the revenue gains in markets x and y are: $R_x = q'_x(p'_x - c_x)$ and $R_y = q'_y(p'_y - c_y)$, and the dead-weight losses are $DWL_x = \frac{1}{2}(p'_x - c_x)(q_x - q'_x)$ and $DWL_y = \frac{1}{2}(p'_y - c_y)(q_y - q'_y)$. The corresponding ratios are:

$$\frac{q'_x(p'_x - c_x)}{\frac{1}{2}(p'_x - c_x)(q_x - q'_x)} = 2\frac{q'_x}{q_x - q'_x}, \qquad \frac{q'_y(p'_y - c_y)}{\frac{1}{2}(p'_y - c_y)(q_y - q'_y)} = 2\frac{q'_y}{q_y - q'_y}.$$

Setting these equal:

$$\frac{q'_x}{q_x - q'_x} = \frac{q'_y}{q_y - q'_y},$$

so that $q'_x(q_y - q'_y) = q'_y(q_x - q'_x)$ or $q'_x q_y = q'_y q_x$, or $\frac{q'_x}{q_x} = \frac{q'_y}{q_y}$. Therefore, the quantities should be reduced in the same proportion.

Exercises for Chapter 6

Exercise 6.1 Let $C(q)$ denote the total cost of producing q units of a good. Show that the average cost is minimized at the point where average and marginal costs coincide.

Exercise 6.2 The price–cost margin (markup), m, for a monopolist is defined as

$$m = \frac{\text{price} - \text{marginal cost}}{\text{price}}.$$

Find the relationship between ϵ, the elasticity of demand and m.

Exercise 6.3 Let demand for a good be given by $p = a - bq$ or equivalently by $q = \left(\frac{a}{b}\right) - \left(\frac{1}{b}\right)p = \alpha - \beta p$, where $\alpha = \left(\frac{a}{b}\right)$ and $\beta = \left(\frac{1}{b}\right)$. The cost of production of the good is $c(q) = cq$, where c is a positive constant. Assume there is only one firm producing the good.

(a) Is the firm's profit function concave?

(b) Find the profit-maximizing level of output of the good.

(c) Suppose that the firm is taxed, paying \$$\tau$ per unit of output. Give the firm's new objective function and find the firm's profit-maximizing level of output.

(d) Suppose that the firm is taxed, such that on each dollar of sales the firm pays the fraction τ in taxes. Give the firm's new objective function and find the profit-maximizing output level.

(e) In cases (c) and (d), analyze the effect of changing τ on the optimal choice of q.

Exercise 6.4 A monopolist faces a demand curve $P(Q) = kQ^{-\alpha}$ $(0 < \alpha < 1)$ and cost function $c(Q) = cQ^2$, $c > 0$.

(a) Find the profit-maximizing level of output, Q^*. Confirm that the second-order condition for a maximum is satisfied.

(b) Now suppose that the cost function is $c(Q) = c_1Q + c_2Q^2$, where $c_1, c_2 > 0$ (the demand function is still $P(Q) = kQ^{-\alpha}$). Confirm that the profit function is concave.

Exercise 6.5 Consider the following supply and demand model:

$$\text{demand} : P_d(Q) = \alpha - \beta Q; \ \text{supply} : P_s(Q) = \gamma - \delta Q.$$

The equilibrium quantity Q_o solves $P_d(Q) = P_s(Q)$, i.e., $P_d(Q_o) = P_s(Q_o)$.

(a) Find the elasticities of the demand and supply functions, η_d and η_s.

(b) Calculate the impact of a small (per) unit tax, t, on the quantity sold.

(c) Relate your answer $\left(\frac{dQ}{dt}\right)$, to the elasticity formula:

$$\frac{dQ}{dt} = -\frac{1}{\left(\frac{P}{Q}\right)}\left\{\frac{1}{\eta_s} - \frac{1}{\eta_d}\right\}$$

(d) Replace the supply function with $P_s(Q) = \gamma - \frac{1}{2}\delta Q^2$ and repeat the calculations for the pair: $P_d(Q) = \alpha - \beta Q$ and $P_s(Q) = \gamma - \frac{1}{2}\delta Q^2$.

Exercise 6.6 Suppose that instead of a unit tax, a proportional tax is imposed so that the after-tax relation between the selling and buying price is given by $P_d(Q) = (1+t)P_s(Q)$. Calculate $\frac{dQ}{dt}$, at $t = 0$.

Exercise 6.7 In a city, electricity is produced by one firm. This firm is given the sole right to produce the city's electricity because it can produce the electricity needed more efficiently than could any other firm or firms. The firm is thus a natural monopoly. The total revenue function for the firm is $R(Q) = P(Q)Q = 24Q - 5Q^2$ $(P(Q) = 24 - 5Q)$ and its total cost is $25 - Q^2$.

One of the problems that the city faces is that since the firm is a natural monopoly, and since electricity has a very inelastic demand, the firm can make large profits. To control these large profits, the city is considering a revenue tax (tR) as a means of regulating the firm. Assuming that the firm is a profit maximizer:

(a) Calculate the optimal price and quantity before regulation, and after the revenue tax is applied.

(b) If the tax rate is set at 20%, will the profits of the firm be decreased? Will the city benefit from this regulation, i.e., will it face lower prices and higher quantities?

(c) Calculate the tax revenue the city will receive.

Exercise 6.8 To control the use of tobacco, the federal government can impose many different types of tax on the producers of tobacco.

Assume that there is only one producer (monopolist), who faces the demand curve $P(Q) = 56 - 8Q$, and whose total cost curve is $C(Q) = 6Q^2 - 10$. If the producer is a profit maximizer:

(a) Determine the profit-maximizing output of the monopolist and the associated price.

(b) Assume now that the government levies a per-unit tax of t on the producer in order to reduce the use of tobacco. What effect does this have on the profit-maximizing price and output?

(c) Alternatively, assume the government levies a sales tax of 50% on the producer. What effect does this have on the profit-maximizing price and output?

(d) If the per-unit tax is set at 7, and the sales tax is set at 50%, which tax is more effective?

Exercise 6.9 Assume that a perfectly competitive producer has a production function for a homogeneous good given by $Q(L) = 20L^{\frac{1}{2}}$. The costs of production are given by a fixed cost of \$50 plus a wage rate of \$8 for each unit of labour employed. If the market price of the good is \$4,

(a) Find the amount of labour needed to maximize profits, and the associated level of profits.

(b) Assume that a payroll tax is imposed, which causes the wage rate to become $w = w(1+t)$. Determine the new amount of labour needed to maximize profits.

(c) If $t = 0.25$, what is the amount of labour needed to maximize profits? What is the associated level of profits?

Exercise 6.10 A competitive firm faces the market price of \$9 per unit. Its total cost is given by $Q^3 - 9Q^2 + 33Q + 10$.

(a) Determine the level of output that maximizes the firm's profits, and the value of profits at this level. Will the firm continue to produce?

(b) Determine the minimum price at which the firm will stay in the industry.

Exercise 6.11 Consider a firm with cost function $C = 6 + Q^2$ facing the market demand function $Q = 18 - 2P$.

(a) If the firm behaves as a monopolist, what will be the price and quantity in equilibrium?

(b) If the firm behaves competitively, what will be the equilibrium price and quantity?

(c) Suppose the monopolist were offered a subsidy S per unit produced. What level of subsidy would induce the monopolist to produce at the competitive equilibrium?

Exercise 6.12 Consider a firm with the following cost and demand functions: $C(Q) = Q^2 + 2Q + 1$, $Q = 16 - 2P$.

(a) Calculate the equilibrium price and quantities:

(i) if the firm acts as a monopolist

(ii) if the firm acts competitively.

(b) Calculate the profits the firm makes:

(i) as a monopolist

(ii) as a competitor.

(c) Suppose the government offered the monopolist a per-unit subsidy, S. What level of S will bring about the competitive solution?

(d) What is the subsidized monopolist's level of profit?

Exercise 6.13 Consider a monopolist with a demand function $P = P(Q)$ and cost function $c = c(Q), c'(Q) > 0$. Show that if the monopolist maximizes profit, the monopolist will operate on the portion of the demand curve where the elasticity exceeds 1.

Exercise 6.14 Consider a monopolist with a demand function $P = P(Q)$ and cost function $c = c(Q), c'(Q) > 0$. Show that with profit maximization, the firm will operate on the portion of the demand curve where the elasticity exceeds 1.

Exercise 6.15 In some market, the demand function is given by $p_d(q) = \frac{a}{\ln q}$ and the supply function is given by $p_s(q) = bq$.

(a) Calculate the elasticities of demand and supply.

(b) The equilibrium quantity satisfies $p_d(q) = p_s(q)$. Starting from a point where there is no tax on the good, suppose that a small tax t is imposed – so that $p_d(q) = p_s(q) + t$. Calculate $\frac{dq}{dt}$, $\frac{dp_d}{dt}$ and $\frac{dp_s}{dt}$, all at $t = 0$.

Exercise 6.16 Suppose that the demand and supply functions in a market are given by $P_d(Q) = Q^{-\alpha}$ and $P_s(Q) = e^{kQ} - 1$. The equilibrium quantity depends on the parameters α and k: $Q(\alpha, k)$.

(a) Show that $\frac{\partial Q}{\partial \alpha} < 0$ and $\frac{\partial Q}{\partial k} < 0$.

(b) Suppose now that the government imposes a "goods and services" style tax at rate t (formally, a value added tax) so that the equilibrium quantity after tax is $Q(t)$ (ignore the parameters α and k from now on). Thus, $Q(t)$ satisfies $P_d(Q(t)) = (1+t)P_s(Q(t))$. Find $\frac{dQ(0)}{dt}$ in terms of elasticities.

7

Systems of equations, differentials and derivatives

7.1 Introduction

In many economic problems, functions of more than one variable arise naturally. Preferences are represented by a utility function that depends on goods consumed, $u(x_1,x_2,\ldots,x_n)$, output in production is represented as a function of inputs, $f(k_1,k_2,\ldots,k_n)$, and so on. As with the single-variable case, differential techniques play an important role in a variety of applications, such as comparative statics and optimization. For example, the impact of variation in one variable on the value of a function routinely arises in considering its marginal impact. In the production case, the marginal productivity of input k_1 is given by $[f(k_1+\Delta,k_2,\ldots,k_n)-f(k_1,k_2,\ldots,k_n)]/\Delta$, and measures the impact on output of a small change in input k_1. Such a calculation is called a partial derivative, since only variation in one of a number of variables is considered. Developing these techniques leads to a number of applications and provides the necessary tools to develop some important results – in particular the implicit function theorem.

Following some introductory discussion, partial derivatives are defined in Section 7.2. This is the natural extension of the derivative of a function of one variable. After this, level contours are introduced (Section 7.3). A level contour identifies the set of x that yields a specific value for the function: in microeconomic theory, indifference curves and isoquants are level contours of utility and production functions respectively. In two dimensions particularly, level contours provide a useful way to visualize the shape of a function. Concepts such as the marginal rate of substitution between goods in consumption or production are explained naturally in terms of level contours and are defined using partial derivatives.

One important use of partial derivatives is in locating maxima and minima of a function: values of $(x_1,\dots,x_n)$ that make the function as large or as small as possible. This technique is illustrated in Section 7.5. The remaining sections consider (non-linear) systems of equations. Section 7.7 outlines some of the basic issues relating to non-linear systems, such as the existence and uniqueness of solutions. Following this, the implicit function is introduced in Section 7.8. This theorem studies how variation of one or more parameters in a system of equations affects each variable in the system.

For a linear function of many variables, the impact of parameter variation is easy to determine. The function $f(x_1,x_2) = ax_1 + bx_2$ is a linear function of two variables and $f(x_1,x_2,\dots,x_n) = a_1x_1 + a_2x_2 + \cdots + a_nx_n$ is a linear function of n variables, $(x_1,x_2,\dots,x_n)$. The impact of a change in x_1 on the value of $f(x_1,x_2) = ax_1 + bx_2$ may be computed directly:

$$f(x_1+\Delta x_1,x_2) - f(x_1,x_2) = (ax_1+\Delta x_1+bx_2)-(ax_1+bx_2) = a\Delta x_1.$$

From this, the change in f per unit change is x_1 may be calculated as:

$$\frac{f(x_1+\Delta x_1,x_2)-f(x_1,x_2)}{\Delta x_1} = a.$$

Similar calculations may be applied for non-linear functions to compute the impact of variation in some variable on the function.

EXAMPLE 7.1: Consider

$$f(x_1,x_2) = (x_1x_2)^2 + \cos(x_1)\sin(x_2),$$

suppose that x_1 is increased to $x_1+\Delta x_1$:

$$\begin{aligned} f(x_1+\Delta,x_2) &= ([x_1+\Delta x_1]\cdot x_2)^2 + \cos(x_1+\Delta x_1)\sin(x_2) \\ &= [x_1^2 + 2x_1\Delta x_1 + (\Delta x_1)^2]\cdot x_2^2 + \cos(x_1+\Delta x_1)\sin(x_2) \\ &= x_1^2\cdot x_2^2 + [2x_1\Delta x_1 + (\Delta x_1)^2]\cdot x_2^2 + \cos(x_1+\Delta x_1)\sin(x_2). \end{aligned}$$

Subtracting $f(x_1,x_2)$, then

$$\begin{aligned} f(x_1+\Delta x_1,x_2) - f(x_1,x_2) = [2x_1\Delta x_1 + (\Delta x_1)^2]\cdot x_2^2 \\ + [\cos(x_1+\Delta x_1) - \cos(x_1)]\sin(x_2). \end{aligned}$$

Dividing by Δx_1,

$$\frac{f(x_1+\Delta x_1,x_2)-f(x_1,x_2)}{\Delta x_1} = 2x_1x_2^2 + \frac{[\cos(x_1+\Delta x_1)-\cos(x_1)]}{\Delta x_1}\sin(x_2) + (\Delta x_1)\cdot x_2^2.$$

As Δx_1 goes to 0, the term $(\Delta x_1)\cdot x_2^2$ goes to 0 and $\frac{[\cos(x_1+\Delta x_1)-\cos(x_1)]}{\Delta x_1}$ approaches the derivative of $\cos(x_1)$, which is $-\sin(x_1)$. Therefore, as Δx_1 approaches 0,

$$\frac{f(x_1+\Delta x_1,x_2)-f(x_1,x_2)}{\Delta x_1} \to 2x_1x_2^2-\sin(x_1)\sin(x_2).$$

◊

This example shows that calculating the impact of parameter or variable changes on a function of many variables or parameters is not significantly different from similar computations in the one-variable case.

The need for such calculations comes up routinely in economics. Consider maximizing utility $u(x_1,x_2)$ subject to a budget constraint, $p_1x_1+p_2x_2=y$, where p_i is the price of good i and y is income. Rearranging the budget constraint: $x_2=\frac{y}{p_2}-\frac{p_1}{p_2}x_1$. Substituting this into the utility function gives

$$\hat{u}(x_1,y,p_1,p_2)=u\left(x_1,\frac{y}{p_2}-\frac{p_1}{p_2}x_1\right).$$

Taking (y,p_1,p_2) as fixed (beyond the control of the consumer), maximizing utility amounts to choosing x_1 to maximize $\hat{u}(x_1,y,p_1,p_2)$. The impact of variations in x_1 on $\hat{u}$ requires calculation of the direct effect via x_1 and the indirect effect via $x_2=\frac{y}{p_1}-\frac{p_1}{p_2}x_1$. When x_1 is chosen optimally, the maximized value of $\hat{u}$ gives the utility of the individual from choosing optimally – let $u^*(y,p_1,p_2)=\max_{x_1}\hat{u}(x_1,y,p_1,p_2)$. The function u^* gives the utility of the individual as a function of (y,p_1,p_2), so one may consider the impact of income and price variations on the individual's welfare. The impact on utility of an increase in income from y to $y+\Delta y$ is given by $u^*(y+\Delta y,p_1,p_2)-u^*(y,p_1,p_2)$, and the per-unit change is given by

$$\frac{u^*(y+\Delta y,p_1,p_2)-u^*(y,p_1,p_2)}{\Delta y}.$$

This expression is a partial derivative – calculating the impact on u^* of a small change in y, with the other parameters affecting u^* held fixed. The *partial derivative* is the analogous concept to the derivative of a function of one variable, for functions of more than one variable. In this example, partial derivatives appear both in the choice of x_1 and the subsequent calculation of the marginal utility of income.

In production theory, a production function $f(k,l)$ associates a level of output $Q=f(k,l)$ to inputs of capital (k) and labor (l). If capital is increased from k to $k+\Delta$, then output increases to $f(k+\Delta,l)$ and the change in output is $f(k+\Delta,l)-f(k,l)$.

Dividing this by Δ gives $[f(k+\Delta,l)-f(k,l)]/\Delta$, the change in output per unit change in the capital input. As Δ goes to 0, assuming f is differentiable with respect to k, the limiting value is called the marginal product of capital.

In demand theory, when a profile of goods $(x_1,\dots,x_n)$ is chosen to maximize utility subject to a budget constraint $p_1x_1+\cdots+p_nx_n=y$, the resulting demands depend on the profile of prices and income. Thus, the demand for good i is $x_i(p_1,\dots,p_n,y)$. When prices other than p_i and income y are constant, changes in p_i lead to changes in x_i, and this connection between x_i and p_i gives the demand function for good i. As p_i varies, the resulting change in x_i is

$$\frac{[x_i(p_1,\dots,p_{i-1},p_i+\Delta p_i,p_{i+1},p_n,y)-x_i(p_1,\dots,p_{i-1},p_i+\Delta p_i,p_{i+1},p_n,y)]}{\Delta p_i}.$$

The partial derivative with respect to one variable depends on the values of the other variables entering the function. For example, the impact of a price change on demand depends on the prices of other goods and income. In Figure 7.1, the partial derivative with respect to p_i at $(p_1,\dots,p_i,\dots,p_n,y)$ is different from the partial derivative at $(\tilde{p}_1,\dots,p_i,\dots,\tilde{p}_n,\tilde{y})$, since $|\Delta\tilde{x}_i|>|\Delta x_i|$.

The next section defines the partial derivative of a function – a natural extension of the derivative of a function of a single variable, to functions of many variables. Following this, level contours are introduced and then, using partial derivatives, the equilibrium output in a duopoly is determined.

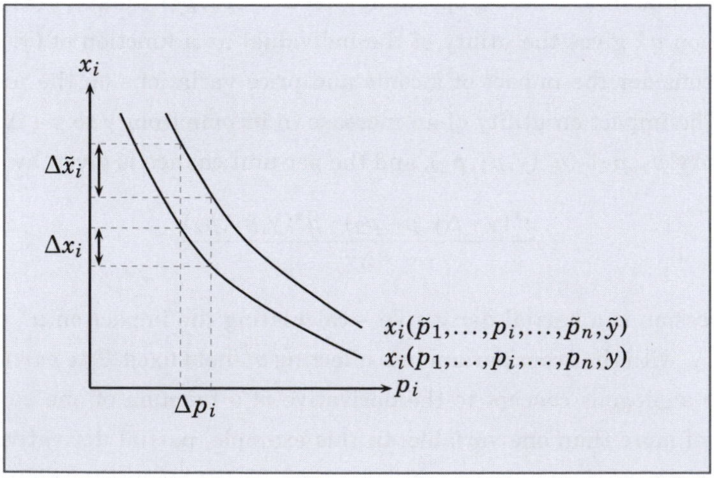

Figure 7.1: The impact of price variations on demand

7.2 Partial derivatives

Let $f : X \to \mathscr{R}$, where $X \subseteq \mathscr{R}^n$, so $f(x) = f(x_1, x_2, \ldots, x_n)$. If all the xs except one are held fixed, we can view f as a function of one variable, and then consider the derivative of f with respect to this variable. This is called the partial derivative. The *partial derivative of f with respect* to x_i is denoted $\frac{\partial f}{\partial x_i}$ and defined as:

$$\frac{\partial f}{\partial x_i} = \lim_{\Delta x_i \to 0} \frac{f(x_1, x_2, \ldots, x_{i-1}, x_i + \Delta x_i, x_{i+1}, \ldots, x_n) - f(x_1, x_2, \ldots, x_n)}{\Delta x_i}.$$

Thus, $\frac{\partial f}{\partial x_i}$ is defined provided this limit exists for all $\Delta x_i \to 0$. The partial derivative of f with respect to x_i measures the change in f due to a small change in x_i, when other variables are held constant.

EXAMPLE 7.2: With $f(x_1, x_2) = (x_1 x_2)^2 + \cos(x_1)\sin(x_2)$,

$$\frac{\partial f}{\partial x_1} = 2(x_1 x_2)x_2 - \sin(x_1)\sin(x_2) = 2x_1 x_2^2 - \sin(x_1)\sin(x_2),$$

$$\frac{\partial f}{\partial x_2} = 2(x_1 x_2)x_1 + \cos(x_1)\cos(x_2) = 2x_1^2 x_2 + \cos(x_1)\cos(x_2).$$

◊

REMARK 7.1: Often, for notational convenience, the partial derivative of f with respect to x_i is denoted f_{x_i} or f_i when no confusion can arise. Also, where there are just two variables in consideration, these may be labeled x and y with partial derivatives f_x and f_y. With more than two variables, it is often more convenient to label by index, $x_i, i = 1, \ldots, n$. □

Note that for $f(x_1, x_2, \ldots, x_n)$, $\frac{\partial f}{\partial x_i}$ is a function of all the x_i. One can take the partial derivative of this function – the partial derivative of the partial derivative. Thus, the partial derivative of $\frac{\partial f}{\partial x_i}$ with respect to x_i is $\frac{\partial}{\partial x_i}\left\{\frac{\partial f}{\partial x_i}\right\}$ and the partial derivative of $\frac{\partial f}{\partial x_i}$ with respect to x_j is $\frac{\partial}{\partial x_j}\left\{\frac{\partial f}{\partial x_i}\right\}$. For ease of notation, these are sometimes written as $\frac{\partial^2 f}{\partial x_i^2}$ and $\frac{\partial^2 f}{\partial x_j \partial x_i}$ or as $f_{x_i x_i}$ and $f_{x_j x_i}$. In the two-variable case with $f(x,y)$, the partial derivatives are $f_x = \frac{\partial f}{\partial x}$ and $f_y = \frac{\partial f}{\partial y}$ and the *second partial derivatives* are written as $f_{xx} = \frac{\partial}{\partial x}\left\{\frac{\partial f}{\partial x}\right\}$, $f_{yy} = \frac{\partial}{\partial y}\left\{\frac{\partial f}{\partial y}\right\}$ $f_{xy} = \frac{\partial}{\partial x}\left\{\frac{\partial f}{\partial y}\right\}$. and $f_{yx} = \frac{\partial}{\partial y}\left\{\frac{\partial f}{\partial x}\right\}$.

EXAMPLE 7.3: To illustrate, consider the two-variable case with $f(x,y) = x^{\alpha}y^{\beta}$. Then:

$$f_x(x,y) = \frac{\partial f}{\partial x} = \alpha x^{\alpha-1}y^{\beta} = \alpha\frac{x^{\alpha}}{x}y^{\beta} = \frac{\alpha}{x}f(x,y)$$
$$f_x(x,y) = \frac{\partial f}{\partial y} = \beta x^{\alpha}y^{\beta-1} = \beta x^{\alpha}\frac{y^{\beta}}{y} = \frac{\beta}{y}f(x,y).$$

Notice that $f_x(x,y) = \frac{\partial f}{\partial x}(x,y) = \frac{\partial\{f(x,y)\}}{\partial x} = \alpha x^{\alpha-1}y^{\beta}$ depends on x and y – it is a function of (x,y). The function $f_x(x,y)$ has a partial derivative with respect to x, which is $\frac{\partial}{\partial x}\left\{\frac{\partial f}{\partial x}\right\} = \alpha(\alpha-1)x^{\alpha-2}y^{\beta}$. Similarly, $f_x(x,y)$ has a partial derivative with respect to y given by $\frac{\partial}{\partial y}\left\{\frac{\partial f}{\partial x}\right\} = \alpha\beta x^{\alpha-1}y^{\beta-1}$.

In Example 7.2, the second partial derivatives are given by:

$$f_{x_1x_1} = \frac{\partial}{\partial x_1}\left\{\frac{\partial f}{\partial x_1}\right\} = 2x_2^2 - \cos(x_1)\sin(x_2)$$
$$f_{x_2x_1} = \frac{\partial}{\partial x_2}\left\{\frac{\partial f}{\partial x_1}\right\} = 4x_1x_2 - \sin(x_1)\cos(x_2)$$
$$f_{x_2x_2} = \frac{\partial}{\partial x_1}\left\{\frac{\partial f}{\partial x_2}\right\} = 2x_1^2 - \cos(x_1)\sin(x_2)$$
$$f_{x_1x_2} = \frac{\partial}{\partial x_1}\left\{\frac{\partial f}{\partial x_2}\right\} = 4x_1x_2 - \sin(x_1)\cos(x_2).$$

◊

Note that in Example 7.3, $f_{x_2x_1} = f_{x_1x_2}$. This is not accidental. When the partial derivatives exist and are continuous, then the second partial derivatives satisfy a special relation known as Young's theorem.

Theorem 7.1: *Suppose the function $f(x_1,\ldots,x_n)$ has continuous partial derivatives, $f_{x_i}(x_1,\ldots,x_n)$, in a neighborhood of $(\bar{x}_1,\ldots,\bar{x}_n)$, and suppose they are differentiable, $\frac{\partial}{\partial x_i}\left\{\frac{\partial f}{\partial x_j}\right\}$ exists for each i,j; then the second partial derivatives are equal:*

$$f_{x_ix_j} = \frac{\partial^2 f}{\partial x_i\partial x_j} = \frac{\partial}{\partial x_i}\left\{\frac{\partial f}{\partial x_j}\right\} = \frac{\partial}{\partial x_j}\left\{\frac{\partial f}{\partial x_i}\right\} = \frac{\partial^2 f}{\partial x_j\partial x_i} = f_{x_jx_i}.$$

EXAMPLE 7.4: Consider $f(x,y) = \ln(xy)\cdot\sin y$, so that $f_x = \frac{y}{xy}\sin y = \frac{1}{x}\sin y$ and $f_{yx} = \frac{1}{x}\cos y$. Differentiating first with respect to y gives: $f_y = \frac{x}{xy}\sin y + \ln(xy)\cos y = \frac{1}{y}\sin y + \ln(xy)\cos y$. And then differentiating with respect to x: $f_{xy} = \frac{y}{xy}\cos y = \frac{1}{x}\cos y$. So, $f_{xy} = f_{yx}$. ◊

7.3 Level contours

Level contours arise frequently in economics, for example, as isoquants in production theory or as indifference curves in consumer theory. Given a function of n variables, $f(x_1,\ldots,x_n)$, the level contour of f at level or height $\bar{f}$ is the set of values of $(x_1,\ldots,x_n)$ such that $f(x_1,\ldots,x_n)=\bar{f}$. If this set is denoted $L_{\bar{f}}$, then

$$L_{\bar{f}} = \{(x_1,\ldots,x_n) \mid f(x_1,\ldots,x_n)=\bar{f}\}$$

Different values of $\bar{f}$ determine different level contours. The level contour is a way of visualizing a function of more than one variable.

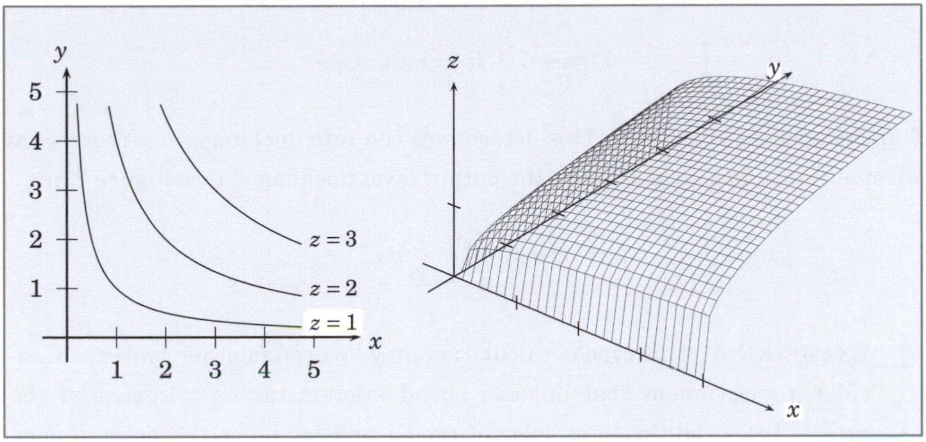

Figure 7.2: The function $f(x,y)=x^{\frac{1}{2}}y^{\frac{1}{2}}$

In Figure 7.2, the function $f(x,y) = x^{\frac{1}{2}}y^{\frac{1}{2}}$ is plotted, and three level contours graphed, giving the (x,y) pairs associated with the level contours $\{(x,y) \mid f(x,y)=1\}$, $\{(x,y) \mid f(x,y)=2\}$ and $\{(x,y) \mid f(x,y)=3\}$.

Given a utility function $u(x,y)$, the indifference curve corresponding to utility level $\bar{u}$ is defined as the set of (x,y) combinations that provide level of utility, $\bar{u}$: $I_{\bar{u}} = \{(x,y) \mid u(x,y) = \bar{u}\}$. The isoquant corresponding to a production function f is the set of input combinations (x,y) that yield a given level of output Q: $I_Q = \{(x,y) \mid f(x,y)=Q\}$. Considering the production function f, since a small change in x has impact f_x on the value of output and a small change in y has impact f_y, the total change in output from small changes (dx,dy) in (x,y) is $df = f_x(x,y)dx + f_y(x,y)dy$. This is called the *total derivative* of f at (x,y). Note that df depends on (x,y) and (dx,dy). An isoquant is a level contour of the function f. Starting from a point (x,y) on the contour, small variations in x and y that leave the value of

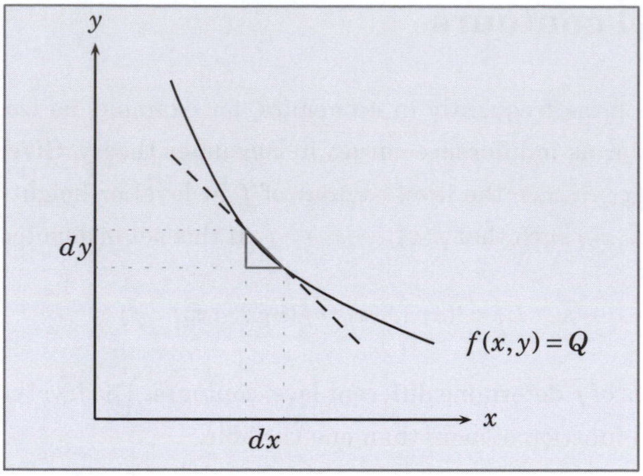

Figure 7.3: Isoquant slope

f unchanged satisfy $df = 0$. This determines the rate of change in y required to offset a change in x so as to leave the output level unchanged (see Figure 7.3):

$$dy = -\frac{f_x(x,y)}{f_y(x,y)}dx.$$

EXAMPLE 7.5: This type of calculation may be used in other contexts. Consider a government that imposes an ad-valorem tax on all goods at the rate τ. Let s_i be the total sales of good i and let p_i be the price of good i, both constant over time. Total tax revenue is $T = \sum_{i=1}^{n}(1+\tau)s_i p_i$. Suppose now that prices increase at the rate $\pi(t)$ so that at time t, the price of good i is $p_i(1+\pi(t))$. Suppose also, that the government adjusts the tax rate τ over time to keep the total tax revenue constant: $T = \sum_{i=1}^{n}(1+\tau(t))(1+\pi(t))s_i p_i = (1+\tau(t))(1+\pi(t))\sum_{i=1}^{n} s_i p_i$. Then, differentiating with respect to t,

$$0 = \left\{\frac{d\tau(t)}{dt}(1+\pi(t)) + (1+\tau(t))\frac{d\pi(t)}{dt}\right\}\sum_{i=1}^{n} s_i p_i.$$

Therefore,

$$0 = \left[\frac{d\tau(t)}{dt}(1+\pi(t)) + (1+\tau(t))\frac{d\pi(t)}{dt}\right].$$

Thus, taxes should be adjusted over time according to the formula:

$$\frac{d\tau(t)}{dt} = -\frac{(1+\tau(t))}{(1+\pi(t))}\frac{d\pi(t)}{dt}.$$ ◊

7.4 The elasticity of substitution

The elasticity of demand measures the slope $\frac{dQ}{dP}$ of a curve (the demand function) relative to the ratio of the variables at that point on the curve, $\frac{Q}{P}$. This is the slope of the curve at a point, divided by the slope of the ray from the origin to that point. The elasticity of substitution measures how that ray varies as the slope varies (all in percentage terms). In principle, this measure can be applied to the level surface of any function, but two particularly important functions in economics are the utility and production functions. Consider Figure 7.4, where the indifference curve u is given and a change in slope from a to b depicted. The marginal rates of substitution at a and b are denoted $MRS(a)$ and $MRS(b)$ and denote the slope of the indifference curve at the respective points. The ray slope from the origin to a is the ratio of consumption of x_2 relative to x_1, $\frac{x_2^a}{x_1^a}$, when the marginal rate of substitution is $MRS(a) = \frac{u_{x_1}(a)}{u_{x_2}(a)}$. When the marginal rate of substitution changes to $MRS(b)$, the slope of the ray becomes $\frac{x_2^b}{x_1^b}$. Therefore, the percentage change in the ratio of consumption resulting from the percentage change in the MRS is given by the following expression, called the elasticity of substitution:

$$\eta = \frac{\frac{\left[\frac{x_2^b}{x_1^b}\right]-\left[\frac{x_2^a}{x_1^a}\right]}{\left[\frac{x_2^a}{x_1^a}\right]}}{\frac{MRS(b)-MRS(a)}{MRS(a)}} = \frac{MRS(a)}{\frac{x_2^a}{x_1^a}} \frac{\left[\frac{x_2^b}{x_1^b}\right]-\left[\frac{x_2^a}{x_1^a}\right]}{MRS(b)-MRS(a)}. \tag{7.1}$$

In Equation 7.1, with u given, the ratio $\frac{x_2}{x_1}$ is viewed as a function of MRS, with the elasticity tracking the change in the ratio resulting from the change in MRS. Considering this measure at a point corresponds to the difference between $MRS(a)$ and $MRS(b)$ becoming small (approaching 0 in the limit). Here, this leads to an expression in terms of the $\ln(x)$ function. Recall that the derivative of $\ln(x)$ is $\frac{1}{x}$ so that $\frac{1}{x} = \frac{d\ln(x)}{dx} \approx \frac{\ln(x+\Delta)-\ln(x)}{\Delta}$. Thus, $\ln(x+\Delta)-\ln(x) \approx \frac{\Delta}{x} = \frac{x'-x}{x}$ where $x' = x+\Delta$. From this perspective, the numerator in Equation 7.1 is $d\ln\left(\frac{x_2}{x_1}\right)$ and the denominator is $d\ln(MRS)$. In differential terms:

$$\eta = \frac{d\ln\left(\frac{x_2}{x_1}\right)}{d\ln(MRS)} = \frac{d\ln\left(\frac{x_2}{x_1}\right)}{d\ln\left(\frac{u_{x_1}}{u_{x_2}}\right)}, \tag{7.2}$$

recalling the fact that MRS is obtained from $du = 0 = u_{x_1}dx_1 + u_{x_2}dx_2$, giving $\frac{dx_2}{dx_1} = -\frac{u_{x_1}}{u_{x_2}}$, giving $MRS = \frac{u_{x_1}}{u_{x_2}}$ (measuring the slope in absolute terms).

Regarding the sign of η, note that, in the movement from a to b, the slope of the indifference curve is always negative but *increases* towards 0. So, in absolute

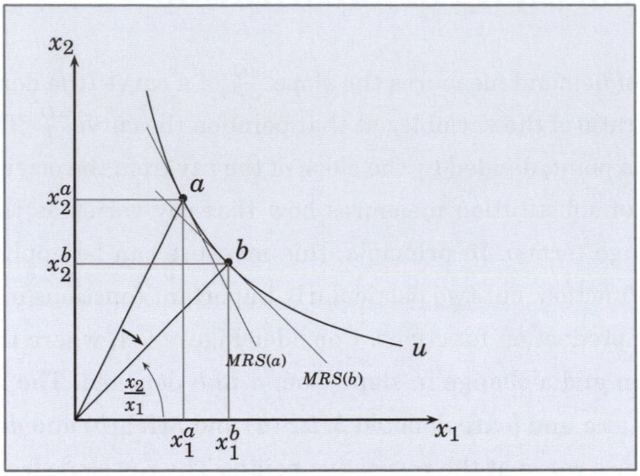

Figure 7.4: Elasticity of substitution

terms, the slope (MRS) declines (is getting smaller). Likewise, the ratio of x_2 to x_1 gets smaller, moving from a to b. Therefore, in moving from a to b both the ratio and MRS decline. Conversely, in the movement from b to a both increase. Consequently, as defined, the elasticity of substitution is non-negative, provided the level surface of the function is convex, as depicted.

EXAMPLE 7.6: The constant elasticity of substitution function is:

$$u(x_1,x_2) = [\alpha x_1^\rho + (1-\alpha)x_2^\rho]^{\frac{1}{\rho}}.$$

Therefore,

$$\begin{aligned} u_{x_1} &= \frac{1}{\rho}[\alpha x_1^\rho + (1-\alpha)x_2^\rho]^{\frac{1}{\rho}-1}\alpha\rho x_1^{\rho-1} \\ u_{x_2} &= \frac{1}{\rho}[\alpha x_1^\rho + (1-\alpha)x_2^\rho]^{\frac{1}{\rho}-1}(1-\alpha)\rho x_2^{\rho-1}. \end{aligned}$$

Therefore, MRS is:

$$\frac{u_{x_1}}{u_{x_2}} = \frac{\alpha\rho x_1^{\rho-1}}{(1-\alpha)\rho x_2^{\rho-1}} = \frac{\alpha}{1-\alpha}\left(\frac{x_1}{x_2}\right)^{\rho-1}.$$

Taking the ln of both sides,

$$\ln\left(\frac{u_{x_1}}{u_{x_2}}\right) = \ln\left(\frac{\alpha}{1-\alpha}\right) + (\rho-1)\ln\left(\frac{x_1}{x_2}\right) = \ln\left(\frac{\alpha}{1-\alpha}\right) + (1-\rho)\ln\left(\frac{x_2}{x_1}\right).$$

Therefore,

$$\ln\left(\frac{x_2}{x_1}\right) = -\frac{1}{1-\rho}\ln\left(\frac{\alpha}{1-\alpha}\right) + \frac{1}{1-\rho}\ln\left(\frac{u_{x_1}}{u_{x_2}}\right)$$

and

$$\eta = \frac{d\ln\left(\frac{x_2}{x_1}\right)}{d\ln\left(\frac{u_{x_1}}{u_{x_2}}\right)} = \frac{1}{1-\rho}.$$

◊

The constant elasticity function can exhibit substantial variation in shape, depending on the value of ρ. Figure 7.5 illustrates three possible values for ρ. The figure suggests some properties of this function as ρ varies. When $\rho = 1$, the function is linear. As ρ decreases to 0, the function approaches the Cobb–Douglas function, $x_1^{\alpha}x_2^{1-\alpha}$, while, as ρ goes to $-\infty$, the function approaches the Leontief function, $\min\{x_1, x_2\}$ (and as ρ tends to ∞, the function tends to $\max\{x_1, x_2\}$).

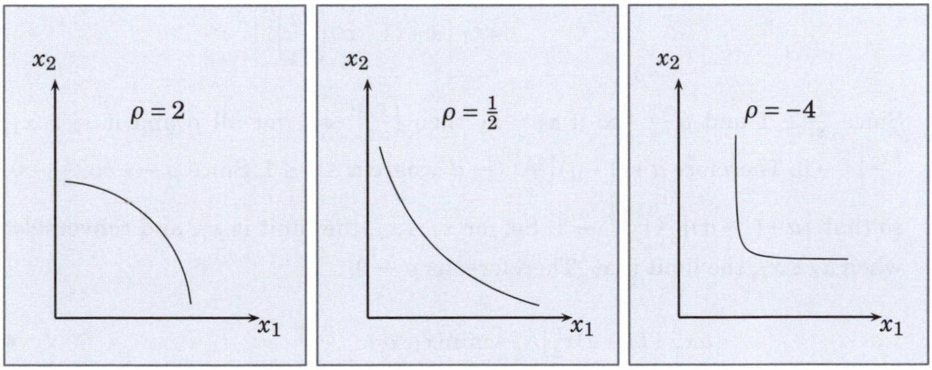

Figure 7.5: Plots of level contours of $u = [x_1^{\rho} + x_2^{\rho}]^{\frac{1}{\rho}}$

Theorem 7.2: *Given* $u(x_1, x_2) = [\alpha x_1^{\rho} + (1-\alpha)x_2^{\rho}]^{\frac{1}{\rho}}$, *then:*

- $[\alpha x_1^{\rho} + (1-\alpha)x_2^{\rho}]^{\frac{1}{\rho}} \to x_1^{\alpha}x_2^{1-\alpha}$ *as* $\rho \to 0$
- $[\alpha x_1^{\rho} + (1-\alpha)x_2^{\rho}]^{\frac{1}{\rho}} \to \min\{x_1, x_2\}$ *as* $\rho \to -\infty$.

PROOF: Write u^{ρ} to make dependence on ρ explicit. Assume at least one of $\{x_1, x_2\}$ is strictly positive, so that $\alpha x_1^{\rho} + (1-\alpha)x_2^{\rho}$ is strictly positive. Let $U^{\rho}(x_1, x_2) = f(u^{\rho}(x_1, x_2)) = \ln u^{\rho}(x_1, x_2) = \frac{\ln[\alpha x_1^{\rho} + (1-\alpha)x_2^{\rho}]}{\rho} = \frac{q(\rho)}{\rho}$. So, f is a continuous monotonic transformation of u. Observe that $\alpha x_1^{\rho} + (1-\alpha)x_2^{\rho} \to 1$ as $\rho \to 0$, so $\ln[\alpha x_1^\rho + (1-\alpha)x_2^\rho] \to 0$ and therefore both numerator and denominator go to 0 as ρ goes to 0. Using l'Hospital's rule, the limit of this ratio is the limit of the derivatives:

$$\lim_{\rho\to 0} \frac{\ln[\alpha x_1^\rho + (1-\alpha)x_2^\rho]}{\rho} = \lim_{\rho\to 0} \frac{\alpha x_1^\rho \ln x_1 + (1-\alpha)x_2^\rho \ln x_2}{\alpha x_1^\rho + (1-\alpha)x_2^\rho} = \alpha \ln x_1 + (1-\alpha)\ln x_2,$$

using the fact that $\frac{d}{d\rho} z^\rho = \frac{d}{d\rho} e^{(\ln z)\rho} = \ln z e^{(\ln z)\rho} = \ln z \cdot z^\rho$ and $\lim_{\rho\to 0} z^\rho = 1$. The function $f^{-1}(z) = e^z$ is continuous, so that $u^\rho = f^{-1}(U^\rho)$ is continuous. Since $U^\rho \to \alpha \ln x_1 + (1-\alpha)\ln x_2$,

$$u^\rho(x_1, x_2) \to e^{\alpha \ln x_1 + (1-\alpha)\ln x_2} = e^{\alpha \ln x_1} e^{(1-\alpha)\ln x_2} = x_1^\alpha x_2^{1-\alpha}.$$

So, as ρ tends to 0, the utility function tend to the Cobb–Douglas function.

In the second case, suppose that $x_1 \le x_2$. Write

$$\begin{aligned}[\alpha x_1^\rho + (1-\alpha)x_2^\rho]^{\frac{1}{\rho}} &= \left[x_1^\rho \left\{\alpha + (1-\alpha)\left(\frac{x_2}{x_1}\right)^\rho\right\}\right]^{\frac{1}{\rho}} \\ &= x_1 \left[\alpha + (1-\alpha)\left(\frac{x_2}{x_1}\right)^\rho\right]^{\frac{1}{\rho}}.\end{aligned}$$

Since $\frac{x_2}{x_1} \ge 1$ and $\rho \to -\infty$ if $x_2 = x_1$ then $\left(\frac{x_2}{x_1}\right)^\rho = 1$ for all ρ and if $x_2 > x_1$, $\left(\frac{x_2}{x_1}\right)^\rho \to 0$. Therefore $\alpha + (1-\alpha)\left(\frac{x_2}{x_1}\right)^\rho \to d$ where $\alpha \le d \le 1$. Since $\rho \to -\infty$, $\frac{1}{\rho} \to 0$, so that $\left[\alpha + (1-\alpha)\left(\frac{x_2}{x_1}\right)^\rho\right]^{\frac{1}{\rho}} \to 1$. So, for $x_1 \le x_2$, the limit is x_1, and conversely; when $x_2 \le x_1$, the limit is x_2. Therefore, as $\rho \to 0$,

$$[\alpha x_1^\rho + (1-\alpha)x_2^\rho]^{\frac{1}{\rho}} \to \min\{x_1, x_2\}. \qquad \blacksquare$$

7.5 Market equilibrium: an application

In the following discussion, partial derivatives are used to find the equilibrium output in an oligopolistic market with n firms. Consider a market where the (inverse) demand is given by $P(Q)$, giving the price at which the quantity Q will sell. Suppose there are n firms, where firm i produces output q_i. Let Q denote aggregate output: $Q = \sum_{i=1}^n q_i$. Suppose also that firm i has cost function $c(q_i)$, giving the total cost of producing q_i units for firm i. Sales of q_i units at price p generate a profit of $pq_i - c(q_i)$ for firm i, and suppose the objective of the firm is to maximize profit. The problem in this environment is that the price depends on total output Q, and Q depends on q_i and the other firms' output levels, $\{q_j\}_{j\neq i}$.

So, profit of firm i depends on the output levels of all the firms. Making this explicit:

$$\pi_i(q_1,\ldots,q_n) = q_i P(Q) - c_i(q_i) = q_i P\left(\sum_{j=1}^{n} q_i\right) - c(q_i).$$

Given the profile of output levels $(q_1,\ldots,q_n)$, a firm, say firm i, may consider the possibility of changing output to some other level, say q'_i. It is in the firm's interest to do so if:

$$\pi_i(q_1,\ldots,q_{i-1},q'_i,q_{i+1},\ldots,q_n) > \pi_i(q_1,\ldots,q_{i-1},q_i,q_{i+1},\ldots,q_n).$$

In that case, changing output from q_i to q'_i raises profit. When firms behave in this way (competing by choice of quantity and picking the profit-maximizing output, taking the actions of others as given), the form of competition is called *Cournot competition*. From the previous expression, if firm i chooses q_i to maximize π_i, at the chosen level of q_i it must be that either increasing or reducing q_i by a small amount provides no increase in firm i's profit:

$$\frac{\partial \pi_i(q_1,\ldots,q_{i-1},q_i,q_{i+1},\ldots,q_n)}{\partial q_i} = 0.$$

Definition 7.1: *Call an output profile $(q_1^*,\ldots,q_n^*)$ a Cournot equilibrium if no firm has an incentive to change output level.*

This condition (that no firm has an incentive to change output) will be true if, for each i,

$$\pi_i(q_1^*,\ldots,q_{i-1}^*,q_i,q_{i+1}^*,\ldots,q_n^*)$$

is maximized at $q_i = q_i^*$, taking q_j^* constant for $j \neq i$.

A *necessary* condition for maximization of $\pi_i(q_1^*,\ldots,q_{i-1}^*,q_i,q_{i+1}^*,\ldots,q_n^*)$ by choice of q_i is that the (partial) derivative with respect to q_i is equal to 0 at $q_i = q_i^*$. The derivative of $\pi_i(q_1^*,\ldots,q_{i-1}^*,q_i,q_{i+1}^*,\ldots,q_n^*)$ with respect to q_i equals 0 at $q_i = q_i^*$:

$$\frac{\partial}{\partial q_i}\pi_i(q_1^*,\ldots,q_{i-1}^*,q_i^*,q_{i+1}^*,\ldots,q_n^*) = 0.$$

(Here, $\frac{\partial}{\partial q_i}\pi_i(q_1^*,\ldots,q_{i-1}^*,q_i^*,q_{i+1}^*,\ldots,q_n^*) = 0$ is understood to mean the derivative

$$\frac{\partial}{\partial q_i}\pi_i(q_1^*,\ldots,q_{i-1}^*,q_i,q_{i+1}^*,\ldots,q_n^*)$$

evaluated at $q_i = q_i^*$.)

Whether this condition is *sufficient* for the output vector to be an equilibrium is taken up later. The next section applies these calculations in the case where demand and cost functions are linear.

7.5.1 Oligopoly: linear demand

To illustrate the calculations, consider the case of linear demand $P(Q) = a - bQ$, and let firm i's cost be given by $c(q_i) = cq_i$, so that every firm faces the same constant marginal cost, c, and 0 fixed cost. Then

$$\begin{aligned}\pi_i &= q_i\left(a - b\sum_{j=1}^{n} q_j\right) - cq_i \\ &= aq_i - bq_i\left(\sum_{j=1}^{n} q_j\right) - cq_i \\ &= aq_i - bq_i\left(\sum_{j\neq i} q_j\right) - bq_i^2 - cq_i.\end{aligned}$$

Taking the partial derivative of π_i with respect to q_i and setting to zero gives:

$$\frac{\partial \pi_i}{\partial q_i} = a - c - 2bq_i - b\sum_{j\neq i} q_j = 0,$$

and rearranging the equality:

$$\frac{(a-c)}{b} = 2q_i + \sum_{j\neq i} q_j.$$

This gives a system of equations:

$$\begin{pmatrix} 2 & 1 & \cdots & 1 \\ 1 & 2 & \cdots & 1 \\ \vdots & \vdots & \ddots & \vdots \\ 1 & 1 & \cdots & 2 \end{pmatrix}\begin{pmatrix} q_1 \\ q_2 \\ \vdots \\ q_n \end{pmatrix} = \begin{pmatrix} \frac{a-c}{b} \\ \frac{a-c}{b} \\ \vdots \\ \frac{a-c}{b} \end{pmatrix},$$

which has the solution: $(q_1^*, \ldots, q_n^*) = \left(\frac{a-c}{(n+1)b}, \ldots, \frac{a-c}{(n+1)b}\right)$. (To see this note that if the q_i are all equal and satisfy any equation, then they satisfy all equations, because with $q_i = q^*$, each equation requires $(n+1)q^* = \frac{a-c}{b}$.) Thus, $q_i^* = q^* = \frac{1}{n+1}\frac{a-c}{b}$ for each i, and

$$\frac{\partial}{\partial q_i}\pi_i(q_1^*, \ldots, q_{i-1}^*, q_i^*, q_{i+1}^*, \ldots, q_n^*) = 0$$

is satisfied. This is called a symmetric equilibrium, since each firm makes the same choice. In fact, the solution might have been determined more easily. From the condition

$$\frac{\partial \pi_i}{\partial q_i} = a - b\left(\sum_{j\neq i} q_i\right) - 2bq_i - c = 0,$$

assuming that the equilibrium is symmetric (each firm makes the same choice), this will be satisfied by the symmetric choice $q_i = q^*$ so that $\sum_{j\neq i} q_i = (n-1)q^*$, giving $a - b(n-1)q^* - 2bq^* - c = 0$, or $a - c - b(n+1)q^* = 0$, or $q^* = \frac{1}{n+1}\frac{a-c}{b}$. In this case, total output is $Q_o = nq^* = \frac{n}{n+1}\frac{a-c}{b}$.

REMARK 7.2: Recall that the competitive quantity, Q_c, is the quantity that leads to price equaling marginal cost: $P(Q_c) = c$. In the linear case, $Q_c = \frac{(a-c)}{b}$, so that Q_o, the equilibrium oligopoly output, is related to Q_c according to $Q_o = \frac{n}{(n+1)}Q_c$. The price associated with the oligopoly output is $P(Q_o) = a - bQ_o = a - b\frac{n}{(n+1)}Q_c$. Substituting in the value of $Q_c : P(Q_o) = a - b\frac{n}{(n+1)}\frac{(a-c)}{b} = a - \frac{n}{(n+1)}[(a-c)] = \frac{a(n+1)-n(a-c)}{(n+1)} = \frac{(a+nc)}{(n+1)} = \frac{a}{(n+1)} + c\frac{n}{(n+1)}$. To summarize:

$$Q_o = \frac{n}{n+1}Q_c, \text{ and } P(Q_o) = a\frac{1}{n+1} + \frac{n}{n+1}c.$$

As the number of firms becomes large ($n \to \infty$), the quantity converges to the competitive quantity, $Q_o \to Q_c$, and the price converges to the competitive price (marginal cost), since $a\frac{1}{n+1} + \frac{n}{n+1}c \to c$. □

7.5.2 Oligopoly: non-linear demand

In the general case of non-linear price and cost functions, the profit function is:

$$\pi_i(q_1,\ldots,q_n) = q_i P\left(\sum_{i=1}^{n} q_i\right) - c(q_i).$$

Repeating the earlier calculations:

$$\frac{\partial \pi_i}{\partial q_i} = P\left(\sum_{i=1}^{n} q_i\right) + q_i \cdot \frac{dP}{dQ} - c'(q_i) = 0.$$

In a symmetric equilibrium, each firm produces the same quantity, say q^*, and this must satisfy the equation:

$$P(nq^*) + q^* P'(nq^*) - c'(q^*) = 0$$

$$P(nq^*)\left\{1 + \frac{1}{n}\left[\frac{nq^*}{P(nq^*)}P'(nq^*)\right]\right\} = c'(q^*),$$

or, using $Q^* = nq^*$,

$$P(Q^*)\left\{1+\frac{1}{n}\left[\frac{Q^*}{P(Q^*)}P'(Q^*)\right]\right\} = c'(q^*)$$

and with $\eta(Q)$ denoting the elasticity of demand at Q, $\eta(Q) = -\frac{P}{Q}\frac{dQ}{dP}$,

$$P(Q^*)\left[1-\frac{1}{n}\frac{1}{\eta(Q^*)}\right] = c'(q^*).$$

Thus, again price converges to marginal cost as $n \to \infty$. This is a common theme in economic models: competition (a large number of firms) leads to efficient levels of output.

7.5.3 Industry concentration indices

The previous discussion suggests that when a small number of firms are supplying a market, there is less competition, output is lower and prices higher. In industrial organization, a market is said to be "concentrated" if there are a small number of suppliers or if a small number of firms dominate the market. A natural question is: when is a market too concentrated, in the sense that market dominance of a few firms leads to large inefficiencies? One measure of concentration is the Herfindahl–Hirschman index. This index is a simple function of market shares of firms. The discussion here relates that index to the Cournot model using the techniques developed earlier.

When there are a small number of firms in an industry, some firms will have a larger share of industry sales. If one or two firms have a large share of total sales, the industry is considered to be concentrated. The Herfindahl–Hirschman index (HHI) of concentration in an industry is defined:

$$HHI = (100s_1)^2 + (100s_2)^2 + \cdots + (100s_n)^2 = 100^2\left(\sum_{i=1}^{n} s_i^2\right),$$

where s_i is the share of firm i of industry sales. This sum is minimized when the shares are all equal, so that larger values of HHI suggest greater industry concentration. (For example, with just two firms, $s_1 + s_2 = 1$, so that $s_1^2 + s_2^2 = s_1^2 + (1-s_1)^2$. Differentiating with respect to s_1 and setting to 0 gives $2s_1 - 2(1-s_1) = 0 = 4s_1 - 2$. So $s_1 = \frac{1}{2}$ and this corresponds to a minimum, since the second derivative of $s_1^2 + (1-s_1)^2$ equals 4, which is positive.) With n firms, the minimum occurs at $s_i = \frac{1}{n}$ for all i. In this case, $s_i^2 = \frac{1}{n^2}$ and $\sum_{i=1}^{n} s_i^2 = \sum_{i=1}^{n} \frac{1}{n^2} = n\frac{1}{n^2} = \frac{1}{n}$. Consequently, the smallest possible value for HHI is $100^2\frac{1}{n}$, when there are n firms. (In fact, viewing the HHI index as a function of the shares: $HHI(s_1, \ldots, s_n)$, the total change is $d(HHI) = (100^2)2\cdot[s_1ds_1 + s_2ds_2 + \cdots + s_nds_n]$, where the share changes must add to 0: $\sum_{i=1}^n ds_i = 0$. If, for some j,k, $s_j > s_k$ then setting $ds_k = -ds_j$ with $ds_j < 0$ decreases the HHI. Thus, $s_j = s_k$ is necessary for a minimum.)

Consider now the firm's optimizing condition in choosing q_i. In what follows, allow different firms to have different cost functions: the cost function of firm i is $c_i(q_i)$. Maximizing profit for i gives the first-order condition:

$$P\left(\sum_{i=1}^n q_i\right) + q_i \cdot \frac{dP}{dQ} - c_i'(q_i) = 0.$$

Write $Q = \sum_i q_i$, so that $s_i = \frac{q_i}{Q}$. Rearrange the condition as:

$$P(Q) - c_i'(q_i) + q_i \cdot P'(Q) = 0.$$

Divide by $P(Q)$:

$$\frac{P(Q) - c_i'(q_i)}{P(Q)} + q_i \cdot \frac{P'(Q)}{P(Q)} = 0.$$

Multiply by $\frac{q_i}{Q}$:

$$\frac{q_i}{Q}\frac{P(Q) - c_i'(q_i)}{P(Q)} + Q \cdot \frac{P'(Q)}{P(Q)}\frac{q_i^2}{Q^2} = 0.$$

Recalling that $s_i = \frac{q_i}{Q}$ and

$$\eta(Q) = -\frac{P}{Q}\frac{dQ}{dP} \Rightarrow \frac{1}{\eta(Q)} = -\frac{Q}{P}\frac{dP}{dQ} = -\frac{Q}{P}P'(Q) = -Q\frac{P'(Q)}{P(Q)},$$

this expression may be written:

$$s_i\frac{P(Q) - c_i'(q_i)}{P(Q)} - \frac{1}{\eta(Q)}s_i^2 = 0.$$

So,

$$\sum_{i=1}^n s_i \frac{P(Q) - c_i'(q_i)}{P(Q)} = \frac{1}{\eta(Q)}\sum_{i=1}^n s_i^2 = \frac{1}{100^2}\frac{1}{\eta(Q)}100^2\sum_{i=1}^n s_i^2 = \frac{HHI}{100^2\,\eta(Q)}$$

Since $P(Q) - c_i'(q_i)$ is the price over cost margin of firm i, $\sum_{i=1}^n s_i \frac{P(Q)-c_i'(q_i)}{P(Q)}$ is the weighted average of these margins. So, the higher is the HHI, the higher is the average price–cost margin or markup.

7.6 Total differentials

The partial derivative of a function $f(x_1,\ldots,x_n)$ with respect to a variable x_i measures the impact of a change in x_i on the value of the function f. If f_{x_i} is the partial derivative of f with respect to x_i, then the impact of a small change in x_i, Δx_i, on f is approximately $f_{x_i}\,\Delta x_i$. One may consider the impact of multiple changes in all the x_i at the same time, to consider a total variation in f resulting from a set of small changes in all the x_i.

Given a function $f(x_1,\ldots,x_n)$, the total differential of f, denoted df, is defined:

$$df = \sum_{i=1}^{n} \frac{\partial f}{\partial x_i} dx_i = \sum_{i=1}^{n} f_{x_i} dx_i. \tag{7.3}$$

Note that df depends on $x = (x_1,\ldots,x_n)$ and $dx = (dx_1,\ldots,dx_n)$. Put $\frac{\partial f}{\partial x} = \left(\frac{\partial f}{\partial x_1}, \frac{\partial f}{\partial x_2},\ldots,\frac{\partial f}{\partial x_n}\right)$ and $dx = (dx_1, dx_2, \ldots dx_n)$. Then $df = \frac{\partial f}{\partial x} \cdot dx$. In some branches of mathematics, this sum is sometimes referred to as a "1-form", consisting of the inner product of $(f_{x_1},\ldots,f_{x_n})$ and the vector direction $(dx_1,\ldots,dx_n)$. Given $(x_1,\ldots,x_n)$, the partial derivatives are determined, the functions $f_{x_i}(x_1,\ldots,x_n)$, and Equation 7.3 is an equation in the changes, the dx_i.

EXAMPLE 7.7: The following functions illustrate the total differential:

1. Let $f(x,y) = x^2 \ln(y)$.

$$df = 2x\ln(y)dx + x^2\frac{1}{y}dy.$$

2. Let $f(x,y) = \sin(x)\cos(y)\ln(xy)$. The total differential is:

$$df = \left[\cos(x)\cos(y)\ln(xy) + \sin(x)\cos(y)\frac{1}{x}\right]dx + \left[-\sin(x)\sin(y)\ln(xy) + \sin(x)\cos(y)\frac{1}{y}\right]dy.$$

3. Let $f(x,y) = e^{\tan(xy)}$. The total differential is:

$$df = e^{\tan(xy)}\left[\left(\frac{y}{\cos^2(xy)}\right)dx + \left(\frac{x}{\cos^2(xy)}\right)dy\right],$$

using the fact that $\frac{d}{dz}\tan(z) = \sec^2(z) = \frac{1}{\cos^2(z)}$.

4. Let $f(x,y) = \left(\frac{\ln(y)}{e^x}\right)^{\alpha}$. The total differential is:

$$df = \alpha\left(\frac{\ln(y)}{e^x}\right)^{\alpha-1}\left[\left(\frac{-\ln y}{e^x}\right)dx + \left(\frac{1}{ye^x}\right)dy\right].$$

◊

The partial derivative of a function $f(x,y)$, say with respect to x, $f_x(x,y)$, measures the change in $f(x,y)$ due to a small change in x, dx. This is approximately

$$f_x \approx \frac{f(x+dx,y)-f(x,y)}{dx}.$$

Thus, $f_x(x,y)dx \approx f(x+dx,y)-f(x,y)$ is the change in f resulting from the change in x. So, $df = f_x(x,y)dx + f_y(x,y)dy$ measures the overall change in f from the combined effects of changing x and y. When the overall change is 0, $df = 0$, and so $0 = f_x(x,y)dx + f_y(x,y)dy$. This is an equation in dx and dy, given (x,y), which may be solved to give $dy = -\frac{f_x(x,y)}{f_y(x,y)}dx$, the change in y necessary to keep the value of the function constant, given the change in x of dx.

For example, consider the function $f(x,y) = ax + by$. Then $df = adx + bdy$, since $f_x = a$ and $f_y = b$. If $df = 0$, then $0 = adx + bdy$ and $dy = -\frac{a}{b}dx$. If x is increased by dx, y must be reduced by $dy = -\frac{a}{b}dx$ to maintain the value of the function constant. The following example utilizes this "trade-off" computation in an economics application (using the marginal rate of substitution).

Sometimes, partial derivatives may be used in the simplification of a problem. The following example reduces a two-variable constrained optimization problem to a one-variable problem, where partial derivatives are then used to locate the optimum. Utility is given as a function of income (y) and leisure (l): $u = u(y,l)$. If current income and leisure levels are (y,l), then small changes in income and leisure, dy and dl, produce a change in utility given by: $du = u_y(y,l)dy + u_l(y,l)dl$, the total differential of u at (y,l). To simplify notation, write $du = u_y dy + u_l dl$. When this is set to 0, the equation $0 = u_y dy + u_l dl$ relates small changes in y to small changes in l that leave utility unchanged: $dy = -\frac{u_l}{u_y}dl$ or $dl = -\frac{u_l}{u_y}dy$. Thus, $\frac{dy}{dl} = -\frac{u_l}{u_y}$ given the rate at which y must be substituted for l at the point (y,l) to keep utility constant: as l (leisure) increases, y must be reduced at the rate $-\frac{u_l}{u_y}$ to keep utility constant. The partial derivatives u_y and u_l are the marginal utility of y and l respectively.

With work defined as available time (24) minus leisure (l), the time spent working is $24 - l$ and at wage ω, income is $y = \omega(24 - l)$. Therefore, $dy = -\omega dl$. Substitute into the total differential: $du = u_y(-\omega dl) + u_l dl = [-\omega \cdot u_y + u_l]dl$. If utility is maximized, then it should not be possible for small changes in l to increase utility. For this to be the case requires that $[-\omega \cdot u_y + u_l] = 0$, or $\omega \cdot u_y = u_l$, or $\omega = \frac{u_l}{u_y}$. Figure 7.6 depicts the solution.

The same condition is obtained directly by optimization. The income constraint, $y = \omega(24-l)$, can be substituted into the utility function: $u(\omega(24-l),l) = u^*(l)$. The problem of maximizing utility reduces to maximizing u^* with respect to l.

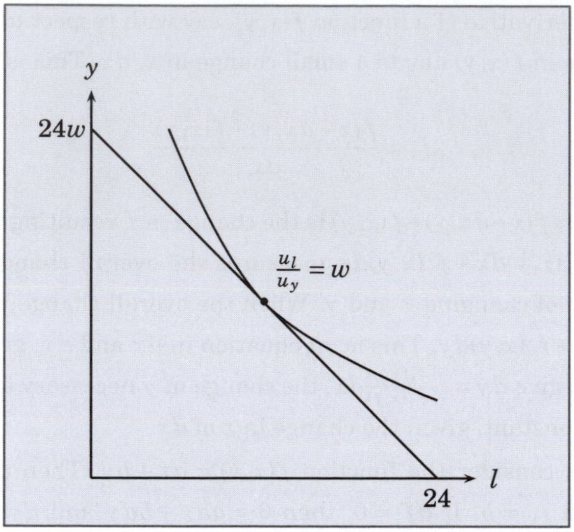

Figure 7.6: Optimal income–work combination

$$\max_l u^* \text{ gives } \frac{du^*}{dl} = 0 = \frac{\partial u}{\partial y}\frac{dy}{dl} + \frac{\partial u}{\partial l}.$$

Now, $\frac{dy}{dl} = -\omega$, from the budget constraint, so the first-order condition becomes

$$-u_y\omega + u_l = 0, \text{ or } \omega = \frac{u_l}{u_y} = \frac{MU_l}{MU_y},$$

where $\frac{MU_l}{MU_y}$ is the ratio of marginal utilities. For a numerical example, suppose that $u(y,l) = y^{\frac{1}{2}}l^{\frac{1}{2}}$ and so:

$$u_y = \frac{1}{2}y^{-\frac{1}{2}}l^{\frac{1}{2}} = \frac{\frac{1}{2}y^{\frac{1}{2}}l^{\frac{1}{2}}}{y};\ u_l = \frac{1}{2}y^{\frac{1}{2}}l^{-\frac{1}{2}} = \frac{\frac{1}{2}y^{\frac{1}{2}}l^{\frac{1}{2}}}{l}.$$

This gives $\omega = \frac{u_l}{u_y} = \frac{y}{l}$ or $\omega l = y$ as the optimizing condition. Recalling that $y = \omega(24-l)$, the optimizing condition may be written $\omega l = \omega(24-l)$. Thus, $2\omega l = 24\omega$ or $l = \frac{24}{2} = 12$ and $y = \omega l = 12\omega$.

EXAMPLE 7.8: This example considers a standard profit maximization problem. Let a firm's profit be given by:

$$\pi(k,l) = Pf(k,l) - rk - wl.$$

The total derivative of π is:

$$d\pi = \pi_k dk + \pi_l dl = (Pf_k - r)dk + (Pf_l - w)dl.$$

For profit to be maximized, it must be that there is no variation in k or l that raises profit. This implies that $(Pf_k - r) = 0$ and $(Pf_l - w) = 0$. If, for example, $(Pf_k - r) < 0$, then reducing k $(dk < 0)$ would raise profit, while if $(Pf_k - r) > 0$, then raising k $(dk > 0)$ would raise profit. This gives the *first-order conditions*. The first-order conditions are:

$$\frac{\partial \pi}{\partial k} = Pf_k - r = 0,\ f_k = \frac{\partial f}{\partial k}$$

and

$$\frac{\partial \pi}{\partial l} = Pf_l - w = 0,\ f_l = \frac{\partial f}{\partial l}.$$

These conditions imply that $f_k = \frac{r}{P}$ and $f_l = \frac{w}{P}$.

Consider a specific functional form: $f(k,l) = k^\alpha l^\beta$, then:

$$f_k = \alpha k^{\alpha-1} l^\beta = \alpha \frac{f}{k}, \text{ and } f_l = \beta k^\alpha l^{\beta-1} = \beta \frac{f}{l}.$$

Rearranging, $\frac{f_k}{f_l} = \frac{\alpha}{\beta}\frac{l}{k} = \frac{r}{w}$, so that $\alpha w l = \beta r k$ and $k = \frac{\alpha w}{\beta r} l$. This determines k as a function of l. To solve for l, use either of the first-order conditions. Taking $f_k = \alpha k^{\alpha-1} l^\beta = \frac{r}{P}$ and substituting for k yields:

$$\alpha \left(\frac{\alpha w}{\beta r}\right)^{\alpha-1} l^{\alpha-1} l^\beta = \frac{r}{P}$$

$$\alpha \left(\frac{\alpha w}{\beta r}\right)^{\alpha-1} l^{(\beta+\alpha-1)} = \frac{r}{P}.$$

Solving for l:

$$l = \left\{\frac{r}{P}\frac{1}{\alpha}\left(\frac{\alpha w}{\beta r}\right)^{1-\alpha}\right\}^{1/(\alpha+\beta-1)}.$$

Similarly, the value of k is then determined:

$$\begin{aligned}
k &= \frac{\alpha w}{\beta r}\left\{\frac{r}{P}\frac{1}{\alpha}\left(\frac{\alpha w}{\beta r}\right)^{1-\alpha}\right\}^{1/(\alpha+\beta-1)} \\
&= \left\{\left(\frac{\alpha w}{\beta r}\right)^{\alpha+\beta-1}\left(\frac{r}{P}\right)\frac{1}{\alpha}\left(\frac{\alpha w}{\beta r}\right)^{1-\alpha}\right\}^{1/(\alpha+\beta-1)} \\
&= \left\{\left(\frac{\alpha w}{\beta r}\right)^{\beta}\left(\frac{r}{P}\right)\frac{1}{\alpha}\right\}^{1/(\alpha+\beta-1)}
\end{aligned}$$

$$= \left\{ \left(\frac{\alpha\omega}{\beta r}\right)^{\beta} \left(\frac{w}{P}\right) \left(\frac{\beta}{\alpha}\right) \left(\frac{r}{\omega}\right) \left(\frac{1}{\beta}\right) \right\}^{1/(\alpha+\beta-1)}$$

$$= \left\{ \left(\frac{\alpha\omega}{\beta r}\right)^{\beta-1} \left(\frac{w}{P}\right) \left(\frac{1}{\beta}\right) \right\}^{1/(\alpha+\beta-1)}$$

$$= \left\{ \frac{w}{P} \frac{1}{\beta} \left(\frac{\beta r}{\alpha\omega}\right)^{1-\beta} \right\}^{1/(\alpha+\beta-1)}.$$

◊

7.6.1 The impact of parameter changes

The following discussion illustrates the impact of parameter shifts on the solution to a pair of equations. A production function $q = f(x, y)$ gives the output, q, associated with inputs x and y. Assume that raising input of either x or y raises output, $f_x > 0$ and $f_y > 0$. An *isoquant* identifies different combinations of x and y that give the same level of output. For example, $I_q = \{(x, y) \mid f(x, y) = q\}$ is the set of (x, y) pairs that give output q. Under standard assumptions on the function f, an isoquant is convex to the origin, as depicted in Figure 7.7. Changing x and y by small amounts, dx and dy, changes output according to $dq = df = f_x dx + f_y dy$. If the changes in x and y leave output unchanged, $0 = f_x dx + f_y dy$, then $(x + dx, y + dy)$ is on the same isoquant as (x, y), as indicated in Figure 7.7, where $a = (x, y)$ and $b = (x + dx, y + dy)$. Now, suppose that the producer must use inputs x and y in the proportion k, so that the choice of (x, y) must satisfy $y = kx$. Thus, output level q is achieved with input usage (x, y) at location a. Implicitly, the choice of q and k determines the values of x and y: $(k, q) \to (x, y)$. To make this explicit, write $x = s(k, q)$ and $y = t(k, q)$. So $f(x, y) = q$ and $y = kx$ implicity determine (x, y).

Consider a variation in k, dk. The total differential of $y = kx$ is $dy = xdk + kdx$, giving small variations in y, k, x that permit the equation $y = kx$ to continue to be satisfied. Next, if x and y must change so as to leave output unchanged, $0 = f_x dx + f_y dy$, and this with the requirement $dy = xdk + kdx$ determines the way in which x and y change. Substituting the requirement $dy = xdk + kdx$ into $0 = f_x dx + f_y dy$ gives:

$$0 = f_x dx + f_y(xdk + kdx) = (f_x + f_y k)dx + f_y xdk.$$

Thus, $(f_x + f_y k)dx = -f_y xdk$ or $dx = -\frac{f_y x}{(f_x + f_y k)} dk$. This measures the change in x due to a change in k, with q unchanged. Dividing through by dk, it has the

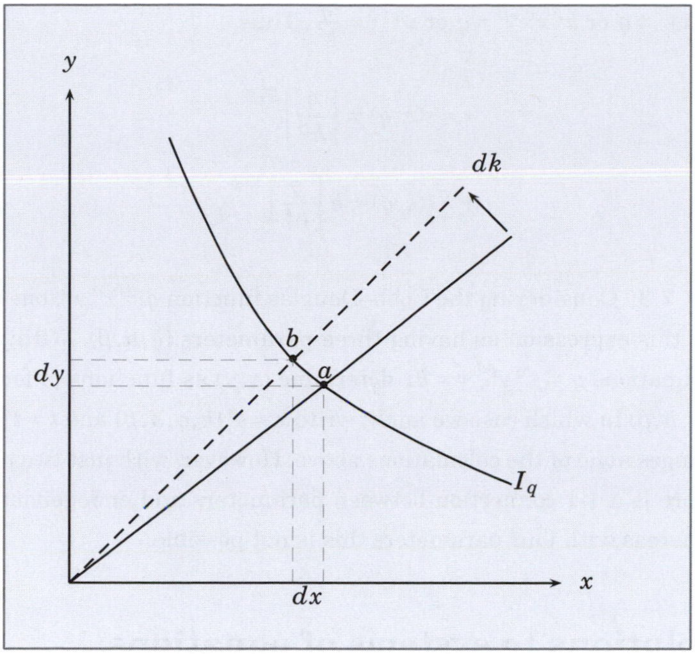

Figure 7.7: A parameter change

interpretation as a partial derivative (since q is held fixed):

$$\frac{\partial x}{\partial k} = -\frac{f_y x}{(f_x + f_y k)}. \tag{7.4}$$

In a similar way, we can determine the impact of a change in k in y. Recall that $dy = xdk + kdx = xdk + k\left[-\frac{f_y x}{(f_x+f_y k)}\right]dk = \left(x + k\left[-\frac{f_y x}{(f_x+f_y k)}\right]\right)dk$. Thus,

$$\frac{\partial y}{\partial k} = \left(x + k\left[-\frac{f_y x}{(f_x + f_y k)}\right]\right). \tag{7.5}$$

In these two Equations 7.4 and 7.5, the values of x and y are understood to be the solution values, located at a.

Recall that when x and y are solved in terms of (k,q), we have $x = s(k,q)$ and $y = t(k,q)$, so the calculation in Equation 7.4 corresponds to $\frac{\partial x}{\partial k} = \frac{\partial s}{\partial k}$. In general, determining the functions s and t may be difficult, depending on the functional form of f. Therefore, the calculations in Equations 7.4 and 7.5 provide insight into the impact of parameter variations, *even though we have not solved for x and y.* In the case where the production function has the Cobb–Douglas form, $f(x,y) = x^{\alpha}y^{\beta}$, the functions may be found easily. Since $x^{\alpha}y^{\beta} = q$ and $y = kx$, substituting gives $x^{\alpha}(kx)^{\beta} = q$ or $k^{\beta}x^{\alpha+\beta} = q$ or $x^{\alpha+\beta} = \frac{q}{k^{\beta}}$. Thus,

$$x = s(k,q) = \left[\frac{q}{k^{\beta}}\right]^{\frac{1}{\alpha+\beta}}$$

$$y = t(k,q) = k\left[\frac{q}{k^{\beta}}\right]^{\frac{1}{\alpha+\beta}}.$$

REMARK 7.3: Considering the Cobb–Douglas function $q = x^{\alpha}y^{\beta}$, one might also consider this expression as having three parameters (q,α,β). With $y = kx$, the pair of equations $q = x^{\alpha}y^{\beta}$, $y = kx$ determine (x,y) as functions of four parameters (k,q,α,β) in which case we might write $x = s^*(k,q,\alpha,\beta)$ and $t = t^*(k,q,\alpha,\beta)$. This changes none of the calculations above. However, with just two parameters (k,q) there is a 1-1 connection between parameters and endogenous variables (x,y), whereas with four parameters this is not possible. □

7.7 Solutions to systems of equations

Consider a system of linear equations $\boldsymbol{A}\boldsymbol{y} = \boldsymbol{x}$, where $\boldsymbol{A}$ is an $n \times n$ matrix, $\boldsymbol{y}$ a vector of variables and $\boldsymbol{x}$ a vector of parameters. Provided $\boldsymbol{A}$ is non-singular, we can solve for $\boldsymbol{y}$: there is a unique solution $\boldsymbol{y} = \boldsymbol{A}^{-1}\boldsymbol{x}$. Furthermore, the impact of variations in $\boldsymbol{x}$ on the solution value of $\boldsymbol{y}$ is easy to determine. If $\boldsymbol{x}$ changes to $\boldsymbol{x} + \boldsymbol{\Delta x}$, the new solution value of $\boldsymbol{y}$ is given by

$$\boldsymbol{y}' = \boldsymbol{A}^{-1}(\boldsymbol{x} + \Delta\boldsymbol{x}) = \boldsymbol{A}^{-1}\boldsymbol{x} + \boldsymbol{A}^{-1}\Delta\boldsymbol{x} = \boldsymbol{y} + \boldsymbol{A}^{-1}\Delta\boldsymbol{x},$$

so the change in $\boldsymbol{y}$ is $\Delta\boldsymbol{y} = \boldsymbol{y}' - \boldsymbol{y} = \boldsymbol{A}^{-1}\Delta\boldsymbol{x}$.

In the non-linear case, similar issues arise. In the two-equation two-unknown system,

$$f_1(y_1, y_2) = x_1$$

$$f_2(y_1, y_2) = x_2,$$

where (y_1, y_2) are the variables and (x_1, x_2) the parameters, the following questions arise:

1. Does the equation system have a solution, (y_1^*, y_2^*)?
2. Is the solution unique?
3. Assuming there is a solution, how does it vary with the parameters (x_1, x_2)?

These questions are considered in detail next, but it possible to indicate the main insights with a brief discussion. To address the first question (existence of a solution), consider the system:

$$f_1(y_1, y_2) = 0y_1 + 1y_2 = 1 \quad (= x_1)$$
$$f_2(y_1, y_2) = e^{-y_1} + y_2 = 1 \quad (= x_2).$$

Rearranging, the first equation gives $y_2 = 1$, the second, $y_2 = 1 - e^{-y_1}$. Regardless of

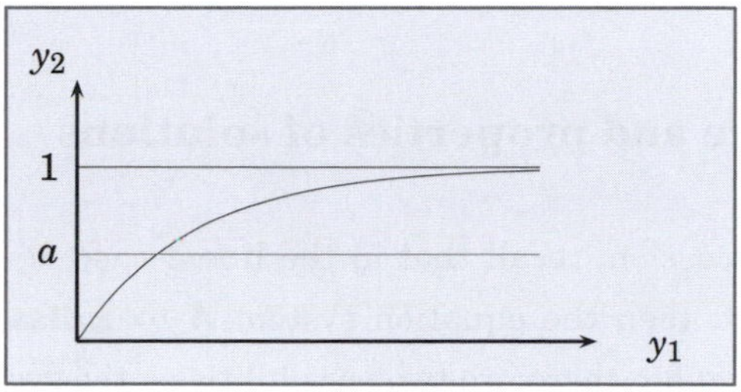

Figure 7.8: No solution

how large y_1 is, $1 - e^{-y_1} < 1$, so the pair of equations has no solutions (Figure 7.8). However, if $x_2 = 1$ is changed to $x_2 = a$, as depicted, then there is a solution and it is unique. Finally, in this case, small variations in (x_1, x_2) from the point $(x_1, x_2) = (1, a)$ lead to small variations in the solution, answering the third question.

The impact of parameter variation on a non-linear system of equations is complicated by the fact that there may be multiple solutions. Even in the univariate

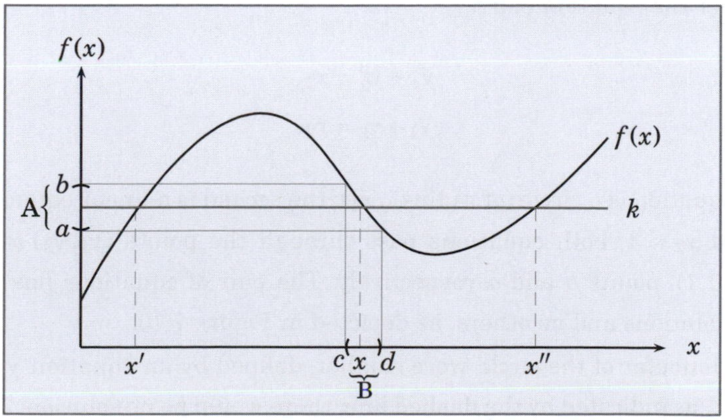

Figure 7.9: Multiple solutions

case, the issue arises, since an equation, $k = f(x)$, may have multiple solutions. Consider the function $f(x)$ depicted in Figure 7.9 and the equation $k = f(x)$, which has multiple solutions, at x', x and x''. However, if we identify a particular solution, x, as we vary k, the solution, x, should vary smoothly with k. As k varies in the region $A = (a,b)$, the solution x varies in the region (c,d). If k increases to b, the solution x moves to c: with $\Delta k = b - k$ and $\Delta x = c - x$, $\frac{\Delta x}{\Delta k} = \frac{c-x}{b-k}$. Noting that $dk = f'(x)dx$, with $f'(x) \neq 0$, $\frac{dx}{dk} = \frac{1}{f'(x)}$. Note that this "works" because $f'(x) \neq 0$, and we have a well-defined (implicit) function $g : (a,b) \to (c,d)$ with $g(k) = x$. The next section explores this idea in the multivariate case.

7.7.1 Existence and properties of solutions

To motivate the discussion, recall that in the linear case, if $\boldsymbol{A}$ is a square $n \times n$ non-singular matrix, then the equation system $\boldsymbol{Ay} = \boldsymbol{x}$ has a unique solution, $\boldsymbol{y} = \boldsymbol{A}^{-1}\boldsymbol{x}$. If $\boldsymbol{A}$ is singular, there are two possibilities – the system has no solution or an infinite number of solutions. Observe that if both y^* and $\bar{y}$ solve $\boldsymbol{Ay} = \boldsymbol{x}$, then for any number θ,

$$\boldsymbol{A}(\theta y^* + (1-\theta)\bar{y}) = \theta \boldsymbol{A} y^* + (1-\theta)\boldsymbol{A}\bar{y} = \theta x + (1-\theta)x = x.$$

So, if y^* and $\bar{y}$ solve the equation $\boldsymbol{Ay} = \boldsymbol{x}$ then so does $(\theta y^* + (1-\theta)\bar{y})$ for any value of θ. So, when there are at least two solutions, there are an infinite number of solutions. Notice also, that if θ is close to 1, $(1-\theta)$ is close to 0 and $(\theta y^* + (1-\theta)\bar{y})$ is close to y^*. So, when there is more than one solution every solution has other solutions arbitrarily close. These properties contrast with the situation in the non-linear case.

Consider the equation pair:

$$y_1^2 + y_2^2 = x_1 \tag{7.6}$$

$$y_1 + y_2 = x_2. \tag{7.7}$$

The first equation is a circle (of radius $\sqrt{x_1}$); the second is a linear equation. When $x_1 = 1$ and $x_2 = 1$, both equations pass through the points $(y_1, y_2) = (1,0)$ and $(y_1, y_2) = (0,1)$, points b and a respectively. The pair of equations has those two points as solutions and no others, as depicted in Figure 7.10.

If the diameter of the circle were smaller, defined by an equation $y_1^2 + y_2^2 = \hat{x}_1$ with $\hat{x}_1 < \frac{1}{2}$, as indicated by the dashed line, there would be no solution. The pair of equations are inconsistent. This example suggests two observations: it is difficult to determine whether or not a non-linear system of equations has a solution; and

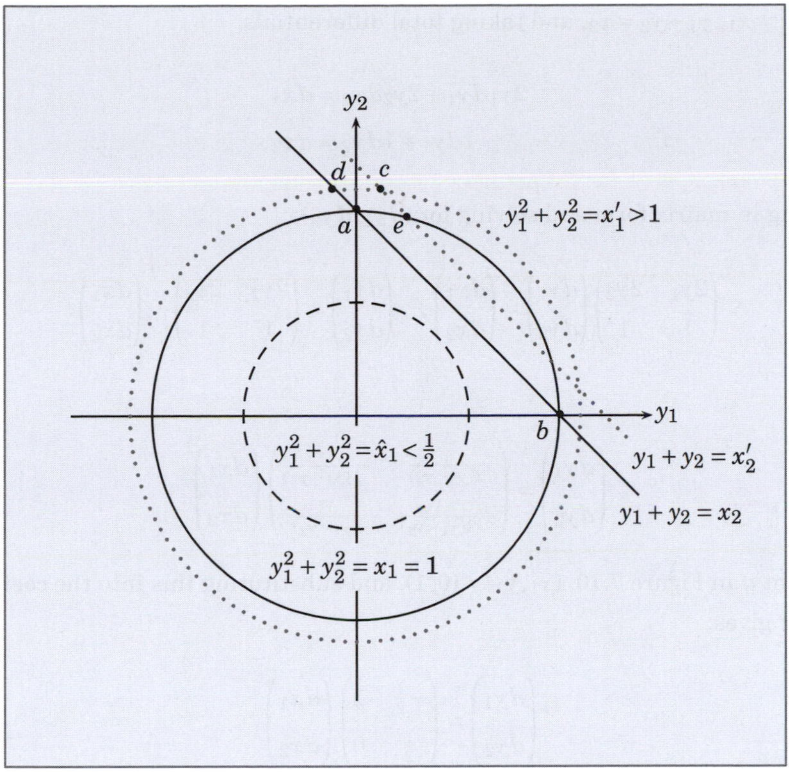

Figure 7.10: A system with two solutions

"typically", if a non-linear system has more than one solution, the solutions are a distance apart: here at $(y_1,y_2)=(1,0)$ and $(y_1,y_2)=(0,1)$. This local uniqueness is exploited in the implicit function theorem.

Given the equation pair,

$$f_1(y_1,y_2)=x_1$$
$$f_2(y_1,y_2)=x_2\,,$$

one may ask if, *given* $x=(x_1,x_2)$, the value of $y=(y_1,y_2)$ is uniquely determined, assuming there is a solution, y. In the two-equation example, $y_1^2+y_2^2=x_1$ and $y_1+y_2=x_2$, when $x_1=1=x_2$, there are two possible solutions for the y values: $(y_1,y_2)=(1,0)$, at a, and $(y_1,y_2)=(0,1)$ at b.

However, there is a unique solution if we restrict attention to a neighborhood of a or of b. In Figure 7.10, if (x_1',x_2') is close to (x_1,x_2), then there is a unique point close to a that satisfies the pair of equations $y_1^2+y_2^2=x_1'$ and $y_1+y_2=x_2'$, the point c. This observation may be used to determine the local variation in (y_1,y_2) resulting from a variation in (x_1,x_2). To see this, consider the pair of equations

$y_1^2 + y_2^2 = x_1$, $y_1 + y_2 = x_2$, and taking total differentials:

$$\begin{aligned} 2y_1 dy_1 + 2y_2 dy_2 &= dx_1 \\ 1dy_1 + 1dy_2 &= dx_2. \end{aligned}$$

Writing in matrix form and solving for (dy_1, dy_2):

$$\begin{pmatrix} 2y_1 & 2y_2 \\ 1 & 1 \end{pmatrix} \begin{pmatrix} dy_1 \\ dy_2 \end{pmatrix} = \begin{pmatrix} dx_1 \\ dx_2 \end{pmatrix}, \quad \begin{pmatrix} dy_1 \\ dy_2 \end{pmatrix} = \begin{pmatrix} 2y_1 & 2y_2 \\ 1 & 1 \end{pmatrix}^{-1} \begin{pmatrix} dx_1 \\ dx_2 \end{pmatrix}.$$

Thus,

$$\begin{pmatrix} dy_1 \\ dy_2 \end{pmatrix} = \begin{pmatrix} \frac{1}{2(y_1-y_2)} & -\frac{y_2}{(y_1-y_2)} \\ -\frac{1}{2(y_1-y_2)} & \frac{y_1}{(y_1-y_2)} \end{pmatrix} \begin{pmatrix} dx_1 \\ dx_2 \end{pmatrix}.$$

At point a in Figure 7.10, $(y_1, y_2) = (0, 1)$, and substituting this into the coefficient matrix gives:

$$\begin{pmatrix} dy_1 \\ dy_2 \end{pmatrix} = \begin{pmatrix} -\frac{1}{2} & 1 \\ \frac{1}{2} & 0 \end{pmatrix} \begin{pmatrix} dx_1 \\ dx_2 \end{pmatrix}.$$

This may be compared with Figure 7.10 at a. First suppose that $dx_2 = 0$ and $dx_1 > 0$ – this corresponds to a dilation of the circle. Then $dy_1 = -\frac{1}{2} dx_1$ and $dy_2 = \frac{1}{2} dx_1$ indicated by the move from point a to point d. If dx_1 is set to 0 and $dx_2 > 0$, this corresponds to a right shift in the straight line. Then (approximately) $dy_1 = dx_2$ and $dy_2 = 0$, which corresponds to a direct move to the right from a along the circle to point e. Note that this is approximate, with dx_2 taken to be very small (so the movement from a to e is horizontal).

In considering the general case, the question of whether a non-linear system has a solution, or not, will not be considered. The entire discussion focuses on the properties of solutions, on the assumption that the equation system has at least one solution. Let $x, y \in \mathcal{R}^n$, and suppose that $x_j = f_j(y_1, y_2, \ldots, y_n)$, $j = 1, \ldots, n$. This gives an equation system:

$$\begin{aligned} x_1 &= f_1(y_1, y_2, \ldots, y_n) \\ x_2 &= f_2(y_1, y_2, \ldots, y_n) \\ \vdots &= \quad \vdots \\ x_n &= f_n(y_1, y_2, \ldots, y_n). \end{aligned} \tag{7.8}$$

Writing the vector of functions as $f = (f_1,\dots,f_n)$, this can be written succinctly as $x = f(y)$, with $x = (x_1, x_2,\dots,x_n)$ and $y = (y_1, y_2,\dots, y_n)$.

Given the equation system $x = f(y)$, consider a variation of y to some point $\tilde{y}$ in some neighborhood of y, $\mathcal{N}(y)$. This determines a new value for x: $\tilde{x} = f(\tilde{y})$. Provided f is continuous, $\tilde{x}$ will be close to x because $\tilde{y}$ is close to y. Now, suppose that for any $\tilde{y} \in \mathcal{N}(y)$, with $\tilde{y} \neq y$, $\tilde{x} = f(\tilde{y}) \neq f(y) = x$. In this case, *the function f is 1-1 in the neighborhood of y*: different values of y determine different values of x. Therefore, when $\tilde{x}$ is close to x, there is a unique $\tilde{y}$ in the neighborhood of y with $\tilde{x} = f(\tilde{y})$. In this case, y is *implicitly defined* as a function of x, and so one can compute the impact on any give solution value y of a variation in x.

Referring to Figure 7.10 at $x = (x_1, x_2) = (1,1)$ there are two values of y that satisfy $x = f(y)$: $y = a$ and $y = b$. Considering the point $y = a$, when x changes to $\tilde{x} = (x_1', x_2')$ the corresponding value of y moves from $y = a$ to $\tilde{y} = c$, the solution in the neighborhood of a. Thus, y is implicitly defined as a function of x in this neighborhood. In this way, we can associate to each point in a neighborhood of (x_1, x_2) a unique point in a neighborhood of a. And, in this case, we can study how y varies in response to variations in x.

Considering equation j, $x_j = f_j(y_1,\dots,y_n)$ in Equation system 7.8,

$$dx_j = \frac{\partial f_j}{\partial y_1} dy_1 + \cdots + \frac{\partial f_j}{\partial y_m} dy_n,$$

and the entire system may be written:

$$\begin{pmatrix} dx_1 \\ \vdots \\ dx_n \end{pmatrix} = \begin{pmatrix} \frac{\partial f_1}{\partial y_1} & \cdots & \frac{\partial f_1}{\partial y_n} \\ \vdots & \ddots & \vdots \\ \frac{\partial f_m}{\partial y_1} & \cdots & \frac{\partial f_m}{\partial y_n} \end{pmatrix} \begin{pmatrix} dy_1 \\ \vdots \\ dy_n \end{pmatrix}.$$

In abbreviated form: $\boldsymbol{dx} = \boldsymbol{J}\boldsymbol{dy}$, and provided $\boldsymbol{J}$ is non-singular, $\boldsymbol{dy} = \boldsymbol{J}^{-1}\boldsymbol{dx}$. The study of such systems is taken up next.

7.8 The implicit function theorem

For the square linear system $\boldsymbol{Ay} = \boldsymbol{x}$, if $\boldsymbol{A}$ has a non-zero determinant (is non-singular), there are two implications: (1) the system has a solution and (2) the solution is unique. Then as we vary $\boldsymbol{x}$ to $\boldsymbol{x}'$, $\boldsymbol{y}$ must vary to $\boldsymbol{y}'$, to satisfy $\boldsymbol{Ay}' = \boldsymbol{x}'$. (Because $\boldsymbol{y} = \boldsymbol{A}^{-1}\boldsymbol{x} = \boldsymbol{Bx}$ uniquely determines $\boldsymbol{y}$ as a function of $\boldsymbol{x}$.) The key requirement is that $\boldsymbol{A}$ is non-singular. For the non-linear equation system, there is an analogous criterion based on a *Jacobian determinant*. This criterion, a non-zero Jacobian determinant, provides information about the *local uniqueness* of a solution but not about the existence of a solution. The Jacobian determinant also provides information about how solutions vary with parameters. Given the systems of equations $x = f(y)$, where the x_i are fixed constants, let y^* be a solution to $x = f(y^*)$. Given a solution y^*, under what circumstances is this the only solution in a neighborhood of y^*? And, if $x = f(y^*)$, under what circumstances does a small change in x lead to a small change in the solution? For both questions, a sufficient condition is that the Jacobian determinant (introduced below) is non-zero at y^*.

The *Jacobian* of the equation system is defined as $\boldsymbol{J} = \left(\frac{\partial x}{\partial y}\right) = \left\{\frac{\partial x_j}{\partial y_i}\right\}, i, j = 1, \dots, n$.

$$\boldsymbol{J} = \begin{pmatrix} \frac{\partial f_1}{\partial y_1} & \frac{\partial f_1}{\partial y_2} & \cdots & \frac{\partial f_1}{\partial y_i} & \cdots & \frac{\partial f_1}{\partial y_n} \\ \frac{\partial f_2}{\partial y_1} & \frac{\partial f_2}{\partial y_2} & \cdots & \frac{\partial f_2}{\partial y_i} & \cdots & \frac{\partial f_2}{\partial y_n} \\ \vdots & \vdots & \ddots & \vdots & \vdots & \vdots \\ \frac{\partial f_i}{\partial y_1} & \frac{\partial f_i}{\partial y_2} & \cdots & \frac{\partial f_i}{\partial y_i} & \cdots & \frac{\partial f_i}{\partial y_n} \\ \vdots & \vdots & \vdots & \vdots & \ddots & \vdots \\ \frac{\partial f_n}{\partial y_1} & \frac{\partial f_n}{\partial y_2} & \cdots & \frac{\partial f_n}{\partial y_i} & \cdots & \frac{\partial f_n}{\partial y_n} \end{pmatrix}$$

Before discussing the Jacobian in detail, an example will illustrate the role of the Jacobian.

EXAMPLE 7.9: Consider the pair of equations:

$$x_1 = 1 = xy = f_1(x, y)$$
$$x_2 = 0 = y - x^2 = f_2(x, y).$$

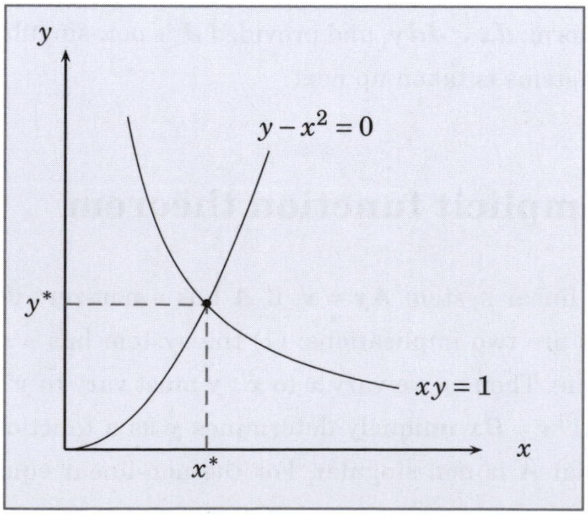

The Jacobian of the system is:

$$\boldsymbol{J} = \begin{pmatrix} \frac{\partial f_1}{\partial y_1} & \frac{\partial f_1}{\partial y_2} \\ \frac{\partial f_2}{\partial y_1} & \frac{\partial f_2}{\partial y_2} \end{pmatrix} = \begin{pmatrix} 1 & 1 \\ -2x & 1 \end{pmatrix}.$$

So, the Jacobian determinant is $|\boldsymbol{J}| = 1+2x$. Note that the two curves "cut", providing a well-defined solution. The non-zero Jacobian determinant confirms this. Furthermore, variations in one of the parameters, (x_1,x_2), will vary the solution a small amount. This also is an implication of the non-zero Jacobian at the solution point (x^*,y^*). ◊

A key property of $\boldsymbol{J}$ is that when there is dependency between the equations $x_i = f_i(y_1,\ldots,y_n)$, $i = 1,\ldots,n$, the determinant of $\boldsymbol{J}$ is zero. ($|\boldsymbol{J}|$ is called the *Jacobian determinant*). To see why this is so, suppose that the nth equation can be written as a function of the other $n-1$ equations: $x_n = \varphi(f_1,f_2,\ldots,f_{n-1})$. Differentiate this equation to get:

$$\frac{\partial x_n}{\partial y'} = \sum_{i=1}^{n-1} \frac{\partial \varphi}{\partial f_i}\frac{\partial f_i}{\partial y'} = \sum_{i=1}^{n-1} \varphi_i \frac{\partial f_i}{\partial y'} = \sum_{i=1}^{n-1} \varphi_i \frac{\partial x_i}{\partial y'}, \; \varphi_i = \frac{\partial \varphi}{\partial f_i}.$$

This shows that the nth row of $\boldsymbol{J}$ is a linear combination of the first $(n-1)$ rows and this implies that $|\boldsymbol{J}| = 0$.

EXAMPLE 7.10: For example, let $x_1 = f_1(y_1,y_2) = y_1+y_2$ and $x_2 = f_2(y_1,y_2) = y_1^2+2y_1y_2+y_2^2$. Observe that $y_1^2+2y_1y_2+y_2^2 = (f_1(y_1,y_2))^2$, so that the equations are dependent (non-linearly). In the notation above, $x_2 = \varphi(f_1)$ or $f_2 = \varphi(f_1)$.

$$\boldsymbol{J} = \begin{pmatrix} \frac{\partial x_1}{\partial y_1} & \frac{\partial x_1}{\partial y_2} \\ \frac{\partial x_2}{\partial y_1} & \frac{\partial x_2}{\partial y_2} \end{pmatrix} = \begin{pmatrix} 1 & 1 \\ 2y_1+2y_2 & 2y_1+2y_2 \end{pmatrix} = \begin{pmatrix} \frac{\partial f_1}{\partial y_1} & \frac{\partial f_1}{\partial y_2} \\ 2\frac{\partial f_1}{\partial y_1} & 2\frac{\partial f_1}{\partial y_2} \end{pmatrix}.$$

Thus, $|\boldsymbol{J}| = 2\left(\frac{\partial f_1}{\partial y_1}\right)\left(\frac{\partial f_1}{\partial y_2}\right) - 2\left(\frac{\partial f_1}{\partial y_1}\right)\left(\frac{\partial f_1}{\partial y_2}\right) = 0$. If $x_2 = x_1^2$, there are an infinite number of solutions and (y_1,y_2) is not determined in terms of (x_1,x_2). If $x_2 \neq x_1^2$, there is no solution. ◊

The condition $|\boldsymbol{J}| \neq 0$ is necessary for a system of equations to be independent. Note, from the example, that this condition detects linear and non-linear dependence. However, the Jacobian determinant does not provide information about the existence of a solution.

EXAMPLE 7.11: Consider again the equation system:

$$y_1^2 + y_2^2 = x_1$$
$$y_1 + y_2 = x_2.$$

The Jacobian is:

$$\boldsymbol{J} = \begin{pmatrix} 2y_1 & 2y_2 \\ 1 & 1 \end{pmatrix}.$$

When x_2 is small, say $x_2 < \frac{1}{2}$, the equation system has no solution, but, for example, at $y_1 = 0$, $y_2 = 1$, $|\boldsymbol{J}| = -2 \neq 0$. In this equation system, whether there is a solution or not depends on (x_1, x_2), but the Jacobian depends only on (y_1, y_2), so it cannot provide information on the existence of a solution. ◊

In the linear case: $\boldsymbol{x} = \boldsymbol{A}\boldsymbol{y}$ so that $\boldsymbol{J} = \boldsymbol{A}$ and so $|\boldsymbol{J}| = |\boldsymbol{A}|$. Hence $|\boldsymbol{J}| \neq 0$ implies that $\boldsymbol{A}$ is invertible and $\boldsymbol{y} = \boldsymbol{A}^{-1}\boldsymbol{x}$. In this case, the solution is unique and the impact of variations in $\boldsymbol{x}$ and $\boldsymbol{y}$ is easy to determine $\boldsymbol{\Delta y} = \boldsymbol{A}^{-1}\boldsymbol{\Delta x}$.

In the non-linear case, $\boldsymbol{J}$ will generally depend on the value of the vector $\boldsymbol{y}$ – see the first row of the example above. To make this explicit, one may write $\boldsymbol{J}(y)$. The following theorem connects the value of $|\boldsymbol{J}|$ to the function f, and considers the impact of a variation in x_j on the values of the y_i. For the latter, one additional piece of notation is required. Let e_j denote a column vector with 1 in the jth row and zeros elsewhere. Write $\boldsymbol{J}_i^j$ for the matrix obtained from $\boldsymbol{J}$ by replacing the ith column with e_j.

Theorem 7.3: *If f is a continuous function with continuous partial derivatives, $f : \mathcal{R}^n \to \mathcal{R}^n$ and $|\boldsymbol{J}(\tilde{y})| \neq 0$ for all $\tilde{y}$ in a neighborhood of y, then f is 1-1 in a neighborhood of y: if y' is close to y, there is a unique $x' = f(y')$ close to $x = f(y)$. Furthermore, at the point y, $\frac{\partial y_i}{\partial x_j} = \frac{|J_j^i|}{|J|}$.*

REMARK 7.4: The motivation for the theorem is relatively easy to see. With $x = f(y)$, and f 1-1 on a neighborhood of y, x' close to x is associated with a unique point y' close to y: there is a function g associating points in the neighborhood of x to points in the neighborhood of y, $(y_1, \ldots, y_n) = (g_1(x_1, \ldots, x_n), \ldots, g_1(x_1, \ldots, x_n))$, such that $x = f(g(x))$, because at x, $g(x)$ gives the value $y = g(x)$ with $x = f(y)$. Then the partial derivatives of the g functions give the impact of variations in the x_i on the y_i: $\frac{\partial y_1}{\partial x_1} = \frac{\partial g_1}{\partial x_1}$ and $\frac{\partial y_1}{\partial x_2} = \frac{\partial g_1}{\partial x_2}$. Considering the two-variable case:

$$x_1 = f_1(g_1(x_1,x_2), g_2(x_1,x_2))$$
$$x_2 = f_2(g_1(x_1,x_2), g_2(x_1,x_2)).$$

Consider a small variation in x_1, holding x_2 fixed. From the equation, $x_1 = f_1(g_1(x_1,x_2), g_2(x_1,x_2))$:

$$dx_1 = \frac{\partial f_1}{\partial y_1}\frac{\partial g_1}{\partial x_1}dx_1 + \frac{\partial f_1}{\partial y_2}\frac{\partial g_2}{\partial x_1}dx_1.$$

From the equation $x_2 = f_2(g_1(x_1,x_2), g_2(x_1,x_2))$:

$$0 = \frac{\partial f_2}{\partial y_1}\frac{\partial g_1}{\partial x_1}dx_1 + \frac{\partial f_2}{\partial y_2}\frac{\partial g_2}{\partial x_1}dx_1.$$

Dividing each of these equations by dx_1:

$$1 = \frac{\partial f_1}{\partial y_1}\frac{\partial g_1}{\partial x_1} + \frac{\partial f_1}{\partial y_2}\frac{\partial g_2}{\partial x_1}$$
$$0 = \frac{\partial f_2}{\partial y_1}\frac{\partial g_1}{\partial x_1} + \frac{\partial f_2}{\partial y_2}\frac{\partial g_2}{\partial x_1}.$$

In matrix form:

$$\begin{pmatrix} 1 \\ 0 \end{pmatrix} = \begin{pmatrix} \frac{\partial f_1}{\partial y_1} & \frac{\partial f_1}{\partial y_2} \\ \frac{\partial f_2}{\partial y_1} & \frac{\partial f_2}{\partial y_2} \end{pmatrix} \begin{pmatrix} \frac{\partial g_1}{\partial x_1} \\ \frac{\partial g_2}{\partial x_1} \end{pmatrix},$$

so, from Cramer's rule:

$$\frac{\partial y_1}{\partial x_1} = \frac{\partial g_1}{\partial x_1} = \frac{\begin{vmatrix} 1 & \frac{\partial f_1}{\partial y_2} \\ 0 & \frac{\partial f_2}{\partial y_2} \end{vmatrix}}{\begin{vmatrix} \frac{\partial f_1}{\partial y_1} & \frac{\partial f_1}{\partial y_2} \\ \frac{\partial f_2}{\partial y_1} & \frac{\partial f_2}{\partial y_2} \end{vmatrix}}.$$

□

The next example applies the theorem to the pair of equations $x_1 = y_1^2 + y_2^2$, $x_2 = y_1 + y_2$.

EXAMPLE 7.12: Consider again the equation pair (depicted in Figure 7.10):

$$x_1 = f_1(y_1, y_2) = y_1^2 + y_2^2$$
$$x_2 = f_2(y_1, y_2) = y_1 + y_2,$$

with $x_1 = x_2 = 1$. Then

$$\begin{pmatrix} \frac{\partial f_1}{\partial y_1} & \frac{\partial f_1}{\partial y_2} \\ \frac{\partial f_2}{\partial y_1} & \frac{\partial f_2}{\partial y_2} \end{pmatrix} = \begin{pmatrix} 2y_1 & 2y_2 \\ 1 & 1 \end{pmatrix}; \quad \begin{pmatrix} 1 & \frac{\partial f_1}{\partial y_2} \\ 0 & \frac{\partial f_2}{\partial y_2} \end{pmatrix} = \begin{pmatrix} 1 & 2y_2 \\ 0 & 1 \end{pmatrix}; \quad \begin{pmatrix} \frac{\partial f_1}{\partial y_1} & 1 \\ \frac{\partial f_2}{\partial y_1} & 0 \end{pmatrix} = \begin{pmatrix} 2y_1 & 1 \\ 1 & 0 \end{pmatrix}. \tag{7.9}$$

So:

$$\frac{\partial y_1}{\partial x_1} = \frac{1}{2(y_1 - y_2)} \quad \text{and} \quad \frac{\partial y_2}{\partial x_1} = \frac{-1}{2(y_1 - y_2)}. \tag{7.10}$$

At $(y_1, y_2) = (0,1)$, $\frac{\partial y_1}{\partial x_1} = -\frac{1}{2}$ and $\frac{\partial y_2}{\partial x_1} = \frac{1}{2}$. The impact of dilating the circle (depicted in Figure 7.10 by the circle formed with *dotted* lines) is to move the solution left from a to d. ◇

REMARK 7.5: Note that when these matrices in Equation 7.9 are evaluated at point (y_1, y_2), the implicit assumption is that the point satisfies the equations. For example, the point $(y_1, y_2) = (4,0)$ lies on neither equation (with $x_1 = x_2 = 1$), but one may nevertheless compute the matrices of partial derivatives: again, these partial derivatives are computed at solution points to the system of equations. □

REMARK 7.6: The calculations in Example 7.12 do not give an explicit solution for (y_1, y_2). However, this is straightforward here. From $y_1 + y_2 = x_2$, $y_1 = x_2 - y_2$, which may be substituted into $y_1^2 + y_2^2 = x_1$ to give $(x_2 - y_2)^2 + y_2^2 = x_1$ or $x_2^2 - 2x_2y_2 + y_2^2 + y_2^2 = x_1$. Collecting terms, $2y_2^2 - 2x_2y_2 + x_2^2 - x_1 = 0$. This quadratic in y_2 has two roots, y_2^a, y_2^b:

$$\begin{aligned} y_2^a, y_2^b &= \frac{2x_2 \pm \sqrt{(2x_2)^2 - 4(2)(x_2^2 - x_1)}}{4} \\ &= \frac{2x_2 \pm \sqrt{4x_2^2 - 8x_2^2 + 8x_1}}{4} \\ &= \frac{2x_2 \pm \sqrt{4(2x_1 - x_2^2)}}{4} \\ &= \frac{2x_2 \pm 2\sqrt{(2x_1 - x_2^2)}}{4} \\ &= \frac{x_2 \pm \sqrt{(2x_1 - x_2^2)}}{2}. \end{aligned}$$

At $x_1 = x_2 = 1$, this gives $y_2^a, y_2^b = \frac{1\pm 1}{2}$, so $y_2^a = 1$ and $y_2^b = 0$. Taking the first root, $y_2^a(x_1, x_2) = \frac{1}{2}x_2 + \frac{1}{2}(2x_1 - x_2^2)^{\frac{1}{2}}$, and differentiating with respect to x_1 (dilating the circle), gives

$$\begin{aligned}\frac{\partial y_2^a}{\partial x_1} &= \frac{1}{2}\left(\frac{1}{2}\right)(2x_1 - x_2^2)^{-\frac{1}{2}}2 \\ &= \frac{1}{2}(2x_1 - x_2^2)^{-\frac{1}{2}} \\ &= \frac{1}{2\sqrt{(2x_1 - x_2^2)}}.\end{aligned}$$

Notice that

$$y_2^a = \frac{1}{2}x_2 + \frac{1}{2}\sqrt{(2x_1 - x_2^2)}, \quad y_1^a = \frac{1}{2}x_2 - \frac{1}{2}\sqrt{(2x_1 - x_2^2)}.$$

Thus, $y_1^a - y_2^a = -\sqrt{(2x_1 - x_2^2)}$ and $\frac{1}{y_1^a - y_2^a} = -\frac{1}{\sqrt{(2x_1 - x_2^2)}}$, so:

$$\frac{\partial y_2^a}{\partial x_1} = \frac{1}{2\sqrt{(2x_1 - x_2^2)}} = -\frac{1}{2(y_1^a - y_2^a)},$$

which compares with Equation 7.10. □

In concluding this section, recall the general system of Equation 7.8, which may be written in the form $x = f(y)$. Then, the total differential may be written, in matrix notation, as $\boldsymbol{dx} = \boldsymbol{J}(y)\boldsymbol{dy}$. If $\boldsymbol{J}(y)$ is non-singular, then we may write $\boldsymbol{dy} = \boldsymbol{J}(y)^{-1}\boldsymbol{dx}$, giving the variations in all y as a result of a variation in the vector x.

7.8.1 General systems of equations

In the previous discussion, equations had the form $x_i = f_i(y_1, \ldots, y_n)$, with the interpretation that the x_i are parameters determining the values of the y_i, just as the values of the x_i in $\boldsymbol{x} = \boldsymbol{Ay}$ determine the y_i. In either case: $x_i = f_i(y_1, \ldots, y_n)$ or $b_i = \sum_{j=1}^n a_{ij}y_j$, the parameters appear "nicely" on one side. If we write $F^i(y_1, y_2, \ldots, y_n, x_i) = x_i - f_i(y_1, \ldots, y_n)$, then $x_i = f_i(y_1, \ldots, y_n)$ for $i = 1, \ldots, n$ is equivalent to $F^i(y_1, y_2, \ldots, y_n, x_i) = 0$, for $i = 1, \ldots, n$.

More generally, each parameter x_i may enter all equations: $F^i(y_1, y_2, \ldots, y_n, x_i, x_2, \ldots, x_m) = 0$. For example,

$$F^1(y_1, y_2, x_1, x_2, x_3) = x_3 - x_1 y_1^2 - x_2 y_2^2$$
$$F^2(y_1, y_2, x_1, x_2, x_3) = 1 - x_3 x_1 y_1^2 - x_2 y_2^2$$

gives a pair of equations determining two variables, (y_1, y_2), with three parameters, (x_1, x_2, x_3). The general system of equations has the form:

$$\begin{aligned} F^1(y_1, \ldots, y_n; x_1, \ldots, x_m) &= 0 \\ F^2(y_1, \ldots, y_n; x_1, \ldots, x_m) &= 0 \\ \vdots \quad &= \vdots \\ F^n(y_1, \ldots, y_n; x_1, \ldots, x_m) &= 0. \end{aligned}$$

In this system, the y variables are viewed as endogenous variables (variables to be determined) and the x variables are viewed as exogenous variables (variables whose values are fixed – parameters). Given the values for the x_i, this is a system of n equations $(F^1, F^2, \ldots, F^n)$, in n unknowns, $(y_1, y_2, \ldots, y_n)$. Note that the number of x_i (m) need not equal the number of equations (n), but the number of y_i (n) must equal the number of equations. This is because the y variables are the variables that must be solved from the n equations, given some fixed set of values for the x variables.

The impact of a parameter change (say x_i changes) on some variable (say y_j) is measured by $\frac{\partial y_j}{\partial x_i}$. As in the previous discussion, this may be computed in terms of the ratio of the determinants of two matrices. The following discussion describes the procedure. For this system, the Jacobian $\boldsymbol{J}$ and an additional matrix $\boldsymbol{J}_i^j$ are defined:

$$\boldsymbol{J} = \begin{pmatrix} \frac{\partial F^1}{\partial y_1} & \frac{\partial F^1}{\partial y_2} & \cdots & \frac{\partial F^1}{\partial y_j} & \cdots & \frac{\partial F^1}{\partial y_n} \\ \frac{\partial F^2}{\partial y_1} & \frac{\partial F^2}{\partial y_2} & \cdots & \frac{\partial F^2}{\partial y_j} & \cdots & \frac{\partial F^2}{\partial y_n} \\ \vdots & \vdots & \ddots & \vdots & \vdots & \vdots \\ \frac{\partial F^i}{\partial y_1} & \frac{\partial F^i}{\partial y_2} & \cdots & \frac{\partial F^i}{\partial y_j} & \cdots & \frac{\partial F^i}{\partial y_n} \\ \vdots & \vdots & \vdots & \vdots & \ddots & \vdots \\ \frac{\partial F^n}{\partial y_1} & \frac{\partial F^n}{\partial y_2} & \cdots & \frac{\partial F^n}{\partial y_i} & \cdots & \frac{\partial F^n}{\partial y_n} \end{pmatrix}$$

$$J_i^j = \begin{pmatrix} \frac{\partial F^1}{\partial y_1} & \frac{\partial F^1}{\partial y_2} & \cdots & \frac{-\partial F^1}{\partial x_i} & \cdots & \frac{\partial F^1}{\partial y_n} \\ \frac{\partial F^2}{\partial y_1} & \frac{\partial F^2}{\partial y_2} & \cdots & \frac{-\partial F^2}{\partial x_i} & \cdots & \frac{\partial F^2}{\partial y_n} \\ \vdots & \vdots & \ddots & \vdots & \vdots & \vdots \\ \frac{\partial F^i}{\partial y_1} & \frac{\partial F^i}{\partial y_2} & \cdots & \frac{-\partial F^i}{\partial x_i} & \cdots & \frac{\partial F^i}{\partial y_n} \\ \vdots & \vdots & \vdots & \vdots & \ddots & \vdots \\ \frac{\partial F^n}{\partial y_1} & \frac{\partial F^n}{\partial y_2} & \cdots & \frac{-\partial F^n}{\partial x_i} & \cdots & \frac{\partial F^n}{\partial y_n} \end{pmatrix}.$$

The matrix $\boldsymbol{J}_i^j$ is obtained from $\boldsymbol{J}$ by replacing the jth column of $\boldsymbol{J}$ with the vector of partial derivatives of $(-F^1, -F^2, \ldots, -F^n)$ with respect to x_i.

Theorem 7.4: *Suppose that F^i has continuous partial derivatives for all i and $|\boldsymbol{J}(\bar{x})| \neq 0$ at $(y, \bar{x})$. Then there exist continuous functions $f^1, \ldots, f^i, \ldots, f^n$, such that $y_j = f^j(x), j = 1, 2, \ldots n$, for x in a neighborhood of $\bar{x}$. Furthermore, the effect on y_j of a change in x_i, $\frac{\partial y_j}{\partial x_i}$, is given by*

$$\frac{\partial y_j}{\partial x_i} = \frac{|\boldsymbol{J}_i^j|}{|\boldsymbol{J}|}.$$

To illustrate, return to the production problem discussed in Section 7.6.1 with a different specification of the production and constraint technology. Let demand, Q, be exogenously given and determining the required output. Suppose that the technology relating inputs x and y to output is $Q = x^\alpha y^\beta$. In addition, a technology constraint requires that $y = kx^\gamma$. We wish to consider the impact of changes in the required production level Q on x and y. The equations relating the variables may be written:

$$F^1(x, y; Q, \alpha, \beta, \gamma, k) = Q - x^\alpha y^\beta = 0 \quad (1)$$

$$F^2(x, y; Q, \alpha, \beta, \gamma, k) = y - kx^\gamma = 0. \quad (2)$$

These equations are depicted in Figure 7.11, where the impact of varying Q is illustrated.

$$\boldsymbol{J} = \begin{pmatrix} \frac{-\alpha Q}{x} & \frac{-\beta Q}{y} \\ \frac{-\gamma y}{x} & 1 \end{pmatrix} \text{ and } \begin{pmatrix} -\frac{\partial F^1}{\partial Q} \\ -\frac{\partial F^2}{\partial Q} \end{pmatrix} = \begin{pmatrix} -1 \\ 0 \end{pmatrix}.$$

Thus, $|\boldsymbol{J}| = -\frac{Q}{x}(\alpha + \beta\gamma)$, and

$$\frac{\partial x}{\partial Q} = \frac{1}{|\boldsymbol{J}|} \begin{vmatrix} -1 & \frac{-\beta Q}{y} \\ 0 & 1 \end{vmatrix} = \frac{1}{(\alpha + \beta\gamma)\frac{Q}{x}}$$

$$\frac{\partial y}{\partial Q} = \frac{1}{|\boldsymbol{J}|} \begin{vmatrix} \frac{-\alpha Q}{x} & -1 \\ \frac{-\gamma y}{x} & 0 \end{vmatrix} = \frac{\gamma\left(\frac{y}{x}\right)}{(\alpha + \beta\gamma)\left(\frac{Q}{x}\right)}.$$

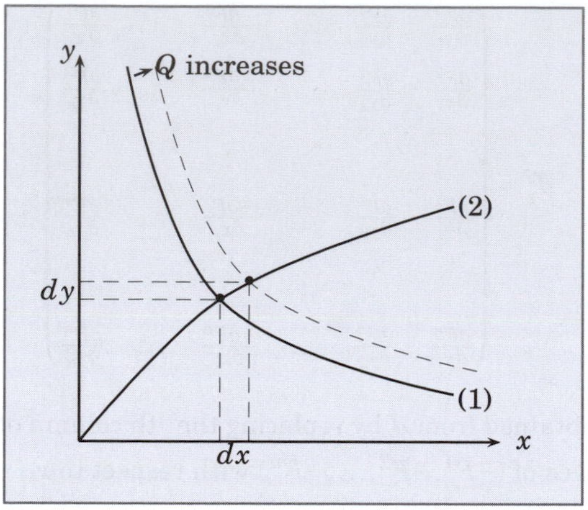

Figure 7.11: The impact of varying output on x and y

EXAMPLE 7.13: Consider a macroeconomic model with two equations:

$$y = c(y - T(y), r) + I(y, r) + G$$
$$M = L(y, r).$$

The first equation is called the IS curve and the second the LM curve in macroeconomics. In this widely studied model, a common task is to determine the impact of changes in policy parameters (government expenditure G and the money supply M) on income y and the interest rate r. These are called policy multipliers and are computed next. For example, $\frac{\partial y}{\partial G}$ is the policy multiplier relating government expenditure to output and measures the impact on output of an increase in government expenditure.

Write c_1 for the partial derivative of c with respect to its first argument, $c_r = \frac{\partial c}{\partial r}$, $c_r = \frac{\partial c}{\partial r}$, $L_r = \frac{\partial L}{\partial r}$ and $L_y = \frac{\partial L}{\partial y}$. Then, differentiating the first equation:

$$dy = c_1(1 - T')dy + c_r dr + I_y dy + I_r dr$$

gives the variation in r resulting from a variation in y when restricted to stay on this curve. Rearranging:

$$(1 - c_1(1 - T') + I_y)dy = (c_r + I_r)dr$$
$$\frac{(1 - c_1(1 - T') + I_y)}{(c_r + I_r)} = \frac{dr}{dy}.$$

A common economic assumption is that $0 < (1 - c_1(1 - T') + I_y) < 1$ along with

$c_r < 0$ and $I_r < 0$, so that $\frac{dr}{dy} < 0$. Similar calculations for the LM curve yield $\frac{dr}{dy} = -\frac{L_y}{L_r}$, which is positive under the assumptions $L_y > 0$ and $L_r < 0$. The curves are depicted in Figure 7.12.

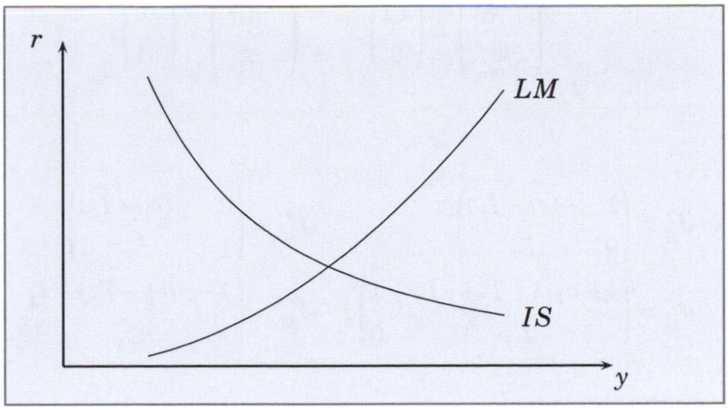

Figure 7.12: The IS–LM curves

In this model, the exogenous variables are government expenditure (G) and the money supply (M). The endogenous variables are income (y) and the rate of interest (r). (As a numerical example, let $T(y) = \frac{1}{2}y$, $c(y - T(y), r) = \left(\sqrt{y - \frac{1}{2}y}\right) \cdot \frac{1}{r}$, $I(y,r) = \frac{y}{r}$ and $L(y,r) = y^{\frac{1}{2}} r^{-1}$.) The first equation asserts that national income (y) equals consumption (c), plus investment I, plus government expenditure, G. The second says that money supply (M) equals money demand L (and demand depends on the interest rate (r) and the level of income). These equations may be rearranged in the general form given previously:

$$F^1(y,r,G,M) = y - c(y - T(y),r) - I(y,r) - G = 0$$
$$F^2(y,r,G,M) = L(y,r) - M = 0.$$

Provided the Jacobian has full rank, we may view y and r as functions of G and M, $y = f^1(G,M)$, $r = f^2(G,M)$, and compute the corresponding partial derivatives according to the rule above. (In the theorem above y and r correspond to y_1 and y_2, while G and M correspond to x_1 and x_2.) With this rearrangement, the Jacobian is obtained:

$$\boldsymbol{J} = \begin{pmatrix} \frac{\partial F^1}{\partial y} & \frac{\partial F^1}{\partial r} \\ \frac{\partial F^2}{\partial y} & \frac{\partial F^2}{\partial r} \end{pmatrix} = \begin{pmatrix} 1 - c_1(1 - T_y) - I_y & -(c_r + I_r) \\ L_y & L_r \end{pmatrix}.$$

The Jacobian has determinant $|\boldsymbol{J}| = [1 - c_1(1 - T_y) - I_y]L_r + L_y(c_r + I_r)$. To calculate $\boldsymbol{J}_G^y$, $\boldsymbol{J}_M^y$, $\boldsymbol{J}_G^r$ and $\boldsymbol{J}_M^r$ (where these correspond to $\boldsymbol{J}_j^i$, with $i = y, r$ and $j = G, M$), observe

$$\begin{pmatrix} -\frac{\partial F^1}{\partial G} \\ -\frac{\partial F^2}{\partial G} \end{pmatrix} = \begin{pmatrix} +1 \\ 0 \end{pmatrix}, \qquad \begin{pmatrix} -\frac{\partial F^1}{\partial M} \\ -\frac{\partial F^2}{\partial M} \end{pmatrix} = \begin{pmatrix} 0 \\ +1 \end{pmatrix}.$$

Thus,

$$\boldsymbol{J}_G^y = \begin{pmatrix} 1 & -(c_r + I_r) \\ 0 & L_r \end{pmatrix}, \qquad \boldsymbol{J}_M^y = \begin{pmatrix} 0 & -(c_r + I_r) \\ 1 & L_r \end{pmatrix}$$

$$\boldsymbol{J}_G^r = \begin{pmatrix} 1 - c_1(1 - T_y) - I_y & 1 \\ L_y & 0 \end{pmatrix}, \quad \boldsymbol{J}_M^r = \begin{pmatrix} 1 - c_1(1 - T_y) - I_y & 0 \\ L_y & 1 \end{pmatrix}.$$

Applying the theorem:

$$\frac{\partial y}{\partial G} = \frac{|\boldsymbol{J}_G^y|}{|\boldsymbol{J}|} = \frac{L_r}{|\boldsymbol{J}|}, \quad \text{and} \quad \frac{\partial y}{\partial M} = \frac{|\boldsymbol{J}_M^y|}{|\boldsymbol{J}|} = \frac{c_r + I_r}{|\boldsymbol{J}|}. \tag{7.11}$$

Similarly,

$$\frac{\partial r}{\partial G} = \frac{|\boldsymbol{J}_G^r|}{|\boldsymbol{J}|} = \frac{-L_y}{|\boldsymbol{J}|}, \quad \text{and} \quad \frac{\partial r}{\partial M} = \frac{|\boldsymbol{J}_M^r|}{|\boldsymbol{J}|} = \frac{-[1 - c_1(1 - T_y) - I_y]}{|\boldsymbol{J}|}. \tag{7.12}$$

◊

REMARK 7.7: It is worth noting that, for a slightly different perspective, these results may be obtained directly by totally differentiating the original equation system.

$$dy = c_1(1 - T_y)dy + c_r dr + I_y dy + I_r dr + dG$$

$$L_y dy + L_r dr = dM,$$

$$[1 - c_1(1 - T_y) - I_y]dy - (c_r + I_r)dr = dG$$

$$L_y dy + L_r dr = dM$$

$$\begin{pmatrix} [1 - c_1(1 - T_y) - I_y] & -(c_r + I_r) \\ L_y & L_r \end{pmatrix} \begin{pmatrix} dy \\ dr \end{pmatrix} = \begin{pmatrix} dG \\ dM \end{pmatrix}.$$

This system has the form: $\boldsymbol{A}\boldsymbol{x} = \boldsymbol{b}$ where

$$\boldsymbol{A} = \begin{pmatrix} [1 - c_1(1 - T_y) - I_y] & -(c_r + I_r) \\ L_y & L_r \end{pmatrix}, \boldsymbol{x} = \begin{pmatrix} dy \\ dr \end{pmatrix}, \boldsymbol{b} = \begin{pmatrix} dG \\ dM \end{pmatrix}.$$

Note that $\boldsymbol{A} = \boldsymbol{J}$ and that $|\boldsymbol{A}| = [1 - c_1(1 - T_y) - I_y]L_r + L_y(G_r + I_r)$. Using Cramer's rule:

$$\begin{aligned} dy &= \frac{1}{|\boldsymbol{A}|}\begin{vmatrix} dG & -(c_r + I_r) \\ dM & L_r \end{vmatrix} \\ &= \frac{L_r dG + (c_r + I_r)dM}{|\boldsymbol{A}|} \\ &= \left(\frac{L_r}{|\boldsymbol{A}|}\right)dG + \left(\frac{(c_r + I_r)}{|\boldsymbol{A}|}\right)dM, \end{aligned}$$

and

$$\begin{aligned} dr &= \frac{1}{|\boldsymbol{A}|}\begin{vmatrix} [1 - c_1(1 - T_y) - I_y] & dG \\ L_y & dM \end{vmatrix} \\ &= \frac{[1 - c_1(1 - T_y) - I_y]dM - L_y dG}{|\boldsymbol{A}|} \\ &= \left(\frac{-L_y}{|\boldsymbol{A}|}\right)dG + \left(\frac{[1 - c_1(1 - T_y) - I_y]}{|\boldsymbol{A}|}\right)dM. \end{aligned}$$

Considering the expression

$$dy = \left(\frac{L_r}{|\boldsymbol{A}|}\right)dG + \left(\frac{(c_r + I_r)}{|\boldsymbol{A}|}\right)dM,$$

if $dM = 0$, the change in y, dy, results from a change in G, dG, with M held fixed. This measures the partial derivative of y with respect to G, and is given in the expression by the coefficient on dG: $\left(\frac{L_r}{|\boldsymbol{A}|}\right)$. Thus:

$$\frac{\partial y}{\partial G} = \left[\frac{L_r}{|\boldsymbol{A}|}\right], \quad \frac{\partial r}{\partial M} = \left[\frac{[1 - c_1(1 - T_y) - I_y]}{|\boldsymbol{A}|}\right],$$

and so on. This is the same result as appears in Equations 7.11 and 7.12. □

The following discussion provides a graphical perspective on the calculations. A highway starts from point $(x, y) = (0,0)$ and follows the path $y = x + \beta x^2 - \gamma x^3$ (given the x coordinate, there is a unique y coordinate determined by this equation). The direction x is east and the direction y is north. The point (x, y) on this highway that is a distance r from the starting point satisfies $x^2 + y^2 = r^2$. (We assume that $\frac{2\beta}{3\gamma} > r$.) This point is uniquely defined by these two equations.

Suppose that we move slightly further along the highway by increasing the distance r; is the direction more northerly or more easterly?

From the starting point at radius r, consider an increase in the radius to r' with $r' - r > 0$ small. This changes the coordinates (x, y) slightly to (x', y').

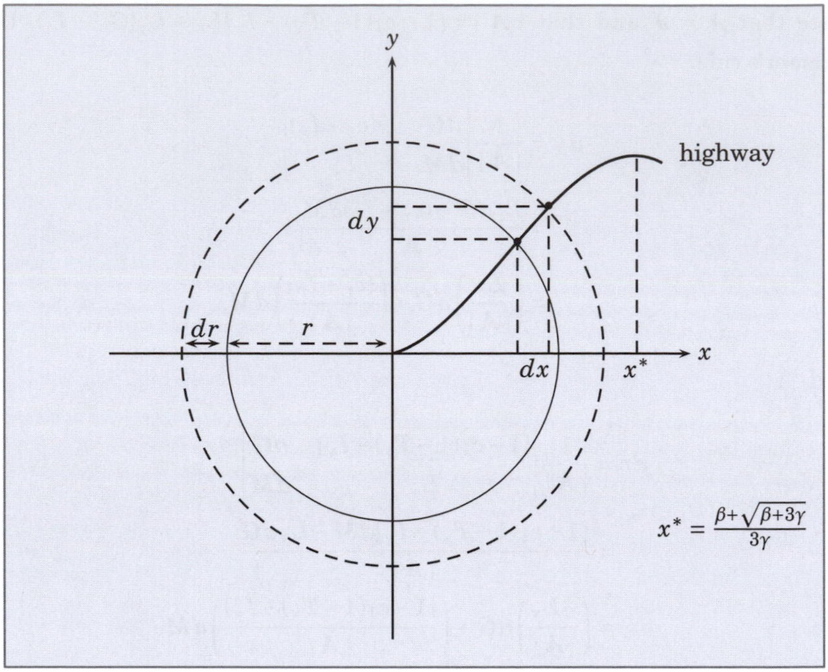

Figure 7.13: The impact of increasing distance

The eastward movement is $x' - x$ and the northward movement is $y' - y$. Then

$$\frac{x'-x}{r'-r} \approx \frac{\partial x}{\partial r} \quad \text{and} \quad \frac{y'-y}{r'-r} \approx \frac{\partial y}{\partial r}.$$

Thus, the northward movement is greater than the eastward movement if:

$$\frac{y'-y}{x'-x} = \frac{\frac{y'-y}{r'-r}}{\frac{x'-x}{r'-r}} \approx \frac{\frac{\partial y}{\partial r}}{\frac{\partial x}{\partial r}} > 1$$

Note that, along the highway,

$$\frac{dy}{dx} = 1 + 2\beta x - 3\gamma x^2 = 1 + (2\beta - 3\gamma x)x = 1 + 3\gamma\left(\frac{2\beta}{3\gamma} - x\right)x.$$

So, $\frac{dy}{dx} > 1$ if $\left(\frac{2\beta}{3\gamma} - x\right) > 0$ or $(2\beta - 3\gamma x) > 0$. At any point (x, y) on the circle $x^2 + y^2 = r^2$, and since $y^2 \geq 0$, $x^2 \leq r^2$, and since the solution is positive, $x < r$, so that $2\beta - 3\gamma x > 2\beta - 3\gamma r$ or $\frac{2\beta}{3\gamma} - x > \frac{2\beta}{3\gamma} - r$. Since $\frac{2\beta}{3\gamma} \geq r$, $\frac{2\beta}{3\gamma} - x \geq 0$. Therefore, at the point on the circle where the curve $y = x + \beta x^2 - \gamma x^3$ cuts, the slope of the curve is greater than 1: at the solution, $\frac{dy}{dx} > 1$, and the northward movement is greater than the eastward movement.

The following example reconsiders this discussion by calculating the partial derivatives of y and x with respect to r, and comparing them.

EXAMPLE 7.14: Recall:

$$x^2 + y^2 = r^2$$
$$y - x - \beta x^2 + \gamma x^3 = 0.$$

$$2xdx + 2ydy = 2rdr$$
$$dy - dx - 2\beta xdx + 3\gamma x^2 dx = 0$$

$$\begin{pmatrix} 2x & 2y \\ 1 + 2\beta x - 3\gamma x^2 & -1 \end{pmatrix} \begin{pmatrix} dx \\ dy \end{pmatrix} = \begin{pmatrix} 2rdr \\ 0 \end{pmatrix},$$

or, since only r is varied,

$$\begin{pmatrix} 2x & 2y \\ 1 + 2\beta x - 3\gamma x^2 & -1 \end{pmatrix} \begin{pmatrix} \frac{\partial x}{\partial r} \\ \frac{\partial y}{\partial r} \end{pmatrix} = \begin{pmatrix} 2r \\ 0 \end{pmatrix}.$$

Using Cramer's rule, the coefficient determinant is $\Delta = -2x - 2y(1 + 2\beta x - 3\gamma x^2)$, while $\Delta_1 = -2r$ and $\Delta_2 = -(1 + 2\beta x - 3\gamma x^2)2r$. Then

$$\frac{\partial x}{\partial r} = \frac{\Delta_1}{\Delta}, \qquad \frac{\partial y}{\partial r} = \frac{\Delta_2}{\Delta} \text{ and } \frac{\frac{\partial y}{\partial r}}{\frac{\partial x}{\partial r}} = \frac{\Delta_2}{\Delta_1} = (1 + 2\beta x - 3\gamma x^2) = (1 + (2\beta - 3\gamma x)x).$$ ◇

The following example reconsiders this discussion by calculating the partial derivatives of y and z with respect to t, and evaluating at $t = 0$.

Example 7.13. Recall

[illegible]

Exercises for Chapter 7

Exercise 7.1 The following simple macroeconomic model has four equations relating government policy variables to aggregate income and investment. Aggregate income equals consumption plus investment and government expenditure:

$$y = c + i + g.$$

Consumption is positively related to the real rate of interest:

$$c = \alpha + \beta y, \ \alpha, \beta > 0.$$

Investment is negatively related to the rate of interest, r:

$$i = \gamma - \delta r, \ \gamma, \delta > 0.$$

The money supply, m_s, equals the money demand, m_d, and this in turn is positively related to the level of income ($\rho > 0$) and negatively related to the rate of interest ($\varphi > 0$):

$$m_s = m_d = \rho y - \varphi r, \ \varphi, k > 0.$$

Find the impact on y of (a) a small change in g, Δg, and (b) a small change in m, Δm.

Exercise 7.2 Consider a market defined by the following supply and demand curves.

$$q^d = a + bP^d + c\bar{Y}$$
$$q^s = d + eP^s,$$

where a, c, d, $e > 0$, and $b < 0$, $a > d$.

(a) (i) Compute the equilibrium price P^* and quantity q^*.

(ii) Sign the comparative statics partial derivatives $\frac{\partial P^*}{\partial \bar{Y}}$, $\frac{\partial q^*}{\partial \bar{Y}}$.

(b) Suppose a tax $t > 0$ is levied on producers:

(i) Compute the new equilibrium price and quantity.

(ii) Sign the derivatives $\frac{\partial q^*}{\partial t}$, $\frac{\partial P^d}{\partial t}$, $\frac{\partial P^s}{\partial t}$.

Exercise 7.3 Consider the following macroeconomic model of national income:

$$C = \alpha + \beta Y, \quad \alpha > 0, 0 < \beta < 1$$
$$Y = C + I + G$$
$$I = \gamma Y, \quad 0 < \gamma < 1.$$

What are the exogenous and endogenous variables? Write this system in matrix form and solve for the reduced form (i.e., write the system in the form $\boldsymbol{Ax} = \boldsymbol{b}$, where $\boldsymbol{x}$ is the vector of endogenous variables and $\boldsymbol{b}$ the vector of exogenous variables and solve as $\boldsymbol{x} = \boldsymbol{A}^{-1}\boldsymbol{b}$). Using the reduced form expressions, find the following comparative statics derivatives:

$$\frac{\partial c}{\partial G}, \quad \frac{\partial Y}{\partial G}, \quad \frac{\partial I}{\partial G}.$$

Exercise 7.4 Consider the following macroeconomic model:

$$y = c(y,r) + i(r) + G$$
$$m = l(r,y).$$

(a) Identify the exogenous and endogenous variables.

(b) (i) What is the effect of an increase in government expenditure G, on output y and the interest rate r?

(ii) What is the effect of an increase in the money supply M, on output and the interest rate?

(c) Consider what happens if we replace the above money market equation with the following: $m = \mu y$. Repeat part (b) with this assumption.

(d) Represent the two different systems in (y,r) space and show the above effect diagrammatically.

Exercise 7.5 Consider the following macroeconomic model:

$$S(Y,r) + T(Y) + IM = I(r) + G + x(e)$$
$$L(Y,r) = M,$$

where Y is national income, S is savings, which is a function of income and the rate of interest, IM is imports, x is exports, I is investment, G is government expenditure, M is money supply and e is the exchange rate. The first equation is the equilibrium condition for the goods market and the second equation is the equilibrium condition for the money market.

Using this model, determine how the equilibrium values of Y and r change with a change in the exchange rate e and with a change in the exogenous level of imports (here imports is an exogenous variable, just like G). Make the usual assumptions: $S_y > 0, S_r > 0, T_y > 0, I_r < 0, x_e < 0, L_y > 0, L_r < 0$.

Exercise 7.6 Consider a production process using input x to produce output y. The relationship between inputs and outputs is somewhat odd: for technological reasons the input–output levels must satisfy two conditions:

$$x^2 + y^2 = \alpha,\ 0 < \alpha < 1$$
$$\beta y = x(1-x),\ 0 < \beta.$$

This pair of equations has a unique non-negative solution ($x, y \geq 0$). Plot these functions in (x, y) space. At this solution, calculate $\frac{dx}{d\alpha}$, $\frac{dy}{d\alpha}$, $\frac{dx}{d\beta}$ and $\frac{dy}{d\beta}$. In relation to your graph, interpret the signs of the derivatives (they are positive or negative).

Exercise 7.7 Repeat the previous question, with input and output required to satisfy the following two conditions:

$$y - \gamma \ln(x+1) = 0$$
$$x^\alpha y^\beta = \gamma.$$

Exercise 7.8 Consider the following IS–LM model with an import sector:

$$y = c(y) + I(i) + G$$
$$m = L(y, i)$$
$$x(\pi) = z(y, \pi).$$

The first two equations are the IS and LM curves, respectively. The third equation asserts that exports (x) depend only on the exchange rate, π, while imports (z) depend on the exchange rate and the level of domestic income.

(a) Calculate the slopes of the IS and LM curves.

(b) Find $\frac{dy}{dG}$, $\frac{di}{dG}$ and $\frac{d\pi}{dG}$. (Differentiate each of the three equations to give a 3×3 equation system in $(dy, di, d\pi)$ and use Cramer's rule.)

Exercise 7.9 An individual has utility function $u(x,y) = x^{\alpha}y^{\beta}$, where $0 < \alpha < 1, 0 < \beta < 1$ and $0 < \alpha+\beta < 1$. The prices of x and y, respectively, are p and q, and the individual has income I. The marginal utility of x is denoted u_x, and here you see that $u_x(x,y) = \alpha x^{\alpha-1}y^{\beta} = \frac{\alpha x^{\alpha}y^{\beta}}{x} = \alpha\frac{u(x,y)}{x}$. Similarly, $u_y(x,y) = \beta x^{\alpha}y^{\beta-1} = \frac{\beta x^{\alpha}y^{\beta}}{y} = \frac{\beta u(x,y)}{y}$. The utility maximizing levels of x and y are determined by the tangency of the indifference curve to the budget constraint: $px+qy = I$. The tangency condition is given by the condition that the marginal rate of substitution equals the price ratio: $MRS = \frac{u_x}{u_y} = \frac{p}{q}$. Thus we have two equations in (x,y): (1) $\frac{u_x}{u_y} = \frac{p}{q}$ and (2) $px+qy = I$. Note that equation (1) has a very simple form: it can be written as a linear equation in x and y after appropriate multiplication and cancellation.

(a) Write the pair of equations in the form

$$F^1(x,y;\alpha,\beta,p,q,I) = 0$$

$$F^2(x,y;\alpha,\beta,p,q,I) = 0$$

and calculate $\frac{\partial x}{\partial I}, \frac{\partial x}{\partial p}, \frac{\partial x}{\partial q}, \frac{\partial x}{\partial \alpha}, \frac{\partial x}{\partial \beta}$.

(b) Solve (1) and (2) directly for x and y (as functions of (α,β,p,q,I)), then take the partial derivatives of x with respect to the parameters, to confirm your answer in (a).

(c) Illustrate, with a graph, how $\frac{\partial x}{\partial I}, \frac{\partial x}{\partial p}, \frac{\partial x}{\partial q}, \frac{\partial x}{\partial \alpha}, \frac{\partial x}{\partial \beta}$ are calculated.

Exercise 7.10 A production possibilities (transformation) curve in two goods, x and y, is given by $T(x,y) = k$, where $T(x,y) = x^2 + y$. The social utility function is given by $u(x,y) = \alpha\ln(x)+\beta\ln(y)$. The efficient choices for x and y are given at the point where the marginal rate of substitution equals the marginal rate of transformation ($MRS = MRT$ or $\frac{u_x}{u_y} = \frac{T_x}{T_y}$). Thus, there are two equations determining the solution: (1) $\frac{u_x}{u_y} = \frac{T_x}{T_y}$ and (2) $x^2 + y = k$.

(a) Rearrange these to get two equations in the form

$$F^1(x,y;\alpha,\beta,k) = 0$$

$$F^2(x,y;\alpha,\beta,k) = 0.$$

(b) Use the implicit function theorem to determine $\frac{\partial x}{\partial \alpha}, \frac{\partial y}{\partial \alpha}, \frac{\partial x}{\partial k}, \frac{\partial y}{\partial k}$.

Exercise 7.11 Consider the following model

$$C = 5+0.85Y$$

$$Y = C+I+G$$

$$I = 0.03Y.$$

Solve for the effects of an increase in government spending on: consumption, income, investment.

Exercise 7.12 Consider the IS–LM model

$$a + br - C\left(\frac{M}{P}\right) = d - er + fY \quad \text{(savings = investment)}$$
$$\frac{M}{P} = AY^{\alpha}r^{-\beta} \quad \text{(money demand = money supply).}$$

(a) Totally differentiate the pair of equations.

(b) Use Cramer's rule to show what effect an increase in M has on r, P (for simplicity let $dY = 0$).

Exercise 7.13 A firm is required to produce output Q using inputs x and y. The production technology is $x^{\alpha}y^{\beta}$. In addition, the firm faces the technology constraint $y = ke^{x}$, $k > 0$. Find $\frac{\partial y}{\partial Q}$, $\frac{\partial x}{\partial Q}$, $\frac{\partial y}{\partial k}$, $\frac{\partial x}{\partial k}$. Also, try finding $\frac{\partial y}{\partial \alpha}$, $\frac{\partial x}{\partial \alpha}$.

Exercise 7.14 Consider the following two equations: $x^2 + y^2 = r^2$ and $y = k\ln(x+1)$, with $r > 0$ and $k > 0$. Plot both equations and indicate on your graph the impact of an increase in r and k on the solution. Find $\frac{dy}{dr}$, $\frac{dx}{dr}$ $\frac{dy}{dk}$ and $\frac{dx}{dk}$.

Exercise 7.15 Consider the following two equations: $xy = k$ and $y = 1 - e^{-\gamma x}$, $k > 0$ and $\gamma > 0$. Plot both equations and indicate on your graph the impact of an increase in k and γ on the solution. Find $\frac{dy}{dk}$, $\frac{dx}{dk}$ $\frac{dy}{d\gamma}$ and $\frac{dx}{d\gamma}$.

Exercise 7.16 An output technology is given by $x + ky^2 = a$, where k and a are positive constants. Government regulations require that outputs of x and y satisfy the relation $e^{\beta y} - e^{\alpha x} = c$, where α, β and c are positive constants. (We assume that $\frac{\ln(c+1)}{\beta} < \sqrt{\frac{a}{k}}$ to ensure there is a solution to the pair of equations.)

(a) The government has decided that it is desirable to raise x and reduce y, by a policy of increasing c. Will the policy have the desired effect? Calculate the impact on x and y of an increase in c.

(b) An economic consultant recommends instead that the government raise β. Would this policy have the desired effect of increasing x and reducing y?

Exercise 7.17 An individual has income or wealth, y, to allocate between present, c, and future, f, consumption. Suppose that the individual has utility function $u(c,f) = c \cdot f$. Income not consumed in the current period is invested at interest rate r to yield the amount available for future consumption: $f = (y-c)(1+r)$. Since y is given, the choice of c determines f. This is a budget constraint relating current and future consumption. Find the impact of a change in income, y, on consumption c.

Exercise 7.18 An individual has preferences given by indifference curves of the form $xy = k$: if the agent is consuming x units of one good, the consumption of $\frac{k}{y}$ units of the other good leads to utility level k. The problem facing the individual is that the goods x and y are bundled so that they cannot be bought in any combination. Specifically, x units of one good come in a package with $y = 1 - e^{-\alpha x}$ units of the other good. Suppose that the individual currently is at utility level k, consuming at levels x and y according to the packaging constraint. Find the impact of an increase in utility on the consumption levels x and y: $\frac{\partial x}{\partial k}$ and $\frac{\partial y}{\partial k}$.

Exercise 7.19 According to a macroeconomic model, the aggregate level of income Y equals consumption C, investment I and government expenditures G ($Y = C + I + G$). Consumption depends on income $C = C(Y)$ while investment depends on the rate of interest r, so $I = I(r)$. Thus, $Y = C(Y) + I(r) + G$. The money supply M equals money demand, which is a function of the level of income and the rate of interest $L = L(Y, r)$. Thus, $M = L(Y, r)$. Finally, the exchange rate π is assumed to depend on income and the interest rate $\pi = \pi(Y, r)$. Together, this gives three equations. The level of government expenditure (G) and the money supply (M) are assumed to be controlled by the government. Given values for G and M, the three equations determine the values of Y, r and π. Suppose that these functions can be written in the form:

$$
\begin{aligned}
Y &= cY^{\alpha} - \gamma r + G \\
M &= \frac{Y}{r} \\
\pi &= Y^{\beta} r^{\delta}.
\end{aligned}
$$

In this model, there are two government policy parameters, government expenditure, G, and money supply M. The impact of changes in these policy variables on output, interest rates and the exchange rate are given by such expressions as $\frac{\partial Y}{\partial G}$, $\frac{\partial r}{\partial M}$, and $\frac{\partial \pi}{\partial G}$. Determine these policy "multipliers" and confirm that $\frac{\partial \pi}{\partial G} = 0$ if $\beta = -\delta$.

Exercise 7.20 Suppose that the supply function for a particular good is given by the equation $p = \alpha + \delta q^{\beta}$, $\alpha, \delta, \beta > 0$. The demand function for the good is given by $\ln(pq) = k$. Shifts in the supply function are caused by changes in α, δ and β. Plot the supply and demand equations and calculate the impact on (equilibrium) price and quantity of supply shifts.

8 Taylor series

8.1 Introduction

Problems such as maximization of a function f or finding the root of an equation $f(x) = 0$ arise frequently in economics. However, often the function may be difficult to work with, motivating the use of approximations to the function f that are easier to study and provide insight regarding f. One relatively simple class of approximating functions are the polynomials. A function h is a polynomial in x if it has the form:

$$h(x) = a_0 + a_1 x + a_2 x^2 + \cdots + a_n x^n,$$

where each a_i is a constant. So, h is the sum of terms where each term is a constant multiplied by x to the power of some integer, such as $a_3 x^3$, where a_3 is the constant and x is raised to the power of 3. The function h is called an nth order polynomial because the highest power is n (assuming $a_n \neq 0$). For example, $h(x) = 2 + x + 5x^2 + 8x^3$ is a third-order polynomial, as is $g(x) = 2x^3$. Under suitable circumstances, a large class of functions can be approximated by a polynomial, so that given the function f, one may find a polynomial such as h approximating f. This makes it possible to discover properties of f by study of the simpler function h and that fact is a primary motivation for studying these approximations. Taylor's theorem concerns the approximation of functions with polynomials and provides the theoretical basis for the polynomial approximations.

Section 8.1.1 motivates the approach by illustrating how a quadratic function may be chosen to approximate a continuous function. Following this, in Section 8.1.2 the main theorem is stated. This asserts that any continuous differentiable function may be approximated by a polynomial function – the Taylor series expansion. Because Taylor series expansions involve approximation, there is a gap between the approximation and the actual function. However, the quality of the approximation improves as the length of the Taylor series expansion increases – the higher the order of the polynomial, the better the approximation. The improvement from lengthening the approximation is considered with an example in Section 8.1.3. Sections 8.2 to 8.4 provide a number of applications. Section 8.2 shows how Taylor series expansions may be used to address a variety of problems – such as relating concavity to the derivatives of a function, finding the roots of an equation, and generating numerical optimization routines to find maxima and minima of a function. In Section 8.3, expected utility is introduced. In this section, the theory is used to provide an approximate measure of utility loss from risk and a monetary measure of the cost incurred from bearing risk. Section 8.4 uses Taylor series to find the approximately optimal solution to a portfolio selection problem. Finally, Section 8.5 describes the multivariate version of Taylor's theorem and Section 8.6 provides a proof of the theorem.

8.1.1 Taylor series approximations: an example

To illustrate the way in which Taylor series may be used to approximate a function, consider the function:

$$f(x) = \ln(x+1)e^{-\frac{x}{2}}.$$

Although simple to describe mathematically, the function has reasonably complex behavior as x varies. Starting from $x = 0$, the function increases initially and then declines, "flattening out" to 0 as x becomes large. Often, functions such as this are difficult to study. For example, the function has a unique maximum, x^*, defined by $f'(x^*) = 0$, but this expression is very difficult to solve. One way to address this difficulty is to approximate f by another function, g, which is "close" to f but simpler to manipulate.

Consider the function, $g(x)$, which is given by the following formula:

$$g(x) = 0.1524 + 0.44285x - 0.17489x^2.$$

Observe that the function g "tracks" the function $f(x)$ to some degree – see Figure 8.1. Both functions, f and g, have the same value at $x = 1$ and are close when x is close to 1. So, g is a good approximation to f when x is close to 1 and g has a simple form.

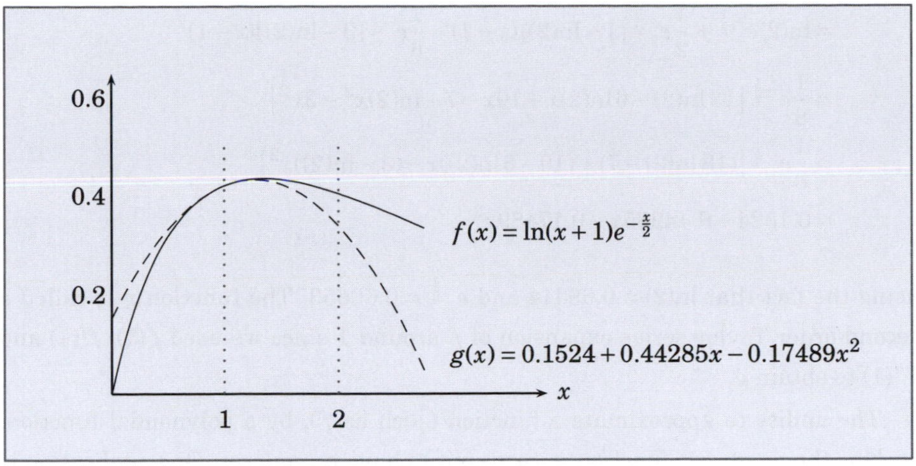

Figure 8.1: Taylor series approximation of $f(x) = \ln(x+1)e^{-\frac{x}{2}}$

In fact, the function g is a second-order Taylor series expansion of f, defined according to the formula:

$$g(x) = f(1) + f'(1)(x-1) + \frac{1}{2}f''(1)(x-1)^2. \tag{8.1}$$

The logic for this is discussed later. The following calculations confirm that this formula in Equation 8.1 yields the expression for g given above. Observe that:

$$\begin{aligned} f'(x) &= \frac{e^{-\frac{1}{2}x}}{(x+1)} - \frac{1}{2}\ln(x+1)e^{-\frac{1}{2}x} \\ f'(1) &= \frac{1}{2}e^{-\frac{1}{2}} - \frac{1}{2}\ln(2)e^{-\frac{1}{2}}, \end{aligned}$$

and

$$\begin{aligned} f''(x) &= -\frac{e^{-\frac{1}{2}x}}{(x+1)^2} - \frac{e^{-\frac{1}{2}x}}{(x+1)} + \frac{1}{4}\ln(x+1)e^{-\frac{1}{2}x} \\ &= -\frac{x+2}{(x+1)^2}e^{-\frac{1}{2}x} + \frac{1}{4}\ln(x+1)e^{-\frac{1}{2}x} \\ f''(1) &= -\frac{3}{4}e^{-\frac{1}{2}} + \frac{1}{4}\ln(2)e^{-\frac{1}{2}}. \end{aligned}$$

With these calculations in place, g may be obtained:

$$\begin{aligned} g(x) &= f(1) + f'(1)(x-1) + \frac{1}{2}f''(1)(x-1)^2 \\ &= \ln(2)e^{-\frac{1}{2}} + \left[\frac{1}{2}e^{-\frac{1}{2}} - \left(\frac{1}{2}\right)\ln(2)e^{-\frac{1}{2}}\right](x-1) + \left[\frac{1}{8}\ln(2)e^{-\frac{1}{2}} - \frac{3}{8}e^{-\frac{1}{2}}\right](x-1)^2 \end{aligned}$$

$$
\begin{aligned}
&= \ln(2)e^{-\frac{1}{2}} + \frac{1}{2}e^{-\frac{1}{2}}[1-\ln(2)](x-1) - \frac{1}{8}e^{-\frac{1}{2}}[3-\ln(2)](x-1)^2 \\
&= \frac{1}{8}e^{-\frac{1}{2}}\left[13\ln(2) - 6\ln(2)x + 10x - 7 + \ln(2)x^2 - 3x^2\right] \\
&= \frac{1}{8}e^{-\frac{1}{2}}\left[(13\ln(2)-7) + (10-6\ln(2))x - (3-\ln(2))x^2\right] \\
&= 0.1524 + 0.44285x - 0.17489x^2,
\end{aligned}
$$

using the fact that $\ln(2) = 0.69314$ and $e^{-\frac{1}{2}} = 0.60653$. The function g is called a second-order Taylor series expansion of f around 1 since we used $f(1)$, $f'(1)$ and $f''(1)$ to obtain g.

The ability to approximate a function (such as f), by a polynomial function, such as the quadratic function g, turns out to have many uses. One application is optimization. If the maximum of f is difficult to determine but g is close to f, then finding the maximum of g may be a good approximation to maximizing f. Note that $f'(x)$ may be written $f'(x) = \left[\frac{1}{(x+1)} - \frac{1}{2}\ln(x+1)\right]e^{-\frac{1}{2}x}$. Since for any x, $e^{-\frac{1}{2}x} \neq 0$, $f'(x) = 0$ implies $\left[\frac{1}{(x+1)} - \frac{1}{2}\ln(x+1)\right] = 0$: x^* satisfies $\left[\frac{1}{(x^*+1)} - \frac{1}{2}\ln(x^*+1)\right] = 0$. The solution for x^* is with $x^* = 1.34575$ and $f'(x^*) = 0$ is satisfied. The function f has a unique maximum at x^*.

The function g has a maximum at $\hat{x} = 1.266$. Assume that, since g approximates f, the maximum of g might be close to the maximum of f, so if we now expand f around $\hat{x}$, we obtain a "new" function, $\hat{g}$; this should be a better approximation of f close to the maximum:

$$
\begin{aligned}
\hat{g}(x) &= f(1.266) + f'(1.266)(x-1.266) + \frac{1}{2}f''(1.266)(x-1.266)^2 \\
&= 0.22901 + 0.307258x - 0.114577x^2.
\end{aligned}
$$

Observe that $\hat{g}$ is maximized at $x' = 1.341$, which is very close to x^*. These calculations suggest a simple procedure for locating the maximum of a function using a simple algorithm, and illustrate one important use of Taylor series expansions.

8.1.2 Taylor series expansions

Taylor's theorem asserts that a continuous function with continuous derivatives may be approximated by a polynomial function. For notational convenience write $f^{(k)}(x) = \frac{d^k f}{dx^k}$ for the kth derivative of the function f. (As usual the lower derivatives are often written $f'(x) = \frac{df}{dx}$, $f''(x) = \frac{d^2 f}{dx^2}$ and $f'''(x) = \frac{d^3 f}{dx^3}$ instead of $f^{(1)}(x)$, $f^{(2)}(x)$ and $f^{(3)}(x)$.)

Theorem 8.1: *Let f be a continuous function on an interval $[a,b]$ with continuous derivatives f', f'', $\dots f^n$. Then given $x, x_0 \in (a,b)$:*

$$f(x) = f(x_0) + f'(x_0)(x-x_0) + \frac{1}{2!}f''(x_0)(x-x_0)^2$$
$$+\frac{1}{3!}f'''(x_0)(x-x_0)^3 + \cdots + \frac{1}{n!}f^n(p)(x-x_o)^n \tag{8.2}$$

for some p between x and x_0, and

$$f(x) \approx f(x_0) + f'(x_0)(x-x_0) + \frac{1}{2!}f''(x_0)(x-x_0)^2$$
$$+\frac{1}{3!}f'''(x_0)(x-x_0)^3 + \cdots + \frac{1}{n!}f^n(x_0)(x-x_0)^n, \tag{8.3}$$

when x is close to x_0.

In Theorem 8.1, the expression on the right side is called a Taylor series expansion of f of order n around x_0. Note that in one of the expansions, Equation 8.2, the last term is $\frac{1}{n!}f^n(p)(x-x_0)^n$ – the point p is chosen to give *equality* of the expansion with the function (and depends on both x and x_0). The second expansion, Equation 8.3, has last term $\frac{1}{n!}f^n(x_0)(x-x_0)^n$, and the expansion is *approximately* equal to $f(x)$. The term $n!$ is "n factorial", defined $n! = n(n-1)(n-2)\cdots 3\cdot 2\cdot 1$, so, for example, $3! = 3\times 2\times 1 = 6$. By convention, $0! = 1$. For example, a third-order expansion is:

$$f(x) = f(x_o) + f'(x_o)(x-x_o) + \frac{1}{2!}f''(x_o)(x-x_o)^2 + \frac{1}{3!}f'''(p)(x-x_o)^3.$$

To illustrate the formula, consider the quadratic function $f(x) = x^2$. The first and second derivatives are $f'(x) = 2x$ and $f''(x) = 2$ for all x, so that for any x_0 and any number p, $f''(x_0) = 2 = f''(p)$ and the second derivative, $f''(x)$ is independent of x. An exact first-order expansion has the form $f(x) = f(x_0) + f'(p)(x-x_0)$ for some p, so that for the function $f(x) = x^2$:

$$f(x) = f(x_0) + f'(p)\cdot(x-x_0)$$
$$x^2 = x_0^2 + 2p\cdot(x-x_0).$$

This equality requires that $p = \frac{x_0+x}{2}$. To see this, rearrange the equation:

$$x^2 - x_0^2 = 2p\cdot(x-x_0)$$
$$(x-x_0)(x+x_0) = 2p\cdot(x-x_0)$$

$$(x_0 + x) = 2p$$
$$\frac{(x_0 + x)}{2} = p.$$

Or, with approximation, $f(x) \approx f(x_0) + f'(x_0)(x - x_0) = [f(x_0) - f'(x_0)x_0] + f'(x_0) \cdot x$, a linear equation with intercept $[f(x_0) - f'(x_0)x_0]$ and slope $f'(x_0)$.

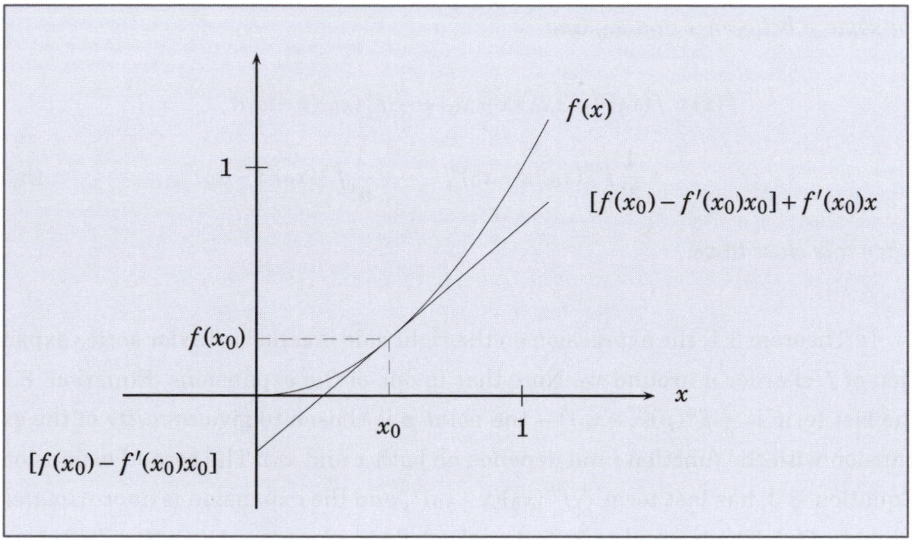

Figure 8.2: First-order Taylor series expansion

One advantage of the approximate expansion is that there is no need to calculate p, since the expansion does not have to be exact.

A second-order Taylor series expansion of x^2 gives:

$$f(x) = x_0^2 + 2x_0(x - x_0) + \frac{1}{2}2(x - x_0)^2 = x_0^2 + 2x_0x - 2x_0^2 + x^2 - 2xx_0 + x_0^2 = x^2.$$

The expansion is exact because $f''(x) = 2$ for all values of x: the second derivative is a constant. Put differently, if $f'''(x) = 0$, so the third derivative is 0, then the second derivative must be constant and so a second-order expansion is exact if the third derivative is 0.

EXAMPLE 8.1: Consider the function $f(x) = \ln(x+1) \cdot x^2$. Note that $f'(x) = \frac{x^2}{x+1} + 2\ln(x+1)x$ and $f''(x) = -\frac{x^2}{(x+1)^2} + \frac{4x}{(x+1)} + 2\ln(x+1)$. At $x = 2$, $f(x) = 4\ln(3)$, $f'(x) = \frac{4}{3} + 4\ln(3)$, and $f''(x) = -\frac{4}{9} + \frac{8}{3} + 2\ln(3) = \frac{20}{9} + 2\ln(3)$. This function has a first-order Taylor series expansion around 2 of: $f(x) \approx 4\ln(3) + (4\ln(3) + \frac{4}{3})$

$(x-2)$ and a second-order expansion $f(x) \approx 4\ln(3) + (4\ln(3) + \frac{4}{3})(x-2) + (\ln(3) + \frac{10}{9})(x-2)^2$. Since $\ln(3) = 1.0986$, $4\ln(3) = 4.3944$, $4\ln(3) + \frac{4}{3} = 5.7274$ and $\ln(3) + \left(\frac{10}{9}\right) = 1.0986 + \left(\frac{10}{9}\right) = 2.2097$, so $f(x) \approx 4.3944 + 5.7274(x-2) + 2.2097(x-2)^2$.

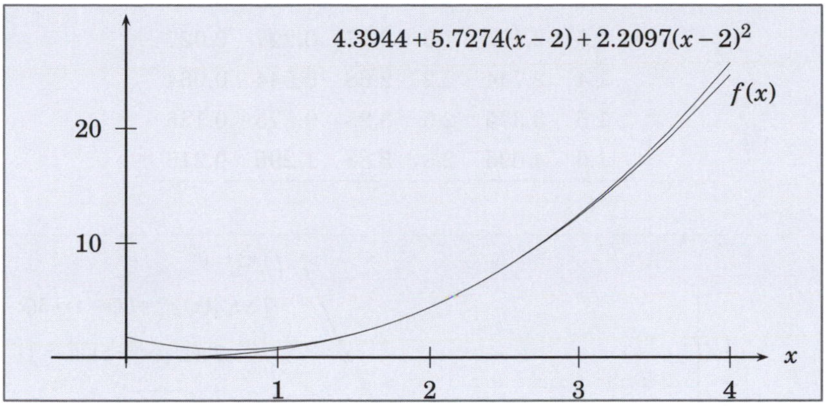

From the figure, the approximation lies above $f(x)$, is tangent at $x = 2$ and provides a good approximation on the interval (1,3). ◊

8.1.3 Approximations and accuracy

As the length of the Taylor series expansion increases, the accuracy of the approximation increases, but the complexity of the approximation also increases with higher-order polynomials. So, there is a trade-off between increasing accuracy and increasing complexity. To see this, consider $f(x) = x^3$ and expansions around $x_0 = 1$. The first-order Taylor series expansion is $\text{TSE}_1(x) = 1 + 3(x-1) = f(1) + f'(1)(x-1)$; and the second-order Taylor series expansion is $\text{TSE}_2(x) = 1 + 3(x-1) + 3(x-1)^2 = f(1) + f'(1)(x-1) + \frac{1}{2}f''(1)(x-1)^2$. The error associated with the first-order Taylor series expansion is $\text{E}_1(x) = \mid f(x) - \text{TSE}_1(x) \mid$, and with the second-order expansion, $\text{E}_2(x) = \mid f(x) - \text{TSE}_2(x) \mid$. The following table compares the errors.

x	x^3	$\text{TSE}_1(x)$	$\text{TSE}_2(x)$	$\text{E}_1(x)$	$\text{E}_2(x)$
0.4	0.064	−0.8	0.28	0.864	0.216
0.5	0.125	−0.5	0.25	0.625	0.125
0.6	0.216	−0.2	0.28	0.416	0.064
0.7	0.343	0.1	0.37	0.243	0.027

0.8	0.512	0.4	0.52	0.112	0.008
0.9	0.729	0.7	0.73	0.029	0.001
1	1	1	1	0	0
1.1	1.331	1.3	1.33	0.031	0.001
1.2	1.728	1.6	1.72	0.128	0.008
1.3	2.197	1.9	2.17	0.297	0.027
1.4	2.744	2.2	2.68	0.544	0.064
1.5	3.375	2.5	3.25	0.875	0.125
1.6	4.096	2.8	3.88	1.296	0.216

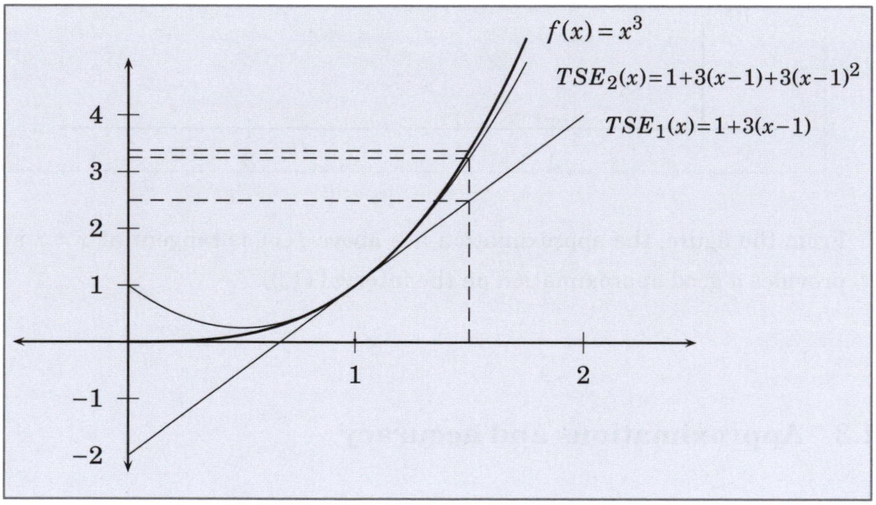

Figure 8.3: Taylor series approximations

In Figure 8.3, $f(0) = 0$, $TSE_1(0) = -2$ and $TSE_2(0) = 1$.

8.2 Applications of Taylor series expansions

The following sections consider some applications of Taylor series expansions. The first application is a proof that concavity of a function implies that the second derivative is non-positive. Following this, numerical optimization routines based on Taylor series are developed. Then some applications on decision-making with risk are described.

8.2.1 Concavity and the second derivative

The following theorem shows that concavity implies a non-positive second derivative; a similar argument may be used to show that convex functions have non-negative second derivatives.

Theorem 8.2: *If f is concave and $f''(x)$ continuous then for any x, $f''(x) \le 0$.*

PROOF: Pick x and let $\Delta x = \tilde{x} - x, \tilde{x}$ close to x. Note $\frac{1}{2}(x+\Delta x) + \frac{1}{2}(x-\Delta x) = x$ and f concave gives:

$$f(x) = f\left(\frac{1}{2}(x+\Delta x) + \frac{1}{2}(x-\Delta x)\right) \ge \frac{1}{2}f(x+\Delta x) + \frac{1}{2}f(x-\Delta x).$$

Therefore

$$2f(x) \ge f(x+\Delta x) + f(x-\Delta x)$$

(or $f(x+\Delta x) + f(x-\Delta x) \le 2f(x)$). Now, Taylor's theorem gives:

$$f(y) = f(y_0) + f'(y_0)(y-y_0) + \frac{1}{2!}f''(\tilde{y})(y-y_0)^2,$$

where $\tilde{y}$ is between y and y_0. First, take $y = x+\Delta x, y_0 = x$, so $y - y_0 = \Delta x$ and then

$$f(x+\Delta x) = f(x) + f'(x)\Delta x + \frac{1}{2}f''(x^*)(\Delta x)^2, \; x^* \text{ between } x \text{ and } x+\Delta x.$$

Next, take $y = x-\Delta x$, $y_0 = x$, so $y - y_0 = -\Delta x$ and then

$$f(x-\Delta x) = f(x) - f'(x)\Delta x + \frac{1}{2}f''(\hat{x})(\Delta x)^2, \; \hat{x} \text{ between } x \text{ and } x-\Delta x.$$

Adding these together:

$$f(x+\Delta x) + f(x-\Delta x) = 2f(x) + \frac{1}{2}[f''(x^*) + f''(\hat{x})](\Delta x)^2.$$

Recall from concavity that $2f(x) \ge f(x+\Delta x) + f(x-\Delta x)$, so

$$0 \ge f(x+\Delta x) + f(x-\Delta x) - 2f(x) = \frac{1}{2}[f''(x^*) + f''(\hat{x})](\Delta x)^2$$

Since $(\Delta x)^2 > 0$, this implies that $0 \ge \frac{1}{2}[f''(x^*) + f''(\hat{x})]$. Since f'' is continuous, as $\Delta x \to 0$, $x^*, \hat{x} \to x$, and so $f''(x^*), f''(\hat{x}) \to f''(x)$. Therefore, $0 \ge f''(x)$. ∎

8.2.2 Roots of a function

Given a function f, a common problem is to find a root of f, a value for x such that $f(\bar{x}) = 0$. Finding a root involves solving the equation $f(x) = 0$. In practice this can be difficult. However, using Taylor series there is a simple procedure that may be used to search for the roots of a function. Consider a first-order Taylor series expansion of f around some x_0:

$$f(x) \approx f(x_0) + f'(x_0)(x - x_0).$$

Given x_0, suppose x is chosen to set $f(x_0)+f'(x_0)(x-x_0) = 0$, with solution x_1. Then

$$f(x_1) \approx f(x_0) + f'(x_0)(x_1 - x_0) = 0,$$

so that x_1 is "approximately" a root of f. If this process is repeated with x_1 replacing x_0, a revised estimate, x_2, for a root is obtained: $f(x_2) \approx f(x_1)+f'(x_1)(x_2-x_1) = 0$. And so on. This process is easy to implement in practice. Assuming $f'(x_0) \neq 0$, the expression can be arranged to give an algorithm:

$$\begin{aligned}
x_1 - x_0 &= -\frac{f(x_0)}{f'(x_0)} \\
x_1 &= x_0 - \frac{f(x_0)}{f'(x_0)} \\
x_2 &= x_1 - \frac{f(x_1)}{f'(x_1)} \\
\vdots \quad & \quad \vdots \\
x_{n+1} &= x_n - \frac{f(x_n)}{f'(x_n)}.
\end{aligned}$$

To illustrate, consider the function $f(x) = -\frac{1}{2} + 4x - 4x^2 + x^3$ (Figure 8.4). Applying this iterative process:

x_i	f	f'	$x_i - \frac{f(x_i)}{f'(x_i)}$
1	0.5	−1	1.5
1.5	−0.125	−1.25	1.4
1.4	0.004	−1.32	1.403
1.403	0.000	−1.318	1.403

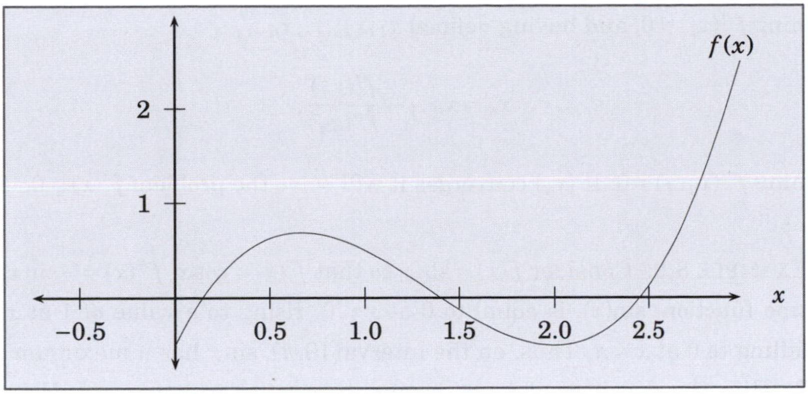

Figure 8.4: Roots of a function: $f(x) = -\frac{1}{2} + 4x - 4x^2 + x^3$

8.2.3 Numerical optimization

Suppose we wish to maximize some differentiable function $f(x)$. A maximum has the property that $f'(x) = 0$. How do we solve $f'(x) = 0$? One approach uses Taylor series expansions,

$$f(x) \simeq f(x_0) + f'(x_0)(x - x_0) + \frac{1}{2} f''(x_0)(x - x_0)^2.$$

When x_0 is "close" to x, a value of x that maximizes the right-hand side (RHS) will also maximize the LHS. Differentiate with respect to x and set to 0. Since the RHS is a quadratic function of x, the derivative is linear in x. This yields

$$f'(x_0) + f''(x_0)(x - x_0) = 0.$$

Solving for x (call the solution x_1)

$$x_1 = x_0 - \frac{f'(x_0)}{f''(x_0)},$$

assuming that $f''(x_0) \neq 0$. Presumably, x_1 will be closer to a maximum of $f(x)$ than x_0. We can continue the iteration

$$x_2 = x_1 - \frac{f'(x_1)}{f''(x_1)},$$

assuming $f''(x_1) \neq 0$, and having defined $x_1, x_2, \ldots, x_{n-1}$,

$$x_n = x_{n-1} - \frac{f'(x_n)}{f''(x_n)},$$

assuming $f''(x_{n-1}) \neq 0$. If $\{x_n\}$ converges it will solve the problem $f'(x) = 0$.

EXAMPLE 8.2: Consider $f(x) = \sin x$ so that $f'(x) = \cos x$, $f''(x) = -\sin x$. The sine function, $\sin(x)$, is equal to 0 at $x = 0$, rising to a value of 1 at $x = \frac{\pi}{2}$, falling to 0 at $x = \pi$. Thus, on the interval $[0, \pi]$, $\sin x$ has a maximum at $\frac{\pi}{2}$. Starting the algorithm at $\frac{1}{2} = x_0$ begins an iteration to estimate $\frac{\pi}{2}$. Using the formula, the iteration has the form:

$$x_n = x_{n-1} + \frac{\cos(x_{n-1})}{\sin(x_{n-1})}.$$

Beginning with $x_0 = \frac{1}{2}$,

$$\begin{aligned}
x_1 &= \frac{1}{2} + \frac{\cos\left(\frac{1}{2}\right)}{\sin\left(\frac{1}{2}\right)} = \frac{1}{2} + \frac{0.878}{0.479} = 2.330 \\
x_2 &= 2.330 + \frac{\cos(2.330)}{\sin(2.330)} = 2.330 + \frac{-0.688}{0.725} = 1.381 \\
x_3 &= 1.381 + \frac{\cos(1.381)}{\sin(1.381)} = 1.381 + \frac{0.189}{0.982} = 1.573 \\
x_4 &= 1.573 + \frac{\cos(1.573)}{\sin(1.573)} = 1.573 + \frac{-(2.20) \times 10^{-3}}{0.999} = 1.5707963.
\end{aligned}$$

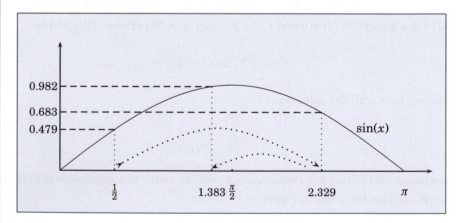

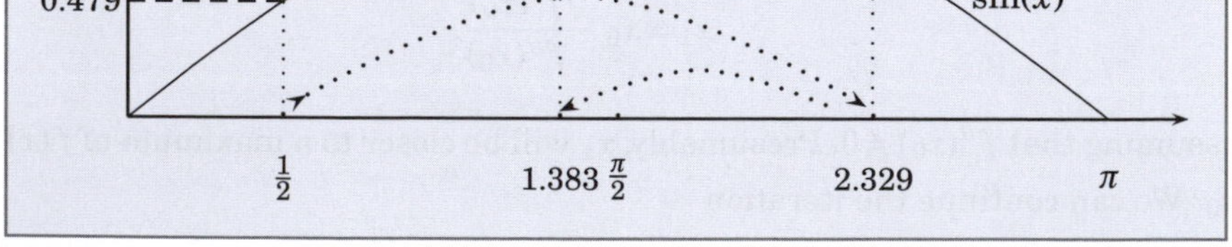

Since $\frac{\pi}{2} = 1.5707963$, even at the fourth iteration we are getting an accurate estimate of $\frac{\pi}{2}$! Starting at $\frac{1}{2}$, observe that the algorithm moves rightward in the direction of increase of the function given by the derivative (to 2.329) overshooting. From this point, again moving in the direction of increase of the function, now leftward, the algorithm moves to 1.383. The next move is rightward, since the direction of increase is rightward – to 1.573. ◇

EXAMPLE 8.3: Consider the task of maximizing profit, where the profit function is:

$$\pi(x) = 10\ln(x+1) - 0.1e^{0.1x} - x^2.$$

First, observe that the derivatives are $\pi'(x) = \frac{10}{x+1} - (0.1)^2 e^{0.1x} - 2x$ and $\pi''(x) = \frac{-10}{(x+1)^2} - (0.1)^3 e^{0.1x} - 2$. Thus $\pi'' < 0$ and the profit function is concave. Therefore, if x^* satisfies $\pi'(x^*) = 0$, then x^* maximizes π. Using a second-order Taylor series expansion:

$$\pi(x) \approx \pi(x_0) + \pi'(x_0)(x - x_0) + \frac{1}{2}\pi''(x_0)(x - x_0)^2,$$

so that

$$0 = \pi'(x) \approx \pi'(x_0) + \pi''(x_0)(x - x_0) \Rightarrow x \approx x_0 - \frac{\pi'(x_0)}{\pi''(x_0)}.$$

There is no obvious point to start the iteration, so choose $x_0 = 0$. The algorithm proceeds according to:

$$x_{i+1} = x_i - \frac{\pi'(x_i)}{\pi''(x_i)} = x_i - \left(\frac{\frac{10}{x+1} - (0.1)^2 e^{0.1x} - 2x}{\frac{-10}{(x+1)^2} - (0.1)^3 e^{0.1x} - 2} \right).$$

The following table gives the sequence of iterations: From this it is clear that if $x_{i+1} = x_i$ and $\pi''(x_i) \neq 0$, then $\pi'(x_i) = 0$ and this implies that x_i maximizes π:

x_i	$\pi'(x)$	$\pi''(x)$	$x_i - \frac{\pi'(x)}{\pi''(x)}$
0	9.99	−12.0010	0.832431
0.832431	3.781503	−4.97923	1.591887
1.591887	0.662694	−3.48974	1.781785
1.781785	0.019295	−3.29346	1.787643
1.787643	1.59×10^{-05}	−3.28804	1.787648
1.787648	1.08×10^{-11}	−3.28803	1.787648
1.787648	2.25×10^{-16}	−3.28803	1.787648
1.787648	2.25×10^{-16}	−3.28803	1.787648
1.787648	2.25×10^{-16}	−3.28803	1.787648

The profit-maximizing choice of x is $x^* = 1.787648$. ◇

EXAMPLE 8.4: A production function has isoquants given by $xy = k$: if inputs are used at the levels x^* and y^*, the output level is x^*y^*. A technological constraint requires that the inputs x and y satisfy the relation $y = 1 - e^{-\alpha x}$. Suppose that $\alpha = 2$, and the firm decides to produce an output of 5 units, so $k = 5$. In this case, the values of x^* and y^* are determined by the two equations $xy = 5$ and $y = 1 - e^{-2x}$. The following calculations show how the values may be calculated using a Taylor series expansion.

The first equation can be written $x = \frac{5}{y}$ and when this is substituted into the second equation we get $y = 1 - e^{-2\frac{5}{y}} = 1 - e^{\frac{-10}{y}}$ or $y - 1 + e^{\frac{-10}{y}} = 0$. Let $f(y) = y - 1 + e^{\frac{-10}{y}}$, so we wish to find y^* such that $f(y^*) = 0$. Now, $f(y^*) \approx f(y_0) + f'(y_0)(y^* - y_0)$, when y^* is "close" to y_0, and since $f(y^*) = 0$, $f(y_0) + f'(y_0)(y^* - y_0) \approx 0$, so that $y^* \approx y_0 - \frac{f(y_0)}{f'(y_0)}$. This provides the basis for the iteration $y_{i+1} \approx y_i - \frac{f(y_i)}{f'(y_i)}$. Since $f(y) = y - 1 + e^{\frac{-10}{y}}$, $f'(y) = 1 + \left(\frac{10}{y^2}\right) e^{\frac{-10}{y}}$:

$$y_{i+1} = y_i - \left(\frac{y_i - 1 + e^{\frac{-10}{y_i}}}{1 + \left(\frac{10}{y_i^2}\right) e^{\frac{-10}{y_i}}} \right).$$

Use a starting value for the iteration of $y_0 = 2$:

x_i	$f(x)$	$f'(x)$	$x_i - \frac{f(x)}{f'(x)}$
2	1.006738	1.016845	1.009939
1.009939	0.00999	1.000496	0.999955
0.999955	2.37×10^{-07}	1.000454	0.999955
0.999955	-4.7×10^{-15}	1.000454	0.999955
0.999955	-5.1×10^{-17}	1.000454	0.999955
0.999955	-5.1×10^{-17}	1.000454	0.999955

Thus, $y^* = 0.999955$ and $x = \frac{5}{0.999955}$ is the required solution. ◊

The numerical optimization procedure generalizes directly to multivariate optimization. The second-order Taylor series approximation in the multivariate case is:

$$f(x) \approx f(x_0) + \nabla f(x_0) \cdot (x - x_0) + \frac{1}{2!}(x - x_0)' H(x_0)(x - x_0),$$

where $\nabla f(x_0)$ is the vector of partial derivatives of f evaluated at x_0, and where $\boldsymbol{H}(x_0)$ is the Hessian matrix of second-order partial derivatives, also evaluated at

x_0. Maximizing the right side by choice of the vector x gives:

$$\nabla f(x_0) + H(x_0)(x - x_0) = 0.$$

Assuming the Hessian matrix is non-singular, the solution x_1 satisfies

$$x_1 - x_0 = -H(x_0)^{-1}\nabla f(x_0),$$

and the resulting iteration is:

$$x_{i+1} - x_i = -H(x_i)^{-1}\nabla f(x_i).$$

EXAMPLE 8.5: Consider the objective function:

$$\pi(x,y) = 2x^{\frac{1}{2}} + 4y^{\frac{1}{2}} + xy - x^2 - y^2.$$

Using a second-order expansion, it is straightforward to develop an algorithm to locate the values of x and y that maximize π. First, observe that a second-order Taylor series expansion of $\pi(x,y)$ around (x_0, y_0) is:

$$\pi(x,y) \approx \pi(x_0,y_0) + \pi_x(x_0,y_0)(x-x_0) + \pi_y(x_0,y_0)(y-y_0) + \frac{1}{2}\{\pi_{xx}(x_0,y_0)(x-x_0)^2 + 2\pi_{xy}(x_0,y_0)(x-x_0)(y-y_0) + \pi_{yy}(x_0,y_0)(y-y_0)^2\}.$$

Maximize the right side by choice of x and y. Differentiating with respect to x and y, respectively, and setting the equations to 0 gives the solution equations. For x:

$$\begin{aligned} 0 &= \pi_x(x_0,y_0) + \frac{1}{2}\{2\pi_{xx}(x_0,y_0)(x-x_0) + 2\pi_{xy}(x_0,y_0)(y-y_0)\} \\ &= \pi_x(x_0,y_0) + \pi_{xx}(x_0,y_0)(x-x_0) + \pi_{xy}(x_0,y_0)(y-y_0). \end{aligned}$$

Similarly, with y:

$$0 = \pi_y(x_0,y_0) + \pi_{xy}(x_0,y_0)(x-x_0) + \pi_{yy}(x_0,y_0)(y-y_0).$$

In matrix terms:

$$\begin{pmatrix} 0 \\ 0 \end{pmatrix} = \begin{pmatrix} \pi_x(x_0,y_0) \\ \pi_y(x_0,y_0) \end{pmatrix} + \begin{pmatrix} \pi_{xx}(x_0,y_0) & \pi_{xy}(x_0,y_0) \\ \pi_{xy}(x_0,y_0) & \pi_{yy}(x_0,y_0) \end{pmatrix} \begin{pmatrix} (x-x_0) \\ (y-y_0) \end{pmatrix}.$$

Rearranging:

$$\begin{pmatrix}(x-x_0)\\(y-y_0)\end{pmatrix} = -\begin{pmatrix}\pi_{xx}(x_0,y_0) & \pi_{xy}(x_0,y_0)\\ \pi_{xy}(x_0,y_0) & \pi_{yy}(x_0,y_0)\end{pmatrix}^{-1}\begin{pmatrix}\pi_x(x_0,y_0)\\ \pi_y(x_0,y_0)\end{pmatrix}.$$

The solution has the form ($z=(x,y)$):

$$z = z_0 - H_\pi(x,y)^{-1}\nabla\pi(x,y),$$

where H_π is the matrix of partial derivatives and $\nabla\pi = (\pi_x, \bar{x}_y)$.

For the specific functional form:

$$\begin{aligned}
\pi(x,y) &= 2x^{\frac{1}{2}} + 4y^{\frac{1}{2}} + xy - x^2 - y^2\\
\pi_x(x,y) &= x^{-\frac{1}{2}} + y - 2x\\
\pi_y(x,y) &= 2y^{-\frac{1}{2}} + x - 2y\\
\pi_{xx}(x,y) &= -\frac{1}{2}x^{-\frac{3}{2}} - 2\\
\pi_{yy}(x,y) &= -y^{-\frac{3}{2}} - 2\\
\pi_{xy}(x,y) &= 1.
\end{aligned}$$

Thus,

$$H(x,y) = \begin{pmatrix}-\frac{1}{2}x^{-\frac{3}{2}} - 2 & 1\\ 1 & -y^{-\frac{3}{2}} - 2\end{pmatrix},$$

and so the algorithm is (beginning at $(x,y)=(x_0,y_0)$):

$$\begin{pmatrix}x_{i+1}\\ y_{i+1}\end{pmatrix} = \begin{pmatrix}x_i\\ y_i\end{pmatrix} - \begin{pmatrix}-\frac{1}{2}x_i^{-\frac{3}{2}} - 2 & 1\\ 1 & -y_i^{-\frac{3}{2}} - 2\end{pmatrix}^{-1}\begin{pmatrix}x_i^{-\frac{1}{2}} + y_i - 2x_i\\ 2y_i^{-\frac{1}{2}} + x_i - 2y_i\end{pmatrix},$$

given initial value (x_0,y_0). A plot of the function:

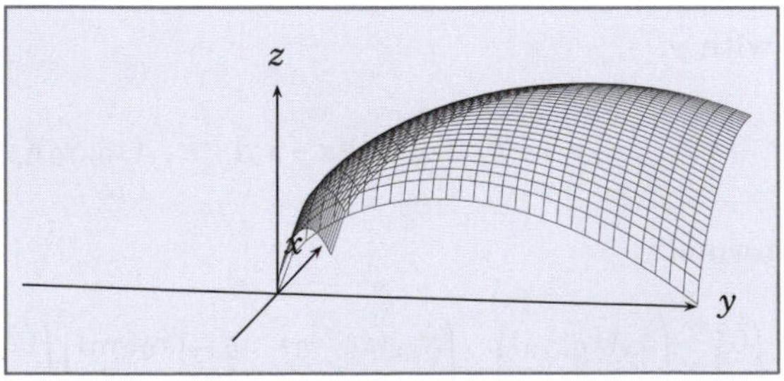

◊

8.3 Expected utility theory and behavior towards risk

The presence of risk (variability) lowers the value of an asset: other things equal, comparing two assets, the one that has greater risk will have a lower value. In this section, Taylor series expansions are used to determine approximate welfare and monetary measures of the negative impact of risk. Before providing these measures, some introduction to utility and risk is necessary.

8.3.1 Expected utility theory

The usual consumer theory model has $u(x_1,x_2,\ldots,x_n)$ = utility from consumption of the $(x_1,\ldots,x_n)$ bundle of goods. In this model, the individual receives and consumes x_1 of good 1, x_2 of good 2 and so on. Suppose that one receives some random return instead of a sure consumption bundle. Then this formulation is inappropriate: when a person receives a random return X, we cannot attach a utility to the return in the same way. Let X be a random return:

X	*Probability*
x_1	p_1
x_1	p_1
$\vdots$	$\vdots$
x_n	p_n

where p_i is the probability that X takes on the value x_i. With this random return, the individual receives income x_i with probability p_i. When you hold a lottery ticket that pays a million dollars if you win (and you win with probability p, say), then the situation is described:

X	*Probability*
0	$(1-p)$
1 million	p

Note that you either get 0 or 1 million – you do not get both! Above, the person receives some return (some x_i). Only *one* x_i is actually received. Note also that we often have to make a decision about whether to purchase a random return or not, *prior* to knowing the outcome. We cannot choose to buy the lottery ticket after we know whether we have won or not. At the time of purchase, we face the possibility of getting either 0 or one million dollars – and have to make a decision. Similarly, when one buys a stock today, the value of the stock tomorrow is unknown. In situations such as these, we need a way to compare the utility associated with different uncertain outcomes, so that we can choose the most preferred one. Expected utility theory is such a way. According to expected utility theory an individual has a utility function, u, defined on income. To calculate the utility associated with the random income or return stream X given previously, calculate the utility of income in each state $i, u(x_i)$, multiply by the probability p_i to get $p_i u(x_i)$ and sum these terms, $\sum p_i u(x_i)$. This is called expected utility and denoted $E\{u(X)\}$. Here, $u(x_i)$ is utility from x_i. (Some outcome x_i will occur – and this is what the individual actually receives.) Now we can compare two random returns X and Y in expected utility terms:

X	*Probability*	Y	*Probability*
x_1	p_1	y_1	q_1
x_2	p_2	y_2	q_2
$\vdots$	$\vdots$	$\vdots$	$\vdots$
x_n	p_n	y_m	q_m

Then the individual prefers the random return y to x if:

$$\sum_{j=1}^{m} q_j u(y_j) > \sum_{i=1}^{n} p_i u(x_i).$$

EXAMPLE 8.6: Let X and Y be two random variables with the following distributions:

X	*Probability*	Y	*Probability*
100	$\frac{1}{3}$	200	$\frac{1}{4}$
196	$\frac{1}{3}$	220	$\frac{1}{4}$
400	$\frac{1}{3}$	244	$\frac{1}{4}$
		264	$\frac{1}{4}$

Suppose that the utility function is the square root function: $u(\cdot) = \sqrt{\cdot}$.

$$E\{u(X)\} = \sum p_i u(x_i) = \frac{1}{3}[10 + 14 + 20] = \frac{44}{3} = 14.67$$

$$E\{u(Y)\} = \sum p_i u(y_i) = \frac{1}{4}[14.14 + 14.83 + 15.62 + 16.25] = \frac{60.84}{4} = 15.21.$$

Thus, y is preferred to x. Why should this be? Some insight is given from the mean and variance.

$$E(X) = \sum p_i x_i = 232 = \frac{696}{3}$$

$$E(Y) = \sum q_i y_i = 232 = \frac{928}{4}$$

$$\text{Var}(X) = \frac{1}{3}[(100-232)^2 + (196-232)^2 + (400-232)^2] = \frac{46944}{3} = 15648$$

$$\text{Var}(Y) = \frac{1}{4}[(200-232)^2 + (220-232)^2 + (244-232)^2 + (264-232)^2]$$

$$= \frac{2336}{4} = 584$$

Thus, while both have the same mean return, Y has a smaller variance – is "less risky". ◊

8.3.2 Risk-averse preferences and welfare loss

The reason that Y is preferred to X in Example 8.6 is related to the shape of u. In general, $u(\cdot)$ is assumed to be concave: this is said to represent risk-averse preferences. An individual is said to be risk averse if u is concave. It is easy to check that $\sqrt{\cdot}$ is a concave function. Loosely speaking, risk-averse individuals prefer low variance for a given mean return. To understand the role of concavity, consider the case where the random variable X has only two possible values, x_1 and x_2, as depicted in Figure 8.5.

X	*Probability*
x_1	p
x_2	$1-p$

The individual prefers the sure amount $\mu = px_1 + (1-p)x_2$ to the uncertain return which gives x_1 with probability p and x_2 with probability $(1-p)$. Thus concavity means that the individual prefers the mean return μ, for sure, rather than the uncertain return X (with mean μ). One can check that the opposite is the case when the utility function is convex. This can be taken further to clarify the role of the variance.

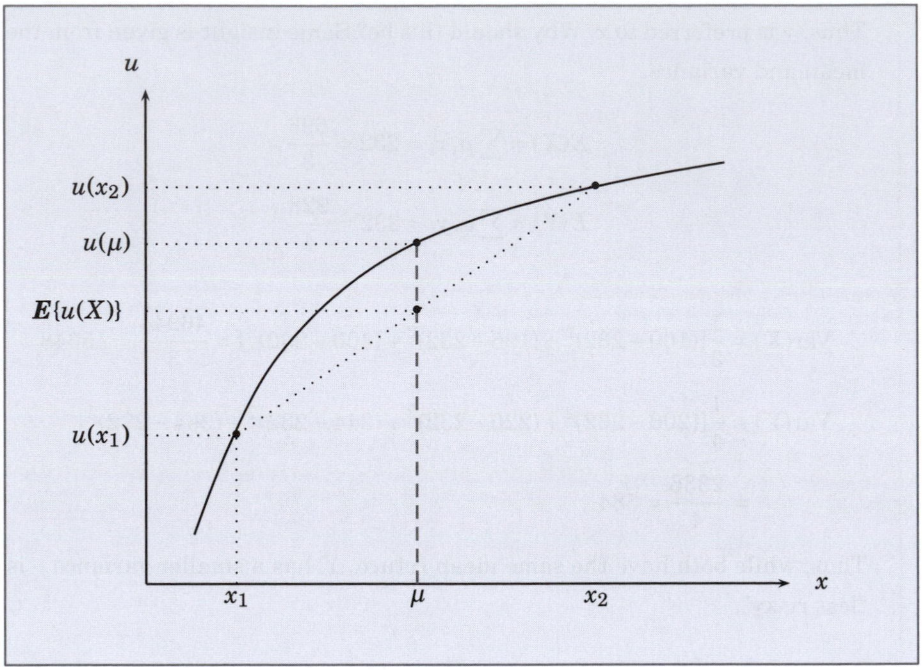

Figure 8.5: Risk aversion

Suppose x_1, x_2 are close to μ, so that we can use a Taylor series approximation:

$$u(x_1) = u(\mu) + u'(\mu)(x_1 - \mu) + \frac{1}{2}u''(\mu)(x_1 - \mu)^2 + \frac{1}{3!}u'''(\mu_1)(x_1 - \mu)^3$$
$$u(x_2) = u(\mu) + u'(\mu)(x_2 - \mu) + \frac{1}{2}u''(\mu)(x_2 - \mu)^2 + \frac{1}{3!}u'''(\mu_2)(x_2 - \mu)^3.$$

We can use these equations to calculate the expected utility, $E(u) = \sum p_i u(x_i)$, as follows:

$$\begin{aligned}E(u) &= p_1 u(x_1) + p_2 u(x_2)\\ &= [p_1 u(\mu) + p_2 u(\mu)] + [p_1 u'(\mu)(x_1 - \mu) + p_2 u'(\mu)(x_2 - \mu)]\\ &\quad + \frac{1}{2}[p_1 u''(\mu)(x_1 - \mu)^2 + p_2 u''(\mu)(x_2 - \mu)^2]\\ &\quad + \frac{1}{3!}[p_1 u'''(\mu_1)(x_1 - \mu)^3 + p_2 u'''(\mu_2)(x_2 - \mu)^3].\end{aligned}$$

If $(x_i - \mu)$ is "small" then the last term is relatively small and can be ignored. Then,

$$\begin{aligned}E\{u(X)\} &\approx [p_1 u(\mu) + p_2 u(\mu)] + [p_1 u'(\mu)(x_1 - \mu) + p_2 u'(\mu)(x_2 - \mu)]\\ &\quad + \frac{1}{2}[p_1 u''(\mu)(x_1 - \mu)^2 + p_2 u''(\mu)(x_2 - \mu)^2].\end{aligned}$$

Adding terms:

$$E\{u(X)\} \approx [p_1 + p_2]u(\mu) + [p_1(x_1 - \mu) + p_2(x_2 - \mu)]u'(\mu) + \frac{1}{2}[p_1(x_1 - \mu)^2 + p_2(x_2 - \mu)^2]u''(\mu).$$

Note that $[p_1 + p_2] = 1$, and $[p_1(x_1 - \mu) + p_2(x_2 - \mu)] = 0$ from the definition of the mean; while $[p_1(x_1 - \mu)^2 + p_2(x_2 - \mu)^2] = \sigma^2$, from the definition of the variance. Thus, we have the expression:

$$E\{u(X)\} \approx u(\mu) + \frac{1}{2}u''(\mu)\sigma^2. \tag{8.4}$$

If we write the last expression as $E\{u(X)\} - u(\mu) \cong \frac{1}{2}u''(\mu)\sigma^2$, we see that when u is concave, then u'' is negative – so that expected utility is less than $u(\mu)$. (See Figure 8.6.) Note also that the difference depends on the variance of the random return. Thus, the welfare loss due to small risk is:

$$\frac{1}{2}u''(\mu)\sigma^2.$$

So, $\frac{1}{2}u''(\mu)\sigma^2$ is the loss in utility or the disutility associated with the risk. Figure 8.6 illustrates the idea. The disutility associated with risk is measured in utility terms but we can also give a monetary measure – discussed next.

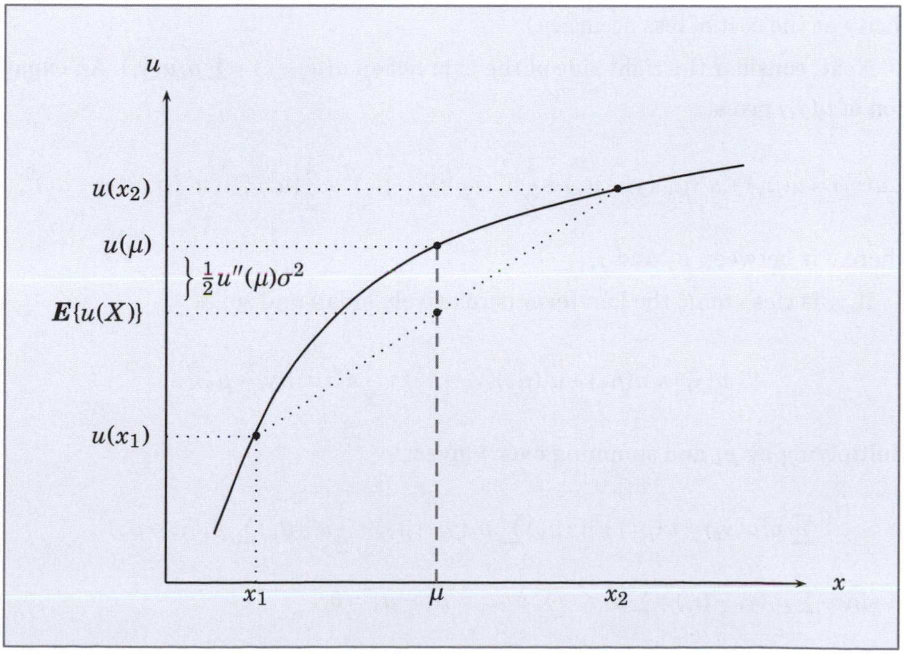

Figure 8.6: Utility loss associated with risk

8.3.3 Risk aversion and the cost of small risk

Let income be represented by a random variable Y, where Y takes on the value y_i with probability p_i (i.e., $\text{Prob}(Y = y_i) = p_i$). Thus $\mu_y = \sum p_i y_i$ and $\sigma_y^2 = \sum p_i(y_i - \mu_y)^2$. Instead of asking how risk affects the level of utility, one can focus on the monetary cost of bearing risk. How much would a person pay to avoid facing the risk and receive a sure income μ_y? If the individual receives μ_y and makes a payment of c, the amount received is $y^* = \mu_y - c$, and the individual is indifferent between the sure income y^* and facing the random return Y. The amount, c, is given as the solution of the equation: $u(\mu_y - c) = \sum p_i u(y_i)$. In general, solving this explicitly for c is very difficult. However, we can use Taylor series expansions to find an approximate solution. This is done now, expanding each side separately. First, consider an expansion of $u(\mu_y - c)$ around μ_y.

$$\begin{aligned} u(\mu_y - c) &= u(\mu_y) - u'(\mu^*)c, \text{ some } \mu^* \in [\mu_y - c, \mu_y] \\ &= u(\mu_y) - u'(\mu_y)c + [u'(\mu_y) - u'(\mu^*)]c \\ &\approx u(\mu_y) - u'(\mu_y)c, \text{ when } c \text{ is small so } [u'(\mu_y) - u'(\mu^*)] \text{ is small.} \end{aligned}$$

By taking an expansion to the first order for $u(\mu_y - c)$, the approximation obtained is linear in c. (A better approximation is given by $u(\mu_y - c) \approx u(\mu_y) - u'(\mu_y)c + \frac{1}{2}u''(\mu_y)c^2$, but that is quadratic in c. Here, the linear approximation gives simplicity at the cost of less accuracy.)

Next, consider the right side of the expression $u(\mu_y - c) = \sum p_i u(y_i)$. An expansion of $u(y_i)$ gives:

$$u(y_i) = u(\mu_y) + u'(\mu_y)(y_i - \mu_y) + \frac{1}{2}u''(\mu_y)(y_i - \mu_y)^2 + \frac{1}{2}[u''(\xi) - u''(\mu_y)](y_i - \mu_y)^2,$$

where ξ is between μ_y and y_i.

If y_i is close to μ_y the last term is relatively small and so:

$$u(y_i) \approx u(\mu_y) + u'(\mu_y)(y_i - \mu_y) + \frac{1}{2}u''(\mu_y)(y_i - \mu_y)^2.$$

Multiplying by p_i and summing over i gives:

$$\sum p_i u(y_i) \simeq u(\mu_y) + u'(\mu_y)\sum p_i(y_i - \mu_y) + \frac{1}{2}u''(\mu_y)\sum p_i(y_i - \mu_y)^2,$$

or, since $\sum p_i(y_i - \mu_y) = \sum p_i y_i - \sum p_i \mu_y = \mu_y - \mu_y = 0$,

$$\sum p_i u(y_i) \simeq u(\mu_y) + \frac{1}{2}u''(\mu_y)\sigma_y^2.$$

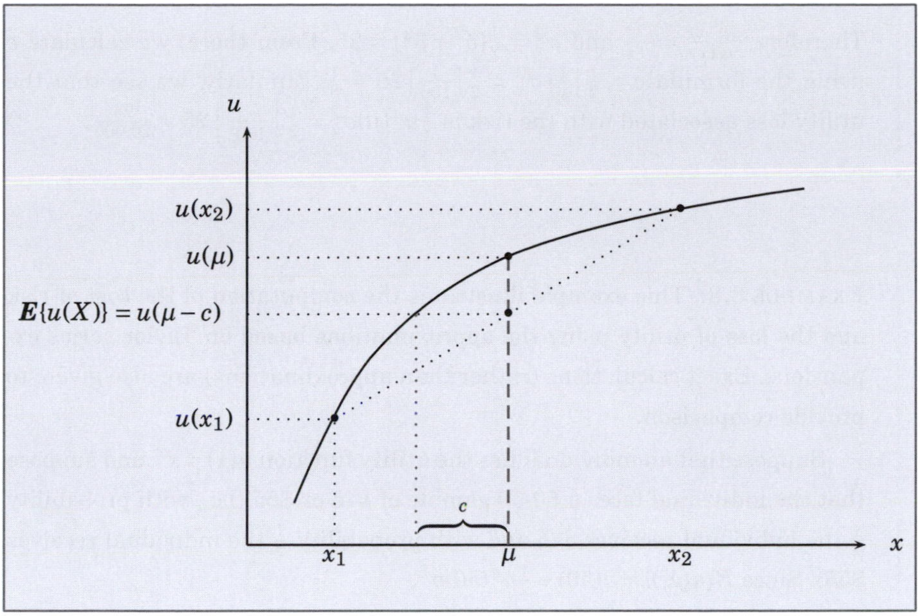

Figure 8.7: The cost of risk

Combining these two approximations:

$$u(\mu_y) - u'(\mu_y)c \approx u(\mu_y - c) = \sum p_i u(y_i) \approx u(\mu_y) + \frac{1}{2}u''(\mu_y)\sigma_y$$

and so

$$u(\mu_y) - u'(\mu_y)c \approx u'(\mu_y) + \frac{1}{2}u''(\mu_y)\sigma_y.$$

Solving for c (approximately) gives:

$$c \approx \frac{1}{2}\frac{-u''(\mu_y)}{u'(\mu_y)}\sigma^2.$$

Thus the "cost" of the risk is proportional to σ_y^2. The proportionality factor $-\frac{u''(\mu_y)}{u'(\mu_y)}$ is called the "degree of absolute risk aversion". Figure 8.7 indicates how c is determined.

EXAMPLE 8.7: Let $u(x) = \ln(x)$, so that $u'(x) = \frac{1}{x}$ and $u''(x) = -\frac{1}{x^2}$. Suppose the random return is:

X	*Probability*
95	$\frac{1}{2}$
105	$\frac{1}{2}$

Therefore, $\frac{u''(x)}{u'(x)} = -\frac{1}{x}$ and $\sigma_x^2 = \frac{1}{2}[5^2+5^2] = 25$. From these, we calculate c using the formula $c = \frac{1}{2}\left[\frac{1}{x}\right]\sigma_x^2 = \frac{1}{2}\left[\frac{1}{100}\right]25 = \frac{1}{8}$. Similarly, we see that the utility loss associated with the risk is $\frac{1}{2}u''(\mu)\sigma_x^2 = \frac{1}{2}\left[\frac{-1}{100^2}\right]25 = \frac{25}{20{,}000}$. ◊

EXAMPLE 8.8: This example illustrates the computation of the cost of risk and the loss of utility using the approximations based on Taylor series expansions. Exact calculations (rather than approximations) are also given, to provide comparison.

Suppose that an individual has the utility function $u(x) = x^{\frac{1}{3}}$ and suppose that the individual faces a 50-50 gamble of \$45 or \$55 (i.e., with probability $\frac{1}{2}$ the individual receives \$45 and with probability $\frac{1}{2}$ the individual receives \$55). Since $E\{u(X)\} \approx u(50) + \frac{1}{2}u''(50)\sigma^2$,

$$u(50) - E\{u(X)\} = -\frac{1}{2}u''(50)\sigma^2 = \frac{2}{9}50^{-\frac{5}{3}} \times 25 = \frac{2}{9}(0.001474)25 = 0.008187.$$

Thus, the approximate loss of utility from facing the risk, relative to receiving the sure amount of 50, is 0.008187.

Similarly, the approximate cost associated with the risk is given by $c \approx -\frac{1}{2}\frac{u''(50)}{u'(50)}\sigma^2$. To compute this, observe that $u(x) = x^{\frac{1}{3}}$, so that $u'(x) = \frac{1}{3}x^{\frac{-2}{3}}$ and $u''(x) = -\frac{2}{9}x^{-\frac{5}{3}}$. Therefore,

$$\frac{u''(x)}{u'(x)} = \frac{-\frac{2}{9}x^{-\frac{5}{3}}}{\frac{1}{3}x^{-\frac{2}{3}}} = -\frac{6}{9}x^{-1} = -\frac{6}{9} \times \frac{1}{x} = -\frac{2}{3} \times \frac{1}{x}.$$

Since $\sigma^2 = \frac{1}{2}(50-45)^2 + \frac{1}{2}(55-45)^2 = \frac{1}{2}5^2 + \frac{1}{2}5^2 = 25$,

$$c \approx -\frac{1}{2}\frac{u''(50)}{u'(50)}\sigma^2 = -\frac{1}{2}\left(-\frac{2}{3} \times \frac{1}{50}\right)25 = \frac{1}{6} = 0.16667.$$

Now, observe that in this example, the exact value of expected utility is:

$$\mathrm{E}(u(X)) = \frac{1}{2}u(45) + \frac{1}{2}u(50) = \frac{1}{2}45^{\frac{1}{3}} + \frac{1}{2}50^{\frac{1}{3}} = 3.620462.$$

Recall that c is chosen to satisfy:

$$u(50-c) = \mathrm{E}\{u(X)\},$$

so that c must satisfy $u(50-c) = (50-c)^{\frac{1}{3}} = 3.679923$, or $50-c = 3.679923^3 = 49.8329$, which gives $c = 50-49.8329 = 0.167101$. Summarizing, the exact cost of the risk is 0.167101, and the approximation of this using the Taylor series expansions is 0.16667. ◊

8.4 Diversification and portfolio selection

Diversification of risk involves combining different random returns that are independent or have low correlation. In that way, the probability of all returns being low (or high) at the same time is small, so that extreme variation in return is less likely. This is briefly discussed and related to the cost of risk. Then approximately optimal portfolio selection using Taylor series expansions is considered.

8.4.1 Diversification

This section illustrates how portfolio diversification reduces risk. Suppose that we have n projects. Assume that the return on each project is *independent* of the return on other projects. Project j has return $Y(j)$, a random variable. This random variable has n_j possible values, $(Y_1(j), Y_2(j), \dots, Y_{n_j}(j))$, with associated probabilities $(p_1(j), p_2(j), \dots, p_{n_j}(j))$. The probability that $Y(j)$ takes the value $Y_i(j)$ is $p_i(j)$:

$$\mu(j) = \sum_{i=1}^{n_j} p_i(j) y_i(j) = E(Y(j)),$$

$$\sigma^2(j) = \sum_{i=1}^{n_j} p_i(j)[Y_i(j) - \mu(j)]^2 = \mathrm{Var}\{Y(j)\}.$$

Let $\mu_j = \mu, \sigma^2(j) = \sigma^2$ – so we assume that each of the random variables has the same mean and variance. If an individual undertakes project j, the cost of risk is:

$$c = \frac{1}{2}\frac{-u''(\mu)}{u'(\mu)}\sigma^2.$$

Suppose instead the individual forms a portfolio containing $\frac{1}{n}$ of each project. Denote this portfolio Z. $Z = \frac{1}{n}\sum_j^n Y_j$ a random variable. The mean and variance are:

$$E(Z) = \mu = \frac{1}{n}\sum_{j=1}^{n}\mu_j = \frac{1}{n}\sum_{i=1}^{n}\mu = \frac{1}{n}n\mu = \mu$$

and $\text{Var}(Z) = \frac{1}{n}\sigma^2$. To see the latter:

$$\text{Var}(Z) = \text{Var}\left(\frac{1}{n}\sum_{j=1}^{n} Y_j\right) = \frac{1}{n^2}\left[\sum_{j=1}^{n}\text{Var}(Y_j)\right],$$

assuming the Y are independent. This is then equal to

$$\frac{1}{n^2}\left[\sum_{j=1}^{n}\sigma_j^2\right] = \frac{1}{n^2}\left[\sum_{j=1}^{n}\sigma^2\right] = \frac{1}{n^2}\left[n\sigma^2\right] = \frac{1}{n}\sigma^2.$$

Using our earlier results, we see that Z is a portfolio that has cost of risk to the ith individual given by:

$$c = \frac{1}{2}\left[\frac{-u''(\mu)}{u'(\mu)}\right]\frac{\sigma^2}{n}.$$

We observe that as the number (n) of projects increases, the cost of risk goes to 0. You can easily see what is happening here – the portfolio Z has the same mean return as each of the $Y(j)$ assets but the variance of Z is $\frac{1}{n}$ the variance of each of the $Y(j)$.

8.4.2 Portfolio selection

This section studies the optimal division of a portfolio into a risky and a safe asset. The purchase of Z units of a risky asset yields an end-of-year gross return Z per unit of risky asset purchased. Let $p_i = \text{Prob}(Z = z_i)$. The asset price is P (price at beginning of year), the individual's wealth at the beginning of the year is W. Thus, if the individual purchases α units of the asset and if at the end of the year the risky asset return turns out to be z_i, then the individual's end-of-year wealth is $W - \alpha P + \alpha z = W + \alpha(z_i - P)$. Therefore, the (end-of-year) expected utility is given by: $E\{u(W + \alpha(Z - P))\} = \sum p_i u(W + \alpha(z_i - P))$.

The individual's problem is to choose α to solve: $\max_\alpha \sum p_i u(W + \alpha(z_i - P))$. Differentiating with respect to α gives

$$\sum p_i u'(W + \alpha(z_i - P))(z_i - P) = 0.$$

Solving this gives the optimal solution for α. However, this may be difficult to solve, so that finding an approximate solution via a Taylor series approximation is useful. If $z_i - P$ is small,

$$u'(W + \alpha(z_i - P)) \simeq u'(W) + u''(W)\alpha(z_i - P).$$

Note that by using a first-order Taylor series expansion of $u'(W+\alpha(z_i-P))$, the expression above is linear in α. So

$$\sum p_i\{u'(W)+u''(W)\alpha(z_i-P)\}(z_i-P)\simeq 0$$

or

$$\sum p_i u'(W)(z_i-P)+\sum p_i u''(W)\alpha(z_i-P)^2\simeq 0.$$

To simplify this expression, observe that

$$\begin{aligned}&\sum p_i u'(W)(z_i-P)+\sum p_i u''(W)\alpha(z_i-P)^2\\ &\qquad = u'(W)\sum p_i(z_i-P)+u''(W)\alpha\sum p_i(z_i-P)^2.\end{aligned}$$

Focus on the terms $\sum p_i(z_i-P)$ and $\sum p_i(z_i-P)^2$. We see that:

$$\sum p_i(z_i-P)=\sum p_i z_i-\sum p_i P=\mu-P.$$

Note that $\sigma_z=\sum p_i(z_i-\mu_z)^2$ is the variance of Z, but the expression above involves $\sum p_i(z_i-P)^2$.

$$\begin{aligned}\sum p_i(z_i-P)^2 &= \sum p_i\left((z_i-\mu)+(\mu-P)\right)^2\\ &= \sum p_i(z_i-\mu)^2+2\sum p_i(\mu-P)(z_i-\mu)+\sum p_i(\mu-P)^2\\ &= \sigma^2+2(\mu-P)\sum p_i(z_i-\mu)+(\mu-P)^2\sum p_i\\ &= \sigma^2+(\mu-P)^2.\end{aligned}$$

(Using the fact that $\sum p_i=1$ and $\sum p_i(z_i-\mu)=0$.)

This calculation gives $\sum p_i(z_i-P)^2=\sigma^2+(\mu-P)^2)$. So,

$$u'(W)\sum p_i(z_i-P)+u''(W)\alpha\sum p_i(z_i-P)^2=u'(W)(\mu-P)+u''(W)\alpha[\sigma^2+(\mu-P)^2].$$

Therefore, the first-order condition is approximately:

$$u'(W)(\mu-P)+u''(W)\alpha[\sigma^2+(\mu-P)^2]\simeq 0.$$

Solving for α:

$$\alpha\approx\frac{1}{-\frac{u''(W)}{u'(W)}}\frac{(\mu-P)}{\sigma^2+(\mu-P)^2}.$$

EXAMPLE 8.9: Let $u(\xi)=\ln(\xi)$, $W=100$, $P=2$. Suppose that the random variable Z has the distribution $\text{Prob}(Z=1)=\text{Prob}(Z=5)=\frac{1}{2}$. Thus, $\mu=3$,

$\sigma^2 = 4$. With these parameters:

$$\frac{-u''(100)}{u'(100)} = -\frac{-1/(100)^2}{1/100} = \frac{1}{100} \Rightarrow \left[1/\frac{u''(W)}{u'(W)}\right] = 100.$$

Using the formula for α, $\alpha \simeq 100\frac{(3-2)}{4+1} = \frac{100}{5} = 20$, so the approximate solution is to purchase roughly 20 units of asset. ◊

8.5 The multivariate version of Taylor's theorem

Continuous functions of many variables can also be approximated by Taylor series expansions. Let $f(x): \mathscr{R}^n \to \mathscr{R}$, be a function of n variables where $\mathscr{R}$ is the real number line and $\mathscr{R}^n = \mathscr{R} \times \cdots \times \mathscr{R}$, the n-times product of $\mathscr{R}$. Recall that the partial derivative of f with respect to x_i is denoted f_{x_i} and the second partial $f_{x_i x_j} = \frac{\partial^2 f}{\partial x_i \partial x_j}$. Write:

$$\boldsymbol{\nabla f} = \begin{pmatrix} f_{x_1} \\ f_{x_2} \\ \vdots \\ f_{x_n} \end{pmatrix}, \qquad \boldsymbol{H} = \begin{pmatrix} f_{x_1x_1} & f_{x_1x_2} & \cdots & f_{x_1x_n} \\ f_{x_2x_1} & f_{x_2x_2} & \cdots & f_{x_1x_n} \\ \vdots & \vdots & \ddots & \vdots \\ f_{x_nx_1} & f_{x_nx_2} & \cdots & f_{x_nx_n} \end{pmatrix} \quad \text{and} \quad \boldsymbol{x} - \boldsymbol{x_0} = \begin{pmatrix} x_1 - x_{01} \\ x_2 - x_{02} \\ \vdots \\ x_n - x_{0n} \end{pmatrix}.$$

Also, write $\boldsymbol{\nabla f}(x)$ and $\boldsymbol{H}(x)$ to denote the matrix of derivatives evaluated at x. A Taylor series expansion of f to the second order around x_0 is

$$f(x) = f(x_0) + \nabla f(x_0) \cdot (x - x_0) + \frac{1}{2!}(x - x_0)' H(p)(x - x_0).$$

Theorem 8.3: *Let $f: \mathscr{R}^n \to \mathscr{R}$, be continuous with continuous first and second partial derivatives. Then given x, x_0:*

$$f(x) = f(x_0) + \nabla f(x_0) \cdot (x - x_0) + \frac{1}{2!}(x - x_0)' H(p)(x - x_0),$$

where p is on a line between x and x_0: for some $\theta \in (0,1)$, $p = \theta x_0 + (1-\theta)x$.

EXAMPLE 8.10: Let $f(x,y) = x^2y^3$. So, $f_x(x,y) = 2xy^3$, $f_y(x,y) = 3x^2y^2$, $f_{xx} = 2y^3$, $f_{yy} = 6x^2y$ and $f_{xy} = 6xy^2$. Then a first-order Taylor series expansion of

f around $x = 1$, $y = 1$ is:

$$\begin{aligned} f(x,y) &\approx f(1,1) + f_x(1,1)(x-1) + f_y(1,1)(y-1) \\ &\approx 1 + 2(x-1) + 3(y-1) \\ &\approx -4 + 3y + 2x. \end{aligned}$$

The second-order Taylor series expansion has the form:

$$\begin{aligned} f(x,y) \approx{}& f(1,1) + f_x(1,1)(x-1) + f_y(1,1)(y-1) \\ &+ \frac{1}{2}[f_{xx}(1,1)(x-1)^2 + 2f_{xy}(1,1)(x-1)(y-1) + f_{yy}(1,1)(y-1)^2]. \end{aligned}$$

Substituting the values for the function:

$$\begin{aligned} x^2y^3 &\approx 1 + 2(x-1) + 3(y-1) + \frac{1}{2}[2(x-1)^2 + 2\cdot 6(x-1)(y-1) + 6(y-1)^2] \\ &\approx 1 + 2(x-1) + 3(y-1) + (x-1)^2 + 6(x-1)(y-1) + 3(y-1)^2 \\ &\approx 6 - 9y - 6x + 3y^2 + x^2 + 6xy. \end{aligned}$$ ◊

8.6 Proof of Taylor's theorem

The proof of Taylor's theorem makes use of Rolle's theorem:

Theorem 8.4: *Let f be a continuous differentiable function on the interval $[a,b]$. Suppose that $f(a) = f(b) = 0$. Then, there is a point $c \in (a,b)$ with $f'(c) = 0$.*

PROOF: If $f(x) = 0$ for all $x \in [a,b]$, then $f'(x) = 0$ on $[a,b]$. Otherwise, there is some point $x \in [a,b]$ with $f(x) > 0$ or $f(x) < 0$. Suppose the former, so $f(x) > f(a) = f(b)$. Since f is continuous, it has a maximum on $[a,b]$ and this cannot be at a or b. Hence there is an interior maximum at some point c with $f'(c) = 0$. If instead $f(x) < f(a) = f(b)$, there is a minimum on $[a,b]$ at some point c with $f'(c) = 0$. ■

Theorem 8.5: *Let f be a continuous function on an interval $[a,b]$ with continuous derivatives f', f'', ..., f^n. Then given $x, x_0 \in (a,b)$:*

$$\begin{aligned} f(x) = f(x_0) + f'(x_0)(x-x_0) + \frac{1}{2!}f''(x_0)(x-x_0)^2 \\ + \frac{1}{3!}f'''(x_0)(x-x_0)^3 + \cdots + \frac{1}{n!}f^n(p)(x-x_0)^n, \end{aligned}$$

where p is between x and x_0, and

$$f(x) \approx f(x_0) + f'(x_0)(x - x_0) + \frac{1}{2!} f''(x_0)(x - x_0)^2 + \frac{1}{3!} f'''(x_0)(x - x_0)^3 + \cdots + \frac{1}{n!} f^n(x_0)(x - x_0)^n$$

when x is close to x_0.

PROOF: Consider the second-order expansion – the generalization to the nth-order case is straightforward. Let

$$g(z) = f(x) - f(z) - f'(z)(x - z) - \frac{K}{2}(x - z)^2.$$

Observe that $g(x) = 0$ and $g(x_0) = f(x) - f(x_0) - f'(x_0)(x - x_0) - \frac{K}{2}(x - x_0)^2$. Choose K to make this 0, so that $g(x_0) = 0$. Then $g(x) = g(x_0)$, so that by Rolle's theorem, there is a point between x_0 and x, say ξ with $g'(\xi) = 0$ ($\min\{x_0, x\} < \xi < \max\{x_0, x\}$). Differentiating $g(z)$:

$$g'(z) = -f'(z) - f''(z)(x - z) + f'(z) + 2\frac{K}{2}(x - z) = -f''(z)(x - z) + K(x - z).$$

At ξ, $g'(\xi) = -f''(\xi)(x - \xi) + K(x - \xi) = 0$. Since $\xi \neq x$, $f''(\xi) = K$.

For the nth-order case:

$$g(z) = f(x) - f(z) - f'(z)(x - z) - \frac{1}{2!} f''(z)(x - z)^2 - \cdots - \frac{1}{(n-1)!} f^{(n-1)}(z)(x - z)^{n-1} - \frac{K}{n!}(x - z)^n.$$

Again, $g(x) = 0$ and choose K so that $g(x_0) = 0$. Rolle's theorem implies that there is some ξ between x_0 and x with $g'(\xi) = 0$. In the derivative of g, all but two terms cancel:

$$g'(z) = -\frac{1}{(n-1)!} f^{(n)}(z)(x - z)^{n-1} + \frac{K}{(n-1)!}(x - z)^{n-1}.$$

At ξ,

$$g'(\xi) = -\frac{1}{(n-1)!} f^{(n)}(\xi)(x - \xi)^{n-1} + \frac{K}{(n-1)!}(x - \xi)^{n-1} = 0.$$

Since $x \neq \xi$, $f^{(n)}(\xi) = K$. ∎

For the multivariate version, the proof is similar.

Theorem 8.6: *Let $f : \mathscr{R}^n \to \mathscr{R}$, be continuous with continuous first and second partial derivatives. Then given x, x_0:*

$$f(x) = f(x_0) + \nabla f(x_0) \cdot (x - x_0) + \frac{1}{2!}(x - x_0)' H(p)(x - x_0),$$

where p is on a line between x and x_0: for some $\theta \in (0,1)$, $p = \theta x_0 + (1-\theta)x$.

PROOF: To see why this is so, given z, Δz, let $g(t) = f(z + t\Delta z)$. Then, g is a function of one variable and a second-order Taylor series expansion around 0 gives:

$$g(t) = g(0) + g'(0)t + \frac{1}{2}g''(s)t^2,\ 0 < s < t.$$

Since

$$g'(t) = \sum_i f_i(z + t\Delta z)\Delta z_i \Rightarrow g'(0) = \sum_i f_i(z)\Delta z_i$$

and

$$g''(t) = \sum_j \sum_i f_{ij}(z + t\Delta z)(\Delta z_i)^2 \Rightarrow g''(s) = \sum_j \sum_i f_{ij}(z + s\Delta z)(\Delta z_i)^2,$$

this gives

$$\begin{aligned} g(t) &= g(0) + g'(0)t + \frac{1}{2}g''(s)t^2 \\ f(z + t\Delta z) &= f(z) + \sum_i f_i(z)t\Delta z_i + \frac{1}{2}\sum_j \sum_i f_{ij}(z + s\Delta z)t^2 \Delta z_i^2. \end{aligned}$$

If $z = x_0$ and $\Delta z = \frac{x - x_0}{t}$, then $t\Delta z = (x - x_0) = \Delta x$ and $s\Delta z = \frac{s}{t}(x - x_0) = \alpha(x - x_0)$, $\alpha = \frac{s}{t} \in (0,1)$.

$$f(x) = f(x_0) + \sum_i f_i(x_0)\Delta x_i + \sum_j \sum_i f_{ij}(x_0 + \alpha\Delta x)(\Delta x_i)^2.$$

Writing $p = x_0 + \alpha(x - x_0) = x_0(1 - \alpha) + \alpha x$, gives:

$$f(x) = f(x_o) + \nabla f(x_0) \cdot (x - x_0) + \frac{1}{2!}(x - x_0)' H(p)(x - x_0).$$

■

Exercises for Chapter 8

Exercise 8.1 Consider the following functions:

$$f(x) = e^x - 1, \tag{8.5}$$

$$g(x) = \ln(x+1), \tag{8.6}$$

$$h(x) = x^2, \tag{8.7}$$

$$u(x) = x^{\frac{1}{2}}, \tag{8.8}$$

$$v(x) = \frac{1}{x}, \tag{8.9}$$

$$w(x) = \sin(x). \tag{8.10}$$

In all cases take $x > 0$.

(a) Plot each function and determine whether each function is concave, convex or neither.

(b) For each function, give the second-order Taylor series expansion.

Exercise 8.2 Give the second-order Taylor series expansions for the following functions,

$$f(x) \approx f(\bar{x}) + f'(\bar{x})(x-\bar{x}) + \frac{1}{2}f''(\bar{x})(x-\bar{x})^2 :$$

$$f(x) = x^{\frac{1}{3}} \tag{8.11}$$

$$f(x) = \ln(x) \tag{8.12}$$

$$f(x) = \sin(x) \tag{8.13}$$

$$f(x) = \cos(x) \tag{8.14}$$

$$f(x) = e^{-rx} \tag{8.15}$$

$$f(x) = x^{\alpha} \tag{8.16}$$

$$f(x) = a^{x}. \tag{8.17}$$

Exercise 8.3 Consider the average cost function $c(Q) = e^{-Q} + \frac{1}{2}Q^2$. Show that the cost function is strictly convex. Find the minimum of the average cost function c: find the value of Q that minimizes c. To do this, first observe that we can write a Taylor series approximation of c:

$$c(Q) \cong c(Q_0) + c'(Q_0)(Q - Q_0) + \frac{1}{2}c''(Q_0)(Q - Q_0)^2.$$

When Q_0 is close to the value of Q that minimizes c, then minimizing $c(Q)$ should give approximately the same answer as minimizing

$$c(Q_0) + c'(Q_0)(Q - Q_0) + \frac{1}{2}c''(Q_0)(Q - Q_0)^2.$$

If we minimize the latter expression by differentiating with respect to Q, we get the first-order condition: $c'(Q_0) + c''(Q_0)(Q - Q_0) = 0$. Solving for Q gives $Q = Q_0 - \left[\frac{c'(Q_0)}{c''(Q_0)}\right]$. This is the basis of an iteration: guess an initial Q_0; this gives $Q_1 = Q_0 - \left[\frac{c'(Q_0)}{c''(Q_0)}\right]$. With Q_1 we can obtain another estimate $Q_2 = Q_1 - \left[\frac{c'(Q_1)}{c''(Q_1)}\right]$, and with Q_2 we can obtain Q_3, and so on. At the nth iteration $Q_{n+1} = Q_n - \left[\frac{c'(Q_n)}{c''(Q_n)}\right]$. The process converges when the change from Q_n to Q_{n+1} becomes small. In this case, we end up with (approximately) $Q_n = Q_{n+1}$. Looking at the iteration rule $Q_{n+1} = Q_n - \left[\frac{c'(Q_n)}{c''(Q_n)}\right]$ yields $0 = Q_{n+1} - Q_n = -\left[\frac{c'(Q_n)}{c''(Q_n)}\right]$. Thus $\left[\frac{c'(Q_n)}{c''(Q_n)}\right] = 0$, and since $[c''(Q_n)] \neq 0$, we get $c'(Q_n) = 0$. Thus, since c is convex, Q_n minimizes c.

Exercise 8.4 Harveys brewery has a monopoly in the beer market. The demand function is $P(Q) = 100 - \left(\frac{1}{3}\right)Q^2$ and the cost function is $C(Q) = Q^2 + e^{0.01Q^2}$.

(a) Is the profit function concave?

(b) Using numerical optimization, find the profit-maximizing level of output. (Hint: start with $Q_0 = 8$ and perform four iterations.)

Exercise 8.5 A monopolist is faced with the following demand curve and cost function

$$P(Q) = 154 - 2e^{Q}, \quad C(Q) = 4Q + 6Q^2.$$

(a) Confirm that the profit function is concave.

(b) Find the profit-maximizing level of output using numerical optimization. Write down the algorithm. (Start iterations at $q_0 = 2$ and perform five iterations.)

Exercise 8.6 A manufacturer of jackets faces the following demand and cost functions:

$$P(q) = 1000 - 10q^2, \qquad C(q) = 100q + q\ln q.$$

(a) Confirm that the profit function is concave.

(b) Find the profit-maximizing level of output by numerical optimization.

Exercise 8.7 Faced with the following demand and cost functions

$$P(q) = 1000 - q^2, \qquad C(q) = 25 + q^2 + q(\ln q)^2,$$

a widget firm wishes to find the profit-maximizing level of output. Find the profit-maximizing level of output.

Exercise 8.8 Suppose a firm's profit function (as a function of output) is given by $\pi(q) = 2\ln(q+1) + \sqrt{q} - q$. Find the profit-maximizing value of q.

Exercise 8.9 Suppose a firm's revenue function (as a function of output) is given by $r(q) = 3(1 - e^{-q}) + 3\ln(q+1) - 2q$. Find the revenue-maximizing value of q.

Exercise 8.10 Consider the function $f(x) = 1 + x + 4x^2 - 3x^3$. This function has two points at which the first derivative of f is 0. Denote these x^* and $\bar{x}$, so that $f'(x^*) = f'(\bar{x}) = 0$. It is easy to see that one of these is 1, say that $\bar{x} = 1$. Find the value of x^* using a Taylor series expansion.

Exercise 8.11 An individual has utility defined on income according to $u(y) = \ln(y+1)$. The individual has initial income $\bar{y}$ and accepts a bet on a coin toss: if heads, the individual receives a dollar, if tails, the individual pays a dollar. The coin is fair (50/50 chance of heads or tails), thus the average utility from the bet is

$$\frac{1}{2}u(\bar{y}-1) + \frac{1}{2}u(\bar{y}+1).$$

If the individual had not accepted the bet, the utility obtained is $u(\bar{y})$ for sure. Compare the average utility of the bet with the sure utility from not betting. Should the individual have accepted the bet?

Exercise 8.12 An individual has the utility function $u(y) = \ln(y+1)$. This gives the utility of having y dollars. The individual faces a random income next period: current income is \$100 and next period income will either fall by \$5 to \$95 or increase by \$5 to \$105. There is a 50% chance that income will rise and a 50% chance that income will fall, so that average income next period is $100 = \frac{1}{2}95 + \frac{1}{2}105$. The risk premium, c, is defined as the amount of money less than \$100 that makes the individual indifferent between accepting that sum for sure and facing the risk associated with receiving only \$95. That is, the individual is indifferent between \$100 − c and the risky return \$95 with 50% chance and \$105 with 50% chance. Thus, $u(100-c) = \frac{1}{2}u(95) + \frac{1}{2}u(105)$, or $\ln(101-c) = \frac{1}{2}\ln(101-5) + \frac{1}{2}\ln(101+5)$. Find the approximate value of c using Taylor series expansions. Plot the utility function and illustrate how c is calculated on your graph.

Exercise 8.13 Repeat the previous question using the utility function $u(y) = e^y - 1$. In this case, note that the utility function is convex (second derivative positive) so that your picture will change significantly!

Exercise 8.14 A student has just received a present of a new computer. The student's yearly income is \$50 and the student has a utility function $u(y) = y^{\frac{1}{2}}$. Unfortunately, there is a 10% chance that the computer will be stolen and the price of a new computer is \$5, so in the event of theft, the replacement will leave the student with only \$45 income for the year. Thus, in the absence of theft, the student's income is \$50, while if the computer is stolen, the income drops to \$45. Computer insurance is available at a cost of \$1. Use the calculations from a Taylor series expansion to decide whether the student should buy insurance or not. Note $u'(y) = \frac{1}{2}y^{-\frac{1}{2}}$ and $u''(y) = -\frac{1}{4}y^{-\frac{3}{2}}$, so that

$$\frac{u''(y)}{u'(y)} = -\frac{1}{4}y^{-\frac{3}{2}}\left[\frac{1}{2}y^{-\frac{1}{2}}\right]^{-1} = -\frac{1}{4}y^{-\frac{3}{2}}[2y^{1/2}] = \frac{1}{2}y^{-1}.$$

Exercise 8.15 Consider an individual with the following utility function defined over wealth: $u(w) = \sqrt{w}$. Suppose the person has the opportunity to take the following gamble: draw a ticket from a hat; a blue ticket yields a prize of \$200, a yellow ticket yields \$50 and an orange ticket yields \$100. There are four orange tickets, two blue, and two yellow.

(a) Calculate the expected payoff from the bet.

(b) Calculate the approximate cost of risk to this individual (using a Taylor series expansion).

(c) How much is the person willing to pay to partake in this gamble?

(d) Suppose instead the utility function is $u(w) = w$. Rework your answers to parts (b) and (c).

(e) Suppose now the utility function is $u(w) = w^2$. Rework your answers to parts (b) and (c).

(f) Explain intuitively why your answers differ in parts (c), (d) and (e).

Exercise 8.16 A risk-averse individual has the following utility function defined over income: $u(w) = \sqrt{w}$. The person is faced with the following lottery with possible outcomes denoted x_i and their associated probabilities $P(x_i)$:

x	$P(x)$
10	$\frac{1}{2}$
15	$\frac{1}{8}$
1	$\frac{1}{8}$
7	$\frac{1}{4}$

(a) What is the expected payoff from this lottery?

(b) Calculate this person's measure of absolute risk aversion.

(c) If tickets for this lottery cost \$7.50, will he buy a ticket? (Use a Taylor series expansion.)

(d) Another individual has utility function $v(x) = kx$, $k > 0$. Will this individual buy a ticket?

Exercise 8.17 An individual owns a house worth \$200,000 that is located next to a river. There is a 1% chance that the river will flood in the spring and cause \$60,000 worth of damage. The utility function of the home owner is $u(w) = \ln w$.

(a) How much is the individual willing to pay for house insurance?

(b) Suppose instead that the homeowner's utility function is $u = w$. How much would he now be willing to pay for insurance? Why is this amount more in part (a)?

Exercise 8.18 A group of graduate students have recently pooled together their life's savings and bought a ski cabin. Apart from its inaccessability, the major drawback with this cabin is that it is located near a hill that is somewhat prone to avalanches. The cabin is worth \$100,000. There are two types of avalanche. The first, which occurs with 20% probability, will cause \$30,000 worth of damage; the second occurs with 2% probability but causes \$90,000 worth of damage. Suppose there are three students in this group with utility functions

$$u_1(w) = w^{\frac{1}{2}}, \quad u_2(w) = w^2, \quad u_3(w) = w.$$

How much will each student be willing to pay to insure the cabin? What is the relationship between these amounts and how is this related to their attitudes towards risk? Is it true that a risk-loving individual will never purchase insurance?

Exercise 8.19 According to expected utility theory, faced with uncertain outcomes x_1 with probability p and x_2 with probability $(1-p)$, an individual associates utility $E(u) = pu(x_1) + (1-p)u(x_2)$. The risk premium, c, associated with the risk is defined: $u(\mu - c) = E(u) = pu(x_1) + (1-p)u(x_2)$, where $\mu = px_1 + (1-p)x_2$, the mean of the random variable x. In the case where $u(x) = \sqrt{x}$, $x_1 = 90$, $x_2 = 110$ and $p = \frac{1}{2}$, find π. An approximation for c is given by $c \approx -\frac{1}{2}\frac{u''(\mu)}{u'(\mu)}\sigma^2$, where $\sigma^2 = p(x_1-\mu)^2 + (1-p)(x_2-\mu)^2$. Estimate c using this approximation and compare it with the exact solution.

Exercise 8.20 (a) Let a risk-averse agent have utility function $u(x) = \ln(x+1)$. Suppose that this individual faces the gamble: win \$95 with probability $\frac{1}{2}$ and win \$105 with probability $\frac{1}{2}$. How much less than \$100 would the individual accept instead of the gamble?

(b) Let a random variable have two possible outcomes x_1 and x_2, with probabilities p_1 and p_2, respectively. Thus, the mean is $\mu = p_1x_1 + p_2x_2$. Let the agent's utility function be u. Suppose that $p_1u(x_1)+p_2u(x_2) = u(z)$ (and both x_1 and x_2 are close to z). Show that:

$$\frac{\mu - z}{p_1(x_1-z)^2 + p_2(x_2-z)^2} \approx -\frac{1}{2}\frac{u''(z)}{u'(z)}.$$

9 Vectors

9.1 Introduction

A vector is an array of numbers, $x = (x_1, x_2, \dots, x_n)$, and may be interpreted in a number of ways: as the values of a set of variables, as a coordinate position in $\mathcal{R}^n$, as a direction (move x_i units in direction i, for $i = 1, \dots, n$), and so on. Depending on context, different interpretations are appropriate. In what follows, (Sections 9.2 and 9.3), some of the key summarizing features of a vector are considered: the length of a vector, the distance between vectors and the angle between vectors. As mentioned, from a geometric perspective, vectors can be interpreted as having direction – in subsequent chapters on optimization, direction of improvement of a function is central to the characterization of an optimal choice. As it turns out, this direction is represented by a vector of partial derivatives of the function, the gradient vector. Vectors implicitly define hyperplanes and halfspaces, key concepts in economic theory – Section 9.4 discusses these concepts. Finally, Section 9.5 describes Farkas' lemma, an old and important result in optimization theory.

9.2 Vectors: length and distance

The vector $\boldsymbol{x} = (x_1, x_2, \dots, x_n)$ is a point in $\mathcal{R}^n$, a list of n real numbers. A vector may also be viewed in terms of direction as: x_1 units along axis 1, x_2 units along axis 2, and so on, as illustrated in Figure 9.1. Expressed that way, the vector $\boldsymbol{x}$ describes a direction from $\mathbf{0}$ to a point.

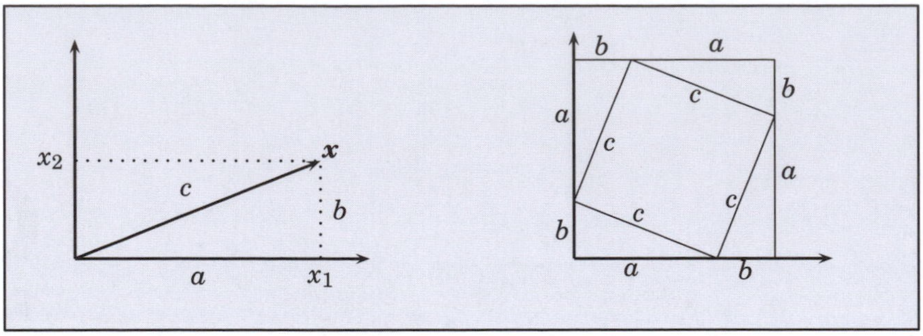

Figure 9.1: Vectors and length

What is the length of the vector $\boldsymbol{x}$, labeled c in Figure 9.1? To answer this, from the vector $\boldsymbol{x}$ form a right-angled triangle where the hypotenuse has length c and the adjacent and opposite have lengths a and b, respectively. The length of the vector $\boldsymbol{x}$ is the length of the hypotenuse, c, and the value of c can be determined from the opposite and adjacent.

On the right side of the figure, the triangle is arranged around a square with side length c. This square with side length c is inscribed inside a square of side length $a+b$. The area of the larger square consists of four occurrences of the triangle and one of the square with side length c. The area of the triangle is $\frac{1}{2}ab$ (half the base by the perpendicular height). Thus, $(a+b)^2 = c^2 + 4\left(\frac{1}{2}\right)ab = c^2 + 2ab$. Since $(a+b)^2 = a^2+b^2+2ab$, $a^2+b^2 = c^2$, so that $c = \sqrt{a^2+b^2}$, which is Pythagoras' theorem. Thus, the length of the vector $\boldsymbol{x}$, denoted $\parallel \boldsymbol{x} \parallel$, is the length of the line c given by $\sqrt{x_1^2+x_2^2}$:

$$\parallel \boldsymbol{x} \parallel = \sqrt{x_1^2+x_2^2}.$$

This extends to n dimensions directly. The length of a vector $\boldsymbol{x} = (x_1, x_2, \ldots, x_n)$ is denoted $\|\boldsymbol{x}\|$ and defined

$$\|\boldsymbol{x}\| = \sqrt{x_1^2+x_2^2+\cdots+x_n^2} = \sqrt{\sum_{i=1}^{n} x_i^2} = \sqrt{\boldsymbol{x}'\boldsymbol{x}}.$$

From the definition, the length of $\boldsymbol{x}$ and $-\boldsymbol{x}$ are the same: $\parallel \boldsymbol{x} \parallel = \parallel -\boldsymbol{x} \parallel$. The sum and difference of two vectors $\boldsymbol{x}$ and $\boldsymbol{y}$ define vectors $\boldsymbol{x}+\boldsymbol{y}$ and $\boldsymbol{x}-\boldsymbol{y}$, as depicted in Figure 9.2.

The distance between two vectors, $\boldsymbol{x}$, $\boldsymbol{y}$, is the length of the vector $\boldsymbol{x}-\boldsymbol{y}$:

$$\parallel \boldsymbol{x}-\boldsymbol{y} \parallel = \sqrt{\sum_{i=1}^{n}(x_i-y_i)^2}.$$

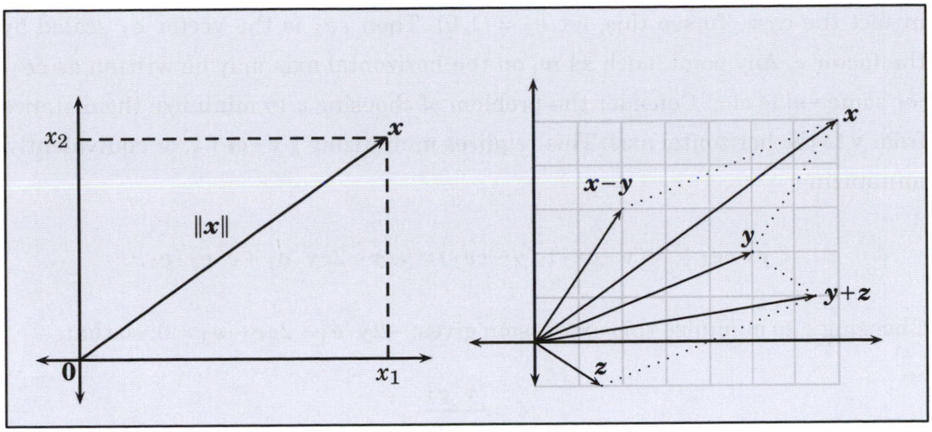

Figure 9.2: Vectors

Given a vector $\boldsymbol{x}$, its length defines a circle or sphere with $\boldsymbol{x}$ in the boundary, so a circle with center $\mathbf{0}$ has radius $r = \| \boldsymbol{x} \|$. Thus, the smaller the length of a vector, the smaller the circle on whose boundary it lies, as illustrated in Figure 9.3, where $\boldsymbol{x}$ and $\boldsymbol{x}'$ have the same length (and $\boldsymbol{y}$, $\boldsymbol{y}'$ also have the same length).

Given a vector $\boldsymbol{z}$, equidistant points from $\boldsymbol{z}$ all lie on a circle centered on $\boldsymbol{z}$ of some fixed radius. In the second diagram in Figure 9.3, the point $\boldsymbol{w}$, closest to $\boldsymbol{y}$ and on the horizontal axis, has the property that $\boldsymbol{y}-\boldsymbol{w}$ is a vector perpendicular (orthogonal) to the horizontal axis. The circle centered on $\boldsymbol{y}$ and tangent to the horizontal axis has only the point $\boldsymbol{w}$ in common with the horizontal axis. The figure suggests that $\boldsymbol{y}-\boldsymbol{w}$ is perpendicular to the horizontal axis and this is

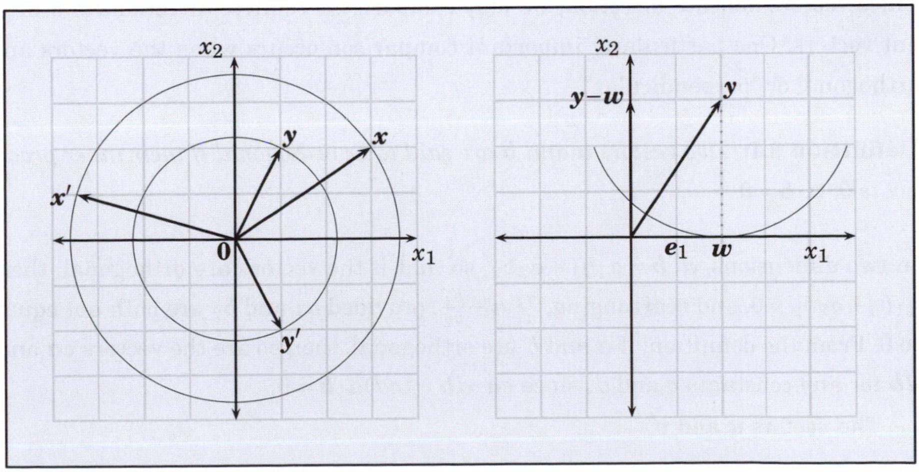

Figure 9.3: Length

in fact the case. To see this, let $\boldsymbol{e}_1 = (1,0)$. Then $c\boldsymbol{e}_1$ is the vector $\boldsymbol{e}_1$ scaled by the factor c. Any point, such as $\boldsymbol{w}$, on the horizontal axis may be written as $c\boldsymbol{e}_1$, for some value of c. Consider the problem of choosing c to minimize the distance from $\boldsymbol{y}$ to the horizontal axis. This requires minimizing $\| \boldsymbol{y} - c\boldsymbol{e}_1 \|$, or equivalently, minimizing:

$$\| \boldsymbol{y} - c\boldsymbol{e}_1 \|^2 = (\boldsymbol{y} - c\boldsymbol{e}_1)\cdot(\boldsymbol{y} - c\boldsymbol{e}_1) = \boldsymbol{y}\cdot\boldsymbol{y} - 2c\boldsymbol{y}\cdot\boldsymbol{e}_1 + c^2\boldsymbol{e}_1\cdot\boldsymbol{e}_1.$$

Choosing c to minimize this expression gives: $-2\boldsymbol{y}\cdot\boldsymbol{e}_1 + 2c\boldsymbol{e}_1\cdot\boldsymbol{e}_1 = 0$, so that

$$\bar{c} = \frac{\boldsymbol{y}\cdot\boldsymbol{e}_1}{\boldsymbol{e}_1\cdot\boldsymbol{e}_1}.$$

So, $\boldsymbol{w} = \bar{c}\boldsymbol{e}_1$ is the point on the horizontal closest to $\boldsymbol{y}$. Observe that

$$(\boldsymbol{y} - \boldsymbol{w})\cdot\boldsymbol{e}_1 = (\boldsymbol{y} - \bar{c}\boldsymbol{e}_1)\cdot\boldsymbol{e}_1 = \boldsymbol{y}\cdot\boldsymbol{e}_1 - \bar{c}\boldsymbol{e}_1\cdot\boldsymbol{e}_1 = \boldsymbol{y}\cdot\boldsymbol{e}_1 - \boldsymbol{y}\cdot\boldsymbol{e}_1 = 0.$$

Since any point, $\boldsymbol{v}$, on the horizontal axis may be written $\boldsymbol{v} = c\boldsymbol{e}_1$, for some c, $(\boldsymbol{y}-\boldsymbol{w})\cdot\boldsymbol{v} = (\boldsymbol{y}-\boldsymbol{w})\cdot c\boldsymbol{e}_1 = c(\boldsymbol{y}-\boldsymbol{w})\cdot\boldsymbol{e}_1 = 0$. In particular, $(\boldsymbol{y}-\boldsymbol{w})\cdot\boldsymbol{w} = 0$. The explanation is simple. The choice of $\boldsymbol{w}$ yields a vector $\boldsymbol{y}-\boldsymbol{w}$ with 0 in the first coordinate. Every vector on the horizontal axis has 0 in the second coordinate, so $(\boldsymbol{y} - \boldsymbol{w})\cdot\boldsymbol{x} = 0$ for any $\boldsymbol{x}$ on the horizontal axis $\boldsymbol{x}$.

9.3 Vectors: direction and angles

Since vectors indicate direction, we may compare the relative directions of different vectors. One particularly important comparison occurs when the vectors are orthogonal or "perpendicular".

Definition 9.1: *Two vectors $\boldsymbol{a}$ and $\boldsymbol{b}$ are said to be orthogonal if their inner product is 0, $\boldsymbol{a}\cdot\boldsymbol{b} = 0$.*

In two dimensions, $\boldsymbol{a}\cdot\boldsymbol{b} = a_1b_1 + a_2b_2$, so that if the vectors are orthogonal, then $a_1b_1 + a_2b_2 = 0$, and rearranging, $\frac{a_2}{a_1} = -\frac{b_1}{b_2}$, provided a_1 and b_2 are both not equal to 0. From the definition, if $\boldsymbol{a}$ and $\boldsymbol{b}$ are orthogonal, then so are the vectors $c\boldsymbol{a}$ and $d\boldsymbol{b}$ for any constants c and d, since $c\boldsymbol{a}\cdot d\boldsymbol{b} = (cd)\boldsymbol{a}\cdot\boldsymbol{b} = 0$.

The vectors $\boldsymbol{u}$ and $\boldsymbol{v}$,

$$\boldsymbol{u} = \begin{pmatrix} 4 \\ 2 \end{pmatrix} \quad \text{and} \quad \boldsymbol{v} = \begin{pmatrix} -1 \\ 2 \end{pmatrix},$$

have inner product $\boldsymbol{u}\cdot\boldsymbol{v} = (4)(-1)+(2)(2) = -4+4 = 0$, so these vectors are orthogonal (Figure 9.4).

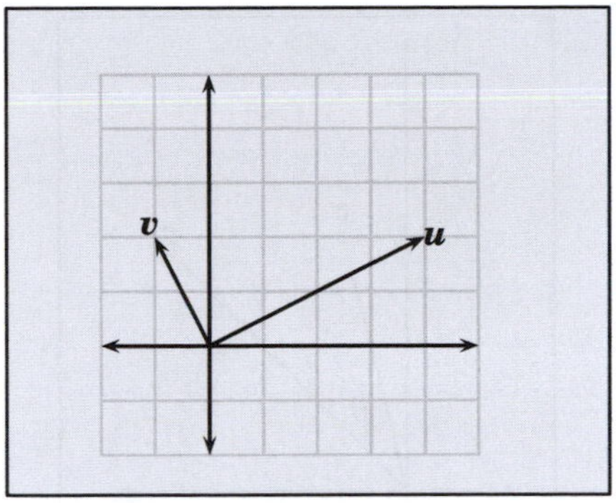

Figure 9.4: Orthogonality

Consider two vectors, $\boldsymbol{v}$ and $\boldsymbol{u}$, depicted in Figure 9.5. Two cases are depicted – where the angle is acute and where the angle is obtuse. In either case, drop a perpendicular line from the point $\boldsymbol{v}$ onto a point lying on the vector $\boldsymbol{u}$, $c\boldsymbol{u}$. Orthogonality requires that this line, $\boldsymbol{v}-c\boldsymbol{u}$, be orthogonal to $\boldsymbol{u}$: $(\boldsymbol{v}-c\boldsymbol{u})\cdot\boldsymbol{u} = 0$. This implies that $\boldsymbol{v}\cdot\boldsymbol{u} - c\boldsymbol{u}\cdot\boldsymbol{u} = 0$, so that

$$c = \frac{\boldsymbol{v}\cdot\boldsymbol{u}}{\boldsymbol{u}\cdot\boldsymbol{u}} = \frac{\boldsymbol{v}\cdot\boldsymbol{u}}{\|\,\boldsymbol{u}\,\|^2}.$$

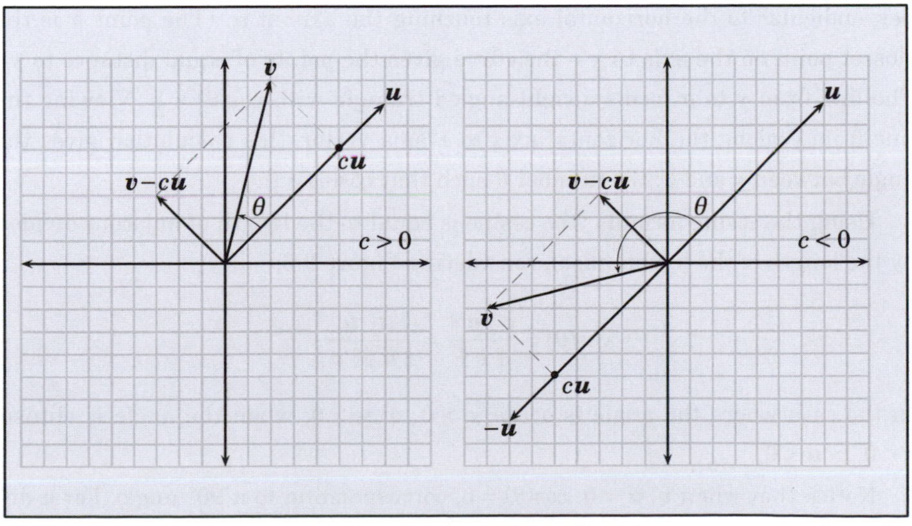

Figure 9.5: Angles between vectors

The point $c\boldsymbol{u}$ is the closest point to $\boldsymbol{v}$ among the set of points $\{\boldsymbol{w} \mid \boldsymbol{w} = \delta \boldsymbol{u}$, for some number $\delta\}$. This is seen in Figure 9.6. Note that the condition determining the

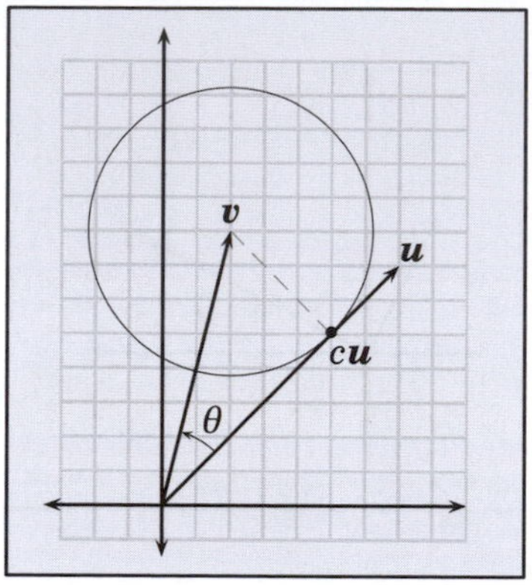

Figure 9.6: Projections

choice of c, $(\boldsymbol{v} - c\boldsymbol{u}) \cdot \boldsymbol{u} = 0$ implies that for any α, $\alpha\boldsymbol{u}$ is orthogonal to $(\boldsymbol{v} - c\boldsymbol{u})$: $(\boldsymbol{v}-c\boldsymbol{u})\cdot\alpha\boldsymbol{u} = \alpha(\boldsymbol{v}-c\boldsymbol{u})\cdot\boldsymbol{u} = 0$. As before, c can be obtained as a distance minimizer: minimizing the distance from $\boldsymbol{v}$ to the set of points $\{\boldsymbol{w} \mid \boldsymbol{w} = \alpha\boldsymbol{u}$ some $\alpha\}$. Thus, c is the solution to $\min_c(\boldsymbol{v} - c\boldsymbol{u})\cdot(\boldsymbol{v} - c\boldsymbol{u})$ and this gives $-2\boldsymbol{v}\cdot\boldsymbol{u} + c\boldsymbol{u}\cdot\boldsymbol{u} = 0$ or $c = \frac{\boldsymbol{u}\cdot\boldsymbol{v}}{\boldsymbol{u}\cdot\boldsymbol{u}}$.

The angle between vectors may be defined using the previous calculations. In Figure 9.7, the standard trigonometric functions are given: sin, cos and tan. So, for example, the cosine of an angle is given as the adjacent divided by the hypotenuse in the right-angled triangle. Given a vector $\boldsymbol{y}$, as depicted, drop a line perpendicular to the horizontal axis touching the axis at α. (The point α is the closest point on the axis to $\boldsymbol{y}$ – the curve gives the points of equal distance to $\boldsymbol{y}$.) The line from $\boldsymbol{y}$ to α forms a right-angled triangle with $\cos(\theta) = \frac{a}{h}$. Viewing the line from 0 along the horizontal axis to $\boldsymbol{z}$ as a vector, this calculation gives the angle between $\boldsymbol{y}$ and $\boldsymbol{z}$: the number θ such that $\cos(\theta) = \frac{a}{h}$.

Using the cosine formula (the cosine is equal to the length of adjacent divided by the length of the hypotenuse), considering Figure 9.6:

$$\cos(\theta) = \frac{c \parallel \boldsymbol{u} \parallel}{\parallel \boldsymbol{v} \parallel} = \frac{\boldsymbol{v}\cdot\boldsymbol{u}}{\parallel \boldsymbol{u} \parallel\parallel \boldsymbol{v} \parallel}.$$

In the case where the angle is acute, $c > 0$, $\boldsymbol{v}\cdot\boldsymbol{u} > 0$; when the angle is obtuse, $c < 0$, $\boldsymbol{v}\cdot\boldsymbol{u} < 0$.

Notice that when $\boldsymbol{v}\cdot\boldsymbol{u} = 0$, $\cos(\theta) = 0$, corresponding to a 90° angle. For a different perspective consider Figure 9.8, where $\boldsymbol{\beta}$ and $\boldsymbol{z}$ are orthogonal: $\boldsymbol{\beta}\cdot\boldsymbol{z} = 0$ or $\beta_1 z_1 + \beta_2 z_2 = 0$ so that $z_2 = -\frac{\beta_1}{\beta_2} z_1$. Thus, the angle between z and the horizontal

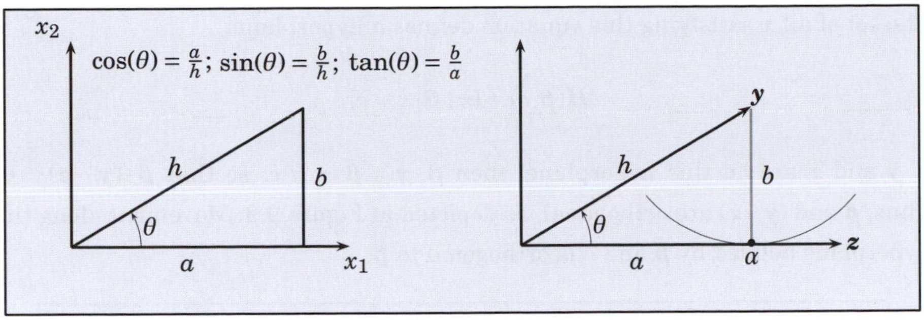

Figure 9.7: Trigonometric functions

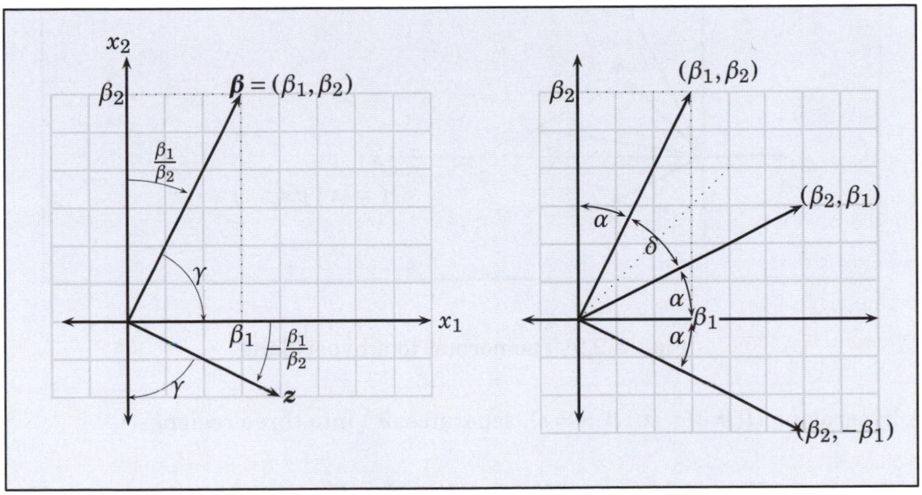

Figure 9.8: Orthogonal vectors

axis is (in absolute terms) $\frac{\beta_1}{\beta_2}$, the same as the angle between $\boldsymbol{\beta}$ and the vertical axis. Since γ and $\frac{\beta_1}{\beta_2}$ combine to form a 90° angle, the angle between $\boldsymbol{\beta}$ and $\boldsymbol{z}$ is 90°. Alternatively, from the second diagram in Figure 9.8, note that $2\alpha + \delta = 90°$, so the angle between $(\boldsymbol{\beta}_1, \boldsymbol{\beta}_2)$ and $(\boldsymbol{\beta}_2, -\boldsymbol{\beta}_1)$ is 90° and $(\boldsymbol{\beta}_1, \boldsymbol{\beta}_2) \cdot (\boldsymbol{\beta}_2, -\boldsymbol{\beta}_1) = 0$.

9.4 Hyperplanes and direction of increase

Let $\boldsymbol{\beta} = (\beta_1, \dots, \beta_n) \neq 0$ be a vector of constants, c a constant and consider the equation:

$$\boldsymbol{\beta x} = \sum_{i=1}^{n} \beta_i x_i = c.$$

The set of all $\boldsymbol{x}$ satisfying this equation defines a hyperplane:

$$H(\boldsymbol{\beta}, c) = \{\boldsymbol{x} \mid \boldsymbol{\beta} \cdot \boldsymbol{x} = c\}.$$

If $\boldsymbol{y}$ and $\boldsymbol{z}$ are on this hyperplane, then $\boldsymbol{\beta} \cdot \boldsymbol{y} = \boldsymbol{\beta} \cdot \boldsymbol{z} = c$, so that $\boldsymbol{\beta} \cdot (\boldsymbol{y} - \boldsymbol{z}) = 0$. Thus, $\boldsymbol{\beta}$ and $(\boldsymbol{y} - \boldsymbol{z})$ are orthogonal, as depicted in Figure 9.9. Movement along the hyperplane defined by $\boldsymbol{\beta}$ and c is orthogonal to $\boldsymbol{\beta}$.

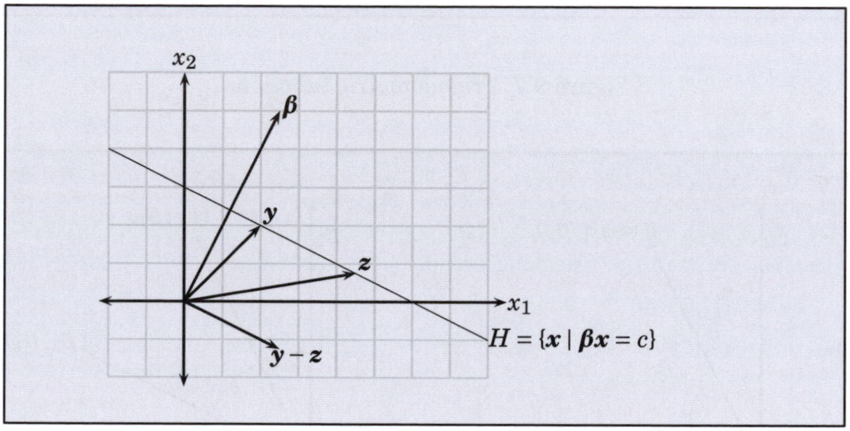

Figure 9.9: The normal to a hyperplane

A hyperplane $H(\boldsymbol{\beta}, c) = \{\boldsymbol{x} \mid \boldsymbol{\beta} \cdot \boldsymbol{x} = c\}$, separates $\mathscr{R}^n$ into three regions:

$$H = \{\boldsymbol{x} \mid \boldsymbol{\beta x} = c\}, \quad H_+ = \{\boldsymbol{x} \mid \boldsymbol{\beta x} > c\}, \quad H_- = \{\boldsymbol{x} \mid \boldsymbol{\beta x} < c\}.$$

On which sides of H do H_+ and H_- lie? Consider the point $\boldsymbol{y}$ and move from $\boldsymbol{y}$ along the direction $\boldsymbol{\beta}$ a fraction $\delta > 0$ to $\boldsymbol{y} + \delta\boldsymbol{\beta}$, as depicted in Figure 9.10. Observe that

$$c' = \boldsymbol{\beta} \cdot (\boldsymbol{y} + \delta\boldsymbol{\beta}) = \boldsymbol{\beta} \cdot \boldsymbol{y} + \delta\boldsymbol{\beta} \cdot \boldsymbol{\beta} = c + \delta\boldsymbol{\beta} \cdot \boldsymbol{\beta} > c$$

Therefore, $\boldsymbol{\beta}$ points into the region H_+: starting at $\boldsymbol{y}$ and moving from $\boldsymbol{y}$ to $\boldsymbol{w} = \boldsymbol{y} + \delta\boldsymbol{\beta}$, $\boldsymbol{w}$ is in H_+.

Equivalently, let $f(\boldsymbol{x}) = \boldsymbol{\beta} \cdot \boldsymbol{x}$, then $\boldsymbol{\beta}$ points in the $\boldsymbol{x}$ direction on which $f(\boldsymbol{x})$ increases. Furthermore, notice that the direction $\boldsymbol{\beta}$ gives the greatest increase among all directions of equal length. For example, $\boldsymbol{r}$ has the same length as $\delta\boldsymbol{\beta}$ (since they both lie on the same circle around $\boldsymbol{y}$), but

$$c'' = \boldsymbol{\beta} \cdot (\boldsymbol{y} + \boldsymbol{r}) < \boldsymbol{\beta} \cdot (\boldsymbol{y} + \delta\boldsymbol{\beta}) = c'.$$

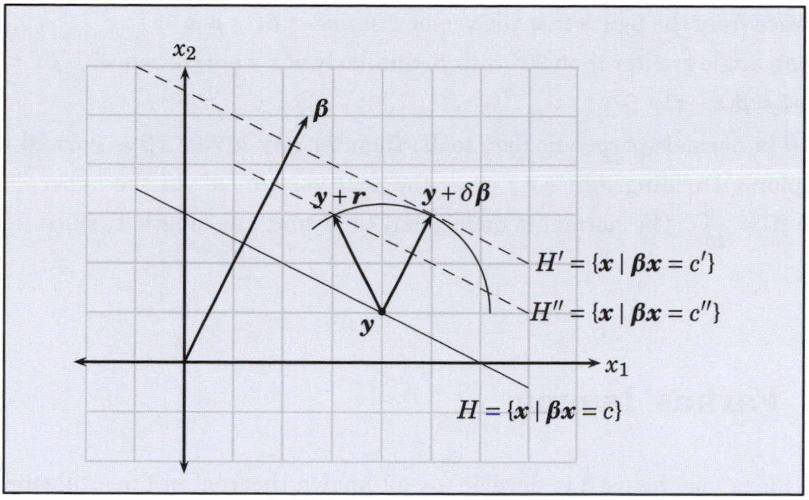

Figure 9.10: The normal to a hyperplane

As noted earlier, the inner product of $\boldsymbol{\beta}$ with any vector $\boldsymbol{\xi}$ may be positive, zero or negative. If $\boldsymbol{\beta\xi} = 0$, they are orthogonal. Informally, if $\boldsymbol{\beta\xi} > 0$, the angle between $\boldsymbol{\beta}$ and $\boldsymbol{\xi}$ is less than 90°; and if $\boldsymbol{\beta\xi} < 0$, the angle between $\boldsymbol{\beta}$ and $\boldsymbol{\xi}$ is greater than 90°. To see this from a different perspective, consider Figure 9.11, where $\boldsymbol{z}$ is chosen so that $\boldsymbol{\beta z} > c$, say $\boldsymbol{\beta z} = c' > c$.

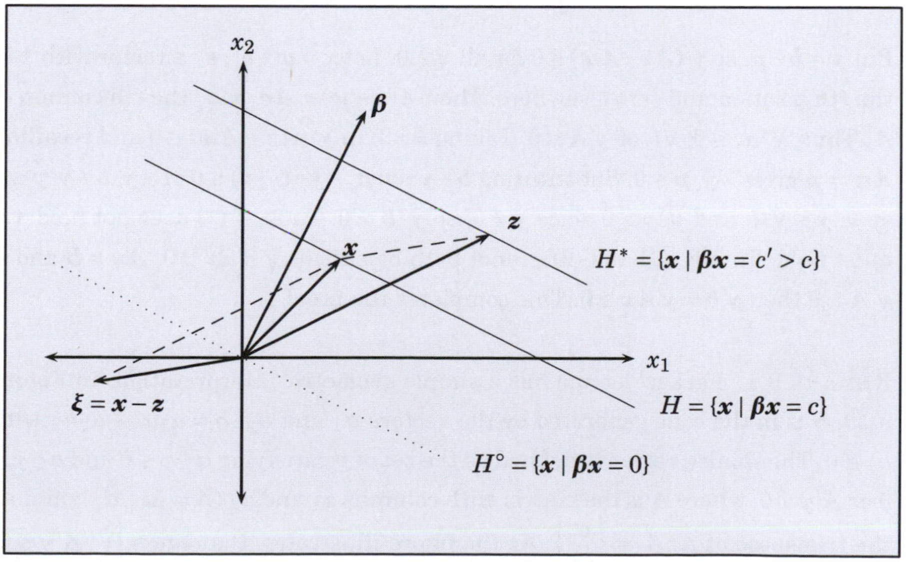

Figure 9.11: The sign of an inner product

Notice from the figure that the vector $\boldsymbol{\xi}$ satisfies $\boldsymbol{\beta\xi} = \boldsymbol{\beta}(\boldsymbol{x}-\boldsymbol{z}) = c - c' < 0$, and $\boldsymbol{\xi}$ forms an angle greater than 90° with $\boldsymbol{\beta}$. Similarly, if $\boldsymbol{z}$ were chosen with $\boldsymbol{\beta z} = \hat{c} < c$, then $\boldsymbol{\beta\xi} = \boldsymbol{\beta}(\boldsymbol{x}-\boldsymbol{z}) > 0$.

If $\boldsymbol{\beta}$ is normal (perpendicular) to H, then for any $\alpha \neq 0$, $\alpha\boldsymbol{\beta}$ is normal to the hyperplane. Dividing β by $\|\boldsymbol{\beta}\|$ gives the *unit normal*, β^*, to the hyperplane H where $\beta^* = \frac{\beta}{\|\boldsymbol{\beta}\|}$. The normal is orthogonal to H and has length 1, since $\beta^{*\prime}\beta^* = \frac{\beta'\beta}{\|\boldsymbol{\beta}\|^2} = 1$.

9.5 Farkas' lemma

These ideas may be used to develop a well-known theorem in linear algebra and linear inequalities called Farkas' lemma. This result is particularly useful in optimization theory. Let $\boldsymbol{A}$ be an $m \times n$ matrix and $\boldsymbol{b}$ an $n \times 1$ vector.

Theorem 9.1: *Either (1)* $\{\boldsymbol{Ax} = \boldsymbol{b},\ \boldsymbol{x} \geq 0\}$*, or (2)* $\{\boldsymbol{A}'\boldsymbol{y} \leq 0$ *and* $\boldsymbol{y}'\boldsymbol{b} > 0\}$ *has a solution, but not both.*

PROOF: Suppose that (1) has no solution: $\boldsymbol{b} \notin K \equiv \{\boldsymbol{Ax} \mid \boldsymbol{x} \geq 0\}$. Let $\boldsymbol{p} \in \{\boldsymbol{Ax} \mid \boldsymbol{x} \geq 0\}$ solve $\min_{z \in K} \|\boldsymbol{b}-\boldsymbol{z}\|$. Since K is convex, $\boldsymbol{p}$ is unique. Let $\boldsymbol{w} \geq 0$ satisfy $\boldsymbol{p} = \boldsymbol{Aw}$. The hyperplane perpendicular to $\boldsymbol{b}-\boldsymbol{p}$ separates $\boldsymbol{b}$ and K: $\forall z \in K$, $(b-p)\cdot(z-p) \leq 0$ (see Figure 9.12.) Equivalently:

$$(\boldsymbol{b}-\boldsymbol{p})\cdot(\boldsymbol{Ax}-\boldsymbol{Aw}) \leq 0,\ \forall x \geq 0.$$

Put $\boldsymbol{y} = \boldsymbol{b}-\boldsymbol{p}$, so $\boldsymbol{y}\cdot(\boldsymbol{Ax}-\boldsymbol{Aw}) \leq 0$ for all $\boldsymbol{x} \geq 0$. Let $\boldsymbol{x} = \boldsymbol{w}+\boldsymbol{e}_i$, $\boldsymbol{e}_i$ a vector with 1 in the ith position and zero elsewhere. Then $\boldsymbol{A}(\boldsymbol{x}-\boldsymbol{w}) = \boldsymbol{Ae}_i = \boldsymbol{a}_i$ the ith column of $\boldsymbol{A}$. Thus, $\boldsymbol{y}\cdot\boldsymbol{a}_i \leq 0$, $\forall i$, or $\boldsymbol{y}'\boldsymbol{A} \leq 0$. Taking $\boldsymbol{x} = 0$ in $\boldsymbol{y}\cdot(\boldsymbol{Ax}-\boldsymbol{Aw}) \leq 0$ and recalling $\boldsymbol{Aw} = \boldsymbol{p}$ gives $-\boldsymbol{y}\cdot\boldsymbol{p} \leq 0$. Substituting $\boldsymbol{b}-\boldsymbol{y}$ for $\boldsymbol{p}$, $-\boldsymbol{y}\cdot(\boldsymbol{b}-\boldsymbol{y}) \leq 0$ or $-\boldsymbol{y}\cdot\boldsymbol{b}+\boldsymbol{y}\cdot\boldsymbol{y} \leq 0$ or $\boldsymbol{y}\cdot\boldsymbol{y} \leq \boldsymbol{y}\cdot\boldsymbol{b}$ and $\boldsymbol{y}\cdot\boldsymbol{y} > 0$ since $\boldsymbol{y} \neq 0$, so $\boldsymbol{y}\cdot\boldsymbol{b} > 0$. Thus, if (1) does not hold, (2) must hold. Finally, (1) and (2) cannot both be satisfied: if $\exists \boldsymbol{x} \geq 0$, $\boldsymbol{Ax} = \boldsymbol{b}$ and if $\boldsymbol{y}'\boldsymbol{A} \geq 0$ then $\boldsymbol{y}'\boldsymbol{b} = \boldsymbol{y}'\boldsymbol{Ax} \geq 0$. This completes the proof. ■

REMARK 9.1: Farkas' lemma has a simple geometric interpretation. In Figure 9.13, $\boldsymbol{b}$ is in the cone generated by the vectors $\boldsymbol{a}_1$ and $\boldsymbol{a}_2$: $\boldsymbol{b} = \alpha_1 a_1 + \alpha_2 a_2$ with $\alpha_i \geq 0$. The shaded region is defined as the set of $\boldsymbol{y}$ satisfying $\boldsymbol{a}_1\cdot\boldsymbol{y} \leq 0$ and $\boldsymbol{a}_2\cdot\boldsymbol{y} \leq 0$ or $\boldsymbol{A}'\boldsymbol{y} \leq 0$, where $\boldsymbol{A}$ is the matrix with columns a_1 and a_2 ($\boldsymbol{A} = \{\boldsymbol{a}_1, \boldsymbol{a}_2\}$) and $\boldsymbol{A}'$ the transpose of $\boldsymbol{A}$: $\boldsymbol{A}' = \begin{Bmatrix} \boldsymbol{a}_1' \\ \boldsymbol{a}_2' \end{Bmatrix}$. As the figure illustrates, the region $\{\boldsymbol{y} \mid \boldsymbol{A}'\boldsymbol{y} \leq 0\}$ is a subset of the region $\{\boldsymbol{y} \mid \boldsymbol{b}'\boldsymbol{y} \leq 0\}$. Thus, $\boldsymbol{Ax} = \boldsymbol{b}$ has a non-negative solution and $\boldsymbol{A}'\boldsymbol{y} \leq 0$ implies $\boldsymbol{b}'\boldsymbol{y} \leq 0$. □

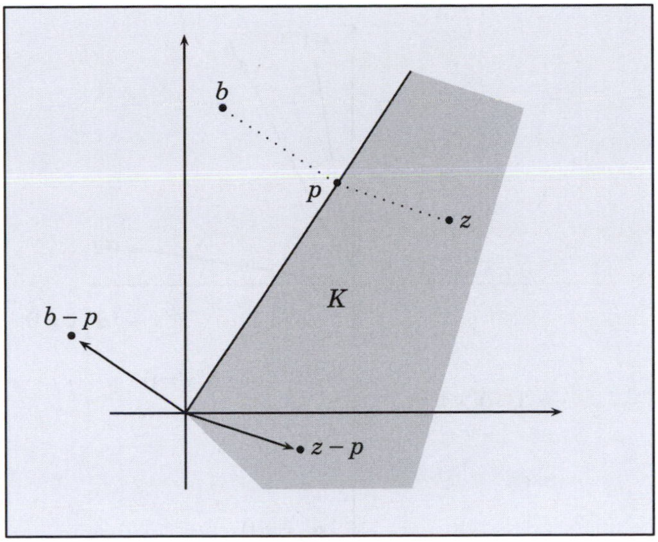

Figure 9.12: Proof of Farkas' lemma

The lemma may be rearranged as follows:

LEMMA 1: The pair $\boldsymbol{A}'\boldsymbol{y} \le 0$ and $\boldsymbol{b}'\boldsymbol{y} > 0$ have a solution if and only if the pair $\boldsymbol{A}'\boldsymbol{y} \ge 0$ and $\boldsymbol{b}'\boldsymbol{y} < 0$ have a solution. Consequently, Farkas' lemma may be written as:

either (1) $\{\boldsymbol{A}x = \boldsymbol{b}, \boldsymbol{x} \ge 0\}$, or (2) $\{\boldsymbol{A}'\boldsymbol{y} \ge 0$ and $\boldsymbol{b}'\boldsymbol{y} < 0\}$ has a solution, but not both. ◇

PROOF: For, if $\boldsymbol{z}$ satisfies $\boldsymbol{A}'\boldsymbol{z} \le 0$ and $\boldsymbol{b}'\boldsymbol{z} > 0$ then $-\boldsymbol{A}'\boldsymbol{z} \ge 0$ and $-\boldsymbol{b}'\boldsymbol{z} < 0$ or $\boldsymbol{A}'(-\boldsymbol{z}) \ge 0$ and $\boldsymbol{b}'(-\boldsymbol{z}) < 0$ or $\boldsymbol{A}'\boldsymbol{y} \ge 0$ and $\boldsymbol{b}'\boldsymbol{y} < 0$ with $\boldsymbol{y} = -\boldsymbol{z}$. ■

Farkas' lemma may be expressed in other ways. One corollary is:

LEMMA 2: Either:

	or	
$\boldsymbol{A}\boldsymbol{x} \le \boldsymbol{b}$		$\boldsymbol{A}'\boldsymbol{y} \ge 0$
$\boldsymbol{x} \ge 0$		$\boldsymbol{b}'\boldsymbol{y} < 0$
		$\boldsymbol{y} \ge 0$
has a solution		has a solution. ◇

PROOF: In Farkas' lemma, let $\bar{\boldsymbol{A}} = \begin{pmatrix} A & I \end{pmatrix}$, so that $\bar{\boldsymbol{A}}' = \begin{pmatrix} A' \\ I \end{pmatrix}$. Then, applying lemma (1), either $[\bar{\boldsymbol{A}}\boldsymbol{z} = \boldsymbol{b},\ \boldsymbol{z} \ge 0]$ has a solution or, $[\bar{\boldsymbol{A}}'\boldsymbol{y} \ge 0$ and $\boldsymbol{b}'\boldsymbol{y} < 0]$ has a

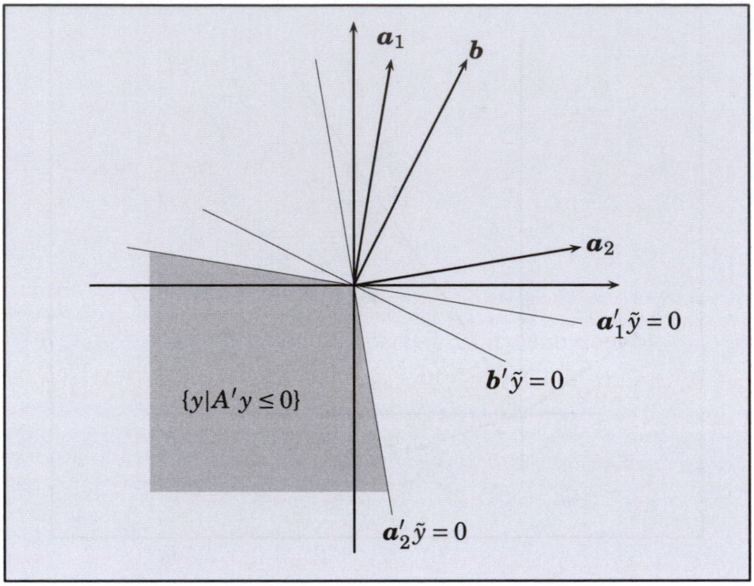

Figure 9.13: Farkas' lemma: a geometric view

solution. Let $\boldsymbol{z} = (\boldsymbol{x}, \boldsymbol{w})$, so that the first expression is equivalent to $[\boldsymbol{Ax} + \boldsymbol{Iw} = \boldsymbol{b}$ and $(\boldsymbol{x}, \boldsymbol{w}) \geq 0]$. There is an $(\boldsymbol{x}, \boldsymbol{w})$ satisfying this if and only if there is an $\boldsymbol{x} \geq 0$ with $\boldsymbol{Ax} \leq \boldsymbol{b}$.

The second expression in lemma (1) is, in this context: $[\begin{pmatrix} \boldsymbol{A}' \\ \boldsymbol{I} \end{pmatrix} \boldsymbol{y} = \begin{pmatrix} \boldsymbol{A}'\boldsymbol{y} \\ \boldsymbol{y} \end{pmatrix} \geq \begin{pmatrix} 0 \\ 0 \end{pmatrix}$ and $\boldsymbol{b}'\boldsymbol{y} < 0]$, and this is equivalent to there being a $\boldsymbol{y}$ with $\boldsymbol{A}'\boldsymbol{y} \geq 0$, $\boldsymbol{b}'\boldsymbol{y} < 0$ and $\boldsymbol{y} \geq 0$. So, either one or the other of the following hold (but not both). Summarizing, there is an $\boldsymbol{x} \geq 0$ with $\boldsymbol{Ax} \leq \boldsymbol{b}$, or there is a $\boldsymbol{y}$ with $\boldsymbol{A}'\boldsymbol{y} \geq 0$, $\boldsymbol{b}'\boldsymbol{y} < 0$ and $\boldsymbol{y} \geq 0$. ■

Exercises for Chapter 9

Exercise 9.1 Find the distance between the vectors $\boldsymbol{x} = (5,3)$ and $\boldsymbol{y} = (4,6)$.

Exercise 9.2 Let $\boldsymbol{x} = (1,3)$, $\boldsymbol{y} = (2,2)$ and $\boldsymbol{z} = (1,-1)$. Find the angle between $\boldsymbol{x}$ and $\boldsymbol{y}$, and that between $\boldsymbol{x}$ and $\boldsymbol{z}$.

Exercise 9.3 Let $\boldsymbol{x} = (4,2)$ and $\boldsymbol{y} = (3,3)$. Find c such that $(\boldsymbol{y}-c\boldsymbol{x})\cdot(\boldsymbol{y}-c\boldsymbol{x}) \le (\boldsymbol{y}-\boldsymbol{v})\cdot(\boldsymbol{y}-\boldsymbol{v})$ for all $\boldsymbol{v} = \tilde{c}\boldsymbol{x}$, for some $\tilde{c}$. Confirm that $(\boldsymbol{y}-c\boldsymbol{x})\cdot\boldsymbol{x} = 0$.

Exercise 9.4

(a) Let $\boldsymbol{\beta} = (1,2)$. Find the hyperplane $\boldsymbol{\beta}\cdot\boldsymbol{x} = 2$. Noting that $(2,0)$ is on this hyperplane, plot $(2,0)+\boldsymbol{\beta}$. Identify the region $\{\boldsymbol{x} \mid \boldsymbol{\beta}\boldsymbol{x} \ge 2\}$.

(b) Let $\boldsymbol{\beta} = (-1,1)$. Find the hyperplane $\boldsymbol{\beta}\cdot\boldsymbol{x} = 2$. Noting that $(-1,1)$ is on this hyperplane, plot $(-1,1)+\boldsymbol{\beta}$. Identify the region $\{\boldsymbol{x} \mid \boldsymbol{\beta}\boldsymbol{x} \ge 2\}$.

(c) Let $\boldsymbol{\beta} = (-1,-1)$. Find the hyperplane $\boldsymbol{\beta}\cdot\boldsymbol{x} = 2$. Noting that $(-1,-1)$ is on this hyperplane, plot $(-1,-1)+\boldsymbol{\beta}$. Identify the region $\{\boldsymbol{x} \mid \boldsymbol{\beta}\boldsymbol{x} \ge 2\}$.

Exercise 9.5 Consider the matrix:

$$\boldsymbol{R}(\theta) = \begin{pmatrix} \cos(\theta) & -\sin(\theta) \\ \sin(\theta) & \cos(\theta) \end{pmatrix},$$

where θ is an angle measured in degrees. For example:

$$\boldsymbol{R}(90) = \begin{pmatrix} \cos(90) & -\sin(90) \\ \sin(90) & \cos(90) \end{pmatrix} = \begin{pmatrix} 0 & -1 \\ 1 & 0 \end{pmatrix}.$$

The transformation $\boldsymbol{R}(\theta)\boldsymbol{x}$ moves the vector $\boldsymbol{x}$ through an angle of θ degrees.

(a) Letting $\boldsymbol{x} = (1,1)$, find $\boldsymbol{x}' = \boldsymbol{R}(45)\boldsymbol{x}$. Using the formulae $c = \frac{\boldsymbol{x}\cdot\boldsymbol{x}'}{\|\boldsymbol{x}\|}$ and $\cos(\theta) = \frac{c\|x\|}{\|\boldsymbol{x}'\|}$, confirm that $\cos(\theta) = \frac{1}{\sqrt{2}} = \frac{c\|x\|}{\|\boldsymbol{x}'\|}$.

(b) Letting $\boldsymbol{y} = (1,3)$, find $\boldsymbol{y}' = \boldsymbol{R}(30)\boldsymbol{y}$. Using the formulae $c = \frac{\boldsymbol{x}\cdot\boldsymbol{y}'}{\|\boldsymbol{y}\|^2}$ and $\cos(\theta) = \frac{c\|\boldsymbol{y}\|}{\|\boldsymbol{y}'\|}$, confirm that $\cos(\theta) = \frac{\sqrt{3}}{2} = \frac{c\|\boldsymbol{y}\|}{\|\boldsymbol{y}'\|}$.

Exercise 9.6 Show that

$$\begin{pmatrix} 1 & 3 & 5 \\ 2 & 1 & 3 \\ 4 & 2 & 1 \end{pmatrix}\begin{pmatrix} y_1 \\ y_2 \\ y_3 \end{pmatrix} \geq \begin{pmatrix} 0 \\ 0 \\ 0 \end{pmatrix}, \quad \begin{pmatrix} 3 & 4 & 3 \end{pmatrix}\begin{pmatrix} y_1 \\ y_2 \\ y_3 \end{pmatrix} < 0$$

has a solution.

Exercise 9.7 Consider the following observations on two random variables, $\boldsymbol{x}$ and $\boldsymbol{y}$:

$$\boldsymbol{x} = \begin{pmatrix} 9 \\ 7 \\ 3 \\ 5 \end{pmatrix}, \quad \boldsymbol{y} = \begin{pmatrix} 12 \\ 8 \\ 2 \\ 6 \end{pmatrix}.$$

The correlation coefficient is defined:

$$r = \frac{\sum(x_i - \bar{x})(y_i - \bar{y})}{\sqrt{\sum(x_i - \bar{x})^2}\sqrt{\sum(y_i - \bar{y})^2}}$$

where $\bar{x}$ and $\bar{y}$ are the respective means. Find the correlation coefficient and relate it to the angle between the deviations around $\boldsymbol{x}$ and $\boldsymbol{y}$ (determine the angle between the vectors of deviations).

10

Quadratic forms

10.1 Introduction

A quadratic form is given by the expression $\boldsymbol{x}'\boldsymbol{A}\boldsymbol{x}$ where $\boldsymbol{x}$ is an $n \times 1$ vector and $\boldsymbol{A}$ an $n \times n$ matrix. Quadratic forms arise in various branches of economics, and particularly in optimization theory to identify a candidate solution as a maximum or minimum. For example, in the study of quadratic functions, $ax^2 + bx + c = (ax + b)x + c$, this function is either "u-shaped" or the opposite, depending on whether a is positive or negative. So, the turning point either locates a minimum or a maximum depending of the sign of a. If, for example, $a > 0$, then large values of x correspond with large values of the function and the turning point corresponds to a minimum value of the function.

Essentially the same situation occurs when the function is a quadratic in many variables: $\boldsymbol{x}'\boldsymbol{A}\boldsymbol{x} + \boldsymbol{b}\boldsymbol{x} + c$. In this case, the matrix $\boldsymbol{A}$ plays an analogous role to a in the scalar case: whether a "turning point" is a maximum or minimum depends on properties of $\boldsymbol{A}$. Letting

$$f(\boldsymbol{x}) = \boldsymbol{x}'\boldsymbol{A}\boldsymbol{x} + \boldsymbol{b}\boldsymbol{x} + c = \sum_{i=1}^{n}\sum_{j=1}^{n} a_{ij}x_i x_j + \sum_{k=1}^{n} b_k x_k + c,$$

in the context of optimization, positive and negative definiteness are the central concepts. Section 10.2 introduces quadratic forms and shows how, in the two-variable case, properties of the $\boldsymbol{A}$ matrix determine the shape of the function f. In Section 10.3, positive and negative definiteness are defined and the properties of positive and negative matrices are characterized in terms of determinant conditions.

When the function $f(\boldsymbol{x})$ is arbitrary (not necessarily quadratic), the criteria developed here extend directly by using a second-order quadratic approximation of the function f to identify local maxima or local minima. Finally, Section 10.4 considers the case where there are linear restrictions on $\boldsymbol{x}$. A matrix $\boldsymbol{A}$ is positive definite if for all $\boldsymbol{x} \neq \boldsymbol{0}$, $\boldsymbol{x}'\boldsymbol{A}\boldsymbol{x} > 0$ and negative definite when $\boldsymbol{x}'\boldsymbol{A}\boldsymbol{x} < 0$ for all non-zero $\boldsymbol{x}$. So, for example, positive definiteness requires that for all non-zero $\boldsymbol{x}$, $\boldsymbol{x}'\boldsymbol{A}\boldsymbol{x} > 0$. Suppose instead, that this requirement applies only to those non-zero $\boldsymbol{x}$ satisfying an additional condition, $\boldsymbol{B}'\boldsymbol{x} = 0$: under what circumstances is $\boldsymbol{x}'\boldsymbol{A}\boldsymbol{x} > 0$ for all non-zero $\boldsymbol{x}$ satisfying $\boldsymbol{B}'\boldsymbol{x} = 0$? This issue is considered in Section 10.4. The criteria developed there appear in the study of constrained optimization problems.

10.2 Quadratic forms

A quadratic form in n variables is given as:

$$\boldsymbol{x}'\boldsymbol{A}\boldsymbol{x} = (x_1, x_2, \ldots, x_n)\begin{pmatrix} a_{11} & \cdots & a_{1n} \\ \vdots & \ddots & \vdots \\ a_{n1} & \cdots & a_{nn} \end{pmatrix}\begin{pmatrix} x_1 \\ \vdots \\ x_n \end{pmatrix} = \sum_i \sum_j a_{ij} x_i x_j,$$

where a_{ij} is the element in the ith row and jth column of $\boldsymbol{A}$.

In the case of two variables:

$$x = \begin{pmatrix} x_1 \\ x_2 \end{pmatrix}, \quad \boldsymbol{A} = \begin{pmatrix} a_{11} & a_{12} \\ a_{21} & a_{22} \end{pmatrix}$$

and

$$\boldsymbol{x}'\boldsymbol{A}\boldsymbol{x} = \begin{pmatrix} x_1 & x_2 \end{pmatrix}\begin{pmatrix} a_{11} & a_{12} \\ a_{21} & a_{22} \end{pmatrix}\begin{pmatrix} x_1 \\ x_2 \end{pmatrix} = a_{11}x_1^2 + (a_{12} + a_{21})x_1 x_2 + a_{22}x_2^2,$$

which is a quadratic form in the variables x_1 and x_2.

An important question concerns the sign of $\boldsymbol{x}'\boldsymbol{A}\boldsymbol{x}$. Under what conditions is the quadratic form always positive or always negative, regardless of the value of $\boldsymbol{x}$? The question arises in a variety of problems, and in particular in the area of optimization.

EXAMPLE 10.1: For example, if $a_{11} = 3$, $a_{22} = 4$ and $a_{12} + a_{21} = 3$, then the associated quadratic form is $q(x_1, x_2) = 3x_1^2 + 4x_2^2 + 3x_1 x_2$. Let $f(\boldsymbol{x})$ be the quadratic function plotted in the figure:

$$z = f(\boldsymbol{x}) = x_1 + x_2 + q(x_1, x_2) = x_1 + x_2 + 3x_1^2 + 4x_2^2 + 3x_1 x_2.$$

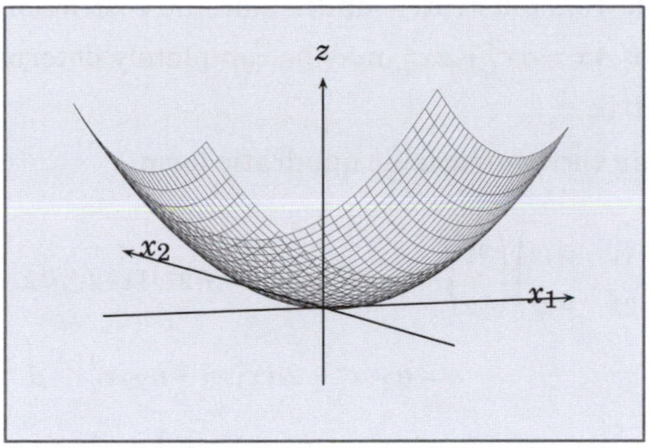

This function is smallest when $(x_1,x_2) = (-\frac{5}{22},-\frac{2}{11}) = (x_1^*,x_2^*)$. That is: $f(x_1^*,x_2^*) \le f(x_1,x_2)$ for all (x_1,x_2).

Notice that the "bowl" shape of the function is crucial: around (x_1^*,x_2^*) the function curves up. This shape property is determined by the quadratic form $q(x_1,x_2)$, which has the property that $q(x_1,x_2) \ge 0$ for all (x_1,x_2), and illustrates the relationship between quadratic forms and optimization. ◊

In general, the sign of the quadratic form $\boldsymbol{x}'\boldsymbol{A}\boldsymbol{x}$ depends on both $\boldsymbol{A}$ and the vector $\boldsymbol{x}$, as Example 10.2 illustrates.

EXAMPLE 10.2: Let $\boldsymbol{A} = \begin{pmatrix} a & 0 \\ 0 & -\hat{a} \end{pmatrix}$ where $a > 0$ and $\hat{a} > 0$. Then

$$\boldsymbol{x}'\boldsymbol{A}\boldsymbol{x} = (x_1,x_2)\begin{pmatrix} a & 0 \\ 0 & -\hat{a} \end{pmatrix}\begin{pmatrix} x_1 \\ x_2 \end{pmatrix} = ax_1^2 - \hat{a}x_2^2.$$

Then

$$\boldsymbol{x}'\boldsymbol{A}\boldsymbol{x} > 0 \quad \text{if} \quad ax_1^2 - \hat{a}x_2^2 > 0 \quad \text{or} \quad x_1^2 > \frac{\hat{a}}{a}x_2^2$$
$$\boldsymbol{x}'\boldsymbol{A}\boldsymbol{x} < 0 \quad \text{if} \quad ax_1^2 - \hat{a}x_2^2 < 0 \quad \text{or} \quad x_1^2 < \frac{\hat{a}}{a}x_2^2.$$

So, the sign of $\boldsymbol{x}'\boldsymbol{A}\boldsymbol{x}$ depends on the value of $\boldsymbol{x}$ and the parameters of $\boldsymbol{A}$. If, instead,

$$\boldsymbol{A} = \begin{pmatrix} a & 0 \\ 0 & \hat{a} \end{pmatrix}$$

then $\boldsymbol{x}'\boldsymbol{A}\boldsymbol{x} = ax_1^2 + \hat{a}x_2^2$, which is never negative since a and $\hat{a}$ are positive and $x_i^2 \ge 0$. ◊

Thus, as Example 10.2 illustrates, under suitable restrictions on $\boldsymbol{A}$, the sign of a quadratic form, $\boldsymbol{x}'\boldsymbol{A}\boldsymbol{x} = ax_1^2 + \hat{a}x_2^2$, may be completely determined by the parameters of the $\boldsymbol{A}$ matrix.

Consider again the two-variable quadratic form:

$$\begin{aligned}(x_1\ x_2)\begin{pmatrix}a_{11} & a_{12}\\ a_{21} & a_{22}\end{pmatrix}\begin{pmatrix}x_1\\ x_2\end{pmatrix} &= a_{11}x_1^2 + (a_{12}+a_{21})x_1x_2 + a_{22}x_2^2\\ &= a_{11}x_1^2 + 2\bar{a}x_1x_2 + a_{22}x_2^2, \quad \bar{a} = \frac{1}{2}(a_{12}+a_{21})\\ &= (x_1\ x_2)\begin{pmatrix}a_{11} & \bar{a}\\ \bar{a} & a_{22}\end{pmatrix}\begin{pmatrix}x_1\\ x_2\end{pmatrix}. \qquad (10.1)\end{aligned}$$

If $a_{11} = 0$, then $\boldsymbol{x}' = (1,0)$ gives $\boldsymbol{x}'\boldsymbol{A}\boldsymbol{x} = 0$, so, for a positive or negative definite matrix, $a_{ii} \neq 0$ for all i. The expression $a_{11}x_1^2 + 2\bar{a}x_1x_2 + a_{22}x_2^2$ may be rearranged (assume $a_{11} \neq 0$ or that $a_{22} \neq 0$ and with a symmetric calculation):

$$\begin{aligned}a_{11}x_1^2 + 2\bar{a}x_1x_2 + a_{22}x_2^2 &= a_{11}\left[x_1 + \frac{\bar{a}}{a_{11}}x_2\right]^2 + \left[a_{22} - \frac{\bar{a}^2}{a_{11}}\right]x_2^2\\ &= a_{11}\left[x_1 + \frac{\bar{a}}{a_{11}}x_2\right]^2 + \frac{1}{a_{11}}\left[a_{11}a_{22} - \bar{a}^2\right]x_2^2.\end{aligned}$$

If $a_{11} > 0$ and $a_{11}a_{22} - \bar{a}^2 > 0$, then this expression is positive if $(x_1,x_2) \neq (0,0)$. Conversely, if $a_{11} < 0$ and $a_{11}a_{22} - \bar{a}^2 > 0$, then the expression is negative for all $(x_1,x_2) \neq (0,0)$. In either of the other two cases $[a_{11} > 0,\ a_{11}a_{22} - \bar{a}^2 < 0]$ and $[a_{11} < 0,\ a_{11}a_{22} - \bar{a}^2 < 0]$, the sign of $\boldsymbol{x}'\boldsymbol{A}\boldsymbol{x}$ can be positive or negative depending on the values of (x_1,x_2). Consequently, the quadratic form is strictly positive for all $(x_1,x_2) \neq (0,0)$ if and only if $a_{11} > 0$ and $a_{11}a_{22} - \bar{a}^2 > 0$; and strictly negative for all $(x_1,x_2) \neq (0,0)$ if and only if $a_{11} < 0$ and $a_{11}a_{22} - \bar{a}^2 > 0$.

EXAMPLE 10.3: This example illustrates how different values of the a_{ij} parameters in the $\boldsymbol{A}$ matrix affect the sign of the quadratic form, Equation 10.1:

$$\boldsymbol{A}_1 = \begin{pmatrix}4 & 3\\ 1 & 2\end{pmatrix} \Rightarrow \boldsymbol{x}'\boldsymbol{A}_1\boldsymbol{x} = 4x^2 + 4xy + 2y^2$$

$$\boldsymbol{A}_2 = \begin{pmatrix}4 & 3\\ 5 & 2\end{pmatrix} \Rightarrow \boldsymbol{x}'\boldsymbol{A}_1\boldsymbol{x} = 4x^2 + 8xy + 2y^2.$$

Considering $\boldsymbol{A}_1$, $a_{11} = 4 > 0$, $a_{11}a_{22} - \bar{a}^2 = 4\cdot 2 - 2^2 = 4 > 0$. Similarly, for $\boldsymbol{A}_2$, $a_{11} = 4 > 0$, $a_{11}a_{22} - \bar{a}^2 = 4\cdot 2 - 4^2 = -8 < 0$. The pair of conditions $a_{11} > 0$ and

$a_{11}a_{22} - \bar{a}^2 > 0$ are satisfied by the matrix $\boldsymbol{A}_1$, but not by $\boldsymbol{A}_2$.

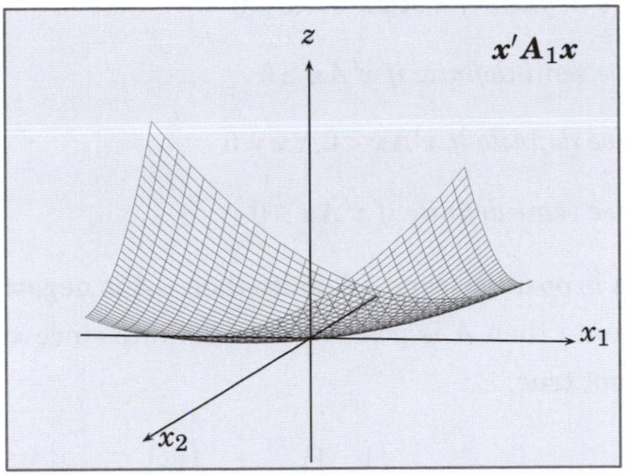

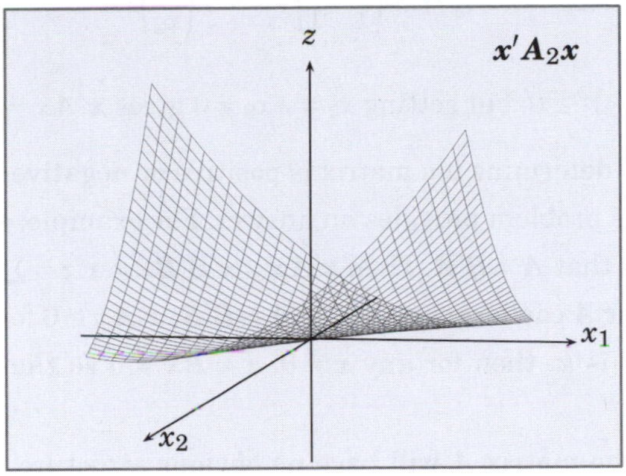

In the graphs, the quadratic form based on $\boldsymbol{A}_1$ is always non-negative and strictly positive for $(x_1, x_2) \neq (0,0)$, whereas the quadratic form based on $\boldsymbol{A}_2$ is negative for some values of (x_1, x_2). For example, at $(x_1, x_2) = (-1, 1)$, $4x^2 + 8xy + 2y^2 = 4(-1)^2 + 8(-1)(1) + 2(1)^2 = -2$. ◊

The next section describes a test to determine if a matrix is positive or negative definite.

10.3 Positive and negative definite matrices

For the following discussion let $\boldsymbol{A}$ be an $n \times n$ square matrix. Any vector $\boldsymbol{x}$ gives a quadratic form, $\boldsymbol{x}'\boldsymbol{A}\boldsymbol{x}$. Depending on the properties of the matrix, $\boldsymbol{A}$, it may be that the expression $\boldsymbol{x}'\boldsymbol{A}\boldsymbol{x}$ is always positive or always negative (whenever $\boldsymbol{x} \neq 0$). Definition 10.1 formalizes this concept and the discussion provides criteria for checking when a matrix is positive or negative definite.

Definition 10.1: *Definite and semi-definite matrices:*

- $\boldsymbol{A}$ *is positive definite if* $\boldsymbol{x}'\boldsymbol{A}\boldsymbol{x} > 0, \forall \boldsymbol{x} \neq 0$
- $\boldsymbol{A}$ *is positive semi-definite if* $\boldsymbol{x}'\boldsymbol{A}\boldsymbol{x} \geq 0$
- $\boldsymbol{A}$ *is negative definite if* $\boldsymbol{x}'\boldsymbol{A}\boldsymbol{x} < 0, \forall \boldsymbol{x} \neq 0$
- $\boldsymbol{A}$ *is negative semi-definite if* $\boldsymbol{x}'\boldsymbol{A}\boldsymbol{x} \leq 0$.

Observe that if $\boldsymbol{A}$ is positive (semi-) definite then $-\boldsymbol{A}$ is negative (semi-)definite. If $\boldsymbol{A}$ is positive definite then $\boldsymbol{A}$ is positive semi-definite since $\boldsymbol{x}'\boldsymbol{A}\boldsymbol{x} > 0 \Rightarrow \boldsymbol{x}'\boldsymbol{A}\boldsymbol{x} \geq 0$. The converse is not true:

$$\boldsymbol{A} = \begin{pmatrix} 1 & 1 \\ 1 & 1 \end{pmatrix}, \quad \boldsymbol{x} = \begin{pmatrix} x_1 \\ x_2 \end{pmatrix},$$

so $\boldsymbol{x}'\boldsymbol{A}\boldsymbol{x} = (x_1 + x_2)^2 \geq 0$, but setting $x_1 = -x_2 \neq 0$ gives $\boldsymbol{x}'\boldsymbol{A}\boldsymbol{x} = 0$.

How can one determine if a matrix is positive or negative definite? Sometimes the nature of the problem provides an answer. For example, suppose that $\boldsymbol{B}$ is an $n \times k$ matrix and that $\boldsymbol{A} = \boldsymbol{B}'\boldsymbol{B}$. Then $\boldsymbol{x}'\boldsymbol{A}\boldsymbol{x} = \boldsymbol{x}'\boldsymbol{B}'\boldsymbol{B}\boldsymbol{x} = \boldsymbol{z}'\boldsymbol{z} = \sum_{i=1}^{n} z_i^2 \geq 0$, where $\boldsymbol{z} = \boldsymbol{B}\boldsymbol{x}$. So, whenever $\boldsymbol{A}$ can be expressed as $\boldsymbol{A} = \boldsymbol{B}'\boldsymbol{B}$, $\boldsymbol{x}'\boldsymbol{A}\boldsymbol{x} \geq 0$ for all $\boldsymbol{x}$. Furthermore, if the rank of $\boldsymbol{B}$ is k, then for any $\boldsymbol{x} \neq 0$, $\boldsymbol{z} = \boldsymbol{B}\boldsymbol{x} \neq 0$ so that $\boldsymbol{x}'\boldsymbol{A}\boldsymbol{x} = \sum_{i=1}^{n} z_i^2 > 0$ whenever $\boldsymbol{x} \neq 0$.

In general, the matrix **A** will have no obvious structure implying the sign of the corresponding quadratic form and motivates the need for a systematic procedure to investigate definiteness of the matrix. The systematic procedure is based on the properties of "minors" derived from the matrix **A**.

Definition 10.2: *Given a square matrix* **A**, *the determinant of any* $k \times k$ *matrix* $(0 < k \leq n)$ *obtained by deleting any* $n-k$ *rows and* $n-k$ *columns is called a minor (or a kth-order minor to identify the dimensions of the matrix on which the determinant is derived).*

- *The* (i,j) *minor of a square matrix* **A**, *denoted* m_{ij}, *is the determinant of the matrix obtained from* **A** *by deleting the ith row and jth column.*
- *A minor obtained from a square matrix* **A** *as the determinant of a* $k \times k$ *matrix given by deleting* $n-k$ *matching rows and columns is called a (kth-order) principal minor.*
- *A principal minor is called a leading principal minor if it is the determinant of a submatrix obtained from* **A** *by deleting the last* $(n-k)$ *rows and columns,* $0 < k \leq n$.

Definition 10.3: *The trace,* $\text{tr}(\boldsymbol{A})$*, of a square matrix,* $\boldsymbol{A}$*, is the sum of the diagonal elements of the matrix:* $\text{tr}(\boldsymbol{A}) = \sum_{i=1}^{n} a_{ii}$.

PROPOSITION 10.1: The following observations are useful:

1. If $\boldsymbol{A}$ is positive definite then $a_{ii} > 0, \forall i$; and if $\boldsymbol{A}$ is positive semi-definite then $a_{ii} \geq 0, \forall i$.
2. If $\boldsymbol{A}$ is positive definite $r(\boldsymbol{A}) = n$.
3. If $\boldsymbol{A}$ is positive definite then $\text{tr}(\boldsymbol{A}) > 0$.
4. The determinant of a symmetric positive definite matrix is positive. ◊

PROOF: Let $x = (0, 0, \ldots, 1, \ldots, 0)$, a vector with 1 in the ith position and 0 elsewhere. Then $\boldsymbol{x}'\boldsymbol{A}\boldsymbol{x} = a_{ii}$. If $\boldsymbol{A}$ is positive definite then $a_{ii} > 0, \forall i$; and if $\boldsymbol{A}$ is positive semi-definite then $a_{ii} \geq 0, \forall i$. To show the second result, note that if $\boldsymbol{A}$ is positive definite, then for all $\boldsymbol{x} \neq 0$, $\boldsymbol{x}'\boldsymbol{A}\boldsymbol{x} \neq 0$. Therefore, for all $\boldsymbol{x} \neq 0$, $\boldsymbol{A}\boldsymbol{x} \neq \boldsymbol{0}$, which implies linear independence of the columns of $\boldsymbol{A}$, so that $\boldsymbol{A}$ has full rank, n. The positive determinant property follows from Theorem 10.1. ■

EXAMPLE 10.4: Consider the matrix $\boldsymbol{A}$:

$$\boldsymbol{A} = \begin{pmatrix} 1 & 0 \\ -4 & 1 \end{pmatrix}.$$

This matrix has positive trace, determinant and leading principal minors, but is not positive definite. (With $\boldsymbol{x}' = (2, 1)$, $\boldsymbol{x}'\boldsymbol{A}\boldsymbol{x} = -3$.) Notice that the matrix $\boldsymbol{A}$ is not symmetric. ◊

10.3.1 Symmetric matrices

Recall that a matrix $\boldsymbol{A}$ is *symmetric* if $\boldsymbol{A} = \boldsymbol{A}'$ (or $a_{ij} = a_{ji}$ for all i, j). The following considerations show that there is no loss of generality in considering only symmetric matrices when checking for positive definiteness. Consider any non-symmetric matrix $\boldsymbol{B}$. Noting that $\boldsymbol{x}'\boldsymbol{B}\boldsymbol{x} = \boldsymbol{x}'\boldsymbol{B}'\boldsymbol{x}$, if $\boldsymbol{B}$ is positive definite then so is $\boldsymbol{B} + \boldsymbol{B}'$ since $2\boldsymbol{x}'\boldsymbol{B}\boldsymbol{x} = \boldsymbol{x}'\boldsymbol{B}\boldsymbol{x} + \boldsymbol{x}'\boldsymbol{B}'\boldsymbol{x} = \boldsymbol{x}'(\boldsymbol{B} + \boldsymbol{B}')\boldsymbol{x}$, so that $\boldsymbol{x}'\boldsymbol{B}\boldsymbol{x} > 0$ if and only if $\boldsymbol{x}'(\boldsymbol{B} + \boldsymbol{B}')\boldsymbol{x} > 0$. A matrix $\boldsymbol{B}$ is positive definite (semi-definite) if and only if the matrix $\boldsymbol{B} + \boldsymbol{B}'$ is positive definite (semi-definite). Note that $\boldsymbol{B} + \boldsymbol{B}'$ is symmetric. Therefore, to determine if $\boldsymbol{B}$ is positive definite (positive semi-definite) etc., one need only confirm that $(\boldsymbol{B}+\boldsymbol{B}')$ is positive definite (positive semi-definite). In discussing (semi-)definite matrices we need only consider symmetric matrices.

10.3.2 Criteria for positive and negative definiteness

In general, it is difficult to check positive or negative definiteness by inspecting a matrix. An alternative criterion is based on considering a sequence of determinants. Let $\boldsymbol{A}$ be a symmetric $n\times n$ matrix and let $\boldsymbol{D}_i$ be an $i\times i$ matrix – the submatrix of $\boldsymbol{A}$ obtained by deleting the last $n-i$ rows and the last $n-i$ columns. Thus $|\boldsymbol{D}_i|$ is the ith leading principal minor. In words, for positive definiteness, the leading principal minors $\{|\boldsymbol{D}_i|\}_{i=1}^n$ must all be positive; for negative definiteness they must alternate in sign, starting negative:

$$\boldsymbol{A}=\begin{pmatrix} a_{11} & a_{12} & \cdots & a_{1j} & \cdots & a_{1n}\\ a_{21} & a_{22} & \cdots & a_{2j} & \cdots & a_{2n}\\ \vdots & \vdots & \ddots & \vdots & \vdots & \vdots\\ a_{j1} & a_{j2} & \cdots & a_{jj} & \cdots & a_{jn}\\ \vdots & \vdots & \vdots & \vdots & \ddots & \vdots\\ a_{n1} & a_{n2} & \cdots & a_{nj} & \cdots & a_{nn}\end{pmatrix},\quad \boldsymbol{D}_i=\begin{pmatrix} a_{11} & a_{12} & \cdots & a_{1i}\\ a_{21} & a_{22} & \cdots & a_{2i}\\ \vdots & \vdots & \ddots & \vdots\\ a_{i1} & a_{i2} & \cdots & a_{ii}\end{pmatrix}.$$

With this notation:

Theorem 10.1: *Let* $\boldsymbol{A}$ *be a symmetric matrix.*
The matrix $\boldsymbol{A}$ *is* positive definite *if and only if* $|\boldsymbol{D}_i|>0, \forall i=1,\dots,n$. *That is:*

$$|\boldsymbol{D}_1|>0,\ |\boldsymbol{D}_2|>0,\ |\boldsymbol{D}_3|>0,\ \dots,\ |\boldsymbol{D}_n|>0.$$

The matrix $\boldsymbol{A}$ *is* negative definite *if and only if* $(-1)^i\,|\boldsymbol{D}_i|>0,\ \forall i=1,\dots,n$. *That is:*

$$|\boldsymbol{D}_1|<0,\ |\boldsymbol{D}_2|>0,\ |\boldsymbol{D}_3|<0,\ \dots,(-1)^n\,|\boldsymbol{D}_n|>0.$$

Thus, with positive definiteness, $|\boldsymbol{D}_i|$ is positive for all i, whereas with negative definiteness, $|\boldsymbol{D}_i|$ alternates in sign, beginning negative.

In the 2×2 case, because the matrix is symmetric, $a_{12}=a_{21}$ and $a_{11}\neq 0$, the quadratic form is:

$$\boldsymbol{x}'\boldsymbol{A}\boldsymbol{x}=a_{11}x_1^2+2a_{12}x_1x_2+a_{22}x_2^2=a_{11}\left(x_1+\frac{a_{12}}{a_{11}}x_2\right)^2+\left(a_{22}-\frac{a_{12}^2}{a_{11}}\right)x_2^2.$$

From this expression, it is clear that if $a_{11} > 0$ and $a_{11}a_{22} - a_{12}^2 > 0$, then provided $x_1 \neq 0$ or $x_2 \neq 0$, $\boldsymbol{x}'\boldsymbol{A}\boldsymbol{x} > 0$. Note that a_{11} is the determinant of the matrix $\boldsymbol{D}_1$, and $a_{11}a_{22} - a_{12}^2$ is the determinant of the matrix $\boldsymbol{D}_2$. From these calculations, note that to check positive or negative definiteness, there are at most n determinants to evaluate.

10.3.3 Positive and negative semi-definiteness

The next theorem gives similar conditions for positive and negative semi-definiteness and uses the following notation. Let $\tilde{\boldsymbol{D}}_i$ be any submatrix of $\boldsymbol{A}$ containing i rows and columns obtained by deleting any $n-i$ rows and the *corresponding* columns (if the kth row is deleted, then the kth column is also deleted). For example, if $\boldsymbol{A}$ is an $n \times n$ matrix:

$$\tilde{\boldsymbol{D}}_i = \begin{pmatrix} a_{kk} & a_{kj} & \cdots & a_{kl} & \cdots & a_{kr} \\ a_{jk} & a_{jj} & \cdots & a_{jl} & \cdots & a_{jr} \\ \vdots & \vdots & \ddots & \vdots & \vdots & \vdots \\ a_{lk} & a_{lj} & \cdots & a_{ll} & \cdots & a_{lr} \\ \vdots & \vdots & \vdots & \vdots & \ddots & \vdots \\ a_{rk} & a_{rj} & \cdots & a_{rl} & \cdots & a_{rr} \end{pmatrix}$$

The determinant of any $\tilde{\boldsymbol{D}}_i$ is an ith-order principal minor.

For example, if $\boldsymbol{A}$ is a 3×3 matrix:

$$\boldsymbol{A} = \begin{pmatrix} a_{11} & a_{12} & a_{13} \\ a_{21} & a_{22} & a_{23} \\ a_{31} & a_{32} & a_{33} \end{pmatrix},$$

then there are three $\tilde{\boldsymbol{D}}_i$ matrices:

$$\begin{pmatrix} a_{11} & a_{12} \\ a_{21} & a_{22} \end{pmatrix}, \quad \begin{pmatrix} a_{22} & a_{23} \\ a_{32} & a_{33} \end{pmatrix}, \quad \begin{pmatrix} a_{11} & a_{13} \\ a_{31} & a_{33} \end{pmatrix},$$

and hence three principal minors of order 2.

Theorem 10.2: *Let* $\boldsymbol{A}$ *be a symmetric matrix.*
The matrix $\boldsymbol{A}$ *is positive semi-definite if and only if:*

$$\textit{all } |\tilde{\boldsymbol{D}}_1| \geq 0, \textit{ all } |\tilde{\boldsymbol{D}}_2| \geq 0, \ldots \textit{ all } |\tilde{\boldsymbol{D}}_i| \geq 0, \ldots \textit{ all } |\tilde{\boldsymbol{D}}_n| \geq 0.$$

The matrix $\boldsymbol{A}$ *is negative semi-definite if and only if:*

$$\textit{all } |\tilde{\boldsymbol{D}}_1| \leq 0, \textit{ all } |\tilde{\boldsymbol{D}}_2| \geq 0, \ \ldots \ \textit{all } (-1)^i\, |\tilde{\boldsymbol{D}}_i| \geq 0, \ldots \ \textit{all } (-1)^n\, |\tilde{\boldsymbol{D}}_n| \geq 0.$$

Observe that the statements for positive and negative semi-definiteness are *not* obtained by replacing strict inequalities by non-strict inequalities. To check for positive semi-definiteness, it is necessary to check that all 1st-order principal minors are positive, all 2nd-order principal minors are positive, all 3rd-order principal minors are positive, and so on.

There are $n = \frac{n!}{(n-1)!1!}$ 1st-order principal minors, $\frac{n!}{(n-2)!2!}$ 2nd-order principal minors, $\frac{n!}{(n-3)!3!}$ 3rd-order principal minors, and so on. There are $\sum_{i=1}^{n}\left[\frac{n!}{(n-i)!i!}\right] = 2^n - 1$ such minors in total.

10.4 Definiteness with equality constraints

In constrained optimization problems, checking for maxima or minima leads to considering definiteness of a quadratic form subject to linear constraints: for example, that $\boldsymbol{x}'\boldsymbol{A}\boldsymbol{x} > 0$ for all $\boldsymbol{x}$ satisfying $\boldsymbol{B}\boldsymbol{x} = 0$, for some $m \times n$ matrix $\boldsymbol{B}$, where

$$\boldsymbol{B} = \begin{pmatrix} b_{11} & b_{12} & \cdots & b_{1n} \\ b_{21} & b_{22} & \cdots & b_{2n} \\ \vdots & \vdots & \vdots & \vdots \\ b_{m1} & b_{m2} & \cdots & b_{mn} \end{pmatrix}.$$

First, consider the simplest case where there is one constraint so $\boldsymbol{B}\boldsymbol{x} = 0$ may be written $\boldsymbol{b}'\boldsymbol{x}$ where $\boldsymbol{b}' = (b_1, b_2, \ldots, b_n)$. Form the bordered matrix:

$$\boldsymbol{M}_{kk} = \begin{pmatrix} 0 & b_1 & b_2 & \cdots & b_k \\ b_1 & a_{11} & a_{12} & \cdots & a_{1k} \\ b_2 & a_{21} & a_{22} & \cdots & a_{2k} \\ \vdots & \vdots & \vdots & \ddots & \vdots \\ b_k & a_{k1} & a_{k2} & \cdots & a_{kk} \end{pmatrix},$$

or, writing $\boldsymbol{b}_k$ for the first k elements of $\boldsymbol{b}$ and $\boldsymbol{A}_{kk}$ for the matrix obtained from $\boldsymbol{A}$ by deleting the last $k+1, k+2, \ldots, n$ rows and columns, $\boldsymbol{M}_{kk}$ may be written $\boldsymbol{M}_{kk} = \begin{pmatrix} 0 & \boldsymbol{b}'_k \\ \boldsymbol{b}_k & \boldsymbol{A}_{kk} \end{pmatrix}$.

Theorem 10.3: *Let* $\boldsymbol{A}$ *be a symmetric matrix and* $b_1 \neq 0$. *Then,*

1. $\boldsymbol{x}'\boldsymbol{A}\boldsymbol{x} > 0$ *for all* $\boldsymbol{x} \neq 0$ *and* $\boldsymbol{b}'\boldsymbol{x} = 0$ *if and only if*

$$|\boldsymbol{M}_{kk}| < 0, \quad k \geq 2$$

2. $\boldsymbol{x}'\boldsymbol{A}\boldsymbol{x} < 0$ *for all* $\boldsymbol{x} \neq 0$ *and* $\boldsymbol{b}'\boldsymbol{x} = 0$ *if and only if*

$$(-1)^k \, |\boldsymbol{M}_{kk}| > 0, \quad k \geq 2$$

In the case where $m > 1$, similar computations apply. Then,

$$\boldsymbol{M}_{kk} = \begin{pmatrix} 0 & 0 & \vdots & 0 & b_{11} & b_{12} & \cdots & b_{1k} \\ 0 & 0 & \vdots & 0 & b_{21} & b_{22} & \cdots & b_{2k} \\ \vdots & \vdots & \ddots & \vdots & \vdots & \cdots & \vdots & \vdots \\ 0 & 0 & \vdots & 0 & b_{m1} & b_{m2} & \cdots & b_{mk} \\ b_{11} & b_{21} & \vdots & b_{m1} & a_{11} & a_{12} & \cdots & a_{1k} \\ b_{12} & b_{22} & \vdots & b_{m2} & a_{21} & a_{22} & \cdots & a_{2k} \\ \vdots & \vdots & \vdots & \vdots & \vdots & \vdots & \ddots & \vdots \\ b_{1k} & b_{2k} & \vdots & b_{mk} & a_{k1} & a_{k2} & \cdots & a_{kk} \end{pmatrix}.$$

The matrix $\boldsymbol{M}_{kk}$ has $m+k$ rows and columns. Writing $\boldsymbol{A}_{kk}$ and $\boldsymbol{B}_{mk}$ for the component matrices,

$$\boldsymbol{M}_{kk} = \begin{pmatrix} \boldsymbol{0} & \boldsymbol{B}_{mk} \\ \boldsymbol{B}'_{mk} & \boldsymbol{A}_{kk} \end{pmatrix}.$$

Theorem 10.4: *Let* $\boldsymbol{A}$ *be a symmetric matrix and let* $|\boldsymbol{B}_{mm}| \neq 0$. *Then,*

1. $\boldsymbol{x}'\boldsymbol{A}\boldsymbol{x} > 0$ *for all* $\boldsymbol{x} \neq 0$ *with* $\boldsymbol{B}\boldsymbol{x} = 0$ *if and only if:*

$$(-1)^m \, |\boldsymbol{M}_{kk}| > 0, \quad k = m+1, \ldots, n.$$

2. $\boldsymbol{x}'\boldsymbol{A}\boldsymbol{x} < 0$ *for all* $\boldsymbol{x} \neq 0$ *with* $\boldsymbol{B}\boldsymbol{x} = 0$ *if and only if:*

$$(-1)^k \, |\boldsymbol{M}_{kk}| > 0, \quad k = m+1, \ldots, n.$$

REMARK 10.1: This characterization may be expressed in different ways. Let $\boldsymbol{I}_p$ be an identity matrix of dimension p. Then:

$$\begin{pmatrix} \mathbf{0} & \boldsymbol{I}_k \\ \boldsymbol{I}_m & \mathbf{0} \end{pmatrix} \begin{pmatrix} \mathbf{0} & \boldsymbol{B}_{mk} \\ \boldsymbol{B}'_{mk} & \boldsymbol{A}_{kk} \end{pmatrix} \begin{pmatrix} \mathbf{0} & \boldsymbol{I}_m \\ \boldsymbol{I}_k & \mathbf{0} \end{pmatrix} = \begin{pmatrix} \boldsymbol{B}'_{mk} & \boldsymbol{A}_{kk} \\ \mathbf{0} & \boldsymbol{B}_{mk} \end{pmatrix} \begin{pmatrix} \mathbf{0} & \boldsymbol{I}_m \\ \boldsymbol{I}_k & \mathbf{0} \end{pmatrix} = \begin{pmatrix} \boldsymbol{A}_{kk} & \boldsymbol{B}'_{mk} \\ \boldsymbol{B}_{mk} & \mathbf{0} \end{pmatrix}.$$

From this, noting that

$$\begin{pmatrix} \mathbf{0} & \boldsymbol{I}_m \\ \boldsymbol{I}_k & \mathbf{0} \end{pmatrix}' = \begin{pmatrix} \mathbf{0} & \boldsymbol{I}_k \\ \boldsymbol{I}_m & \mathbf{0} \end{pmatrix},$$

$$\mid \boldsymbol{M}_{kk} \mid = \left(\begin{vmatrix} \mathbf{0} & \boldsymbol{I}_k \\ \boldsymbol{I}_m & \mathbf{0} \end{vmatrix} \right)^2 \times \begin{vmatrix} \boldsymbol{A}_{kk} & \boldsymbol{B}'_{mk} \\ \boldsymbol{B}_{mk} & \mathbf{0} \end{vmatrix} = \begin{vmatrix} \boldsymbol{A}_{kk} & \boldsymbol{B}'_{mk} \\ \boldsymbol{B}_{mk} & \mathbf{0} \end{vmatrix}.$$

Therefore, the determinental conditions given previously can also be expressed in terms of matrices in the form:

$$\begin{pmatrix} \boldsymbol{A}_{kk} & \boldsymbol{B}'_{mk} \\ \boldsymbol{B}_{mk} & \mathbf{0} \end{pmatrix}.$$

□

REMARK 10.2: If there is just one constraint, $m = 1$, then constrained positiveness requires $\mid \boldsymbol{M}_{kk} \mid < 0$, $k = m+1, \ldots, n$. More generally, with $m > 1$ m odd gives $\mid \boldsymbol{M}_{kk} \mid < 0$, $k = m+1, \ldots, n$, while with m even the inequality is reserved so that $\mid \boldsymbol{M}_{kk} \mid > 0$, $k = m+1, \ldots, n$. In contrast, for constrained negatives $\mid \boldsymbol{M}_{kk} \mid$ alternates in sign starting positive at $k = m+1$ if m is odd and starting negative if m is even. □

Exercises for Chapter 10

Exercise 10.1 Consider the following symmetric matrices:

$$\boldsymbol{A} = \begin{pmatrix} 3 & -1 & 1 \\ -1 & 3 & -2 \\ 1 & -2 & 2 \end{pmatrix}, \qquad \boldsymbol{B} = \begin{pmatrix} 3 & 2 & 1 \\ 2 & 12 & 4 \\ 1 & 4 & 2 \end{pmatrix}.$$

Confirm that both $\boldsymbol{A}$ and $\boldsymbol{B}$ are positive definite.

Exercise 10.2 Consider the following matrices:

$$\boldsymbol{A} = \begin{pmatrix} -3 & 2 & 5 \\ 2 & -6 & 1 \\ 5 & 1 & -13 \end{pmatrix}, \qquad \boldsymbol{B} = \begin{pmatrix} -5 & 2 & 1 \\ 2 & -1 & -1 \\ 1 & -1 & -7 \end{pmatrix}.$$

Confirm that both $\boldsymbol{A}$ and $\boldsymbol{B}$ are negative definite.

Exercise 10.3 Consider the matrix:

$$\boldsymbol{A} = \begin{pmatrix} \frac{1}{4} & -\frac{1}{3} \\ -\frac{1}{3} & \frac{2}{9} \end{pmatrix}.$$

Show that $\boldsymbol{A}$ is positive definite for all $\boldsymbol{x} = (x_1, x_2)$ satisfying $\left(\frac{1}{2}, \frac{2}{3}\right) \cdot (x_1, x_2) = 0$.

Exercise 10.4 Consider the matrix:

$$\boldsymbol{A} = \begin{pmatrix} \frac{1}{4} & -\frac{4}{3} \\ -\frac{4}{3} & \frac{32}{9} \end{pmatrix}.$$

Show that $\boldsymbol{A}$ is positive definite for all $\boldsymbol{x} = (x_1, x_2)$ satisfying $\left(2, \frac{8}{3}\right) \cdot (x_1, x_2) = 0$.

Exercise 10.5 Consider the matrix:

$$\boldsymbol{Q}(\alpha,\beta,\gamma)=\begin{pmatrix}\alpha(\alpha-1) & \alpha\beta & \alpha\gamma\\ \alpha\beta & \beta(\beta-1) & \beta\gamma\\ \alpha\gamma & \beta\gamma & \gamma(\gamma-1).\end{pmatrix}.$$

Let $\boldsymbol{x}'\boldsymbol{Q}(\alpha,\beta,\gamma)\boldsymbol{x}$ be a quadratic form. Also, let $\boldsymbol{b}=(w,r,p)$ be a vector of strictly positive entries. Suppose that $\alpha=\beta=\gamma=\frac{1}{4}$. Show that $\boldsymbol{x}'\boldsymbol{Q}(\alpha,\beta,\gamma)\boldsymbol{x}<0$ for all $\boldsymbol{x}$ satisfying $\boldsymbol{b}'\boldsymbol{x}=0$.

11

Multivariate optimization

11.1 Introduction

This chapter is concerned with multivariate optimization, finding values for $(x_1, x_2, \dots, x_n)$ to make the function $f(x_1, x_2, \dots, x_n)$ as large or small as possible. In Section 11.2, the two-variable case is described; Section 11.3 considers the n-variable case. The extension from two to n variables is natural. Section 11.2 provides sufficient conditions for a local optimum and provides motivation for those conditions. Also, in this section, the envelope theorem is described. Section 11.3 provides sufficient conditions for both local and global optima. These conditions are developed in terms of concavity of the objective function and the properties of Hessian matrices in Section 11.4.

11.2 The two-variable case

Consider the problem of choosing (x, y) to maximize $z = f(x, y)$, a function of two variables. The first requirement is that no unilateral variations in x or y improve the value of f, the *first-order conditions*:

$$\frac{\partial f(x,y)}{\partial x} = f_x(x,y) = 0$$
$$\frac{\partial f(x,y)}{\partial y} = f_y(x,y) = 0. \tag{11.1}$$

The total differential is $dz = f_x dx + f_y dy$. This gives the change in the value of the function due to the variation (dx, dy) in x and y from the value (x, y).

Note that dz depends on (x,y) since $dz = f_x(x,y)dx + f_y(x,y)dy$. When the choice of x and y maximizes f, there can be no changes dx and dy that raise the value of f, as required by the partial derivative conditions above – that f_x and f_y are both 0. Consider the function plotted in Figure 11.1, $f(x,y) = 1 - x^2 - y^2$. There, at

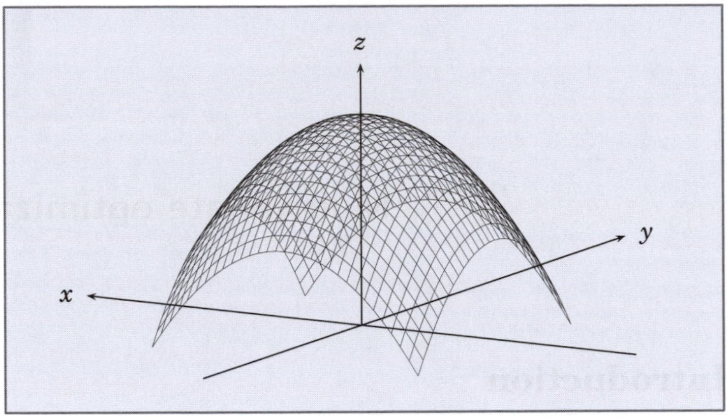

Figure 11.1: The function $f(x,y) = 1 - x^2 - y^2$

$(0,0)$, $f_x = f_y = 0$, and it is clear from the figure that this point locates the highest point of the function. Confirming that a point is the maximizing value requires a comparison similar to the univariate test $f''(x) < 0$. The next theorem summarizes the key result for local optimality of a point $(\bar{x},\bar{y})$.

Theorem 11.1: *Suppose that* $(\bar{x},\bar{y})$ *satisfy*

1. *(a)* $f_x(\bar{x},\bar{y}) = f_y(\bar{x},\bar{y}) = 0$

   (b) $f_{xx}(\bar{x},\bar{y}) < 0, f_{yy}(\bar{x},\bar{y}) < 0$ *and* $\left[f_{xx}(\bar{x},\bar{y})f_{yy}(\bar{x},\bar{y}) - f_{xy}^2(\bar{x},\bar{y})\right] > 0$,

   then $(\bar{x},\bar{y})$ *is a local maximum.*

2. *(a)* $f_x(\bar{x},\bar{y}) = f_y(\bar{x},\bar{y}) = 0$

   (b) $f_{xx}(\bar{x},\bar{y}) > 0, f_{yy}(\bar{x},\bar{y}) > 0$ *and* $\left[f_{xx}(\bar{x},\bar{y})f_{yy}(\bar{x},\bar{y}) - f_{xy}^2(\bar{x},\bar{y})\right] > 0$.

   then $(\bar{x},\bar{y})$ *is a local minimum.*

The following examples illustrate the use of the theorem.

EXAMPLE 11.1: Suppose that a firm has a product that is produced at unit cost:

$$\min c(x,y) = e^x e^y - \ln x - \ln y.$$

What are the unit cost minimizing levels of x and y? The first-order conditions are:

$$c_x = e^x e^y - \frac{1}{x} = 0$$
$$c_y = e^x e^y - \frac{1}{y} = 0.$$

Thus, $\frac{1}{x} = \frac{1}{y}$ and $x = y$. From this, $e^{2x} = \frac{1}{x}$, so that solving $xe^{2x} = 1$ gives the solution ($x = 0.4263$). The second-order conditions are:

$$c_{xx} = e^x e^y + \frac{1}{x^2}$$
$$c_{yy} = e^x e^y + \frac{1}{y^2}$$
$$c_{xy} = e^x e^y.$$

So, $c_{xx} > 0$ and

$$c_{xx}c_{yy} - c_{xy}^2 = \left[e^x e^y + \frac{1}{x^2}\right]\left[e^x e^y + \frac{1}{y^2}\right] - [e^x e^y]^2$$
$$= e^x e^y \frac{1}{y^2} + e^x e^y \frac{1}{x^2} + \frac{1}{x^2}\frac{1}{y^2} > 0.$$

◊

The next example considers a profit maximization problem.

EXAMPLE 11.2: Suppose that a firm has profit function

$$f(x,y) = \sqrt{x} + \ln(y+1) - x - \frac{1}{2}y.$$

What values of x and y maximize profit?

In this example, the computations are simplified because the function may be written $f(x,y) = f_1(x) + f_2(y)$: $f(x,y) = \sqrt{x} + \ln(y+1) - x - \frac{1}{2}y = [\sqrt{x} - x] + [\ln(y+1) - \frac{1}{2}y]$, and the cross-partial derivatives vanish.

The function may be plotted:

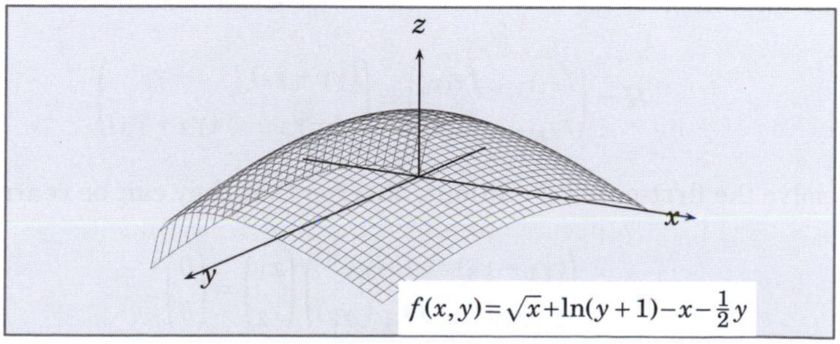

The first-order conditions are:

$$f_x = \frac{1}{2}\frac{1}{\sqrt{x}} - 1 = 0$$

$$f_y = \frac{1}{y+1} - \frac{1}{2} = 0.$$

So, $x^{\frac{1}{2}} = \frac{1}{2}$ and $2 = y+1$ giving $x = \frac{1}{4}$, $y = 1$. The second derivatives are: $f_{xx} = -\frac{1}{4}x^{-\frac{3}{2}}$, $f_{yy} = -(y+1)^{-2}$ and $f_{xy} = 0$. So, f_{xx} and f_{yy} are both negative in a neighborhood of $(\bar{x},\bar{y}) = \left(\frac{1}{4},1\right)$, $f(x,y) < f\left(\frac{1}{4},1\right)$, for $(x,y) \neq \left(\frac{1}{4},1\right)$. ◊

REMARK 11.1: In Example 11.2, the function $f(x,y)$ has cross-partial derivatives that are 0, so that $f_{x,y}(x,y) = 0$ for all (x,y). Later, in Equation 11.2, with $f_x(\bar{x},\bar{y}) = f_y(\bar{x},\bar{y}) = 0$ and $f_{xy}(\bar{x},\bar{y}) = f_{yx}(\bar{x},\bar{y}) = 0$, the Taylor series expansion reduces to:

$$f(x,y) = f(\bar{x},\bar{y}) + [f_{xx}(x^*,y^*)(\Delta x)^2 + f_{yy}(x^*,y^*)(\Delta y)^2].$$

So, if $f_{xx}(x^*,y^*) < 0$ and $f_{yy}(x^*,y^*) < 0$, since $(\Delta x)^2$ and $(\Delta y)^2$ are both positive, $f(x,y) < f(\bar{x},\bar{y})$. If $f_{xx}(\bar{x},\bar{y}) < 0$ and $f_{yy}(\bar{x},\bar{y}) < 0$, then this will also hold for (x^*,y^*) close to $(\bar{x},\bar{y})$, so that for all (x,y) close to $(\bar{x},\bar{y})$, $f(x,y) < f(\bar{x},\bar{y})$. □

EXAMPLE 11.3: A problem in physics leads to the following objective. Minimize $f(x_1,x_2)$, where f is defined:

$$f(x_1,x_2) = \frac{1}{2}(\gamma_1 x_1^2 + \gamma_3(x_2-x_1)^2 + \gamma_2 x_2^2) - bx_2,\ \gamma_i > 0, \forall i = 1,2,3.$$

Differentiating with respect to x_1 and x_2 gives:

$$f_{x_1} = \gamma_1 x_1 - \gamma_3(x_2 - x_1) = 0$$

$$f_{x_2} = \gamma_3(x_2 - x_1) + \gamma_2 x_2 - b = 0.$$

The second derivatives are: $f_{x_1x_1} = \gamma_1 + \gamma_3$, $f_{x_1x_2} = -\gamma_3$, $f_{x_2x_2} = \gamma_2 + \gamma_3$. This gives the matrix $\boldsymbol{H}$ as

$$\boldsymbol{H} = \begin{pmatrix} f_{x_1x_1} & f_{x_1x_2} \\ f_{x_1x_2} & f_{x_2x_2} \end{pmatrix} = \begin{pmatrix} (\gamma_1+\gamma_3) & -\gamma_3 \\ -\gamma_3 & (\gamma_2+\gamma_3) \end{pmatrix}.$$

To solve the first-order conditions, observe that they can be rearranged as:

$$\begin{pmatrix} (\gamma_1+\gamma_3) & -\gamma_3 \\ -\gamma_3 & (\gamma_2+\gamma_3) \end{pmatrix} \begin{pmatrix} x_1 \\ x_2 \end{pmatrix} = \begin{pmatrix} 0 \\ b \end{pmatrix}.$$

Cramer's rule may be used to solve for x_1 and x_2. The determinant of H is

$$|\boldsymbol{H}| = (\gamma_1+\gamma_3)(\gamma_2+\gamma_3)-\gamma_3^2 = \gamma_1\gamma_2+\gamma_1\gamma_3+\gamma_2\gamma_3+\gamma_3^2-\gamma_3^2 = \gamma_1\gamma_2+\gamma_1\gamma_3+\gamma_2\gamma_3.$$

So,

$$x_1 = \frac{1}{|\boldsymbol{H}|}\begin{vmatrix} 0 & -\gamma_3 \\ b & (\gamma_2+\gamma_3) \end{vmatrix}, \quad x_2 = \frac{1}{|\boldsymbol{H}|}\begin{vmatrix} (\gamma_1+\gamma_3) & 0 \\ -\gamma_3 & b \end{vmatrix}.$$

Or,

$$x_1 = \frac{b\gamma_3}{\gamma_1\gamma_2+\gamma_1\gamma_3+\gamma_2\gamma_3}, \quad x_2 = \frac{b(\gamma_1+\gamma_3)}{\gamma_1\gamma_2+\gamma_1\gamma_3+\gamma_2\gamma_3}.$$

Since

$$f_{x_1x_1} = \gamma_1+\gamma_2 > 0, \quad f_{x_1x_1}f_{x_2x_2} - f_{x_1x_2}^2 = \gamma_1\gamma_2+\gamma_1\gamma_3+\gamma_2\gamma_3 > 0,$$

the second-order condition for a minimum is satisfied. The figure plots f for $\gamma_i = 2, \forall i,\ b = 1$.

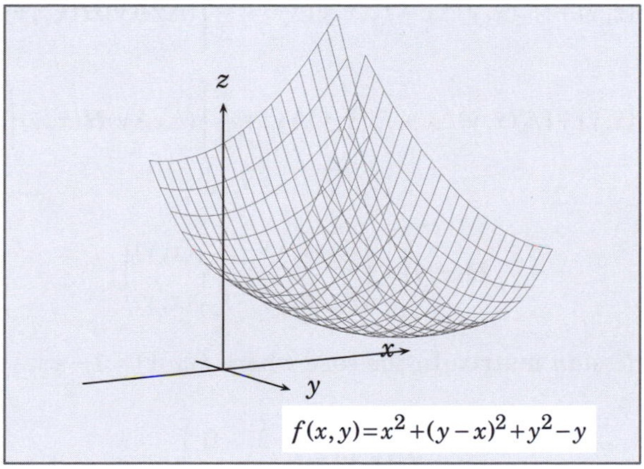

◊

11.2.1 Motivation for the second-order conditions

Recall that in the univariate case, a Taylor series expansion gave:

$$f(x) = f(\bar{x}) + f'(\bar{x})(x-\bar{x}) + \frac{1}{2!}f''(p)(x-\bar{x})^2$$

for some p between x and $\bar{x}$. So, with $f'(\bar{x}) = 0$, $f(x) = f(\bar{x}) + \frac{1}{2}f''(p)(x-\bar{x})^2$, so that if $f''(p) < 0$ for all p close to $\bar{x}$, then $f(x) < f(\bar{x})$ and $\bar{x}$ is a local maximum. Essentially the same ideas apply to the multivariate case. In the two-variable case, a Taylor series expansion of f around $(\bar{x},\bar{y})$, $(x,y)=(\bar{x}+\Delta x,\bar{y}+\Delta y)$ is given by:

$$\begin{aligned} f(x,y) &= f(\bar{x},\bar{y})+[f_x(\bar{x},\bar{y})\Delta x+f_y(\bar{x},\bar{y})\Delta y] \\ &\quad +[f_{xx}(x^*,y^*)(\Delta x)^2+2f_{xy}(x^*,y^*)\Delta x\Delta y+f_{yy}(x^*,y^*)(\Delta y)^2] \qquad (11.2)\\ &\approx f(\bar{x},\bar{y})+[f_x(\bar{x},\bar{y})\Delta x+f_y(\bar{x},\bar{y})\Delta y] \\ &\quad +[f_{xx}(\bar{x},\bar{y})(\Delta x)^2+2f_{xy}(\bar{x},\bar{y})\Delta x\Delta y+f_{yy}(\bar{x},\bar{y})(\Delta y)^2], \qquad (11.3) \end{aligned}$$

where (x^*,y^*) is on the line segment joining the points (x,y) and $(\bar{x},\bar{y})$:

$$(x^*,y^*)\in\{(\tilde{x},\tilde{y})\mid(\tilde{x},\tilde{y})=\theta(x,y)+(1-\theta)(\bar{x},\bar{y}),\theta\in[0,1]\},$$

and where $\Delta x = x-\bar{x}$ and $\Delta y = y-\bar{y}$. The Taylor series expansion is exact with (x^*,y^*) (Equation 11.2) or approximate with $(\bar{x},\bar{y})$ (Equation 11.3). The expression may be written in matrix terms:

$$f(x,y)=f(\bar{x},\bar{y})+[f_x(\bar{x},\bar{y})\Delta x+f_y(\bar{x},\bar{y})\Delta y]+\frac{1}{2}\left[(\Delta x\,\Delta y)\boldsymbol{H}(x^*,y^*)\begin{pmatrix}\Delta x\\ \Delta y\end{pmatrix}\right] \qquad (11.4)$$

$$\approx f(\bar{x},\bar{y})+[f_x(\bar{x},\bar{y})\Delta x+f_y(\bar{x},\bar{y})\Delta y]+\frac{1}{2}\left[(\Delta x\Delta y)\boldsymbol{H}(\bar{x},\bar{y})\begin{pmatrix}\Delta x\\ \Delta y\end{pmatrix}\right], \qquad (11.5)$$

where

$$\boldsymbol{H}(x,y)=\begin{pmatrix} f_{xx}(x,y) & f_{xy}(x,y)\\ f_{yx}(x,y) & f_{yy}(x,y)\end{pmatrix},$$

is called the *Hessian* matrix. In the case where $f(x,y)=1-x^2-y^2$,

$$\boldsymbol{H}(x,y)=\begin{pmatrix}-2 & 0\\ 0 & -2\end{pmatrix}$$

so, the expression in Equation 11.4 reduces to

$$f(x,y)=f(\bar{x},\bar{y})+[f_x(\bar{x},\bar{y})\Delta x+f_y(\bar{x},\bar{y})\Delta y]-2[(\Delta x)^2+(\Delta y)^2],$$

and at $(\bar{x},\bar{y})$, if $f_x(\bar{x},\bar{y})=f_y(\bar{x},\bar{y})=0$ the expression reduces to $f(x,y)=-2[(\Delta x)^2+(\Delta y)^2]$. Since this is less than or equal to 0 for all values of Δx and Δy, the value of f at $(\bar{x},\bar{y})$ is at least as large as at any other (x,y) value. In this case, and in general, the properties of the Hessian matrix are crucial in determining whether a point satisfying the first-order conditions (Equations 11.1) determine a local maximum (or minimum).

Note that the entries (partial derivatives) in the Hessian matrix depend on x and y: writing $\boldsymbol{H}(x,y)$ emphasizes that the entries are evaluated at (x,y). If the first-order conditions hold at $(\bar{x},\bar{y})$, then $f_x(\bar{x},\bar{y}) = f_y(\bar{x},\bar{y}) = 0$. In this case, Equations 11.2 and 11.3 become:

$$f(x,y) = f(\bar{x},\bar{y}) + [f_{xx}(x^*,y^*)(\Delta x)^2 + 2f_{xy}(x^*,y^*)\Delta x\Delta y + f_{yy}(x^*,y^*)(\Delta y)^2] \quad (11.6)$$

$$\approx f(\bar{x},\bar{y}) + [f_{xx}(\bar{x},\bar{y})(\Delta x)^2 + 2f_{xy}(\bar{x},\bar{y})\Delta x\Delta y + f_{yy}(\bar{x},\bar{y})(\Delta y)^2], \quad (11.7)$$

where (x^*,y^*) is on the line segment joining (x,y) and $(\bar{x},\bar{y})$. In terms of the Hessian matrix:

$$f(x,y) = f(\bar{x},\bar{y}) + (\Delta x\,\Delta y)\boldsymbol{H}(x^*,y^*)\begin{pmatrix}\Delta x\\ \Delta y\end{pmatrix} \quad (11.8)$$

$$\approx f(\bar{x},\bar{y}) + (\Delta x\,\Delta y)\boldsymbol{H}(\bar{x},\bar{y})\begin{pmatrix}\Delta x\\ \Delta y\end{pmatrix}. \quad (11.9)$$

The sign of this expression depends on properties of the matrix:

$$\boldsymbol{H}(x,y) = \begin{pmatrix} f_{xx}(x,y) & f_{xy}(x,y)\\ f_{xy}(x,y) & f_{yy}(x,y)\end{pmatrix}.$$

The key property is called negative definiteness.

Definition 11.1: *A square symmetric matrix* $\boldsymbol{A} = \begin{pmatrix} a & b\\ b & c\end{pmatrix}$ *is said to be negative definite if for all* $\boldsymbol{z} = (z_1 z_2) \neq (0,0)$, $\boldsymbol{z}'\boldsymbol{A}\boldsymbol{z} < 0$. *The matrix is said to be positive definite if* $\boldsymbol{z}'\boldsymbol{A}\boldsymbol{z} > 0$ *for all such* $\boldsymbol{z}$.

Recall the criteria for negative and positive definiteness:

PROPOSITION 11.1: In terms of the elements of the matrix, positive definiteness and negative definiteness are characterized as follows:

1. $\boldsymbol{A}$ is negative definite if and only if $a < 0$ and $ac - b^2 > 0$, and
2. $\boldsymbol{A}$ is positive definite if and only if $a > 0$ and $ac - b^2 > 0$. ◇

REMARK 11.2: This implies that if $\boldsymbol{A}$ is negative (positive) definite and $\tilde{\boldsymbol{A}}$ is (sufficiently) close to $\boldsymbol{A}$, then $\tilde{\boldsymbol{A}}$ is negative (positive) definite. For example, if $a < 0$ and $ac - b^2 > 0$ and $\tilde{a},\tilde{b},\tilde{c}$ are close to a,b,c, respectively, then $\tilde{a} < 0$ and $\tilde{a}\tilde{c} - \tilde{b}^2 > 0$, so that $\tilde{\boldsymbol{A}}$ is negative definite. □

Consequently, $\boldsymbol{H}(x,y)$ is negative definite at (x,y) if:

$$f_{xx}(x,y) < 0 \text{ and } f_{xx}(x,y)f_{yy}(x,y) - f_{xy}^2(x,y) > 0.$$

Likewise, $\boldsymbol{H}(x,y)$ is positive definite at (x,y) if:

$$f_{xx}(x,y) > 0 \text{ and } f_{xx}(x,y)f_{yy}(x,y) - f_{xy}^2(x,y) > 0.$$

REMARK 11.3: As noted earlier, if $f_{xx}(\bar{x},\bar{y}) < 0$, then for $[f_{xx}(\bar{x},\bar{y})f_{yy}(\bar{x},\bar{y}) - f_{xy}^2(\bar{x},\bar{y})] > 0$ to hold, it must be that $f_{yy}(\bar{x},\bar{y}) < 0$, since $f_{xy}^2(\bar{x},\bar{y}) \equiv (f_{xy}(\bar{x},\bar{y}))^2 \geq 0$. So, condition 1(b) in Theorem 11.1 is equivalent to $f_{xx}(\bar{x},\bar{y}) < 0$ and $\left[f_{xx}(\bar{x},\bar{y})f_{yy}(\bar{x},\bar{y}) - f_{xy}^2(\bar{x},\bar{y})\right] > 0$. An analogous equivalence applies in the local minimum case. □

Given the previous discussion, the main theorem, Theorem 11.1, is straightforward to prove.

Theorem 11.1 *Suppose that* $(\bar{x},\bar{y})$ *satisfy*

1. (a) $f_x(\bar{x},\bar{y}) = f_y(\bar{x},\bar{y}) = 0$

   (b) $f_{xx}(\bar{x},\bar{y}) < 0, f_{yy}(\bar{x},\bar{y}) < 0$ *and* $\left[f_{xx}(\bar{x},\bar{y})f_{yy}(\bar{x},\bar{y}) - f_{xy}^2(\bar{x},\bar{y})\right] > 0.$

   then $(\bar{x},\bar{y})$ *is a local maximum.*

2. (a) $f_x(\bar{x},\bar{y}) = f_y(\bar{x},\bar{y}) = 0$

   (b) $f_{xx}(\bar{x},\bar{y}) > 0, f_{yy}(\bar{x},\bar{y}) > 0$ *and* $\left[f_{xx}(\bar{x},\bar{y})f_{yy}(\bar{x},\bar{y}) - f_{xy}^2(\bar{x},\bar{y})\right] > 0.$

   then $(\bar{x},\bar{y})$ *is a local minimum.*

PROOF: Suppose that at $(\bar{x},\bar{y})$, $f_x(\bar{x},\bar{y}) = f_y(\bar{x},\bar{y}) = 0$, and suppose also that condition 1(b) is satisfied, so that $\boldsymbol{H}(\bar{x},\bar{y})$ is negative definite. Given a small number η, let $-\eta < \Delta x < \eta$ and $-\eta < \Delta y < \eta$. Since (x^*,y^*) lies between $(\bar{x},\bar{y})$ and $(\bar{x}+\Delta x, \bar{y}+\Delta y)$ in Equation 11.8, (x^*,y^*) is within η of $(\bar{x},\bar{y})$. In view of the preceding observation, it must be that $\boldsymbol{H}(x^*,y^*)$ is negative definite for small η (since $\boldsymbol{H}(x^*,y^*)$ is close to $\boldsymbol{H}(\bar{x},\bar{y})$). Therefore,

$$f(x,y) = f(\bar{x},\bar{y}) + (\Delta x\, \Delta y)\boldsymbol{H}(x^*,y^*)\begin{pmatrix}\Delta x\\ \Delta y\end{pmatrix} \quad \text{and} \quad (\Delta x\, \Delta y)\boldsymbol{H}(x^*,y^*)\begin{pmatrix}\Delta x\\ \Delta y\end{pmatrix} < 0,$$

so it must be that $f(x,y) < f(\bar{x},\bar{y})$. Similar reasoning applies to the case of a local minimum. ■

As mentioned in Remark 11.3, the conditions $f_{xx}(x,y) < 0$ and $f_{xx}(x,y)f_{yy}(x,y) - f_{xy}^2(x,y) > 0$ together imply that $f_{yy}(x,y) < 0$ so that the pair of conditions

$[f_{xx}(x,y) < 0$ and $f_{xx}(x,y)f_{yy}(x,y) - f_{xy}^2(x,y) > 0]$ are equivalent to $[f_{xx}(x,y) < 0$, $f_{yy}(x,y) < 0$ and $f_{xx}(x,y)f_{yy}(x,y) - f_{xy}^2(x,y) > 0]$. However, as a practical matter, it is often useful to check the simpler conditions $f_{xx}(x,y) < 0$ and $f_{yy}(x,y) < 0$ first: if either fails, then the condition $f_{xx}(x,y)f_{yy}(x,y) - f_{xy}^2(x,y) > 0$ cannot be satisfied.

REMARK 11.4: In fact, since the Taylor series expansion $f(x,y) = f(\bar{x},\bar{y}) + (\Delta x\, \Delta y)\boldsymbol{H}(x^*,y^*)\begin{pmatrix}\Delta x \\ \Delta y\end{pmatrix}$ holds generally (given any (x,y), and $(\bar{x},\bar{y})$ with $f_x(\bar{x},\bar{y}) = f_y(\bar{x},\bar{y}) = 0$, there is (x^*,y^*) such that this equation is satisfied). So, for example, if $\boldsymbol{H}(x,y)$ were negative definite for all (x,y), then $f(x,y) < f(\bar{x},\bar{y})$ for all $(x,y) \neq (\bar{x},\bar{y})$ and the first-order conditions give a global maximum. □

With a view to generalization to the many-variable case, the second-order conditions can be expressed directly in terms of submatrices of the Hessian matrix $\boldsymbol{H}$. Considering $\boldsymbol{H}$, write $\boldsymbol{H}_1$ for the matrix consisting of just the top left entry $\boldsymbol{H}_1 = (f_{xx})$, and $\boldsymbol{H}_2$ for the matrix $\boldsymbol{H}$. Because the determinant of a matrix with one row and column is just the single entry of that matrix (if $\boldsymbol{A} = [a]$, $|\boldsymbol{A}| = a$),

$$|\boldsymbol{H}_1| = f_{xx}$$

and

$$|\boldsymbol{H}_2| = f_{xx}f_{yy} - f_{xy}^2.$$

Consequently:

REMARK 11.5: The second-order conditions for a maximum are

$$|\boldsymbol{H}_1| < 0 \quad \text{and} \quad |\boldsymbol{H}_2| > 0.$$

The second-order conditions for a minimum are

$$|\boldsymbol{H}_1| > 0 \quad \text{and} \quad |\boldsymbol{H}_2| > 0.$$

In the case of two variables, the benefit from expressing the second-order conditions in terms of determinants derived from the Hessian matrix is not significant. However, with three or more variables, the matrix presentation is more succinct. (See Section 11.3.) □

EXAMPLE 11.4: Let $f(x,y) = -2x^2 - 7y^2 + xy + 12x + 37y$. Then, $f_x = -4x + y + 12$, $f_{xx} = -4$, $f_y = -14y + x + 37$, $f_{yy} = -14$, $f_{xy} = 1$. Solving for $f_x = 0$, $f_y = 0$

gives $x = \frac{41}{11}$, $y = \frac{32}{11}$.

$$H = \begin{pmatrix} -4 & 1 \\ 1 & -14 \end{pmatrix}, \quad |H_1| = -4 < 0, \quad |H_2| = \begin{vmatrix} -4 & 1 \\ 1 & -14 \end{vmatrix} = 55 > 0,$$

so the Hessian is negative definite. Consequently, the solution to the first-order conditions is a local maximum. ◇

The next example involves somewhat more complex computation.

EXAMPLE 11.5: Consider the problem:

$$\max f(x,y) = \sqrt{x}[1 + \ln(y+1)] - x - y^2.$$

A graph of the function is:

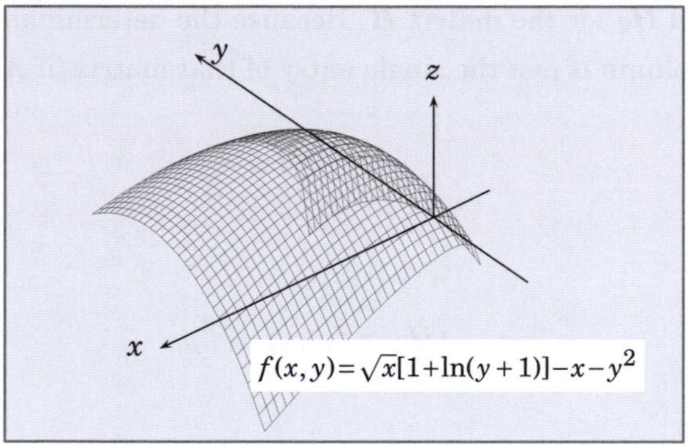

The first-order conditions are:

$$f_x = \frac{1}{2}(x)^{-\frac{1}{2}}[1 + \ln(y+1)] - 1 = 0.$$
$$f_y = \frac{\sqrt{x}}{(y+1)} - 2y = 0.$$

From the first-order conditions, $x^{\frac{1}{2}} = \frac{1}{2}[1 + \ln(y+1)]$. Substituting into the equation $f_y = 0$ gives $\frac{1}{2}[1 + \ln(y+1)] = 2y(y+1)$. Solving for y gives $y = 0.2448129$. Thus, $x = \frac{1}{4}[1 + \ln(1.2448129)]^2$. Next, consider the second-order conditions,

$$f_{xx} = -\frac{1}{4}(x)^{-\frac{3}{2}}[1 + \ln(y+1)] < 0$$
$$f_{yy} = -\frac{\sqrt{x}}{(y+1)^2} - 2$$
$$f_{xy} = \frac{1}{2}(x)^{-\frac{1}{2}}\frac{1}{(y+1)}.$$

Thus,

$$\begin{aligned}
f_{xx}f_{yy} - f_{xy}^2 &= \left(-\frac{1}{4}(x)^{-\frac{3}{2}}[1+\ln(y+1)]\right)\left[-\frac{\sqrt{x}}{(y+1)^2} - 2\right] - \left(\frac{1}{4}(x)^{-1}\frac{1}{(y+1)^2}\right) \\
&= \frac{1}{4x(y+1)^2}([1+\ln(y+1)]-1) + 2\left(\frac{1}{4}\right)x^{-\frac{3}{2}}(1+\ln(y+1)) \\
&= \frac{\ln(y+1)}{4x(y+1)^2} + 2\left(\frac{1}{4}\right)x^{-\frac{3}{2}}(1+\ln(y+1)) > 0.
\end{aligned}$$

Since $\ln(y+1) = \ln(1.2448129) > 0$, the solution is a local maximum. ◇

The final example in this section considers a problem with general parameter specification.

EXAMPLE 11.6: A firm uses two inputs, gas (x) and electricity (z), to heat a building during winter. The problem is to decide on the appropriate combinations of the inputs to minimize cost. The cost function is given by:

$$c(x,z) = px^{-\alpha} + \beta\frac{x}{f(z)} + \gamma z, \quad \alpha,\beta,\gamma > 0,$$

where $f(z)$ is an increasing function ($f'(z) > 0$). The first-order conditions are:

$$\begin{aligned}
c_x(x,z) &= -\alpha p x^{-\alpha-1} + \frac{\beta}{f(z)} = 0, \\
c_z(x,z) &= -\beta\frac{x}{f(z)^2}f'(z) + \gamma = 0.
\end{aligned}$$

And the second derivatives are given by:

$$\begin{aligned}
c_{xx}(x,z) &= \alpha(\alpha+1)px^{-\alpha-2} \\
c_{zz}(x,z) &= -\beta x\left[-2\frac{1}{f(z)^3}(f'(z))^2 + \frac{1}{f(z)^2}f''(z)\right] \\
c_{xz}(x,z) &= -\frac{\beta}{f(z)^2}f'(z).
\end{aligned}$$

Thus, the second-order conditions are:

$$\begin{aligned}
c_{xx} &= \alpha(\alpha+1)px^{-\alpha-2} > 0 \\
c_{zz} &= -\beta x\left[-2\frac{1}{f(z)^3}(f'(z))^2 + \frac{1}{f(z)^2}f''(z)\right] > 0
\end{aligned}$$

$$c_{xx}c_{zz} - c_{xz}^2 = -(\alpha(\alpha+1)px^{-\alpha-2})\left(\beta x\left[-2\frac{1}{f(z)^3}(f'(z))^2 + \frac{1}{f(z)^2}f''(z)\right]\right)$$
$$-\frac{\beta}{f(z)^2}^2(f'(z))^2 > 0.$$

These conditions are impossible to confirm without a specific functional form for f. Consider the case where $f(z) = e^z$, so that $f(z) = f'(z) = f''(z) = e^z$. Then $\frac{1}{f(z)^3}(f'(z))^2 = \frac{1}{f(z)} = \frac{1}{f(z)^2}f''(z)$, so that $c_{zz}(x,z) = -\beta x\left[-2\frac{1}{f(z)} + \frac{1}{f(z)}\right] = -\beta x\frac{1}{f(z)}(-1) = \frac{\beta x}{f(z)}$. Also, $c_{xz}(x,z) = -\frac{\beta}{f(z)^2}f'(z) = -\frac{\beta}{f(z)}$. Thus, the second-order conditions reduce to:

$$c_{xx} = \alpha(\alpha+1)px^{-\alpha-2} > 0$$
$$c_{zz} = \frac{\beta x}{f(z)} > 0$$
$$c_{xx}c_{zz} - c_{xz}^2 = [\alpha(\alpha+1)px^{-\alpha-1}]\left[\frac{\beta}{f(z)}\right] - \left(\frac{\beta}{f(z)}\right)^2 > 0.$$

The first-order condition gives $-\alpha px^{-\alpha-1} + \frac{\beta}{f(z)} = 0$ or $\alpha px^{-\alpha-1} = \frac{\beta}{f(z)}$. Substitute $\frac{\beta}{f(z)}$ for $-\alpha px^{-\alpha-1}$ in the last condition to obtain

$$\left[(\alpha+1)\frac{\beta}{f(z)}\right]\left[\frac{\beta}{f(z)}\right] - \left(\frac{\beta}{f(z)}\right)^2 = (\alpha)\left(\frac{\beta}{f(z)}\right)^2 > 0,$$

which verifies that the first-order condition is satisfied. To solve for x and z, recall the first-order conditions:

$$c_x(x,z) = -\alpha px^{-\alpha-1} + \frac{\beta}{f(z)} = 0, \quad c_z(x,z) = -\beta\frac{x}{f(z)} + \gamma = 0.$$

From $c_x(x,z) = 0$, $\alpha px^{-\alpha-1} = \frac{\beta}{f(z)}$, so $x^{-(\alpha+1)} = \frac{1}{\alpha p}\frac{\beta}{f(z)}$. Rearranging,

$$x = \left(\frac{1}{\alpha p}\right)^{\frac{-1}{(\alpha+1)}}\left(\frac{\beta}{f(z)}\right)^{\frac{-1}{(\alpha+1)}} = (\alpha p)^{\frac{1}{(\alpha+1)}}\left(\frac{f(z)}{\beta}\right)^{\frac{1}{(\alpha+1)}},$$

and from $c_z(x,z) = 0$, $\beta\frac{x}{f(z)} = \gamma$ so that, substituting for x,

$$\left(\frac{\beta}{f(z)}\right)\left(\frac{1}{\alpha p}\right)^{\frac{-1}{(\alpha+1)}}\left(\frac{\beta}{f(z)}\right)^{\frac{-1}{(\alpha+1)}} = \gamma.$$

Rearranging,

$$\gamma = \left(\frac{1}{\alpha p}\right)^{\frac{-1}{(\alpha+1)}}\left(\frac{\beta}{f(z)}\right)\left(\frac{\beta}{f(z)}\right)^{\frac{-1}{(\alpha+1)}} = \left(\frac{1}{\alpha p}\right)^{\frac{-1}{(\alpha+1)}}\left(\frac{\beta}{f(z)}\right)^{1+\frac{-1}{(\alpha+1)}} = \left(\frac{1}{\alpha p}\right)^{\frac{-1}{(\alpha+1)}}\left(\frac{\beta}{f(z)}\right)^{\frac{\alpha}{(\alpha+1)}}.$$

So,

$$\left(\frac{\beta}{f(z)}\right)^{\frac{\alpha}{(\alpha+1)}} = \gamma\left(\frac{1}{\alpha p}\right)^{\frac{1}{(\alpha+1)}} \quad \text{and then} \quad \left(\frac{\beta}{f(z)}\right) = \gamma^{\frac{\alpha+1}{\alpha}}\left(\frac{1}{\alpha p}\right)^{\frac{1}{\alpha}}.$$

Thus,

$$f(z) = \beta(\gamma)^{\frac{-(\alpha+1)}{\alpha}}\left(\frac{1}{\alpha p}\right)^{\frac{-1}{\alpha}} \quad \text{or} \quad z = f^{-1}\left(\beta(\gamma)^{\frac{-(\alpha+1)}{\alpha}}\left(\frac{1}{\alpha p}\right)^{\frac{-1}{\alpha}}\right).$$

When $f(z) = e^z$, $q = f(z)$ gives $z = f^{-1}(q) = \ln(q)$. Thus,

$$z = \ln\left[\beta(\gamma)^{\frac{-(\alpha+1)}{\alpha}}\left(\frac{1}{\alpha p}\right)^{\frac{-1}{\alpha}}\right].$$

◊

11.2.2 Failure of the second-order conditions

The second-order conditions confirm that at a candidate optimum, in a neighborhood of that point the function "bends down" in all directions in the case of maximization, and "bends up" in all directions in the case of minimization. There are a variety of ways in which this condition may fail. The following examples illustrate two common cases. Consider the function

$$f(x,y) = x^2 - y^2.$$

The function is depicted in Figure 11.2 and is sometimes called a saddle function in view of its shape.

As the figure indicates, the point $(0,0)$ satisfies the first-order conditions but from there variations in x raise f while variations in y reduce f. Observe that $f_x = 2x$, $f_y = -2y$, $f_{xx} = 2$, $f_{yy} = -2$ and $f_{xy} = 0$, so the Hessian matrix is:

$$\begin{pmatrix} 2 & 0 \\ 0 & -2 \end{pmatrix}.$$

Thus, $|\boldsymbol{H}_1| = 2$ and $|\boldsymbol{H}_2| = -2$ so that neither the conditions for a local maximum or a local minimum are satisfied.

As a second example, consider the function:

$$f(x,y) = x^2 y.$$

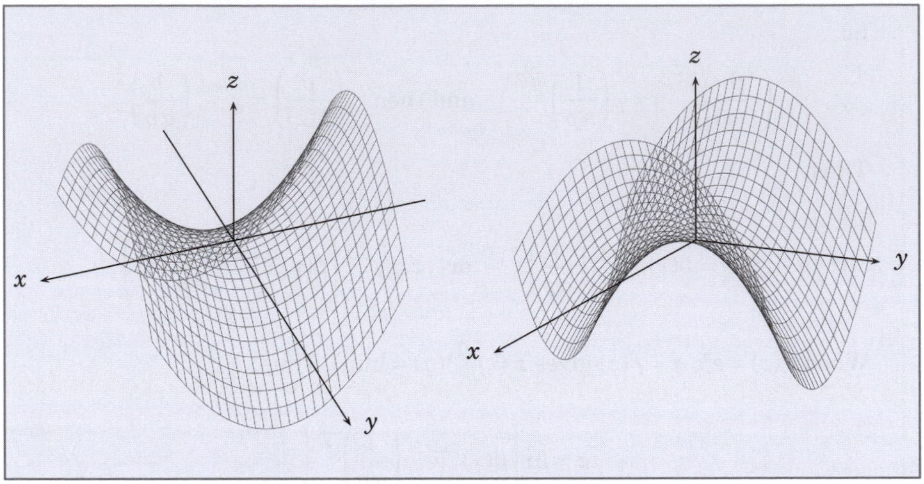

Figure 11.2: The function $f(x,y) = x^2 - y^2$

This function is plotted in Figure 11.3. Again, the point (0,0) satisfies the first-order conditions since $f_x = 2xy$, $f_y = x^2$, but the function twists at (0,0) . The Hessian matrix is:

$$\boldsymbol{H} = \begin{pmatrix} 2y & 2x \\ 2x & 0 \end{pmatrix}.$$

So $|\boldsymbol{H}_1| = 2y$ and $|\boldsymbol{H}_2| = -4x^2$. Both of these are 0 at $(x,y) = 0$.

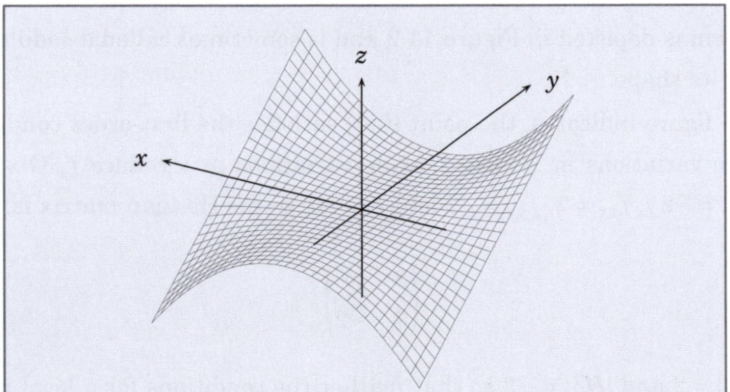

Figure 11.3: The function $f(x,y) = x^2 y$

11.2.3 The envelope theorem

Consider the problem of maximizing a function of x when the function depends on some exogenous parameter α : $f(x,\alpha)$. For example: $f(x,\alpha) = x^{\frac{1}{2}} - \alpha x$. The usual procedure is to set the first derivative to zero: $f_x(x,\alpha) = 0$. This then gives the solution $x^* = x(\alpha)$ as a function of α, chosen to satisfy $f_x(x(\alpha),\alpha) = 0$. Note that if α varies, so does $x(\alpha)$ to maintain the equality of the partial derivative with 0. As α varies, so will the optimal choice of x to maintain the identity $f_x(x(\alpha),\alpha) \equiv 0$. In the example, $f(x,\alpha) = x^{\frac{1}{2}} - \alpha x$ gives $f_x(x,\alpha) = \frac{1}{2}x^{-\frac{1}{2}} - \alpha = 0$. Rearranging, $\frac{1}{2\alpha} = x^{\frac{1}{2}}$ or $x(\alpha) = \left(\frac{1}{2\alpha}\right)^2$. If the optimal choice of x given α, $x(\alpha)$, is substituted into the function f, then $f(x(\alpha),\alpha)$ is the value of the function being maximized, at the solution. Note that $f(x(\alpha),\alpha)$ depends only on α; the value of the optimized function may be written $f^*(\alpha) \equiv f(x(\alpha),\alpha)$. In the example:

$$f^*(\alpha) = f(x(\alpha),\alpha) = \left[\left(\frac{1}{2\alpha}\right)^2\right]^{\frac{1}{2}} - \alpha\left(\frac{1}{2\alpha}\right)^2 = \left(\frac{1}{2\alpha}\right) - \left(\frac{1}{4\alpha}\right) = \left(\frac{1}{4\alpha}\right).$$

Now consider the following question. If α varies, how much will the value of the *maximized* function vary? That is, what is the value of $\frac{df^*(\alpha)}{d\alpha}$? Differentiating:

$$\frac{df^*(\alpha)}{d\alpha} = f_x(x(\alpha),\alpha)\frac{dx(\alpha)}{d\alpha} + f_\alpha(x(\alpha),\alpha).$$

Note now that from the first-order condition $f_x(x(\alpha),\alpha) = 0$ at each value of α, so that for each α,

$$\frac{df^*(\alpha)}{d\alpha} = f_\alpha(x(\alpha),\alpha).$$

In words, the change in the maximized value of the function, $f^*(\alpha)$, as the parameter (α) changes, is equal to the derivative of the (unoptimized) function, $f(x,\alpha)$, with respect to the parameter, with the derivative evaluated at the solution $(x(\alpha))$: $f_\alpha(x^*(\alpha),\alpha)$. In the example, $f^*(\alpha) = \left(\frac{1}{4\alpha}\right)$, so that

$$\frac{df^*(\alpha)}{d\alpha} = -\left(\frac{1}{4\alpha^2}\right) = -\left(\frac{1}{2\alpha}\right)^2 = -x(\alpha).$$

If we differentiate $f(x,\alpha) = x^{\frac{1}{2}} - \alpha x$ with respect to α and evaluate at the solution $x(\alpha)$, we get $-x(\alpha)$. To maximize $f(x,\alpha) = x^{\frac{1}{2}} - \alpha x$, the optimal choice of x is that which makes the gap between $x^{\frac{1}{2}}$ and αx as large as possible. In Figure 11.4, this is illustrated with the gap maximized at $x(\alpha)$ and with resulting value $x(\alpha)^{\frac{1}{2}} - \alpha x(\alpha) = a - b$.

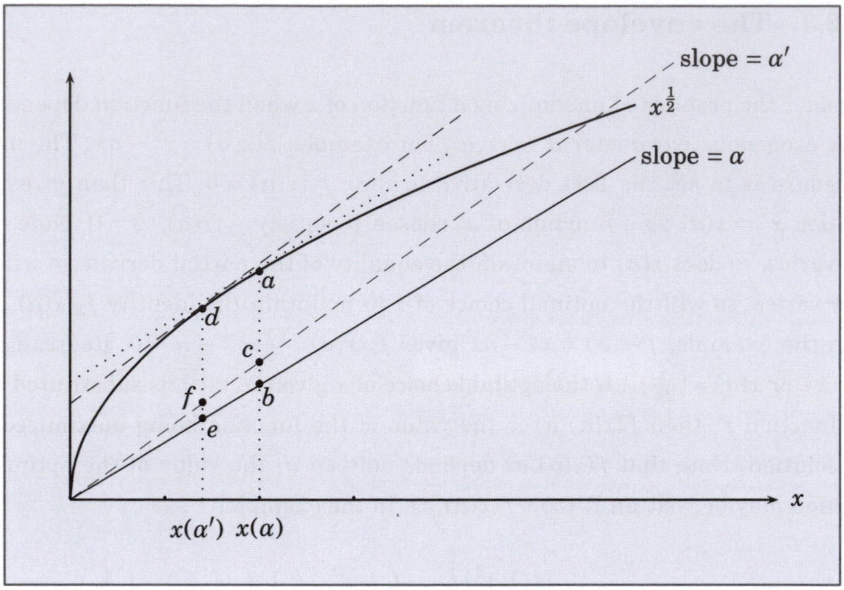

Figure 11.4: The impact of parameter variation on the objective value

In the figure, profit at α is $(a-b)$ and at α' profit is $(d-f)$. Observe that

$$\pi(\alpha') = d - f \approx (a-c) = (a-b) - (c-b) = \pi(\alpha) - (\alpha' - \alpha)x(\alpha),$$

so

$$\frac{\pi(\alpha') - \pi(\alpha)}{(\alpha' - \alpha)} \approx -x(\alpha).$$

This observation extends to the general case. Let $f(x_1, x_2, \cdots, x_n, \alpha_1, \alpha_2, \cdots, \alpha_m)$. For ease of notation write $x = (x_1, x_2, \cdots, x_n)$ and $\alpha = (\alpha_1, \alpha_2, \cdots, \alpha_m)$. The first-order conditions are $f_{x_i}(x, \alpha) = 0, i = 1, 2, \cdots, n$. Solving these gives $x(\alpha) = (x_1(\alpha), x_2(\alpha), \cdots, x_n(\alpha))$. Then the optimized value of the function is $f^*(\alpha) = f(x(\alpha), \alpha)$. To calculate the change in f^* due to a change in one α, say α_i, take the partial derivative of $f^*(\alpha)$ with respect to α_i. This gives

$$\frac{\partial f^*(\alpha)}{\partial \alpha_i} = \sum_{j=1}^{n} \frac{\partial f(x(\alpha), \alpha)}{\partial x_i} \frac{\partial x_i(\alpha)}{\partial \alpha_i} + \frac{\partial f(x(\alpha), \alpha)}{\partial \alpha_i}.$$

Note now that from the first-order conditions, the partial derivative of $f(x, \alpha)$ evaluated at $x = x(\alpha)$ is equal to 0 for each $j = 1, 2, \cdots, n$: $\frac{\partial f(x(\alpha), \alpha)}{\partial x_j} = 0$, $\forall j$. Thus,

$$\frac{\partial f^*(\alpha)}{\partial \alpha_i} = \frac{\partial f(x(\alpha), \alpha)}{\partial \alpha_i}.$$

The following examples illustrate this result.

EXAMPLE 11.7: The profit function of a firm depends on output Q and advertising A. Profit is given by

$$\pi(Q,A) = 500 - 4Q^2 - 8Q + 6AQ - 10A^2 + 100A.$$

So, the first-order conditions are:

$$\begin{aligned}\pi_Q &= -8Q + 6A - 8 \quad = 0\\ \pi_A &= 6Q - 20A + 100 = 0.\end{aligned}$$

Thus,

$$\begin{pmatrix}-8 & 6\\ 6 & -20\end{pmatrix}\begin{pmatrix}Q\\ A\end{pmatrix} = \begin{pmatrix}8\\ -100\end{pmatrix}.$$

Solving these gives $Q = \frac{110}{31}$ and $A = \frac{188}{31}$. Since $\pi_{QQ} = -8$, $\pi_{AA} = -20$ and $\pi_{QA} = 6$, $\pi_{QQ}\pi_{AA} - \pi_{QA}^2 = 160 - 36 = 124$. So, the second-order conditions for a maximum are satisfied. ◊

Example 11.8 develops Example 11.7 to illustrate the envelope theorem.

EXAMPLE 11.8: Suppose that the profit function is now $\pi(Q,A) = 500 - \alpha Q^2 - \beta Q + \gamma AQ - 5A^2 + 50A$, where α, β and γ are parameters. In this case, the first-order conditions become:

$$\pi_Q = -2\alpha Q - \beta + \gamma A = 0, \quad \pi_A = \gamma Q - 10A + 50 = 0.$$

Rearranging:

$$\begin{pmatrix}-2\alpha & \gamma\\ \gamma & -10\end{pmatrix}\begin{pmatrix}Q\\ A\end{pmatrix} = \begin{pmatrix}\beta\\ -50\end{pmatrix}.$$

Using Cramer's rule:

$$\Delta = \begin{vmatrix}-2\alpha & \gamma\\ \gamma & -10\end{vmatrix} = 20\alpha - \gamma^2,$$

$$\Delta_1 = \begin{vmatrix}\beta & \gamma\\ -50 & -10\end{vmatrix} = -10\beta + 50\gamma, \quad \Delta_2 = \begin{vmatrix}-2\alpha & \beta\\ \gamma & -50\end{vmatrix} = 100\alpha - \gamma\beta.$$

Thus,

$$Q(\alpha,\beta,\gamma)=\frac{\Delta_1}{\Delta}=\frac{-10\beta+50\gamma}{20\alpha-\gamma^2},\quad A(\alpha,\beta,\gamma)=\frac{\Delta_2}{\Delta}=\frac{100\alpha-\gamma\beta}{20\alpha-\gamma^2}.$$

For the second-order conditions:

$$\pi_{QQ}=-2\alpha<0,\quad \text{and}\quad \pi_{QQ}\pi_{AA}-\pi_{QA}^2=20\alpha-\gamma^2>0\ \left(\text{if } \alpha>\frac{\gamma^2}{20}\right).$$

Finally, using the envelope theorem, the impact of parameter changes on the value of the objective may be calculated:

$$\frac{\partial\pi^*}{\partial\alpha}=-Q(\alpha,\beta,\gamma)^2,\quad \frac{\partial\pi^*}{\partial\beta}=-Q(\alpha,\beta,\gamma),\quad \frac{\partial\pi^*}{\partial\gamma}=Q(\alpha,\beta,\gamma)A(\alpha,\beta,\gamma).$$

For example, $\frac{\partial\pi^*}{\partial\alpha}$ is the change in the firm's profit (at the optimal solution) due to a change in the parameter α. ◇

The next example considers a standard economic model: profit maximization by a price-taking firm with two inputs. Profit maximization leads to optimal choices, which in turn determine the firm's profit. How does this vary as input prices vary? The envelope theorem provides a simple way to answer this question. In addition, the example illustrates the impact of parameter changes on the choice variables.

EXAMPLE 11.9: A firm produces output, Q, with two inputs, x and y, according to the production function $Q=x^\alpha y^\alpha$, $0<\alpha<\frac{1}{2}$. The cost of input x is $r>0$ and the cost of input y is $w>0$. The output Q sells at unit price p.

The following discussion considers the problem of profit maximization and consideration of the impact of parameter changes on the optimal values of the choice variables and the level of profit at the solution. The profit function is

$$\pi(x,y)=px^\alpha y^\alpha-rx-wy.$$

Maximizing this determines solution values for x and y, say x^* and y^*, which depend on the parameters (p,r,w,α), $x^*=x(p,\alpha,r,w)$ and $y^*=y(p,\alpha,r,w)$. Substituting these into the profit function gives the level of profit at the optimal choices of (x,y):

$$\pi^*(p,w,r)=p(x^*)^\alpha(y^*)^\alpha-rx^*-wy^*,$$

with x^* and y^* functions of (p,α,r,w).

How does *maximized* profit vary with a small change in r? According to the envelope theorem, the answer may be obtained by differentiating the unoptimized profit function with respect to r and evaluating the derivative at the solution. Here, the derivative of the unoptimized function with respect to r is $-x$ and evaluating at the solution gives $-x^*$. (Likewise, the impact of an output price change is obtained by differentiating the unoptimized function with respect to p and evaluating the derivative at the solution.) The following computations confirm these observations.

SOLVING FOR THE OPTIMAL SOLUTION

The first-order conditions for profit maximization are:

$$\begin{aligned}\pi_x(x,y) &= \alpha p x^{\alpha-1} y^{\alpha} - r = \alpha p Q x^{-1} - r = 0\\ \pi_y(x,y) &= \alpha p x^{\alpha} y^{\alpha-1} - w = \alpha p Q y^{-1} - w = 0,\end{aligned}$$

where $Q = x^{\alpha} y^{\alpha}$. Therefore, $rx = wy$ or $x = \frac{w}{r} y$, and

$$\begin{aligned}0 &= \alpha p x^{\alpha-1} y^{\alpha} - r\\ &= \alpha p \left(\frac{w}{r} y\right)^{\alpha-1} y^{\alpha} - r\\ &= \alpha p \left(\frac{w}{r}\right)^{\alpha-1} y^{\alpha-1} y^{\alpha} - r\\ &= \alpha p \left(\frac{w}{r}\right)^{\alpha-1} y^{2\alpha-1} - r.\end{aligned}$$

Therefore, $\frac{\alpha p}{r}\left(\frac{w}{r}\right)^{\alpha-1} = y^{1-2\alpha}$, so that

$$y = \frac{\alpha p}{r}^{\frac{1}{1-2\alpha}} \left(\frac{w}{r}\right)^{\frac{\alpha-1}{1-2\alpha}}$$

$$y = (\alpha p)^{\frac{1}{1-2\alpha}} w^{\frac{\alpha-1}{1-2\alpha}} r^{\frac{-\alpha}{1-2\alpha}}. \tag{11.10}$$

Similarly,

$$x = (\alpha p)^{\frac{1}{1-2\alpha}} r^{\frac{\alpha-1}{1-2\alpha}} w^{\frac{-\alpha}{1-2\alpha}}. \tag{11.11}$$

The second-order conditions are satisfied, since:

$$\begin{aligned}\pi_{xx} &= \alpha(\alpha-1) p Q x^{-2} < 0\\ \pi_{yy} &= \alpha(\alpha-1) p Q y^{-2} < 0\end{aligned}$$

$$\pi_{xx}\pi_{yy} - \pi_{xy}^2 = \alpha^2(\alpha-1)^2 p^2 \frac{Q^2}{y^2x^2} - \alpha^2\alpha^2 p^2 \frac{Q^2}{x^2y^2} = \alpha^2 p^2 \frac{Q^2}{y^2x^2}[(\alpha-1)^2 - \alpha^2] > 0,$$

using $\pi_{xy} = \alpha\alpha pQx^{-1}y^{-1}$ and the fact that $[(\alpha-1)^2-\alpha^2] = [\alpha^2-2\alpha+1-\alpha^2] = (1-2\alpha)$.

THE IMPACT OF PARAMETER CHANGES ON CHOICE VARIABLES

From the solutions, the impact of parameter changes on the optimal values can be computed directly. For example:

$$\begin{aligned}\frac{\partial x}{\partial p} &= \left(\frac{1}{1-2\alpha}\right)(\alpha p)^{\frac{1}{1-2\alpha}-1} r^{\frac{\alpha-1}{1-2\alpha}} w^{\frac{-\alpha}{1-2\alpha}} \\ &= \left(\frac{1}{1-2\alpha}\right)\left(\frac{1}{p\alpha}\right)x\end{aligned}$$

and

$$\begin{aligned}\frac{\partial y}{\partial r} &= \frac{-\alpha}{1-2\alpha}(\alpha p)^{\frac{1}{1-2\alpha}} w^{\frac{\alpha-1}{1-2\alpha}} r^{\frac{-\alpha}{1-2\alpha}-1} \\ &= \left(\frac{-\alpha}{1-2\alpha}\right)\left(\frac{1}{r}\right)y.\end{aligned}$$

THE IMPACT OF PARAMETER CHANGES ON PROFIT (ENVELOPE CALCULATIONS) At the optimal solution, the profit is $\pi^*(p,\alpha,r,w) = \pi(x(p,\alpha,r,w),\ y(p,\alpha,r,w))$. According to the envelope theorem, the impact of a parameter change on the objective may be computed by differentiating the unoptimized objective with respect to the parameter and evaluating the derivative at the solution. In the present context, for example:

$$\begin{aligned}\frac{\partial \pi^*}{\partial r} &= \frac{\partial \pi(x(p,\alpha,r,w), y(p,\alpha,r,w))}{\partial r} = -x(p,\alpha,r,w) \\ \frac{\partial \pi^*}{\partial w} &= \frac{\partial \pi(x(p,\alpha,r,w), y(p,\alpha,r,w))}{\partial w} = -y(p,\alpha,r,w).\end{aligned}$$

The following discussion confirms this. Recall that

$$\begin{aligned}\pi(x(p,\alpha,r,w), y(p,\alpha,r,w)) = px(p,\alpha,r,w)^{\alpha} y(p,\alpha,r,w)^{\alpha} \\ -rx(p,\alpha,r,w) - wy(p,\alpha,r,w).\end{aligned}$$

Then

$$\begin{aligned}&\frac{\partial \pi(x(p,\alpha,r,w), y(p,\alpha,r,w))}{\partial r} \\ &= \alpha px(p,\alpha,r,w)^{\alpha-1} y(p,\alpha,r,w)^{\alpha}\frac{\partial x(p,\alpha,r,w)}{\partial r} \\ &\quad + \alpha px(p,\alpha,r,w)^{\alpha} y(p,\alpha,r,w)^{\alpha-1}\frac{\partial y(p,\alpha,r,w)}{\partial r}\end{aligned}$$

$$
\begin{aligned}
&-r\frac{\partial x(p,\alpha,r,w)}{\partial r} - w\frac{\partial y(p,\alpha,r,w)}{\partial r} - x(p,\alpha,r,w)\\
&= [\alpha p x(p,\alpha,r,w)^{\alpha-1} y(p,\alpha,r,w)^{\alpha} - r]\frac{\partial x(p,\alpha,r,w)}{\partial r}\\
&\quad + [\alpha p x(p,\alpha,r,w)^{\alpha} y(p,\alpha,r,w)^{\alpha-1} - w]\frac{\partial y(p,\alpha,r,w)}{\partial r} - x(p,\alpha,r,w).
\end{aligned}
$$

We know that $x(p,\alpha,r,w)$ and $y(p,\alpha,r,w)$ were chosen to satisfy the first-order conditions, so that

$$
\begin{aligned}
&[\alpha p x(p,\alpha,r,w)^{\alpha-1} y(p,\alpha,r,w)^{\alpha} - r] = 0,\\
&[\alpha p x(p,\alpha,r,w)^{\alpha} y(p,\alpha,r,w)^{\alpha-1} - w] = 0.
\end{aligned}
$$

Therefore,

$$\frac{\partial \pi(x(p,\alpha,r,w), y(p,\alpha,r,w))}{\partial r} = -x(p,\alpha,r,w).$$

So, the derivative of the optimized function with respect to the parameter r is equal to the derivative of the unoptimized function with respect to the parameter, with the derivative evaluated at the optimal solution. Similarly,

$$\frac{\partial \pi(x(p,\alpha,r,w), y(p,\alpha,r,w))}{\partial w} = -y(p,\alpha,r,w).$$

These calculations are closely related to dual prices in linear programming. The partial derivative of $\pi(x(p,\alpha,r,w), y(p,\alpha,r,w))$ with respect to r measures the change in the value of the objective at the optimal solution due to a change in the cost of one of the inputs (x).

THE IMPACT OF PARAMETER CHANGES ON PROFIT (DIRECT CALCULATIONS)

Finally, the optimized function can be differentiated directly. These tedious calculations confirm the result obtained directly from the envelope theorem. Observe, using Equations 11.11 and 11.10, $x = (\alpha p)^{\frac{1}{1-2\alpha}} r^{\frac{\alpha-1}{1-2\alpha}} w^{\frac{-\alpha}{1-2\alpha}}$, and $y = (\alpha p)^{\frac{1}{1-2\alpha}} w^{\frac{\alpha-1}{1-2\alpha}} r^{\frac{-\alpha}{1-2\alpha}}$, so:

$$x^{\alpha} = (\alpha p)^{\frac{\alpha}{1-2\alpha}} r^{\frac{(\alpha-1)\alpha}{1-2\alpha}} w^{\frac{-\alpha^2}{1-2\alpha}} \tag{11.12}$$

$$y^{\alpha} = (\alpha p)^{\frac{\alpha}{1-2\alpha}} w^{\frac{(\alpha-1)\alpha}{1-2\alpha}} r^{\frac{-\alpha^2}{1-2\alpha}}. \tag{11.13}$$

Because $\frac{(\alpha-1)\alpha}{1-2\alpha} + \frac{-\alpha^2}{1-2\alpha} = \frac{-\alpha}{1-2\alpha}$,

$$
\begin{aligned}
x^{\alpha} y^{\alpha} &= (\alpha p)^{\frac{2\alpha}{1-2\alpha}} r^{\frac{-\alpha}{1-2\alpha}} w^{\frac{-\alpha}{1-2\alpha}}\\
&= \alpha^{\frac{2\alpha}{1-2\alpha}} p^{\frac{2\alpha}{1-2\alpha}} r^{\frac{-\alpha}{1-2\alpha}} w^{\frac{-\alpha}{1-2\alpha}}
\end{aligned}
$$

and

$$px^{\alpha}y^{\alpha} = \alpha^{\frac{2\alpha}{1-2\alpha}} p^{\frac{1}{1-2\alpha}} r^{\frac{-\alpha}{1-2\alpha}} w^{\frac{-\alpha}{1-2\alpha}}$$

since $\frac{2\alpha}{1-2\alpha} + 1 = \frac{2\alpha+(1-2\alpha)}{1-2\alpha} = \frac{1}{1-2\alpha}$.

From the solution values of x and y,

$$x = (\alpha p)^{\frac{1}{1-2\alpha}} r^{\frac{\alpha-1}{1-2\alpha}} w^{\frac{-\alpha}{1-2\alpha}}$$

$$y = (\alpha p)^{\frac{1}{1-2\alpha}} w^{\frac{\alpha-1}{1-2\alpha}} r^{\frac{-\alpha}{1-2\alpha}}$$

and

$$rx = (\alpha p)^{\frac{1}{1-2\alpha}} r^{(\frac{\alpha-1}{1-2\alpha}+1)} w^{\frac{-\alpha}{1-2\alpha}} = \alpha^{\frac{1}{1-2\alpha}} p^{\frac{1}{1-2\alpha}} r^{\frac{-\alpha}{1-2\alpha}} w^{\frac{-\alpha}{1-2\alpha}}$$

$$wy = (\alpha p)^{\frac{1}{1-2\alpha}} w^{(\frac{\alpha-1}{1-2\alpha}+1)} r^{\frac{-\alpha}{1-2\alpha}} = \alpha^{\frac{1}{1-2\alpha}} p^{\frac{1}{1-2\alpha}} w^{\frac{-\alpha}{1-2\alpha}} r^{\frac{-\alpha}{1-2\alpha}}.$$

Thus,

$$rx + wy = 2\alpha^{\frac{1}{1-2\alpha}} p^{\frac{1}{1-2\alpha}} w^{\frac{-\alpha}{1-2\alpha}} r^{\frac{-\alpha}{1-2\alpha}}.$$

So, $\pi^* = p(x^*)^{\alpha}(y^*)^{\alpha} - (rx^* + wy^*)$,

$$\begin{aligned} p(x^*)^{\alpha}(y^*)^{\alpha} - (rx^* + wy^*) &= \alpha^{\frac{2\alpha}{1-2\alpha}} p^{\frac{1}{1-2\alpha}} r^{\frac{-\alpha}{1-2\alpha}} w^{\frac{-\alpha}{1-2\alpha}} - 2\alpha^{\frac{1}{1-2\alpha}} p^{\frac{1}{1-2\alpha}} w^{\frac{-\alpha}{1-2\alpha}} r^{\frac{-\alpha}{1-2\alpha}} \\ &= \left[\alpha^{\frac{2\alpha}{1-2\alpha}} - 2\alpha^{\frac{1}{1-2\alpha}}\right] p^{\frac{1}{1-2\alpha}} r^{\frac{-\alpha}{1-2\alpha}} w^{\frac{-\alpha}{1-2\alpha}} \\ &= \left[\alpha^{\frac{1}{1-2\alpha}} \alpha^{-1} - 2\alpha^{\frac{1}{1-2\alpha}}\right] p^{\frac{1}{1-2\alpha}} r^{\frac{-\alpha}{1-2\alpha}} w^{\frac{-\alpha}{1-2\alpha}} \\ &= \alpha^{\frac{1}{1-2\alpha}} \left[\alpha^{-1} - 2\right] p^{\frac{1}{1-2\alpha}} r^{\frac{-\alpha}{1-2\alpha}} w^{\frac{-\alpha}{1-2\alpha}} \\ &= \alpha^{\frac{1}{1-2\alpha}} \alpha^{-1}[1-2\alpha] p^{\frac{1}{1-2\alpha}} r^{\frac{-\alpha}{1-2\alpha}} w^{\frac{-\alpha}{1-2\alpha}}, \end{aligned}$$

using the fact that $\frac{2\alpha}{1-2\alpha} = \frac{1}{1-2\alpha} - 1$. Differentiating with respect to r gives:

$$\begin{aligned} \frac{\partial \pi^*}{\partial r} &= \left[\frac{-\alpha}{1-2\alpha}\right] \alpha^{\frac{1}{1-2\alpha}} \alpha^{-1}[1-2\alpha] p^{\frac{1}{1-2\alpha}} r^{\frac{-\alpha}{1-2\alpha}-1} w^{\frac{-\alpha}{1-2\alpha}} \\ &= -\alpha^{\frac{1}{1-2\alpha}} p^{\frac{1}{1-2\alpha}} r^{\frac{\alpha-1}{1-2\alpha}} w^{\frac{-\alpha}{1-2\alpha}} \\ &= -x^*, \end{aligned}$$

using $\frac{-\alpha}{1-2\alpha} - 1 = \frac{-\alpha-(1-2\alpha)}{1-2\alpha} = \frac{\alpha-1}{1-2\alpha}$ and $\left[\frac{-\alpha}{1-2\alpha}\right] \alpha^{-1}[1-2\alpha] = -1$. ◊

11.3 Optimization: *n* variables

Consider the problem of maximizing or minimizing a function f: $\max f(x_1,\ldots,x_n)$ or $\min f(x_1,\ldots,x_n)$. For convenience, write $f(x)$, where $x=(x_1,\ldots,x_n)$. The first derivatives of f are $f_i=\frac{\partial f}{\partial x_i}, i=1,\ldots,n$. Set $\nabla f=(f_1,f_2,\ldots,f_n)$. If it is the case that at a point x^*, the partial derivative of f with respect to x_i is not equal to 0, say $f_i(x^*)=\frac{\partial f(x^*)}{\partial x_i}>0$, this means that at the point x^* the value of f is increased if x_i is increased from x_i^*. This implies that for x^* to be a local maximum of the function f, it must be the case that every partial derivative is 0. A *necessary* condition for x^* to be a local maximum is that partial derivatives at x^* are 0:

$$\frac{\partial f(x^*)}{\partial x_i}=0,\ i=1,\ldots,n.$$

Under what conditions can one be sure that x^* is a local maximum? In other words, what are necessary and sufficient conditions for x^* to be a local maximum? Let $\boldsymbol{H}(p)$ be the *Hessian* matrix evaluated at p:

$$\boldsymbol{H}=\left\{\frac{\partial^2 f}{\partial x_i\partial x_j}\right\}_{ij}=\begin{pmatrix}\frac{\partial^2 f}{\partial^2 x_1} & \frac{\partial^2 f}{\partial x_1\partial x_2} & \cdots & \frac{\partial^2 f}{\partial x_1\partial x_n}\\ \frac{\partial^2 f}{\partial x_2\partial x_1} & \frac{\partial^2 f}{\partial^2 x_2} & \cdots & \frac{\partial^2 f}{\partial x_2\partial x_n}\\ \vdots & \vdots & \ddots & \vdots\\ \frac{\partial^2 f}{\partial x_n\partial x_1} & \frac{\partial^2 f}{\partial x_n\partial x_2} & \cdots & \frac{\partial^2 f}{\partial x_n\partial x_n}\end{pmatrix}=\begin{pmatrix}f_{11} & f_{12} & \cdots & f_{1n}\\ f_{21} & f_{22} & \cdots & f_{2n}\\ \vdots & \vdots & \ddots & \vdots\\ f_{n1} & f_{n2} & \cdots & f_{nn}\end{pmatrix}.$$

In general, $\boldsymbol{H}$ will depend on x. This dependence is made explicit by writing $\boldsymbol{H}(x)$. Determining if $\boldsymbol{H}$ is positive or negative definite is possible by checking a set of determinental conditions. Define $\boldsymbol{H}_j$ to be the $j\times j$ matrix obtained from $\boldsymbol{H}$ by deleting the last $n-j$ rows and columns:

$$\boldsymbol{H}_j=\begin{pmatrix}f_{11} & f_{12} & \cdots & f_{1j}\\ f_{21} & f_{22} & \cdots & f_{2j}\\ \vdots & \vdots & \ddots & \vdots\\ f_{j1} & f_{j2} & \cdots & f_{jj}\end{pmatrix}.$$

Theorem 11.2: *Suppose that x^* satisfies:*

$$\frac{\partial f(x^*)}{\partial x_i}=0,\ i=1,\ldots,n.$$

If

$$|\boldsymbol{H}_1| < 0,\ |\boldsymbol{H}_2| > 0,\ |\boldsymbol{H}_3| < 0,\ \ldots, \quad \text{or} \quad (-1)^j |\boldsymbol{H}_j| > 0,\ j = 1,\ldots,n,$$

then x^* is a local maximum.

If

$$|\boldsymbol{H}_1| > 0, |\boldsymbol{H}_2| > 0, |\boldsymbol{H}_3| > 0, \ldots, \quad \text{or} \quad |\boldsymbol{H}_j| > 0,\ j = 1,\ldots,n,$$

then x^* is a local minimum.

The following example illustrates the calculations.

EXAMPLE 11.10: Let $f(x,y,z) = 3x^2 + 12y^2 + 5z^2 - 13x + 12y + 2xz - 3yz$.

Then: $f_x = 6x - 13 + 2z$, $f_{xx} = 6$, $f_{xy} = 0$, $f_{xz} = 2$, $f_y = 24y + 12 - 3z$, $f_{yy} = 24$, $f_{yz} = -3$, $f_z = 10z + 2x - 3y$, $f_{zz} = 10$, and

$$\boldsymbol{H} = \begin{pmatrix} 6 & 0 & 2 \\ 0 & 24 & -3 \\ 2 & -3 & 10 \end{pmatrix},$$

The first-order conditions solve to give: $x = \frac{205}{86}$, $y = -\frac{25}{43}$ and $z = -\frac{28}{43}$.

$$|\boldsymbol{H}_1| = 6 > 0, \quad |\boldsymbol{H}_2| = \begin{vmatrix} 6 & 0 \\ 0 & 24 \end{vmatrix} = 144 > 0, \quad |\boldsymbol{H}_3| = \begin{vmatrix} 6 & 0 & 2 \\ 0 & 24 & -3 \\ 2 & -3 & 10 \end{vmatrix} = 1290 > 0,$$

so the Hessian is positive definite. ◊

The next example is the standard profit-maximization, now with three inputs.

EXAMPLE 11.11: Consider the production function $q = f(x,y,z) = x^{\alpha} y^{\beta} z^{\gamma}$. A price-taking firm must maximize $\pi(x,y,z) = pf(x,y,z) - rx - sy - tz$. The first-order conditions are $\pi_x = pf_x - r = 0$, $\pi_y = pf_y - s = 0$ and $\pi_z = pf_z - t = 0$. The first- and second-order partial derivatives are:

$$\begin{aligned} f_x &= \alpha x^{\alpha-1} y^{\beta} z^{\gamma} \\ f_y &= \beta x^{\alpha} y^{\beta-1} z^{\gamma} \\ f_z &= \gamma x^{\alpha} y^{\beta} z^{\gamma-1} \end{aligned}$$

and

$$f_{xx} = \alpha(\alpha-1)x^{\alpha-2}y^{\beta}z^{\gamma} = \alpha(\alpha-1)\frac{q}{x^2}$$
$$f_{yy} = \beta(\beta-1)x^{\alpha}y^{\beta-2}z^{\gamma} = \beta(\beta-1)\frac{q}{y^2}$$
$$f_{zz} = \gamma(\gamma-1)x^{\alpha}y^{\beta}z^{\gamma-2} = \gamma(\gamma-1)\frac{q}{z^2}$$
$$f_{xy} = \alpha\beta x^{\alpha-1}y^{\beta-1}z^{\gamma} = \alpha\beta\frac{q}{xy}$$
$$f_{xz} = \alpha\gamma x^{\alpha-1}y^{\beta}z^{\gamma-1} = \alpha\gamma\frac{q}{xz}$$
$$f_{yz} = \beta\gamma x^{\alpha}y^{\beta-1}z^{\gamma-1} = \beta\gamma\frac{q}{yz}.$$

From these calculations, the Hessian is:

$$\boldsymbol{H} = \begin{pmatrix} \pi_{xx} & \pi_{xy} & \pi_{xz} \\ \pi_{yx} & \pi_{yy} & \pi_{yz} \\ \pi_{zx} & \pi_{zy} & \pi_{zz} \end{pmatrix} = \begin{pmatrix} f_{xx} & f_{xy} & f_{xz} \\ f_{yx} & f_{yy} & f_{yz} \\ f_{zx} & f_{zy} & f_{zz} \end{pmatrix}$$
$$= \begin{pmatrix} \alpha(\alpha-1)\frac{q}{x^2} & \alpha\beta\frac{q}{xy} & \alpha\gamma\frac{q}{xz} \\ \alpha\beta\frac{q}{xy} & \beta(\beta-1)\frac{q}{y^2} & \beta\gamma\frac{q}{yz} \\ \alpha\gamma\frac{q}{xz} & \beta\gamma\frac{q}{yz} & \gamma(\gamma-1)\frac{q}{z^2} \end{pmatrix}.$$

From this, $\boldsymbol{H}_1 = \left(\alpha(\alpha-1)\frac{q}{x^2}\right)$, $\boldsymbol{H}_2 = \begin{pmatrix} \alpha(\alpha-1)\frac{q}{x^2} & \alpha\beta\frac{q}{xy} \\ \alpha\beta\frac{q}{xy} & \beta(\beta-1)\frac{q}{y^2} \end{pmatrix}$ and $\boldsymbol{H}_3 = \boldsymbol{H}$.

Thus, $|\boldsymbol{H}_1| = \alpha(\alpha-1)\frac{q}{x^2} < 0$. Also,

$$|\boldsymbol{H}_2| = \alpha(\alpha-1)\beta(\beta-1)\frac{q}{x^2}\frac{q}{y^2} - \alpha^2\beta^2\left(\frac{q}{xy}\right)^2$$
$$= [\alpha(\alpha-1)\beta(\beta-1) - \alpha^2\beta^2]\frac{q}{x^2}\frac{q}{y^2}$$
$$= \alpha\beta[1-\alpha-\beta]\frac{q}{x^2}\frac{q}{y^2}.$$

Since $1-\alpha-\beta > 0$, this expression is positive. Finally,

$$|\boldsymbol{H}_3| = \alpha(\alpha-1)\frac{q}{x^2}\begin{vmatrix} \beta(\beta-1)\frac{q}{y^2} & \beta\gamma\frac{q}{yz} \\ \beta\gamma\frac{q}{yz} & \gamma(\gamma-1)\frac{q}{z^2} \end{vmatrix} - \alpha\beta\frac{q}{xy}\begin{vmatrix} \alpha\beta\frac{q}{xy} & \beta\gamma\frac{q}{yz} \\ \alpha\gamma\frac{q}{xz} & \gamma(\gamma-1)\frac{q}{z^2} \end{vmatrix}$$
$$+ \alpha\gamma\frac{q}{xz}\begin{vmatrix} \alpha\beta\frac{q}{xy} & \beta(\beta-1)\frac{q}{y^2} \\ \alpha\gamma\frac{q}{xz} & \beta\gamma\frac{q}{yz} \end{vmatrix}$$

$$= \alpha(\alpha-1)\frac{q}{x^2}\left[\beta(\beta-1)\gamma(\gamma-1)-\beta^2\gamma^2\right]\frac{q}{y^2z^2}$$
$$-\alpha\beta\frac{q}{xy}\left[\alpha\beta\gamma(\gamma-1)-\alpha\gamma\beta\gamma\right]\frac{q}{xyz^2}$$
$$+\alpha\gamma\frac{q}{xz}[\alpha\beta\beta\gamma-\alpha\gamma\beta(\beta-1)]\frac{q}{xy^2z}$$
$$= -\alpha\beta\gamma(1-\alpha-\beta-\gamma)\frac{q^2}{x^2y^2z^2},$$

so $|\boldsymbol{H}_3| < 0$, since $(1-\alpha-\beta-\gamma) > 0$. ◊

11.3.1 Motivation for the Hessian conditions

To motivate the previous discussion, consider a Taylor series expansion of f. The matrix version of Taylor's theorem gives:

$$f(x) = f(x^*) + \nabla f(x^*)\cdot(x-x^*) + (\boldsymbol{x}-\boldsymbol{x}^*)'\boldsymbol{H}(p)(\boldsymbol{x}-\boldsymbol{x}^*),$$

where p is between x and x^*. If x^* satisfies: $f_i(x^*) = 0, \forall i$, then

$$f(\tilde{x}) = f(x^*) + (\tilde{\boldsymbol{x}}-\boldsymbol{x}^*)'\boldsymbol{H}(p)(\tilde{\boldsymbol{x}}-\boldsymbol{x}^*). \tag{11.14}$$

From this expression, whether $f(\tilde{x})$ is larger or smaller than $f(x^*)$ depends entirely on $\boldsymbol{H}(p)$. If, for all vectors, $\boldsymbol{z} \neq 0$, $\boldsymbol{z}'\boldsymbol{H}(p)\boldsymbol{z} < 0$, then $\boldsymbol{H}(p)$ is negative definite. From Equation 11.14 this implies $f(\tilde{x}) < f(x^*)$. If, for all vectors, $\boldsymbol{z} \neq 0$, $\boldsymbol{z}'\boldsymbol{H}(p)\boldsymbol{z} > 0$ the matrix $\boldsymbol{H}(p)$ is positive definite and Equation 11.14 implies $f(\tilde{x}) > f(x^*)$. So the sufficiency conditions depend on properties of the Hessian matrix $\boldsymbol{H}$, and the associated quadratic form, characterized in terms of positive and negative definiteness. Theorem 11.3 gives criteria for negative and positive definiteness.

Theorem 11.3: *The matrix $\boldsymbol{H}(x)$ is negative definite if:*

$$|\boldsymbol{H}_1(x)| < 0, |\boldsymbol{H}_2(x)| > 0, |\boldsymbol{H}_3(x)| < 0, \ldots, \text{ or } (-1)^j|\boldsymbol{H}_j(x)| > 0, \text{ for } j = 1, \ldots, n.$$

The matrix $\boldsymbol{H}(x)$ is positive definite if:

$$|\boldsymbol{H}_1(x)| > 0, |\boldsymbol{H}_2(x)| > 0, |\boldsymbol{H}_3(x)| > 0, \ldots \text{ or } |\boldsymbol{H}_j(x)| > 0, \text{ for } j = 1, \ldots, n.$$

Note that if $\boldsymbol{H}(x^*)$ satisfies the conditions for positive or negative definiteness, then provided p is close to x^*, $\boldsymbol{H}(p)$ will also satisfy those conditions. If $\tilde{x}$ is close to x^* in Equation 11.14, then so is p. So, for example, if at x^*, $f_i(x^*) = 0$, for all i, and $\boldsymbol{H}(x^*)$ is negative definite, then so is $\boldsymbol{H}(p)$ and $f(\tilde{x}) < f(x^*)$.

11.3.2 Definiteness and second-order conditions

Recall that the Hessian matrix is the matrix of second partial derivatives of the function. Because the order of differentiation does not matter, $f_{x_i x_j} = f_{x_j x_i}$, the Hessian matrix is symmetric and the expression $\boldsymbol{z}'\boldsymbol{H}(p)\boldsymbol{z}$ is a quadratic form. The main theorem giving second-order conditions in terms of the Hessian matrix is:

Theorem 11.4: *Let x^* satisfy $f_i(x^*) = 0, i = 1, \ldots, n$. Then*

1. *x^* is a local maximum if $\boldsymbol{H}(\tilde{x})$ is negative semi-definite, $\forall \tilde{x}$ close to x^*.*
2. *x^* is a unique local maximum if $\boldsymbol{H}(x^*)$ negative definite.*
3. *x^* is a global maximum if $\boldsymbol{H}(x)$ is negative semi-definite $\forall x$.*
4. *x^* is a unique global maximum if $\boldsymbol{H}(x)$ is negative definite $\forall x$.*
5. *x^* is a local minimum if $\boldsymbol{H}(x)$ is positive semi-definite.*
6. *x^* is a unique local minimum if $\boldsymbol{H}(x^*)$ is positive definite, $\forall \tilde{x}$ close to x^*.*
7. *x^* is a global minimum if $\boldsymbol{H}(x)$ positive semi-definite $\forall x$.*
8. *x^* is a unique global minimum if $\boldsymbol{H}(x)$ positive definite $\forall x$.*

PROOF: These results are easy to verify by considering the Taylor series expansion:

$$f(x) = f(x^*) + \nabla f(x^*) \cdot (x - x^*) + (\boldsymbol{x} - \boldsymbol{x}^*)'\boldsymbol{H}(p)(\boldsymbol{x} - \boldsymbol{x}^*)$$

and if at x^*, $f_i(x^*) = 0$ so $\nabla f(x^*) = 0$, then the expansion gives:

$$f(x) = f(x^*) + (\boldsymbol{x} - \boldsymbol{x}^*)'\boldsymbol{H}(p)(\boldsymbol{x} - \boldsymbol{x}^*).$$

For example, if $\boldsymbol{H}(\tilde{x})$ is negative semi-definite, $\forall \tilde{x}$ close to x^*, then this will also be satisfied by p in the Taylor series expansion, since f is between $\tilde{x}$ and x^*: $\boldsymbol{H}(\tilde{x})$ is negative semi-definite when $\tilde{x}$ is close to x^*. In this case, $(\tilde{\boldsymbol{x}} - \boldsymbol{x}^*)'\boldsymbol{H}(p)(\tilde{\boldsymbol{x}} - \boldsymbol{x}^*) \leq 0$, so that $f(\tilde{x}) \leq f(x^*)$. This confirms that x^* is a local maximum. A similar discussion applies to the other results. ∎

The next examples illustrate three-variable optimization and the envelope theorem. Following this, a variation of the example is used to illustrate the envelope theorem.

EXAMPLE 11.12: A monopolist can sell a product in three different countries. The demand in each country is $Q_1 = 45 - \frac{1}{2}P_1$, $Q_2 = 50 - \frac{1}{3}P_2$ and $Q_3 = 75 - \frac{1}{4}P_3$. The cost function for the monopolist is $C = 30 + Q^2$ where $Q = Q_1 + Q_2 + Q_3$. The profit function for the monopolist is:

$$\pi = TR - TC = P_1Q_1 + P_2Q_2 + P_3Q_3 - 30 - (Q_1 + Q_2 + Q_3)^2.$$

Using the inverse demand functions, $P_1 = 90 - 2Q_1$, $P_2 = 150 - 3Q_2$, $P_3 = 300 - 4Q_3$, profit is:

$$\pi = (90 - 2Q_1)Q_1 + (150 - 3Q_2)Q_2 + (300 - 4Q_3)Q_3 - 30 - (Q_1 + Q_2 + Q_3)^2.$$

The first-order conditions for profit maximizing output in each country are:

$$\pi_{Q_1} = 90 - 4Q_1 - 2(Q_1 + Q_2 + Q_3) = 0$$

$$\pi_{Q_2} = 150 - 6Q_2 - 2(Q_1 + Q_2 + Q_3) = 0$$

$$\pi_{Q_3} = 300 - 8Q_3 - 2(Q_1 + Q_2 + Q_3) = 0.$$

In matrix form, these conditions give:

$$\begin{pmatrix} 90 \\ 150 \\ 300 \end{pmatrix} - \begin{pmatrix} 6 & 2 & 2 \\ 2 & 8 & 2 \\ 2 & 2 & 10 \end{pmatrix} \begin{pmatrix} Q_1 \\ Q_2 \\ Q_3 \end{pmatrix} = \begin{pmatrix} 0 \\ 0 \\ 0 \end{pmatrix},$$

and solving:

$$\begin{pmatrix} 6 & 2 & 2 \\ 2 & 8 & 2 \\ 2 & 2 & 10 \end{pmatrix}^{-1} = \begin{pmatrix} \frac{19}{100} & -\frac{1}{25} & -\frac{3}{100} \\ -\frac{1}{25} & \frac{7}{50} & -\frac{1}{50} \\ -\frac{3}{100} & -\frac{1}{50} & \frac{11}{100} \end{pmatrix};$$

$$\begin{pmatrix} Q_1 \\ Q_2 \\ Q_3 \end{pmatrix} = \begin{pmatrix} \frac{19}{100} & -\frac{1}{25} & -\frac{3}{100} \\ -\frac{1}{25} & \frac{7}{50} & -\frac{1}{50} \\ -\frac{3}{100} & -\frac{1}{50} & \frac{11}{100} \end{pmatrix} \begin{pmatrix} 90 \\ 150 \\ 300 \end{pmatrix} = \begin{pmatrix} \frac{21}{10} \\ \frac{57}{5} \\ \frac{273}{10} \end{pmatrix}.$$

Since

$$\pi_{Q_1Q_1} = -6, \quad \pi_{Q_1Q_2} = -2, \quad \pi_{Q_1Q_3} = -2, \quad \pi_{Q_2Q_2} = -8,$$
$$\pi_{Q_2Q_3} = -2, \quad \pi_{Q_3Q_3} = -10,$$

$$\boldsymbol{H} = \begin{pmatrix} -6 & -2 & -2 \\ -2 & -8 & -2 \\ -2 & -2 & -10 \end{pmatrix}, \quad \boldsymbol{H}_1 = -6 < 0, \quad \boldsymbol{H}_2 = \begin{vmatrix} -6 & -2 \\ -2 & -8 \end{vmatrix} = 44 > 0,$$

$$\boldsymbol{H} = \boldsymbol{H}_3; \quad |\boldsymbol{H}| = \begin{vmatrix} -6 & -2 & -2 \\ -2 & -8 & -2 \\ -2 & -2 & -10 \end{vmatrix} = -400 < 0,$$

and the solution is a maximum. ◊

EXAMPLE 11.13: Suppose that in Example 11.12 the cost function is $C = 30 + \alpha Q^2$, where α is some positive constant. Then the profit function becomes

$$\pi = (90 - 2Q_1)Q_1 + (150 - 3Q_2)Q_2 + (300 - 4Q_3)Q_3 - 30 - \alpha(Q_1 + Q_2 + Q_3)^2,$$

and the first-order conditions are:

$$\pi_{Q_1} = 90 - 4Q_1 - 2\alpha(Q_1 + Q_2 + Q_3) = 0$$
$$\pi_{Q_2} = 150 - 6Q_2 - 2\alpha(Q_1 + Q_2 + Q_3) = 0$$
$$\pi_{Q_3} = 300 - 8Q_3 - 2\alpha(Q_1 + Q_2 + Q_3) = 0.$$

Thus,

$$\begin{pmatrix} 90 \\ 150 \\ 300 \end{pmatrix} - \begin{pmatrix} 4+2\alpha & 2\alpha & 2\alpha \\ 2\alpha & 6+2\alpha & 2\alpha \\ 2\alpha & 2\alpha & 8+2\alpha \end{pmatrix} \begin{pmatrix} Q_1 \\ Q_2 \\ Q_3 \end{pmatrix} = \begin{pmatrix} 0 \\ 0 \\ 0 \end{pmatrix}.$$

Some calculation yields:

$$\begin{pmatrix} 4+2\alpha & 2\alpha & 2\alpha \\ 2\alpha & 6+2\alpha & 2\alpha \\ 2\alpha & 2\alpha & 8+2\alpha \end{pmatrix}^{-1} = \begin{pmatrix} \frac{7\alpha+12}{4(13\alpha+12)} & -\frac{\alpha}{(13\alpha+12)} & -\frac{3\alpha}{4(13\alpha+12)} \\ -\frac{\alpha}{(13\alpha+12)} & \frac{3\alpha+4}{2(13\alpha+12)} & -\frac{\alpha}{2(13\alpha+12)} \\ -\frac{3\alpha}{4(13\alpha+12)} & -\frac{\alpha}{2(13\alpha+12)} & \frac{5\alpha+6}{4(13\alpha+12)} \end{pmatrix}$$

and

$$\begin{pmatrix} \frac{7\alpha+12}{4(13\alpha+12)} & -\frac{\alpha}{(13\alpha+12)} & -\frac{3\alpha}{4(13\alpha+12)} \\ -\frac{\alpha}{(13\alpha+12)} & \frac{3\alpha+4}{2(13\alpha+12)} & -\frac{\alpha}{2(13\alpha+12)} \\ -\frac{3\alpha}{4(13\alpha+12)} & -\frac{\alpha}{2(13\alpha+12)} & \frac{5\alpha+6}{4(13\alpha+12)} \end{pmatrix} \begin{pmatrix} 90 \\ 150 \\ 300 \end{pmatrix} = \begin{pmatrix} -\left(\frac{15}{2}\right)\left(\frac{-36+29\alpha}{12+13\alpha}\right) \\ -\frac{15(\alpha-20)}{12+13\alpha} \\ \left(\frac{15}{2}\right)\frac{31\alpha+60}{12+13\alpha} \end{pmatrix} = \begin{pmatrix} Q_1 \\ Q_2 \\ Q_3 \end{pmatrix}.$$

For the second-order conditions, the Hessian matrix is:

$$\begin{pmatrix} -(4+2\alpha) & -2\alpha & -2\alpha \\ -2\alpha & -(6+2\alpha) & -2\alpha \\ -2\alpha & -2\alpha & -(8+2\alpha) \end{pmatrix},$$

so $|\,\boldsymbol{H}_1\,| = -(4+2\alpha) < 0$, $|\,\boldsymbol{H}_2\,| = [(4+2\alpha)(6+2\alpha)] - (2\alpha)^2 = 24+20\alpha > 0$, and $|\,\boldsymbol{H}_3\,| = -(208\alpha+192) < 0$.

This example may be used to illustrate the envelope theorem. If the optimized function is denoted $\pi^*(\alpha)$, then

$$\frac{d\pi^*}{d\alpha} = -(Q_1^* + Q_2^* + Q_3^*)^2,$$

where Q_i^* is the optimal value of Q_i. The following calculations confirm this. Observe that with the cost function $C = 30 + \alpha Q^2$, the optimal values of the Q_i depend on α, as calculated above. Write these functions: $Q_1(\alpha), Q_2(\alpha), Q_3(\alpha)$. Then

$$\begin{aligned} R_1(Q_1(\alpha)) &= P_1(\alpha)Q_1(\alpha) = [90 - 2Q_1(\alpha)]Q_1(\alpha) \\ R_2(Q_2(\alpha)) &= P_2(\alpha)Q_2(\alpha) = [150 - 3Q_2(\alpha)]Q_2(\alpha) \\ R_3(Q_3(\alpha)) &= P_3(\alpha)Q_3(\alpha) = [300 - 4Q_3(\alpha)]Q_3(\alpha) \end{aligned}$$

and

$$\pi^*(\alpha) = R_1(Q_1(\alpha)) + R_2(Q_2(\alpha)) + R_3(Q_3(\alpha)) - 30 - \alpha(Q_1(\alpha) + Q_2(\alpha) + Q_3(\alpha))^2.$$

Thus,

$$\begin{aligned} \frac{d\pi^*}{d\alpha} = {} & \frac{dR_1(Q_1(\alpha))}{dQ_1}\frac{dQ_1}{d\alpha} + \frac{dR_2(Q_2(\alpha))}{dQ_2}\frac{dQ_2}{d\alpha} + \frac{dR_3(Q_3(\alpha))}{dQ_3}\frac{dQ_3}{d\alpha} \\ & - 2\alpha(Q_1(\alpha) + Q_2(\alpha) + Q_3(\alpha))\left[\frac{dQ_1}{d\alpha} + \frac{dQ_2}{d\alpha} + \frac{dQ_3}{d\alpha}\right] \\ & - (Q_1(\alpha) + Q_2(\alpha) + Q_3(\alpha))^2 \end{aligned}$$

$$= \left[\frac{dR_1(Q_1(\alpha))}{dQ_1} - 2\alpha(Q_1(\alpha)+Q_2(\alpha)+Q_3(\alpha))\right]\frac{dQ_1}{d\alpha}$$
$$+ \left[\frac{dR_2(Q_2(\alpha))}{dQ_2} - 2\alpha(Q_1(\alpha)+Q_2(\alpha)+Q_3(\alpha))\right]\frac{dQ_2}{d\alpha}$$
$$+ \left[\frac{dR_3(Q_3(\alpha))}{dQ_3} - 2\alpha(Q_1(\alpha)+Q_2(\alpha)+Q_3(\alpha))\right]\frac{dQ_3}{d\alpha}$$
$$-(Q_1(\alpha)+Q_2(\alpha)+Q_3(\alpha))^2.$$

Since

$$\left[\frac{dR_i(Q_i(\alpha))}{dQ_i} - 2\alpha(Q_1(\alpha)+Q_2(\alpha)+Q_3(\alpha))\right] = 0, i = 1,2,3,$$

$\frac{d\pi^*}{d\alpha} = -(Q_1(\alpha)+Q_2(\alpha)+Q_3(\alpha))^2$. Alternatively, recall that the profit function (before optimization) is:

$$\pi = (90-2Q_1)Q_1 + (150-3Q_2)Q_2 + (300-4Q_3)Q_3 - 30 - \alpha(Q_1+Q_2+Q_3)^2.$$

Differentiating with respect to the parameter α gives $-(Q_1+Q_2+Q_3)^2$, and evaluating this at the solution yields: $(Q_1(\alpha)+Q_2(\alpha)+Q_3(\alpha))^2$. Thus, the derivative of the unoptimized function with respect to the parameter α evaluated at the solution is the same as the derivative of the optimized function with respect to the parameter – an illustration of the envelope theorem. ◊

REMARK 11.6: When f is *additively separable*, the second order conditions are simpler. The function f is additively separable if $f(x_1,\ldots,x_n) = \sum_{i=1}^n f^i(x_i) = f^1(x_1)+\cdots+f^n(x_n)$. The Hessian has the form:

$$\boldsymbol{H} = \begin{pmatrix} \frac{\partial^2 f^1}{\partial^2 x_1^2} & 0 & \cdots & 0 \\ 0 & \frac{\partial^2 f^2}{\partial^2 x_2^2} & \cdots & 0 \\ \vdots & \vdots & \ddots & \vdots \\ 0 & 0 & 0 & \frac{\partial^2 f^n}{\partial^2 x_n^2} \end{pmatrix}.$$

Then f is negative definite if and only if $\frac{\partial^2 f^i}{\partial x_i^2} < 0, \forall i$. □

EXAMPLE 11.14 (PRICE DISCRIMINATION IN THREE MARKETS): Price and revenue in market 1 are given by $P_1 = P_1(Q_1)$ and $R_1(Q_1) = P_1(Q_1)Q_1$, respectively. Similarly, in markets 2 and 3, price and revenue are, respectively, $P_2 = P_2(Q_2)$, $P_3 = P_3(Q_3)$, $R_2(Q_2) = P_2(Q_2)Q_2$ and $R_3(Q_3) = P_3(Q_3)Q_3$. Production cost is given by $c(Q_1,Q_2,Q_3)$. The profit function is then:

$$\pi(Q_1,Q_2,Q_3) = R_1(Q_1) + R_2(Q_2) + R_3(Q_3) - c(Q_1,Q_2,Q_3).$$

(If the cost function is additively separable, so that $c(Q_1,Q_2) = \gamma_1(Q_1) + \gamma_2(Q_2) + \gamma_3(Q_3)$, then the problem of maximizing profit reduces to choosing Q_1, Q_2 and Q_3 independently. In general, if the level of production of one good affects the cost of producing the other good, then the cost function is not additively separable.)

The first-order conditions are:

$$\pi_1 = R_1'(Q_1) - c_1 = 0, \left(\pi_1 = \frac{\partial \pi}{\partial Q_1}, c_1 = \frac{\partial c}{\partial Q_1}\right)$$

$$\pi_2 = R_2'(Q_2) - c_2 = 0, \left(\pi_2 = \frac{\partial \pi}{\partial Q_2}, c_2 = \frac{\partial c}{\partial Q_2}\right)$$

$$\pi_3 = R_3'(Q_3) - c_3 = 0, \left(\pi_3 = \frac{\partial \pi}{\partial Q_3}, c_3 = \frac{\partial c}{\partial Q_3}\right).$$

The second derivatives are $\pi_{11} = R_1''(Q_1) - c_{11}$, $\pi_{22} = R_2''(Q_2) - c_{22}$ and $\pi_{33} = R_3''(Q_3) - c_{33}$. The cross derivatives are $\pi_{ij} = -c_{ij}$, when $i \neq j$. ($\pi_{ij} = \frac{\partial^2 \pi}{\partial Q_i \partial Q_j}$, $c_{ij} = \frac{\partial^2 c}{\partial Q_i \partial Q_j}$). The first-order conditions give:

$$MR_1(Q_1) = R_1'(Q_1) = P_1(Q_1) + Q_1 P_1'(Q_1) = c_1(Q_1,Q_2,Q_3) = MC_1(Q_1,Q_2,Q_3).$$

So

$$P_1(Q_1)\left[1 + \frac{Q_1}{P_1}\frac{dP_1}{dQ_1}\right] = c_1$$

or

$$P_1(Q_1)\left[1 + \frac{1}{\eta_1}\right] = c_1,$$

$\eta_1 < 0$. Similarly, $P_2(Q_2)\left[1 + \frac{1}{\eta_2}\right] = c_2$, with $\eta_2 < 0$ and $P_3(Q_3)\left[1 + \frac{1}{\eta_3}\right] = c_3$, with $\eta_3 < 0$. If η_i is a large negative number, then market i is sensitive to price and the price in this market should be set close to marginal cost.

If the cost is linear: $c(Q_1,Q_2,Q_3) = \gamma_1 Q_1 + \gamma_2 Q_2 + \gamma_3 Q_3$, and the Hessian matrix has a simple form, since $c_{ij} = 0$ for $i \neq j$. Taking each market to have diminishing marginal revenue: $R_i'' < 0$, $i = 1,2,3$. For the second-order conditions, observe that the Hessian matrix is:

$$\begin{pmatrix} \pi_{11}(Q_1) & 0 & 0 \\ 0 & \pi_{22}(Q_2) & 0 \\ 0 & 0 & \pi_{33}(Q_3) \end{pmatrix} = \begin{pmatrix} R_1''(Q_1) & 0 & 0 \\ 0 & R_2''(Q_2) & 0 \\ 0 & 0 & R_3''(Q_3) \end{pmatrix}.$$

Therefore, the second-order conditions are: $\pi_{11} = R_1''(Q_1) < 0, \forall Q_1$, $\pi_{11}\pi_{22} = R_1''(Q_1)R_2''(Q_2) > 0$, and $\pi_{11}\pi_{22}\pi_{33} = R_1''(Q_1)R_2''(Q_2)R_3''(Q_3) < 0$. ◊

11.4 Concavity, convexity and the Hessian matrix

The next theorem relates concavity and convexity to the "definiteness" of the Hessian.

Theorem 11.5:

1. *If f is strictly concave, then f is concave; and concavity of f is equivalent to negative semi-definiteness of the Hessian matrix $\boldsymbol{H}(x)$, for all x.*

2. *If the Hessian matrix $\boldsymbol{H}(x)$ is negative definite for all x, then f is strictly concave.*

3. *If f is strictly convex then f is convex; and convexity of f implies that the Hessian matrix $\boldsymbol{H}(x)$ is positive semi-definite for all x.*

4. *If the Hessian matrix $\boldsymbol{H}(x)$ is positive definite for all x, then f is strictly convex.*

PROOF: The following discussion proves 1. From the definitions, strict concavity implies concavity. The remaining calculations show that concavity is equivalent to negative semi-definiteness. Recall that a function, f, is concave if given x and y, $f(\theta x + (1-\theta)y) \geq \theta f(x) + (1-\theta)f(y)$, $\theta \in [0,1]$. Concavity is determined by comparing any two points with a third point on the line defined by those two points. Given any two points x, y the line defined by these two points is $L_{xy} = \{w \mid w = (1-\theta)x + \theta y, \text{ some } \theta \in \mathscr{R}\}$. Let $f^{L_{xy}}$ be the function f defined on L_{xy} by: $f^{L_{xy}}(\theta) = f(\theta x + (1-\theta)y)$, $\forall \theta$. Observe that for every line L_{xy}, $f^{L_{xy}}$ is concave in θ if and only if f is concave.

$$
\begin{aligned}
f^{L_{xy}}(\lambda\theta+(1-\lambda)\theta') &= f\big((\lambda\theta+(1-\lambda)\theta')x+[1-(\lambda\theta+(1-\lambda)\theta')]y\big) \\
&= f\big(\lambda\theta x+(1-\lambda)\theta' x+[(\lambda+(1-\lambda))-(\lambda\theta+(1-\lambda)\theta')]y\big) \\
&= f\big(\lambda\theta x+(1-\lambda)\theta' x+(\lambda(1-\theta)y+(1-\lambda)(1-\theta')y\big) \\
&= f\big(\lambda[\theta x+(1-\theta)y]+(1-\lambda)[\theta' x+(1-\theta')y]\big) \\
&\geq \lambda f(\theta x+(1-\theta)y)+(1-\lambda)f(\theta' x+(1-\theta')y) \\
&= \lambda f^{L_{xy}}(\theta)+(1-\lambda)f^{L_{xy}}(\theta'),
\end{aligned}
$$

where the inequality holds if and only if f is concave. Thus, $f^{L_{xy}}(\theta) = f(\theta x+(1-\theta)y) = f(y+\theta(x-y))$ is concave for any x, y. Let $z = x-y$ so $f^{L_{xy}}(\theta) = f(y+\theta z)$. By assumption, $f^{L_{xy}}$ is concave so that $f^{L_{xy}}_{\theta\theta} \leq 0$. Now,

$$f^{L_{xy}}_{\theta} = \sum_i f_i(y+\theta z)z_i$$

and

$$f^{L_{xy}}_{\theta\theta}(\theta) = \sum_j\sum_i f_{ji}(y+\theta z)z_j z_i = \boldsymbol{z}'\boldsymbol{H}(y+\theta z)\boldsymbol{z}.$$

So, concavity of $f^{L_{xy}}$ in θ for every L_{xy} is equivalent to $\boldsymbol{z}'\boldsymbol{H}(y+\theta z)\boldsymbol{z} \leq 0$ for all y, z and θ. ■

REMARK 11.7: Note that strict concavity of a function f does not imply that $\boldsymbol{H}(x)$ is negative definite for all x; and strict convexity does not imply that $\boldsymbol{H}(x)$ is positive definite for all x. For example, consider $f(x) = -(x-1)^4$, a strictly concave function. Differentiating gives $f''(x) = -12(x-1)^2$. Note that $f''(1) = 0$ so that f'' is only negative semi-definite despite the fact that the function f is strictly concave. □

Exercises for Chapter 11

Exercise 11.1 The total weekly revenue (in dollars) that a company obtains from selling goods x and y is:

$$R(x,y) = -\left(\tfrac{1}{4}\right)x^2 - \left(\tfrac{3}{8}\right)y^2 - \left(\tfrac{1}{4}\right)xy + 300x + 240y.$$

The total weekly cost attributed to production is

$$C(x,y) = 180x + 140y + 5000.$$

Find the profit-maximizing levels of x and y and check the second-order conditions.

Exercise 11.2 A television relay station will serve three towns, A, B and C, located in the (x,y) plane at coordinates $A = (30,20)$, $B = (-20,10)$ and $C = (10,-10)$. Plot the location of the three towns in the (x,y) plane and find the location of a transmitter, chosen to minimize the sum of the squares of the distances of the transmitter from each of the towns. For two points (x_1,y_1) and (x_2,y_2), the distance between them is $[(x_1-x_2)^2+(y_1-y_2)^2]^{\frac{1}{2}}$.

Exercise 11.3 Consider the following duopoly model. There are two firms supplying a market where demand is given by $p(Q) = a - bQ$. Firm i produces q_i units of output and so the total level of production is $Q = q_1 + q_2$. Both firms face the same constant marginal cost, so the cost of producing q_i for firm i is cq_i. Thus the profit functions of firms 1 and 2 respectively, are given by:

$$\pi_1(q_1,q_2) = p(q_1+q_2)q_1 - cq_1 = (a - b[q_1+q_2])q_1 - cq_1$$
$$\pi_2(q_1,q_2) = p(q_1+q_2)q_2 - cq_2 = (a - b[q_1+q_2])q_2 - cq_2.$$

(a) Suppose that each firm takes the output of the other firm as fixed, and calculates its own best choice of quantity. In this case, firm i solves $\frac{\partial \pi_i}{\partial q_i} = 0$, taking as given the value of q_j. Calculate for the linear demand case the equations:

$$\begin{pmatrix} \frac{\partial \pi_1}{\partial q_1} \\ \frac{\partial \pi_2}{\partial q_2} \end{pmatrix} = \begin{bmatrix} 0 \\ 0 \end{bmatrix}.$$

Solve for q_1 and q_2 and denote the solutions q_1^d and q_2^d, where d denotes duopoly. Compare the value of $Q^d = q_1^d + q_2^d$, with the monopoly level of output.

(b) Suppose now that we have n firms instead of two. This case is called oligopoly. Let the output of the ith firm be q_i^o, where o denotes oligopoly, and see if you can solve for $Q^o = q_1^o + q_2^o + \ldots + q_n^o$.

(c) Suppose now that a government tax per unit of output is imposed on each firm, with tax t_i per unit on firm i. Then the profit functions become:

$$\pi_1(q_1, q_2; t_1) = p(q_1 + q_2)q_1 - cq_1 - t_1 q_1 = (a - b[q_1 + q_2])q_1 - cq_1 - t_1 q_1$$

$$\pi_2(q_1, q_2; t_2) = p(q_1 + q_2)q_2 - cq_2 - t_2 q_2 = (a - b[q_1 + q_2])q_2 - cq_2 - t_2 q_2.$$

Repeat the exercise in (a), and solve for q_1 and q_2. In this case, q_1 and q_2 will depend on t_1 and t_2 (and the other parameters, a, b, c), so we can write $q_1(t_1, t_2)$ and $q_2(t_1, t_2)$. The total tax revenue of the government is $T(t_1, t_2) = t_1 q_1(t_1, t_2) + t_2 q_2(t_1, t_2)$. Maximize T with respect to t_1 and t_2, and check that the second-order conditions for a maximum are satisfied.

Exercise 11.4 Consider a firm that sells its output, y, at a fixed price p. In the production process, the firm creates pollution. To control pollution, the firm can engage in pollution abatement, a. The technology is as follows: the cost of producing y units of output when operating at abatement level a is given by $c(y, a)$, and the level of pollution emission is $e(y, a)$. We assume that for the cost function, $c_y > 0, c_{yy} > 0, c_a > 0$ and $c_{aa} > 0$. Similarly, for the emission process, e, we assume that $e_y > 0, e_{yy} > 0, e_a < 0$ and $e_{aa} > 0$. In addition, we assume that $c_{ay} (= c_{ya})$ and $e_{ay} (= e_{ya})$ are both positive. Finally, the government imposes a (unit) tax, τ, on the level of pollution emission. The profit function for the firm is then given by:

$$\pi(y, a) = py - c(y, a) - \tau e(y, a).$$

(a) Find the first-order conditions for profit maximization with respect to choice of (y, a).

(b) Determine the second-order conditions.

(c) In the case where $c(y,a) = y^2 a + \gamma a y$ and $e(y,a) = y - a^2 + \beta a y$, γ, β parameters, find the profit-maximizing levels of y and a. Check the second-order conditions. From the solution to the first-order conditions and calculate $\partial y/\partial \tau$ and $\partial a/\partial \tau$.

Exercise 11.5 A firm's costs of plant operation are a function of two variables, x and y. The cost function is given by:

$$c(x,y) = 3x^2 - 2xy + y^2 + 4x + 3y.$$

Give the solution to the first-order conditions and confirm that the solution minimizes the cost function.

Exercise 11.6 A box with edge lengths x, y and z has volume $V = xyz$. The surface area of a box with these dimensions (open at the top) is: $S = xy + 2yz + 2xz$. (The base of the box has length x and width y. Since there is no top, xy occurs only once in the definition of S.) Suppose you have to choose a box size with volume 32; what choice of x, y and z is optimal to minimize surface area? Note that, in this question, there are only two choice variables once the volume is fixed. The variable z can be substituted out: $z = 32/xy$. Find the optimal choices of the variables to minimize surface area S, and check that the second-order conditions for a minimum are satisfied.

Exercise 11.7 Consider a firm that sells its output, y, at a fixed price p. In the production process, the firm creates pollution. To control pollution, the firm can engage in pollution abatement, a. The technology is as follows: the cost of producing y units of output when operating at abatement level a is given by $c(y,a)$, and the level of pollution emission is $e(y,a)$. Finally, the government imposes a (unit) tax, τ, on the level of pollution emission. The profit function for the firm is then given by:

$$\pi(y,a) = py - c(y,a) - \tau e(y,a).$$

Give the first- and second-order conditions for a maximum. Find the profit-maximizing levels of output and abatement when the cost function is given by $c(y,a) = y^2 + \frac{1}{2}a^2$, and $e(y,a) = y(10 - a)$. Check that the second-order conditions are satisfied. We assume that the tax rate is less than $1 (0 < \tau < 1)$, and $p > 10$.

Exercise 11.8 Find the maximizing values of (x,y) for the following function, and confirm that the second-order conditions are satisfied:

$$f(x,y) = \sqrt{x} + \ln y - x^2 - 3y.$$

Exercise 11.9 Find the maximizing values of (x,y) for the following function, and confirm that the second-order conditions are satisfied:

$$f(x,y) = a\ln x + b\ln y - \frac{1}{2}c(x^2+y^2).$$

Exercise 11.10 Find the maximizing values of (x,y) for the following function, and confirm that the second-order conditions are satisfied:

$$f(x,y) = \ln(ax+by) - \frac{1}{2}c(x^2+y^2).$$

Exercise 11.11 Consider the problem of maximizing a function, $f(x,y;a)$, by choice of x and y, where a is some parameter. Suppose that the solution values for x and y are $x(a)$ and $y(a)$, respectively. The optimized value of the function is then $f^*(a) = f(x(a), y(a); a)$.

(a) Explain why $\frac{df^*(a)}{da} = \frac{\partial f(x(a),y(a);a)}{\partial a}$.

(b) Illustrate this in the one-variable case, where $f(x;a) = a\ln(x) - \frac{1}{a}x$.

Exercise 11.12 Consider the following problem. Minimize $f(x,y) = (x-a)^2 + (y-b)^2 - \frac{1}{2}cxy$.

(a) Find the *minimizing* values for x and y. (Assume that $c < 1$.)

(b) Confirm that the second-order conditions are satisfied. Let $f^*(a,b,c)$ be the value of the function at the solution.

(c) Find $\frac{\partial f^*(a,b,c)}{\partial a}$ and $\frac{\partial f^*(a,b,c)}{\partial c}$.

Exercise 11.13 A firm maximizes profit, where profit is equal to revenue minus cost. Output (Q) depends on two inputs, K and L, related according to the production function $Q = f(K,L)$. Output sells at price p and the costs of the inputs K and L are r and w, respectively. So the profit maximization problem is:

$$\max \pi(K,L) = pf(K,L) - rK - wL.$$

Give the first- and second-order conditions for profit maximization. In the particular case where f is $f(K,L) = K^\alpha L^\beta$ with $\alpha > 0$, $\beta > 0$ and $0 < \alpha + \beta < 1$, verify that the second-order conditions are satisfied and solve for the optimal choice of (K,L).

Exercise 11.14 A standard econometric problem is determining the ordinary least squares estimator (OLS). In the underlying model, the ith observation on one variable y is denoted y_i and the corresponding observation on another variable x is denoted x_i. It is postulated that y and x are linearly related with some error, $y_i = \beta_0 + \beta_1 x_i + \epsilon_i$. If a and b are chosen as the intercept and slope, respectively, then the estimated error associated with the ith observation is $e_i = y_i - a - bx_i$. The OLS estimator is obtained by minimizing the sum of squares of the estimated errors, $\sum_{i=1}^n e_i^2$, by choice of a and b. Thus, the OLS estimator is obtained by solving the problem:

$$\min_{a,b} S = \sum_{i=1}^n e_i^2 = \min \sum_{i=1}^n (y_i - a - bx_i)^2 = \min \sum_{i=1}^n (y_i - a - bx_i)^2.$$

Find the OLS estimator and confirm the second-order conditions for a minimum.

Exercise 11.15 A monopolist can sell a product in three different countries. The demand in each country is:

$$Q_1 = 50 - \frac{1}{2}P_1$$

$$Q_2 = 50 - \frac{1}{3}P_2$$

$$Q_3 = 75 - \frac{1}{4}P_3.$$

The cost function for the monopolist is

$$C(Q) = 30 + Q^2, \qquad \text{where } Q = Q_1 + Q_2 + Q_3.$$

Calculate the amount of the product that the profit-maximizing monopolist should sell in each country and check the second-order conditions for a maximum.

Exercise 11.16 For the following functions, calculate the Hessians and determine if they are positive or negative definite:

(a) $$f(x,y) = -2x^2 - 7y^2 + xy + 12x + 37y$$

(b) $$f(x,y) = 10x^2 + 5y^2 + 3x - 12y + 9xy$$

(c) $$f(x,y,z) = 3x^2 + 12y^2 + 5z^2 - 13x + 12y + 2xz - 3yz$$

(d) $$f(x,y,z) = 2x + 3y - 5xz + 2xy - 6x^2 - 8y^2 - 11z^2.$$

Exercise 11.17 For the following function, find all extreme points and check if they are maxima or minima:

$$f(x,y) = 3y^3 + 2x^2 + 4xy + 8y^2 + 2y.$$

Exercise 11.18 Given the three-input Cobb–Douglas production function,

$$Y = AL^{\alpha}K^{\beta}M^{\gamma},$$

where L is the amount of labour, K the amount of capital and M the amount of other materials used, prove that the second-order conditions for a profit maximization hold if and only if $\alpha+\beta+\gamma<1$ $(\alpha,\beta,\gamma>0)$. The good sells for a price of a dollar per unit. The costs of labour, capital and materials are, respectively, w, r and p. Find the Hessian matrix and confirm that it is negative definite.

Exercise 11.19 The geometric average of three (positive) numbers x, y and z is given by $(xyz)^{\frac{1}{3}}$. Suppose that $x+y+z=k$. Show that the geometric average is maximized when $x=y=z$.

Exercise 11.20 A firm uses two inputs x,y to produce Q according to $Q=A(x^{\alpha}+y^{\beta})$, with $0<\alpha,\beta<1$. The production cost is given by $c(x,y)$, so that profit is $\pi(x,y)=pA(x^{\alpha}+y^{\beta})-c(x,y)$, where p is the price of Q. Give the first- and second-order conditions for a maximum. In the case where $c(x,y)=ax+by+dxy$, $a,b,d>0$, find the maximizing values of x,y and show that the conditions for a maximum are satisfied.

Exercise 11.21 For the function

$$f(x,y)=10x^2+10y^2+3x-12y+10xy,$$

show that the Hessian is positive definite.

Exercise 11.22 For the function

$$f(x,y,z)=2x+3y-5xz+2xy-6x^2-9y^2-12z^2,$$

show that the Hessian is negative definite.

12

Equality-constrained optimization

12.1 Introduction

Optimization of a function $f(x_1,\ldots,x_n)$ involves choosing the x_i variables to make the function f as large or as small as possible. With constrained optimization, permissible choices for the x_i are restricted or constrained. In what follows, such constraints are represented by a function, g, whereby the x_i variables must satisfy a condition of the form $g(x_1,\ldots,x_n)=c$. For example, in the standard utility maximization problem, the task is to maximize utility $u(x_1,\ldots,x_n)$ subject to a budget constraint $p_1x_1+\cdots+p_nx_n=c$, where p_i is the price of good i and c is the income available.

Constraints limit the choices that are feasible. In the case of two goods, the budget constraint is $p_1x_1+p_2x_2=c$ so that the feasible pairs lie on a line. In this case, there is one constraint with two choice variables. The choice of one variable determines the choice of the other: given x_1, x_2 is determined as $x_2=\frac{c}{p_2}-\frac{p_1}{p_2}x_1$. Suppose that in addition to the budget constraint there is a second constraint – that the two goods are complements that must be consumed in a set ratio: $\frac{x_2}{x_1}=r$ or $x_2=rx_1$. Given this, from $p_1x_1+p_2x_2=c$ one has $p_1x_1+p_2rx_1=c$ or $(p_1+p_2r)x_1=c$ or $x_1=\frac{c}{(p_1+p_2r)}$ and $x_2=\frac{cr}{(p_1+p_2r)}$. Thus, there is a unique pair satisfying the two constraints. This observation is generally true; if the number of constraints equals the number of choice variables, there are generally at most a finite number of solutions to the set of constraints. For this reason, the set of constraints is generally less than the number of choice variables.

The next section provides a motivating example showing how the Lagrangian technique relates to the familiar first-order condition (marginal rate of substitution equals the price ratio) for an optimum in the consumer choice problem.

Section 12.2 considers the case of two variables and one constraint. The case of n variables and one constraint is discussed in Section 12.4. It turns out that the Lagrange multiplier appearing in the definition of the Lagrangian has a significant economic interpretation: it provides a measure of the value of the constraint. Specifically, it measures the marginal change in the value of the objective resulting from a small change in the constraint. This is examined in Section 12.5. In the case of unconstrained optimization, identification of sufficient conditions for a global optimum is simplified when the objective has a specific shape, particularly concavity or convexity. Similarly, with constrained optimization, under suitable conditions, discussed in Section 12.6, the local conditions for an optimum are sufficient for the solution to be a global optimum.

12.1.1 A motivating example

The optimization techniques described next are based on the use of Lagrangian functions. The discussion here shows how that formulation may be related to optimization ideas introduced earlier. Consider the problem of maximizing $u(y,l)$ where y is income and l is leisure. Work is time not used for leisure, $24-l$, and income is given by the wage w times the amount of time spent working, so income and leisure are connected according to the equation or constraint $y = w(24-l)$. Assume that the marginal benefit of both income and leisure is positive: $u_y > 0$ and $u_l > 0$. Maximizing $u(y,l)$ subject to the constraint $y = w(24-l)$, is an equality-constrained optimization problem. Preferences and the leisure–income constraint are depicted in Figure 12.1. Note that the condition $y = w(24-l)$ implies that there is only one variable to be chosen – once l is chosen, the value of y is determined. Using this fact, y can be eliminated from the problem – the constraint can be substituted into the objective to give $u^*(l) = u(w(24-l),l)$ and the function u^* can be maximized with respect to l as an unconstrained problem. The function u^* is depicted in Figure 12.2. Differentiating and setting to 0:

$$u^{*'} = 0 \Rightarrow -u_y w + u_l = 0 \Rightarrow \frac{u_l}{u_y} = w.$$

For example, if $u(y,l) = \ln y + \ln l$, then $u^*(l) = \ln(w(24-l)) + \ln(l)$. Differentiating $u^*(l)$ and setting the derivative to 0:

$$u^{*'}(l) = 0 = \frac{-w}{w(24-l)} + \frac{1}{l} = 0 \Rightarrow -l + 24 - l = 0 = 24 - 2l.$$

This has solution $l = \frac{24}{2} = 12$, $y = w(24-12) = 12w$. The second-order condition for a maximum is easy to check: $u^{*'}(l) = -\left(\frac{1}{(24-l)}\right)^2 - \left(\frac{1}{l}\right)^2 < 0$. Thus, using the con-

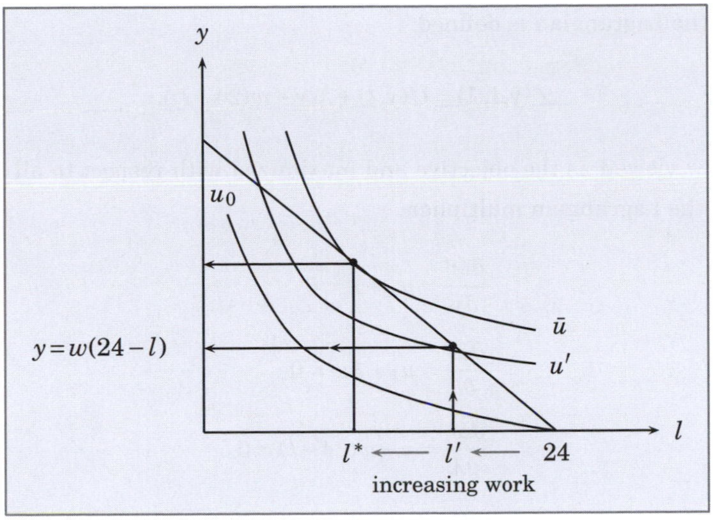

Figure 12.1: Constrained optimal choice

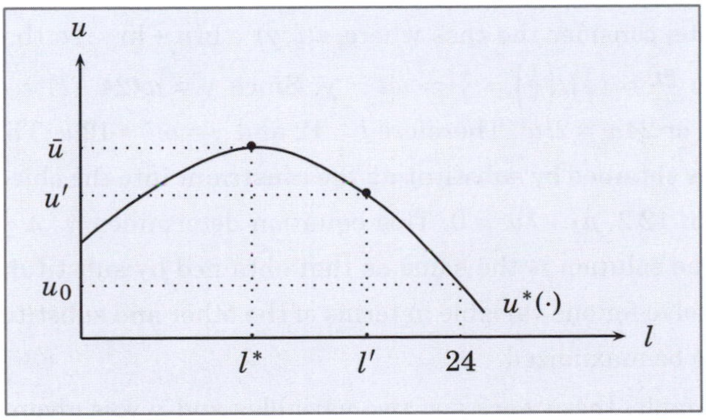

Figure 12.2: Optimal leisure choice

straint to solve for one variable in terms of the other leads to an unconstrained problem in one variable that can be solved as the unconstrained maximization of a function of one variable. With just two choice variables, this approach is often satisfactory, but becomes less manageable as the number of choice variables increases.

An alternative approach to constrained optimization is to define a function called a Lagrangian. The Lagrangian is denoted $\mathscr{L}$. It is the original function plus the constraint multiplied by a "Lagrange multiplier". The arguments of the function $\mathscr{L}$ are the choice variables *and* a Lagrange multiplier, λ. In the present example, the Lagrangian is defined:

$$\mathscr{L}(y,l,\lambda) = U(y,l) + \lambda(y - w(24-l)).$$

This is now viewed as the objective and maximized with respect to all variables, including the Lagrangian multiplier.

$$\frac{\partial \mathscr{L}}{\partial y} = u_y + \lambda = 0 \tag{12.1}$$

$$\frac{\partial \mathscr{L}}{\partial l} = u_l + \lambda w = 0 \tag{12.2}$$

$$\frac{\partial \mathscr{L}}{\partial \lambda} = y - w(24-l) = 0. \tag{12.3}$$

The first two conditions give $\frac{u_l}{u_y} = w$, and this condition imposes the tangency of the indifference curve to the income–leisure constraint in Figure 12.1. The third equation gives $y - w(24-l) = 0$. Solving these determines the optimal values for l and y.

To illustrate, consider the case where $u(l,y) = \ln u + \ln y$, so that $u_y = \frac{1}{y}, u_l = \frac{1}{l}$, and so $w = \frac{u_l}{u_y} = \left(\frac{1}{l}\right)/\left(\frac{1}{y}\right) = \frac{y}{l} \Rightarrow wl = y$. Since $y = w(24-l) = wl$, this gives $wl = 24w - lw$ or $24w = 2lw$. Therefore $l = 12$ and $y = wl = 12w$. This is the same solution as was obtained by substituting the constraint into the objective function. From Equation 12.2, $u_l + \lambda w = 0$. This equation determines $\lambda : \lambda = -\frac{u_l}{w} = -\frac{1}{wl} = -\frac{1}{12w}$. Thus, the solution is the same as that obtained by substitution, using the constraint to solve for one variable in terms of the other and substituting that into the function to be maximized.

In this example, there were just two variables and y was given as an explicit function of l. More generally, with many variables, more complicated constraint functions and possibly multiple constraints, solution by substitution becomes infeasible and use of Lagrangians is the standard approach. Before discussing the general case, the two-variable case is considered in more detail.

12.2 The two-variable case

The following discussion considers in general terms the case where there are two choice variables and one equality constraint. Let the objective function be $f(x,y)$ and let the constraint be $c = g(x,y)$. Throughout, assume that $(x,y) \in Z$ where Z is an open set in $\mathscr{R}^2$. Thus, if $(\bar{x},\bar{y})$ are in Z, all points (x',y') sufficiently close to $(\bar{x},\bar{y})$ are in Z. Furthermore, f and g are assumed to be continuously differentiable (continuous with continuous derivatives). The task is to maximize or minimize $f(x,y)$ subject to (x,y), satisfying the constraint $c = g(x,y)$.

Definition 12.1: *A point* (x^*,y^*) *is a constrained local maximum if* $g(x^*,y^*)=0$ *and* $f(x^*,y^*) \geq f(x,y)$ *for all* (x,y) *in a neighborhood of* (x^*,y^*) *and satisfying* $g(x,y)=0$.

A constrained local minimum is similarly defined. To identify necessary and sufficient conditions for (x^*,y^*) to be a constrained local maximum or minimum it is useful to formulate the Lagrangian, $\mathscr{L}$:

$$\mathscr{L}(x,y,\lambda) = f(x,y) + \lambda[c - g(x,y)]. \tag{12.4}$$

In addition, define the *bordered Hessian*, $\bar{\boldsymbol{H}}$:

$$\bar{\boldsymbol{H}} = \begin{pmatrix} 0 & g_x & g_y \\ g_x & f_{xx} - \lambda g_{xx} & f_{xy} - \lambda g_{xy} \\ g_y & f_{xy} - \lambda g_{xy} & f_{yy} - \lambda g_{yy} \end{pmatrix}. \tag{12.5}$$

Sometimes, the bordered Hessian is written with a negative sign before the partial derivatives of the constraint ($-g_x$ and $-g_y$). This is equivalent to multiplying the first row and column by -1, a transformation that leaves the determinant of $\bar{\boldsymbol{H}}$ unchanged. With this notation:

Theorem 12.1: *Consider the problem of maximizing or minimizing the function* $f(x,y)$ *subject to a constraint* $g(x,y) = c$. *Let* $\mathscr{L}(x,y,\lambda) = f(x,y,\lambda) + \lambda[c - g(x,y)]$. *Suppose* (x^*,y^*,λ^*) *satisfies:*

$$\mathscr{L}_x = f_x(x^*,y^*) - \lambda^* g_x(x^*,y^*) = 0 \tag{12.6}$$

$$\mathscr{L}_y = f_y(x^*,y^*) - \lambda^* g_y(x^*,y^*) = 0 \tag{12.7}$$

$$\mathscr{L}_\lambda = c - g(x^*,y^*) = 0. \tag{12.8}$$

Then,

- *if* $|\bar{\boldsymbol{H}}| > 0$, (x^*,y^*) *is a constrained local maximum;*
- *if* $|\bar{\boldsymbol{H}}| < 0$, (x^*,y^*) *is a constrained local minimum.*

Assuming the derivatives are non-zero, conditions 12.6 and 12.7 imply that

$$\frac{f_x(x^*,y^*)}{f_y(x^*,y^*)} = \frac{\lambda^* g_x(x^*,y^*)}{\lambda^* g_y(x^*,y^*)} = \frac{g_x(x^*,y^*)}{g_y(x^*,y^*)}.$$

This, along with the constraint, $c - g(x^*, y^*) = 0$, yields two equations in the two unknowns, (x^*, y^*):

$$\frac{f_x(x^*, y^*)}{f_y(x^*, y^*)} = \frac{g_x(x^*, y^*)}{g_y(x^*, y^*)} \tag{12.9}$$

$$g(x^*, y^*) = c. \tag{12.10}$$

These may be solved for (x^*, y^*) and λ^* solved residually from Equations 12.6 or 12.7.

EXAMPLE 12.1: A cylinder has circumference length $2\pi r$, where r is the radius. The area of a circle of radius r is πr^2. Thus, the surface area of the cylinder of radius r and height h is: $a(r,h) = 2\pi r^2 + 2\pi rh$ (top and bottom + wall). The cylinder has a volume of $v(r,h) = \pi r^2 h$.

Consider the problem of maximizing volume subject to total surface area being A. This leads to the Lagrangian:

$$\begin{aligned}
\mathscr{L}(r,h,\lambda) &= v(r,h) + \lambda(A - a(r,h)) \\
\mathscr{L} &= \pi r^2 h + \lambda(A - 2\pi r^2 - 2\pi rh): \\
\frac{\partial \mathscr{L}}{\partial r} &= 2\pi rh - \lambda 4\pi r - \lambda 2\pi h = 0 && (1) \\
\frac{\partial \mathscr{L}}{\partial h} &= \pi r^2 - \lambda 2\pi r = 0 && (2) \\
\frac{\partial \mathscr{L}}{\partial \lambda} &= A - 2\pi r^2 - 2\pi rh = 0, && (3)
\end{aligned}$$

and $\frac{\partial \mathscr{L}}{\partial h}$ gives $\lambda = \frac{1}{2}r$. Substitute into (1): $2\pi rh - 2\pi r^2 - \pi rh = 0$, or $\pi r(2h - 2r - h) = 0$, or $h = 2r$. Substitute into (3): $A - 2\pi r^2 - 4\pi r^2 = 0$, which gives $A = 6\pi r^2$ or:

$$r^2 = \frac{A}{6\pi}, \quad r = \sqrt{\frac{A}{6\pi}}, \qquad h = 2r = 2\sqrt{\frac{A}{6\pi}} = \sqrt{\frac{4A}{6\pi}}.$$

Thus

$$\mathscr{L}^* = \pi \cdot r^2 \cdot h = \pi \frac{A}{6\pi} \times \sqrt{\frac{4A}{6\pi}} = \sqrt{\frac{4A^3}{36 \times 6\pi}} = \sqrt{\frac{A^3}{54\pi.}}$$

For example, if $A = 24\pi$ then $r = 2, h = 4, \lambda = 1, \mathscr{L}^* = 16\pi$. To check the second-order condition (with $g(h,r) = a(h,r)$, the constraint function is the area):

$$|\bar{\boldsymbol{H}}| = \begin{vmatrix} 0 & g_r & g_h \\ g_r & \mathscr{L}_{rr} & \mathscr{L}_{hr} \\ g_h & \mathscr{L}_{rh} & \mathscr{L}_{hh} \end{vmatrix} = \begin{vmatrix} 0 & 4\pi r + 2\pi h & 2\pi r \\ 4\pi r + 2\pi h & 2\pi h - \lambda 4\pi & 2\pi r - \lambda 2\pi \\ 2\pi r & 2\pi r - \lambda 2\pi & 0 \end{vmatrix}.$$

$$|\bar{\boldsymbol{H}}| = 32\pi^3 r^3 - 16\pi^3 r^2 \lambda + 8\pi^3 r^2 h - 16\pi^3 rh\lambda.$$

With $\lambda = \frac{1}{2}r$, this reduces to:

$$|\bar{\boldsymbol{H}}| = 24\pi^3 r^3.$$

This is sufficient for a maximum (local). The area and volume functions may be depicted:

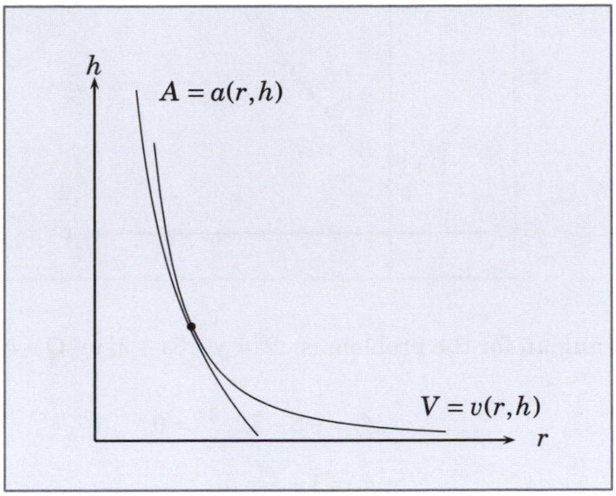

◊

REMARK 12.1: In fact, the problem in Example 12.1 is easily solved by substitution. For fixed surface area A, radius and height are connected by the formula: $A = 2\pi rh + 2\pi r^2$, so that

$$h(r) = \frac{A - 2\pi r^2}{2\pi r} = \frac{A}{2\pi r} - r.$$

Volume is given as a function of r and h by $v(r,h) = \pi r^2 h$. Substituting for h gives

$$v^*(r) = v(r, h(r)) = \pi r^2 h = \pi r^2 \left[\frac{A}{2\pi r} - r\right] = \frac{A}{2} r - \pi r^3.$$

From this, $v^{*\prime}(r) = \frac{A}{2} - 3\pi r^2$ and $v^{*\prime\prime}(r) = -6\pi r$, so the function v^* is concave and solving for r in $v^{*\prime}(r) = \frac{A}{2} - 3\pi r^2 = 0$ gives $r^2 = \frac{A}{6\pi}$ and $r = \sqrt{\frac{A}{6\pi}}$. □

EXAMPLE 12.2: A production technology gives combinations of x and y that yield a fixed level of output Q. The technology is described by the equation $y = Q + e^{2x}$. The cost of using these inputs is given by the equation $C = y - 3x$

(so that raising x reduces cost!). The task is to minimize cost subject to the production constraint.

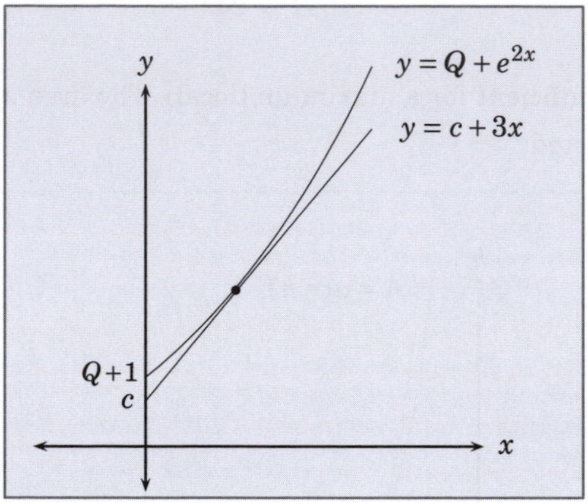

The Lagrangian for the problem is: $\mathscr{L} = y - 3x + \lambda[y - Q - e^{2x}]$. Therefore,

$$\mathscr{L}_x = -3 - \lambda 2e^{2x} = 0$$
$$\mathscr{L}_y = 1 + \lambda = 0$$
$$\mathscr{L}_\lambda = [y - Q - e^{2x}] = 0.$$

From the second equation, $\lambda = -1$ and substituting into the first equation gives: $3 = 2e^{2x}$. Solving for x, $x = \frac{1}{2}\ln\left(\frac{3}{2}\right)$. Since $y = Q + e^{2x}$, we get $y = Q + e^{2\frac{1}{2}\ln(\frac{3}{2})} = Q + e^{\ln(\frac{3}{2})} = Q + \frac{3}{2}$.

For the second-order condition, $\mathscr{L}_{xx} = -\lambda 4e^{2x}$, $\mathscr{L}_{yy} = 0$, $\mathscr{L}_{xy} = 0$. The constraint is $0 = g(x,y)$, where $g(x,y) = -y + Q + e^{2x}$, so that $g_x = +2e^{2x}$ and $g_y = -1$. The bordered Hessian is

$$\bar{\boldsymbol{H}} = \begin{pmatrix} 0 & 2e^{2x} & -1 \\ 2e^{2x} & -\lambda 4e^{2x} & 0 \\ -1 & 0 & 0 \end{pmatrix} \Rightarrow |\bar{\boldsymbol{H}}| = -1 \begin{vmatrix} 2e^{2x} & -\lambda 4e^{2x} \\ -1 & 0 \end{vmatrix} = \lambda 4e^{2x} = -4e^{2x} < 0.$$

Thus, the solution to the first-order conditions gives a local minimum. ◊

EXAMPLE 12.3: In this example, the task is to find a pair (x,y) that minimize the radius of a circle, subject to (x,y) being on a line:

$$\min f(x,y) = x^2 + (y-1)^2,$$

subject to $\alpha = y - kx$, $0 < \alpha < 1$, $0 < k$. The Lagrangian is

$$\mathscr{L} = x^2 + (y-1)^2 + \lambda(\alpha - y + kx).$$

This leads to first-order conditions:

$$\mathscr{L}_x = 2x\lambda k = 0$$
$$\mathscr{L}_y = 2(y-1) - \lambda = 0$$
$$\mathscr{L}_\lambda = (\alpha - y + kx) = 0.$$

From these, $y = \alpha + kx$, $y - 1 = (\alpha - 1) + kx$ so $2[(\alpha-1)+kx] - \lambda = 0$. Then $2x + 2[(\alpha-1)+kx]k = 0$ or $(1+k^2)x = (1-\alpha)k$, so $x = \frac{(1-\alpha)k}{1+k^2}$. Solving for y: $y = \alpha + \frac{(1-\alpha)k^2}{1+k^2} = \frac{\alpha+k^2}{1+k^2}$ and $\lambda = -\left(\frac{1}{k}\right)2x = -\frac{2(1-\alpha)}{1+k^2}$. Next, $\mathscr{L}_{xx} = 2$, $\mathscr{L}_{yy} = 2$, $\mathscr{L}_{xy} = 0$, and $g_x = -k$, $g_y = 1$. The bordered Hessian is

$$\bar{\boldsymbol{H}} = \begin{pmatrix} 0 & -k & 1 \\ -k & 2 & 0 \\ 1 & 0 & 2 \end{pmatrix},$$

and $|\bar{\boldsymbol{H}}| = k(-2k) + 1(-2) = -2(1+k^2) < 0$. A graph of the problem is given in the figure.

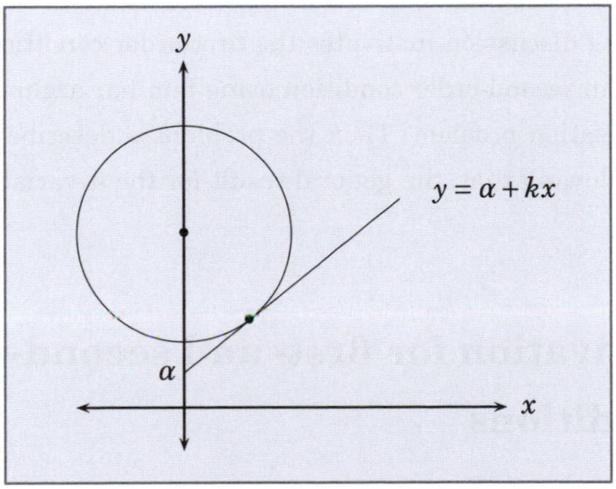

◇

REMARK 12.2: Theorem 12.1 gives *sufficient* conditions for a local maximum. But a local (or global) maximum may fail to satisfy these conditions. Suppose that $f(x,y) = xy$ and $g(x,y) = (x+y-1)^2$. Consider the problem of maximizing f subject to the constraint $g(x,y) = 0$. The constraint is satisfied if and only if $x+y-1=0$ or $x+y=1$ or $y = 1-x$. Substituting $y = 1-x$ into the objective and maximizing $\hat{f}(x) = x(1-x)$ gives $\hat{f}'(x) = 1-2x$ and $\hat{f}''(x) = -2$ so that the local and global maxima are found at $x^* = y^* = \frac{1}{2}$. But observe that $g_x(x,y) = g_y(x,y) = (x+y-1)$ so that at the solution $(x^*,y^*) = \left(\frac{1}{2},\frac{1}{2}\right)$, $g_x(x^*,y^*) = g_y(x^*,y^*) = 0$. However, $(f_x,f_y) = (y,x)$ so that $(f_x(x^*,y^*),f_y(x^*,y^*)) = \left(\frac{1}{2},\frac{1}{2}\right)$. So, for example, in Theorem 12.1, the condition $f_x(x^*,y^*) = \lambda^* g_x(x^*,y^*)$ fails, although (x^*,y^*) is the global optimum. Because $(g_x(x^*,y^*),g_y(x^*,y^*)) = (0,0)$, it is impossible to write $f_x(x^*,y^*) = \lambda^* g_x(x^*,y^*)$. □

The issue here is that at the solution, $(g_x(x^*,y^*),g_y(x^*,y^*)) = (0,0)$. Theorem 12.2 confirms this observation.

Theorem 12.2: *Let f and g be continuously differentiable functions on an open set Z in $\mathcal{R}^2$. Suppose that (x^*,y^*) is a constrained local maximum or minimum of f, subject to the constraint $g(x,y) = c$. Suppose that $(g_x(x^*,y^*),g_y(x^*,y^*)) \neq (0,0)$. Then there is a λ^* such that*

$$\begin{pmatrix} f_x(x^*,y^*) \\ f_y(x^*,y^*) \end{pmatrix} = \lambda^* \begin{pmatrix} g_x(x^*,y^*) \\ g_y(x^*,y^*) \end{pmatrix}.$$

Therefore, when $(g_x(x^*,y^*),g_y(x^*,y^*)) \neq (0,0)$, the conditions $f_x = g_x$, $f_y = g_y$ and $g = 0$ give three equations in three unknowns, (x,y,λ), which may be solved to identify a candidate solution to the optimization problem.

The following discussion motivates the first-order conditions and derives the bordered Hessian second-order condition using familiar arguments for the unconstrained optimization problem. Then the problem is described from a geometric perspective. Following that, the general result for the n-variable case is given, in Theorem 12.4.

12.3 Motivation for first- and second-order conditions

Observe that the constraint $g(x,y) = c$ may be used to solve for y as a function of x. Assume that either $\frac{\partial g}{\partial x}$ or $\frac{\partial g}{\partial y} \neq 0$, so that we can write one variable as a function of the other. Recall the discussion on implicit functions. Let $F(x,y) = 0 = g(x,y) - c$ and assume that F has continuous partial derivatives. If $\frac{\partial F}{\partial y} \neq 0$, then y can be written as a function, h, of x while if $\frac{\partial F}{\partial x} \neq 0$, then x can be written as a function φ, of y. Since $\frac{\partial F}{\partial y} = -\frac{\partial g}{\partial y}$, if $\frac{\partial g}{\partial y} \neq 0$, then there is a function with $y = h(x)$. Similarly, since $\frac{\partial F}{\partial x} = -\frac{\partial g}{\partial x}$, if $\frac{\partial g}{\partial x} \neq 0$ then there is a function φ with $x = \varphi(y)$. (These are local properties.) For example, if $g(x,y) = xy = c$, then $y = \frac{c}{x} = h(x)$; if $g(x,y) = \ln(xy) = c$ then $xy = e^c$, and $y = \left(\frac{1}{x}\right)e^c = h(x)$; if $g(x,y) = px + qy = c$, then $y = \frac{c}{q} - \frac{p}{q}x = h(x)$. Note that the presumption here is that $g_y \neq 0$. (In particular, this must hold at the candidate solution to the optimization problem, so that the issue raised in Remark 12.2 is assumed not to arise.)

12.3.1 The first-order condition

Assume that we can write y as a function of x: $y = h(x)$. Then, by substituting for y the problem reduces to $\max_x f(x,h(x))$ (or $\max_y f(\varphi(y),y)$ if x is substituted out). The original problem with two choice variables and a constraint has been reduced to a one-variable unconstrained problem – maximizing the function $\hat{f}(x) \stackrel{\text{def}}{=} f(x,h(x))$. With just two variables, this procedure is, in principle, straightforward. With more than two variables, the approach is more complex and in general it is more convenient to use Lagrangians.

The following calculations show that the solution from $\max_x f(x,h(x))$ and then determining y from x ($y = h(x)$) is the same as that obtained from the Lagrangian method, solving $\max f(x,y)$ subject to $g(x,y) = c$. Specifically, solving $\max_x f(x,h(x))$ leads to the same first-order conditions for (x,y) as obtained earlier, and the second-order condition is shown to be equivalent to the bordered Hessian condition. Maximizing $f(x,h(x))$ with respect to x, the first-order condition is

$$0 = \frac{d\hat{f}}{dx} = f_x + f_y h' = 0, \text{ where } h' = \frac{dy}{dx} = -\frac{g_x}{g_y}.$$

Thus,

$$0 = f_x + f_y\left(-\frac{g_x}{g_y}\right), \text{ so that } \frac{f_x}{g_x} = \frac{f_y}{g_y}.$$

This condition and $c = g(x,y)$ give the same conditions as obtained from the Lagrangian first-order conditions.

12.3.2 The second-order condition

Maximizing the function $\hat{f}(x) = f(x,h(x))$ gives a first-order condition for an optimum. The second-order condition, $\hat{f}''(x) < 0$ for a maximum or $\hat{f}''(x) > 0$ for a minimum, may also be related to the bordered Hessian condition: the discussion to follow shows that $\hat{f}'' < 0$ or $\hat{f}'' > 0$ is equivalent to the corresponding bordered Hessian conditions for a local constrained optimum.

For a maximum we require that $\frac{d^2f}{dx^2} < 0$, which is $\frac{d}{dx}\left\{\frac{df}{dx}\right\} = \frac{d}{dx}\{f_x + f_y h'\} < 0$. Evaluating this,

$$\begin{aligned}\frac{d}{dx}\{f_x + f_y h'\} &= \frac{d}{dx}\{f_x(x,h(x)) + f_y(x,h(x))h'(x)\}\\ &= f_{xx} + f_{xy}h' + f_{xy}h' + f_{yy}(h')^2 + f_y h''\\ &= f_{xx} + 2f_{xy}h' + f_{yy}(h')^2 + f_y h''.\end{aligned}$$

Since

$$\begin{aligned}h'(x) &= -\frac{g_x(x,h(x))}{g_y(x,h(x))},\\ h''(x) &= -\frac{g_y[g_{xx} + g_{xy}h'] - g_x[g_{xy} + g_{yy}h']}{g_y^2}\\ h'' &= -\frac{g_{xx}g_y - g_{xy}g_x - g_{xy}g_x + g_{yy}g_x^2/g_y}{g_y^2} \quad (\text{using } -h' = \frac{g_x}{g_y})\\ &= -\frac{g_{xx}g_y^2 - 2g_{xy}g_x g_y + g_{yy}g_x^2}{g_y^3}.\end{aligned}$$

Using $h' = \frac{g_x}{g_y}$, h'' as above and $\lambda = \frac{f_y}{g_y}$ gives the second derivative as:

$$\begin{aligned}\frac{d^2f}{dx} &= f_{xx} + 2f_{xy}\left[-\frac{g_x}{g_y}\right] + f_{yy}\left[\frac{g_x^2}{g_y^2}\right] + f_y h''\\ &= \frac{1}{g_y^2}\left\{f_{xx}g_y^2 - 2f_{xy}g_x g_y + f_{yy}g_x^2 - \frac{f_y}{g_y}(g_{xx}g_y^2 - 2g_{xy}g_x g_y + g_{yy}g_x^2)\right\}\\ &= \frac{1}{g_y^2}\{(f_{xx} - \lambda g_{xx})g_y^2 - 2(f_{xy} - \lambda g_{xy})g_x g_y + (f_{yy} - \lambda g_{yy})g_x^2\},\end{aligned}$$

so

$$\frac{d^2f}{dx} \times (g_y^2) = \{(f_{xx} - \lambda g_{xx})g_y^2 - 2(f_{xy} - \lambda g_{xy})g_x g_y + (f_{yy} - \lambda g_{yy})g_x^2\}.$$

Now, consider the determinant:

$$|\bar{\boldsymbol{H}}| = \begin{pmatrix} 0 & g_x & g_y \\ g_x & f_{xx} - \lambda g_{xx} & f_{xy} - \lambda g_{xy} \\ g_y & f_{xy} - \lambda g_{xy} & f_{yy} - \lambda g_{yy} \end{pmatrix}.$$

Evaluating the determinant:

$$\begin{aligned}|\bar{\boldsymbol{H}}| &= -g_x\{g_x(f_{yy}-\lambda g_{yy})-g_y(f_{xy}-\lambda g_{xy})\} - g_y\{-g_x(f_{xy}-\lambda g_{xy})+g_y(f_{xx}-\lambda g_{xx})\} \\ &= -\{(f_{xx}-\lambda g_{xx})g_y^2 + 2(f_{xy}-\lambda g_{xy})g_x g_y + (f_{yy}-\lambda g_{yy})g_x^2\} \\ &= -\frac{d^2 f}{dx^2}\times(g_y^2).\end{aligned}$$

Therefore, $|\bar{\boldsymbol{H}}|$ has the opposite sign to $\frac{\partial^2 f}{\partial x^2}$.

For a maximum, the condition $\frac{\partial^2 f}{\partial x^2} < 0$ is sufficient and this is equivalent to the condition:

$$\begin{vmatrix} 0 & g_x & g_y \\ g_x & f_{xx}-\lambda g_{xx} & f_{xy}-\lambda g_{xy} \\ g_y & f_{xy}-\lambda g_{xy} & f_{yy}-\lambda g_{yy} \end{vmatrix} > 0.$$

Since multiplying a row and a column by -1 does not change either the value or sign of the determinant, the determinant may also be written:

$$\begin{vmatrix} 0 & -g_x & -g_y \\ -g_x & f_{xx}-\lambda g_{xx} & f_{xy}-\lambda g_{xy} \\ -g_y & f_{xy}-\lambda g_{xy} & f_{yy}-\lambda g_{yy} \end{vmatrix}.$$

Conversely, in a minimization problem the determinant should be negative, i.e.,

$$\begin{vmatrix} 0 & g_x & g_y \\ g_x & f_{xx}-\lambda g_{xx} & f_{xy}-\lambda g_{xy} \\ g_y & f_{xy}-\lambda g_{xy} & f_{yy}-\lambda g_{yy} \end{vmatrix} < 0.$$

Finally, since $\mathcal{L}_{xx} = f_{xx} - \lambda g_{xx}$, etc., the determinant may be written succinctly:

$$|\bar{\boldsymbol{H}}| = \begin{vmatrix} 0 & g_x & g_y \\ g_x & \mathcal{L}_{xx} & \mathcal{L}_{xy} \\ g_y & \mathcal{L}_{yx} & \mathcal{L}_{yy} \end{vmatrix}.$$

12.3.3 A geometric perspective

With just one constraint, g, and one objective, f, constrained optimization involves locating a point where the objective is tangent to the constraint and then confirming that this is an optimum. The following discussion provides a different perspective. The slope of the constraint may also be measured in terms of a line perpendicular to the constraint. Figure 12.3 shows the perpendicular, P, to the line L and the curve $f(x,y)=c$ in each panel. In fact, the perpendicular to a curve $f(x,y)=c$ is proportional to the vector of partial derivatives, $(f_x(x,y), f_y(x,y))$. Furthermore, this vector P points in the direction of greatest local increase in the function and therefore has a direct connection to the optimization problem. To motivate this perspective, first consider the case where the curve is a line such as L in Figure 12.3A. From the figure, the slope of the line is $-\frac{y}{x}$ and the angle between L and the horizontal axis is α. With the vertical dashed line as the baseline, the slope of the angle β is $\frac{x}{y}$. Because α and β are in two corners of a right-angled triangle, these two angles sum to 90°. Consider the line P, drawn perpendicular to L. Note that the angles α and α^* are equal, and because the angles α^* and β^* sum to 90°, β^* is the same angle as β, and so has the same slope, $\frac{x}{y}$. Noting that the slope of α times the slope of β is $-\frac{y}{x}\times\frac{x}{y}=-1$, two lines, L and P, are perpendicular if and only if their slopes multiply to -1. Now, consider

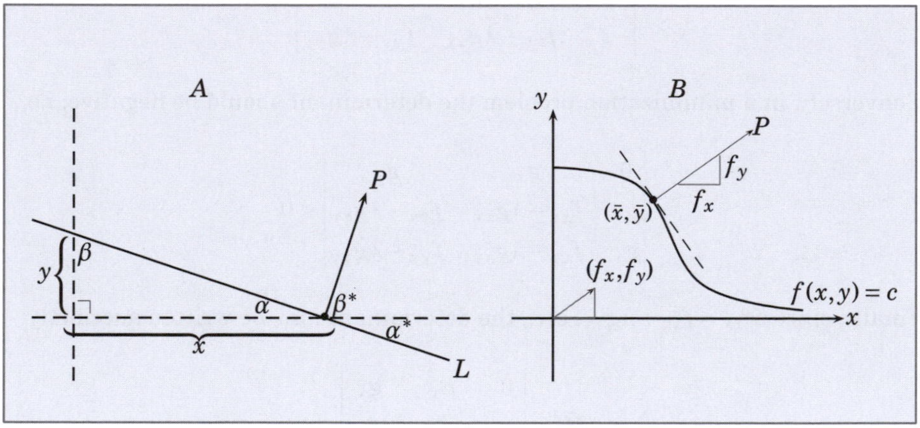

Figure 12.3: Orthogonal lines and normal vectors

Figure 12.3B. Along the level surface $f(x,y)=c$, and totally differentiating and setting to 0 gives: $0=f_x(x,y)dx+f_y(x,y)dy$, so that the slope along the level surface is $-\frac{f_x}{f_y}$. The slope at $(\bar{x},\bar{y})$ given by the dashed line is $\frac{dy}{dx}=-\frac{f_x(\bar{x},\bar{y})}{f_y(\bar{x},\bar{y})}$. The line P is chosen perpendicular to the dashed line. Since the slope of P, s_P, multiplied by the slope of the dashed line equals -1: $-1=-\frac{f_x(\bar{x},\bar{y})}{f_y(\bar{x},\bar{y})}\times s_P$, $s_P=\frac{f_y(\bar{x},\bar{y})}{f_x(\bar{x},\bar{y})}$. Thus, the direction P, starting from $(\bar{x},\bar{y})$, is proportional to $(f_x(\bar{x},\bar{y}), f_y(\bar{x},\bar{y}))$. The line perpendicular to the tangent to the level surface is proportional to $(f_x(\bar{x},\bar{y}), f_y(\bar{x},\bar{y}))$.

To illustrate these ideas in the economic context, consider utility maximization where the utility function is $u(x,y) = x^{\alpha} y^{\beta}$. The partial derivatives of u with respect to x and y are $u_x = \alpha x^{\alpha-1} y^{\beta} = \alpha \frac{u}{x}$ and $u_y = \beta x^{\alpha} y^{\beta-1} = \beta \frac{u}{y}$. Thus, $\nabla u(x,y) = (u_x, u_y) = \left(\alpha \frac{u}{x}, \beta \frac{u}{y}\right)$. Notice, for example, that at (x_1, y_1), x_1 is small relative to y_1, so that $\alpha \frac{u}{x_1}$ is large relative $\beta \frac{u}{y_1}$ and so the vector $\nabla u(x_1, y_1) = (\alpha \frac{u}{x_1}, \beta \frac{u}{y_1})$ has first component large relative to the second. In Figure 12.4, the dashed line

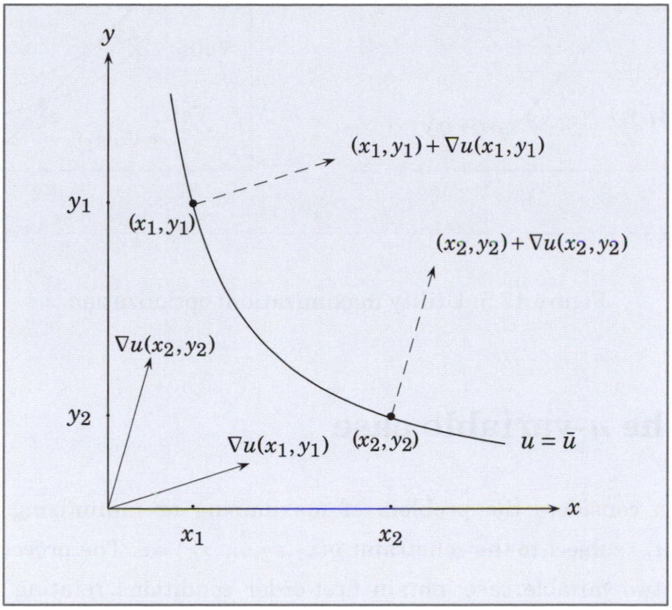

Figure 12.4: Utility maximization: the indifference curve

starting from (x_1, y_1) is the vector $\nabla u(x_1, y_1)$. Similarly, the dashed line starting from (x_2, y_2) is the vector $\nabla u(x_2, y_2)$. Notice that these dashed lines are perpendicular to the level surface of the function u (the set of points $\{(x,y) \mid u(x,y) = \bar{u}\}$). Also, notice that for $i = 1,2$, starting from (x_i, y_i) and moving in the direction $\nabla u(x_i, y_i)$ increases the value of the function. For utility maximization, the constraint has the form $g(x,y) = px + qy = c$, where p and q are the prices of x and y, respectively. In this case, $g_x = p$ and $g_y = q$, so that for all (x,y), $\nabla g(x,y) = (p,q)$ and the gradient of $g(x,y)$ is constant, independent of (x,y). This is depicted in Figure 12.5a. In Figure 12.5b, at $z_1 = (x_1, y_1)$ there is no number λ such that $\nabla u(z_1) = \lambda \nabla g(z_1)$ and the indifference curve cuts the budget constraint. The situation is similar at z_2. Neither z_1 nor z_2 is optimal. Whenever this is the case, $\nabla u(z) \neq \lambda \nabla g(z)$ for any λ, the slope of the indifference curve and the budget constraint differ so that z cannot be optimal. In contrast, at point z_3, $\nabla u(z_3) = \lambda \nabla g(z_3)$, where $\nabla u(z_3)$ is proportional to $\nabla g(z_3)$ and λ scales $\nabla g(z_3)$ to equal $\nabla u(z_3)$. This is the first-order condition.

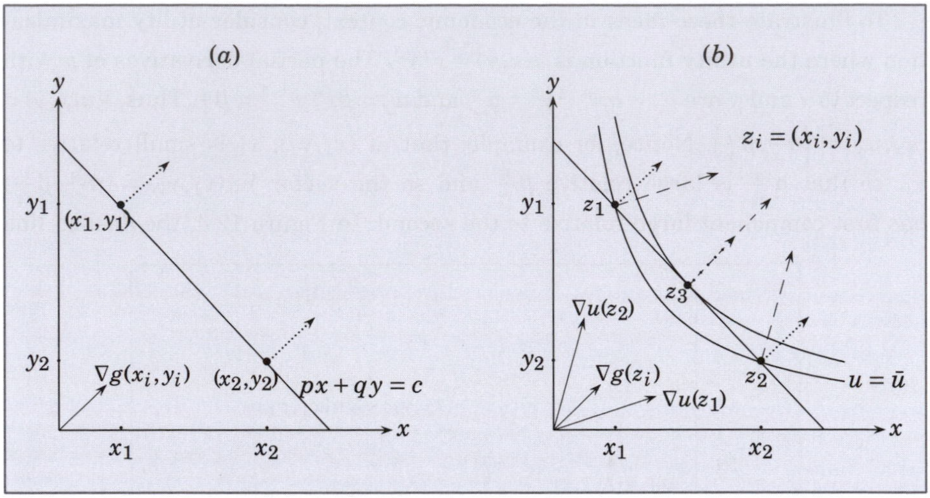

Figure 12.5: Utility maximization: optimization

12.4 The n-variable case

This section considers the problem of maximizing or minimizing a function $f(x_1,x_2,\dots,x_n)$ subject to the constraint $g(x_1,x_2,\dots,x_n)=c$. The procedure is similar to the two-variable case: obtain first-order conditions relating the partial derivatives of the objective to those of the constraint and then confirm that the solution to these equations actually is a local maximum or local minimum. The generalization of the Lagrangian to n variables is straightforward. Let

$$\mathcal{L} = f(x_1,x_2,\dots,x_n) + \lambda(c - g(x_1,\dots,x_n)).$$

The first-order conditions are:

$$\frac{\partial \mathcal{L}}{\partial x_i} = \mathcal{L}_i = f_i(x) - \lambda g_i(x) = 0$$
$$\frac{\partial \mathcal{L}}{\partial \lambda} = \mathcal{L}_\lambda = c - g(x) = 0.$$

Let (x^*,λ^*) be a solution to these $n+1$ equations. Whether this is a local maximum or local minimum depends on second-order conditions. These in turn are derived in terms of bordered matrices and associated quadratic forms, developed in the next section. After that, the main result is developed. This gives sufficiency conditions for a solution to the first-order conditions to be a local maximum or minimum.

12.4.1 Quadratic forms and bordered matrices

This section recalls some material on quadratic forms and matrix definiteness. A matrix $\boldsymbol{A}$ determines a quadratic form according to the expression $\boldsymbol{x}'\boldsymbol{A}\boldsymbol{x}$. The matrix $\boldsymbol{A}$ is said to be negative or positive definite if $\boldsymbol{x}'\boldsymbol{A}\boldsymbol{x} > 0$ or $\boldsymbol{x}'\boldsymbol{A}\boldsymbol{x} < 0$ for all non-zero vectors $\boldsymbol{x}$. The positive and negative definiteness criteria play a central role in identifying local optima in the study of unconstrained optimization problems. In constrained optimization, there is a comparable criterion based on bordered matrices. The following theorem presents a result on quadratic forms. Its application to constrained optimization is developed subsequently.

Theorem 12.3: *Let* $\mathbf{A}$ *be an* $n \times n$ *symmetric matrix and let* $\mathbf{b}$ *be an* $n \times 1$ *vector with* $\mathbf{b} \neq \mathbf{0}$. *Also, let* $\mathbf{C} = \mathbf{C}_n = \begin{pmatrix} 0 & \mathbf{b}' \\ \mathbf{b} & \mathbf{A} \end{pmatrix}$

$$\mathbf{C} = \begin{pmatrix} 0 & b_1 & b_2 & \cdots & b_n \\ b_1 & a_{11} & a_{12} & \cdots & a_{1n} \\ b_2 & a_{21} & a_{22} & \cdots & a_{2n} \\ \vdots & \vdots & \vdots & \ddots & \vdots \\ b_n & a_{n1} & a_{n2} & \cdots & a_{nn} \end{pmatrix}, \qquad \mathbf{C}_j = \begin{pmatrix} 0 & b_1 & b_2 & \cdots & b_j \\ b_1 & a_{11} & a_{12} & \cdots & a_{1j} \\ b_2 & a_{21} & a_{22} & \cdots & a_{2j} \\ \vdots & \vdots & \vdots & \ddots & \vdots \\ b_j & a_{j1} & a_{j2} & \cdots & a_{jj} \end{pmatrix}.$$

Then

- $\boldsymbol{\xi}'\mathbf{A}\boldsymbol{\xi} < 0$, *for all* $\boldsymbol{\xi} \neq 0$ *with* $\mathbf{b}'\boldsymbol{\xi} = 0$, *if and only if* $(-1)^j|\mathbf{C}_j| > 0$, *for all* $j = 2,3,\ldots,n$.
- $\boldsymbol{\xi}'\mathbf{A}\boldsymbol{\xi} > 0$ *for all* $\boldsymbol{\xi} \neq 0$ *with* $\mathbf{b}'\boldsymbol{\xi} = 0$, *if and only if* $|\mathbf{C}_j| < 0$, *for all* $j = 2,3,\ldots,n$.

The logic for this can be seen in the two-variable case. Let

$$\boldsymbol{b} = \begin{pmatrix} \alpha \\ \beta \end{pmatrix}, \quad \boldsymbol{A} = \begin{pmatrix} a_{11} & a_{12} \\ a_{21} & a_{22} \end{pmatrix}, \quad \boldsymbol{C} = \begin{pmatrix} 0 & \alpha & \beta \\ \alpha & a_{11} & a_{12} \\ \beta & a_{21} & a_{22} \end{pmatrix}.$$

Consider a quadratic form in $\boldsymbol{A}$, with $\boldsymbol{x}' = (x_1, x_2)$, $\boldsymbol{x}'\boldsymbol{A}\boldsymbol{x} = x_1^2 a_{11} + 2a_{12}x_1x_2 + a_{22}x_2^2$ (note $a_{12} = a_{21}$). Suppose that the constraint $(\alpha,\beta)\cdot(x_1,x_2) = 0$ is imposed. Then $x_2 = -\frac{\alpha}{\beta}x_1$ and imposing this condition on the quadratic form gives:

$$\boldsymbol{x}'\boldsymbol{A}\boldsymbol{x} = x_1^2 a_{11} + 2a_{12}x_1\left(-\frac{\alpha}{\beta}x_1\right) + a_{22}\left(-\frac{\alpha}{\beta}x_1\right)^2$$

$$= x_1^2\left[a_{11} + 2a_{12}\left(-\frac{\alpha}{\beta}\right) + a_{22}\left(-\frac{\alpha}{\beta}\right)^2\right]$$
$$= x_1^2\left[\beta^2 a_{11} - 2a_{12}\alpha\beta + a_{22}\alpha^2\right]\frac{1}{\beta^2}.$$

This expression is positive if and only if $[\beta^2 a_{11} - 2a_{12}\alpha\beta + a_{22}\alpha^2]$ is positive.

Now, consider the determinant of $\boldsymbol{C}$:

$$\begin{aligned}|\boldsymbol{C}| &= -\alpha(\alpha a_{22} - \beta a_{12}) + \beta(\alpha a_{21} - \beta a_{11}) \\ &= -\alpha^2 a_{22} + \alpha\beta a_{12} + \beta\alpha a_{21} - \beta^2 a_{11} \\ &= -\alpha^2 a_{22} + 2\alpha\beta a_{12} - \beta^2 a_{11} \\ &= -\left[\beta^2 a_{11} - 2a_{12}\alpha\beta + a_{22}\alpha^2\right].\end{aligned}$$

For $\boldsymbol{x} \neq 0$, $\boldsymbol{x}'\boldsymbol{A}\boldsymbol{x} > 0$ for all $\boldsymbol{x}$ satisfying $(\alpha,\beta)\cdot\boldsymbol{x} = 0$ if and only if $|\boldsymbol{C}| < 0$.

12.4.2 Bordered Hessians and optimization

In the present context of n-variable optimization, $(x_1,\ldots,x_n)$ are chosen to maximize the objective $f(x_1,\ldots,x_n)$, subject to the constraint $g(x_1,\ldots,x_n) = c$. Define the bordered Hessian matrices:

$$\bar{\boldsymbol{H}} = \begin{pmatrix} 0 & g_1 & g_2 & \cdots & g_n \\ g_1 & \mathcal{L}_{11} & \mathcal{L}_{12} & \cdots & \mathcal{L}_{1n} \\ g_2 & \mathcal{L}_{21} & \mathcal{L}_{22} & \cdots & \mathcal{L}_{2n} \\ \vdots & \vdots & \vdots & \ddots & \vdots \\ g_n & \mathcal{L}_{n1} & \mathcal{L}_{n2} & \cdots & \mathcal{L}_{nn} \end{pmatrix}, \qquad \bar{\boldsymbol{H}}_j = \begin{pmatrix} 0 & g_1 & g_2 & \cdots & g_j \\ g_1 & \mathcal{L}_{11} & \mathcal{L}_{12} & \cdots & \mathcal{L}_{1j} \\ g_2 & \mathcal{L}_{21} & \mathcal{L}_{22} & \cdots & \mathcal{L}_{2j} \\ \vdots & \vdots & \vdots & \ddots & \vdots \\ g_j & \mathcal{L}_{j1} & \mathcal{L}_{j2} & \cdots & \mathcal{L}_{jj} \end{pmatrix}.$$

Comparing notation, note that $\boldsymbol{b}$ corresponds to $\frac{\partial g}{\partial x}$. The matrix $\boldsymbol{A}$ corresponds to the matrix of second derivatives of the Lagrangian, and $\boldsymbol{C}_j$ corresponds to the bordered Hessian principal minor $\bar{\boldsymbol{H}}_j$. With this notation, the main result on constrained optimization is:

Theorem 12.4: *Let $\mathcal{L}(x_1,\ldots,x_n,\lambda) = f(x_1,\ldots,x_n) + \lambda(c - g(x_1,\ldots,x_n))$ and suppose that at (x^*,λ^*) the following are satisfied:*

$$\frac{\partial \mathcal{L}}{\partial x_i} = \mathcal{L}_i = 0,\ i = 1,\ldots,n$$
$$\frac{\partial \mathcal{L}}{\partial \lambda} = \mathcal{L}_\lambda = 0.$$

If, in addition, the bordered Hessian $\bar{\boldsymbol{H}}$ satisfies (1) at x^*, then x^* is a local maximum and if $\bar{\boldsymbol{H}}$ satisfies (2) at x^*, then x^* is a local minimum, where

1. $|\bar{\boldsymbol{H}}_2| > 0$, $|\bar{\boldsymbol{H}}_3| < 0$, $|\bar{\boldsymbol{H}}_4| > 0$, …, $(-1)^n |\bar{\boldsymbol{H}}_n| > 0$

2. $|\bar{\boldsymbol{H}}_2| < 0$, $|\bar{\boldsymbol{H}}_3| < 0$, …, $|\bar{\boldsymbol{H}}_n| < 0$.

Note that the theorem asserts that x^* is a *local* maximum or minimum; x^* may not be a global maximum or minimum. Some motivation for this result is as follows. The Lagrangian first-order necessary conditions for an optimum are:

$$\frac{\partial \mathscr{L}}{\partial x_i} = f_i(x) - \lambda g_i(x) = 0,\ i = 1,\ldots,n \tag{12.11}$$

$$\frac{\partial \mathscr{L}}{\partial \lambda} = c - g(x) = 0. \tag{12.12}$$

Figure 12.6 illustrates failure of the tangency conditions (Equation 12.11). The figure shows that when there is no λ such that condition $f_i = \lambda g_i$ for all i, then generally there is an objective improving change that satisfies the constraint. There is an n-variable analogy to Theorem 12.2.

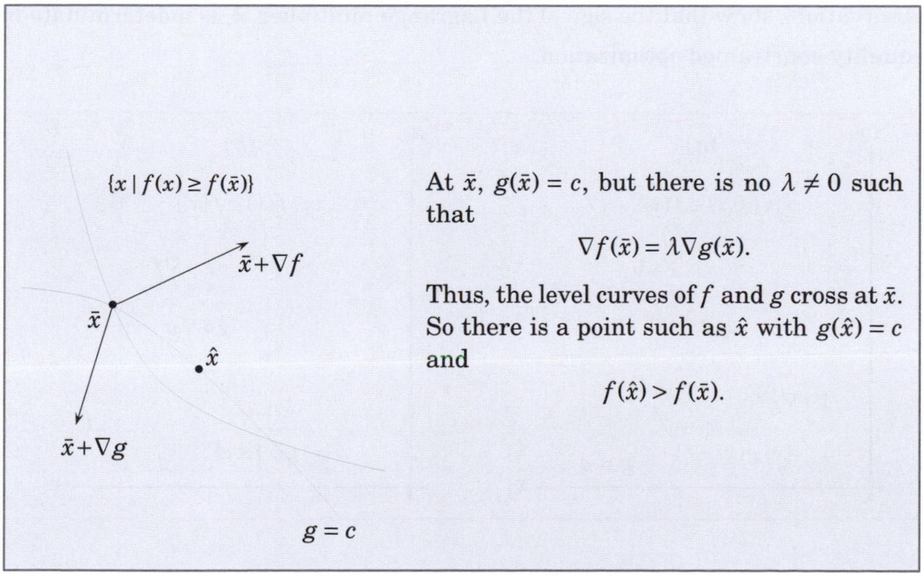

Figure 12.6: Tangency condition fails

Theorem 12.5: *Let f and g be continuously differentiable functions on an open set Z in $\mathscr{R}^n$. Suppose that (x^*) is a constrained local maximum or minimum of*

f, subject to the constraint $g(x) = c$. Suppose that $\nabla g(x^*) \neq 0$. Then there is a λ^* such that

$$\nabla f(x^*) = \lambda^* \nabla g(x^*).$$

12.4.2.1 The sign of the Lagrange multiplier

Sometimes it is useful to know the sign of the Lagrangian multiplier at the solution. However, with equality-constrained optimization, the multiplier may be positive or negative, depending on the way the constraint is expressed. Figure 12.7 illustrates the case where the first-order tangency conditions are satisfied. There are two possibilities. In the first, ∇f and ∇g "point" in the same direction, so that for some $\lambda > 0$, $\nabla f = \lambda \nabla g$. This occurs when the upper contour sets $\{x \mid g(x) \geq g(\bar{x})\}$ and $\{x \mid f(x) \geq f(\bar{x})\}$ are on the same side of the g and f curves through $\bar{x}$. This is depicted in (b).

When ∇f and ∇g "point" in opposite directions, then for some $\lambda < 0$, $\nabla f = \lambda \nabla g$. This occurs when the upper contour sets $\{x \mid g(x) \geq g(\bar{x})\}$ and $\{x \mid f(x) \geq f(\bar{x})\}$ are on opposite sides of the g and f curves through $\bar{x}$. This is depicted in (a). These observations show that the sign of the Lagrange multiplier, λ, is indeterminate in equality-constrained optimization.

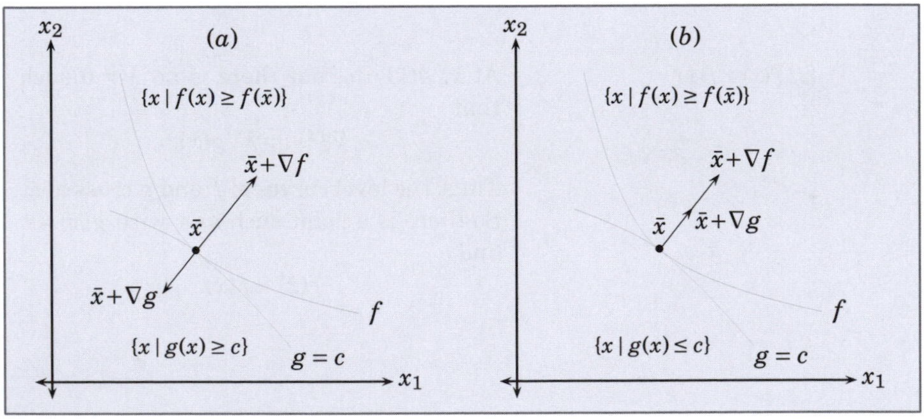

Figure 12.7: Tangencies at the solution

To illustrate, suppose that $f(x_1, x_2) = x_1 x_2$ and $g(x_1, x_2) = x_1^2 + x_2^2 = c$. Then

$$\nabla f = (f_{x_1}, f_{x_2}) = (x_2, x_1) \quad \text{and} \quad \nabla g = (g_{x_1}, g_{x_2}) = (2x_1, 2x_2),$$

so that $\nabla f = \frac{1}{2}\nabla g$, and $\lambda = \frac{1}{2}$. Both ∇f and ∇g point into the positive orthant with the vector ∇f twice the length of ∇g. Now, suppose that $\hat{g}(x_1,x_2) = -x_1^2 - x_2^2$ so that the constraint is unchanged if $\hat{g}(x_1,x_2) = -c$. Since $\nabla \hat{g} = (-2x_1, -2x_2) = -\nabla g$, $\nabla f = -\frac{1}{2}\nabla \hat{g}$ and $\hat{\lambda} = -\frac{1}{2}$. The sign of the Lagrange multiplier depends on the way in which the constraint is expressed.

12.4.3 The second-order condition reconsidered

The following discussion elaborates on the connection between the bordered Hessian conditions and the optimality of x^*, obtained as a solution to the first-order conditions. In particular, the discussion provides additional perspective on the connection between the bordered Hessian and the second-order condition.

The Lagrangian equals the objective function at any x satisfying the constraint:

$$\mathscr{L}(x,\lambda) = f(x) + \lambda[c - g(x)] = f(x).$$

At a constrained maximum, (x^*,λ^*), $f(x^*+v) - f(x^*) \leq 0$, for any v with $g(x^*+v) - c = 0$. Consider a Taylor series expansion of $\mathscr{L}(x^*+v,\lambda)$ around x^*. This gives:

$$\mathscr{L}(x^*+v,\lambda) = \mathscr{L}(x^*,\lambda) + \sum_i \frac{\partial \mathscr{L}(x^*,\lambda)}{\partial x_i} v_i + \frac{1}{2}\boldsymbol{v}'\boldsymbol{H}_{\mathscr{L}}(\hat{x},\lambda)\boldsymbol{v}$$
$$\mathscr{L}(x^*+v,\lambda) = \mathscr{L}(x^*,\lambda) + \frac{1}{2}\boldsymbol{v}'\boldsymbol{H}_{\mathscr{L}}(\hat{x},\lambda)\boldsymbol{v},$$

where $\boldsymbol{H}_{\mathscr{L}}(\hat{x},\lambda)$ is the matrix $\frac{\partial^2 \mathscr{L}}{\partial x_i \partial x_j}$ evaluated at $(\hat{x},\lambda)$ where $\hat{x}$ is a point between x^* and x^*+v, and where the second equality follows from the first-order conditions, $\frac{\partial \mathscr{L}}{\partial x_i} = 0$, $\forall i$.

Since x^* and x^*+v are feasible, $(c - g(x^*) = 0,\ c - g(x^*+v) = 0)$, $\mathscr{L}(x^*+v,\lambda) = f(x^*+v)$ and $\mathscr{L}(x^*,\lambda) = f(x^*)$, so from the previous expression

$$f(x^*+v) = f(x^*) + \frac{1}{2}\boldsymbol{v}'\boldsymbol{H}_{\mathscr{L}}(\hat{x},\lambda)\boldsymbol{v}.$$

Therefore $f(x^*+v) < f(x^*)$ if $\boldsymbol{v}'\boldsymbol{H}_{\mathscr{L}}(\hat{x},\lambda)\boldsymbol{v} < 0$.

The constraint function satisfies, using a first-order Taylor series,

$$0 = g(x^*+v) - c = g(x^*) - c + g_x(\tilde{x})\cdot v = g_x(\tilde{x})\cdot v,$$

where $\tilde{x}$ is (on the line) between x^* and x^*+v.

Recall that given a vector $\boldsymbol{b}$ and matrix $\boldsymbol{A}$, $\boldsymbol{z}'\boldsymbol{A}\boldsymbol{z} < 0$ for all $\boldsymbol{z}$ satisfying $\boldsymbol{b}'\boldsymbol{z} = 0$ if and only if the matrix

$$\boldsymbol{Q} = \begin{pmatrix} 0 & \boldsymbol{b}' \\ \boldsymbol{b} & \boldsymbol{A} \end{pmatrix}$$

satisfies the conditions $(-1)^j|\boldsymbol{Q}_j| > 0$, $j = 2,\ldots,n$. Equivalently, $|\boldsymbol{Q}_2| > 0$, $|\boldsymbol{Q}_3| < 0$, and so on. (Here, $\boldsymbol{Q}_j$ is the northwest principal minor of $\boldsymbol{Q}$, of order $j+1$.) Returning to the optimization problem, consider the matrices:

$$\boldsymbol{Q}^* = \begin{pmatrix} 0 & g_x(x^*)' \\ g_x(x^*) & H_{\mathscr{L}}(x^*,\lambda) \end{pmatrix} \quad \text{and} \quad \boldsymbol{Q}(\hat{x},\tilde{x}) = \begin{pmatrix} 0 & g_x(\tilde{x})' \\ g_x(\tilde{x}) & H_{\mathscr{L}}(\hat{x},\lambda) \end{pmatrix}.$$

Since these functions are continuous, for v small, $\boldsymbol{Q}(\hat{x},\tilde{x}) \approx \boldsymbol{Q}^*$ so that if $\boldsymbol{Q}^*$ satisfies $(-1)^j|\boldsymbol{Q}_j^*| > 0$ for $j \geq 2$, then $(-1)^j|\boldsymbol{Q}_j(\hat{x},\tilde{x})| > 0$ for $j \geq 2$ also. Therefore, since $\boldsymbol{g}_x(\tilde{x})\cdot\boldsymbol{v} = 0$, $\boldsymbol{v}'\boldsymbol{H}_{\mathscr{L}}(\hat{x},\lambda)\boldsymbol{v} < 0$ and so $f(x^*+v) < f(x^*)$.

12.5 Interpretation of the Lagrange multiplier

The Lagrangian is $\mathscr{L} = f(x_1,\ldots,x_n) + \lambda(c - g(x_1,\ldots,x_n))$. This has first-order conditions:

$$f_i(x_1^*,\ldots,x_n^*) - \lambda g_i(x_1^*,\ldots,x_n^*) = 0$$
$$c = g(x_1^*,\ldots,x_n^*) = 0.$$

Let (x^*,λ) be a solution. Note that x^* and λ depend on c: as c varies they will also. Thus x^* and λ may be written as functions of $c : x^*(c), \lambda(c)$. These functions give a solution at *each* value of c. In particular, the first-order conditions hold at each value of c:

$$f_i(x_1^*(c),\cdots,x_n^*(c)) - \lambda(c)g_i(x_1^*(c),\cdots,x_n^*(c)) = 0$$
$$c - g(x_1^*(c),\ldots,x_n^*(c)) = 0.$$

The value of the objective at c is: $\mathscr{L}(c) = f(x_1^*(c),\ldots,x_n^*(c)) + \lambda(c)[c - g(x_1^*(c),\ldots, x_n^*(c))]$. This function may be differentiated with respect to c to give:

$$\frac{d\mathscr{L}(c)}{dc} = f_1\frac{dx_1^*}{dc} + f_2\frac{dx_2^*}{dc} + \cdots + f_n\frac{dx_n^*}{dc} - \lambda(c)g_1\frac{dx_1^*}{dc} - \lambda(c)g_2\frac{dx_2^*}{dc} - \cdots - \lambda(c)g_n\frac{dx_n^*}{dc}$$
$$+ \frac{d\lambda}{dc}[c - g(x_1^*(c),\ldots,x_n^*(c))] + \lambda(c).$$

Rearranging:

$$\begin{aligned}\frac{d\mathscr{L}(c)}{dc} &= \sum_{i=1}^{n}[f_i(x_1^*(c)),\ldots,x_n^*(c)) - \lambda(c)g_i(x_1^*(c),\ldots,x_n^*(c))]\frac{dx_i^*}{dc} \\ &+ \frac{d\lambda}{dc}[c - g(x_1^*(c),\ldots,x_n^*(c)] + \lambda(c) \\ &= \lambda(c).\end{aligned}$$

This follows using the first-order conditions. Thus, $\frac{d\mathscr{L}}{dc} = \lambda$, the marginal impact on the value of the objective due to changing the constraint "a little", is equal to the Lagrange multiplier.

The next example uses the Lagrange multiplier to calculate a derivative: the change in a function due to a small change in its argument.

EXAMPLE 12.4: Consider the problem: $\max f(x) = x^2$ subject to $x = c$. The Lagrangian is:

$$\mathscr{L}(x,\lambda) = x^2 + \lambda(c - x),$$

with first-order conditions:

$$\begin{aligned}\mathscr{L}_x &= 2x - \lambda = 0, \Rightarrow x = \frac{\lambda}{2} \\ \mathscr{L}_\lambda &= c - x = 0 \Rightarrow x = c,\end{aligned}$$

so $\lambda(c) = 2c$, and $x(c) = c$.

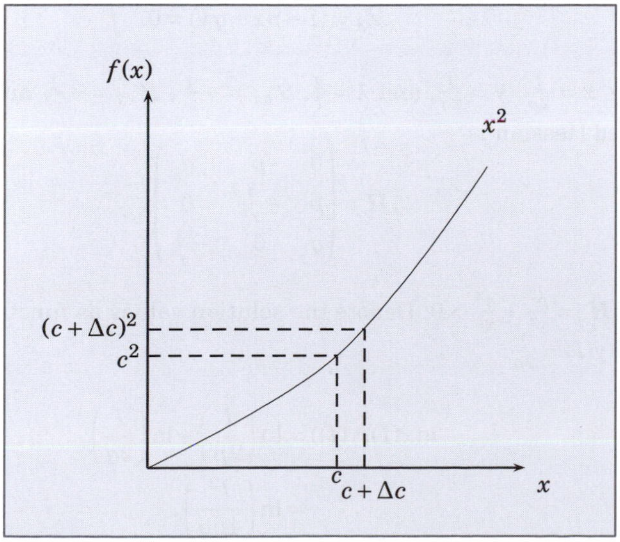

Thus, $\mathcal{L}^*(c) \stackrel{\text{def}}{=} \mathcal{L}(x(c),\lambda(c)) = x(c)^2 = c^2$ so that $\frac{d\mathcal{L}^*}{dc} = \lambda = 2c$.

The figure illustrates this. Note that varying c "a little" amounts to calculating the slope of the function, x^2, which at c is equal to $2c$. Since $\mathcal{L}^*(c) = \mathcal{L}(x(c),\lambda(c)) = f(x(c))$, the Lagrangian at the solution is identically equal to the value of the objective at the solution, $\mathcal{L}^{*'}(c) = f'(x(c))\frac{dx}{dc} = 2x(c)\frac{dx}{dc} = 2c$. ◊

Example 12.5 shows that in the standard utility maximization problem, the Lagrange multiplier measures the marginal utility of income: the utility gain an optimizing individual would achieve from a small increase in income.

EXAMPLE 12.5: Consider the standard utility maximization problem:

$$\max u(x,y) = \ln x + \ln y,$$

subject to $px + qy = I$. The Lagrangian is:

$$\mathcal{L} = \ln x + \ln y + \lambda(I - px - qy).$$

So, the first-order conditions are:

$$\begin{aligned}
\mathcal{L}_x &= \frac{1}{x} - \lambda p = 0 \\
\mathcal{L}_y &= \frac{1}{y} - \lambda q = 0 \\
\mathcal{L}_\lambda &= (I - px + qy) = 0.
\end{aligned}$$

Solving, $x = \frac{I}{2p}$, $y = \frac{I}{2q}$, and $\lambda = \frac{2}{I}$. $\mathcal{L}_{xx} = -\frac{1}{x^2}$, $\mathcal{L}_{yy} = -\frac{1}{y^2}$ and $\mathcal{L}_{xy} = 0$. The bordered Hessian is

$$\bar{\boldsymbol{H}} = \begin{pmatrix} 0 & p & q \\ p & -\frac{1}{x^2} & 0 \\ q & 0 & -\frac{1}{y^2} \end{pmatrix}.$$

Thus, $|\bar{\boldsymbol{H}}| = \frac{p^2}{y^2} + \frac{q^2}{x^2} > 0$. Denote the solution values as functions of I, $x(I) = \frac{I}{2p}$ and $y(I) = \frac{I}{2q}$:

$$\begin{aligned}
u(x(I),y(I)) &= \ln\left(\frac{I}{2p}\right) + \ln\left(\frac{I}{2q}\right) \\
&= \ln\left(\frac{I^2}{4pq}\right).
\end{aligned}$$

Thus,

$$\frac{du(x(I),y(I))}{dI} = \frac{1}{\frac{I^2}{4pq}} \cdot \frac{I}{2pq} = \frac{2}{I} = \lambda.$$

Observe that if we differentiate $\mathscr{L}$ with respect to I we get λ and evaluating at the solution value for λ gives $\frac{2}{I}$.

The same reasoning allows us to conclude that the marginal impact on maximum utility of a change in price p is given by $-\lambda(I)x(I) = -\frac{2}{I}\frac{I}{2p} = -\frac{1}{p}$, differentiating the Lagrangian with respect to p and evaluating at the solution. Differentiating $u(x(I),y(I))$ with respect to p gives $-\frac{1}{\frac{I^2}{4pq}} \cdot \frac{I^2}{4p^2q} = -\frac{1}{p}$, so that the marginal impact of an increase in the price of x is $-\frac{1}{p}$. ◊

Example 12.6 (Cost minimization subject to a production quota): The cost function is $c = rK + wL$ and the production function is $Q = K^\alpha L^\beta$. The problem is:

$$\min \mathscr{L} = rK + wL + \lambda(Q - K^\alpha L^\beta).$$

The first-order conditions are:

$$\mathscr{L}_K = r - \lambda\alpha\frac{K^\alpha L^\beta}{K} = 0 \Rightarrow r - \lambda\alpha\frac{Q}{K} = 0 \tag{1}$$

$$\mathscr{L}_L = w - \lambda\beta\frac{K^\alpha L^\beta}{L} = 0 \Rightarrow w - \lambda\beta\frac{Q}{L} = 0 \tag{2}$$

$$\mathscr{L}_\lambda = Q - K^\alpha L^\beta = 0. \tag{3}$$

Solving using (1) and (2), $\frac{r}{w} = \frac{\alpha}{\beta}\frac{L}{K}$, which implies that $K = \frac{\alpha w}{\beta r}L$. Then

$$Q = K^\alpha L^\beta = \left(\frac{\alpha w}{\beta r}\right)^\alpha L^\alpha L^\beta = \left(\frac{\alpha w}{\beta r}\right)^\alpha L^{\alpha+\beta}.$$

Thus

$$L = \left(\frac{\alpha w}{\beta r}\right)^{\frac{-\alpha}{\alpha+\beta}} Q^{\frac{1}{\alpha+\beta}}.$$

So

$$K = \frac{\alpha w}{\beta r}L = \left(\frac{\alpha w}{\beta r}\right)\left(\frac{\alpha w}{\beta r}\right)^{\frac{-\alpha}{\alpha+\beta}} Q^{\frac{1}{\alpha+\beta}} = \left(\frac{\alpha w}{\beta r}\right)^{\frac{\beta}{\alpha+\beta}} Q^{\frac{1}{\alpha+\beta}}.$$

The solutions can also be written:

$$K = \left(\frac{r}{\alpha}\right)^{\frac{-\beta}{\alpha+\beta}} \left(\frac{w}{\beta}\right)^{\frac{\beta}{\alpha+\beta}} Q^{\frac{1}{\alpha+\beta}} = K(Q)$$

$$L = \left(\frac{r}{\alpha}\right)^{\frac{\alpha}{\alpha+\beta}} \left(\frac{w}{\beta}\right)^{\frac{-\alpha}{\alpha+\beta}} Q^{\frac{1}{\alpha+\beta}} = L(Q).$$

As required, these satisfy $K(Q)^{\alpha}L(Q)^{\beta} = Q, \forall Q$. To solve for λ, observe that

$$\lambda = \left(\frac{r}{\alpha}\right)\frac{K}{Q} = \left(\frac{r}{\alpha}\right)\left(\frac{r}{\alpha}\right)^{\frac{-\beta}{\alpha+\beta}} \left(\frac{w}{\beta}\right)^{\frac{\beta}{\alpha+\beta}} Q^{\frac{1}{\alpha+\beta-1}} = \left(\frac{r}{\alpha}\right)^{\frac{\alpha}{\alpha+\beta}} \left(\frac{w}{\beta}\right)^{\frac{\beta}{\alpha+\beta}} Q^{\frac{1-(\alpha+\beta)}{\alpha+\beta}}.$$

Compare this with $\frac{dc}{dQ}$. Note that $c(Q) = rK(Q) + wL(Q)$,

$$\begin{aligned}
c(Q) &= rK(Q) + wL(Q) \\
&= \alpha\left(\frac{r}{\alpha}\right)K(Q) + \beta\left(\frac{w}{\beta}\right)L(Q) \\
&= \alpha\left(\frac{r}{\alpha}\right)^{\frac{\alpha}{\alpha+\beta}} \left(\frac{w}{\beta}\right)^{\frac{\beta}{\alpha+\beta}} Q^{\frac{1}{(\alpha+\beta)}} + \beta\left(\frac{r}{\alpha}\right)^{\frac{\alpha}{\alpha+\beta}} \left(\frac{w}{\beta}\right)^{\frac{\beta}{\alpha+\beta}} Q^{\frac{1}{(\alpha+\beta)}} \\
&= (\alpha+\beta)\left(\frac{r}{\alpha}\right)^{\frac{\alpha}{\alpha+\beta}} \left(\frac{w}{\beta}\right)^{\frac{\beta}{\alpha+\beta}} Q^{\frac{1}{(\alpha+\beta)}}.
\end{aligned}$$

Therefore,

$$\begin{aligned}
\frac{dc}{dQ} &= \left(\frac{r}{\alpha}\right)^{\frac{\alpha}{\alpha+\beta}} \left(\frac{w}{\beta}\right)^{\frac{\beta}{\alpha+\beta}} Q^{\frac{1-(\alpha+\beta)}{\alpha+\beta}} \\
&= \lambda(Q).
\end{aligned}$$

◇

EXAMPLE 12.7: A production technology relating output Q to inputs x and y is given by the equation $y = Q + e^{kx}$. The cost function is given by $C = y - \delta x$, where $\delta > k$. Consider minimizing cost by choice of x and y.

In this case the Lagrangian is: $\mathscr{L} = y - \delta x + \lambda[y - Q - e^{kx}]$. Therefore,

$$\begin{aligned}
\mathscr{L}_x &= -\delta - \lambda k e^{kx} = 0 \\
\mathscr{L}_y &= 1 + \lambda = 0 \\
\mathscr{L}_\lambda &= [y - Q - e^{kx}] = 0.
\end{aligned}$$

From the second equation, $\lambda = -1$ and substituting into the first equation gives: $\delta = ke^{kx}$. Solving for x, $x = \frac{1}{k}\ln\left(\frac{\delta}{k}\right)$. Since $y = Q + e^{kx}$, we get $y = Q + e^{k\frac{1}{k}\ln(\frac{\delta}{k})} = Q + e^{\ln(\frac{\delta}{k})} = Q + \frac{\delta}{k}$.

For the second-order condition, $\mathscr{L}_{xx} = -\lambda k^2 e^{kx}$, $\mathscr{L}_{yy} = 0$, $\mathscr{L}_{xy} = 0$. The constraint is $0 = g(x,y)$, where $g(x,y) = -y + Q + e^{kx}$, so that $g_x = +ke^{kx}$ and $g_y = -1$. The bordered Hessian is

$$\bar{\boldsymbol{H}} = \begin{pmatrix} 0 & ke^{kx} & -1 \\ ke^{kx} & -\lambda k^2 e^{2x} & 0 \\ -1 & 0 & 0 \end{pmatrix} \Rightarrow |\bar{\boldsymbol{H}}| = -1 \begin{vmatrix} ke^{kx} & -\lambda k^2 e^{kx} \\ -1 & 0 \end{vmatrix}$$

$$= \lambda k^2 e^{kx} = -k^2 e^{kx} < 0.$$

Thus, the solution to the first-order conditions gives a local minimum.

Write the solution values of x and y as $x(k,\delta)$ and $y(k,\delta)$, and the solution value of the Lagrange multiplier as $\lambda(k,\delta)$. Then the cost is given by $C(k,\delta) = y(k,\delta) - \delta x(k,\delta)$. Using the envelope theorem

$$\frac{\partial C}{\partial k} = -\lambda(k,\delta)x(k,\delta)e^{kx(k,\delta)}, \quad \frac{\partial C}{\partial \delta} = -x(k,\delta).$$

Note the answers are equal to the derivative of the Lagrangian with respect to the appropriate variable, evaluated at the solution. ◊

EXAMPLE 12.8: Recall Example 12.3: $\max f(x,y) = x^2 + (y-1)^2$ subject to $\alpha = y - kx$, $0 < \alpha < 1$ and $k > 0$. The solution: $x^* = \frac{(1-\alpha)k}{1+k^2}$, $y^* = \frac{\alpha+k^2}{1+k^2}$, and $\lambda^* = -\frac{2(1-\alpha)}{1+k^2}$. Thus, $y^* - 1 = \frac{\alpha+k^2-(1+k^2)}{1+k^2} = \frac{\alpha-1}{1+k^2}$, so that

$$\begin{aligned} \mathscr{L}^* &= (x^*)^2 + (y^*-1)^2 \\ &= \left(\frac{(1-\alpha)k}{1+k^2}\right)^2 + \left(\frac{\alpha-1}{1+k^2}\right)^2 \\ &= \left(\frac{1}{1+k^2}\right)^2 [((1-\alpha)k)^2 + (\alpha-1)^2] \\ &= \left(\frac{1}{1+k^2}\right)^2 [(1-\alpha)^2(1+k^2)] \\ &= \left(\frac{1}{1+k^2}\right)(1-\alpha)^2. \end{aligned}$$

Therefore,

$$\frac{\partial \mathscr{L}^*}{\partial \alpha} = -\left(\frac{1}{1+k^2}\right)2(1-\alpha) = \lambda^*.$$

Thus, the impact on the constrained objective is obtained by differentiating the Lagrangian with respect to the parameter and evaluating the derivative at the solution. ◊

EXAMPLE 12.9: Recall the optimal design of the cylinder in Example 12.1. There $\mathscr{L}^* = \left(\frac{A^3}{54\pi}\right)^{\frac{1}{2}}$. Differentiating with respect to A,

$$\frac{\partial \mathscr{L}^*}{\partial A} = \frac{1}{2}\left[\left(\frac{A^2}{9}\right)r^2\right]^{-\frac{1}{2}}\left(\frac{3A^2}{54\pi}\right)$$

Using the fact that $r^2 = \left(\frac{A}{6\pi}\right)$,

$$\frac{\partial \mathscr{L}^*}{\partial A} = \frac{1}{2}r = \lambda \,(\text{or } \lambda^*).$$

Again, this is the derivative of the Lagrangian with respect to A, evaluated at the solution. ◊

The discussion this far has focused on the identification of a local optimum. Section 12.6 provides conditions under which a local optimum is also a global optimum.

12.6 Optimization: concavity and convexity

In a constrained optimization problem, the shape of the objective function and the constraints play an important role in characterizing global optima – just as in the unconstrained case. For this discussion, assume that the domain of the function is convex.

Definition 12.2: *A set* $\boldsymbol{X}$ *is convex if*

$$\forall x, y \in \boldsymbol{X} \Rightarrow \theta x + (1-\theta)y \in \boldsymbol{X}, \forall \theta \in [0,1].$$

The main result below relies on quasiconcavity of the objective function. Definition 12.3 introduces a variety of notions of quasiconcavity.

Definition 12.3: *Concavity and quasiconcavity.*

1. *A function f is concave if* $f(\theta x+(1-\theta)y) \geq \theta f(x)+(1-\theta)f(y)$, $\forall x, y \in \boldsymbol{X}$, $\theta \in [0,1]$.

2. *A function f is strictly concave if* $f(\theta x+(1-\theta)y) > \theta f(x)+(1-\theta)f(y)$, $\forall x \neq y \in \boldsymbol{X}$, $\theta \in (0,1)$.

3. *A function f is quasiconcave if* $f(\theta x+(1-\theta)y) \geq \min\{f(x), f(y)\}$, $\forall x, y \in \boldsymbol{X}$, $\forall \theta \in [0,1]$.

4. *A function f is strictly quasiconcave if* $f(\theta x+(1-\theta)y) > \min\{f(x), f(y)\}, \forall x \neq y, \theta \in (0,1)$.

5. *A function f is explicitly quasiconcave if it is quasiconcave and* $f(x) > f(y) \Rightarrow f(\theta x+(1-\theta)y) > f(y)$, $\forall \theta \in (0,1)$.

Concavity is a more demanding requirement than quasiconcavity.

PROPOSITION 12.1: If f is (strictly) concave then f is (strictly) quasiconcave.◊

PROOF: If f is concave then $f(\theta x+(1-\theta)y) \geq \theta f(x)+(1-\theta)f(y) \geq \min\{f(x), f(y)\}$. With strict quasiconcavity and $x \neq y$, $f(\theta x+(1-\theta)y) > \theta f(x)+(1-\theta)f(y) \geq \min\{f(x), f(y)\}$. ■

Furthermore,

PROPOSITION 12.2: If f is strictly quasiconcave then f is explicitly quasiconcave. ◊

PROOF: If f is strictly quasiconcave and $f(x) > f(y)$ then $f(\theta x+(1-\theta)y) > \min\{f(x), f(y)\} > f(y)$ and this confirms the proposition. ■

The relations are depicted in Figure 12.8. Notice that explicit quasiconcavity allows flat sections at the top of the function (a property excluded by strict quasiconcavity).

Recall that concavity is related to the Hessian matrix $\boldsymbol{H}(x) = \left\{\frac{\partial^2 f(x)}{\partial x \partial x'}\right\}$,

$$\boldsymbol{H}(x) = \begin{pmatrix} f_{11}(x) & f_{12}(x) & \cdots & f_{1n}(x) \\ f_{21}(x) & f_{22}(x) & \cdots & f_{2n}(x) \\ \vdots & \vdots & \ddots & \vdots \\ f_{n1}(x) & f_{n2}(x) & \cdots & f_{nn}(x) \end{pmatrix},$$

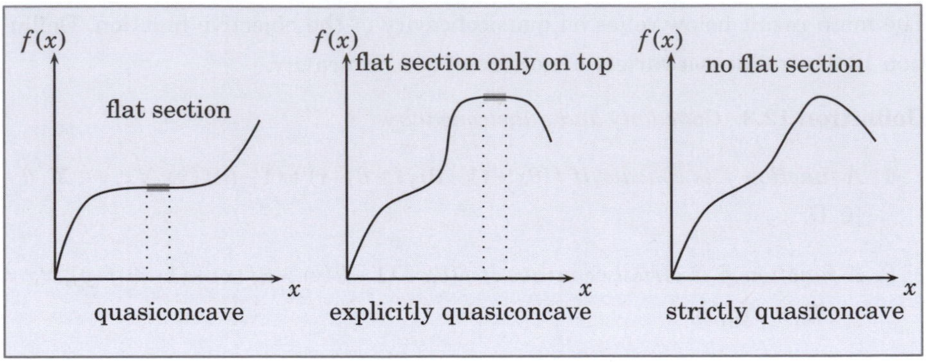

Figure 12.8: Quasiconcavity

where the dependence of the partial derivatives on $x = (x_1, x_2, \ldots, x_n)$ is made explicit.

As discussed earlier, if f is concave then the Hessian matrix $\boldsymbol{H}(x)$ is negative semi-definite for all x. In fact, f is concave if and only if $\boldsymbol{H}(x)$ is negative semi-definite for all x. If f is strictly concave then $\boldsymbol{H}(x)$ is negative semi-definite for all x; and, conversely, if $\boldsymbol{H}(x)$ is negative definite for all x then f is strictly concave. Notice that strict concavity of f does not imply that $\boldsymbol{H}(x)$ is negative definite for all x. This observation is clarified in the following remark.

REMARK 12.3: While an everywhere negative definite Hessian matrix implies strict concavity of the associated function, the converse is not the case: a strictly concave function need not have an everywhere negative definite Hessian matrix. The following example confirms the latter observation. Consider $f(x) = -x^4$, with $f'(x) = -4x^3$ and $f''(x) = -12x^2$. Thus, $f''(x) \leq 0$ with $f''(0) = 0$, so viewing $f''(x)$ as the Hessian matrix (with just one entry), this is not negative definite for all x, since the requirement fails at $x = 0$. However, this function is strictly concave as the following discussion demonstrates. To do so, it is necessary to show that when $0 < \theta < 1$ and $y \neq x$, then $f(\theta x + (1-\theta)y) > \theta f(x) + (1-\theta)f(y)$. With $f(x) = -x^4$,

$$\begin{aligned} f(\theta x + (1-\theta)y) &= -(\theta x + (1-\theta)y)^4 \\ &= -\{4\theta^3 x^3 y + 6\theta^2 x^2 y^2 - 4\theta^4 x^3 y + 12\theta^3 x y^3 + 6\theta^4 x^2 y^2 + 4\theta x y^3 \\ &\quad - 4\theta^4 x y^3 - 12\theta^2 x y^3 - 12\theta^3 x^2 y^2 + \theta^4 x^4 - 4\theta y^4 + y^4 \\ &\quad + 6\theta^2 y^4 - 4\theta^3 y^4 + \theta^4 y^4\}. \end{aligned}$$

Let

$$\delta = f(\theta x+(1-\theta)y)-[\theta f(x)+(1-\theta)f(y)] = -(\theta x+(1-\theta)y)^4 - \{-\theta x^4-(1-\theta)y^4\}.$$

The function f is strictly concave if $\delta > 0$ whenever $\theta \in (0,1)$ and $x \neq y$.

$$\begin{aligned}\delta &= -\{4\theta^3x^3y+6\theta^2x^2y^2-4\theta^4x^3y+12\theta^3xy^3+6\theta^4x^2y^2+4\theta xy^3-4\theta^4xy^3\\ &\quad -12\theta^2xy^3-12\theta^3x^2y^2+\theta^4x^4-3\theta y^4+6\theta^2y^4-4\theta^3y^4+\theta^4y^4-\theta x^4\}\\ &= -\{\theta(x-y)^2(\theta-1)\left[\theta^2y^2-2\theta^2xy+\theta^2x^2+2\theta xy-3\theta y^2+\theta x^2+3y^2+2xy+x^2\right]\}\\ &= \theta(1-\theta)(x-y)^2\left[\theta^2y^2+2\theta(1-\theta)xy+\theta^2x^2+3(1-\theta)y^2+\theta x^2+2xy+x^2\right]\\ &= \theta(1-\theta)(x-y)^2\left[\theta^2y^2+(2\theta(1-\theta)+1)xy+\theta^2x^2+3(1-\theta)y^2+\theta x^2+x^2\right].\end{aligned}$$

From this expression, it is clear that with $x \neq y$ and $\theta \in (0,1)$, $\delta > 0$ if $xy \geq 0$. It remains to show that $\delta > 0$ for $xy < 0$. Suppose that $xy < 0$. Take $x < 0$ and $y > 0$ (the reasoning is the same for the case $y < 0$ and $x > 0$), and set $z = \theta x + (1-\theta)y$ so that $x < z < y$. Regarding z there are two possibilities: $z < 0$ or $z > 0$. The analysis is the same in both cases, so consider the case where $z < 0$. Then there is a β, $0 < \beta < 1$ with $z = \beta x+(1-\beta)0 = \beta x$. So, $z = \beta x+(1-\beta)0 = \theta x+(1-\theta)y$ and thus $(\beta-\theta)x = (1-\theta)y$. Since $y > 0$ and $x < 0$, $\beta < \theta$. Therefore,

$$f(z) = f(\beta x+(1-\beta)0) > \beta f(x)+(1-\beta)f(0) > \theta f(x)+(1-\theta)f(0) > \theta f(x)+(1-\theta)f(y),$$

where the first inequality follows because $f(\beta x) = -\beta^4x^4 > -\beta x^4 = \beta f(x)$ and $f(0) = 0$; the second because $\beta < \theta$ and $f(0) = 0$; and the third because $f(0) > f(y)$. An identical argument applies when $z > 0$. □

So far the results apply only to finding local maxima or minima. For a global result it is necessary to impose additional assumptions. The following result assumes that the constraint set is convex: $\{x \mid g(x) = c\}$ is convex. Observe that $\{x \mid g(x) = c\}$ convex implies $g(x) = a_0 + a_1x_1 + \cdots + a_nx_n$, $(g(\theta x+(1-\theta)y) = \theta g(x) + (1-\theta)g(y)$, $\forall (x,y)$.) In Figure 12.9, the set of x_i that satisfy $g(x) = c$ lie on the straight line.

Theorem 12.6: *Let* $\mathscr{L}(x_1,\ldots,x_n,\lambda) = f(x_1,\ldots,x_n)+\lambda(c-g(x_1,\ldots,x_n))$, *where*

(a) f *is explicitly quasiconcave*

(b) $\{x \mid g(x) = c\}$ *is convex.*

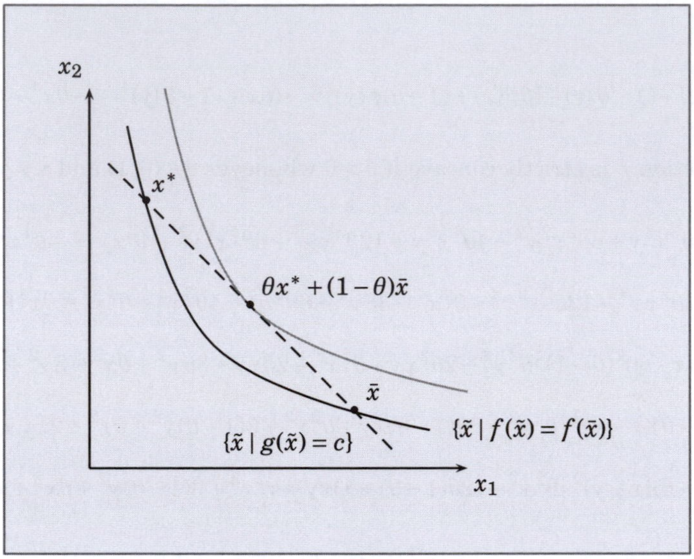

Figure 12.9: A linear constraint

If (x^*, λ^*) *satisfy,*

1. $\mathscr{L}_i(x^*, \lambda^*) = 0$, $i = 1, \ldots, n$
   $\mathscr{L}_\lambda(x^*, \lambda^*) = 0$

2. $\bar{\boldsymbol{H}}_2(x^*) > 0$, $\bar{\boldsymbol{H}}_3(x^*) < 0, \ldots, (-1)^n \bar{\boldsymbol{H}}_n(x^*) > 0$,

then x^* *is a global maximum of the constrained program.*

PROOF: Suppose (x^*, λ^*) satisfies the conditions (1) and (2) but is not a global maximum. Since x^* is not a global maximum, $\exists y, g(y) = c$ and $f(y) > f(x^*)$. Since g is linear, $g(\theta y + (1-\theta)x^*) = c$, or $g(x^* + \theta(y - x^*)) = c$. So $\theta y + (1-\theta)x^* = x^* + \theta(y - x^*)$ is feasible and since f is explicitly quasiconcave, $f(x^* + \theta(y - x^*)) > f(x^*), \forall \theta \in (0,1)$. However, since (x^*, λ^*) satisfy (1) and (2), from Theorem 12.4, x^* is a local maximum so that in some neighborhood of x^*, all $\tilde{x}$ satisfying $g(\tilde{x}) = c$ also satisfy $f(\tilde{x}) \le f(x^*)$. However, θ can be chosen arbitrarily small and $f(x^* + \theta(y - x^*)) > f(x^*)$. Thus, there are feasible points arbitrarily close to x^* with a higher value of f; a contradiction. ■

The theorem is not valid without the assumption that (2) holds, as the next example illustrates.

EXAMPLE 12.10: Define the function $f_1(x)$:

$$\begin{aligned} f_1(x) &= (x-1)^3,\ x \le 2 \\ &= (x-1)^3 - (x-2)^4,\ x > 2. \end{aligned}$$

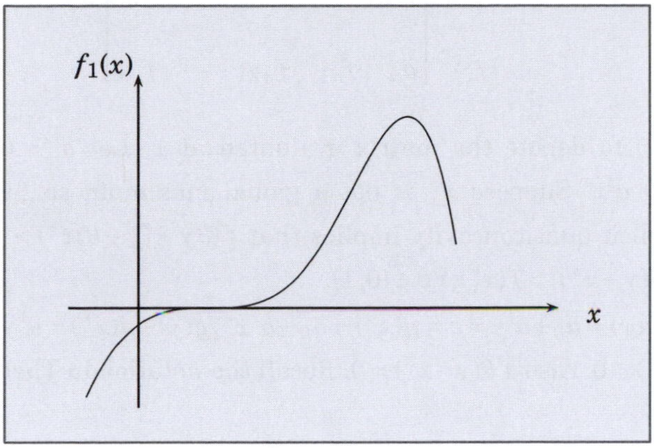

Set $f(x,y) = f_1(x)$, $g(x,y) = x - y$ and $c = 0$. Let $(x^*,y^*,\lambda^*) = (1,-1,0)$. Then $f_x(x,y) = 3(x-1)^2$ if $x < 2$ and $f_x(x,y) = 3(x-1)^2 - 4(x-2)^3$ if $x > 2$. Thus $f_x(1,-1) = 0, f_y(x,y) = 0, g_x(x,y) = 1$ and $g_y = -1$. Therefore,

$$\begin{aligned} f_x(1,-1) - 0 \cdot g_x(1,-1) &= 0 \\ f_y(1,-1) - 0 \cdot g_y(1,-1) &= 0 \\ 0 - (1-1) &= 0, \end{aligned}$$

so (1) is satisfied at $(x^*,y^*,\lambda^*) = (1,-1,0)$. Next,

$$|\bar{\boldsymbol{H}}_2(x^*,y^*,\lambda^*)| = \begin{vmatrix} 0 & g_x & g_y \\ g_x & f_{xx} & f_{xy} \\ g_y & f_{yx} & f_{yy} \end{vmatrix} = 0.$$

Thus (2) is not satisfied. Observe from the graph that $(x^*,y^*,\lambda^*) = (1,-1,0)$ does not correspond to a maximum. ◊

REMARK 12.4: A variation on the proof of Theorem 12.6 may be given using Theorem 12.3 directly, as follows. Since $\{x \mid g(x) = c\}$ is convex, $g(x) = a_0 + a_1x_1 + \ldots + a_nx_n$ because if x, y are in this set then so is $\theta x + (1-\theta)y$. Then $g(\theta x + (1-\theta)y) = c = \theta g(x) + (1-\theta)g(y)$ so that g is linear (affine). Therefore

$g_i = a_i$ and $\mathscr{L}_{ij} = f_{ij}$. Thus

$$\bar{\boldsymbol{H}} = \begin{bmatrix} 0 & a_1 & a_2 & \cdots & a_n \\ a_1 & f_{11} & f_{12} & \cdots & f_{1n} \\ a_2 & f_{21} & f_{22} & \cdots & f_{2n} \\ \vdots & \vdots & \vdots & \ddots & \vdots \\ a_n & f_{n1} & f_{n2} & \cdots & f_{nn} \end{bmatrix}.$$

Write $\bar{\boldsymbol{H}}(x)$ to denote the matrix evaluated at x. Let $a' = (a_1, a_2, \ldots, a_n)$, so $g(x) = a_0 + a'x$. Suppose x^* is not a global maximum so $\exists y, g(y) = c, f(y) > f(x^*)$. Explicit quasiconcavity implies that $f(\theta y + (1-\theta)x^*) > f(x^*)$, $\forall \theta \in (0,1)$, or $f[x^* + \theta(y - x^*)] > f(x^*), \forall \theta \in (0,1)$.

Since $g(y) = a_0 + a'y = c = g(x^*) = a_0 + a'x^*, g(y) - g(x^*) = a'y - a'x^* = c - c = 0$, or $a'(y - x^*) = 0$. Also $a'\theta(y - x^*) = 0$. Recall the notation in Theorem 12.3 and let

$$\boldsymbol{A}(x) = \begin{bmatrix} f_{11}(x) & f_{12}(x) & \cdots & f_{1n}(x) \\ f_{21}(x) & f_{22}(x) & \cdots & f_{2n}(x) \\ \vdots & \vdots & \ddots & \vdots \\ f_{n1}(x) & f_{n2}(x) & \cdots & f_{nn}(x) \end{bmatrix}, \; \boldsymbol{b} = a, \; \boldsymbol{C}_j(x) = \bar{\boldsymbol{H}}_j(x).$$

Given (2) in the statement of Theorem 12.6, $(-1)^j \mid \boldsymbol{C}_j(x^*) \mid > 0, j = 2, \ldots, n$, so from Theorem 12.3, $\boldsymbol{\xi}'\boldsymbol{A}(x^*)\boldsymbol{\xi} < 0$, for all $\boldsymbol{\xi} \neq 0$, with $\boldsymbol{a}'\boldsymbol{\xi} = 0 (= \boldsymbol{b}'\boldsymbol{\xi})$.

As $f(x^* + \theta(y - x^*)) > f(x^*)$, take θ sufficiently small for a Taylor series expansion:

$$f(x^* + \theta(y - x^*)) = f(x^*) + f'(x^*)\theta(y - x^*) + \frac{1}{2}\theta(y - x^*)'\boldsymbol{A}(\tilde{x})\theta(y - x^*) > f(x^*), \tag{12.13}$$

where $\tilde{x}$ is close to x^*. Cancel $f(x^*)$ to get $\theta f'(x^*)(y - x^*) + \frac{1}{2}\theta^2(y - x^*)'\boldsymbol{A}(\tilde{x}) \times (y - x^*) > 0$. Consider the term $(y - x^*)'\boldsymbol{A}(\tilde{x})(y - x^*)$. At x^*, $(y - x^*)'\boldsymbol{A}(x^*)(y - x^*) < 0$, since $a'(y - x^*) = 0$ and $(y - x^*) \neq 0$.

For $\tilde{x}$ sufficiently close to x^*, $(-1)^j \mid \boldsymbol{C}_j(\tilde{x}) \mid > 0, j = 2, \ldots, n$, and since $a'(y - x^*)$ and $(y - x^*) \neq 0$, it must be that $(y - x^*)'\boldsymbol{A}(\tilde{x})(y - x^*) < 0$. Thus to satisfy the inequality in expression 12.13, $\theta f'(x^*)(y - x^*) > 0$, so $f'(x^*)(y - x^*) > 0$. However,

$$0 < f'(x^*)(y - x^*) = \lim_{\theta \to 0} \left\{ \frac{f(x^* + \theta(y - x^*)) - f(x^*)}{\theta} \right\}.$$

This implies that for θ small, $f(x^*+\theta(y-x^*)) > f(x^*)$ and since $x^*+\theta(y-x^*)$ is feasible ($g(x^*+\theta(y-x^*)) = c$) this contradicts the fact that x^* satisfies the conditions for a local maximum. □

12.6.1 Quasiconcavity and bordered Hessians

Just as there is a close connection between concavity (and convexity) and the Hessian matrix, there is a similar connection for quasiconcavity (quasiconcavity), except that the Hessian matrix is replaced by a bordered Hessian matrix. Augment the Hessian matrix with a border of first derivatives as follows. Let

$$\boldsymbol{Q} = \begin{pmatrix} 0 & f_1 & f_2 & \cdots & f_n \\ f_1 & f_{11} & f_{12} & \cdots & f_{1n} \\ f_2 & f_{21} & f_{22} & \cdots & f_{2n} \\ \vdots & \vdots & \vdots & \ddots & \vdots \\ f_n & f_{n1} & f_{n2} & \cdots & f_{nn} \end{pmatrix}, \qquad \boldsymbol{Q}_j = \begin{pmatrix} 0 & f_1 & f_2 & \cdots & f_j \\ f_1 & f_{11} & f_{12} & \cdots & f_{1j} \\ f_2 & f_{21} & f_{22} & \cdots & f_{2j} \\ \vdots & \vdots & \vdots & \ddots & \vdots \\ f_j & f_{j1} & f_{j2} & \cdots & f_{jj} \end{pmatrix}.$$

Write $\boldsymbol{Q}(x)$ and $\boldsymbol{Q}_j(x)$ to denote matrices evaluated at x. These matrices are used to characterize quasiconcavity.

Theorem 12.7:

1. *If f is quasiconcave then,*

$$|\boldsymbol{Q}_2(x)| \geq 0, |\boldsymbol{Q}_3(x)| \leq 0, \ldots, (-1)^n |\boldsymbol{Q}_n(x)| \geq 0, \ \forall x.$$

   ($\boldsymbol{Q}_1 \leq 0$ always holds).

2. *If*

$$|\boldsymbol{Q}_1(x)| < 0, |\boldsymbol{Q}_2(x)| > 0, \ldots, (-1)^n |\boldsymbol{Q}_n(x)| > 0, \ \forall x,$$

   then f is quasiconcave.

3. *If f is quasiconvex, then,*

$$|\boldsymbol{Q}_2(x)| \leq 0, |\boldsymbol{Q}_3(x)| \leq 0, \ldots, |\boldsymbol{Q}_n(x)| \leq 0, \ \forall x.$$

4. *If*

$$|\boldsymbol{Q}_1(x)| < 0, |\boldsymbol{Q}_2(x)| < 0, \ldots, |\boldsymbol{Q}_n(x)| < 0, \ \forall x,$$

   then f is quasiconvex.

These results can be related to constrained optimization as follows. Consider the problem:

$$\max f(x_1,x_2,\ldots,x_n) \quad \text{subject to} \quad g(x_1,x_2,\ldots,x_n) = c.$$

Suppose that $g(x_1,\ldots,x_n) = a_0 + a_1x_1 + a_2x_2 + \ldots + a_nx_n$. Then $g_i = a_i$. Let

$$\mathscr{L} = f(x_1,\ldots,x_n) + \lambda(c - g(x_1,\ldots,x_n)).$$

Then $\mathscr{L}_i = f_i - \lambda g_i$ and $\mathscr{L}_{ij} = f_{ij}$. The first-order conditions are $\mathscr{L}_i = 0, \forall i$, or $f_i = \lambda g_i, \forall i$, or $g_i = \frac{1}{\lambda} f_i, \forall i$. Theorem 12.4 gives second-order conditions in terms of the Hessian matrix. In the present context,

$$\boldsymbol{H}_n = \overline{\boldsymbol{H}} = \begin{bmatrix} 0 & -g_1 & -g_2 & \cdots & -g_n \\ -g_1 & \mathscr{L}_{11} & \mathscr{L}_{12} & \cdots & \mathscr{L}_{1n} \\ -g_2 & \mathscr{L}_{21} & \mathscr{L}_{22} & \cdots & \mathscr{L}_{2n} \\ \vdots & \vdots & \vdots & \ddots & \vdots \\ -g_n & \mathscr{L}_{n1} & \mathscr{L}_{n2} & \cdots & \mathscr{L}_{nn} \end{bmatrix} = \begin{bmatrix} 0 & -f_1/\lambda & -f_2/\lambda & \cdots & -f_n/\lambda \\ -f_1/\lambda & \mathscr{L}_{11} & \mathscr{L}_{12} & \cdots & \mathscr{L}_{1n} \\ -f_2/\lambda & \mathscr{L}_{21} & \mathscr{L}_{22} & \cdots & \mathscr{L}_{2n} \\ \vdots & \vdots & \vdots & \ddots & \vdots \\ -f_n/\lambda & \mathscr{L}_{n1} & \mathscr{L}_{n2} & \cdots & \mathscr{L}_{nn} \end{bmatrix}.$$

Thus

$$\bar{\boldsymbol{H}}_j = \frac{1}{\lambda^2} \begin{bmatrix} 0 & f_1 & f_2 & \cdots & f_j \\ f_1 & f_{11} & f_{12} & \cdots & f_{1j} \\ f_2 & f_{21} & f_{22} & \cdots & f_{2j} \\ \vdots & \vdots & \vdots & \ddots & \vdots \\ f_j & f_{j1} & \mathscr{L}_{j2} & \cdots & \mathscr{L}_{jj} \end{bmatrix},$$

since $\mathscr{L}_{ij} = f_{ij}$ in the linear constraint case. Noting that $\overline{\boldsymbol{H}}_j$ corresponds to $\boldsymbol{Q}_j$,

Theorem 12.8: *If the bordered Hessian satisfies the conditions for f to be quasi-concave, then the bordered Hessian satisfies the sufficient conditions for a local maximum.*

EXAMPLE 12.11: Consider the problem: $\max U(x_1,\ldots,x_n)$ subject to: $p_1x_1 + p_2x_2 + \cdots + p_nx_n = Y$, or $p \cdot x = Y = g(x)$. Thus, $-g_i = -p_i$ and $\mathscr{L}_{ij} = U_{ij}$. Therefore, a sufficient condition for x^* to be a local maximum is

$$\begin{bmatrix} 0 & -p_1 & -p_2 \\ -p_1 & U_{11} & U_{12} \\ -p_2 & U_{21} & U_{22} \end{bmatrix} > 0, \quad \begin{bmatrix} 0 & -p_1 & -p_2 & -p_3 \\ -p_1 & U_{11} & U_{12} & U_{13} \\ -p_2 & U_{21} & U_{22} & U_{23} \\ -p_3 & U_{31} & U_{32} & U_{33} \end{bmatrix} < 0, \text{ etc.}$$

◊

12.7 Optimization with many constraints

Consider a square $n \times n$ matrix $\boldsymbol{A}$, and let $\boldsymbol{B}$ be an $m \times n$ matrix with $m < n$. Thus,

$$\boldsymbol{A} = \begin{pmatrix} a_{11} & a_{12} & \cdots & a_{1n} \\ a_{21} & a_{22} & \cdots & a_{2n} \\ \vdots & \vdots & \ddots & \vdots \\ a_{n1} & a_{n2} & \cdots & a_{nn} \end{pmatrix}, \qquad \boldsymbol{B} = \begin{pmatrix} b_{11} & a_{12} & \cdots & a_{1n} \\ \vdots & \vdots & \ddots & \vdots \\ b_{m1} & a_{m2} & \cdots & a_{mn} \end{pmatrix}.$$

Let $\boldsymbol{B}_m$ denote the square matrix obtained from $\boldsymbol{B}$ by deleting the last $n-m$ columns. Define the matrix:

$$\boldsymbol{C} = \begin{pmatrix} 0 & \boldsymbol{B} \\ \boldsymbol{B}' & \boldsymbol{A} \end{pmatrix}.$$

The matrix $\boldsymbol{C}$ and a secondary family of matrices $(\boldsymbol{C}_r)$ may be defined more explicitly:

$$\boldsymbol{C} = \begin{pmatrix} 0 & \dots & 0 & b_{11} & b_{12} & \cdots & b_{1n} \\ \vdots & \ddots & \vdots & \vdots & \vdots & \vdots & \vdots \\ 0 & \dots & 0 & b_{m1} & b_{m2} & \cdots & b_{mn} \\ b_{11} & \cdots & b_{m1} & a_{11} & a_{12} & \cdots & a_{1n} \\ b_{12} & \cdots & b_{m2} & a_{21} & a_{22} & \cdots & a_{2n} \\ \vdots & \vdots & \vdots & \vdots & \vdots & \ddots & \vdots \\ b_{1n} & \cdots & b_{mn} & a_{n1} & a_{n2} & \cdots & a_{nn} \end{pmatrix},$$

$$\boldsymbol{C}_r = \begin{pmatrix} 0 & \dots & 0 & b_{11} & b_{12} & \cdots & b_{1r} \\ \vdots & \ddots & \vdots & \vdots & \vdots & \vdots & \vdots \\ 0 & \dots & 0 & b_{m1} & b_{m2} & \cdots & b_{mr} \\ b_{11} & \cdots & b_{m1} & a_{11} & a_{12} & \cdots & a_{1r} \\ b_{12} & \cdots & b_{m2} & a_{21} & a_{22} & \cdots & a_{2r} \\ \vdots & \vdots & \vdots & \vdots & \vdots & \ddots & \vdots \\ b_{1r} & \cdots & b_{mr} & a_{r1} & a_{r2} & \cdots & a_{rr} \end{pmatrix}.$$

Recall the matrices from $\boldsymbol{C}$ and $\boldsymbol{C}_r$ from the earlier discussion on quadratic forms. These are used in Theorem 12.9 to provide the key result for the multiconstraint optimization problem.

Theorem 12.9: *Suppose that **A** is symmetric and $\boldsymbol{B}_m$ is non-singular.*

1. *Then $\boldsymbol{x}'\boldsymbol{A}\boldsymbol{x} > 0$ for every $\boldsymbol{x}$ with $\boldsymbol{B}\boldsymbol{x} = 0$, if and only if:*

$$(-1)^m \mid \boldsymbol{C}_r \mid < 0, \quad r = m+1, \dots, n.$$

2. *Then $x'Ax < 0$ for every x with $Bx = 0$, if and only if:*

$$(-1)^r \mid C_r \mid > 0, \quad r = m+1, \ldots, n.$$

Notice that condition 1 requires that all determinants have the same sign (negative if m is odd, positive if m is even), whereas in condition 2 the sign alternates with the size of the matrix. This result may be used to give a sufficiency condition for local optimality of a solution x^* to a constrained optimization problem.

Let $f(x_1, \ldots, x_n)$ be the objective function and $g_k(x_1, \ldots, x_n) = c_k$, $k = 1, \ldots, m$ a collection of m constraints. Assume that $m < n$. Define the Lagrangian:

$$\mathscr{L} = f(x_1, \ldots, x_n) + \lambda_1[c_1 - g_1(x_1, \ldots, x_n)] + \cdots + \lambda_m[c_1 - g_m(x_1, \ldots, x_n)].$$

Define the matrix D_g as:

$$D_g = \begin{pmatrix} \frac{\partial g_1}{\partial x_1} & \frac{\partial g_1}{\partial x_2} & \cdots & \frac{\partial g_1}{\partial x_n} \\ \frac{\partial g_2}{\partial x_1} & \frac{\partial g_2}{\partial x_2} & \cdots & \frac{\partial g_2}{\partial x_n} \\ \vdots & \vdots & \vdots & \vdots \\ \frac{\partial g_m}{\partial x_1} & \frac{\partial g_m}{\partial x_2} & \cdots & \frac{\partial g_m}{\partial x_n} \end{pmatrix},$$

and define

$$C_r = \begin{pmatrix} 0 & D_g \\ D_g' & \nabla^2 \mathscr{L} \end{pmatrix} = \begin{pmatrix} 0 & \cdots & 0 & \frac{\partial g_1}{\partial x_1} & \frac{\partial g_1}{\partial x_2} & \cdots & \frac{\partial g_1}{\partial x_r} \\ \vdots & \ddots & \vdots & \vdots & \vdots & \vdots & \vdots \\ 0 & \cdots & 0 & \frac{\partial g_m}{\partial x_1} & \frac{\partial g_m}{\partial x_2} & \cdots & \frac{\partial g_m}{\partial x_r} \\ \frac{\partial g_1}{\partial x_1} & \cdots & \frac{\partial g_m}{\partial x_1} & \mathscr{L}_{11} & \mathscr{L}_{12} & \cdots & \mathscr{L}_{1r} \\ \frac{\partial g_1}{\partial x_2} & \cdots & \frac{\partial g_m}{\partial x_2} & \mathscr{L}_{21} & \mathscr{L}_{22} & \cdots & \mathscr{L}_{2r} \\ \vdots & \vdots & \vdots & \vdots & \vdots & \ddots & \vdots \\ \frac{\partial g_1}{\partial x_r} & \cdots & \frac{\partial g_m}{\partial x_r} & \mathscr{L}_{r1} & \mathscr{L}_{r2} & \cdots & \mathscr{L}_{rr} \end{pmatrix}.$$

Where $\nabla^2 \mathscr{L}$ is the matrix $\{\mathscr{L}_{ij}\}_{1 \le i \le r, 1 \le j \le r}$, as indicated in the expression. Assume that the rank of D_g is m, and that the x_i are labeled such that the first m columns of D_g are linearly independent at x^*. Thus

Theorem 12.10: *Suppose that x^* is feasible (satisfies all the constraints). If $(x^*, \lambda_1^*, \ldots, \lambda_m^*) = (x^*, \lambda^*)$ satisfies:*

1. $\mathcal{L}_i(x^*, \lambda^*) = 0,\ i = 1, \dots, n$

   $\mathcal{L}_{\lambda_k}(x^*, \lambda^*) = 0,\ k = 1, \dots, m$

2. $(-1)^r \mid C_r \mid > 0,\ r = m+1, \dots n,$

then x^* is a constrained local maximum.

If condition 2 is replaced by $(-1)^m \mid C_r \mid > 0\ r = m+1, \dots, n$, then x^* is a constrained local minimum.

So, for example, if there is one constraint so that $m = 1$, the local maximum condition requires that $(-1)^2 \mid C_2 \mid > 0$ or $\mid C_2 \mid > 0$, $(-1)^3 \mid C_3 \mid > 0$ or $\mid C_2 \mid > 0$, and so on. Similarly, the local minimum condition requires $(-1)^1 \mid C_2 \mid > 0$ or $\mid C_2 \mid < 0$, $(-1)^1 \mid C_3 \mid > 0$ or $\mid C_3 \mid < 0$, and so on. This is the same set of conditions as appears in Theorem 12.4.

Exercises for Chapter 12

Exercise 12.1 A firm has the production function $Y = 30L^{\frac{1}{2}}K^{\frac{2}{3}}$.

(a) What is the cost-minimizing level of L and K if the firm must produce 1000 units? Check the second-order conditions.

(b) By how much will cost increase if the firm must increase output by one unit?

(c) Given this production function, can we find a profit-maximizing level of output? Why or why not?

Exercise 12.2 Assume that a monopolist producing one good can sell to two distinct markets and price discriminate. The demand function in each market is

$$D_1 = 300 - \frac{1}{3}P_1$$
$$D_2 = 300 - \frac{1}{2}P_2.$$

The total cost to the monopolist is $C = 2D_1^2 + 3D_1D_2 + D_2^2$. Also, owing to a government restriction, the total output of the firm must be 100 units. What is the profit-maximizing output in each market? Check the second-order conditions.

Exercise 12.3 Consider a firm that produces beer and pretzels. It must sell a bag of pretzels for every four bottles of beer. If its profit function is

$$\pi = 150B - 4B^2 + 150P - 3P^2 + 10BP + 100$$

(P = # of bags of pretzels, B = # beers), what is the profit-maximizing level of output?

Exercise 12.4 The geometric average of two positive numbers, x and y, is $(xy)^{\frac{1}{2}}$. Consider the task of minimizing the geometric average, subject to the condition $x+y=1$. Show, using the method of Lagrange multipliers, that the optimal choice of (x,y) is $\left(\frac{1}{2},\frac{1}{2}\right)$ and confirm that this gives the constrained minimum.

Exercise 12.5 A swimming pool is 25 meters times 100 meters. In coordinates, a position (x,y) gives length and distance traveled from (0,0). Thus (15,20) denotes the position of a swimmer who swam 20 meters along the x axis and 15 meters along the y axis from (0,0). There are four corners (0,0), (0,25), (100,0) and (100,25). The diagonal line connecting corners (0,25) and (100,0) is given by $4y+x=100$. The distance from corner (0,0) to coordinate (x,y) is $(x^2+y^2)^{\frac{1}{2}}$. A swimmer wishes to swim from the (0,0) corner, and cross the diagonal line in the shortest distance possible. What coordinate position should the person swim to? Use the method of Lagrange multipliers to determine the answer.

Exercise 12.6 A firm wishes to produce a level of output Q at minimum cost. Two inputs, capital (K) and labour (L), are used to produce Q with the production function given by $Q=K^{\alpha}+L^{\alpha}$, where $0<\alpha<1$. The wage rate is ω and the cost of capital (per unit) is r, so that total cost given (K,L) is $C=rK+\omega L$.

(a) Set up the Lagrangian to solve the problem of finding the cost-minimizing levels of K and L required to produce Q.

(b) Solve for K,L and the Lagrange multiplier, λ.

(c) Check that the second-order conditions for a minimum are satisfied.

(d) Observe that in your solution to part (b), both K and L depend on Q. To emphasize this dependency, write $K(Q)$ and $L(Q)$. Thus the cost of producing the output level Q is $C(Q)=rK(Q)+\omega L(Q)$. Find $dC(Q)/dQ$ and compare your answer with λ.

Exercise 12.7 Dr. Who lives at coordinates $(x,y)=(3,12)$. A road in the (x,y) plane follows the path: $k=x^2-6x+y$, where $0<k<3$. On the bus journey home, each traveller must give the driver "drop-off" coordinates. Unfortunately for Dr. Who, the bus does not travel by his house (since $(x,y)=(3,12)$ is not on the line $k=x^2-6x+y$).

(a) What drop-off coordinates should Dr. Who give the driver so that Dr. Who is as close to his house as possible? Hint: the distance between two points, (x_1,y_1) and (x_2,y_2), is $d=[(x_1-x_2)^2+(y_1-y_2)^2]^{\frac{1}{2}}$. However, since minimizing this is equivalent to minimizing $d^*=[(x_1-x_2)^2+(y_1-y_2)^2]$, take d^* as the measure of distance.

(b) Confirm that the second-order conditions are satisfied.

(c) The solution depends on k. Let $d^*(k)$ be the distance at the solution from the drop-off point to Dr. Who's house. Calculate $\frac{d[d^*(k)]}{dk}$ and relate it to the Lagrange multiplier in (a).

Exercise 12.8 A firm sets both price and quantity in a market. It is known that the demand function for this market is $P = k/Q$. By law, the firm is required to choose a price quantity combination lying on the demand curve. The cost of setting (P,Q) combinations is $C(P,Q) = P + \alpha Q^2$.

(a) Minimize $C(P,Q)$ subject to the demand constraint.

(b) Confirm that the second-order conditions are satisfied.

(c) The optimal choices of P and Q depend on $(\alpha,k) : P(\alpha,k), Q(\alpha,k)$, so that the cost at the solution is $C^*(\alpha,k) = C(P(\alpha,k), Q(\alpha,k))$. Find $\partial C^*(\alpha,k)/\partial k$ and compare this with the Lagrange multiplier obtained in (a).

Exercise 12.9 Maximize xy subject to the constraint $x^2 + y = k$, where k is some positive constant. Confirm that the second-order conditions for a constrained optimum are satisfied. Interpret the Lagrange multiplier.

Exercise 12.10 Minimize $(x-1)^2 + y^2$ subject to the constraint $x^2 + k = y$, where k is some positive constant. Confirm that the second-order conditions for a constrained optimum are satisfied. Interpret the Lagrange multiplier.

Exercise 12.11 Consider the following constrained optimization problem: maximize $f(x,y) = y - bx^2$ subject to the constraint: $y = k + a\ln x$. Here k, a and b are positive constants.

(a) Plot the problem.

(b) Find the solution values for x and y.

(c) Confirm that the second-order conditions are satisfied.

(d) Let the solution be $x(a,b,k)$ and $y(a,b,k)$. Then the value of the objective at the solution is $f^*(a,b,k) = f(x(a,b,k), y(a,b,k))$. Find $\frac{\partial f^*(a,b,k)}{\partial a}$. Find $\frac{\partial f^*(a,b,k)}{\partial k}$ and interpret this in terms of the Lagrange multiplier.

Exercise 12.12 An individual faces the problem of minimizing expenditure subject to reaching some fixed level of utility u^*. There are two goods, x and y, and the utility function is given by $u(x,y) = x^\alpha y^\alpha$, $0 < 2\alpha < 1$. The cost of x (per unit) is p and the cost of y (per unit) is q, so the cost of the bundle (x,y) is $px + qy$.

(a) Set up the Lagrangian for this problem. Find the optimal values of x and y and the value of the Lagrange multiplier. Check the second-order conditions.

(b) The solution depends on p, q and u^* so write $x(p,q,u^*)$, $y(p,q,u^*)$, $\lambda(p,q,u^*)$. The expenditure required to achieve utility level u^* given prices p and q is therefore $e(p,q,u^*) = px(p,q,u^*) + qy(p,q,u^*)$. Then

$$\frac{\partial e(p,q,u^*)}{\partial p} \text{ and } \frac{\partial e(p,q,u^*)}{\partial q}$$

give the change in mimimum income required to reach utility level u^* when the price of x and y change. Show that

$$\frac{\partial e(p,q,u^*)}{\partial p} = x(p,q,u^*) \text{ and } \frac{\partial e(p,q,u^*)}{\partial q} = y(p,q,u^*).$$

13

Inequality-constrained optimization

13.1 Introduction

This chapter considers the problem of maximizing or minimizing a function $f : \mathcal{R}^n \to \mathcal{R}$ subject to inequality constraints such as $g(x) \le c$ or $g(x) \ge c$. This type of problem arises frequently in economics. For example, maximizing output $y = f(x_1, x_2)$ subject to expenditure constraint, $g(x_1, x_2) = w_1 x_1 + w_2 x_2 \le c$, is a typical problem.

In Section 13.2, the constrained optimization is formally defined and the main result (Theorem 13.1) is given. All the results in this chapter are derived from this theorem and the constraint qualification. Section 13.2.1 discusses the constraint qualification, a (mild) condition that the derivatives of the constraints must satisfy to guarantee that the "standard" first-order conditions are necessary for a local optimum. In Section 13.2.2, the complementary slackness condition is discussed: on non-binding constraints we can take the Lagrange multipliers to be 0. Section 13.2.3 discusses the logical equivalence of the maximization and minimization programs. In Section 13.2.4, conditions are given under which a local optimum is also a global optimum. In Section 13.2.5, problems with non-negativity constraints are considered. In this case, the necessary conditions for a local optimum may be expressed in terms of the Kuhn–Tucker conditions, and this is done in Theorems 13.5 and 13.6, which are derived directly from the main result, Theorem 13.1. The remaining sections deal with a few miscellaneous topics.

First, Section 13.1.1 gives an introductory example that motivates the key characterizing condition for a solution: at the solution, the gradient of the objective may be written as a linear combination of the gradients of the constraints.

13.1.1 A motivating example

To develop some of the issues, consider the following optimization example.

EXAMPLE 13.1:

$$\begin{aligned} \max \quad & x_1 x_2 \\ \text{subject to:} \quad & \\ & x_1^2 + x_2^2 \leq 1 \\ & x_2 \leq \alpha x_1. \end{aligned}$$

Thus, the objective is $f(x_1, x_2) = x_1 x_2$, and these constraints may be written $g_1(x_1, x_2) = x_1^2 + x_2^2 - 1 \leq 0$ and $g_2(x_1, x_2) = -\alpha x_1 + x_2^2 \leq 0$. The figure depicts the problem for two different values of the parameter α.

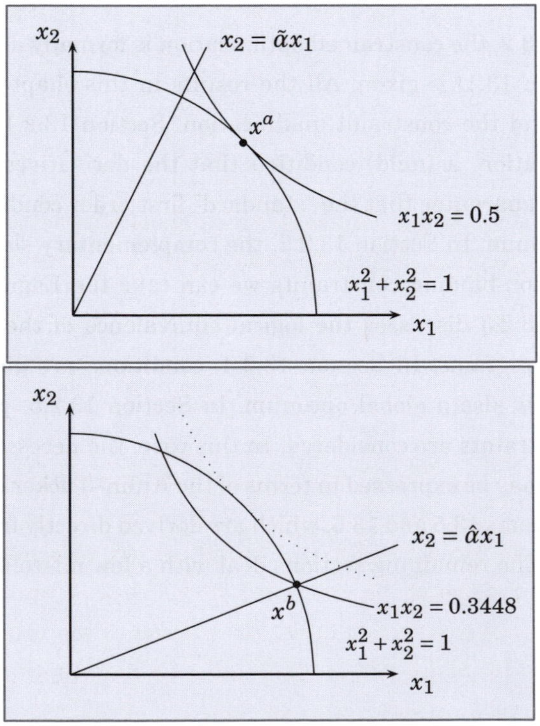

In the figures, constraint g_2 is depicted for two possible values for α: $\tilde{\alpha} = 2 > 1 > \frac{4}{10} = \hat{\alpha}$. At parameter value $\tilde{\alpha}$, only the first constraint is binding and the solution occurs at $x^a = \left(\sqrt{\frac{1}{2}}, \sqrt{\frac{1}{2}}\right)$. At parameter value $\hat{\alpha}$, both constraints are binding and the solution occurs at x^b. Let $\tilde{g}_2(x_1,x_2) = -\tilde{\alpha}x_1 + x_2$, and $\hat{g}_2(x_1,x_2) = -\hat{\alpha}x_1 + x_2$.

At parameter value $\tilde{\alpha}$, the constraints are $g_1(x_1,x_2) \le 0$ and $\tilde{g}_2(x_1,x_2) \le 0$ and the solution, x^a, satisfies $g_1(x^a) = 0$, $\tilde{g}_2(x^a) < 0$. Constraint g_1 is *active* or *binding* at x^a but constraint $\tilde{g}_2$ is not *active* at x^a, because the constraint is not binding at x^a. At parameter value $\hat{\alpha}$, the constraints are $g_1(x_1,x_2) \le 0$ and $\hat{g}_2(x_1,x_2) \le 0$, and, at the solution, x^b satisfies $g_1(x^b) = 0$, $\hat{g}_2(x^b) = 0$. Both constraints g_1 and $\hat{g}_2$ are active at x^b. Depending on circumstance, a constraint may or may not be binding, so that the equations defining the solution must reflect this possibility.

In the case where $\alpha = \tilde{\alpha}$, observe that the slope of the objective is equal to the slope of the active constraint. In the case where $\alpha = \hat{\alpha}$, the slope of the objective lies between the slopes of the constraints. The point x^b is located at the solution to the two constraints $x_2 = 0.4x_1$ and $x_1^2 + x_2^2 = 1$: $x^b = (x_1,x_2) = (0.9285, 0.3714)$. The slope of constraint 1 is 0.4 and the slope of constraint 2 is $-\frac{x_1}{x_2} = -\frac{0.9285}{0.3714} = -2.5$. The slope of the objective is $-\frac{0.27}{x_1^2} = -\frac{0.27}{0.9285} = -0.2908$. Thus, since the slope of the objective lies between the slopes of the two constraints, the slope of the objective may be written as a weighted average of the slopes of the constraints. ◊

That the slope of the objective lies between the slopes of the constraints is an important point. Remark 13.1 clarifies this point.

REMARK 13.1: Rather than express conditions in terms of slopes, it is standard practice to use the normal to a curve to represent the slope. Thus, with $f(x_1,x_2) = x_1x_2$ evaluated at x^b,

$$\nabla f \stackrel{\text{def}}{=} (f_{x_1}, f_{x_2}) = (x_2, x_1) = (0.3714, 0.9285)$$

is the normal to the curve $f(x_1,x_2)$ at (x_1,x_2). As a vector, this has slope $\frac{x_1}{x_2}$, which is the negative of the reciprocal of the slope of the objective $f(x_1,x_2) = x_1x_2$, given by $\frac{dx_2}{dx_1} = -\frac{x_2}{x_1}$. Similarly, the normal of the linear constraint g_2 is $\nabla g_2 = \left(\frac{\partial g_2}{\partial x_1}, \frac{\partial g_2}{\partial x_2}\right) = (-0.4, 1)$ and the normal of the quadratic constraint g_1 is $\nabla g_1 = \left(\frac{\partial g_1}{\partial x_1}, \frac{\partial g_1}{\partial x_2}\right) = (2x_1, 2x_2) = (1.857, 0.7428)$.

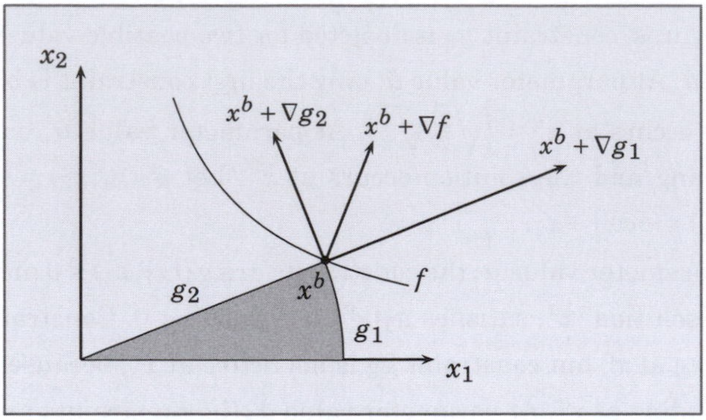

The slopes of the normals of the constraints are, respectively, $\frac{x_2}{x_1} = 0.4$ (g_1) and $-\frac{1}{0.4} = -2.5$ (g_2). The slope of the objective normal is $\frac{x_1}{x_2} = 2.5$. An arc starting with 0 slope first reaches the normal to g_1 then the normal to f and finally the normal to g_1. Thus, the normal of the objective may be expressed as a positive linear combination of the constraint normals:

$$\nabla f = \lambda_1 \nabla g_1 + \lambda_2 \nabla g_2$$

$$\begin{pmatrix} 0.3714 \\ 0.9285 \end{pmatrix} = 0.3448 \begin{pmatrix} 1.8570 \\ 0.7428 \end{pmatrix} + 0.6724 \begin{pmatrix} -0.4 \\ 1 \end{pmatrix}. \tag{13.1}$$

□

The observations in Example 13.1 apply for the general case.

13.1.2 Overview

Given a function $h(x) = h(x_1, \ldots, x_n)$, the normal to the function at x is the gradient of the function at x, $\nabla h = (h_{x_1}, h_{x_2}, \ldots, h_{x_n})$. The key characterizing condition for an optimum is that the gradient (normal) of the objective may be written as a weighted sum of the gradients of the constraints. With k constraints, $\{g_j\}_{j=1}^k$, $\nabla f = \sum_{j=1}^k \lambda_j \nabla g_j$, with $\lambda_j \geq 0$ for all j, as in Theorem 13.1.

Provided a constraint qualification is satisfied, this condition, $\nabla f = \sum_{j=1}^k \lambda_j \nabla g_j$, with $\lambda_j \geq 0$ for all j, along with complementary slackness conditions, determines a system of equations that may be solved to provide necessary conditions for a local optimum. All results here are derived from these conditions.

The constraint qualification is discussed in Section 13.2.1; complementary slackness is discussed in Section 13.2.2. Sufficiency conditions are discussed in Section 13.2.4: concavity of the objective and convexity of the set of feasible points imply that, given the constraint qualification, a solution to the necessary conditions yields a global maximum for the constrained problem. Finally, the necessary conditions can be specialized to include non-negativity conditions, and this is done in Section 13.2.5.

13.2 Optimization

Since a constraint of the form $g(x) \leq c$ may be written as $g(x) - c \leq 0$ and constraints of the form $g(x) \geq c$ may be written as $g(x) - c \geq 0$ or as $-g(x) \leq -c$, we can express all such constraints in the form $g(x) \leq 0$ (or $g(x) \geq 0$). Equality constraints may be dealt with in the same way since $g(x) = c$ is equivalent to $g(x) \leq c$ and $g(x) \geq c$, but at the cost of having two constraints instead of one. Similarly, non-negativity constraints may be expressed in this way; setting $g(x) = -x$, the condition $g(x) \leq 0$ implies that $x \geq 0$.

An inequality-constrained maximization problem is given as:

$$\max f(x) \text{ subject to } g_j(x) \leq 0,\ j = 1, \ldots, r. \tag{13.2}$$

The function $f(x)$ is the objective function and $g_j, j = 1, \ldots, r$, the set of constraints. Feasible points and active constraints are defined next.

Definition 13.1: *A point x^* is feasible if it satisfies all of the constraints: $g_j(x^*) \leq 0, \forall j$. A constraint, g_j is active at a feasible point x^* if $g_j(x^*) = 0$, and otherwise it is inactive. Given the feasible point x^*, let $\mathscr{A}(x^*)$ be the set of active constraints at x^*: $j \in \mathscr{A}(x^*)$ if and only if $g_j(x^*) = 0$.*

A point x^* is a constrained local maximum if there is no alternative point "close" to x^* that satisfies the constraints and gives a higher value for the objective. Formally:

Definition 13.2: *A point x^* is a constrained local maximum of f if there is a neighborhood of x^* (an open set in $\mathscr{R}^n$ containing x^*), $\mathscr{N}(x^*)$, such that $f(x^*) \geq f(\bar{x})$ for all $\bar{x} \in \mathscr{N}(x^*) \cap \{x \mid g_l(x) \leq 0,\ l = 1, \ldots, r\}$.*

The following discussion identifies necessary conditions for a point x^* to be a constrained local maximum. The discussion makes use of Farkas' lemma (see Remark 13.2), applied to the following equation system:

$$\begin{pmatrix} \frac{\partial g_1(x^*)}{\partial x_1} & \frac{\partial g_2(x^*)}{\partial x_1} & \cdots & \frac{\partial g_r(x^*)}{\partial x_1} \\ \vdots & \vdots & \ddots & \vdots \\ \frac{\partial g_1(x^*)}{\partial x_n} & \frac{\partial g_2(x^*)}{\partial x_n} & \cdots & \frac{\partial g_r(x^*)}{\partial x_n} \end{pmatrix} \begin{pmatrix} \lambda_1 \\ \vdots \\ \lambda_r \end{pmatrix} = \begin{pmatrix} \frac{\partial f(x^*)}{\partial x_1} \\ \vdots \\ \frac{\partial f(x^*)}{\partial x_n} \end{pmatrix} \tag{13.3}$$

Each of the i rows may be read as:

$$\frac{\partial f(x^*)}{\partial x_i} = \sum_{j=1}^{r} \lambda_j \frac{\partial g_j(x^*)}{\partial x_i},\ i = 1, \ldots, n. \tag{13.4}$$

This system of equations is central to the characterization of necessary conditions for an optimum.

For ease of notation, write $\nabla f(x^*)$ for the term on the right side of equality 13.3, ∇g for the $n \times r$ matrix on the left side and $\nabla g_l(x^*)$ for the lth column of $\nabla g(x^*)$. With $\lambda = (\lambda_1, \ldots, \lambda_r)$, equation system 13.3 may be written:

$$\nabla g(x^*) \cdot \lambda = \nabla f(x^*). \tag{13.5}$$

This equation system plays a central role in inequality-constrained optimization. Subject to a qualification, if the equation system does not have a non-negative solution, then the point x^* is not a local optimum. Theorem 13.1 gives Equation 13.5 as a necessary condition for optimality, provided a "constraint qualification", $\nabla g_j(x^*) \cdot z < 0$, for any binding constraint, j, is satisfied.

Theorem 13.1: *Given the program 13.2, suppose that at x^* there is some z with $\nabla g_j(x^*) \cdot z < 0$ for each active constraint j. If x^* is a constrained local maximum then*

$$\frac{\partial f(x^*)}{\partial x_i} = \sum_{j=1}^{r} \lambda_j \frac{\partial g_j(x^*)}{\partial x_i}, \quad i = 1, \ldots, n$$

must have a non-negative solution, $\lambda = (\lambda_1, \ldots, \lambda_r)$, $\lambda_j \geq 0$, $j = 1, \ldots, r$.

PROOF: If Equation 13.3 has no non-negative solution, then from Farkas' lemma (Remark 13.2), $\exists z^* \in \mathcal{R}^n$ such that:

$$\sum_{i=1}^{n} \frac{\partial g_j(x^*)}{\partial x_i} \cdot z_i^* \leq 0, \quad \text{for} \quad j = 1, \ldots, r, \quad \text{and} \quad \sum_{i=1}^{n} \frac{\partial f(x^*)}{\partial x_i} \cdot z_i^* > 0. \tag{13.6}$$

More succinctly, $\nabla g_j(x^*) \cdot z^* \leq 0$, $\forall j = 1, \ldots, r$, and $\nabla f(x^*) \cdot z^* > 0$. By assumption, there is some z with $\nabla g_j(x^*) \cdot z < 0$ for each $j \in \mathcal{A}(x^*)$ (that is $\sum_{i=1}^{n} \frac{\partial g_j(x^*)}{\partial x_i} z_i < 0$ for all $j \in \mathcal{A}(x^*)$). Let $z_\alpha = (1-\alpha)z + \alpha z^*$, so that $\nabla g_j(x^*) \cdot z_\alpha < 0$ for each $j \in \mathcal{A}(x^*)$. Consider $x^* + \delta z_\alpha$ where $\delta > 0$ is small, and note that $\lim_{\delta \to 0} \frac{1}{\delta}[f(x^* + \delta z_\alpha) - f(x^*)] = \nabla f(x^*) \cdot z_\alpha > 0$ when α is close to 1. Similarly, for each $j \in \mathcal{A}(x^*)$, $\lim_{\delta \to 0} \frac{1}{\delta}[g_j(x^* + \delta z_\alpha) - g_j(x^*)] = \nabla g_j(x^*) \cdot z_\alpha < 0$. For the inactive constraints, $j \notin \mathcal{A}(x^*)$, $g_j(x^*) < 0$ so that for δ close to 0, $g_j(x^* + \delta z_\alpha) < 0$, $j = 1, \ldots, r$. Consequently, there is an α close to 1, and a δ close to 0 such that $f(x^* + \delta z_\alpha) - f(x^*) > 0$ and $g_j(x^* + \delta z_\alpha) < 0$, $j = 1, \ldots, r$. Summarizing, given that at x^* there is some z with $\nabla g_j(x^*) \cdot z < 0$ for each active constraint j, if Equation 13.3 does not have a non-negative solution, then x^* cannot be a constrained local maximum. ∎

REMARK 13.2: According to Farkas' lemma, if $\boldsymbol{A}$ is an $n \times r$ matrix then exactly one of the conditions holds:

1. $\boldsymbol{Ax} = \boldsymbol{b}$ has a solution $\boldsymbol{x} \geq 0$, or
2. There exists $\boldsymbol{y}$ with $\boldsymbol{A}'\boldsymbol{y} \leq 0$ and $\boldsymbol{b}'\boldsymbol{y} > 0$.

That is, if (1) is satisfied, (2) cannot be satisfied; and if (1) is not satisfied, then (2) must be satisfied. □

The following remarks clarify a few points regarding this result, Theorem 13.1. Remark 13.3 explains how non-negativity constraints are imposed; Remark 13.4 discusses the determinacy of the system (in terms of equations and unknowns) and Remark 13.5 provides some insight into the sign of the multiplier.

REMARK 13.3: Observe that the program given by expression 13.2,

$$\max f(x) \text{ subject to } g_j(x) \leq 0,\ j = 1, \ldots, r,$$

includes the possibility of imposing non-negativity constraints. For example, if $g_k(x) = -x_k$, then $g_k(x) \leq 0$ is equivalent to $-x_k \leq 0$ or $x_k \geq 0$. In Section 13.2.5, a special case is considered – where some of the constraints are non-negativity conditions. There, the Lagrangian together with complementary slackness conditions are used to define the necessary conditions for an optimum. This gives the Kuhn–Tucker conditions for constrained optimization with inequality constraints. □

REMARK 13.4: Regarding the solution of these equation systems, note that in Equation 13.3 (or Equations 13.4 or 13.5), there are n equations in $n + r$ variables (x, λ). In addition, there are r inequalities $g_j(x) \leq 0$, $j = 1, \ldots, r$ to be satisfied at a solution. Complementary slackness, discussed in Section 13.2.2, asserts that the λ_j may be chosen to satisfy $\lambda_j g_j = 0$ for each j, giving an additional r equations. Provided the constraint qualification is satisfied, a necessary condition for an optimum is that the system $\nabla g(x) \cdot \lambda = \nabla f(x)$ and $\lambda_j g_j(x) = 0$, $j = 1, \ldots, r$ of $n+r$ equations in $n+r$ unknowns has a solution $(x_1, \ldots x_n; \lambda_1, \ldots, \lambda_r)$ with $\lambda_j \geq 0$, $\forall j = 1, \ldots, r$. □

REMARK 13.5: A motivation for the sign of the Lagrange multiplier may be given by considering Figure 13.1. At the solution $x_1^* = x_2^*$, $\nabla f(x^*) = (x_2^*, x_1^*)$, and $\nabla g(x^*) = (2x_1^*, 2x_2^*)$, so that $\nabla f(x^*) = \lambda \nabla g(x^*)$ with $\lambda = \frac{1}{2}$. Notice that ∇g points *out* from the feasible region, as does ∇f. Thus, the direction of improvement of f is the *same* as the direction of infeasible choices. This is necessary for the

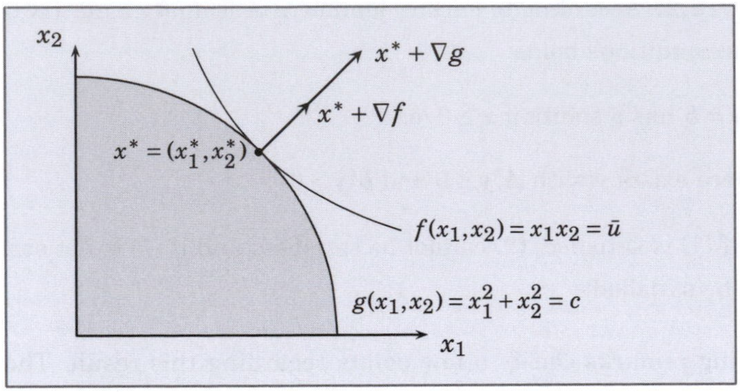

Figure 13.1: Sign of the Lagrange multiplier

candidate solution to be an optimum. This fact is reflected in the expression $\nabla f = \lambda \nabla g$ with $\lambda > 0$. □

The following example illustrates the application of Theorem 13.1 by showing how the necessary condition for an optimum is satisfied at the solution.

EXAMPLE 13.2: Consider the following problem:

$$\max_x \quad f(x_1, x_2) = x_1 x_2$$

subject to:

$$g_1(x_1, x_2) = x_2 + x_1^2 - c \leq 0$$
$$g_2(x_1, x_2) = x_2 - \alpha\sqrt{x_1} \leq 0.$$

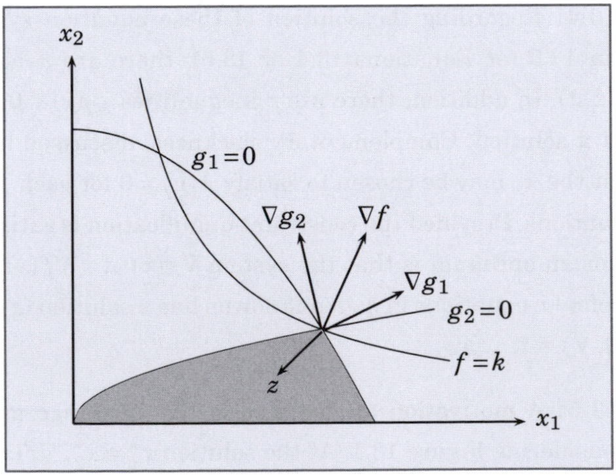

Taking the parameter values as $c = 4$ and $\alpha = 1$, the constraints intersect at $(x_1, x_2) = (1.648, 1.284)$. Let $f(x_1, x_2) = x_1 x_2$, so that at $(x_1, x_2) = (1.648, 1.284)$,

$$\nabla f = \begin{pmatrix} x_2 \\ x_1 \end{pmatrix} = \begin{pmatrix} 1.284 \\ 1.648 \end{pmatrix}; \quad \nabla g_1 = \begin{pmatrix} 2x_1 \\ 1 \end{pmatrix} = \begin{pmatrix} 3.296 \\ 1 \end{pmatrix}; \quad \nabla g_2 = \begin{pmatrix} -\frac{1}{2}\frac{1}{\sqrt{x_1}} \\ 1 \end{pmatrix} = \begin{pmatrix} -0.389 \\ 1 \end{pmatrix}.$$

Therefore,

$$\nabla f = \lambda_1 \nabla g_1 + \lambda_2 \nabla g_2 = 0.5224 \nabla g_1 + 1.1256 \nabla g_2.$$

So, $(x_1, x_2; \lambda_1, \lambda_2) = (1.648, 1.284, 0.5224, 1.1256)$ satisfies the necessary conditions for a maximum. ◊

REMARK 13.6: Note that the vector $\boldsymbol{z}' = (-0.3, -0.6)$ satisfies $\nabla g_1 \cdot \boldsymbol{z} = -1.589$, $\nabla g_1 \cdot \boldsymbol{z} = -0.483$. Thus, the constraint qualification is satisfied. The constraint qualification if satisfied, ensures that the requirement that $\nabla \boldsymbol{g}(x^*) \cdot \lambda = \nabla f(x^*)$ has a non-negative solution, where λ is a necessary condition for a local optimum. □

The next Example, 13.3, considers the quasilinear utility function and maximization of utility on a constraint set that is not convex. The example illustrates the key necessary condition, satisfied at the solution to the program. However, this example is reconsidered later (Example 13.11), where the non-convexity permits the necessary conditions to be satisfied at a non-optimal point.

EXAMPLE 13.3: Consider the problem of maximizing the function $f(x, y) = \alpha \ln x + y$, subject to the constraints $g_1(x, y) = y^2 + (x-1)^2 \leq 2$ and $g_2(x, y) = xy \leq k = 1.12$. For the calculations, take $\alpha = \frac{6}{10}$.

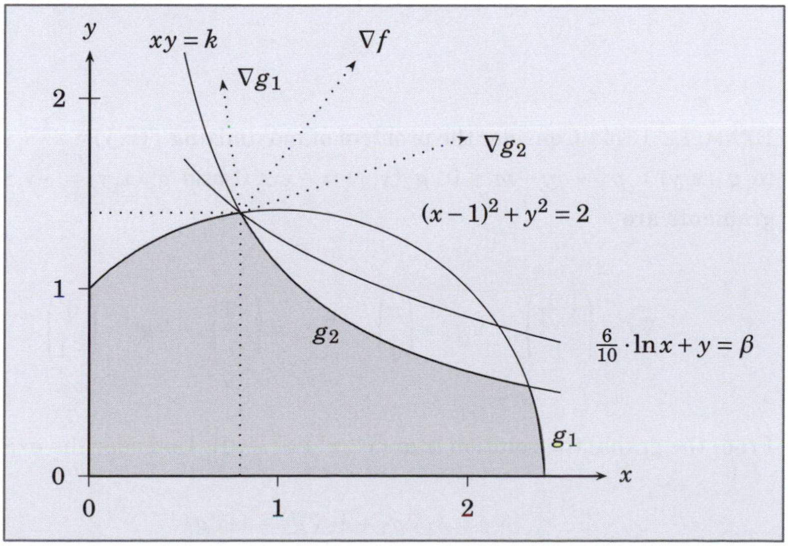

The solution occurs at $x = \frac{8}{10}$ and $y = \frac{14}{10}$. At these values, $xy = k = 1.12$ and $\frac{6}{10}\ln(x) + y = \beta = 1.266$.

Note that

$$\nabla f(x,y) = \begin{pmatrix} \frac{6}{10}\frac{1}{x} \\ 1 \end{pmatrix}, \quad \nabla g_1(x,y) = \begin{pmatrix} 2(x-1) \\ 2y \end{pmatrix}, \quad \nabla g_2(x,y) = \begin{pmatrix} y \\ x \end{pmatrix}.$$

At $x = \frac{8}{10}$, $y = \frac{14}{10}$, forming Equation 13.3 gives:

$$\begin{pmatrix} \frac{6}{8} \\ 1 \end{pmatrix} = \lambda_1 \begin{pmatrix} -\frac{4}{10} \\ \frac{28}{10} \end{pmatrix} + \lambda_2 \begin{pmatrix} \frac{14}{10} \\ \frac{4}{10} \end{pmatrix},$$

which has solution $\lambda_1 = \frac{55}{204}$ and $\lambda_2 = \frac{125}{204}$.

This example is considered further in Section 13.2.6, where a second tangency point is identified, which fails to be a local optimum. ◊

One feature of quasilinear preferences is that demand for one of the goods (the one entering the utility function non-linearly) depends only on the price ratio. If income is not sufficiently large, demand as a function of the price ratio may exceed income. This in turn implies that demand for the other good is negative. To avoid this outcome, it is necessary to impose non-negativity constraints on the variable choices, as is done in Example 13.4.

EXAMPLE 13.4: Consider the problem of maximizing $f(x,y) = x^{\frac{1}{2}} + y$ subject to $g_1(x,y) = px + qy - m \leq 0$, $g_2(x,y) = -x \leq 0$ and $g_3(x,y) = -y \leq 0$. The gradients are

$$\nabla f = \begin{pmatrix} \frac{1}{2}\frac{1}{\sqrt{x}} \\ 1 \end{pmatrix}, \quad \nabla g_1 = \begin{pmatrix} p \\ q \end{pmatrix}, \quad \nabla g_2 = \begin{pmatrix} -1 \\ 0 \end{pmatrix}, \quad \nabla g_3 = \begin{pmatrix} 0 \\ -1 \end{pmatrix}.$$

From the graph, the solution is at $(x^*, y^*) = \left(\frac{m}{p}, 0\right)$. Consider the expression:

$$\nabla f = \lambda_1 \nabla g_1 + \lambda_2 \nabla g_2 + \lambda_3 \nabla g_3$$

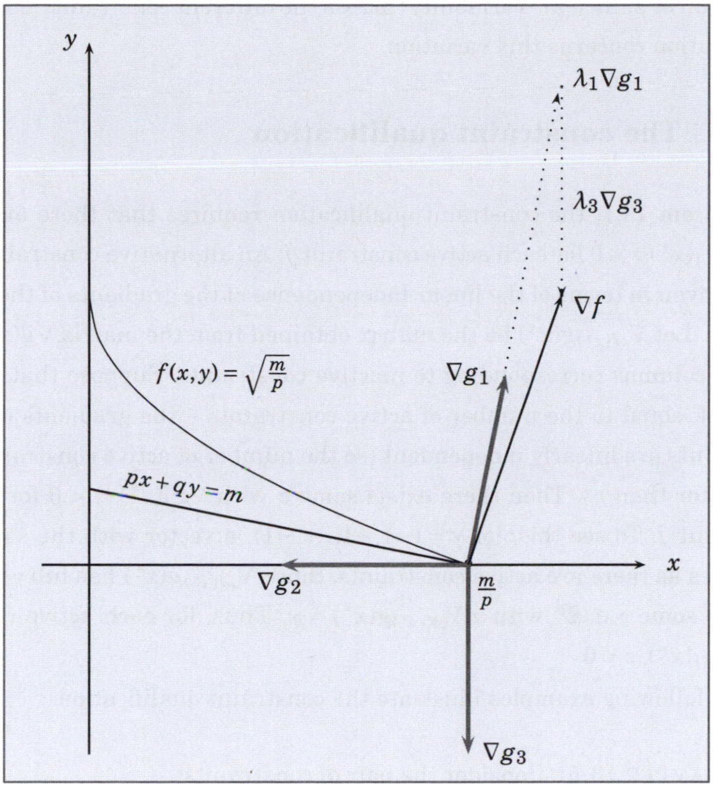

or

$$\begin{pmatrix} \frac{1}{2}\frac{1}{\sqrt{\frac{m}{p}}} \\ 1 \end{pmatrix} = \lambda_1 \begin{pmatrix} p \\ q \end{pmatrix} + \lambda_2 \begin{pmatrix} -1 \\ 0 \end{pmatrix} + \lambda_3 \begin{pmatrix} 0 \\ -1 \end{pmatrix}.$$

Because the normal to the budget constraint is steeper than the normal to the indifference curve, the second coordinate of $\lambda_1 \cdot (p,q)$ will exceed the second coordinate of ∇f (which is 1), as depicted in the figure. Choose λ_3 to reduce the second coordinate to 1. Thus, $\frac{1}{2}\frac{1}{\sqrt{\frac{m}{p}}} = \lambda_1 p$, or $\lambda_1 = \frac{1}{2}\frac{1}{\left(\sqrt{\frac{m}{p}}\right)p} = \frac{1}{2}\frac{1}{\sqrt{mp}}$. Then $\lambda_1 q = \frac{1}{2}\frac{1}{\sqrt{mp}}q = \frac{1}{2}\frac{1}{\sqrt{\frac{m}{p}}}\frac{q}{p}$. Ignoring the non-negativity condition, the optimal x value would be $\frac{1}{4}\frac{q^2}{p^2}$ and this is greater than $\frac{m}{p}$. Thus, $\frac{1}{2}\frac{q}{p} > \sqrt{\frac{m}{p}}$ and so $\frac{1}{2}\frac{q}{p}\frac{1}{\sqrt{\frac{m}{p}}} > 1$. Putting $\lambda_3 = \frac{1}{2}\frac{q}{p}\frac{1}{\sqrt{\frac{m}{p}}} - 1$ and $\lambda_2 = 0$ completes the determination of the λ_i. This example is considered further in Example 13.10. ◊

The constraint qualification is discussed next. If the gradient of the objective is to be written as a weighted average of the gradients of the constraints, $\nabla f = \sum \lambda_j \nabla g_j$, this requires sufficient "variability" across the different constraints. The constraint qualification concerns this variation.

13.2.1 The constraint qualification

In Theorem 13.1, the constraint qualification requires that there exists some z with $\nabla g_j(x^*)\cdot z < 0$ for each active constraint j. An alternative constraint qualification is given in terms of the linear independence of the gradients of the active constraints. Let $\nabla_{\mathscr{A}(x^*)}g(x^*)$ be the matrix obtained from the matrix $\nabla g(x^*)$ by deleting the columns corresponding to inactive constraints. Suppose that $\nabla_{\mathscr{A}(x^*)}g(x^*)$ has rank equal to the number of active constraints – the gradients of the active constraints are linearly independent (so the number of active constraints must be no greater than n). Then there exists some z with $\nabla g_j(x^*)\cdot z < 0$ for each active constraint j. To see this, let $\gamma = (-1,-1,\dots,-1)$, a vector with the same number of entries as there are active constraints. Since $\nabla_{\mathscr{A}(x^*)}g(x^*)$ has full column rank, there is some $z \in \mathscr{R}^n$ with $z'\nabla_{\mathscr{A}(x^*)}g(x^*) = \gamma$. Thus, for each active constraint j, $-1 = \Delta g_j(x^*)\cdot z < 0$.

The following examples illustrate the constraint qualification.

EXAMPLE 13.5: Consider the pair of constraints:

$$g_1(x_1,x_2) = x_2 - \sqrt{x_1} \le 0$$
$$g_2(x_1,x_2) = x_1^2 - 1 - x_2 \le 0.$$

The two curves, $g_1(x_1,x_2) = 0$, $g_2(x_1,x_2) = 0$ intersect at $(x_1,x_2) = (1.49, 1.22)$.

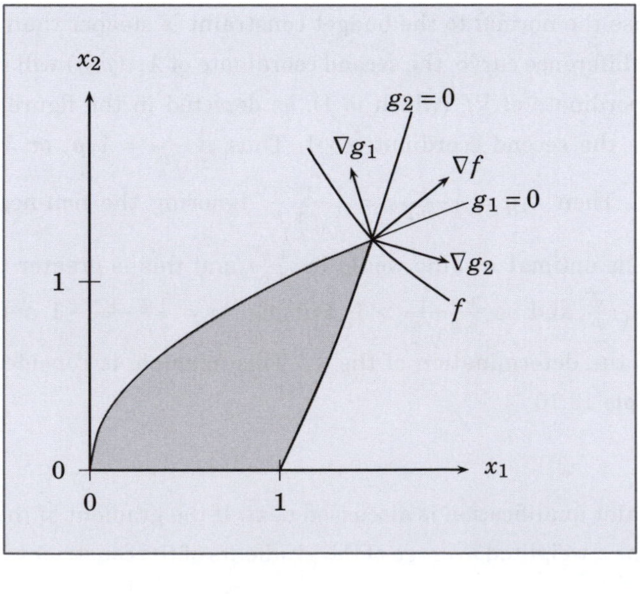

The shaded area depicts the set of points in the non-negative orthant satisfying the two constraints. The gradient of f, ∇f, lies in the cone determined by ∇g_1 and ∇g_2 so that ∇f may be written

$$\nabla f = \lambda_1 \nabla g_1 + \lambda_2 \nabla g_2,$$

with λ_1, λ_2 non-negative. ◊

13.2.1.1 Failure of the constraint qualification

When the constraint qualification fails, the representation in Equation 13.4 (or 13.5) may be impossible to satisfy: at the solution to the problem, x^*, Equation 13.4 cannot be satisfied. The following example illustrates.

EXAMPLE 13.6: The figure plots the constraints determined by two circles of radius $\sqrt{8}$:

$$g_1(x) = (x_1 - 3)^2 + (x_2 - 3)^2 \le 8$$
$$g_2(x) = (x_1 - 7)^2 + (x_2 - 7)^2 \le 8.$$

They have only one point in common: $x = (5,5)$.

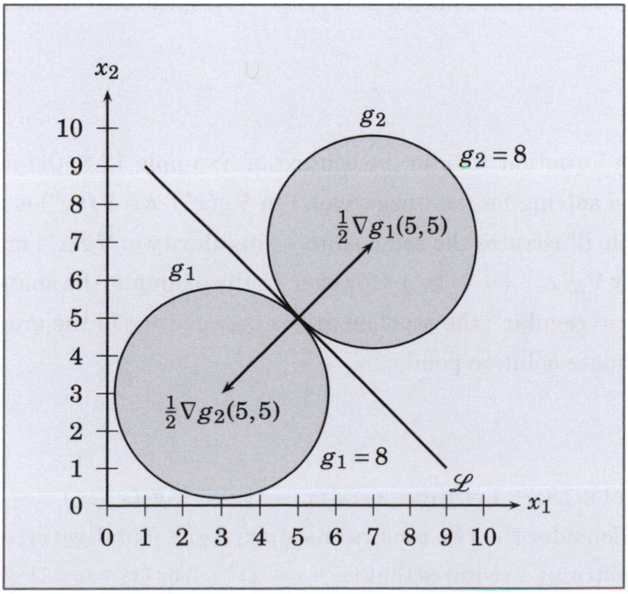

The constraints have gradients:

$$\nabla g_1(x) = \begin{pmatrix} 2(x_1-3) \\ 2(x_2-3) \end{pmatrix}, \quad \nabla g_2(x) = \begin{pmatrix} 2(x_1-7) \\ 2(x_2-7) \end{pmatrix}.$$

The figure depicts these for $x = (5,5)$:

$$\nabla g_1(5,5) = \begin{pmatrix} 4 \\ 4 \end{pmatrix}, \quad \nabla g_2(5,5) = \begin{pmatrix} -4 \\ -4 \end{pmatrix}.$$

Thus, for any λ_1, λ_2,

$$\lambda_1 \nabla g_1(5,5) + \lambda_2 \nabla g_2(5,5) = (\lambda_1 - \lambda_2)\begin{pmatrix} 4 \\ 4 \end{pmatrix} = \beta \begin{pmatrix} 4 \\ 4 \end{pmatrix}.$$

The line $\mathscr{L}$ passes through (5,5). From the figure, note that $\{(x_1,x_2) \mid \nabla g_1(5,5)\cdot(x_1,x_2) \le 0\}$ is the set of points on or below the line $\mathscr{L}$. Likewise $\{(x_1,x_2) \mid \nabla g_2(5,5)\cdot(x_1,x_2) \le 0\}$ is the set of points on or above the line $\mathscr{L}$. There is no point z with $\nabla g_i(5,5)\cdot z < 0$ for $i = 1,2$. Note also that $\nabla g_1(5,5)$ and $\nabla g_2(5,5)$ are linearly dependent: $\nabla g_1(5,5) = (4,4)$ and $\nabla g_2(5,5) = (-4,-4)$. Unless $\nabla f(5,5) = (\alpha,\alpha)$ for some number α, it is impossible to write $\nabla f(5,5) = \lambda_1 \nabla g_1(5,5) + \lambda_2 \nabla g_2(5,5)$, even though the maximization problem has a solution at $(x_1^*, x_2^*) = (5,5)$, the only feasible point. ◇

Considering Equation 13.5, in the context of Example 13.6, the matrix $\boldsymbol{\nabla} g(x^*)$ is singular and solving for (non-negative) λ in $\boldsymbol{\nabla} g(x^*)\cdot\lambda = \nabla f(x^*)$ is impossible. The next example illustrates the same point – singularity of $\boldsymbol{\nabla} g(x^*)$ makes it impossible to satisfy $\boldsymbol{\nabla} g(x^*)\cdot\lambda = \nabla f(x^*)$. However, in the example, the shape of the feasible region seems "regular"; the problem arises because one of the gradients vanishes at the candidate solution point.

EXAMPLE 13.7: Let $g_1(x_1,x_2) = (x_1+x_2-1)^3$, $g_2(x_1,x_2) = x_2$ and $f(x_1,x_2) = x_1x_2$. Consider the two constraints $g_1(x_1,x_2) \le 0$ and $g_2(x_1,x_2) \le \frac{1}{2}$. Satisfying constraint 1 requires that $(x_1+x_2-1)^3 \le 0$ or $(x_1+x_2-1) \le 0$ or $x_1+x_2 \le 1$.

The gradients are:

$$\nabla f(x_1,x_2) = \begin{pmatrix} x_2 \\ x_1 \end{pmatrix}, \quad \nabla g_1(x_1,x_2) = \begin{pmatrix} 3(x_1+x_2-1)^2 \\ 3(x_1+x_2-1)^2 \end{pmatrix}, \quad \nabla g_2(x_1,x_2) = \begin{pmatrix} 0 \\ 1 \end{pmatrix}.$$

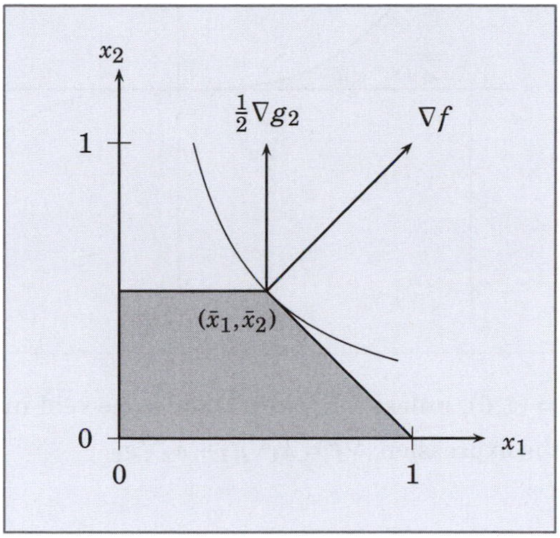

So, the set of (x_1,x_2) pairs satisfying both constraints are those pairs with $x_1+x_2 \leq 1$ and with $x_2 \leq \frac{1}{2}$. The solution occurs at $(\bar{x}_1,\bar{x}_2) = \left(\frac{1}{2},\frac{1}{2}\right)$, where the gradients are

$$\nabla f\left(\frac{1}{2},\frac{1}{2}\right) = \begin{pmatrix} \frac{1}{2} \\ \frac{1}{2} \end{pmatrix}, \quad \nabla g_2\left(\frac{1}{2},\frac{1}{2}\right) = \begin{pmatrix} 0 \\ 0 \end{pmatrix}, \quad \nabla g_2\left(\frac{1}{2},\frac{1}{2}\right) = \begin{pmatrix} 0 \\ 1 \end{pmatrix}.$$

It is impossible to write $\nabla f\left(\frac{1}{2},\frac{1}{2}\right)$ as a linear combination of $\nabla g_1\left(\frac{1}{2},\frac{1}{2}\right)$ and $\nabla g_1\left(\frac{1}{2},\frac{1}{2}\right)$. Here, again, the matrix $\mathbf{\nabla} g(x^*)$ is singular at the candidate solution point. One might remark that a re-specification of the constraint $g_1(x_1,x_2) \leq 1$ to the equivalent $\hat{g}_1(x_1,x_2) = x_1+x_2-1 \leq 0$ provides two linearly independent constraint gradient vectors. ◊

EXAMPLE 13.8: The following example illustrates yet another case where the necessary conditions are not satisfied. The constraints are $g_1(x,y) = y-(1-x)^3 \leq 0$ and $g_2(x,y) = -y \leq 0$. Thus, $\nabla g_1 = \left(\frac{\partial g_1}{\partial x}, \frac{\partial g_1}{\partial y}\right) = (-3(1-x)^2, 1)$ and $\nabla g_2 = \left(\frac{\partial g_2}{\partial x}, \frac{\partial g_2}{\partial y}\right) = (0,-1)$.

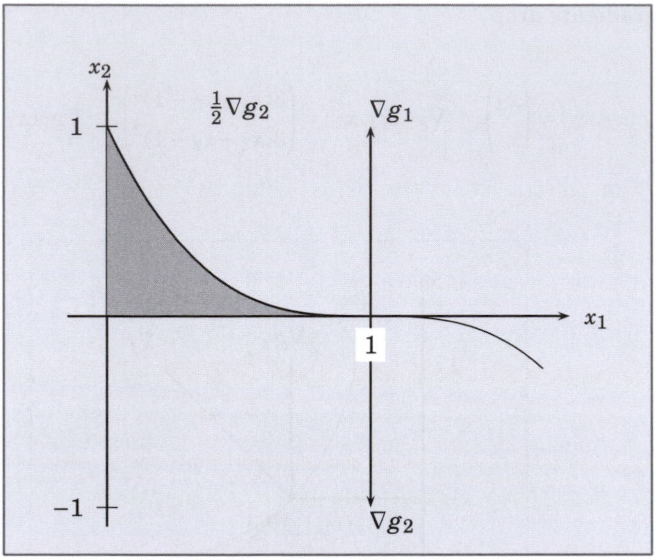

So, at $(x,y) = (1,0)$, unless $\nabla f = \alpha(0,1)$ for some real number α, there is no solution to the expression: $\nabla f = \lambda_1 \nabla g_1 + \lambda_2 \nabla g_2$. ◊

From Theorem 13.1, subject to the constraint qualification, a necessary condition for x^* to be a solution is that

$$\frac{\partial f(x^*)}{\partial x_i} = \sum_{j=1}^{r} \lambda_j \frac{\partial g_j(x^*)}{\partial x_i}, \quad i = 1,\ldots,n$$

must have a non-negative solution, $\lambda = (\lambda_1,\ldots,\lambda_r)$, $\lambda_j \geq 0$, $j = 1,\ldots,r$. Remark 13.4 notes that the Lagrange multipliers and associated constraints may be chosen to satisfy $\lambda_j g_j = 0$: at the solution, if the multiplier is positive, the corresponding constraint is binding ($\lambda_j > 0$ implies $g_j = 0$); and, if the constraint is not binding, the corresponding multiplier is 0 ($g_j < 0$ implies $\lambda_j = 0$).

Imposing this requirement yields a system of $n+k$ equations ($\nabla f = \sum_{j=1}^{k} \lambda_j \nabla g_j$ gives n equations and $\lambda_j g_j = 0$, $j = 1,\ldots,k$ gives k equations) in $n+k$ unknowns, $(\{x_i\}_{i=1}^{n}, \{\lambda_j\}_{j=1}^{k})$. The following discussion confirms the validity of imposing the complementary slackness conditions.

13.2.2 Complementary slackness

Consider the program given in Equation 13.2 and let x^* be a constrained local maximum satisfying the necessary conditions of Theorem 13.1 with Lagrange multipliers $\{\lambda_j\}_{j=1}^k$. At x^*, the active constraints are $\mathscr{A}(x^*) = \{l \mid g_l(x^*) = 0\}$ and the inactive constraints $\mathscr{A}(x^*)^c = \{l \mid g_l(x^*) < 0\}$. There is an open neighborhood of x^*, $\mathscr{O}(x^*)$ with $g_l(x) < 0$ for any $l \in \mathscr{A}(x^*)^c$ and $x \in \mathscr{O}(x^*)$.

Consider the modified program obtained from Equation 13.2 by deleting the constraints not binding at x^*:

$$\max f(x) \text{ subject to } g_j(x) \leq 0,\ j \in \mathscr{A}(x^*). \tag{13.7}$$

For this program, x^* is a constrained local maximum – otherwise, in any neighborhood of x^* there is an x' satisfying the constraints $\mathscr{A}(x^*)$ and with $f(x') > f(x^*)$, violating the assumption that x^* is a constrained local maximum of the original program. Furthermore, for this program the constraint qualification is satisfied at x^* (if the qualification was satisfied at x^* for the original program).

Therefore (applying Theorem 13.1 to the modified program), there exists $\tilde{\lambda}_j \geq 0$, $j \in A(x^*)$ with:

$$\frac{\partial f(x^*)}{\partial x_i} = \sum_{j \in \mathscr{A}(x^*)} \tilde{\lambda}_j \frac{\partial g_j(x^*)}{\partial x_i}.$$

For $j \notin A(x^*)$, let $\tilde{\lambda}_l = 0$ so that:

$$\frac{\partial f(x^*)}{\partial x_i} = \sum_{j=1}^{r} \tilde{\lambda}_j \frac{\partial g_j(x^*)}{\partial x_i}.$$

The condition $\tilde{\lambda}_j g_j = 0$ for all $j = 1,\ldots,r$ is called the complementary slackness condition. Consequently, we can refine Theorem 13.1 with complementary slackness conditions:

Theorem 13.2: *Given the program 13.2, suppose that at x^* there is some z with $\Delta g_l(x^*) \cdot z < 0$ for each active constraint l, then if x^* is a constrained local maximum there exists $\lambda_j \geq 0$, $j = 1,\ldots,r$, with:*

$$\frac{\partial f(x^*)}{\partial x_i} = \sum_{j=1}^{r} \lambda_j \frac{\partial g_j(x^*)}{\partial x_i},\ i = 1,\ldots,n, \tag{13.8}$$

and where $\lambda_j = 0$ if $g_j(x^*) < 0$ (equivalently $\lambda_j g_j(x^*) = 0$, $\forall j = 1,\ldots,r$).

So, in expression 13.8 there are n equations, and the equations $\lambda_j g_j(x^*) = 0$, $\forall j = 1,\ldots,r$, give $n + r$ equations in $n + r$ unknowns (see Remark 13.4).

13.2.3 Minimization and maximization

The case where the objective is to minimize a function subject to constraints may be considered by adapting Theorem 13.2. Let $f(x)$ be the function to be minimized, and specify the j constraints in the form $g_j(x) \geq 0$, $j = 1,\ldots,r$. Observe that minimizing $f(x)$ is the same as maximizing $\hat{f}(x) = -f(x)$; and that the constraint $g_j(x) \geq 0$ is equivalent to $-g_j(x) \leq 0$, so let $\hat{g}_j(x) = -g_j(x)$ for each j. Then the program:

$$\text{(A)} \quad \min f(x) \text{ subject to } g_j(x) \geq 0,\ j = 1,\ldots,r$$

is equivalent to the program:

$$\text{(B)} \quad \max \hat{f}(x) \text{ subject to } \hat{g}_j(x) \leq 0,\ j = 1,\ldots,r.$$

Considering program B, according to Theorem 13.2, if at x^*, $\nabla \hat{g}_l(x^*) \cdot z < 0$ for each active constraint l, then if x^* is a local maximum of $\hat{f}$, there are $\lambda_j \geq 0$, $j = 1,\ldots,r$, such that:

$$\frac{\partial \hat{f}(x^*)}{\partial x_i} = \sum_{j=1}^{r} \lambda_j \frac{\partial \hat{g}_j(x^*)}{\partial x_i},\ i = 1,\ldots,n \tag{13.9}$$

and $\lambda_j \hat{g}_j(x^*) = 0$ for all $j = 1,\ldots,r$.

These conditions may be written in terms of f and g_j. Since $f = -\hat{f}$ and $g = -\hat{g}_j$, Equation 13.9 is unchanged by multiplying by -1 to give

$$\frac{\partial f(x^*)}{\partial x_i} = \sum_{j=1}^{r} \lambda_j \frac{\partial g_j(x^*)}{\partial x_i},\ i = 1,\ldots,n \tag{13.10}$$

and $\lambda_j \hat{g}_j(x^*) = 0$ is equivalent to $\lambda_j g_j(x^*) = 0$. Finally, note that $\nabla \hat{g}_l(x^*) \cdot z < 0$ is equivalent to $\nabla g_l(x^*) \cdot z > 0$. Consequently,

Theorem 13.3: *Suppose that at x^*, there is some z such that $\nabla g_l(x^*) \cdot z > 0$ for each active constraint, l. Then if x^* is a local minimum in program A, there exists $\lambda_j \geq 0$, $j = 1,\ldots,r$ such that Equation 13.10 is satisfied and $\lambda_j g_j(x^*) = 0$, for all j.*

13.2.4 Global and local optima

Theorem 13.1 gives necessary conditions for a constrained local optimum. This leaves open the possibility that there is some alternative feasible x' satisfying the constraints, for which $f(x') > f(x^*)$.

Theorem 13.4: *Suppose that f is concave and the feasible region convex, then, given the constraint qualification, any x^* satisfying Equation 13.3 solves the optimization problem 13.2.*

PROOF: To see this observe that if f is concave then for any $\alpha \in [0,1]$, $f((1-\alpha)x^* + \alpha x') \geq (1-\alpha)f(x^*) + \alpha f(x')$. Rearranging gives

$$f(x^* + \alpha(x' - x^*)) \geq f(x^*) + \alpha[f(x') - f(x^*)]$$

or

$$f(x^* + \alpha(x' - x^*)) - f(x^*) \geq \alpha[f(x') - f(x^*)].$$

Dividing by α and taking limits gives $\nabla f(x^*)\cdot(x' - x^*) \geq f(x') - f(x^*)$. If there is some feasible x' with $f(x') > f(x^*)$, then $\nabla f(x^*)\cdot(x' - x^*) > 0$. Because the feasible set is convex, for every binding constraint, $g_j(x^* + \alpha(x' - x^*)) = g_j((1-\alpha)x^* + \alpha x') \leq g_j(x^*) = 0$ so that $\nabla g_j(x^*)\cdot(x' - x^*) \leq 0$ for every binding constraint. Put $z = (x' - x^*)$. So, $0 < \nabla f(x^*)\cdot z$ and for every binding constraint, j, $\nabla g_j(x^*)\cdot z \leq 0$. According to the necessary condition for an optimum, $\nabla f(x^*) = \sum_{j=1}^{r} \lambda_j \Delta g_j(x^*)$ with $\lambda_j = 0$ for every non-binding constraint. Combining these inequalities:

$$0 < \Delta f(x^*)\cdot z = \left[\sum_{j=1}^{r} \lambda_j \Delta g_j(x^*)\right]\cdot z = \sum_{j=1}^{r} \lambda_j(\Delta g_j(x^*)\cdot z) \leq 0.$$

This contradiction implies there is no such x'. ■

An analogous result holds in the minimization case with f convex.

13.2.5 Non-negativity constraints

Non-negativity constraints may be accommodated in this framework by adapting Theorems 13.1 and 13.2. These calculations lead to the well-known Kuhn–Tucker conditions. The constraint $x_i \geq 0$ may be written $g_i^\circ(x) = -x_i \leq 0$. Consider

$$\max f(x) \text{ subject to } g_j(x) \leq 0,\ j = 1, \ldots, r;\ x_i \geq 0,\ i = 1, \ldots, n, \tag{13.11}$$

or equivalently,

$$\max f(x) \text{ subject to } g_j(x) \leq 0,\ j = 1, \ldots, r;\ g_i^0(x) \leq 0,\ i = 1, \ldots, n. \tag{13.12}$$

With β_i denoting the multiplier on constraint g_i°, Equation 13.4 becomes:

$$\frac{\partial f(x^*)}{\partial x_i} = \sum_{j=1}^{r} \lambda_j \frac{\partial g_j(x^*)}{\partial x_i} + \sum_{k=1}^{n} \beta_k \frac{\partial g_k^\circ(x^*)}{\partial x_i}, \; i = 1,\ldots,n, \tag{13.13}$$

with $\lambda_j \geq 0$ and $\beta_k \geq 0$. Noting that $\frac{\partial g_i^\circ(x^*)}{\partial x_i} = -1$ and $\frac{\partial g_k^\circ(x^*)}{\partial x_i} = 0$, $k \neq i$, this is

$$\frac{\partial f(x^*)}{\partial x_i} = \sum_{j=1}^{r} \lambda_j \frac{\partial g_j(x^*)}{\partial x_i} - \beta_i, \; i = 1,\ldots,n. \tag{13.14}$$

The complementary slackness conditions give: $\lambda_j g_j(x^*) = 0$ and $\beta_i g_i^\circ(x^*) = -\beta_i x_i^* = 0$. From Equation 13.14, the conditions $\beta_i \geq 0$ and $\beta_i x_i^* = 0$, along with $x_i^* \geq 0$, give:

$$\frac{\partial f(x^*)}{\partial x_i} - \sum_{j=1}^{r} \lambda_j \frac{\partial g_j(x^*)}{\partial x_i} \leq 0, \;\; x_i^* \left(\frac{\partial f(x^*)}{\partial x_i} - \sum_{j=1}^{r} \lambda_j \frac{\partial g_j(x^*)}{\partial x_i} \right) = 0, \;\; x_i^* \geq 0, \quad \forall i. \tag{13.15}$$

Similarly,

$$g_j(x^*) \leq 0, \;\; \lambda_j g_j(x^*) = 0, \;\; \lambda_j \geq 0, \quad \forall j. \tag{13.16}$$

Provided the constraint qualification is satisfied, the conditions in Equations 13.15 and 13.16 are necessary for a local maximum of Equation 13.11.

These conditions may be expressed in terms of a Lagrangian function.

Theorem 13.5: *Consider the program:*

$$\max f(x) \text{ subject to } g_j(x) \leq 0, \quad j = 1, \ldots, r; \quad x_i \geq 0, \;\; i = 1,\ldots,n. \tag{13.17}$$

Let

$$\mathscr{L} = f(x) - \sum_{j=1}^{r} \lambda_j g_j(x).$$

Provided the constraint qualification is satisfied, a local maximum in Equation 13.17 satisfies:

$$\begin{aligned} &\frac{\partial \mathscr{L}}{\partial x_i} \leq 0; \quad x_i \frac{\partial \mathscr{L}}{\partial x_i} = 0; \quad x_i \geq 0 \;\; \textit{for} \;\; i = 1,\ldots,n \\ &\frac{\partial \mathscr{L}}{\partial \lambda_j} \geq 0; \quad \lambda_j \frac{\partial \mathscr{L}}{\partial \lambda_j} = 0; \quad \lambda_j \geq 0 \;\; \textit{for} \;\; j = 1,\ldots,r. \end{aligned} \tag{13.18}$$

For a local minimum, consider the program:

$$\min f(x) \text{ subject to } g_j(x) \geq 0,\ j=1,\ldots,r;\ x_i \geq 0,\ i=1,\ldots,n. \tag{13.19}$$

Using the earlier reasoning, the constraint qualification is modified (as in Theorem 13.3). Noting that minimizing f is equivalent to maximizing $-f$ and that the constraint $g_j(x) \geq 0$ is equivalent to $-g_j(x) \leq 0$, the minimization problem may be written as:

$$\max[-f(x)] \text{ subject to } -g_j(x) \leq 0,\ j=1,\ldots,r;\ x_i \geq 0,\ i=1,\ldots,n, \tag{13.20}$$

so the Lagrangian has the form $\mathscr{L} = -f - \sum_{j=1}^{r} \lambda_j[-g_j]$. With $\hat{\mathscr{L}} = -\mathscr{L}$, $\frac{\partial \hat{\mathscr{L}}}{\partial x_i} = -\frac{\partial \mathscr{L}}{\partial x_i}$, $\frac{\partial \hat{\mathscr{L}}}{\partial \lambda_j} = -\frac{\partial \mathscr{L}}{\partial \lambda_j}$, the conditions in Equation 13.18 translate directly to identify necessary conditions for x^* to be a local solution in Equation 13.20.

Theorem 13.6: *Consider the program:*

$$\min f(x) \text{ subject to } g_j(x) \geq 0, \quad j=1,\ldots,r; \quad x_i \geq 0, \quad i=1,\ldots,n. \tag{13.21}$$

Let

$$\mathscr{L} = f(x) - \sum_{j=1}^{r} \lambda_j g_j(x).$$

Provided the constraint qualification is satisfied, a local minimum in Equation 13.21 satisfies:

$$\begin{aligned} \frac{\partial \mathscr{L}}{\partial x_i} &\geq 0; \quad x_i \frac{\partial \mathscr{L}}{\partial x_i} = 0; \quad x_i \geq 0 \quad \text{for} \quad i=1,\ldots,n \\ \frac{\partial \mathscr{L}}{\partial \lambda_j} &\leq 0; \quad \lambda_j \frac{\partial \mathscr{L}}{\partial \lambda_j} = 0; \quad \lambda_j \geq 0 \quad \text{for} \quad j=1,\ldots,r. \end{aligned} \tag{13.22}$$

The following example revisits an earlier example with inequality constraints using complementary slackness explicitly.

EXAMPLE 13.9: Recall Example 13.2: $\max_x f(x_1,x_2) = x_1x_2$ subject to the constraints $g_1(x_1,x_2) = x_2^2 + x_1^2 - 4 \leq 0$ and $g_2(x_1,x_2) = x_2 - \sqrt{x_1} \leq 0$. Formulating this problem with the Lagrangian and imposing the non-negativity constraints gives:

$$\begin{aligned} \mathscr{L} &= x_1x_2 - \lambda_1[x_2^2 + x_1^2 - 4] - \lambda_2[x_2 - \sqrt{x_1}\,] \\ \frac{\partial \mathscr{L}}{\partial x_1} &= x_2 - \lambda_1 2x_1 + \lambda_2 \frac{1}{2}\frac{1}{\sqrt{x_1}} \end{aligned} \tag{13.23}$$

$$\frac{\partial \mathcal{L}}{\partial x_2} = x_1 - \lambda_1 2x_2 - \lambda_2 \tag{13.24}$$

$$\frac{\partial \mathcal{L}}{\partial \lambda_1} = -[x_2^2 + x_1^2 - 4] \tag{13.25}$$

$$\frac{\partial \mathcal{L}}{\partial \lambda_2} = -[x_2 - \sqrt{x_1}\,]. \tag{13.26}$$

Since $(x_1,x_2) = (1,1)$ satisfies the constraints, in any solution, $x_i > 0$, otherwise $x_1x_2 = 0 < 1\cdot 1$. If both x_1 and x_2 are strictly positive, complementary slackness implies that $x_2 - \lambda_1 2x_1 + \lambda_2 \frac{1}{2}\frac{1}{\sqrt{x_1}} = 0$ and $x_1 - \lambda_1 2x_2 - \lambda_2 = 0$. Next, observe that λ_1 and λ_2 are strictly positive. To see this, suppose that $\lambda_1 = 0$. Then Equation 13.23 implies that $x_2 = -\lambda_2 \frac{1}{2}\frac{1}{\sqrt{x_1}}$ and Equation 13.24 gives $x_1 = \lambda_2$. Since $x_1 > 0$, $\lambda_2 > 0$, so that $x_2 = -\lambda_2 \frac{1}{2}\frac{1}{\sqrt{x_1}}$ implies that x_1 and x_2 have opposite signs. Hence $\lambda_1 > 0$. If $\lambda_2 = 0$, then Equation 13.23 gives $x_2 = \lambda_1 2x_1$ and Equation 13.24 gives $x_1 = \lambda_1 2x_2$. Therefore, $\frac{x_2}{x_1} = \frac{2\lambda_1 x_1}{2\lambda_1 x_2} = \frac{x_1}{x_2}$, and so $x_2^2 = x_1^2$ or $x_1 = x_2$, since both are positive. Equation 13.25 then implies that $x_1 > 1$ and $x_2 > 1$ but Equation 13.26 requires $x_2 = \sqrt{x_1}$, which is impossible. Therefore, in the solution, x_1, x_2 are located at the intersection of the constraints and the multipliers are determined as in Example 13.2. ◇

EXAMPLE 13.10: A utility function that arises frequently in economics is the quasilinear function:

$$u(x,y) = v(x) + y,$$

where x and y are the two goods consumed. Denote income by m, so that the budget constraint is $px + qy = m$. Consider the problem of maximizing utility subject to the budget constraint. Utility is maximized at the tangency of the indifference curve and the budget constraint. Suppose that $v(x) = \sqrt{x}$ so that $v'(x) = \frac{1}{2}\frac{1}{\sqrt{x}}$. The indifference curve has slope $\frac{dy}{dx} = -\frac{1}{2}\frac{1}{\sqrt{x}}$ and the budget constraint has slope $-\frac{p}{q}$. Setting these equal, $\frac{1}{2}\frac{1}{\sqrt{x}} = \frac{p}{q}$, or $\sqrt{x} = \frac{1}{2}\frac{q}{p}$, or $x = \frac{1}{4}\frac{q^2}{p^2}$. Thus, expenditure on good x is $px = \frac{1}{4}\frac{q^2}{p}$. Since $px + qy = m$, if $y \ge 0$, then $px = \frac{1}{4}\frac{q^2}{p} \le m$. Otherwise, the situation is as depicted, with a solution where $y < 0$.

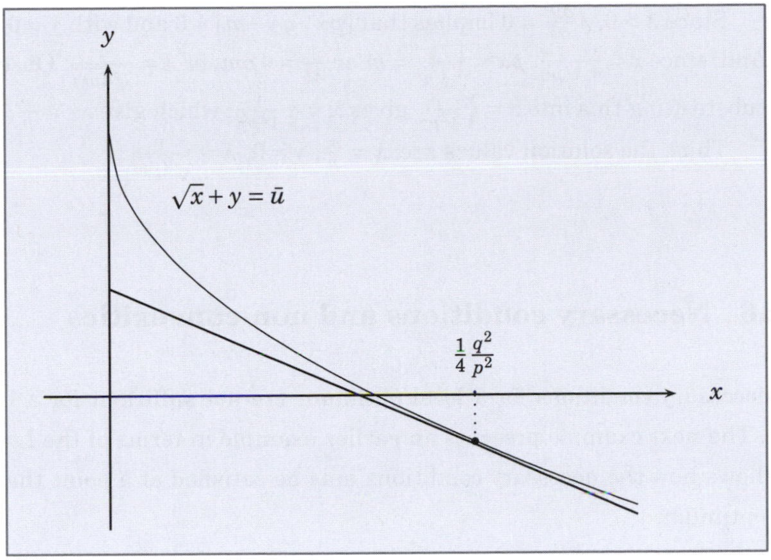

Write the budget constraint as $g(x,y) = px + qy - m \leq 0$. The problem is then to maximize $f(x,y) = \sqrt{x} + y$ subject to $g(x,y) = px + qy - m \leq 0$, $x \geq 0$ and $y \geq 0$. Assume that $\frac{1}{4}\frac{q^2}{p} > m$, so that, in the absence of non-negativity constraints, the value of x demanded would exceed m, as depicted.

The Lagrangian for this problem is $\mathscr{L} = \sqrt{x}+y-\lambda[px+qy-m]$. Applying the conditions in Equation 13.22.

$$\frac{\partial \mathscr{L}}{\partial x} = \frac{1}{2}\frac{1}{\sqrt{x}} - \lambda p \leq 0; \quad x\frac{\partial \mathscr{L}}{\partial x} = 0; \quad x \geq 0$$

$$\frac{\partial \mathscr{L}}{\partial y} = 1 - \lambda q \leq 0; \quad y\frac{\partial \mathscr{L}}{\partial y} = 0; \quad y \geq 0$$

$$\frac{\partial \mathscr{L}}{\partial \lambda} = -[px + qy - m] \geq 0; \quad \lambda\frac{\partial \mathscr{L}}{\partial \lambda} = 0; \quad \lambda \geq 0.$$

To solve this system, first observe that $\frac{\partial \mathscr{L}}{\partial y} \leq 0$ implies that $1 - \lambda q \leq 0$, so that $\lambda \geq \frac{1}{q}$. Next, $\frac{\partial \mathscr{L}}{\partial x} \leq 0$ implies that $\frac{1}{2}\frac{1}{\sqrt{x}} - \lambda p \leq 0$ and rearranging gives $\sqrt{x} \geq \frac{1}{2}\frac{1}{\lambda p}$. So, $x > 0$ and therefore $x\frac{\partial \mathscr{L}}{\partial x} = 0$ implies $\frac{\partial \mathscr{L}}{\partial x} = 0$, giving $\sqrt{x} = \frac{1}{2}\frac{1}{\lambda p}$ or $x = \frac{1}{4}\frac{1}{\lambda^2 p^2}$.

To see that $y = 0$, suppose otherwise, so that $y\frac{\partial \mathscr{L}}{\partial y} = 0$ implies that $\frac{\partial \mathscr{L}}{\partial y} = 0$ and so $1 - \lambda q = 0$, implying that $\lambda = \frac{1}{q}$. In that case, $\sqrt{x} \geq \frac{1}{2}\frac{1}{\lambda p}$ gives $\sqrt{x} \geq \frac{1}{2}\frac{q}{p}$ and $x \geq \frac{1}{4}\frac{q^2}{p^2}$, implying that $px \geq \frac{1}{4}\frac{q^2}{p} > m$.

Since $\lambda > 0$, $\lambda \frac{\partial \mathcal{L}}{\partial \lambda} = 0$ implies that $[px+qy-m]=0$ and with $y=0$, $px=m$. And, since $x = \frac{1}{4}\frac{1}{\lambda^2 p^2}$, $px = \frac{1}{4}\frac{1}{\lambda^2 p} = m$ or $\frac{1}{\lambda^2} = 4pm$, or $\lambda = \frac{1}{\sqrt{4pm}}$. Observe that substituting this into $x = \frac{1}{4}\frac{1}{\lambda^2 p^2}$ gives $x = \frac{1}{4}\frac{1}{\lambda^2 p^2}$, which gives $x = \frac{m}{p}$.

Thus, the solution values are: $x = \frac{m}{p}$, $y = 0$, $\lambda = \frac{1}{\sqrt{4pm}}$. ◊

13.2.6 Necessary conditions and non-convexities

The necessary conditions for a local optimum are not sufficient for a local optimum. The next example presents an earlier example in terms of the Lagrangian and shows how the necessary conditions may be satisfied at a point that is not a local optimum.

EXAMPLE 13.11: Recall Example 13.3: maximize the function $f(x,y) = \alpha \ln x + y$, subject to the constraints $g_1(x,y) = y^2 + (x-1)^2 \le 2$ and $g_2(x,y) = xy \le k = 1.12$. The Lagrangian is:

$$\mathcal{L} = \alpha \ln x + y - \lambda_1(y^2 + (x-1)^2 - 2) - \lambda_2(xy - k).$$

So, the necessary conditions become:

$$\frac{\partial \mathcal{L}}{\partial x} = \alpha \frac{1}{x} - 2\lambda_1(x-1) - \lambda_2 y \le 0; \quad x\frac{\partial \mathcal{L}}{\partial x} = 0; \quad x \ge 0$$

$$\frac{\partial \mathcal{L}}{\partial y} = 1 - 2\lambda_1 y - \lambda_2 x \le 0; \quad y\frac{\partial \mathcal{L}}{\partial y} = 0, \quad y \ge 0$$

$$\frac{\partial \mathcal{L}}{\partial \lambda_1} = -(y^2 + (x-1)^2 - 2) \ge 0; \quad \lambda_1\frac{\partial \mathcal{L}}{\partial \lambda_1} = 0, \ \lambda_1 \ge 0$$

$$\frac{\partial \mathcal{L}}{\partial \lambda_2} = -(xy - k) \ge 0; \quad \lambda_2\frac{\partial \mathcal{L}}{\partial \lambda_2} = 0, \ \lambda_2 \ge 0.$$

Suppose that there is a solution with $\lambda_1 = 0$. Then $\frac{\partial \mathcal{L}}{\partial y} \le 0$ implies $\lambda_2 x > 0$. With $\lambda_1 = 0$, $\frac{\partial \mathcal{L}}{\partial x} \le 0$ gives $\alpha\frac{1}{x} - \lambda_2 y \le 0$, which implies that $y > 0$. Next, $\lambda_2 \frac{\partial \mathcal{L}}{\partial \lambda_2} = 0$ gives $xy = k$ so $\frac{1}{x} = \frac{y}{k}$. Substituting into $\frac{\partial \mathcal{L}}{\partial x} = 0$ ($\alpha\frac{1}{x} - \lambda_2 y = 0$) gives $\alpha\frac{y}{k} - \lambda_2 y = 0$ or $(\frac{\alpha}{k} - \lambda_2)y = 0$ and since $y > 0$, $\lambda_2 = \frac{\alpha}{k}$. Turning to $\frac{\partial \mathcal{L}}{\partial y} = 0$ gives $1 - \lambda_2 x = 0$ or $1 - \frac{\alpha}{k}x = 0$ so that $x = \frac{k}{\alpha}$, so that $xy = k$ implies $y = \alpha$. Thus, the pair $(x', y') = (\frac{\alpha}{k}, \alpha)$ satisfy the necessary conditions for an optimum. Since $\alpha = 0.6$ and $k = 1.12$ in the discussion, here $y' = 0.6$ and $x' = \frac{1.12}{0.6} = 1.867$. The value of the objective at (x', y') is $\beta' = 0.9745 = 0.6\ln(1.866) + 0.6$.

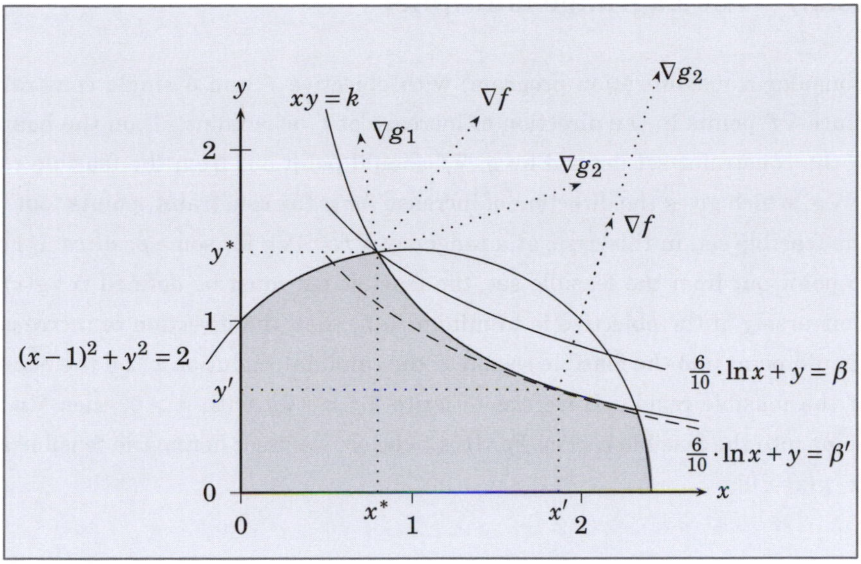

The solution to the constrained optimization problem is at (x^*, y^*) where the necessary conditions for a constrained maximum are satisfied. But the necessary conditions are also satisfied at (x', y') and this point is not a local maximum.

As the figure illustrates, there are points in the neighborhood of (x', y') satisfying the constraint and yielding higher payoff. Along the constraint curve $xy = k$, $y = \frac{k}{x}$ and so the value of the objective along the constraint is $\bar{f}(x) = \alpha \ln x + \frac{k}{x}$. A second-order Taylor series expansion gives:

$$\bar{f}(x) = \bar{f}\left(\frac{k}{\alpha}\right) + \bar{f}'\left(\frac{k}{\alpha}\right)\left(x - \frac{k}{\alpha}\right) + \frac{1}{2}\bar{f}''(p)\left(x - \frac{k}{\alpha}\right)^2,$$

where p is between x and $\frac{k}{\alpha}$. Since $f'(x) = \alpha\frac{1}{x} - k\frac{1}{x^2}$, at $x = \frac{k}{\alpha}$, $f'\left(\frac{k}{\alpha}\right) = \alpha\frac{1}{\frac{k}{\alpha}} - k\frac{1}{\left(\frac{k}{\alpha}\right)^2} = \frac{\alpha^2}{k} - \frac{\alpha^2}{k} = 0$. Also, $f''(x) = -\alpha\frac{1}{x^2} + 2k\frac{1}{x^3} = \left[-\alpha + 2k\frac{1}{x}\right]\frac{1}{x^2}$, so $f''(p) = \left[-\alpha + 2k\frac{1}{p}\right]\frac{1}{p^2}$. When x is close to $\frac{k}{\alpha}$, $p \approx \frac{k}{\alpha}$, and $\left[-\alpha + 2k\frac{1}{p}\right] \approx \alpha > 0$, so $f''(p) \approx \alpha\left(\frac{\alpha}{k}\right)^2 > 0$ when x is close to $\frac{k}{\alpha}$. Therefore,

$$\bar{f}(x) \approx \bar{f}\left(\frac{k}{\alpha}\right) + \frac{1}{2}\alpha\left(\frac{\alpha}{k}\right)^2 \cdot \left(x - \frac{k}{\alpha}\right)^2.$$

So, $f(x) > \bar{f}(\frac{k}{\alpha})$ for $x \neq \frac{k}{\alpha}$ in a neighborhood of $x' = \frac{k}{\alpha}$. Finally, observe that the constraint set is not convex (so Theorem 13.4 does not apply). ◊

13.2.7 The Lagrange multiplier

Consider a maximization program, with objective f and a single constraint g. Since ∇f points in the direction of increase of f, at a point x^*, on the boundary of the constraint set defined by g, ∇f should point out from the feasible region. If ∇g, which gives the direction of increase for g the constraint, points "out" from the feasible set, in this case, at a tangency, $\nabla f = \lambda \nabla g$ for some *positive* λ. For ∇g to point out from the feasible set, the feasible set must be defined $\{x \mid g(x) \leq 0\}$. Conversely, if the objective is to minimize f, then the direction of *increase* of f should point into the feasible region at the candidate solution x^* on the boundary of the feasible region. If we are to write $\nabla f = \lambda \nabla g$ with $\lambda > 0$, then ∇g must point into the feasible region. For this to be so, we must define the feasible set as $\{x \mid g(x) \geq 0\}$.

13.2.8 Equality and inequality constraints

Finally, consider the case where there are both equality and inequality constraints. Given the family of constraints:

$$h_l(x) = 0,\ l = 1, \dots m; \qquad g_j(x) \leq 0,\ j = 1 \dots, r.$$

Consider the problem of minimizing $f(x)$ subject to these constraints.

Theorem 13.7: *Suppose that f, h_l, g_j are continuous with continuous partial derivatives. Let X be an open set in $\mathcal{R}^n$ and $x^* \in X$ and suppose that $f(x^*) \leq f(x), \forall x \in X$. Then there are numbers $\lambda, \mu_1, \mu_2, \dots, \mu_k, \lambda_1, \lambda_2, \dots, \lambda_r$, not all 0, such that:*

$$\lambda \frac{\partial f(x^*)}{\partial x_i} + \sum_{l=1}^{m} \mu_l \frac{\partial h_l(x^*)}{\partial x_i} + \sum_{j=1}^{r} \lambda_r \frac{\partial g_j(x^*)}{\partial x_j} = 0, \quad i = 1, \dots, n. \tag{13.27}$$

In addition,

1. *$\lambda \geq 0$ and $\lambda_j \geq 0$*

2. *For each j such that $g_j(x^*) < 0$, $\lambda_j = 0$*

3. *If the g_j and h_l satisfy a constraint qualification: the gradients of all h_l and the active g_j (those g_j with $g_j(x^*) = 0$) are linearly independent vectors, then λ can be chosen equal to 1: $\lambda = 1$.*

A result similar to Theorem 13.4 also holds with both equality and inequality constraints.

Theorem 13.8: *Suppose that f, h_l, g_j are continuous with continuous partial derivatives. Suppose that at x^*, Equations 13.27 are satisfied, the constraint qualification is satisfied, f is convex and the constraint set convex. The x^* minimizes f subject to the constraints.*

Exercises for Chapter 13

Exercise 13.1 Let

$$f(x,y) = y - \alpha x - x^{\beta}, \quad \alpha, \beta > 1$$
$$g(x,y) = y^2 - 1 - \ln(x+1).$$

Solve the problem $\max f(x,y)$ subject to $g(x,y) \le 0$ and $x,y \ge 0$, using the Kuhn–Tucker conditions.

Exercise 13.2 Let

$$f(x,y) = xy$$
$$g_1(x,y) = ax + by - 2a, \quad a = b > 0$$
$$g_2(x,y) = y + cx^{\gamma} - 1, \quad c = \frac{1}{10}, \gamma = 2.$$

Solve the problem $\max f(x,y)$ subject to $g_i(x,y) \le 0$ for $i = 1,2$ and $x,y \ge 0$, using the Kuhn–Tucker conditions.

Exercise 13.3 Let

$$f(x,y) = x^2 y$$
$$g_1(x,y) = ax + by - 2a, \quad a = b > 0$$
$$g_2(x,y) = y + cx^{\gamma} - 1, \quad c = \frac{1}{10}, \gamma = 2.$$

Solve the problem $\max f(x,y)$ subject to $g_i(x,y) \leq 0$ for $i = 1,2$ and $x,y \geq 0$, using the Kuhn–Tucker conditions.

Exercise 13.4 Let

$$f(x,y) = 14 - x^2 - y^2 - 2x + 2y$$
$$g(x,y) = x + y - 4.$$

Solve the problem $\max f(x,y)$ subject to $g(x,y) \leq 0$ and $x,y \geq 0$, using the Kuhn–Tucker conditions.

Exercise 13.5 Let

$$f(x,y) = xy$$
$$g_1(x,y) = 1 - y + (x-1)^3$$
$$g_2(x,y) = 2y - 2 + 4(x-1)^3.$$

Consider the program: $\max f(x,y)$ subject to $g_i(x,y) \leq 0$, $i = 1,2$. Confirm that the solution to this program occurs at $(x,y) = (1,1)$, but at this point, there is no non-negative solution ((λ_1, λ_2) pair) to the equation:

$$\nabla f = \lambda_1 \nabla g_1 + \lambda_2 \nabla g_2.$$

Explain why the Kuhn–Tucker conditions fail.

14

Integration

14.1 Introduction

The integral of a function f defines a new function, which may have the interpretation of an area or as the "reverse" of differentiation. For example, the computation of consumer surplus and dead-weight loss involve the calculation of area. Given a demand function $P(Q)$ and market price p, the consumer surplus is defined as the area under the demand curve above price and is used as a measure of the welfare generated by the existence of that market. The technique of integration provides a means of determining this area. In a second type of problem, it is useful to identify a function as the derivative of another function. For example, the rate of return on an asset – the change in the value of the asset per unit of time – may be viewed as the derivative of the value of the asset. Thus, knowing the derivative of the value of the asset with respect to time, say $v'(t)$, it is necessary to determine the value of the asset, $v(t)$, from this derivative – given $v'(t)$, $v(t)$ must be found. These problems are solved by integration.

In what follows, the integral of a function is formally defined. Following this, a variety of applications are developed: Section 14.4 considers the welfare measures of producer and consumer surplus. Section 14.5 considers the calculation of cost and profit for a firm. Taxation and dead-weight loss are examined in Section 14.6. In the remaining sections (Sections 14.7, 14.8, 14.9 and 14.10) present-value calculations, Leibnitz's rule, measures of inequality and Ramsey pricing are discussed. Finally, Section 14.11 explores in detail various measures of welfare using these techniques.

14.2 The integral

Let f be a continuous function defined on an interval $[a,b]$. The integral of f over the interval $[a,b]$ is denoted $\int_a^b f(x)dx$. Considering this expression as the area under the function f, when a and b are very close, then on the interval $[a,b]$ the value of the function f does not change much ($f(b) \approx f(a)$), so the area is roughly $f(a)(b-a)$. To extend this idea to any interval (where b is not close to a), partition the interval $[a,b]$ into subintervals $[x_1,x_2],[x_2,x_3],\ldots,[x_{k-1},x_k]$ with $a = x_1$ and $b = x_k$. A partition may be defined by the points determining the intervals: $\mathscr{P} = \{x_1,\ldots,x_n\}$, with $x_i < x_{i+1}$. The norm or mesh of the partition is the largest of these subintervals. Write $|\mathscr{P}|$ for the norm of the partition, define $|\mathscr{P}| = \max_i(x_{i+1}-x_i)$, giving a measure of the widest interval in the partition. For each interval $[x_i,x_{i+1}]$, let s_i be a point in $[x_i,x_{i+1}]$, $x_i \le s_i \le x_{i+1}$, $i = 1,\ldots,k-1$, $s = (s_1,\ldots,s_{k-1})$ and consider the expression:

$$R(\mathscr{P},s) = \sum_{i=1}^{k-1} f(s_i)(x_{i+1}-x_i). \tag{14.1}$$

Definition 14.1: *The function f is (Riemann) integrable on $[a,b]$ if there is a number I, such that for every $\epsilon > 0$ there is a $\delta > 0$, such that if $|\mathscr{P}| < \delta$, then*

$$|R(\mathscr{P},s) - I| < \epsilon,$$

for any $s = (s_1,\ldots,s_k)$, $x_i \le s_i \le x_{i+1}$, $i = 1,\ldots,k-1$. In this case, I is said to be the integral of f over $[a,b]$ and the integral is denoted $\int_a^b f(x)dx = I$.

Assuming the integral exists, the expression, $\int_a^b f(x)dx$ denotes the area between the function f and the horizontal axis. Figure 14.1 depicts the function $f(x) = \sin(\pi \cdot x) + \ln(x+1) + 1$, where x is measured in radians.

In Figure 14.2, the Riemann approximation is illustrated. As the mesh size becomes smaller, the rectangles better approximate the area under the curve. So, the integral is defined as the limit of a sequence of approximations to the area under the curve between a and b. In the figure, the area represented by the rectangles is $\sum_{j=1}^{6} f(s_i)(x_{j+1}-x_j)$. Note that if $a < c < d$, then

$$\int_a^c f(x)dx + \int_c^d f(x)dx = \int_a^d f(x)dx.$$

A different perspective may be given where the integral is seen as the "reverse" of differentiation. With a fixed, varying b varies the integral and we may write the

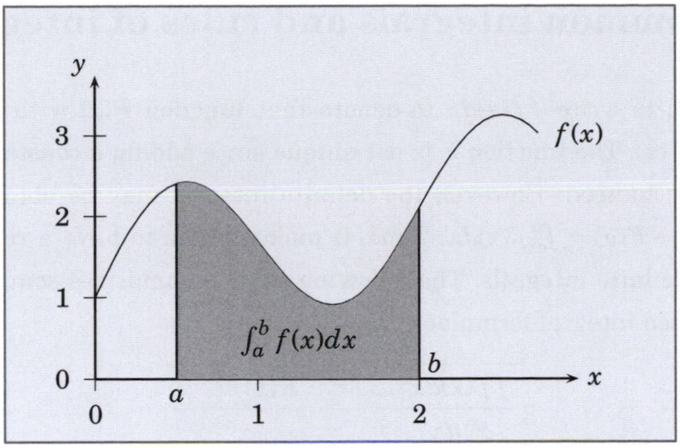

Figure 14.1: The integral of f from a to b

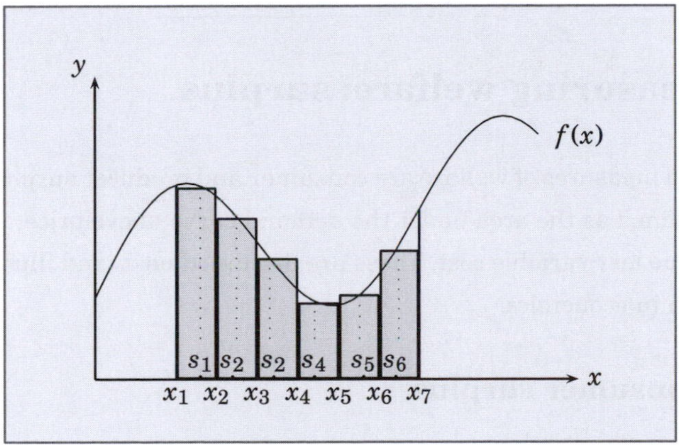

Figure 14.2: The integral of f: definition

integral as a function of b. Define

$$F(b) = \int_a^b f(x)dx.$$

Then $F(b+\epsilon) - F(b) = \int_b^{b+\epsilon} f(x)dx$. If ϵ is small, with f continuous, f is approximately constant on $[b, b+\epsilon]$ and equal to $f(b)$. Then $F(b+\epsilon) - F(b) \cong f(b)[(b+\epsilon) - b] = f(b)\epsilon$. Thus $\frac{F(b+\epsilon)-F(b)}{\epsilon} \cong f(b)$. As $\epsilon \to 0$, $\frac{F(b+\epsilon)-F(b)}{\epsilon} \to F'(b)$ and the approximation becomes exact so that $F'(b) = f(b)$. From this perspective, integration recovers a function F whose derivative is f.

14.3 Common integrals and rules of integration

It is common to write $\int f(x)dx$ to denote that function $F(x)$ with the property that $F'(x) = f(x)$. The function F is not unique since adding a constant leaves the derivative unchanged. However, the definite integral may be obtained directly from F: $F(b) - F(a) = \int_a^b f(x)dx$. Thus, it makes sense to have a record of some standard indefinite integrals. The following table summarizes some of the more frequently used integral formulae.

$\int f(x)dx$		$F(x)$
$\int x^n dx$	$=$	$\frac{1}{n+1}x^{n+1}$
$\int \frac{1}{x}dx$	$=$	$\ln x, x > 0$
$\int e^{ax}dx$	$=$	$\frac{1}{a}e^{ax}$
$\int \sin(ax)dx$	$=$	$-\frac{1}{a}\cos(ax)$
$\int \cos(ax)dx$	$=$	$\frac{1}{a}\sin(ax)$

14.4 Measuring welfare: surplus

Two standard measures of welfare are consumer and producer surplus. Consumer surplus is defined as the area under the demand curve above price; producer surplus is revenue less variable cost. These are discussed next, and illustrate the use of integration in economics.

14.4.1 Consumer surplus

Suppose that there are six consumers of a good. Each consumer requires one unit of the good. Let wp_i denote willingness of i to pay for a unit and suppose that, $wp_1 = 8$, $wp_2 = 6$, $wp_3 = 5$, $wp_4 = 4$, $wp_5 = 3$ and $wp_6 = 1$. With this information we can plot aggregate demand as a function of price (Figure 14.3).

Each consumer pays the price $p^* = 3$. Consumer 1 is willing to pay 8, 5 more than 1 must pay and so has \$5 "surplus". Similarly, consumer 2 has \$3 of surplus, and so on. The total surplus for the consumers is:

$$CS = (8-3)+(6-3)+(5-3)+(4-3)+(3-3) = 11.$$

If $D(p)$ is the number of units demanded at price p, demand, or $p(q)$ inverse demand, then

$$CS = \int_3^8 D(p)dp = \int_0^4 [p(q)-3]dq.$$

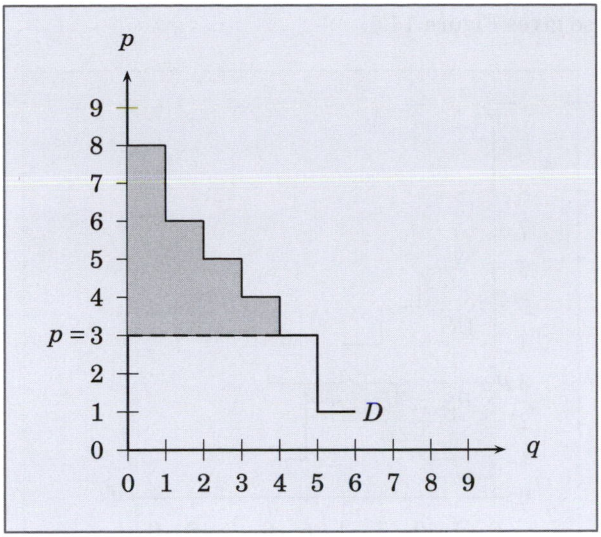

Figure 14.3: Consumer surplus

14.4.2 Producer surplus

Firm supply is given by marginal cost. Let c_i denote the extra cost of producing the ith unit. And set $c_1 = \frac{1}{2}$, $c_2 = \frac{2}{2}$, $c_3 = \frac{3}{2}$, $c_4 = \frac{4}{2}$, $c_5 = \frac{5}{2}$ and $c_6 = \frac{6}{2}$. The total producer surplus (Figure 14.4) is the sum of price less the cost for each additional unit:

$$PS = \left(3 - \frac{1}{2}\right) + \left(3 - \frac{2}{2}\right) + \left(3 - \frac{3}{2}\right) + \left(3 - \frac{4}{2}\right) + \left(3 - \frac{5}{2}\right) = 7\frac{1}{2}.$$

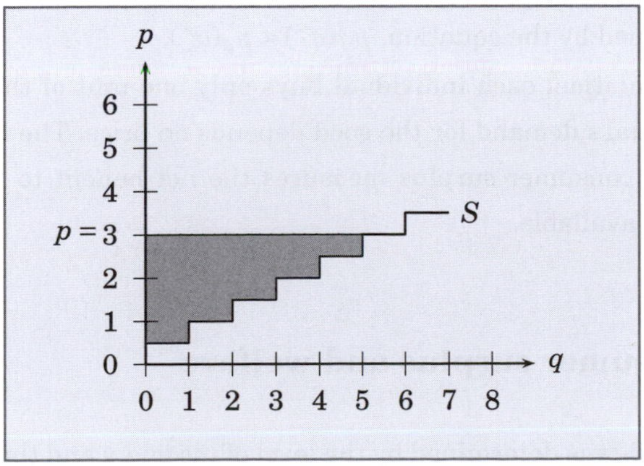

Figure 14.4: Producer surplus

Combining these gives Figure 14.5.

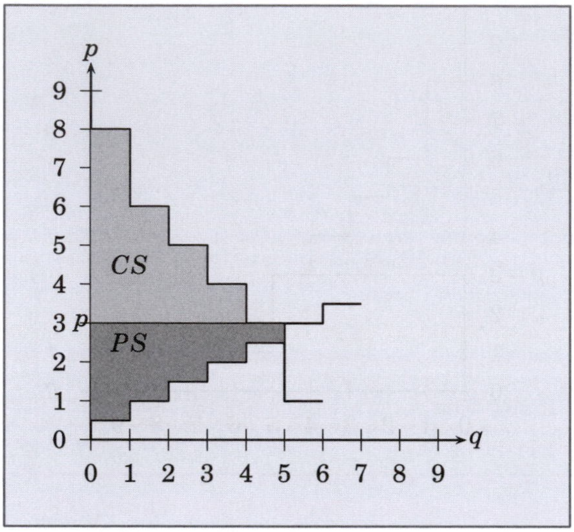

Figure 14.5: Consumer and producer surplus

With $S(p)$ the supply function, we can write total surplus as

$$\int_{\frac{1}{2}}^{8} [D(p) - S(p)]dp.$$

Alternatively, if we write the demand and supply curves in inverse form, $p_d(q)$ and $p_s(q)$, the total surplus may be evaluated as:

$$\int_{0}^{q^*} [p_d(q) - p_s(q)]dq,$$

where q^* is defined by the equation, $p_d(q^*) = p_s(q^*)$.

In this formulation, each individual buys only one unit of the good. Suppose that the individual's demand for the good depends on price. The following discussion shows how consumer surplus measures the net benefit to the consumer of having the good available.

14.4.3 Consumer surplus and welfare

Suppose that utility is determined by the level of income y and the amount of good x consumed. The individual has income y and in the absence of consumption of the good y, has utility $u = y$, so that a dollar of income gives one unit of utility. The unit price of x is p, so that x units cost px and the remaining income is then $y-px$. The utility from x is increasing, but marginal utility from consumption of x is decreasing. To represent this, let the utility from gross income y and consumption x be given according to the formula

$$u(x,y) = x^{\frac{1}{2}} + y - px.$$

Maximizing utility by choice of x gives: $\frac{1}{2}x^{-\frac{1}{2}} - p = 0$, or $x^{\frac{1}{2}} = \frac{1}{2p}$, giving demand for x, $x(p) = \frac{1}{4}\frac{1}{p^2}$, and the inverse demand is $p(x) = \frac{1}{2\sqrt{x}}$. In the absence of any purchase of good x, the individual has welfare $u^0 = y$. With x available at price p per unit, demand is $x(p) = \frac{1}{4}\frac{1}{p^2}$ and so total expenditure on x is $px(p) = p \cdot \frac{1}{4}\frac{1}{p^2} = \frac{1}{4}\frac{1}{p}$. Welfare with x available is

$$\begin{aligned} u^* &= x(p)^{\frac{1}{2}} + y - px(p) \\ &= \frac{1}{2p} + y - \frac{1}{4}\frac{1}{p} \\ &= y + \frac{1}{4}\frac{1}{p}. \end{aligned}$$

Comparing with u^0, the gain from the presence of market x is $\frac{1}{4}\frac{1}{p}$: $u^* - u^0 = \frac{1}{4}\frac{1}{p}$. The individual would be willing to give up (or pay) income equal to $\frac{1}{4}\frac{1}{p}$ for access to the market for good x at price p.

Consider now the computation of consumer surplus associated with good x. From the demand curve $x(p)$ (Figure 14.6), consumer surplus is given by:

$$CS = \int_p^\infty x(t)dt = \frac{1}{4}\int_p^\infty t^{-2}dt = \frac{1}{4}t^{-1}\frac{1}{-1}\Big|_p^\infty = -\frac{1}{4}t^{-1}\Big|_p^\infty = \frac{1}{4}\frac{1}{p}.$$

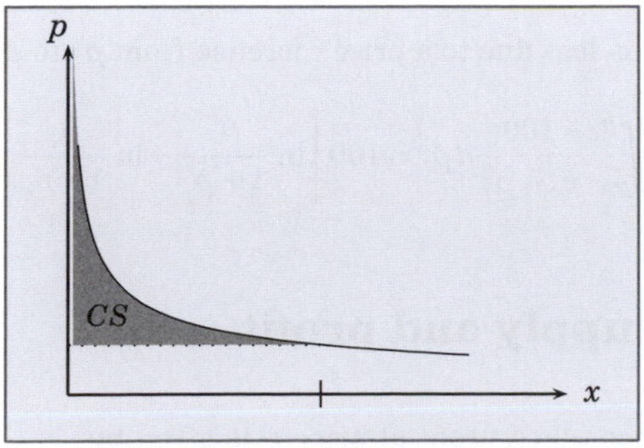

Figure 14.6: Cost and profit

Thus,

$$u^* - u^0 = CS.$$

In this example, there is an exact connection between welfare measured by variation in utility and welfare measured by consumer surplus. This arises because the utility function is quasilinear and, as a result, the level of income y does not appear in the demand function for x. If, for example, $u(x,y) = x^{\frac{1}{2}} y^{\frac{1}{2}}$, then the exact connection breaks down but consumer surplus may still be considered a welfare measure. This is discussed further later.

In general, with income level y and m goods available to purchase at prices $p_1, \ldots, p_m$, demand for good j will have the form $x_j(p_1, \ldots, p_m, y)$. In that case, to compute consumer surplus, we take the prices of other goods as constant and let $\hat{x}_j(p_j) = x_j(p_1, \ldots, p_{j-1}, p_j, p_{j+1}, \ldots, p_m)$. Then, for example, the consumer surplus lost as a result of a price increase from p'_j to p''_j is given by $\int_{p'_j}^{p''_j} \hat{x}_j(p_j) dp_j$. Aggregation to many consumers is straightforward, using the aggregate demand function $X_j(p_j) = \sum_{i=1}^n \hat{x}^i_j(p_j)$, where $\hat{x}^i_j(p_j)$ is the demand for good j by individual i.

Consider the constant elasticity of substitution (CES) utility function $u(x_1, x_2) = (x_1^\rho + x_2^\rho)^{\frac{1}{\rho}}$, where $0 < \rho < 1$. Suppose the individual has income y and faces prices p_i, p_j. Then, with utility maximization, the demand for good i is:

$$x_i(p_i, p_j, y) = \frac{p_i^{r-1} y}{p_i^r + p_j^r}, \quad r = \frac{\rho}{\rho - 1}.$$

For example, letting $\rho = \frac{1}{2}$, so that $r = -1$, this becomes

$$x_i(p_i, p_j, y) = \frac{p_i^{-2} y}{p_i^{-1} + p_j^{-1}} = \frac{y}{p_i + p_i^2 p_j^{-1}}.$$

Suppose $p_j = 1$ and $y = 100$; this then becomes $\hat{x}_i(p_i) = x_i(p_i, 1, 100) = \frac{100}{p_i + p_i^2}$. Then the consumer surplus loss due to a price increase from p_i to p'_i is:

$$\int_{p_i}^{p'_i} \frac{100}{\tilde{p}_i + \tilde{p}_i^2} d\tilde{p}_i = 100 \left[\ln \frac{p'_i}{1 + p'_i} - \ln \frac{p_i}{1 + p_i} \right]$$

14.5 Cost, supply and profit

When a firm is a price taker, profit at price p is given by:

$$\pi(q) = pq - c(q),$$

where $c(q)$ is the total cost (fixed and variable) of producing q units of output. Thus, $F = c(0)$ denotes the fixed cost, independent of whether output is produced or not. Define the variable cost function as $c_v(q) = c(q) - c(0)$. Thus, $c_v'(q) = c'(q)$. If price is fixed, maximizing profit gives

$$\pi'(q) = p - c'(q) = 0$$

or

$$p = c'(q) = mc(q).$$

Thus, quantity is chosen to set marginal cost equal to price. Assuming $c'(q)$ is increasing, this equation defines the firm's supply curve (Figure 14.7). At price p, to maximize profit, q should be chosen to satisfy $p = c'(q) = mc(q)$. Or, $q = mc^{-1}(p) = s(p)$, where $s(p)$ is the firm's supply curve. At price p, the firm will

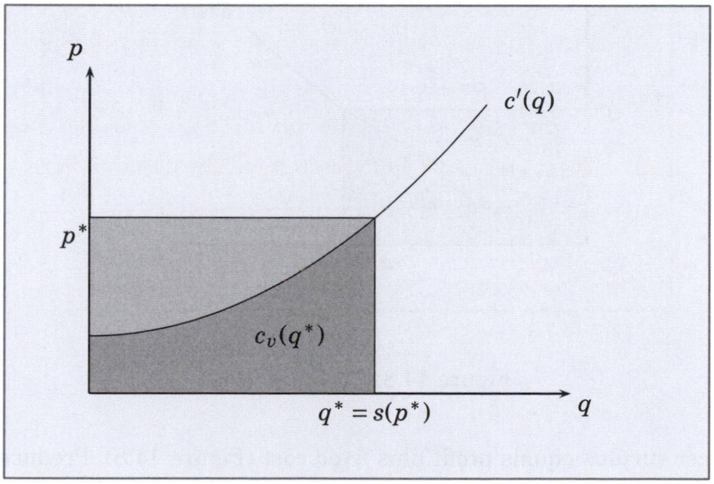

Figure 14.7: The supply curve

supply the quantity $mc^{-1}(p)$ to maximize profit. On the assumption that marginal cost is increasing, supply is upward sloping. Total cost is the sum of variable and fixed cost: $c(q) = c_v(q) + F$ and so $c'(q) = c_v'(q)$. The area under the marginal cost curve gives variable cost:

$$\int_0^{q^*} c'(q)dq = \int_0^{q^*} c_v'(q)dq,$$

which restates to:

$$c(q^*) - c(0) = c_v(q^*) - c_v(0) = c_v(q^*),$$

since $c_v(0) = 0$. Profit for the firm is

$$\pi(q^*) = p^*q^* - c(q^*) = p^*q^* - c_v(q^*) - F.$$

Rearranging:

$$\pi(q^*) + F = p^*q^* - c_v(q^*) = p^*q^* - \int_0^{q^*} c_v'(q)dq.$$

Thus,

$$[\pi(q^*) + F] + \int_0^{q^*} c_v'(q)dq = p^*q^*.$$

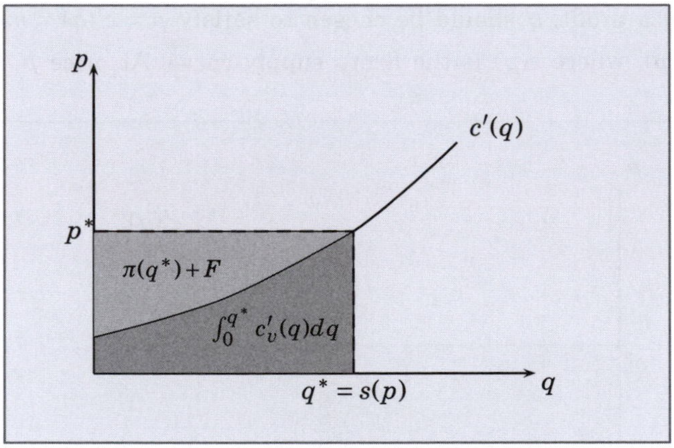

Figure 14.8: Profit and cost

So, producer surplus equals profit plus fixed cost (Figure 14.8). Producer surplus depends only on the marginal cost curve, and profit may be negative if F is large.

14.5.1 Average and marginal cost

These observations may also be seen in terms of average cost:

$$AC(q) = \frac{c(q)}{q} = \frac{c_v(q) + F}{q} = \frac{c_v(q)}{q} + \frac{F}{q}.$$

Then,

$$AC'(q) = \frac{qc'(q) - c(q)}{q^2} = \frac{1}{q}[c'(q) - AC(q)].$$

From this equation, one sees two things. First, average cost is minimized at the point where marginal cost equals average cost ($AC'(q) = 0$ gives $c'(q) = AC(q)$). Second, average cost is declining when $MC < AC$; and increasing when $MC > AC$. These observations are depicted in Figure 14.9. Note that, as depicted, $p^* < AC(q^*)$. From earlier discussion, $p^*q^* = \pi + c(0) + c_v(q)$, (with $c(0) = F$). Therefore,

$$p^* = \frac{\pi + c(0) + c_v(q)}{q^*} < AC(q^*) = \frac{c(0) + c_v(q)}{q^*},$$

and so $\pi < 0$.

Figure 14.9 implicitly assumes that there is positive fixed cost. For, consider the initial point where $q = 0$, and consider an increment of output to Δ. Then total cost is $c(\Delta)$, and average cost satisfies

$$AC(\Delta) = \frac{c(\Delta)}{\Delta} > \frac{c(\Delta) - c(0)}{\Delta} = MC(\Delta),$$

where the strict inequality follows provided $c(0) > 0$.

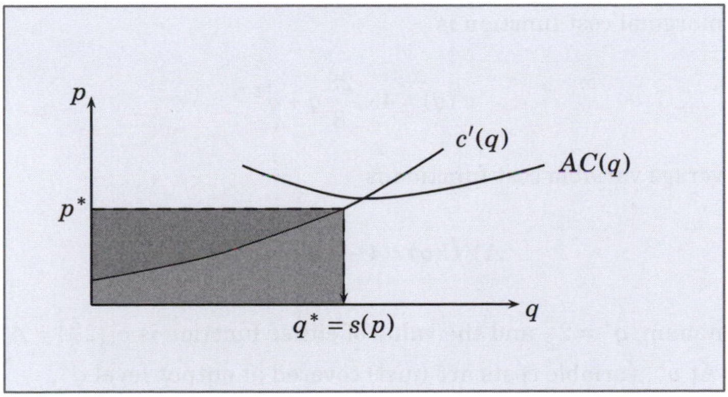

Figure 14.9: Average cost curves and marginal cost

Similarly, considering average variable cost $AVC(q) = \frac{c_v(q)}{q}$, it has derivative $AVC'(q) = \frac{qc_v'(q) - c_v(q)}{q^2}$, so that the minimum AVC occurs where average variable cost coincides with marginal cost. In the case of increasing marginal cost, average variable cost is also increasing and equals marginal cost at 0.

More generally, observe that, since $c_v(0) = 0$:

$$\lim_{q \to 0} \frac{AVC(q)}{q} = \lim_{q \to 0} \frac{c_v(q)}{q} = \lim_{q \to 0} \frac{c_v(q) - c_v(0)}{q} = c_v'(0) = MC(0),$$

so that average variable cost and marginal cost always coincide at $q = 0$. Figure 14.10 depicts the case where marginal cost declines initially. Consider the

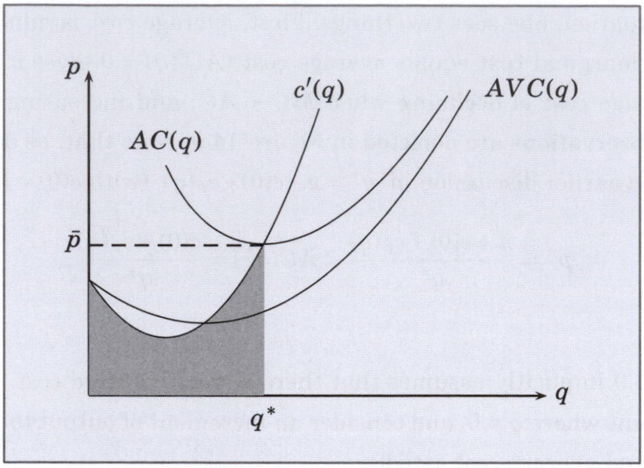

Figure 14.10: Cost curves

total cost function $c(q) = F + 4q - \frac{7}{4}q^2 + \frac{1}{3}q^3$, where $F \geq 0$ is a constant, the fixed cost. The marginal cost function is

$$c'(q) = 4 - \frac{28}{8}q + q^2$$

and the average variable cost function is

$$AVC(q) = 4 - \frac{7}{4}q + \frac{1}{3}q^2.$$

At the minimum, $q^* = 2\frac{5}{8}$ and the value of either function is $c'\left(2\frac{5}{8}\right) = AVC\left(2\frac{5}{8}\right) = \frac{109}{64} = p^*$. At p^*, variable costs are (just) covered at output level q^*,

$$p^* = AVC(q^*) = \frac{c_v(q^*)}{q^*}.$$

Thus, in Figure 14.11,

$$B + D = p^*q^* = c_v(q^*) = \int_0^{q^*} c'(q)dq = A + D.$$

The logic for this is clear; marginal cost is high initially and at price p^* exceeds unit revenue. As output increases beyond q', price exceeds marginal revenue and at point q^* revenue from sales fully covers marginal cost. There is no sales (or production) below price p^*. Above that price, supply follows the marginal cost curve. The supply curve is given by the heavy line (Figure 14.12). No output level between 0 and q^* is ever chosen: output is either 0 or at least as large as q^*.

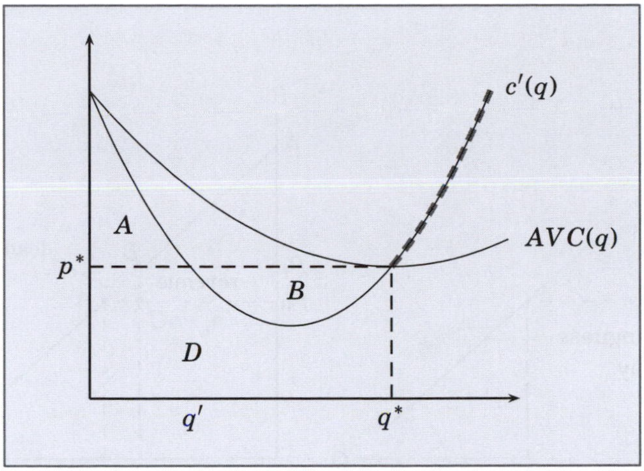

Figure 14.11: Cost curves

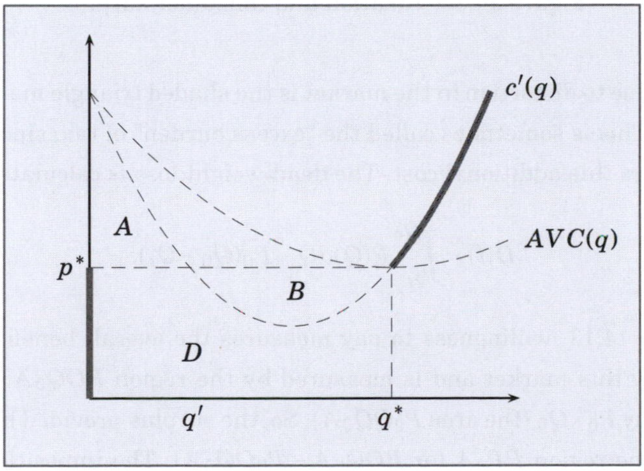

Figure 14.12: Supply

14.6 Taxation and consumer surplus

While taxation is necessary to raise government revenue, it imposes a burden on the consumer and changes consumer behavior. This change in behavior leads to inefficiency and welfare loss, measured in terms of *dead-weight loss* of consumer surplus. The discussion here explains how this arises. A standard demand function is depicted in Figure 14.13, along with the impact of taxation on price and quantity. Denote the demand function by $P(Q)$, and let $P_t = P_0(1+t)$, where t is an ad-valorem tax at rate t.

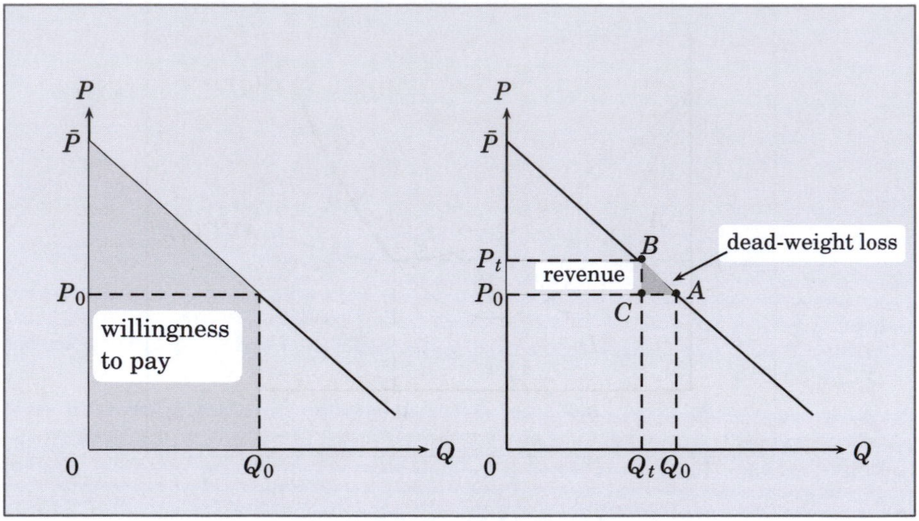

Figure 14.13: Taxation and consumer surplus

The loss due to distortion in the market is the shaded triangle marked as dead-weight loss. This is sometimes called the "excess burden" of tax, since raising the revenue incurs this additional cost. The dead-weight loss is calculated:

$$D(t) = \int_{Q_t}^{Q_0} P(Q)dQ - P_0(Q_0 - Q_t). \tag{14.2}$$

In Figure 14.13, willingness to pay measures the overall benefit provided to consumers by this market and is measured by the region $\bar{P}OQ_0A$. At price P_0, consumers pay $P_0 \times Q_0$ (the area P_0OQ_0A). So, the surplus provided by the market is given by the region $\bar{P}P_0A$ (or $\bar{P}OQ_0A - P_0OQ_0A$). The imposition of the tax raises price to P_t, reducing surplus by P_tP_0AB and generating revenue $(P_t - P_0) \times Q_t$. The surplus reduction exceeds the revenue raised by the amount P_tP_0AB less $(P_t - P_0) \times Q_t$, which is the region ABC, the dead-weight loss associated with the tax.

EXAMPLE 14.1: Let $P(Q) = \frac{1}{Q}$ and suppose that P_0 is given exogenously (horizontal supply). Consider the impact of an ad-valorem tax. One may use the formula in Equation 14.2 to determine the dead-weight loss, D. The area under the demand curve between Q_0 and Q_t is given by:

$$\int_{Q_t}^{Q_0} P(Q)dQ = \int_{Q_t}^{Q_0} \frac{1}{Q}dQ = \ln(Q)\,|_{Q_t}^{Q_0} = \ln(Q_0) - \ln(Q_t) = \ln\left[\frac{Q_0}{Q_t}\right].$$

Next, Q_0 and Q_t are determined by the initial price and the tax:

$$P_0 = \frac{1}{Q_0} \text{ so that } Q_0 = \frac{1}{P_0}; \text{ and } P_0(1+t) = \frac{1}{Q_t} \text{ so that } Q_t = \frac{1}{P_0(1+t)}.$$

These give $\frac{Q_0}{Q_t} = 1+t$ so that $\int_{Q_t}^{Q_0} P(Q)dQ = \ln\left[\frac{Q_0}{Q_t}\right] = \ln(1+t)$. Next, observe that

$$Q_0 - Q_t = \frac{1}{P_0}\left(1 - \frac{1}{1+t}\right) = \frac{1}{P_0}\frac{t}{1+t},$$

so $P_0(Q_0 - Q_t) = \frac{t}{1+t}$. Therefore,

$$D(t) = \int_{Q_t}^{Q_0} p(Q)dq - P_0(Q_0 - Q_t) \quad (14.3)$$

$$= \ln(1+t) - \frac{t}{1+t}. \quad (14.4)$$

With this formula, the impact of varying the tax on the dead-weight loss can be calculated:

$$\begin{aligned} D'(t) &= \frac{1}{1+t} - \left[\frac{(1+t)-t}{(1+t)^2}\right] \\ &= \left[\frac{1}{1+t}\right]\left[1 - \frac{1}{1+t}\right] = \left[\frac{1}{1+t}\right]\left[\frac{t}{1+t}\right] \\ &= \frac{t}{(1+t)^2} > 0. \end{aligned}$$

$$\begin{aligned} D''(t) &= \frac{[(1+t)^2 - 2t(1+t)]}{(1+t)^4} \\ &= (1+t)\frac{[(1+t)-2t]}{(1+t)^4} \\ &= \frac{(1-t)}{(1+t)^3}. \end{aligned}$$

Therefore, the dead-weight loss is increasing ($D'(t) > 0$) at an increasing rate ($D''(t) > 0$)). ◊

EXAMPLE 14.2: In Example 14.1, the supply price was exogenous. This example considers the calculation of producer and consumer surplus where supply is upward sloping. Consider the following supply and demand functions:

$$P_d(Q) = kQ^{-\alpha}, 0 < \alpha < 1, \qquad P_s(Q) = Q^{\beta}.$$

Together these determine the market-clearing quantity, Q_0, and both the consumer and producer surplus. The following discussion shows that the sum of producer and consumer surplus (overall surplus), is:

$$\left[\frac{\beta+\alpha}{(1-\alpha)(1+\beta)}\right] k^{\frac{\alpha+\beta}{\alpha+1}}.$$

Equating supply and demand, $P_d(Q) = kQ^{-\alpha} = Q^{\beta} = P_s(Q)$ gives $k = Q^{\alpha+\beta}$, so the solution is $Q_0 = k^{\frac{1}{\alpha+\beta}}$. The associated price is $P_0 = k^{\frac{\beta}{\alpha+\beta}}$. Note, therefore, that $R_0 = P_0 Q_0 = k^{\frac{\beta+1}{\alpha+\beta}}$. The consumer and producer surplus (CS and PS) are

$$\begin{aligned} CS &= \int_0^{Q_0} P_d(Q)dQ - R_0 \\ &= \int_0^{Q_0} kQ^{-\alpha}dQ - R_0 \\ &= \frac{k}{1-\alpha}(Q_0)^{1-\alpha} - R_0, \end{aligned}$$

$$\begin{aligned} PS &= R_0 - \int_0^{Q_0} P_s(Q)dQ \\ &= R_0 - \int_0^{Q_0} Q^{\beta}dQ \\ &= R_0 - \frac{1}{1+\beta}(Q_0)^{\beta+1}. \end{aligned}$$

Since R_0 cancels out when consumer and producer surplus are combined:

$$CS + PS = \int_0^{Q_0} [P_d(Q) - P_s(Q)]dQ.$$

Considering CS, from the solution to Q_0, $(Q_0)^{1-\alpha} = k^{\frac{1-\alpha}{\alpha+\beta}}$ so that $\frac{k}{1-\alpha}(Q_0)^{1-\alpha} = \frac{1}{1-\alpha}kk^{\frac{1-\alpha}{\alpha+\beta}} = \frac{1}{1-\alpha}k^{1+\frac{1-\alpha}{\alpha+\beta}}$. Note that $1 + \frac{1-\alpha}{\alpha+\beta} = \frac{\alpha+\beta+1-\alpha}{\alpha+\beta} = \frac{\beta+1}{\alpha+\beta}$, so that

$$CS = \frac{1}{1-\alpha}k^{\frac{\beta+1}{\alpha+\beta}} - k^{\frac{\beta+1}{\alpha+\beta}} = \frac{\alpha}{1-\alpha}k^{\frac{\beta+1}{\alpha+\beta}}.$$

For the producer surplus, $(Q_0)^{\beta+1} = k^{\frac{\beta+1}{\alpha+\beta}}$, so that

$$PS = k^{\frac{\beta+1}{\alpha+\beta}} - \frac{1}{1+\beta}k^{\frac{\beta+1}{\alpha+\beta}} = \frac{\beta}{1+\beta}k^{\frac{\beta+1}{\alpha+\beta}}.$$

Therefore,

$$CS+PS=\left[\frac{\alpha}{1-\alpha}+\frac{\beta}{1+\beta}\right]k^{\frac{\beta+1}{\alpha+\beta}}=\left[\frac{\alpha+\beta}{(1-\alpha)(1+\beta)}\right]k^{\frac{\beta+1}{\alpha+\beta}},$$

using $\left[\frac{\alpha}{1-\alpha}+\frac{\beta}{1+\beta}\right]=\left[\frac{\alpha(1+\beta)}{(1-\alpha)(1+\beta)}+\frac{\beta(1-\alpha)}{(1+\beta)(1-\alpha)}\right]=\left[\frac{\alpha(1+\beta)+\beta(1-\alpha)}{(1+\beta)(1-\alpha)}\right]=\left[\frac{\alpha+\beta}{(1+\beta)(1-\alpha)}\right]$.

◊

EXAMPLE 14.3: Suppose that a value added tax t is imposed on the market given by:

$$P_d(Q)=kQ^{-\alpha}, 0<\alpha<1, \quad P_s(Q)=Q^{\beta}.$$

The following calculations determine the dead-weight loss associated with the tax. The after-tax equilibrium quantity is determined by: $P_d(Q)=kQ^{-\alpha}=(1+t)Q^{\beta}=(1+t)P_s(Q)$. This can be rearranged to give $\frac{k}{(1+t)}Q^{-\alpha}=Q^{\beta}$, and letting $k'=\frac{k}{(1+t)}$ gives $k'Q^{-\alpha}=Q^{\beta}$. Repeating earlier calculations gives the after-tax equilibrium price and quantity $Q_t=(k')^{\frac{1}{\alpha+\beta}}$, and before-tax equilibrium is at $Q^*=(k)^{\frac{1}{\alpha+\beta}}$. The after-tax equilibrium price is $P_t=(k')^{\frac{\beta}{\alpha+\beta}}$. The dead-weight loss is:

$$\begin{aligned}
DWL &= \int_{Q_t}^{Q^*}[P_d(Q)-P_s(Q)]dQ \\
&= \int_{0}^{Q^*}[P_d(Q)-P_s(Q)]dQ-\int_{Q^*}^{Q_t}[P_d(Q)-P_s(Q)]dQ \\
&= \left[\frac{\alpha+\beta}{(1-\alpha)(1+\beta)}\right]k^{\frac{\beta+1}{\alpha+\beta}}-\left[\frac{\alpha+\beta}{(1-\alpha)(1+\beta)}\right](k')^{\frac{\beta+1}{\alpha+\beta}} \\
&= \left[\frac{\alpha+\beta}{(1-\alpha)(1+\beta)}\right]\left(k^{\frac{\beta+1}{\alpha+\beta}}-(k')^{\frac{\beta+1}{\alpha+\beta}}\right) \\
&= \left[\frac{\alpha+\beta}{(1-\alpha)(1+\beta)}\right]\left(k^{\frac{\beta+1}{\alpha+\beta}}\left[1-\left(\frac{1}{1+t}\right)^{\frac{\beta+1}{\alpha+\beta}}\right]\right).
\end{aligned}$$

◊

14.7 Present value

A firm operating a plant has revenue flow: $R(t)$ at time t. The cost of operating the plant in present-value terms to time T is $c(T)$ and the discount rate is r.

From these functions, the present value (profit) from operating to period T is

$$\pi(T) = \int_0^T R(t)e^{-rt}dt - c(T).$$

As time goes on (T increases) cost $c(T)$ increases and the increment to revenue decreases if $R(t)e^{-rt}$ is declining. For how long should the firm operate: what value of T maximizes $\pi(T)$?

A special case of this problem occurs when $R(t) = kt$ and $c(T) = ce^{rt}$, with $k, c > 0$. In this case

$$\pi(T) = k\int_0^T te^{-rt}dt - ce^{rT}.$$

To calculate $\int_0^T te^{-rt}dt$ use the integration by parts rule, $\int_0^T udv = uv\,|_0^T - \int_0^T vdu$. In the present case, let $u = t, v = e^{-rt}$ so $dv = (-r)e^{-rt}dt$.

Rewrite $\int_0^T te^{-rt}dt$ as $-(\frac{1}{r})\int_0^T t(-r)e^{-rt}dt$ and consider $\int_0^T t(-r)e^{-rt}dt$. Integrating by parts,

$$\int_0^T t(-r)e^{-rt}dt = te^{-rt}\,|_0^T - \int_0^T e^{-rt}dt = Te^{-rT} - \int_0^T e^{-rt}dt.$$

The term $\int_0^T e^{-rt}dt$ can be integrated:

$$\int_0^T e^{-rt}dt = \left(-\frac{1}{r}\right)\int_0^T (-r)e^{-rt}dt = \left(-\frac{1}{r}\right)e^{-rt}\,|_0^T = \left(-\frac{1}{r}\right)[e^{-rT} - 1].$$

Thus,

$$\int_0^T te^{-rt}dt = -\left(\frac{1}{r}\right)\left[Te^{-rT} + \frac{1}{r}e^{-rT} - \frac{1}{r}\right] = \left[\frac{1}{r^2} - \frac{1}{r^2}e^{-rT} - \frac{1}{r}Te^{-rT}\right].$$

$$\pi(T) = k\left[\frac{1}{r^2} - \frac{1}{r^2}e^{-rT} - \frac{1}{r}Te^{-rT}\right] - ce^{rT}.$$

The first-order condition is:

$$\pi'(T) = 0 = k\left[\frac{1}{r}e^{-rT} - \frac{1}{r}e^{-rT} + Te^{-rT}\right] - cre^{rT} = 0 = kTe^{-rT} - cre^{rT} = 0.$$

The second-order sufficient condition for a maximum is:

$$\pi''(T) = k\left[e^{-rT} - rTe^{-rT}\right] - cr^2e^{rT} < 0.$$

One feature of this example is that the variable T appears in the range of the integral. For the particular functional form, completing the integration and then computing the derivative was simple. Sometimes it is more convenient to differentiate directly and Leibnitz's rule provides a means to do so.

14.8 Leibnitz's rule

Let $f(x,k)$ be a continuous function of x and t and let

$$g(k) = \int_{a(k)}^{b(k)} f(x,k)dx, \quad c \le k \le d. \tag{14.5}$$

Leibnitz's rule provides a formula for differentiating g.

Theorem 14.1: *Suppose that $f(x,k)$ and $\frac{\partial f(x,k)}{\partial k}$ are continuous functions of (x,t) on a region $[a,b]\times[c,d]$ and suppose that $a(k)$ and $b(k)$ are continuous with continuous derivatives on $[a,b]$ (continuously differentiable on $[a,b]$), both with range in $[c,d]$, then*

$$\frac{dg}{dk} = \int_{a(k)}^{b(k)} \frac{\partial f}{\partial k}dx + f(b(k),k)\frac{db(k)}{dk} - f(a(k),k)\frac{da(k)}{dk}.$$

Applied to the present-value calculation:

$$\pi(T) = \int_0^T R(t)e^{-rt}dt - c(T) \Rightarrow \frac{d\pi}{dT} = \frac{d}{dT}\left[\int_0^T R(t)e^{-rt}dt\right] - c'(T).$$

Here $f(t,T) = R(t)e^{-rt}dt$. Applying Leibnitz's rule:

$$\frac{d\pi}{dT} = \left[\int_0^T \frac{\partial f}{\partial T}dt + f(T,T)\frac{dT}{dT} - f(0,T)\frac{d0}{dT}\right] - c'(T).$$

Since $\frac{\partial f}{\partial T} = 0$ and $\frac{d0}{dT} = 0$, $\frac{d\pi}{dT} = f(T,T) - c'(T) = R(T)e^{-rT} - c'(T)$. Setting this to zero gives $R(T)e^{-rT} - c'(T) = 0$ (in the example $kTe^{-rT} - cre^{rT} = 0$).

EXAMPLE 14.4: A student has discovered the function that relates exam score to time spent studying. The key components of the function are as follows. Time spent studying improves exam performance at the rate a^t. However, with more study, fatigue reduces ability to comprehend interesting math quickly, so effort over time should be discounted at some rate: $e^{-r(t+1)}$. Note $e^{-r(t+1)}$ and not e^{-rt}, because it takes some time to "get going". Time spent studying is deducted from sleeping time. One hour of sleep is worth c

dollars, so studying for T hours costs cT. Putting all this together, the "profit from study" function is given by

$$\pi(T) = \int_0^T e^{-r(t+1)} a^t - cT,$$

where T is the number of hours spent studying. Assume that $0 < c < e^{-r}$ and $0 < a < e^r$. A natural question is to find the optimal choice of T.

Here, using Leibnitz's rule will help. The first-order condition is:

$$\pi'(T) = e^{-r(T+1)} a^T - c = 0.$$

Now, $a^T = e^{T \ln a}$, so the first-order condition may be written $e^{-r(T+1)} e^{T \ln a} = c$ or $e^{-r}[e^{T(\ln a - r)}] = c$ or $[e^{T(\ln a - r)}] = e^r c$ or $T(\ln a - r) = r + \ln c$. Thus,

$$T^* = \frac{r + \ln c}{(\ln a - r)}.$$

Now, $0 < c < e^{-r}$ so $\ln c < -r$ or $r + \ln c < 0$ and $0 < a < e^r$ gives $\ln a < r$ or $\ln a - r < 0$. Therefore, $T^* > 0$. Finally, to check the second-order condition,

$$\pi'(T) = e^{-r(T+1)} a^T - c = e^{-r} e^{T(\ln a - r)} - c,$$

$$\pi''(T) = (\ln a - r) e^{-r}[e^{T(\ln a - r)}] = (\ln a - r) e^{-r(T+1)} a^T < 0.$$

A more complex version of this has a depend on T: $a(T)$. In this case, the integral becomes:

$$\pi(T) = \int_0^T e^{-r(t+1)} a(T)^t - cT.$$

In this case, applying Leibnitz's rule:

$$\begin{aligned}
\pi'(T) &= \int_0^T \frac{\partial}{\partial T}\{e^{-r(t+1)} a(T)^t\} dt + e^{-r(T+1)} a(T)^T - c \\
&= \int_0^T e^{-r(t+1)} \frac{\partial}{\partial T}\{a(T)^t\} dt + e^{-r(T+1)} a(T)^T - c \\
&= \int_0^T e^{-r(t+1)} \left\{ t \frac{1}{a(T)} a'(T) a(T)^t \right\} dt + e^{-r(T+1)} a(T)^T - c \\
&= \frac{1}{a(T)} a'(T) \int_0^T t\, e^{-r(t+1)} a(T)^t dt + e^{-r(T+1)} a(T)^T - c.
\end{aligned}$$

Setting this to 0 gives the first-order condition for T. ◊

14.9 Inequality measures

The Lorenz curve is often used as an indicator of the distribution of income in a population. A Lorenz curve L identifies the share of national income belonging to different percentiles of the population, ordered by relative income. In particular, it is often used as an indicator of income inequality. The Lorenz curve is a function attaching a value $L(p)$ to each percentile p. The following discussion defines these curves.

Let F be a distribution function of income in a population: $F(y)$ = proportion of population with income $\leq y$. Assume F is differentiable and let $F'(y) = f(y)$, where f is called the density of the distribution function. Assume $f(y) > 0$ for all y. For $\hat{y} > y$, $\hat{y}$ close to y,

$$F(\hat{y}) - F(y) \cong \int_y^{\hat{y}} f(y)dy = f(y)(\hat{y} - y)$$

gives the proportion of the population with income between y and $\hat{y}$. Given y, if $p = F(y) \leq 1$, then $100p\%$ is the percentage of the population with income $\leq y$. Alternatively, given the fraction p, if y is defined by $p = F(y)$ then y is the level of income such that $100p\%$ have income $\leq y$. To each p we can associate a $y, y(p)$ such that $100p\%$ of population have income in the range $[0, y(p)]$. Thus, $F(y(p)) = p$. The Lorenz curve, $L(p)$, gives for each fraction, p, the total income of the poorest $100p\%$ divided by the total income of the population (Figure 14.14).

The following discussion defines the Lorenz curve, first for the finite population case, and then for a continuous population. Consider first the case where there is a finite number of people in the population. Suppose that the incomes earned in the population are $\{y_1, y_2, \ldots, y_k\}$ with $y_i < y_{i+1}$ and n_i people have income y_i. The total number of people in the population is $n = \sum_{i=1}^k n_i$, the total number with income no greater than y_q is $\sum_{j=1}^q n_j$, and the fraction of the population with income no larger than y_q is $\frac{\sum_{j=1}^q n_j}{\sum_{j=1}^k n_j}$. The total income of people with income no larger than y_q is $\sum_{i=1}^q y_i n_i$. The total income in the population is $\sum_{i=1}^k y_i n_i$. So, the fraction of income held by the population members with income no larger than y_q is:

$$L = \frac{\sum_{i=1}^q y_i n_i}{\sum_{i=1}^k y_i n_i} = \frac{\frac{1}{n}\sum_{i=1}^q y_i n_i}{\frac{1}{n}\sum_{i=1}^k y_i n_i} = \frac{\sum_{i=1}^q y_i \left(\frac{n_i}{n}\right)}{\sum_{i=1}^k y_i \left(\frac{n_i}{n}\right)} = \frac{\sum_{i=1}^q y_i f(y_i)}{\sum_{i=1}^k y_i f(y_i)} = \frac{\sum_{y=y_1}^{y_q} y f(y)}{\sum_{y=y_1}^{y_k} y f(y)},$$

where $f(y_i) = \frac{n_i}{n}$, the fraction of the population with income y_i.

By analogy with this expression, in the continuous case,

$$L(p) = \frac{\int_0^{y(p)} xf(x)dx}{\int_0^{\infty} xf(x)dx} = \frac{\int_0^{y(p)} xf(x)dx}{\mu} = \frac{F(y(p))\int_0^{y(p)} x\left[\frac{f(x)}{F(y(p))}\right]dx}{\int_0^{\infty} xf(x)dx}$$
$$= p\frac{\mu_p}{\mu}$$

where μ_p is the mean income of the poorest fraction p of the population. Thus $L(p)$ gives the percentage of total income held by the poorest $100p\%$ of the population. Here μ is the mean income of the population. To clarify:

$$\int_0^{y(p)} xf(x)dx = F(y(p))\int_0^{y(p)} x\frac{f(x)}{F(y(p))}dx.$$

Note that $\int_0^{y(p)} \frac{f(x)}{F(y(p))}dx = 1$ (since $\int_0^{y(p)} f(x)dx = F(y(p))$ and thus $\frac{f(x)}{F(y(p))}$ is a distribution – the distribution of income in the group with income in the range $[0, y(p)]$. If $\hat{x} > x$ and $\hat{x}$ close to x, $\frac{f(x)(\hat{x}-x)}{F(y(p))} \approx$ "proportion" of the income group $[0, y(p)]$ whose income is in the range $[x, \hat{x}]$. Therefore, $\int_0^{y(p)} x\frac{f(x)}{F(y(p))}dx = \mu_p$ – the mean income of the poorest $100p\%$ (i.e., mean income to pth percentile). Also $F(y(p)) = p$, by definition, so the Lorenz curve may be written $L(p) = p\frac{\mu_p}{\mu}$. Since $\mu_p \leq \mu$, $L(p) \leq p$.

The following calculations show that the Lorenz curve is convex, as depicted in Figure 14.14.

$$L(p) = \frac{1}{\mu}\int_0^{y(p)} xf(x)dx,$$

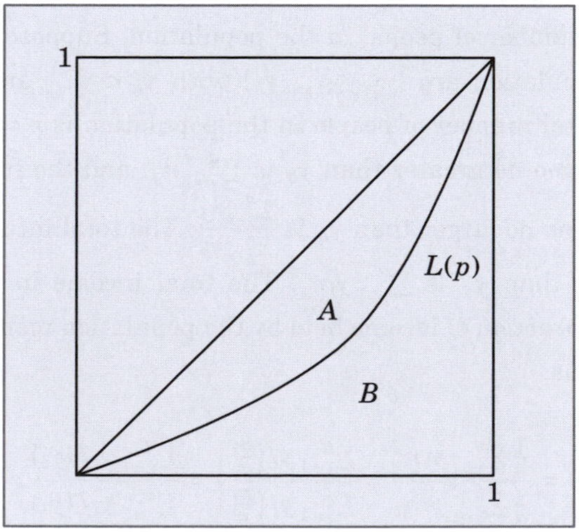

Figure 14.14: The Lorenz curve

so (using Leibnitz's rule),

$$L'(p) = \frac{1}{\mu}[y(p)f(y(p))]\frac{dy(p)}{dp}.$$

Now $p = F(y(p))$, so

$$1 = F'(y(p))\frac{dy(p)}{dp} = f(y(p))\frac{dy(p)}{dp} \Rightarrow \frac{dy(p)}{dp} = \frac{1}{f(y(p))}.$$

Thus,

$$L'(p) = \frac{y(p)}{\mu}, L''(p) = \frac{1}{\mu}\cdot\frac{dy(p)}{dp} = \frac{1}{\mu}\cdot\frac{1}{f(y(p))} > 0.$$

A second measure of inequality is given by the *Gini* coefficient. This is a single number, G, defined:

$$G = \frac{\frac{1}{2} - \int_0^1 L(p)dp}{\frac{1}{2}} = 1 - 2\int_0^1 L(p)dp = \frac{A}{A+B}. \tag{14.6}$$

In Figure 14.14, this is the area A divided by the area $A+B$. The area A is $\frac{1}{2} - \int_0^1 L(p)dp$ (the area under the 45° line and above $L(p)$), while the area $A+B$ is $\frac{1}{2}$. While the Gini coefficient contains less information than the Lorenz curve, it has the virtue of being a single number, so that Gini coefficients for different countries are comparable.

14.10 Ramsey pricing

A standard rule of thumb for efficient pricing is that price be set equal to marginal cost: output should be increased up to the point where the cost of providing an additional unit is equal to the amount a consumer is willing to pay. Where there are fixed costs of production, this is not viable in the long term since marginal cost pricing does not cover fixed cost. In the case of a single good, price must be at least as large as average cost for long-run profit to be non-negative. With more than one good, where there are fixed costs and joint production, the problem is more complicated. The following discussion considers efficient pricing (to maximize welfare) where more than one market is supplied, and where a fixed cost must be recovered by pricing above marginal cost.

A firm supplies two markets, X and Y. The variable cost of joint production is $c(x,y)$, fixed cost is F, and inverse demand in the markets is p_x and

p_y, respectively. At output levels (x^*, y^*), surplus is given by:

$$S(x^*, y^*) = \int_0^{x^*} p_x(x)dx + \int_0^{y^*} p_y(y)dy - c(x^*, y^*).$$

The problem is to maximize surplus subject to revenue from sales covering cost. For example, $c(x,y) = c_x x + c_y y + F$ is a cost function with constant marginal costs c_x and c_y and a fixed cost F. Thus, prices must be set above marginal cost to cover fixed cost.

Define the Lagrangian:

$$\mathcal{L} = \int_0^{x^*} p_x(x)dx + \int_0^{y^*} p_y(y)dy - c(x^*, y^*) + \lambda[p_x(x)x + p_y(y)y - c(x^*, y^*) - F].$$

At a solution (x^*, y^*),

$$\frac{\partial \mathcal{L}}{\partial x^*} = p_x(x^*) - c_x(x^*, y^*) + \lambda[p_x(x^*) + p'_x(x^*)x^* - c_x(x^*, y^*)] = 0$$

$$\frac{\partial \mathcal{L}}{\partial y^*} = p_y(y^*) - c_y(x^*, y^*) + \lambda[p_y(y^*) + p'_y(y^*)y^* - c_y(x^*, y^*)] = 0$$

$$\frac{\partial \mathcal{L}}{\partial \lambda} = p_x(x^*)x^* + p_y(y^*)y^* - c(x^*, y^*) - F = 0.$$

Rearranging these equations,

$$[p_x(x^*) - c_x(x^*, y^*)](1+\lambda) + \lambda p'_x(x^*)x^* = 0$$
$$[p_y(y^*) - c_y(x^*, y^*)](1+\lambda) + \lambda p'_y(y^*)y^* = 0$$

or

$$[p_x(x^*) - c_x(x^*, y^*)] = -\frac{\lambda}{(1+\lambda)} p'_x(x^*)x^* = -p'_x(x^*)x^* k$$

$$[p_y(y^*) - c_y(x^*, y^*)] = -\frac{\lambda}{(1+\lambda)} p'_y(y^*)y^* = -p'_y(y^*)y^* k,$$

where $k = \frac{\lambda}{(1+\lambda)}$. Dividing by $p_x(x^*)$ and $p_y(y^*)$, respectively, and noting that

$$\frac{-p'_x(x^*)x^*}{p_x(x^*)} = \frac{1}{\eta_x(x^*)},$$

$$\frac{[p_x(x^*) - c_x(x^*, y^*)]}{p_x(x^*)} = k\frac{1}{\eta_x(x^*)}$$

$$\frac{[p_y(y^*) - c_y(x^*, y^*)]}{p_y(y^*)} = k\frac{1}{\eta_y(y^*)}.$$

Thus, the markup of price in excess of marginal cost in each market should be proportional to the reciprocal of the elasticity in that market (with the same constant of proportionality, k, for all markets).

REMARK 14.1: In this discussion, differentiation involved differentiating with respect to a variable appearing in the range of an integral. This can be done using Leibnitz's rule, but also by direct computation. The derivative of $g(x^*) = \int_0^{x^*} p_x(x)dx$ with respect to x^* is $p_x(x^*)$, and this can be established with a simple variational argument:

$$\begin{aligned}\frac{1}{\Delta}[g(x^*+\Delta)-g(x^*)] &= \frac{1}{\Delta}\left[\int_0^{x^*+\Delta} p_x(x)dx - \int_0^{x^*} p_x(x)dx\right]\\ &= \frac{1}{\Delta}\left[\int_0^{x^*} p_x(x)dx + \int_{x^*}^{x^*+\Delta} p_x(x)dx - \int_0^{x^*} p_x(x)dx\right]\\ &= \frac{1}{\Delta}\int_{x^*}^{x^*+\Delta} p_x(x)dx\\ &\approx \frac{1}{\Delta}\int_{x^*}^{x^*+\Delta} p_x(x^*)dx,\end{aligned}$$

the approximation following with $p_x(x)$ approximately constant on $[x^*, x^*+\Delta]$. Since $\int_{x^*}^{x^*+\Delta} p_x(x^*)dx = p_x(x^*)\Delta$,

$$\frac{1}{\Delta}[g(x^*+\Delta)-g(x^*)] \approx \frac{1}{\Delta}p_x(x^*)\Delta = p_x(x^*).$$

□

EXAMPLE 14.5: Consider $p_x(x) = a - bx$, $p_y(y) = \alpha - \beta y$, and suppose that cost is $c(x,y) = c_x x + c_y y + F$. Since the reciprocal of the elasticity is $\frac{1}{\eta_x} = -\frac{x}{p_x}p'_x(x)$, the formula:

$$\frac{[p_x(x^*) - c_x(x^*, y^*)]}{p_x(x^*)} = k\frac{1}{\eta_x(x^*)}$$

becomes

$$\frac{p_x(x^*) - c_x}{p_x(x^*)} = k \times \left[-\frac{x^*}{p_x(x^*)}p'_x(x^*)\right].$$

Cancelling $p_x(x^*)$ and with $p'_x(x) = -b$,

$$a - bx^* - c_x = k \times bx^*.$$

Thus, $(a - c_x) = (1+k)bx^*$ so that $x^* = \frac{1}{1+k}\frac{a-c_x}{b} = \delta\frac{a-c_x}{b}$, with $\delta = \frac{1}{1+k}$. Let x_{mc} be the quantity at which price is equal to marginal cost: $x_{mc} = \frac{a-c_x}{b}$ and, with similar notation, $y_{mc} = \frac{\alpha - c_y}{\beta}$. Thus, $x^* = \delta x_{mc}$ and $y^* = \delta y_{mc}$, so that relative to marginal cost pricing, both goods are scaled down by the same proportion. Thus, when $x_{mc} = y_{mc}$, $x^* = y^*$. Revenue over marginal cost in market x is:

$$\begin{aligned}[p_x(x^*) - c_x]x^* &= [(a - c_x) - bx^*]x^* = [(a - c_x) - bx^*]x^* \\ &= [(a - c_x) - \delta(a - c_x)]x^* = [(1 - \delta)(a - c_x)]x^*,\end{aligned}$$

and substituting for x^* gives revenue over marginal cost as: $(1-\delta)\delta\frac{(a-c_x)^2}{b}$. The analogous calculation in market y gives revenue over cost as $(1-\delta)\delta\frac{(\alpha-c_y)^2}{\beta}$. Therefore, setting total revenue over marginal cost equal to fixed cost gives:

$$(1-\delta)\delta\left[\frac{(a-c_x)^2}{b} + \frac{(\alpha-c_y)^2}{\beta}\right] = F$$

This expression (assuming a solution) will have two roots – select the larger root, corresponding to the larger output of x and y. ◊

14.11 Welfare measures

When the price of a good changes, the welfare of a consumer is affected. How should the welfare change be measured? There are a number of answers to this question – here three are considered: consumer surplus, compensating variation and equivalent variation.

14.11.1 Consumer surplus

Maximizing utility $u(x_1,\ldots,x_n)$ subject to the budget constraint $\sum p_i x_i = m$ yields demand functions $x_i(p_1,\ldots,p_n,m)$, $i = 1,\ldots,n$. The program is written:

$$\begin{aligned} \max \quad & u(x_1,\ldots,x_n) \\ \text{subject to} \quad & \sum_k p_k x_k = m. \end{aligned} \tag{14.7}$$

For ease of notation, write the demand function as $x_i(p_i, p_{-i}, m)$ to highlight "own" price, p_i. The optimal choice occurs at the tangency of the budget constraint and the indifference curve as depicted in Figure 14.15 for the case of two goods, (x_i,x_j). A natural question arises: how does a price change affect welfare? For example, if p_i increases to p'_i, how is the individual's welfare affected? Considering Figure 14.15, utility has fallen from u^0 to u^1 so that the loss in welfare might be measured by $u^0 - u^1$. However, utility is an ordinal measure of welfare ($u(x_1,\dots,x_n)$ and $5u(x_1,\dots,x_n)$ have exactly the same indifference curves so the distance $u^0 - u^1$ is arbitrary).

A natural *monetary* measure of welfare loss uses the consumer surplus associated with demand for good i, and this measure is commonly used. However, a change in the price of one good affects demand for other goods through cross-price effects, so that consumer surplus variations resulting from a change in consumption of good i may not fully describe the impact on welfare.

When the price of good x_i increases to p'_i, demand for good x_i decreases from $x_i(p_i,p_j,m)$ to $x_i(p'_i,p_j,m)$ (assuming that the good is not a "Giffen" good). This gives a downward sloping demand curve – as p_i increases, x_i decreases.

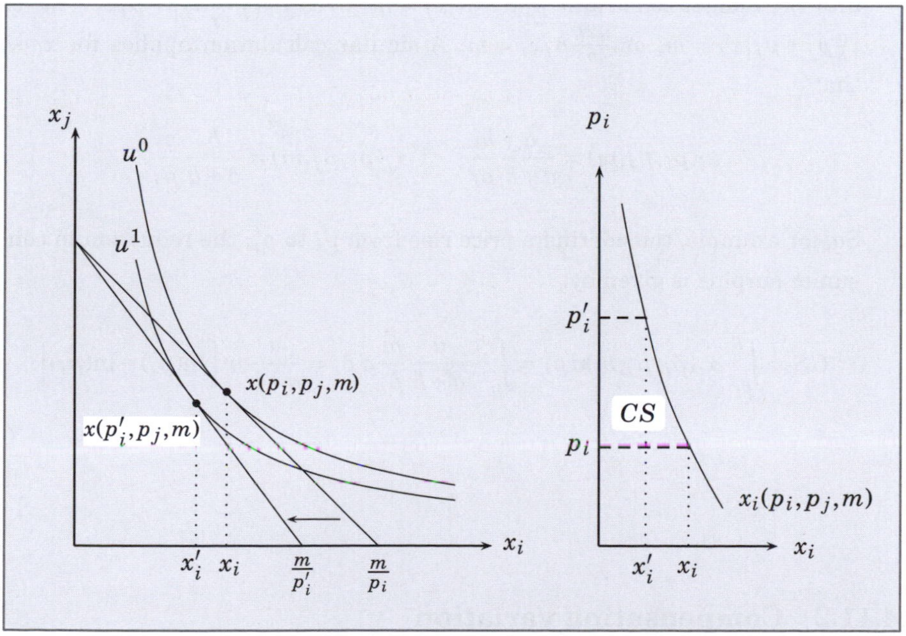

Figure 14.15: Utility maximization and consumer surplus

In Figure 14.15, a price increase from p_i to p'_i moves the budget constraint inward, leading to a tangency at $x_i(p'_i,p_j,m)$ on the new budget constraint: $\{(x_i,x_j) \mid p'_i x_i + p_j x_j = m\}$. Demand for good x_i falls from $x_i = x_i(p_i,p_j,m)$ to $x'_i = x_i(p'_i,p_j,m)$.

The welfare cost measured in terms of consumer surplus is given by the area under the demand curve between p_i and p'_i:

$$CS = \int_{p_i}^{p'_i} x_i(\tilde{p}_i, p_j, m) d\tilde{p}_i. \tag{14.8}$$

The demand function $x_i = x_i(p_i, p_j, m)$ is sometimes referred to as the *Marshallian* demand function, to distinguish it from the *Hicksian* (or "compensated") demand function that is discussed later.

EXAMPLE 14.6: Suppose that $u(x_i, x_j) = a\ln x_i + b\ln x_j$. Optimality involves setting the marginal rate of substitution equal to the price ratio. Thus,

$$\frac{\frac{a}{x_i}}{\frac{b}{x_j}} = \frac{p_i}{p_j}.$$

Rearranging this expression: $\frac{ax_j}{bx_i} = \frac{p_i}{p_j}$, so that $x_i = \frac{a}{b}\frac{p_j}{p_i}x_j$. Substituting this into the budget constraint $p_i x_i + p_j x_j = m$ gives $p_i\left(\frac{a}{b}\frac{p_j}{p_i}x_j\right) + p_j x_j = m$, or $\left[\frac{a}{b}p_j + p_j\right]x_j = m$, or $\frac{a+b}{b}p_j x_j = m$. A similar calculation applies for x_i so that

$$x_i(p_i, p_j, m) = \frac{a}{a+b}\frac{m}{p_i}, \qquad x_j(p_i, p_j, m) = \frac{b}{a+b}\frac{m}{p_j}.$$

So, for example, considering a price rise from p_i to p'_i, the reduction in consumer surplus is given by:

$$CS = \int_{p_i}^{p'_i} x_i(\tilde{p}_i, p_j, m) d\tilde{p}_i = \int_{p_i}^{p'_i} \frac{a}{a+b}\frac{m}{\tilde{p}_i} d\tilde{p}_i = \frac{a}{a+b} m\left[\ln(p'_i) - \ln(p_i)\right].$$

◊

14.11.2 Compensating variation

Two important alternative measures of welfare variation are compensating and equivalent variation. In this section, compensating variation is discussed in detail. The measure is illustrated in Figure 14.16. Suppose there are two goods, i and j, and the initial price–income profile is (p_i, p_j, m). Consumption is at A.

Suppose the price of good i increases from p_i to p'_i. Then, consumption moves

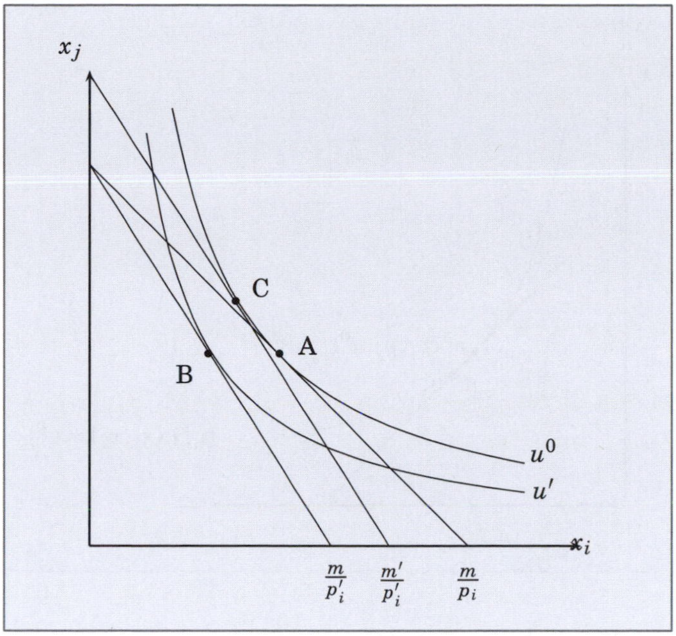

Figure 14.16: Compensating variation

from A to B, and the person has a lower level of welfare. Suppose now that the individual is compensated for the price increase with an increase in income – just enough to restore the person's initial level of utility: increase income from m to m'.

The amount $m'-m$ is exactly the monetary compensation required to offset the welfare-reducing impact of the price increase. After this increase in income $m'-m$, the person ends up consuming at C. This is called the *compensating variation*:

$$CV = \text{compensating variation} = m' - m.$$

The following discussion develops an expression for the compensating variation using the expenditure function and the Hicksian demand curve. Given income m and prices (p_i, p_{-i}), utility maximization leads to demand $x^* = x(p_i, p_{-i}, m)$ and utility $u^0 = u(x(p_i, p_{-i}, m))$. Now, consider the problem where u^0 is fixed and the task is minimizing the cost of achieving the level of utility u^0:

$$\min \quad p_1x_1 + p_2x_2 \qquad \text{subject to} \quad u(x_1, x_2) = u^0. \tag{14.9}$$

This problem involves moving the budget constraint in as far as possible subject to maintaining tangency with the indifference curve associated with utility level u^0. The solution is depicted in Figure 14.17.

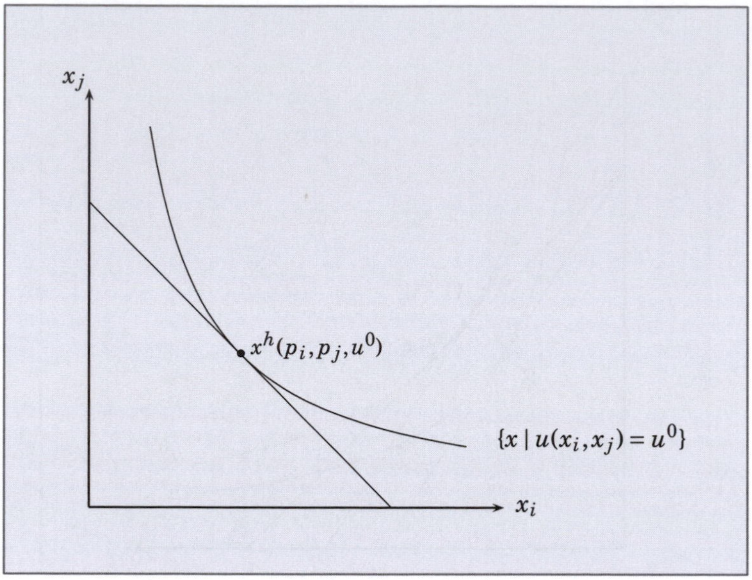

Figure 14.17: The optimal cost-minimizing choice

Notice that the solution to this program depends on the prices and level of utility u^0: the demand function for good i is $x_i^h(p_i, p_j, u^0)$. The superscript "h" denotes the fact that this is the Hicksian (or compensated) demand function with demand dependent on prices and the level of utility, u^0. Example 14.7 computes these demand functions in a simple case.

EXAMPLE 14.7: With $u(x_i, x_j) = a \ln x_i + b \ln x_j$, minimizing cost subject to attaining some utility level u requires that the marginal rate of substitution equals the price ratio. Thus, $\frac{a x_j}{b x_i} = \frac{p_i}{p_j}$, or $x_i = \frac{a}{b}\frac{p_j}{p_i} x_j$. The choice of (x_i, x_j) must satisfy $u = a \ln x_i + b \ln x_j = \ln\left(x_i^a x_j^b\right)$ or $e^u = x_i^a x_j^b$. Substituting the value for x_i in terms of x_j,

$$e^u = x_i^a x_j^b = \left(\frac{a}{b}\frac{p_j}{p_i} x_j\right)^a x_j^b = \left(\frac{a}{b}\frac{p_j}{p_i}\right)^a x_j^{a+b}.$$

So, $x_j^{a+b} = \left(\frac{b}{a}\frac{p_i}{p_j}\right)^a e^u$ and $x_i^{a+b} = \left(\frac{a}{b}\frac{p_j}{p_i}\right)^b e^u$. Thus,

$$x_i^h(p_i, p_j, u) = \left[\left(\frac{a}{b}\frac{p_j}{p_i}\right)^b e^u\right]^{\frac{1}{a+b}}, \qquad x_j^h(p_i, p_j, u) = \left[\left(\frac{b}{a}\frac{p_i}{p_j}\right)^a e^u\right]^{\frac{1}{a+b}}.$$

◊

This contrasts with the *Marshallian* demand curve, $x_i(p_i,p_j,m)$, where demand depends on prices and the level of income and is found by maximizing utility subject to a budget constraint. From Figure 14.17 one can see that minimizing expenditure at given prices defines a budget constraint that is tangent to the indifference curve determined by u^0. Conversely, if this budget level is set, utility maximization will give a solution at the same tangency. Thus,

$$x_i(p_i,p_{-i},m) = x_i^h(p_i,p_j,u^0), \text{ for all } i. \quad (14.10)$$

In words, starting with income m and maximizing utility gives demand $x(p_i,p_{-i},m)$ and corresponding utility level u^0. If this utility level, u^0, is then fixed and expenditure minimized subject to attaining this utility level, the resulting expenditure is m and the corresponding (Hicksian) demand is $x_i^h(p_i,p_j,u^0)$. This connection is illustrated in Example 14.9.

Minimizing cost subject to achieving a given utility level leads to the *expenditure function*. If prices are (p_i,p_{-i}) and a utility level u^0 is given, write $e(p_i,p_{-i},u^0)$ for the (minimum) expenditure required to reach that utility level at those prices. If the price of good i rises to say p_i', the relevant expenditure becomes $e(p_i',p_{-i},u^0)$. Thus, to compensate for the price rise in a way that would allow the individual to maintain the same utility level, the individual would require the additional amount

$$e(p_i',p_{-i},u^0) - e(p_i,p_{-i},u^0).$$

This is called the *compensating variation*. The following discussion shows how the Hicksian demand function can be used to compute the compensating variation.

The Lagrangian associated with the cost minimizing program is:

$$\mathscr{L}(x_i,x_j,\lambda) = p_ix_i + p_jx_j + \lambda(u^0 - u(x_i,x_j)),$$

with first-order conditions:

$$\mathscr{L}_i = p_i - \lambda u_i(x_i,x_j) = 0,$$
$$\mathscr{L}_\lambda = u^0 - u(x_i,x_j) = 0, \quad \left(u_i = \frac{\partial u}{\partial x_i}\right), \; i = 1,2$$

where $u_i = \frac{\partial u}{\partial x_i}$. They determine a solution: $x^h(p_i,p_j,u^0) = (x_i^h(p_i,p_j,u^0), x_j^h(p_i,p_j,u^0))$. Define the expenditure function:

$$e(p_i,p_j,u^0) = p_ix_i^h(p_i,p_j,u^0) + p_jx_j^h(p_i,p_j,u^0).$$

Note that

$$\mathscr{L}(x_i^h(p_i,p_j,u^0),x_j^h(p_i,p_j,u^0),\lambda(p_i,p_j,u^0)) = e(p_i,p_j,u^0),$$

for all (p_i,p_j,u^0) since at any (p_i,p_j,u^0), $x^h(p_i,p_j,u^0)$ satisfies $u^0 - u(x^h(p_i,p_j,u^0)) = 0$. Therefore, we may differentiate with respect to prices or utility. In particular:

$$\frac{\partial \mathscr{L}}{\partial p_i} = \frac{\partial \mathscr{L}}{\partial x_i}\frac{\partial x_i^h}{\partial p_i} + \frac{\partial \mathscr{L}}{\partial x_j}\frac{\partial x_j^h}{\partial p_i} = \frac{\partial e}{\partial p_i}$$

and

$$\begin{aligned}\frac{\partial e}{\partial p_i} = \frac{\partial \mathscr{L}}{\partial p_i} &= x_i^h(p_i,p_j,u^0) + \sum_{k=1}^{2} p_k \frac{\partial x_k^h}{\partial p_i} - \lambda \sum_{k=1}^{2} \frac{\partial u}{\partial x_k}\frac{\partial x_k^h}{\partial p_i} \\ &= x_i^h(p_i,p_j,u^0) + \sum_{k=1}^{2}\left[p_k - \lambda\frac{\partial u}{\partial x_k}\right]\frac{\partial x_k^h}{\partial p_i}.\end{aligned}$$

From the first-order conditions, the second term is zero. Therefore,

$$\frac{\partial e(p_i,p_j,u^0)}{\partial p_i} = x_i^h(p_i,p_j,u^0). \tag{14.11}$$

Thus, the derivative of the expenditure function with respect to the price of a good gives the Hicksian demand function for that good. Suppose that p_i rises to p_i' with p_j remaining constant. If the individual's income were set at $e(p_i,p_j,u^0)$, then optimal consumption would occur at $\bar{x}$ (see Figure 14.18), yielding the lower level of utility u'. How much compensation does the consumer need to be as well off as before (i.e., to maintain utility level u^0)? From the definition of the cost function, $e(p_i',p_j,u^0)$ is the amount of money required to achieve the level of utility u^0 at prices (p_i',p_j) whereas $e(p_i,p_j,u^0)$ is the amount required at prices (p_i,p_j). Therefore, after the price increase, the individual needs $e(p_i',p_j,u^0) - e(p_i,p_j,u^0)$ to be as well off as before. In Figure 14.18, the impact of a price rise is given by the inner dotted line. The outer dotted line results from the compensating income. Let $(x_i^h,x_j^h) = x^h(p_i,p_j,u^0)$ be the bundle chosen at prices (p_i,p_j) to achieve utility u^0, and let $(x_i'^h,x_j'^h) = x^h(p_i',p_j,u^0)$ be the bundle chosen at prices (p_i',p_j) to achieve utility u^0. Thus

$$e(p_i,p_j,u^0) = p_i x_i^h + p_j x_j^h, \quad e(p_i',p_j,u^0) = p_i' x_i'^h + p_j x_j'^h. \tag{14.12}$$

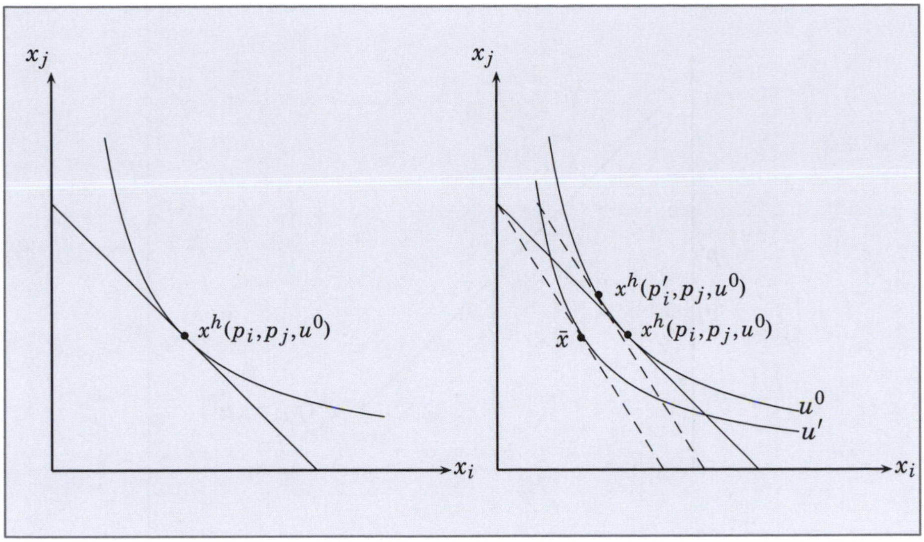

Figure 14.18: The impact of price changes on quantities

The compensation required to make the individual as well off as before the price increase is then:

$$CV = e(p'_i, p_j, u^0) - e(p_i, p_j, u^0). \tag{14.13}$$

This amount can be given an alternative representation. Recall that $F(b) - F(a) = \int_a^b F'(x)dx$. In the present context this can be applied to give:

$$CV = e(p'_i, p_j, u^0) - e(p_i, p_j, u^0) = \int_{p_i}^{p'_i} \frac{\partial e}{\partial p_i} dp_i = \int_{p_i}^{p'_1} x_i^h(\tilde{p}_i, p_j, u^0) d\tilde{p}_i. \tag{14.14}$$

Figure 14.19 illustrates the calculation.

EXAMPLE 14.8:

$$\begin{aligned}
u(x_1, x_2) &= x_1^{\alpha} x_2^{1-\alpha} \\
\mathscr{L}(x_1, x_2, \lambda) &= p_1 x_1 + p_2 x_2 + \lambda\left(u^0 - x_1^{\alpha} x_2^{1-\alpha}\right) \\
\mathscr{L}_{x_1} &= p_1 - \frac{\lambda \alpha x_1^{\alpha} x_2^{1-\alpha}}{x_1} = 0 \\
\mathscr{L}_{x_2} &= p_2 - \frac{\lambda (1-\alpha) x_1^{\alpha} x_2^{1-\alpha}}{x_2} = 0 \\
\mathscr{L}_{\lambda} &= u^0 - x_1^{\alpha} x_2^{1-\alpha} = 0.
\end{aligned}$$

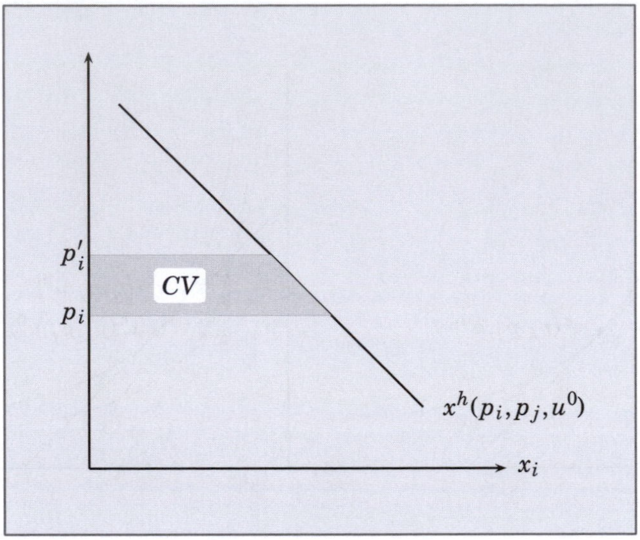

Figure 14.19: Compensating variation

Thus, $\frac{p_1}{p_2} = \frac{\alpha}{1-\alpha}\frac{x_2}{x_1} \Rightarrow x_1 = \frac{p_2}{p_1}\frac{\alpha}{1-\alpha}x_2$. The third equation implies

$$u^0 = \left[\frac{\alpha p_2}{(1-\alpha)p_1}\right]^{\alpha} x_2^{\alpha} x_2^{1-\alpha} = \left[\frac{\alpha p_2}{(1-\alpha)p_1}\right]^{\alpha} x_2 \Rightarrow x_2 = \left[\frac{(1-\alpha)p_1}{\alpha p_2}\right]^{\alpha} u^0.$$

Solving for x_1:

$$x_1 = \left[\frac{(1-\alpha)p_1}{\alpha p_2}\right]^{-1} x_2 = \left[\frac{(1-\alpha)p_1}{\alpha p_2}\right]^{-(1-\alpha)} \left[\frac{(1-\alpha)p_1}{\alpha p_2}\right]^{-\alpha} \left[\frac{(1-\alpha)p_1}{\alpha p_2}\right]^{\alpha} u^0$$

$$= \left[\frac{\alpha p_2}{(1-\alpha)p_1}\right]^{1-\alpha} u^0 = u^0 \left[\frac{(1-\alpha)p_1}{\alpha p_2}\right]^{-(1-\alpha)} = u^0 \left[\frac{(1-\alpha)}{\alpha p_2}\right]^{-(1-\alpha)} p_1^{-(1-\alpha)},$$

$$CV = \int_{p_1}^{p_1'} u^0 \left[\frac{(1-\alpha)}{\alpha p_2}\right]^{-(1-\alpha)} p_1^{-(1-\alpha)} d\tilde{p}_1 = u^0 \left[\frac{(1-\alpha)}{\alpha p_2}\right]^{-(1-\alpha)} \int_{p_1}^{p_1'} p_1^{-1-\alpha} d\tilde{p}_1$$

$$= u^0 \left[\frac{(1-\alpha)}{\alpha p_2}\right]^{-(1-\alpha)} \frac{1}{\alpha}(p_1^{-\alpha} - (p_1')^{-\alpha}) = u^0 \left[\frac{(1-\alpha)}{\alpha p_2}\right]^{-(1-\alpha)} \frac{1}{\alpha}\left(\frac{1}{p_1^{\alpha}} - \frac{1}{(p_1')^{\alpha}}\right).$$

◊

14.11.3 Equivalent variation

Where compensating variation asks what monetary compensation is required to "neutralize" a price increase from lowering utility, equivalent variation asks what income reduction would have the same impact on utility as the price increase. As with compensating variation, equivalent variation has a simple expression in terms of the expenditure function. This is explained next.

Suppose that the price of i increases from p_i to p_i' (see Figure 14.20). The

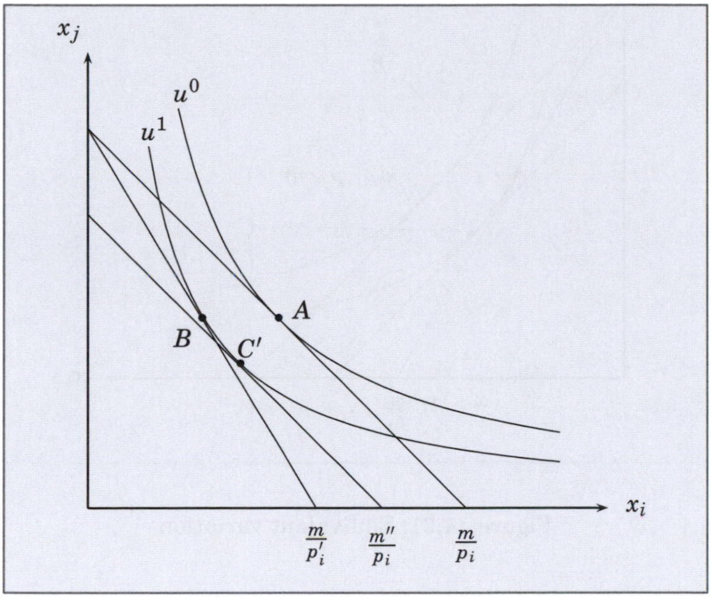

Figure 14.20: Equivalent variation

impact of the price increase is to move consumption from A to B. Welfare drops from u^0 to u^1. Suppose instead that price remains unchanged, but income is reduced to achieve the same loss of welfare. At initial prices, income is reduced from m to m''. In this case, the individual ends up consuming at C'. The loss of income $m - m''$ has the equivalent effect on welfare as a price increase from p_i to p_i'. This is called the *equivalent variation*:

$$EV = \text{equivalent variation} = m - m''.$$

The following discussion provides detailed calculations for EV.

Suppose that income is m and prices $(p_1, \dots, p_n)$. Utility maximization leads to demand $x(p_i, p_{-i}, m)$ and corresponding utility u^0. If now, starting from utility of u^0 the consumption bundle is chosen to minimize expenditure, this is achieved at consumption bundle $x(p_i, p_{-i}, m)$ at a cost of m. In terms of the expenditure function $e(p_i, p_{-i}, u^0) = m$. See Figure 14.21. Now suppose that the price of good

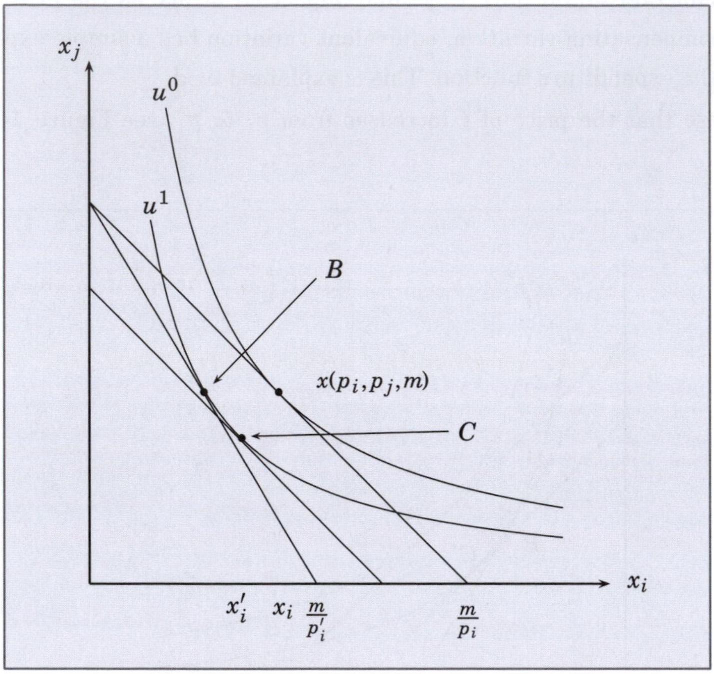

Figure 14.21: Equivalent variation

i increases to p_i'. With income unchanged, utility maximization leads to demand $x(p_i', p_{-i}, m)$, at B in Figure 14.21. So,

$$p_i' x_i(p_i', p_{-i}, m) + \sum_{k \neq i} p_k x_i(p_i', p_{-i}, m) = m. \tag{14.15}$$

Conversely, with u^1 given, at price profile (p_i', p_{-i}) expenditure minimization subject to attaining utility level u^1 leads to demand at B. That is, solving the program:

$$\min \left\{ p_i' x_i + \sum_{j \neq 1} p_j x_j \right\} \text{ subject to } u_i(x) \geq u^1 \tag{14.16}$$

gives the compensated or Hicksian demand $x^h(p_i', p_{-i}, u^1)$, at B in Figure 14.21, with corresponding expenditure $e(p_i', p_{-i}, u^1)$. So,

$$x_i^h(p_i', p_{-i}, u^1) = x_i(p_i', p_{-i}, m), \text{ for all } i. \tag{14.17}$$

Thus,

$$e(p_i', p_{-i}, u^1) = p_i' x_i^h(p_i', p_{-i}, u^1) + \sum_{k \neq i} p_k x_k^h(p_i', p_{-i}, u^1) \qquad (14.18)$$
$$= p_i' x_i(p_i', p_{-i}, m) + \sum_{k \neq i} p_k x_k(p_i', p_{-i}, m)$$
$$= m.$$

Next, at the original prices (p_i, p_{-i}), expenditure minimization, subject to achieving utility level u^1, leads to demand at C. Solving the program:

$$\min \left\{ p_i x_i + \sum_{j \neq 1} p_j x_j \right\} \text{ subject to } u_i(x) \geq u^1 \qquad (14.19)$$

gives the compensated or Hicksian demand $x^h(p_i, p_{-i}, u^1)$, at C in Figure 14.21, with corresponding expenditure $e(p_i, p_{-i}, u^1)$. The drop in welfare, from u^0 to u^1, resulting from the price increase p_i to p_i' generates the same utility loss as would a fall in income from m to $e(p_i, p_{-i}, u^1)$, because both B and C are on the same indifference curve. Since $m = e(p_i', p_{-i}, u^1)$, the income equivalent of the price rise, called the *equivalent variation*, is given by:

$$EV = e(p_i', p_{-i}, u^1) - e(p_i, p_{-i}, u^1) = \int_{p_i}^{p_i'} \frac{\partial e(\tilde{p}_i, p_{-i}, u^1)}{\partial p_i} d\tilde{p}_i \qquad (14.20)$$
$$= \int_{p_i}^{p_i'} x^h(\tilde{p}_i, p_{-i}, u^1) d\tilde{p}_i. \qquad (14.21)$$

14.11.4 Comparison of welfare measures

These welfare measures (CV and EV) may be compared using the Hicksian and Marshallian demand curves $x^h(p_i, p_{-i}, u)$ and $x_i(p_i, p_{-i}, m)$. Consider the demand for good i, $x_i(p_i, p_{-i}, m)$ evaluated at price utility profile (p_i, p_{-i}, m). At these prices and income, the level of utility is u^0. Conversely, at prices (p_i, p_{-i}), the minimum expenditure required to achieve utility u^0 is $e(p_i, p_{-i}, u^0) = m$, and this occurs at the consumption bundle $x_i^h(p_i, p_{-i}, u^0)$. As p_i varies to, say, p_i', the value of $e(p_i', p_{-i}, u^0) = m'$ gives the amount of money required to maintain utility u^0. At profile (p_i', p_{-i}, m') the same observations apply: maximizing utility gives Marshallian demand $x_i(p_i', p_{-i}, m')$ and utility u^0, while minimizing expenditure subject to achieving utility u^0 gives compensated demand $x_i^h(p_i', p_{-i}, u^0)$ and expenditure m'. These observations are depicted in Figure 14.22.

Consider the function $x_i(p_i, p_{-i}, e(p_i, p_{-i}, u^0))$ and the impact of a price change on its value. As price p_i changes to p_i', the initial impact is to push consumption toward the point A. As income is adjusted through $e(p_i, p_{-i}, u^0)$ to allow the individual to maintain the same utility level, this moves demand to $x(p_i', p_j, m') = x^h(p_i', p_{-i}, u^0)$ (at price p_i', $m' = e(p_i', p_j, u^0)$).

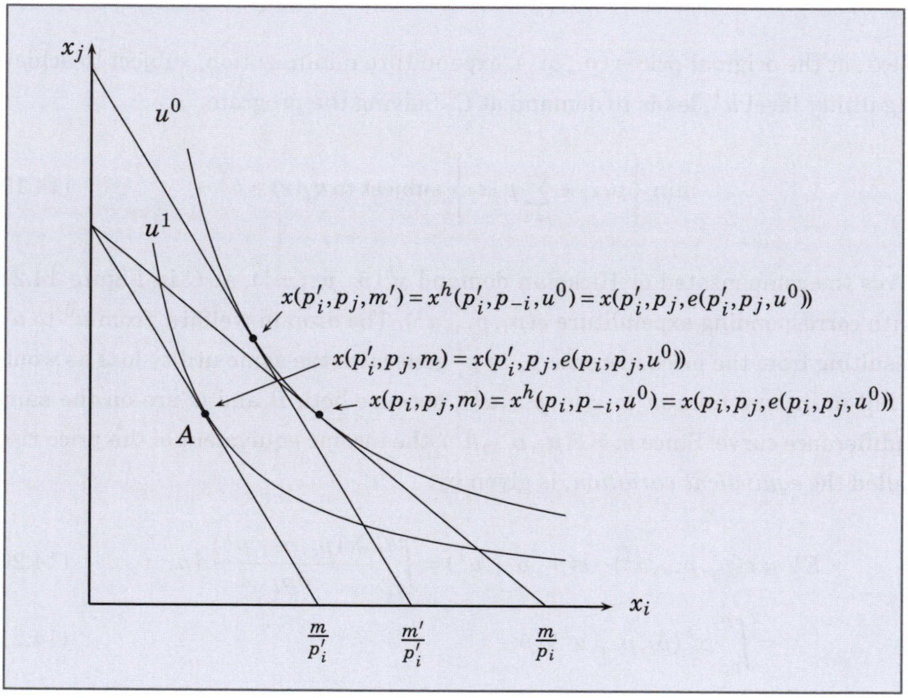

Figure 14.22: Demand variation

Notice that

$$x_i^h(p_i, p_{-i}, u^0) = x_i(p_i, p_j, e(p_i, p_j, u^0)) \tag{14.22}$$

for each value of p_i. This expression is an identity in p_i and connects the Hicksian and Marshallian demand curves.

EXAMPLE 14.9: This example illustrates the connection between Marshallian and Hicksian demand functions that appears in Equations 14.10 and 14.22. Recall the Marshallian demands in Example 14.6: $x_i(p_i, p_j, m) = \frac{a}{a+b}\frac{m}{p_i}$ and $x_j(p_i, p_j, m) = \frac{b}{a+b}\frac{m}{p_j}$. The corresponding utility attained at this level of income is u^0, say:

$$u^0 = a\ln[x_i(p_i,p_j,m)] + b\ln[x_j(p_i,p_j,m)] = a\ln\left[\frac{a}{a+b}\frac{m}{p_i}\right] + b\ln\left[\frac{b}{a+b}\frac{m}{p_j}\right]$$
$$= \ln\left\{\left[\frac{a}{a+b}\frac{m}{p_i}\right]^a\left[\frac{b}{a+b}\frac{m}{p_j}\right]^b\right\}.$$

Therefore, $e^{u^0} = \left[\frac{a}{a+b}\frac{m}{p_i}\right]^a\left[\frac{b}{a+b}\frac{m}{p_j}\right]^b$. Consider:

$$\begin{aligned}
x_i^h(p_i,p_j,u^0) &= \left[\left(\frac{a}{b}\frac{p_j}{p_i}\right)^b e^{u^0}\right]^{\frac{1}{a+b}} \\
&= \left(\frac{a}{b}\frac{p_j}{p_i}\right)^{\frac{b}{a+b}}\left\{\left[\frac{a}{a+b}\frac{m}{p_i}\right]^a\left[\frac{b}{a+b}\frac{m}{p_j}\right]^b\right\}^{\frac{1}{a+b}} \\
&= \left(\frac{a}{b}\right)^{\frac{b}{a+b}}\left(\frac{p_j}{p_i}\right)^{\frac{b}{a+b}}\left\{\left(\frac{a}{a+b}\right)^a m^a\left(\frac{1}{p_i}\right)^a\left(\frac{b}{a+b}\right)^b m^b\left(\frac{1}{p_j}\right)^b\right\}^{\frac{1}{a+b}} \\
&= \left(\frac{a}{b}\right)^{\frac{b}{a+b}}\left(\frac{a}{a+b}\right)^{\frac{a}{a+b}}\left(\frac{b}{a+b}\right)^{\frac{b}{a+b}}\left(\frac{p_j}{p_i}\right)^{\frac{b}{a+b}}\left(\frac{1}{p_i}\right)^{\frac{a}{a+b}}\left(\frac{1}{p_j}\right)^{\frac{b}{a+b}}(m^{a+b})^{\frac{1}{a+b}} \\
&= \left(\frac{a}{a+b}\right)^{\frac{a}{a+b}}\left(\frac{b}{a+b}\right)^{\frac{b}{a+b}}\left(\frac{a}{b}\right)^{\frac{b}{a+b}}p_j^{\frac{b}{a+b}}\left(\frac{1}{p_i}\right)^{\frac{b}{a+b}}\left(\frac{1}{p_i}\right)^{\frac{a}{a+b}}\left(\frac{1}{p_j}\right)^{\frac{b}{a+b}}m \\
&= \left(\frac{a}{a+b}\right)^{\frac{a}{a+b}}\left(\frac{a}{a+b}\right)^{\frac{b}{a+b}}\left(\frac{1}{p_i}\right)m \\
&= \left(\frac{a}{a+b}\right)\left(\frac{1}{p_i}\right)m \\
&= x_i(p_i,p_j,m).
\end{aligned}$$

This gives the equality in Equation 14.10. Conversely, starting with a fixed utility level u^0, the corresponding Hicksian demand functions imply the expenditure:

$$\begin{aligned}
m &= p_i x_i^h(p_i,p_j,u^0) + p_j x_j^h(p_i,p_j,u^0) \\
&= p_i\left[\left(\frac{a}{b}\frac{p_j}{p_i}\right)^b e^{u^0}\right]^{\frac{1}{a+b}} + p_j\left[\left(\frac{b}{a}\frac{p_i}{p_j}\right)^a e^{u^0}\right]^{\frac{1}{a+b}}.
\end{aligned}$$

So,

$$m = \left[p_i\left(\frac{a}{b}\right)^{\frac{b}{a+b}}\left(\frac{p_j}{p_i}\right)^{\frac{b}{a+b}} + p_j\left(\frac{b}{a}\right)^{\frac{a}{a+b}}\left(\frac{p_i}{p_j}\right)^{\frac{a}{a+b}}\right]e^{\frac{u^0}{a+b}}.$$

The Marshallian demand is:

$$\begin{aligned}
x_i(p_i,p_j,m) &= \frac{a}{a+b}\frac{m}{p_i} \\
&= \frac{a}{a+b}\frac{1}{p_i}\left[p_i\left(\frac{a}{b}\right)^{\frac{b}{a+b}}\left(\frac{p_j}{p_i}\right)^{\frac{b}{a+b}} + p_j\left(\frac{b}{a}\right)^{\frac{a}{a+b}}\left(\frac{p_i}{p_j}\right)^{\frac{a}{a+b}}\right]e^{\frac{u^0}{a+b}} \\
&= \frac{a}{a+b}\left[\left(\frac{a}{b}\right)^{\frac{b}{a+b}}\left(\frac{p_j}{p_i}\right)^{\frac{b}{a+b}} + \frac{p_j}{p_i}\left(\frac{b}{a}\right)^{\frac{a}{a+b}}\left(\frac{p_j}{p_i}\right)^{-\frac{a}{a+b}}\right]e^{\frac{u^0}{a+b}} \\
&= \frac{a}{a+b}\left[\left(\frac{a}{b}\right)^{\frac{b}{a+b}} + \left(\frac{b}{a}\right)^{\frac{a}{a+b}}\right]\left(\frac{p_j}{p_i}\right)^{\frac{b}{a+b}}e^{\frac{u^0}{a+b}} \\
&= \frac{a}{a+b}\left[\left(\frac{a}{b}\right)^{\frac{b}{a+b}} + \left(\frac{b}{a}\right)^{1-\frac{b}{a+b}}\right]\left(\frac{p_j}{p_i}\right)^{\frac{b}{a+b}}e^{\frac{u^0}{a+b}} \\
&= \frac{a}{a+b}\left[\left(\frac{a}{b}\right)^{\frac{b}{a+b}} + \left(\frac{a}{b}\right)^{\frac{b}{a+b}}\left(\frac{b}{a}\right)\right]\left(\frac{p_j}{p_i}\right)^{\frac{b}{a+b}}e^{\frac{u^0}{a+b}} \\
&= \frac{a}{a+b}\left(\frac{a}{b}\right)^{\frac{b}{a+b}}\left[1+\left(\frac{b}{a}\right)\right]\left(\frac{p_j}{p_i}\right)^{\frac{b}{a+b}}e^{\frac{u^0}{a+b}} \\
&= \frac{a}{a+b}\left(\frac{a}{b}\right)^{\frac{b}{a+b}}\left[\left(\frac{a+b}{a}\right)\right]\left(\frac{p_j}{p_i}\right)^{\frac{b}{a+b}}e^{\frac{u^0}{a+b}} \\
&= \left[\left(\frac{a}{b}\right)^{b}\left(\frac{p_j}{p_i}\right)^{b}e^{u^0}\right]^{\frac{1}{a+b}} \\
&= x_i^h(p_i,p_j,u^0).
\end{aligned}$$

This completes the calculations. ◊

Returning to Equation 14.22 and differentiating,

$$\frac{\partial x_i^h(p_i,p_{-i},u^0)}{\partial p_i} = \frac{\partial x_i(p_i,p_j,e(p_i,p_j,u^0))}{\partial p_i} + \frac{\partial x_i(p_i,p_j,e(p_i,p_j,u^0))}{\partial m}\frac{\partial e(p_i,p_j,u^0)}{\partial p_i}.$$

Using Equation 14.11,

$$\frac{\partial x_i^h(p_i,p_{-i},u^0)}{\partial p_i} = \frac{\partial x_i(p_i,p_j,e(p_i,p_j,u^0))}{\partial p_i} + \frac{\partial x_i(p_i,p_j,e(p_i,p_j,u^0))}{\partial m}x^h(p_i,p_{-i},u^0), \tag{14.23}$$

an expression called the *Slutsky* equation, which connects the slopes of the compensated and uncompensated demand functions. Rearranging the expression,

$$\frac{\partial x_i(p_i,p_j,e(p_i,p_j,u^0))}{\partial p_i} = \frac{\partial x_i^h(p_i,p_{-i},u^0)}{\partial p_i} - \frac{\partial x_i(p_i,p_j,e(p_i,p_j,u^0))}{\partial m}x^h(p_i,p_{-i},u^0). \tag{14.24}$$

For a normal good (demand increases when income increases), $\frac{\partial x_i}{\partial m} > 0$. Also, $\frac{\partial x_i^h}{\partial p_i}$ is negative, since it involves a rotation along the indifference curve. So, for normal goods,

$$\frac{\partial x_i}{\partial p_i} < \frac{\partial x_i^h}{\partial p_i} < 0. \tag{14.25}$$

When p_i is plotted on the vertical axis, this means that the Marshallian demand function is less steep than the Hicksian demand function (Figure 14.23).

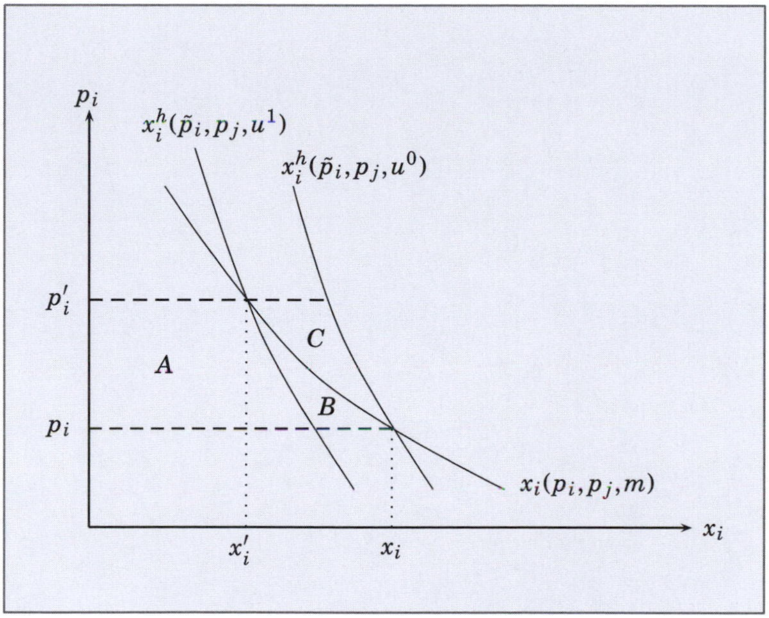

Figure 14.23: Different welfare measures

Recalling the definitions in Equations 14.8, 14.14 and 14.20,

$$\begin{aligned} EV &= \int_{p_i}^{p_1'} x_i^h(\tilde{p}_i, p_j, u^1) d\tilde{p}_i = A \\ CS &= \int_{p_i}^{p_i'} x_i(\tilde{p}_i, p_j, m) d\tilde{p}_i \ = A + B \\ CV &= \int_{p_i}^{p_1'} x_i^h(\tilde{p}_i, p_j, u^0) d\tilde{p}_i = A + B + C. \end{aligned} \tag{14.26}$$

Thus, for normal goods, a price increase gives $EV < CS < CV$. Conversely, a price reduction leads to a reversal of these inequalities.

In the special case where there are no income effects, all three coincide. For example, if $u(x_i, x_j) = v(x_i) + x_j$, the quasilinear case, then the optimizing condition $\frac{u_{x_i}}{u_{x_j}} = \frac{v'(x_i)}{1} = \frac{p_i}{p_j}$ implies that x_i depends only on the price ratio so that variations in m have no effect on x_i.

Exercises for Chapter 14

Exercise 14.1 Suppose that the economy is composed of a single consumer and one productive sector producing commodity Q. The consumer's demand function is

$$Q = \left(\frac{p}{\epsilon}\right)^{1/(\epsilon-1)} \qquad 0 < \epsilon < 1$$

and commodity Q is produced under constant marginal cost c. Calculate the consumer surplus of the consumer if Q is a perfectly competitive industry, in terms of c and ϵ. (Perfectly competitive is understood to mean that price is equal to marginal cost in equilibrium.)

Exercise 14.2 With demand $Q = \left(\frac{p}{\epsilon}\right)^{\frac{1}{\epsilon-1}}$, $0 < \epsilon < 1$, calculate the consumer surplus of the consumer if the government places a per unit tax on Q at the rate t. (Markets are competitive so that price equals marginal cost, c.)

Exercise 14.3 Compare consumer surplus in the two cases considered in Exercises 14.1 and 14.2.

Exercise 14.4 The demand curve for a product is given by: $q = a + bp, b < 0$. If the supplier is able to discriminate prices perfectly, find the total revenue function, $RD(q)$.

Exercise 14.5 Given demand and supply functions $P_d(Q)$ and $P_s(Q)$, respectively, the equilibrium quantity is Q_0, where Q_0 satisfies $P_d(Q_0) = P_s(Q_0)$. The corresponding price is $P_0 = P_d(Q_0) = P_s(Q_0)$. A value-added tax is imposed, leading to a new quantity, Q_t, defined by $P_d(Q_t) = (1+t)P_s(Q_t)$. In the case where

$P_d(Q) = \frac{a}{Q}$ and $P_s(Q) = c + dQ$, where $a = \frac{2c^2}{d}$, solve for Q_0 and Q_t and calculate the dead-weight loss associated with the tax.

Exercise 14.6 Repeat Exercise 14.5, with demand and supply functions $P_d(Q) = a - bQ^2$ and $P_s(Q) = c + dQ^2$. Here, a may not be related to c and d, but $a > c$.

Exercise 14.7 Suppose that an agent has utility function $u(x,y) = \alpha\ln(x) + \alpha\ln(y)$. To achieve utility level u_0, the agent minimizes $px + qy$ subject to the constraint that $u_0 = u(x,y)$, where p and q are the prices of x and y, respectively. Suppose now that the price of x increases from p to $p^* > p$. How much more money does the individual need to maintain the utility level u_0?

Exercise 14.8 An individual has utility function $u(x,y) = \alpha\ln x + \beta\ln y$. The current prices of x and y, respectively, are p and q, the person has income I and, with utility-maximizing behavior, the person currently achieves utility level u'.

Suppose that the price of good x rises from p to p^*. How much more income would the person need to be as well off as before (i.e. achieve utility level u')?

15

Eigenvalues and eigenvectors

15.1 Introduction

Eigenvalues and eigenvectors are scalars and vectors associated with any square matrix and identify key properties of the matrix. They have widespread use in economics, as illustrated by the following examples. Consider a dynamic system of equations, $\boldsymbol{y}_{t+1} = \boldsymbol{A}\boldsymbol{y}_t$, where $\boldsymbol{A}$ is an arbitrary matrix with real entries. Over time, the evolution of the system is governed by powers of $\boldsymbol{A}$: $\boldsymbol{y}_{t+j} = \boldsymbol{A}^j\boldsymbol{y}_t$. This formulation leads naturally to the study of powers of a matrix $\boldsymbol{A}$ with real entries and this task is greatly simplified by using eigenvectors and eigenvalues of $\boldsymbol{A}$. As a second example, Markov processes evolve according to a transition matrix $\boldsymbol{T}$ according to $x_{t+1} = x_t\boldsymbol{T}$. In that case the matrix consists of non-negative entries and each row sums to 1. A steady state of this system is a vector $\bar{\boldsymbol{x}}$ with $\bar{\boldsymbol{x}} = \bar{\boldsymbol{x}}\boldsymbol{T}$. It turns out that this steady state is associated with a specific eigenvalue of the matrix $\boldsymbol{T}$. In least squares estimation, the estimated errors, $\boldsymbol{e}$, are related to the dependent variable, $\boldsymbol{y}$, according to a formula $\boldsymbol{e} = \boldsymbol{M}\boldsymbol{y}$. Given the statistical distribution of $\boldsymbol{y}$, the distribution of $\boldsymbol{e}$ is determined by $\boldsymbol{M}$, and in the study of $\boldsymbol{M}$, its eigenvalues and eigenvectors play a useful role. Finally, in certain optimization problems, location of an optimum is closely related to a particular symmetric matrix having the property that $\boldsymbol{x}\boldsymbol{A}\boldsymbol{x} > 0$ for all $\boldsymbol{x}$. The study of each of these problems is facilitated through the use of eigenvalues and eigenvectors.

In Section 15.2 eigenvalues and eigenvectors are defined and some of the key properties introduced. Section 15.3 describes two dynamic models, of price and population growth, that illustrate the use of eigenvectors and eigenvalues in the study of the time path of these variables. A key theorem, the Perron–Frobenius theorem, is summarized in Section 15.4. Section 15.5 describes two applications in detail: a simplified model of Internet webpage ranking and a generational model of population evolution. When the underlying matrix is symmetric, the associated eigenvalues and eigenvectors have additional properties and these features are discussed in Section 15.6. Eigenvalues are found as the roots of a polynomial equation and so may be real or complex. Therefore, some discussion of complex numbers in this context is provided in Section 15.7. Section 15.8 considers the diagonalization of a matrix (so that a matrix $\boldsymbol{A}$ may be written in the form $\boldsymbol{A} = \boldsymbol{P}\boldsymbol{\Lambda}\boldsymbol{P}^{-1}$, with $\boldsymbol{\Lambda}$ a diagonal matrix.) This key property simplifies the computation of powers of $\boldsymbol{A}$. A number of important properties of eigenvalues are described in Section 15.9.

15.2 Definitions and basic properties

Consider a square $n \times n$ matrix $\boldsymbol{A}$. Given a vector $\boldsymbol{x}$, this is transformed by $\boldsymbol{A}$ to the vector $\boldsymbol{y} = \boldsymbol{A}\boldsymbol{x}$ by multiplication. In general, $\boldsymbol{y}$ and $\boldsymbol{x}$ will bear no particular relation, but in the special case where $\boldsymbol{y}$ is proportional to $\boldsymbol{x}$, $\boldsymbol{y} = \lambda\boldsymbol{x}$, $\boldsymbol{x}$ is called an eigenvector of the matrix $\boldsymbol{A}$.

EXAMPLE 15.1: Let

$$\boldsymbol{A} = \begin{pmatrix} 4 & 6 \\ 2 & 8 \end{pmatrix}, \quad \boldsymbol{x}_1 = \begin{pmatrix} 1 \\ 1 \end{pmatrix}, \quad \boldsymbol{x}_2 = \begin{pmatrix} -3 \\ 1 \end{pmatrix}, \quad \lambda_1 = 10, \quad \lambda_2 = 2.$$

Notice that

$$\boldsymbol{A}\boldsymbol{x}_1 = \begin{pmatrix} 10 \\ 10 \end{pmatrix} = 10\begin{pmatrix} 1 \\ 1 \end{pmatrix}, \qquad \boldsymbol{A}\boldsymbol{x}_2 = \begin{pmatrix} -6 \\ 2 \end{pmatrix} = 2\begin{pmatrix} -3 \\ 1 \end{pmatrix}.$$

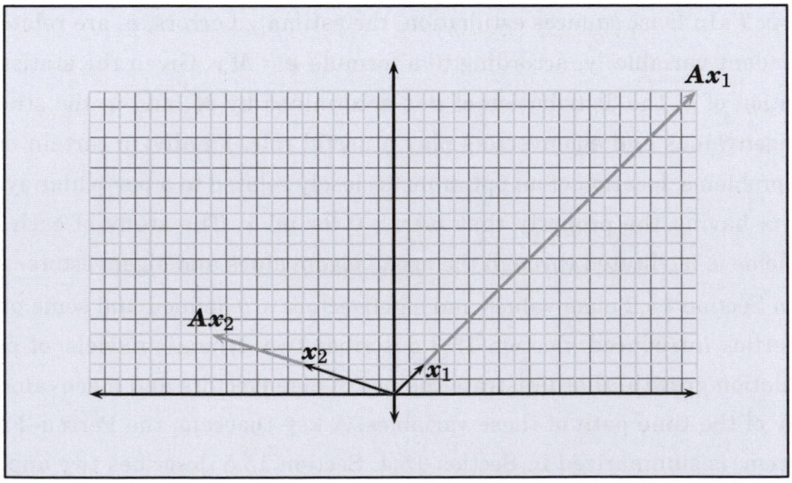

> In the first case, applying $\boldsymbol{A}$ to $\boldsymbol{x}_1$ scales $\boldsymbol{x}_1$ up by a factor of 10; applying $\boldsymbol{A}$ to $\boldsymbol{x}_2$ scales $\boldsymbol{x}_2$ up by a factor of 2. For both vectors, applying $\boldsymbol{A}$ has the effect of scaling the vector but the direction is unchanged. ◇

Let $\boldsymbol{A}$ be a square matrix of order n. Let $\boldsymbol{x}$ be an n-dimensional vector and λ a scalar. If $\boldsymbol{A}\boldsymbol{x} = \lambda\boldsymbol{x}, \boldsymbol{x} \neq 0$, then λ is called a *characteristic root* or an *eigenvalue* of $\boldsymbol{A}$ and $\boldsymbol{x}$ its associated *characteristic vector* or *eigenvector*. Formally:

Definition 15.1: *Given the square matrix* $\mathbf{A}$*, a non-zero vector* $\boldsymbol{x}$ *is called an eigenvector of* $\mathbf{A}$ *if there is some number* λ*, such that* $\mathbf{A}\boldsymbol{x} = \lambda\boldsymbol{x}$*. The scalar* λ *is called the eigenvalue associated with* $\boldsymbol{x}$ *(and may be 0). Different eigenvectors may have the same eigenvalue.*

Note that there is not a unique eigenvector associated with an eigenvalue, because $\boldsymbol{A}\boldsymbol{x} = \lambda\boldsymbol{x} \Leftrightarrow \boldsymbol{A}(c\boldsymbol{x}) = \lambda(c\boldsymbol{x})$ so that if $\boldsymbol{x}$ is an eigenvector of $\boldsymbol{A}$, then so is $c\boldsymbol{x}$, where c is any real non-zero number. Conventionally, this is resolved by some normalization.

Since $\boldsymbol{A}\boldsymbol{x} = \lambda\boldsymbol{x}$ implies $(\boldsymbol{A} - \lambda\boldsymbol{I})\boldsymbol{x} = 0$, if $\boldsymbol{x} \neq 0$, then $(\boldsymbol{A} - \lambda\boldsymbol{I})$ must be singular so that $|\boldsymbol{A} - \lambda\boldsymbol{I}| = 0$. The determinental equation $|\boldsymbol{A} - \lambda\boldsymbol{I}| = 0$, defines an nth-order polynomial in λ called the *characteristic equation* or *characteristic polynomial*.

Definition 15.2: *The characteristic polynomial of* $\mathbf{A}$*,* $p(\lambda)$ *is defined:*

$$p(\lambda) = |\boldsymbol{A} - \lambda\boldsymbol{I}|.$$

In the case where $\boldsymbol{A}$ is a 2×2 matrix,

$$\boldsymbol{A} = \begin{pmatrix} a_{11} & a_{12} \\ a_{21} & a_{22} \end{pmatrix}, \quad |\boldsymbol{A} - \lambda\boldsymbol{I}| = \begin{vmatrix} a_{11} - \lambda & a_{12} \\ a_{21} & a_{22} - \lambda \end{vmatrix} = (a_{11} - \lambda)(a_{22} - \lambda) - a_{21}a_{12}.$$

So,

$$\begin{aligned} |\boldsymbol{A} - \lambda\boldsymbol{I}| &= a_{11}a_{22} - (a_{11} + a_{22})\lambda + \lambda^2 - a_{21}a_{12} \\ &= \lambda^2 - (a_{11} + a_{22})\lambda + a_{11}a_{22} - a_{21}a_{12} \\ &= \lambda^2 - \text{tr}(\boldsymbol{A})\lambda + \det(\boldsymbol{A}), \end{aligned} \tag{15.1}$$

where $\text{tr}(\boldsymbol{A})$ is the sum of the diagonal elements of $\boldsymbol{A}$ and $\det(\boldsymbol{A})$ is the determinant of $\boldsymbol{A}$. From this, in the 2×2 case,

$$\lambda_1, \lambda_2 = \frac{\text{tr}(\boldsymbol{A}) \pm \sqrt{\text{tr}(\boldsymbol{A})^2 - 4\det(\boldsymbol{A})}}{2}.$$

EXAMPLE 15.2:

$$A = \begin{pmatrix} 4 & 6 \\ 3 & 1 \end{pmatrix}, \quad A - \lambda I = \begin{pmatrix} 4 & 6 \\ 3 & 1 \end{pmatrix} - \begin{pmatrix} \lambda & 0 \\ 0 & \lambda \end{pmatrix} = \begin{pmatrix} 4-\lambda & 6 \\ 3 & 1-\lambda \end{pmatrix},$$

so that $|A - \lambda I| = (4-\lambda)(1-\lambda) - 18 = \lambda^2 - 5\lambda - 14 = (\lambda - 7)(\lambda + 2)$. So, $\lambda_1 = 7$ and $\lambda_2 = -2$ are the two eigenvalues. Finding the eigenvector x_2 associated with, say, λ_2 requires solving:

$$\begin{pmatrix} 4-\lambda_2 & 6 \\ 3 & 1-\lambda_2 \end{pmatrix} \begin{pmatrix} x_{21} \\ x_{22} \end{pmatrix} = \begin{pmatrix} 0 \\ 0 \end{pmatrix} \Leftrightarrow \begin{pmatrix} 6 & 6 \\ 3 & 3 \end{pmatrix} \begin{pmatrix} x_{21} \\ x_{22} \end{pmatrix} = \begin{pmatrix} 0 \\ 0 \end{pmatrix}.$$

So, $x_{21} = -x_{22}$ solves this equation. With a normalization, $x_{21} = 1$ gives $x_{22} = -1$ and $x_2 = (1, -1)$. A similar calculation gives $x_1 = (2, 1)$. ◊

REMARK 15.1: Sometimes it is useful to define *left* eigenvalues and eigenvectors. If $y'A = \lambda y'$ for some λ and $y \neq 0$, then y is called a left eigenvector of A. With this terminology, $Ax = \lambda x$ is a right eigenvector, or eigenvalue pair. The left eigenvalues are defined as the characteristic roots of the polynomial $| A' - \lambda I |$ and since $| A' - \lambda I | = | A - \lambda I |$ (a matrix and its transpose have the same determinant), the eigenvalues are unchanged: the set of left eigenvalues is the same as the set of right eigenvalues. However, the eigenvectors generally differ. For example:

$$A = \begin{pmatrix} 4 & 8 \\ 4 & 4 \end{pmatrix}, \quad \lambda_1 = 4(1+\sqrt{2}), \lambda_2 = 4(1-\sqrt{2}),$$

$$\text{left eigenvectors}: y_1 = \begin{pmatrix} \frac{1}{2}\sqrt{2} \\ 1 \end{pmatrix}, y_2 = \begin{pmatrix} -\frac{1}{2}\sqrt{2} \\ 1 \end{pmatrix},$$

$$\text{right eigenvectors}: x_1 = \begin{pmatrix} \sqrt{2} \\ 1 \end{pmatrix}, x_2 = \begin{pmatrix} -\sqrt{2} \\ 1 \end{pmatrix}.$$

However, they satisfy orthogonality conditions. Consider left and right eigenvalue–eigenvector pair (y_i, λ_i) and (x_j, λ_j). Then

$$y_i' A = \lambda_i y_i' \;\Rightarrow\; y_i' A x_j = \lambda_i y_i' x_j \text{ and } A x_j = \lambda_j x_j \;\Rightarrow\; y_j' A x_j = \lambda_j y_j' x_j.$$

Therefore, $\lambda_i y_i' x_j = \lambda_j y_j' x_j$ or $(\lambda_i - \lambda_j) y_i' x_j = 0$. If all the eigenvalues are distinct, then writing $\boldsymbol{Y}$ for a matrix of left eigenvectors, each eigenvector occupying a column of $\boldsymbol{Y}$, and $\boldsymbol{X}$ for the matrix consisting of columns of right eigenvectors, these calculations imply that $\boldsymbol{Y}'\boldsymbol{X}$ is a diagonal matrix. □

For $\boldsymbol{A}$ an $n \times n$ matrix, the equation $| \boldsymbol{A} - \lambda \boldsymbol{I} | = 0$ is a polynomial with n roots, $\lambda_1, \ldots, \lambda_n$. The roots may be real or complex and may be repeated. This polynomial, $p(\lambda)$, may be written:

$$\begin{aligned} p(\lambda) &= a_n \lambda^n + a_{n-1} \lambda^{n-1} + \cdots a_1 \lambda + a_0 \\ &= a_n (\lambda - \lambda_1)(\lambda - \lambda_2) \cdots (\lambda - \lambda_n). \end{aligned}$$

(Take the second form, multiply out the right side and match the coefficients on powers of λ to connect the roots to the coefficients of $p(\lambda)$.) For example,

$$p(\lambda) = \lambda^2 - 3\lambda + 2 = (\lambda - \lambda_1)(\lambda - \lambda_2) = \lambda^2 - (\lambda_1 + \lambda_2)\lambda + \lambda_1 \lambda_2,$$

so that $\lambda_1 + \lambda_2 = 3$ and $\lambda_1 \lambda_1 = 2$, giving $\lambda_1 = 1$, $\lambda_2 = 2$. From the definition, if $\boldsymbol{A}$ is singular, then $\lambda = 0$ is a root of $\boldsymbol{A}$. The fact that $| \boldsymbol{A} - \lambda \boldsymbol{I} | = 0$ has n roots follows from the fundamental theorem of algebra, which states:

Theorem 15.1: *Let $p(z)$ be an nth-order polynomial:*

$$p(z) = a_n z^n + a_{n-1} z^{n-1} + \cdots a_1 z + a_0$$

with $a_n \neq 0$ (and where each a_i may be real or complex). Then p has n roots, $z_1, z_2, \ldots, z_n$: $p(z_i) = 0$, $i = 1, \ldots, n$.

Note that even if the entries of the matrix are real numbers (or even positive real numbers), eigenvalues and eigenvectors may be complex.

EXAMPLE 15.3: Let

$$\boldsymbol{A} = \begin{pmatrix} 1 & 2 \\ -3 & 2 \end{pmatrix}.$$

Then $| \boldsymbol{A} - \lambda \boldsymbol{I} | = (1 - \lambda)(2 - \lambda) + 6 = \lambda^2 - 3\lambda + 8$. From this, the roots are:

$$\lambda_1, \lambda_2 = \frac{3 \pm \sqrt{9 - 32}}{2} = \frac{3 \pm \sqrt{-23}}{2}.$$

Therefore, the eigenvalues are complex. ◊

REMARK 15.2: In the case of a 3×3 matrix, the characteristic polynomial from the determinant of $\boldsymbol{A} - \lambda \boldsymbol{I}$ suggests some useful features of the function. The polynomial is:

$$\begin{aligned} p(\lambda) = -\lambda^3 + (a_{11} + a_{22} + a_{33})\lambda^2 + (-a_{11}a_{22} - a_{11}a_{33} - a_{22}a_{33} + a_{23}a_{32} + a_{21}a_{12} \\ + a_{31}a_{13})\lambda + (a_{11}a_{22}a_{33} - a_{11}a_{23}a_{32} + a_{21}a_{32}a_{13} - a_{21}a_{12}a_{33} \\ + a_{31}a_{12}a_{23} - a_{31}a_{13}a_{22}). \end{aligned}$$

Notice that the coefficient on λ^3 is $(-1)^3$, the coefficient on λ^2 is $\text{tr}(\boldsymbol{A})$ and the constant is $|\,\boldsymbol{A}\,|$. For the n-dimensional case, this becomes:

$$p(\lambda) = (-1)^n \lambda^n + (-1)^{n-1}\text{tr}(A)\lambda^{n-1} + [\text{terms in } \lambda^{n-2}, \lambda^{n-3}, \cdots, \lambda] + \det(A).$$

For example, recalling that $p(\lambda) = |\,\boldsymbol{A} - \lambda \boldsymbol{I}\,|$, $p(0) = |\,\boldsymbol{A}\,| = \det(\boldsymbol{A})$. □

One important observation is that if the eigenvalues are all distinct, then the eigenvectors are linearly independent.

Theorem 15.2: *Let* **A** *have eigenvalues, $\lambda_1, \lambda_2, \ldots, \lambda_n$ all distinct, and let $\boldsymbol{x}_1, \boldsymbol{x}_2, \ldots, \boldsymbol{x}_n$ be associated eigenvectors. Then $\boldsymbol{x}_1, \boldsymbol{x}_2, \ldots, \boldsymbol{x}_n$ are linearly independent.*

PROOF: Consider the case where $\boldsymbol{A}$ is a 3×3 matrix. Suppose that $\boldsymbol{x}_1, \boldsymbol{x}_2, \boldsymbol{x}_3$ are not linearly independent, so there exist α_1, α_2 and α_3, at least one non-zero, such that

$$\alpha_1 \boldsymbol{x}_1 + \alpha_2 \boldsymbol{x}_2 + \alpha_3 \boldsymbol{x}_3 = 0.$$

Multiplying by $(\boldsymbol{A} - \lambda_2 \boldsymbol{I})(\boldsymbol{A} - \lambda_3 \boldsymbol{I})$ gives:

$$(\boldsymbol{A} - \lambda_2 \boldsymbol{I})(\boldsymbol{A} - \lambda_3 \boldsymbol{I})[\alpha_1 \boldsymbol{x}_1 + \alpha_2 \boldsymbol{x}_2 + \alpha_3 \boldsymbol{x}_3] = 0.$$

Using the fact that $(\boldsymbol{A} - \lambda_i \boldsymbol{I})\boldsymbol{x}_j = \boldsymbol{A}\boldsymbol{x}_j - \lambda_i \boldsymbol{x}_j = \lambda_j \boldsymbol{x}_j - \lambda_i \boldsymbol{x}_j = (\lambda_j - \lambda_i)\boldsymbol{x}_j$ gives:

$$\begin{aligned} 0 &= \alpha_1(\boldsymbol{A} - \lambda_2 \boldsymbol{I})(\boldsymbol{A} - \lambda_3 \boldsymbol{I})\boldsymbol{x}_1 + \alpha_2(\boldsymbol{A} - \lambda_2 \boldsymbol{I})(\boldsymbol{A} - \lambda_3 \boldsymbol{I})\boldsymbol{x}_2 + \alpha_3(\boldsymbol{A} - \lambda_2 \boldsymbol{I})(\boldsymbol{A} - \lambda_3 \boldsymbol{I})\boldsymbol{x}_3 \\ &= \alpha_1(\boldsymbol{A} - \lambda_2 \boldsymbol{I})(\lambda_1 - \lambda_3)\boldsymbol{x}_1 + \alpha_2(\boldsymbol{A} - \lambda_2 \boldsymbol{I})(\lambda_2 - \lambda_3)\boldsymbol{x}_2 + \alpha_3(\boldsymbol{A} - \lambda_2 \boldsymbol{I})(\lambda_3 - \lambda_3)\boldsymbol{x}_3 \\ &= \alpha_1(\lambda_1 - \lambda_3)(\boldsymbol{A} - \lambda_2 \boldsymbol{I})\boldsymbol{x}_1 + \alpha_2(\lambda_2 - \lambda_3)(\boldsymbol{A} - \lambda_2 \boldsymbol{I})\boldsymbol{x}_2 + \alpha_3(\lambda_3 - \lambda_3)(\boldsymbol{A} - \lambda_2 \boldsymbol{I})\boldsymbol{x}_3 \\ &= \alpha_1(\lambda_1 - \lambda_3)(\lambda_1 - \lambda_2)\boldsymbol{x}_1 + \alpha_2(\lambda_2 - \lambda_3)(\lambda_2 - \lambda_2)\boldsymbol{x}_2 + \alpha_3(\lambda_3 - \lambda_3)(\lambda_3 - \lambda_2)\boldsymbol{x}_3 \\ &= \alpha_1(\lambda_1 - \lambda_3)(\lambda_1 - \lambda_2)\boldsymbol{x}_1. \end{aligned}$$

Since $\boldsymbol{x}_1 \neq 0$ and $(\lambda_1 - \lambda_3)(\lambda_1 - \lambda_2) \neq 0$, it must be that $\alpha_1 = 0$. Similar computations show that $\alpha_2 = \alpha_3 = 0$. For the n-variable case, the calculations generalize directly. ■

REMARK 15.3: Thus, distinct eigenvalues imply independent eigenvectors. Two real vectors $\boldsymbol{x}$ and $\boldsymbol{y}$ are said to be orthogonal if $\boldsymbol{x}'\boldsymbol{y} = 0$. Although distinct eigenvalues imply that the eigenvectors are independent, they may not be orthogonal. For example, the matrix $\boldsymbol{A}$ has eigenvectors $\boldsymbol{v}_1$ and $\boldsymbol{v}_2$ and eigenvalues λ_1, λ_2. The eigenvectors are linearly independent but not orthogonal, $\boldsymbol{v}_1 \cdot \boldsymbol{v}_2 = 9$:

$$\boldsymbol{A} = \begin{pmatrix} 0 & 1 \\ -8 & 6 \end{pmatrix}, \; \boldsymbol{v}_1 = \begin{pmatrix} 1 \\ 2 \end{pmatrix}, \; \boldsymbol{v}_2 = \begin{pmatrix} 1 \\ 4 \end{pmatrix}, \; \lambda_1 = 2, \; \lambda_2 = 4. \qquad \square$$

REMARK 15.4: Even when the eigenvalues are not distinct, it may be possible to find linearly independent eigenvectors. For example, the identity matrix:

$$\boldsymbol{A} = \begin{pmatrix} 1 & 0 \\ 0 & 1 \end{pmatrix}$$

has characteristic equation $(1-\lambda)^2 = 0$, so that $\lambda_1 = \lambda_2 = 1$. Nevertheless,

$$\begin{pmatrix} 1 & 0 \\ 0 & 1 \end{pmatrix}\begin{pmatrix} 1 \\ 0 \end{pmatrix} = \lambda_1 \begin{pmatrix} 1 \\ 0 \end{pmatrix}, \quad \begin{pmatrix} 1 & 0 \\ 0 & 1 \end{pmatrix}\begin{pmatrix} 0 \\ 1 \end{pmatrix} = \lambda_2 \begin{pmatrix} 0 \\ 1 \end{pmatrix},$$

so that one can find two linearly independent eigenvectors associated with the common eigenvectors $\lambda_1 = \lambda_2$.

However, this is not always possible. Consider the matrix:

$$\boldsymbol{B} = \begin{pmatrix} 1 & 0 \\ 1 & 1 \end{pmatrix}.$$

The matrix $\boldsymbol{B}$ has eigenvalues $\lambda_1 = \lambda_2 = 1$. Thus, any eigenvector must satisfy:

$$\boldsymbol{B} = \begin{pmatrix} 1 & 0 \\ 1 & 1 \end{pmatrix}\begin{pmatrix} x_1 \\ x_2 \end{pmatrix} = \begin{pmatrix} x_1 \\ x_2 \end{pmatrix}$$

and this requires that $x_1 + x_2 = x_2$ or $x_1 = 0$, and so any eigenvector $\boldsymbol{x} = (x_1, x_2)$ has $x_1 = 0$. All eigenvectors of $\boldsymbol{B}$ have the form $\boldsymbol{x} = (x_1, x_2) = (0, z)$. $\square$

REMARK 15.5: One interesting property of the characteristic equation $p(\lambda) = 0$ is that the matrix satisfies it own characteristic equation in the sense that:

$$P(\boldsymbol{A}) = a_n \boldsymbol{A}^n + a_{n-1}\boldsymbol{A}^{n-1} + \cdots + a_1 \boldsymbol{A} + a_0 = 0.$$

This is called the Cayley–Hamilton theorem. In the case where the eigenvalues are all distinct, this is easy to see. Let $\boldsymbol{x}_i$ be an eigenvector. Then

$$\begin{aligned}
P(\boldsymbol{A})\boldsymbol{x}_i &= (a_n\boldsymbol{A}^n + a_{n-1}\boldsymbol{A}^{n-1} + \cdots + a_1\boldsymbol{A} + a_0\boldsymbol{I})\boldsymbol{x}_i \\
&= (a_n\boldsymbol{A}^n\boldsymbol{x}_i + a_{n-1}\boldsymbol{A}^{n-1}\boldsymbol{x}_i + \cdots + a_1\boldsymbol{A}\boldsymbol{x}_i + a_0\boldsymbol{I}\boldsymbol{x}_i) \\
&= (a_n\lambda_i^n\boldsymbol{x}_i + a_{n-1}\lambda_i^{n-1}\boldsymbol{x}_i + \cdots + a_1\lambda_i\boldsymbol{x}_i + a_0\boldsymbol{I}\boldsymbol{x}_i) \\
&= (a_n\lambda_i^n + a_{n-1}\lambda_i^{n-1} + \cdots + a_1\lambda_i + a_0)\boldsymbol{x}_i \\
&= p(\lambda_i)\boldsymbol{x}_i \\
&= 0.
\end{aligned}$$

So, $P(\boldsymbol{A})\boldsymbol{x}_i = 0$ for each eigenvector $\boldsymbol{x}_i$. Since the eigenvalues are distinct, the eigenvectors are linearly independent and with the matrix $\boldsymbol{X} = (\boldsymbol{x}_1,\ldots,\boldsymbol{x}_n)$, $P(\boldsymbol{A})\boldsymbol{X} = 0$ and inverting $\boldsymbol{X}$, $P(\boldsymbol{A}) = 0$. □

Because $\boldsymbol{A}\boldsymbol{x}_i = \lambda_i\boldsymbol{x}_i$, if $\boldsymbol{\Lambda}$ is the diagonal matrix with the eigenvalues on the diagonal and $\boldsymbol{X} = \{\boldsymbol{x}_1,\ldots,\boldsymbol{x}_n\}$ the matrix of eigenvectors, then $\boldsymbol{A}\boldsymbol{X} = \boldsymbol{X}\boldsymbol{\Lambda}$. If $\boldsymbol{X}$ is non-singular, then $\boldsymbol{X}^{-1}\boldsymbol{A}\boldsymbol{X} = \boldsymbol{\Lambda}$ and $\boldsymbol{A} = \boldsymbol{X}\boldsymbol{\Lambda}\boldsymbol{X}^{-1}$. In this case, $\boldsymbol{X}$ is said to diagonalize $\boldsymbol{A}$.

Theorem 15.3: *A square $n \times n$ matrix $\boldsymbol{A}$ is diagonalizable if and only if it has n linearly independent eigenvectors. A sufficient condition for this is that the eigenvalues of $\boldsymbol{A}$ all be distinct.*

15.3 An application: dynamic models

Consider a vector, $\boldsymbol{y}_t$ that evolves over time according to the equation:

$$\boldsymbol{y}_t = \boldsymbol{A}\boldsymbol{y}_{t-1}.$$

Then, starting from $t = 1$, this may be written as $\boldsymbol{y}_t = \boldsymbol{A}^t\boldsymbol{y}_0$, so that powers of $\boldsymbol{A}$ are central to describing the evolution over time.

One particularly useful application of eigenvalues and eigenvectors is in computing powers of a matrix $\boldsymbol{A}$. Suppose that $\boldsymbol{A}$ has n distinct eigenvalues, $\lambda_1,\ldots,\lambda_n$ with eigenvectors $\boldsymbol{x}_1,\ldots,\boldsymbol{x}_n$, so that $\boldsymbol{A}\boldsymbol{x}_i = \lambda_i\boldsymbol{x}_i$, for each i. Arranging these n equations in matrix form:

$$\boldsymbol{A}\boldsymbol{X} = \boldsymbol{A}(\boldsymbol{x}_1,\boldsymbol{x}_2,\cdots,\boldsymbol{x}_n) = (\boldsymbol{x}_1,\boldsymbol{x}_2,\cdots,\boldsymbol{x}_n)\boldsymbol{\Lambda} = \boldsymbol{X}\boldsymbol{\Lambda}$$

where $\boldsymbol{X}$ is the square matrix with $\boldsymbol{x}_i$ in column i and $\boldsymbol{\Lambda}$ is the diagonal matrix with λ_i in the ith row and column position. Because $\boldsymbol{X}$ consists of linearly independent vectors, $\boldsymbol{X}^{-1}$ exists, so that

$$\boldsymbol{A} = \boldsymbol{X}\boldsymbol{\Lambda}\boldsymbol{X}^{-1}, \text{ and } \boldsymbol{X}^{-1}\boldsymbol{A}\boldsymbol{X} = \boldsymbol{\Lambda}$$

Then,

$$\begin{aligned}\boldsymbol{A}^2 &= \boldsymbol{A}\boldsymbol{A} = \boldsymbol{X}\boldsymbol{\Lambda}\boldsymbol{X}^{-1}\boldsymbol{X}\boldsymbol{\Lambda}\boldsymbol{X}^{-1} = \boldsymbol{X}\boldsymbol{\Lambda}^2\boldsymbol{X}^{-1}\\ \boldsymbol{A}^k &= \boldsymbol{X}\boldsymbol{\Lambda}^k\boldsymbol{X}^{-1}.\end{aligned}$$

So, the behavior of $\boldsymbol{A}^k$ can be considered in terms of $\boldsymbol{\Lambda}^k$. Therefore, if a system evolves according to $\boldsymbol{y}_{t+1} = \boldsymbol{A}\boldsymbol{y}_t$, then

$$\boldsymbol{y}_t = \boldsymbol{A}^t\boldsymbol{y}_0 = \boldsymbol{X}\boldsymbol{\Lambda}^k\boldsymbol{X}^{-1}\boldsymbol{y}_0.$$

EXAMPLE 15.4: Let

$$\boldsymbol{A} = \begin{pmatrix} -\frac{3}{2} & 3 \\ -\frac{3}{2} & \frac{11}{4} \end{pmatrix}.$$

Then

$$\boldsymbol{X} = \begin{pmatrix} 3 & 4 \\ 2 & 3 \end{pmatrix}, \quad \boldsymbol{X}^{-1} = \begin{pmatrix} 3 & -4 \\ -2 & 3 \end{pmatrix}, \quad \boldsymbol{\Lambda} = \begin{pmatrix} \frac{1}{2} & 0 \\ 0 & \frac{3}{4} \end{pmatrix}.$$

So, for example,

$$\boldsymbol{A}^t = \begin{pmatrix} 3 & 4 \\ 2 & 3 \end{pmatrix} \begin{pmatrix} \frac{1}{2} & 0 \\ 0 & \frac{3}{4} \end{pmatrix}^t \begin{pmatrix} 3 & -4 \\ -2 & 3 \end{pmatrix},$$

$$\boldsymbol{A}^2 = \begin{pmatrix} -\frac{9}{4}, & \frac{15}{4} \\ -\frac{15}{8}, & \frac{49}{16} \end{pmatrix}, \quad \boldsymbol{A}^3 = \begin{pmatrix} -\frac{9}{4} & \frac{57}{16} \\ -\frac{57}{32} & \frac{179}{64} \end{pmatrix}, \quad \ldots, \quad \boldsymbol{A}^{10} = \begin{pmatrix} -0.4417 & 0.6640 \\ -0.3320 & 0.4990 \end{pmatrix}. \quad \diamond$$

Since $\boldsymbol{\Lambda}$ is a diagonal matrix, $\boldsymbol{\Lambda}^t$ is easy to calculate. Letting $\boldsymbol{w}'_j$ denote the jth row of $\boldsymbol{X}^{-1}$, $\boldsymbol{X}^{-1}\boldsymbol{y}_0$ is a vector with jth element $\boldsymbol{w}'_j\boldsymbol{y}_0$, and $\boldsymbol{\Lambda}^t\boldsymbol{X}^{-1}\boldsymbol{y}_0$ a vector with jth element $\lambda_j^t\boldsymbol{w}'_j\boldsymbol{y}_0$. Consequently,

$$\boldsymbol{y}_t = \sum_{j=1}^{n} \boldsymbol{x}_j\lambda_j^t\boldsymbol{w}'_j\boldsymbol{y}_0 = \sum_{j=1}^{n}(\boldsymbol{w}'_j\boldsymbol{y}_0)\boldsymbol{x}_j\lambda_j^t. \tag{15.2}$$

Thus $\boldsymbol{y}_t$ is a combination of the eigenvectors with the relative importance of each eigenvector $\boldsymbol{x}_j$, determined by its corresponding eigenvalue, λ_j. For example, suppose that the system is two-dimensional so that:

$$\boldsymbol{y}_t = (\boldsymbol{w}_1'\boldsymbol{y}_0)\boldsymbol{x}_1\lambda_1^t + (\boldsymbol{w}_2'\boldsymbol{y}_0)\boldsymbol{x}_2\lambda_2^t \tag{15.3}$$

$$= \lambda_1^t \left[(\boldsymbol{w}_1'\boldsymbol{y}_0)\boldsymbol{x}_1 + (\boldsymbol{w}_2'\boldsymbol{y}_0)\boldsymbol{x}_2 \left(\frac{\lambda_2}{\lambda_1}\right)^t \right]. \tag{15.4}$$

So, if λ_1 is larger than λ_2, then for large t, $\left(\frac{\lambda_2}{\lambda_1}\right)^t \approx 0$ and $\boldsymbol{y}_t \approx \lambda_1^t(\boldsymbol{w}_1'\boldsymbol{y}_0)\boldsymbol{x}_1$. Thus, $\boldsymbol{y}_t$ is approximately proportional to the eigenvector with the largest eigenvalue. In this case, starting from the initial position $\boldsymbol{y}_0$, $\boldsymbol{y}_t$ moves towards the line determined by $\boldsymbol{x}_1$ and then along this eigenvector (towards 0 or away from 0, depending on whether λ_1 is smaller or larger than 1).

Equation 15.2 may be seen directly from a slightly different perspective. If the eigenvectors are linearly independent, then for any initial condition $\boldsymbol{y}_0$ there are scalars $\alpha_1,\ldots,\alpha_n$, such that $\boldsymbol{y}_0 = \alpha_1\boldsymbol{x}_1 + \cdots + \alpha_n\boldsymbol{x}_n$. Therefore,

$$\boldsymbol{y}_t = \boldsymbol{A}^t\boldsymbol{y}_0 = \boldsymbol{A}^t[\alpha_1\boldsymbol{x}_1 + \cdots + \alpha_n\boldsymbol{x}_n] = \alpha_1\lambda_1^t\boldsymbol{x}_1 + \cdots + \alpha_n\lambda_n^t\boldsymbol{x}_n,$$

since $\boldsymbol{A}^t\boldsymbol{x}_k = \lambda_k^t\boldsymbol{x}_k$.

15.3.1 Supply and demand dynamics

The following discussion develops a model of price adjustment determined by the excess of demand over supply: the greater the excess demand, the faster the price rises. In this framework, it is easy to consider the long-run dynamic behavior of prices. Consider the two-market demand and supply model:

$$\begin{aligned} &\text{Market 1:} \quad D_1 = a_0 + a_1p_1 + a_2p_2, \quad S_1 = b_0 + b_1p_1 + b_2p_2 \\ &\text{Market 2:} \quad D_2 = \alpha_0 + \alpha_1p_1 + \alpha_2p_2, \quad S_2 = \beta_0 + \beta_1p_1 + \beta_2p_2. \end{aligned}$$

The excess demand in market i is $D_i - S_i$. The vector of excess demands is then:

$$\begin{aligned} D_1 - S_1 &= a_0 + a_1p_1 + a_2p_2 - (b_0 + b_1p_1 + b_2p_2) \\ &= (a_0 - b_0) + (a_1 - b_1)p_1 + (a_2 - b_2)p_2 \\ D_2 - S_2 &= \alpha_0 + \alpha_1p_1 + \alpha_2p_2 - (\beta_0 + \beta_1p_1 + \beta_2p_2) \\ &= (\alpha_0 - \beta_0) + (\alpha_1 - \beta_1)p_1 + (\alpha_2 - \beta_2)p_2. \end{aligned}$$

Rearranging:

$$\begin{pmatrix} D_1 - S_1 \\ D_2 - S_2 \end{pmatrix} = \begin{pmatrix} (a_0 - b_0) \\ (\alpha_0 - \beta_0) \end{pmatrix} + \begin{pmatrix} (a_1 - b_1) & (a_2 - b_2) \\ (\alpha_1 - \beta_1) & (\alpha_2 - \beta_2) \end{pmatrix} \begin{pmatrix} p_1 \\ p_2 \end{pmatrix}. \tag{15.5}$$

In matrix notation, this may be written:

$$\boldsymbol{e} = \boldsymbol{d} + \boldsymbol{A}\boldsymbol{p}.$$

When demand and supply are equal, excess demand is 0 and the markets clear, giving the market equilibrium price. This is found as $\bar{\boldsymbol{p}}$, the solution to $0 = \boldsymbol{d} + \boldsymbol{A}\bar{\boldsymbol{p}}$. Thus, $\boldsymbol{d}$ and the equilibrium price are related as $\boldsymbol{d} = -\boldsymbol{A}\bar{\boldsymbol{p}}$ (or $\bar{\boldsymbol{p}} = -\boldsymbol{A}^{-1}\boldsymbol{d}$). If the price vector at time t is $\boldsymbol{p}_t$, the associated excess demand is $\boldsymbol{e}_t = \boldsymbol{d} + \boldsymbol{A}\boldsymbol{p}_t$. Since $\boldsymbol{d} = -\boldsymbol{A}\bar{\boldsymbol{p}}$, this may be written $\boldsymbol{e}_t = \boldsymbol{d} + \boldsymbol{A}\boldsymbol{p}_t = -\boldsymbol{A}\bar{\boldsymbol{p}} + \boldsymbol{A}\boldsymbol{p}_t = \boldsymbol{A}(\boldsymbol{p}_t - \bar{\boldsymbol{p}})$. Now consider a price adjustment model, when price moves in the direction of excess demand:

$$\begin{aligned} \boldsymbol{p}_{t+1} - \boldsymbol{p}_t &= \gamma \boldsymbol{e}_t \\ &= \gamma \boldsymbol{A}(\boldsymbol{p}_t - \bar{\boldsymbol{p}}). \end{aligned}$$

Noting that $\boldsymbol{p}_{t+1} - \boldsymbol{p}_t = (\boldsymbol{p}_{t+1} - \bar{\boldsymbol{p}}) - (\boldsymbol{p}_t - \bar{\boldsymbol{p}})$,

$$\begin{aligned} \boldsymbol{p}_{t+1} - \bar{\boldsymbol{p}} &= (\boldsymbol{p}_t - \bar{\boldsymbol{p}}) + \gamma \boldsymbol{A}(\boldsymbol{p}_t - \bar{\boldsymbol{p}}) \\ &= (\boldsymbol{I} + \gamma \boldsymbol{A})(\boldsymbol{p}_t - \bar{\boldsymbol{p}}) \\ &= \boldsymbol{B}(\boldsymbol{p}_t - \bar{\boldsymbol{p}}). \end{aligned}$$

Assuming that $\boldsymbol{B}$ has distinct eigenvalues λ_1 and λ_2, with corresponding eigenvectors, $\boldsymbol{q}_1$ and $\boldsymbol{q}_2$, $\boldsymbol{B}$ may be written: $\boldsymbol{B} = \boldsymbol{Q}\boldsymbol{\Lambda}\boldsymbol{Q}^{-1}$, where $\boldsymbol{\Lambda}$ is the diagonal matrix with λ_1 and λ_2 on the diagonal and where $\boldsymbol{Q} = (\boldsymbol{q}_1, \boldsymbol{q}_2)$, the matrix consisting of the two eigenvectors. Writing $\boldsymbol{v}_t = \boldsymbol{p}_t - \bar{\boldsymbol{p}}$, the equation system may be written:

$$\boldsymbol{v}_{t+1} = \boldsymbol{B}\boldsymbol{v}_t = \boldsymbol{Q}\boldsymbol{\Lambda}\boldsymbol{Q}^{-1}\boldsymbol{v}_t.$$

To illustrate, suppose that Equation 15.5 has the form:

$$\begin{pmatrix} D_1 - S_1 \\ D_2 - S_2 \end{pmatrix} = \begin{pmatrix} -8 \\ 7 \end{pmatrix} + \begin{pmatrix} \frac{1}{2} & \frac{3}{2} \\ -\frac{1}{2} & -\frac{5}{4} \end{pmatrix} \begin{pmatrix} p_1 \\ p_2 \end{pmatrix}.$$

When excess demands are 0, the corresponding price is the equilibrium price.

$$\begin{pmatrix} 0 \\ 0 \end{pmatrix} = \begin{pmatrix} -8 \\ 7 \end{pmatrix} + \begin{pmatrix} \frac{1}{2} & \frac{3}{2} \\ -\frac{1}{2} & -\frac{5}{4} \end{pmatrix} \begin{pmatrix} \bar{p}_1 \\ \bar{p}_2 \end{pmatrix},$$

so that

$$\begin{pmatrix}\bar{p}_1\\ \bar{p}_2\end{pmatrix} = -\begin{pmatrix}\frac{1}{2} & \frac{3}{2}\\ -\frac{1}{2} & -\frac{5}{4}\end{pmatrix}^{-1}\begin{pmatrix}-8\\ 7\end{pmatrix} = -\begin{pmatrix}-10 & -12\\ 4 & 4\end{pmatrix}\begin{pmatrix}-8\\ 7\end{pmatrix} = \begin{pmatrix}4\\ 4\end{pmatrix}.$$

Suppose that $\gamma = 1$, so that in the previous calculations,

$$\boldsymbol{B} = \boldsymbol{I} + \boldsymbol{A} = \begin{pmatrix}1 & 0\\ 0 & 1\end{pmatrix} + \begin{pmatrix}\frac{1}{2} & \frac{3}{2}\\ -\frac{1}{2} & -\frac{5}{4}\end{pmatrix} = \begin{pmatrix}\frac{3}{2} & \frac{3}{2}\\ -\frac{1}{2} & -\frac{1}{4}\end{pmatrix}.$$

The matrix $\boldsymbol{B}$ has eigenvalues $\lambda_1 = \frac{2}{4}$, $\lambda_2 = \frac{3}{4}$ and corresponding eigenvectors $\boldsymbol{q}'_1 = (6,-4)$, $\boldsymbol{q}'_2 = (4,-2)$. So,

$$\boldsymbol{B} = \begin{pmatrix}\frac{3}{2} & \frac{3}{2}\\ -\frac{1}{2} & -\frac{1}{4}\end{pmatrix} = \begin{pmatrix}6 & -4\\ -4 & 2\end{pmatrix}\begin{pmatrix}\frac{2}{4} & 0\\ 0 & \frac{3}{4}\end{pmatrix}\begin{pmatrix}6 & -4\\ -4 & 2\end{pmatrix}^{-1} = \boldsymbol{Q}\boldsymbol{\Lambda}\boldsymbol{Q}^{-1}.$$

The price adjustment equation may be written ($\boldsymbol{v}_t = \boldsymbol{p}_t - \bar{\boldsymbol{p}}$):

$$\boldsymbol{v}_{t+1} = \boldsymbol{B}^t\boldsymbol{v}_t = \boldsymbol{Q}\boldsymbol{\Lambda}\boldsymbol{Q}^{-1}\boldsymbol{v}_0.$$

Starting with $\boldsymbol{v}_0 = (4,4)$, this generates the sequence

$$\boldsymbol{v}_1 = \begin{pmatrix}12.0\\ -3.0\end{pmatrix}, \boldsymbol{v}_2 = \begin{pmatrix}13.50\\ -5.25\end{pmatrix}, \boldsymbol{v}_3 = \begin{pmatrix}12.37\\ -5.43\end{pmatrix}, \boldsymbol{v}_4 = \begin{pmatrix}10.40\\ -4.82\end{pmatrix}, \boldsymbol{v}_5 = \begin{pmatrix}8.37\\ -3.99\end{pmatrix},$$
$$\boldsymbol{v}_6 = \begin{pmatrix}6.56\\ -3.18\end{pmatrix}, \boldsymbol{v}_7 = \begin{pmatrix}5.06\\ -2.48\end{pmatrix}, \boldsymbol{v}_8 = \begin{pmatrix}3.86\\ -1.90\end{pmatrix}, \boldsymbol{v}_9 = \begin{pmatrix}2.93\\ -1.45\end{pmatrix}, \boldsymbol{v}_{10} = \begin{pmatrix}2.21\\ -1.10\end{pmatrix}.$$

Plotting these shows the time path of $\boldsymbol{v}_t = \boldsymbol{p}_t - \bar{\boldsymbol{p}}$ for $t = 1,\dots,10$ in Figure 15.1. Notice the trajectory of the price path as it moves quickly to be proportional to $\boldsymbol{q}_2$, the eigenvector (indicated by the gray arrow) corresponding to the dominant eigenvalue. In this example, at $\boldsymbol{v}_0$, each price exceeds the equilibrium price by 4: $p_{i0} - \bar{p}_i = 4$.

15.3.2 A model of population growth

Consider a population that grows from an initial level of 1: an individual is born in period 1. The individual matures in period 2 and has one offspring, which enters life in period 3. Therefore, in period 3, there are two individuals, one mature, one new. In each period of time, a mature individual produces one offspring, and individuals never die. In period 4, the mature individual produces one offspring adding one individual to the two present in period 3, so that there are

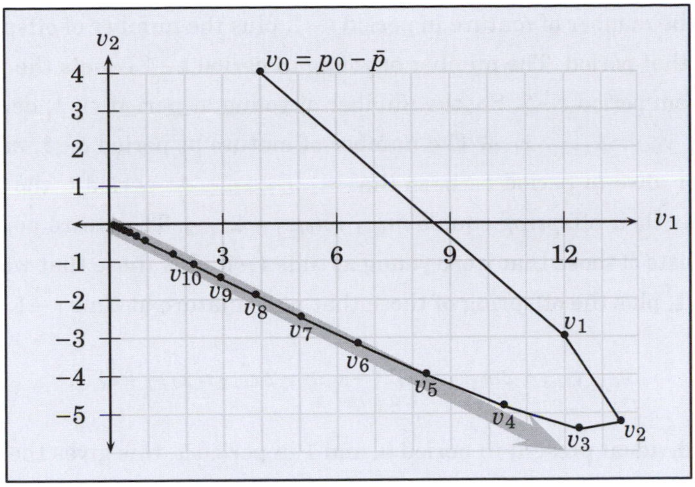

Figure 15.1: Price dynamics

three individuals present in period 4, two of whom are now mature – the original mature individual and the individual that was young the previous period. In period 5, these two mature individuals each produce one offspring and the individual young in period 4 become mature in period 5. And so on. Schematically, this may be shown as:

$$\{y\} \to \{m\} \to \{m, y\} \to \{m, m, y\} \to \{m, m, m, y, y\} \to \{m, m, m, m, m, y, y, y\} \to \cdots$$

Tracking the cohorts of young and mature, the number of young next period equals the number of offspring this period which equals the current number of mature: $y_{t+1} = m_t$. The number of mature next period equals the number of current mature and current young: $m_{t+1} = m_t + y_t$. Thus,

$$\begin{pmatrix} m_{t+1} \\ y_{t+1} \end{pmatrix} = \begin{pmatrix} 1 & 1 \\ 1 & 0 \end{pmatrix} \begin{pmatrix} m_t \\ y_t \end{pmatrix}.$$

This formulation tracks both young and mature over time. Note that $(m_1, y_1) = (0, 1)$, $(m_2, y_2) = (1, 0)$, and so on. The model may also be presented in terms of the growth of the entire population. This formulation is developed next. The dynamics for the whole population provide essentially the same details as the study of both groups separately. Considering the dynamics when written in terms of the evolution of the whole population, $x_t = m_t + y_t$, this leads to the famous Fibonacci equation. Let x_t be the number of individuals (young and mature) in the population at time t. Take any time period t and consider how the population in that period is determined: it will equal the number of young in the previous period,

$t-1$, plus the number of mature in period $t-1$, plus the number of offspring of the mature in that period. The number of young in period $t-1$ equals the population increase from period $t-2$. So, the number of young at period $t-1$, denoted y_{t-1}, is given by $y_{t-1} = x_{t-1} - x_{t-2}$. The number of mature at period $t-1$, m_{t-1} equals the number alive in period $t-2$, so that $m_{t-1} = x_{t-2}$. At period t, the number of mature plus their offspring equals $m_{t-1} + m_{t-1} = 2x_{t-2}$. The entire population at time t consists of those that were young at time $t-1$, plus those that were mature at time $t-1$, plus the offspring of those that were mature at time $t-1$. Thus:

$$x_t = y_{t-1} + 2m_{t-1} = x_{t-1} - x_{t-2} + 2x_{t-2} = x_{t-1} + x_{t-2}.$$

With no individual present in period 0, and 1 in period 1, this gives the recursion:

$$x_t = x_{t-1} + x_{t-2},\ t \geq 2,\ x_0 = 0,\ x_1 = 1.$$

This is a second-order difference equation, but the dynamics may be expressed in first-order terms in a matrix formulation. Define $z_t = x_{t-1}$ so that $x_{t+1} = x_t + z_t$ and $z_{t+1} = x_t$:

$$\begin{pmatrix} x_{t+1} \\ z_{t+1} \end{pmatrix} = \begin{pmatrix} 1 & 1 \\ 1 & 0 \end{pmatrix} \begin{pmatrix} x_t \\ z_t \end{pmatrix} = \boldsymbol{A} \begin{pmatrix} x_t \\ z_t \end{pmatrix}.$$

Because $x_1 = 1$ and $y_1 = x_0 = 0$, this system proceeds according to:

$$\begin{pmatrix} x_{t+1} \\ z_{t+1} \end{pmatrix} = \begin{pmatrix} 1 & 1 \\ 1 & 0 \end{pmatrix}^t \begin{pmatrix} 1 \\ 0 \end{pmatrix} = \boldsymbol{A}^t \begin{pmatrix} 1 \\ 0 \end{pmatrix}. \tag{15.6}$$

To proceed, express the coefficient matrix in terms of its eigenvectors and eigenvalues: $\boldsymbol{A} = \boldsymbol{Q\Lambda Q}^{-1}$:

$$\begin{vmatrix} 1-\lambda & 1 \\ 1 & 0-\lambda \end{vmatrix} = (1-\lambda)(-\lambda) - 1 = \lambda^2 - \lambda - 1.$$

The eigenvalues are the roots of $\lambda^2 - \lambda - 1 = 0$: $\lambda_1, \lambda_2 = \frac{1 \pm \sqrt{1+4}}{2} = \frac{1 \pm \sqrt{5}}{2}$. Considering λ_1, the corresponding eigenvector satisfies:

$$\begin{pmatrix} 1 & 1 \\ 1 & 0 \end{pmatrix} \begin{pmatrix} \xi_1 \\ \xi_2 \end{pmatrix} = \lambda_1 \begin{pmatrix} \xi_1 \\ \xi_2 \end{pmatrix}.$$

From the second line, $\xi_1 = \lambda_1 \xi_2$: normalizing with $\xi_2 = 1$ gives $\xi_1 = \lambda_1$ and the eigenvector is $q_1' = (\xi_1, \xi_2) = (\lambda_1, 1)$. Exactly the same reasoning gives the second eigenvector as $q_2' = (\lambda_2, 1)$. Let $\boldsymbol{Q}$ be the matrix of eigenvectors, so that

$$\boldsymbol{A} = \begin{pmatrix} 1 & 1 \\ 1 & 0 \end{pmatrix} = \boldsymbol{Q}\boldsymbol{\Lambda}\boldsymbol{Q}^{-1} = \begin{pmatrix} \lambda_1 & \lambda_2 \\ 1 & 1 \end{pmatrix} \begin{pmatrix} \lambda_1 & 0 \\ 0 & \lambda_2 \end{pmatrix} \begin{pmatrix} \lambda_1 & \lambda_2 \\ 1 & 1 \end{pmatrix}^{-1}$$

$$= \begin{pmatrix} \lambda_1 & \lambda_2 \\ 1 & 1 \end{pmatrix} \begin{pmatrix} \lambda_1 & 0 \\ 0 & \lambda_2 \end{pmatrix} \begin{pmatrix} 1 & -\lambda_2 \\ -1 & \lambda_1 \end{pmatrix} \frac{1}{\lambda_1 - \lambda_2}.$$

Thus,

$$\boldsymbol{A}^t = \begin{pmatrix} \lambda_1 & \lambda_2 \\ 1 & 1 \end{pmatrix} \begin{pmatrix} \lambda_1 & 0 \\ 0 & \lambda_2 \end{pmatrix}^t \begin{pmatrix} 1 & -\lambda_2 \\ -1 & \lambda_1 \end{pmatrix} \frac{1}{\lambda_1 - \lambda_2}$$

$$= \begin{pmatrix} \lambda_1 & \lambda_2 \\ 1 & 1 \end{pmatrix} \begin{pmatrix} \lambda_1^t & 0 \\ 0 & \lambda_2^t \end{pmatrix} \begin{pmatrix} 1 & -\lambda_2 \\ -1 & \lambda_1 \end{pmatrix} \frac{1}{\lambda_1 - \lambda_2}$$

$$= \begin{pmatrix} \lambda_1 & \lambda_2 \\ 1 & 1 \end{pmatrix} \begin{pmatrix} \lambda_1^t & -\lambda_2\lambda_1^t \\ -\lambda_2^t & \lambda_1\lambda_2^t \end{pmatrix} \frac{1}{\lambda_1 - \lambda_2}$$

$$= \begin{pmatrix} \lambda_1^{t+1} - \lambda_2^{t+1} & -\lambda_2\lambda_1\lambda_1^t + \lambda_1\lambda_2\lambda_2^t \\ \lambda_1^t - \lambda_2^t & -\lambda_2\lambda_1^t + \lambda_1\lambda_2^t \end{pmatrix} \frac{1}{\lambda_1 - \lambda_2}$$

$$= \begin{pmatrix} \lambda_1^{t+1} - \lambda_2^{t+1} & \lambda_1^t - \lambda_2^t \\ \lambda_1^t - \lambda_2^t & \lambda_1^{t-1} - \lambda_2^{t-1} \end{pmatrix} \frac{1}{\lambda_1 - \lambda_2},$$

using the fact that $\lambda_1\lambda_2 = \frac{1+\sqrt{5}}{2}\frac{1-\sqrt{5}}{2} = \frac{1-5}{4} = -1$. From Equation 15.6 and the preceding calculations,

$$\begin{pmatrix} x_{t+1} \\ z_{t+1} \end{pmatrix} = \boldsymbol{A}^t \begin{pmatrix} 1 \\ 0 \end{pmatrix} = \frac{1}{\lambda_1 - \lambda_2} \begin{pmatrix} \lambda_1^{t+1} - \lambda_2^{t+1} & \lambda_1^t - \lambda_2^t \\ \lambda_1^t - \lambda_2^t & \lambda_1^{t-1} - \lambda_2^{t-1} \end{pmatrix} \begin{pmatrix} 1 \\ 0 \end{pmatrix}, \tag{15.7}$$

so that

$$x_{t+1} = \frac{1}{\lambda_1 - \lambda_2}\left[\lambda_1^{t+1} - \lambda_2^{t+1}\right]$$

$$= \frac{1}{\sqrt{5}}\left[\left(\frac{1+\sqrt{5}}{2}\right)^{t+1} - \left(\frac{1-\sqrt{5}}{2}\right)^{t+1}\right]$$

Returning to the equation governing the evolution of mature and young:

$$\begin{pmatrix} m_{t+1} \\ y_{t+1} \end{pmatrix} = \begin{pmatrix} 1 & 1 \\ 1 & 0 \end{pmatrix} \begin{pmatrix} m_t \\ y_t \end{pmatrix}. \tag{15.8}$$

The properties of the coefficient matrix have already been developed. Letting $\boldsymbol{s}'_t = (m_{t+1}, y_{t+1})$ and $\boldsymbol{s}'_2 = (1,0)$, so the system evolves as $\boldsymbol{s}_{t+1} = \boldsymbol{A}\boldsymbol{s}_t$, $\boldsymbol{s}_2 = \boldsymbol{A}\boldsymbol{s}_1$, $\boldsymbol{s}_t = \boldsymbol{A}^{t-1}\boldsymbol{s}_1$, $\boldsymbol{s}_{t+1} = \boldsymbol{A}^t\boldsymbol{s}_1$. (Recall the population evolves (m,y) as $\boldsymbol{s}_1 = (0,1)$, $\boldsymbol{s}_2 = (1,0)$.) The eigenvalues of $\boldsymbol{A}$ are λ_1, $\lambda_2 = \frac{1\pm\sqrt{5}}{2}$ with eigenvectors $\boldsymbol{q}_1 = \begin{pmatrix}\lambda_1\\1\end{pmatrix}$ and $\boldsymbol{q}_2 = \begin{pmatrix}\lambda_2\\1\end{pmatrix}$ and these are linearly independent. Therefore, $\boldsymbol{s}_2$ may be expressed as $\boldsymbol{s}_2 = \alpha_1\boldsymbol{q}_1 + \alpha_2\boldsymbol{q}_2$ (with solution $\alpha_1 = \frac{1}{\sqrt{5}}$, $\alpha_2 = -\frac{1}{\sqrt{5}}$). Thus,

$$\begin{aligned}\boldsymbol{s}_{t+1} &= \boldsymbol{A}^{t-1}\boldsymbol{s}_2 = \boldsymbol{A}^{t-1}(\alpha_1\boldsymbol{q}_1 + \alpha_2\boldsymbol{q}_2)\\ &= \alpha_1\boldsymbol{A}^{t-1}\boldsymbol{q}_1 + \alpha_2\boldsymbol{A}^{t-1}\boldsymbol{q}_2\\ &= \alpha_1\lambda_1^{t-1}\boldsymbol{q}_1 + \alpha_2\lambda_2^{t-1}\boldsymbol{q}_2.\end{aligned}$$

Expanding terms:

$$\begin{pmatrix}m_{t+1}\\ y_{t+1}\end{pmatrix} = \frac{1}{\sqrt{5}}\left(\frac{1+\sqrt{5}}{2}\right)^{t-1}\begin{pmatrix}\frac{1+\sqrt{5}}{2}\\ 1\end{pmatrix} - \frac{1}{\sqrt{5}}\left(\frac{1-\sqrt{5}}{2}\right)^{t-1}\begin{pmatrix}\frac{1-\sqrt{5}}{2}\\ 1\end{pmatrix}. \tag{15.9}$$

Thus,

$$\begin{aligned}m_{t+1} &= \frac{1}{\sqrt{5}}\left[\left(\frac{1+\sqrt{5}}{2}\right)^t - \left(\frac{1-\sqrt{5}}{2}\right)^t\right]\\ y_{t+1} &= \frac{1}{\sqrt{5}}\left[\left(\frac{1+\sqrt{5}}{2}\right)^{t-1} - \left(\frac{1-\sqrt{5}}{2}\right)^{t-1}\right].\end{aligned}$$

So, for example, comparing x_t and m_t, recall that the number of mature at time t equals the population at time $t-1$: $m_{t+1} = x_t$, as given by the equations governing the evolution of x_t and m_t.

Definition 15.3: *The spectral radius of* $\boldsymbol{A}$, *denoted* $\rho(\boldsymbol{A})$ *is defined:*

$$\rho(\boldsymbol{A}) = \max\{\ |\lambda|\ \big|\ \lambda \text{ is an eigenvalue of } \boldsymbol{A}\}.$$

That is, $\rho(\boldsymbol{A})$ *is the largest eigenvalue in terms of absolute value.*

15.4 The Perron–Frobenius theorem

The Perron–Frobenius theorem gives properties of the eigenvalues and eigenvectors of square matrices with real non-negative components or with strictly positive components. Thus $\boldsymbol{A} = \{a\}_{ij}$ with a_{ij} real, $a_{ij} \geq 0$ for all i,j; or the stronger condition that components are strictly positive for all i,j: $a_{ij} > 0$, $\forall i,j$. For the following discussion, a positive matrix or vector is a real matrix or vector with all elements strictly positive; a non-negative matrix or vector is a real matrix or vector with all elements non-negative.

Theorem 15.4: *Let* $\boldsymbol{A}$ *be a positive square matrix,* $a_{ij} > 0$ *(*$\boldsymbol{A} \gg 0$*). Then:*

1. *There is a real positive eigenvalue,* λ^* *with* $\lambda^* = \rho(\boldsymbol{A})$*. This eigenvalue is simple (it is not repeated as a root of* $|\boldsymbol{A} - \lambda \boldsymbol{I}| = 0$*).*

2. *All other eigenvalues* λ' *satisfy* $|\lambda'| < |\lambda| = \lambda^*$*.*

3. *The eigenvalue* λ^* *has a real positive eigenvector,* $\boldsymbol{x}^*$*,* $\boldsymbol{x}_i^* > 0$ *for all* i*.*

4. *The eigenvector* $\boldsymbol{x}^*$ *is the only real non-negative eigenvector: any other eigenvector if real has a negative element or else is complex.*

5. *The eigenvalue* λ^* *satisfies:* $\min_i \sum_j a_{ij} \le \lambda^* \le \max_i \sum_j a_{ij}$.

So, $\boldsymbol{A} \gg 0$ is a sufficient condition for $\boldsymbol{A}$ to have a strictly positive eigenvalue λ^* whose corresponding eigenvector is real and strictly positive.

EXAMPLE 15.5: In this example, the matrix is positive, with one positive eigenvalue and corresponding eigenvector that is non-negative.

$$\boldsymbol{A} = \begin{pmatrix} 1 & 4 \\ 3 & 2 \end{pmatrix}, \quad \lambda_1 = 5, \lambda_2 = -2, \quad \boldsymbol{x}_1 = \begin{pmatrix} 1 \\ 1 \end{pmatrix}, \boldsymbol{x}_2 = \begin{pmatrix} -\frac{4}{3} \\ 1 \end{pmatrix}.$$

If $\boldsymbol{A}$ is rescaled by a factor γ, $\gamma \boldsymbol{A} \boldsymbol{x}_i = \gamma \lambda_i \boldsymbol{x}_i$, so that the eigenvalues of $\gamma \boldsymbol{A}$ are $\gamma\lambda_1$ and $\gamma\lambda_2$. When $\gamma = \frac{1}{5}$, $\gamma \boldsymbol{A}$ is a Markov matrix (non-negative with each row sum equal to 1). ◊

EXAMPLE 15.6: In this example, the matrix is positive and has one real eigenvalue and two complex eigenvalues. Corresponding to the real eigenvalue is a positive eigenvector:

$$\boldsymbol{A} = \begin{pmatrix} 0.1 & 0.1 & 2 \\ 0.8 & 0.7 & 0.7 \\ 0.1 & 2 & 0.1 \end{pmatrix}$$

$$\lambda_1 = 2.2, \lambda_2 = -0.65 + 0.952628i, \lambda_3 = -0.65 - 0.952628i$$

$$x_1 = \begin{pmatrix} 0.577 \\ 0.577 \\ 0.577 \end{pmatrix}, x_2 = \begin{pmatrix} -0.808+0.0i \\ 0.119+0.291i \\ 0.297-0.399i \end{pmatrix}, x_3 = \begin{pmatrix} -0.808-0.0i \\ 0.119-0.291i \\ 0.297+0.399i \end{pmatrix}.$$

◇

REMARK 15.6: The conclusions of Theorem 15.4 are unchanged if the condition that $A \gg 0$ is replaced by the condition that there is some $\bar{k}$ such that for all $k \geq \bar{k}$, $A^k \gg 0$. For example,

$$A = \begin{pmatrix} 1 & 2 \\ 3 & 0 \end{pmatrix}; \quad A^2 = \begin{pmatrix} 7 & 2 \\ 3 & 6 \end{pmatrix}.$$

□

The condition that $A \gg 0$ may be relaxed to $A \geq 0$ ($a_{ij} \geq 0$) if the matrix is assumed to be irreducible. (A matrix is irreducible if, given i, j, there is a sequence $a_{it_1}, a_{t_1 t_2}, \ldots, a_{t_k t_{k+1}}$ all positive and with $t_{k+1} = j$.)

However, in this case, more than one eigenvalue may attain $\rho(A)$.

Theorem 15.5: *Let A be a non-negative irreducible square matrix. Then:*

1. *There is a real positive eigenvalue, λ^* with $\lambda^* = \rho(A)$. This eigenvalue is simple (it is not repeated as a root of $|A - \lambda I| = 0$).*

2. *The eigenvalue λ^* has real positive eigenvector, x^*, $x_i^* > 0$ for all i. This eigenvector is unique (up to a scalar multiple).*

3. *No other eigenvalue $\lambda' \neq \lambda^*$ has a real non-negative eigenvector.*

4. *For any other eigenvalue λ', $|\lambda'| \leq |\lambda| = \lambda^*$.*

5. *The eigenvalue λ^* satisfies:* $\min_i \sum_j a_{ij} \leq \lambda^* \leq \max_i \sum_j a_{ij}$.

The next example gives an irreducible matrix with two eigenvalues having the same norm of 1: $|\lambda_1| = |\lambda_2| = 1$.

EXAMPLE 15.7: Let

$$A = \begin{pmatrix} 0 & 1 \\ 1 & 0 \end{pmatrix}, \lambda_1 = 1, \lambda_2 = -1, x_1 = \begin{pmatrix} 1 \\ 1 \end{pmatrix}, x_2 = \begin{pmatrix} -1 \\ 1 \end{pmatrix}.$$

◇

Note that if $\boldsymbol{A}$ is a non-negative matrix with all entries of $\boldsymbol{A}^n$ positive for some n, then $\boldsymbol{A}$ is irreducible.

REMARK 15.7: If $\boldsymbol{A}$ is arbitrary (real with positive or negative entries) any eigenvalue λ satisfies $|\lambda| \le \max_i \sum_j |a_{ij}|$. □

EXAMPLE 15.8: Let $\boldsymbol{A}$ be a non-negative matrix:

$$\boldsymbol{A} = \begin{pmatrix} a & b \\ c & d \end{pmatrix} \quad \text{and} \quad \boldsymbol{A} - \lambda \boldsymbol{I} = \begin{pmatrix} a-\lambda & b \\ c & d-\lambda \end{pmatrix}.$$

The characteristic equation is $\lambda^2 - (a+d)\lambda + (ad - cb)$. This has roots:

$$\begin{aligned} \lambda_1, \lambda_2 &= \frac{(a+d) \pm \sqrt{(a+d)^2 - 4(ad-cb)}}{2} \\ &= \frac{(a+d) \pm \sqrt{a^2 + d^2 + 2ad - 4ad + 4cb}}{2} \\ &= \frac{(a+d) \pm \sqrt{(a-d)^2 + 4cb}}{2}. \end{aligned}$$

so the roots of $\boldsymbol{A}$ are real. ◊

For non-negative (or positive) matrices larger than 2×2, eigenvalues may be complex, but there is always a real positive root with corresponding real eigenvector.

15.5 Some applications

This section discusses two applications of the theory of eigenvectors. The first is a model of webpage ranking, based essentially on the transition process governing linkages from one webpage to another. The second application concerns population growth in a multigenerational model.

15.5.1 Webpage ranking

The Internet can be viewed as a collection of interconnected nodes; each node is a website. Figure 15.2 illustrates a six-node system. An arrow indicates a link (or

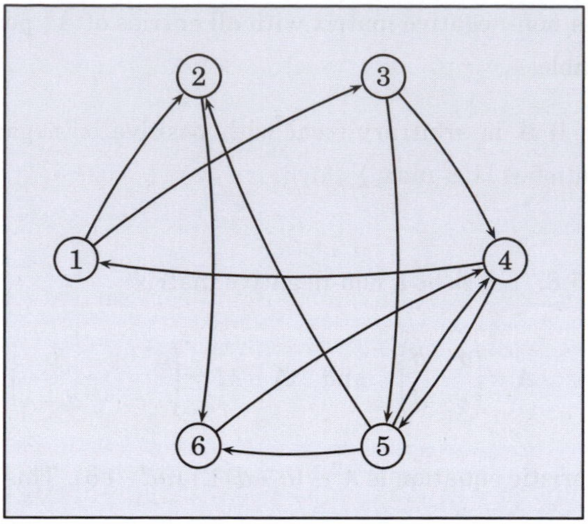

Figure 15.2: Internet nodes

outlink). From Figure 15.2, webpage 1 links to webpage 2 and 3, webpage 4 links to pages 1 and 5. The number of pages that a webpage links to is called the "out-degree", the number of arrows leaving a webpage. This network structure may be represented in matrix terms as an adjacency matrix (see Figure 15.3). Denote the adjacency matrix $\boldsymbol{A} = \{l_{ij}\}$ with $l_{ij} = 0$ if there is no link from i to j, and $l_{ij} = 1$ if there is a link from i to j. In the matrix, a "1" in position (i,j) indicates a link from page i to page j. Let $h(i)$ denote the number of outlinks from i and define t_{ij}:

$$t_{ij} = \begin{cases} \frac{1}{h(i)} & \text{if } h(i) > 0 \text{ and } l_{ij} = 1 \\ 0 & \text{otherwise.} \end{cases}.$$

	1	2	3	4	5	6
1		1	1			
2						1
3				1	1	
4	1				1	
5		1		1		1
6				1		

$= \boldsymbol{A}$

$$\begin{array}{c} \\ 1 \\ 2 \\ 3 \\ 4 \\ 5 \\ 6 \end{array} \begin{array}{c} \begin{array}{cccccc} 1 & 2 & 3 & 4 & 5 & 6 \end{array} \\ \begin{pmatrix} 0 & \frac{1}{2} & \frac{1}{2} & 0 & 0 & 0 \\ 0 & 0 & 0 & 0 & 0 & 1 \\ 0 & 0 & 0 & \frac{1}{2} & \frac{1}{2} & 0 \\ \frac{1}{2} & 0 & 0 & 0 & \frac{1}{2} & 0 \\ 0 & \frac{1}{3} & 0 & \frac{1}{3} & 0 & \frac{1}{3} \\ 0 & 0 & 0 & 1 & 0 & 0 \end{pmatrix} \end{array} = \boldsymbol{T}$$

Figure 15.3: Adjacency matrix and corresponding transition process

Provided $h(i) > 0$ for all i, the corresponding matrix $\boldsymbol{T} = \{t_{ij}\}$ has each row sum to 1. The page rank assigned to i, r_i, is measured by how many pages link to i and the rank of those pages.

- The contribution of page j to the rank of i is positive only if j links to i, $t_{ji} > 0$.
- If j links to many pages, its contribution to i is diluted (the larger $h(j)$, the less the impact on i of a link from j).
- If j has a high rank and links to i, this contributes positively to i's rank.

This process can be expressed in terms of distributions over webpages. If r_j is viewed as the probability of being in state j, $p(j)$, and t_{ij} as the probability of going from state i to state j, $p(j \mid i)$ (so that each link from i is viewed as equally likely), then the aggregate page-to-page movement may be represented in probabilistic terms by:

$$p(i) = \sum_j p(i \mid j)p(j)$$

or

$$r_i = \sum_{\{j \mid l_{ji}=1\}} \frac{r_j}{h(j)} = \sum_{\{j \mid l_{ji}=1\}} t_{ji}r_j = \sum_j t_{ji}r_j.$$

Writing $\boldsymbol{r}' = (r_1, r_2, \ldots, r_n)$ for the vector or rankings, the ranking vector must satisfy:

$$\boldsymbol{r}'\boldsymbol{T} = \boldsymbol{r}'.$$

Transposing, we seek a solution to $\boldsymbol{T}'\boldsymbol{r} = \boldsymbol{r}$. According to the Perron–Frobenius theorem, there is a unique eigenvalue–eigenvector pair $(\lambda^*, \boldsymbol{r}^*)$ with $\lambda^* > 0$ and all entries of $\boldsymbol{r}^*$ non-negative and satisfying

$$\boldsymbol{T}'\boldsymbol{r}^* = \lambda^*\boldsymbol{r}^*.$$

Note that $\boldsymbol{T\iota} = \boldsymbol{\iota}$, $\boldsymbol{\iota}'\boldsymbol{T}' = \boldsymbol{\iota}'$ where $\boldsymbol{\iota}$ is a column vector of 1s. Therefore

$$\lambda^*\boldsymbol{\iota}'\boldsymbol{r}^* = \boldsymbol{\iota}'\boldsymbol{T}'\boldsymbol{r}^* = \boldsymbol{\iota}'\boldsymbol{r}^*;$$

hence, $\lambda^* = 1$ (and in fact, with the eigenvectors normalized to sum to 1), $\boldsymbol{\iota}'\boldsymbol{r}^* = 1$. For the matrix $\boldsymbol{T}'$ (with $\boldsymbol{T}$ in Figure 15.3), the eigenvalues are (to three places of decimals):

$$(\lambda_1,\lambda_2,\lambda_3,\lambda_4,\lambda_5,\lambda_6) = (1,-0.769,-0.099+0.735i,-0.099-0.735i,-0.017 +0.313i,-0.017-0.313i).$$

So, two eigenvalues are real, λ_1 and λ_2. The eigenvectors corresponding to λ_1 and λ_2 ($\boldsymbol{T}'\boldsymbol{y}_i = \lambda_i \boldsymbol{y}_i$) are:

$$y_1 = (0.143,0.131,0.072,0.285,0.179,0.190),$$
$$y_2 = (-0.343,0.435,0.223,0.528,-0.489,-0.354)$$

The remaining eigenvectors are complex.

(As an aside, the corresponding eigenvectors for $\boldsymbol{T}$ ($\boldsymbol{T}\boldsymbol{x}_i = \lambda_i \boldsymbol{x}_i$) are $\boldsymbol{x}_1$ and $\boldsymbol{x}_2$ (again to three decimal places):

$$\boldsymbol{x}_1 = (1,1,1,1,1,1), \quad \boldsymbol{x}_2 = (0.356,-0.646,0.098,-0.382,0.230,0.497).$$

Again, the remaining eigenvectors are complex. The values in the first eigenvector, $\boldsymbol{x}_1$, just reflect the fact that the rows of $\boldsymbol{T}$ sum to 1.)

REMARK 15.8: Notice that iterates of $\boldsymbol{T}$ converge to a matrix with all rows equal. At 20 iterations, $\boldsymbol{T}^{20}$ is given by (to four places of decimals):

$$\boldsymbol{T}^{20} = \begin{pmatrix} 0.1428 & 0.1305 & 0.0713 & 0.2840 & 0.1791 & 0.1921 \\ 0.1405 & 0.1316 & 0.0716 & 0.2897 & 0.1763 & 0.1904 \\ 0.1435 & 0.1307 & 0.0713 & 0.2853 & 0.1791 & 0.1901 \\ 0.1425 & 0.1327 & 0.0724 & 0.2857 & 0.1776 & 0.1892 \\ 0.1428 & 0.1303 & 0.0711 & 0.2852 & 0.1790 & 0.1916 \\ 0.1449 & 0.1290 & 0.0702 & 0.2849 & 0.1807 & 0.1903 \end{pmatrix}.$$

In particular, each row of $\boldsymbol{T}^{20}$ equals x_1. What is the explanation for this? Because the eigenvectors are linearly independent, we can write any vector $\boldsymbol{v}$ in the form:

$$\boldsymbol{v} = \sum \alpha_i^v \boldsymbol{r}_i,$$

a linear combination of the eigenvectors of $\boldsymbol{T}'$ (left eigenvectors of $\boldsymbol{T}$). Notice that the coefficients α_i^v depend on the vector $\boldsymbol{v}$. For each eigenvector, $\boldsymbol{r}_i$ of $\boldsymbol{T}'$

$$\boldsymbol{r}_i'\boldsymbol{T} = \lambda_i \boldsymbol{r}_i' \text{ and } \boldsymbol{r}_i'\boldsymbol{T}^k = \lambda_i^k \boldsymbol{r}_i'.$$

Therefore,

$$\begin{aligned}\boldsymbol{v}'\boldsymbol{T}^k &= \sum \alpha_i^v \boldsymbol{r}'_i \boldsymbol{T}^k = \sum \alpha_i^v \lambda_i^k \boldsymbol{r}'_i \\ &= \lambda_1^k \alpha_1^v \boldsymbol{r}'_1 + \sum_{i\neq 1} \alpha_i^v \left(\frac{\lambda_i}{\lambda_1}\right)^k \boldsymbol{r}'_i.\end{aligned}$$

In view of Remark 15.6, $|\lambda_1| > |\lambda_j|$, $j \neq 1$, $\left(\frac{\lambda_i}{\lambda_1}\right)^k \to 0$, so for k large,

$$\boldsymbol{v}'\boldsymbol{T}^k \approx \lambda_1^k \alpha_1^v \boldsymbol{r}'_1.$$

If $\boldsymbol{v} = \boldsymbol{e}_j$ for any j, where $\boldsymbol{e}_j$ is a vector with 1 in the jth position and 0 elsewhere,

$$\boldsymbol{e}'_j\boldsymbol{T}^k \approx \lambda_1^k \alpha_1^v \boldsymbol{r}'_1.$$

Any row, j, of $\boldsymbol{T}^k$ is approximately proportional to $\boldsymbol{r}'_i$. (That all rows are equal arises from the fact that each row must sum to 1.) □

REMARK 15.9: Not all such matrices ($\boldsymbol{T}$) converge under iteration ($\boldsymbol{T}^k$). A sufficient condition is that the matrix $\boldsymbol{T}$ be irreducible (it is possible to move from any state or location to any other state or location) and aperiodic (returns to the state at irregular times).

The period d_i of the state i is defined as $d_i = \gcd\{n \geq 1 : t_{ii}^{(n)} > 0\}$, with "gcd" denoting greatest common divisor. (Set $d_i = \infty$ if $t_{ii}^{(n)} = 0$). For irreducible matrices, all states have the same period. An irreducible matrix is called aperiodic if its period is 1. A matrix $\boldsymbol{Q}$ is strictly positive if all entries are strictly positive $\boldsymbol{Q} \gg 0$ or $q_{ij} > 0$, $\forall i, j$. A transition matrix, $\boldsymbol{T}$, is irreducible and aperiodic if and only if there is some $\bar{k}$ such that $\boldsymbol{T}^k \gg 0$ for all $k \geq \bar{k}$. □

15.5.2 Leslie matrices

Consider a species that has a maximum life span of n periods. A representative member of the population aged j gives birth to f_j offspring, and survives to the next period with probability s_j.

Thus, if there are x_i individuals aged i in the population, then the total number of offspring produced is $f_1x_1 + f_2x_2 + \cdots + f_nx_n$. The number aged $i > 1$ in the population is the number aged $i-1$ in the previous period that survive to age i:

$x_i = s_{i-1}x_{i-1}$, $i > 1$. Arranging these equations in matrix form:

$$\boldsymbol{x}_{t+1} = \begin{pmatrix} x_{1,t+1} \\ x_{2,t+1} \\ \vdots \\ x_{n-1,t+1} \\ x_{n,t+1} \end{pmatrix} = \begin{pmatrix} f_1 & f_2 & \cdots & f_{n-1} & f_n \\ s_1 & 0 & \cdots & 0 & 0 \\ 0 & s_2 & \cdots & 0 & 0 \\ \vdots & \vdots & \ddots & 0 & 0 \\ 0 & 0 & \cdots & s_{n-1} & 0 \end{pmatrix} \begin{pmatrix} x_{1t} \\ x_{2t} \\ \vdots \\ x_{n-1,t} \\ x_{nt} \end{pmatrix} = \boldsymbol{L}\boldsymbol{x}_t.$$

Thus, the long-run evolution of the population is given by $\boldsymbol{x}_{t+1} = \boldsymbol{L}\boldsymbol{x}_t$: $\boldsymbol{x}_t = \boldsymbol{L}^t\boldsymbol{x}_0$, where $\boldsymbol{x}_0$ is the initial population by age cohort. If a long-run steady-state distribution exists, it satisfies:

$$\boldsymbol{x} = \boldsymbol{L}\boldsymbol{x},$$

so that $\boldsymbol{x}$ is an eigenvector of $\boldsymbol{L}$. This occurs when 1 is an eigenvalue of $\boldsymbol{L}$, although in general this will not be the case.

REMARK 15.10: Note that the Leslie matrix can exhibit periodicity. Consider:

$$\boldsymbol{L} = \begin{pmatrix} 0 & 2 \\ \frac{1}{2} & 0 \end{pmatrix}, \quad \boldsymbol{L}^2 = \begin{pmatrix} 1 & 0 \\ 0 & 1 \end{pmatrix}, \quad \boldsymbol{L}^3 = \begin{pmatrix} 0 & 2 \\ \frac{1}{2} & 0 \end{pmatrix}.$$

This matrix has two eigenvalues, $\lambda_1 = 1$ and $\lambda_2 = -1$, with corresponding (and linearly independent) eigenvectors $\boldsymbol{x}_1' = (2,1)$, $\boldsymbol{x}_2' = (-2,1)$.

Any vector $\boldsymbol{z}$ (initial state) may be written $\boldsymbol{z} = \alpha_1\boldsymbol{x}_1 + \alpha_2\boldsymbol{x}_2$, where the α_i depend on $\boldsymbol{z}$. Since $\boldsymbol{L}\boldsymbol{x}_i = \lambda_i\boldsymbol{x}_i$ and $\boldsymbol{L}^k\boldsymbol{x}_i = \lambda_i^k\boldsymbol{x}_i$,

$$\begin{aligned} \boldsymbol{L}^k\boldsymbol{z} &= \alpha_1\boldsymbol{L}^k\boldsymbol{x}_1 + \alpha_2\boldsymbol{L}^k\boldsymbol{x}_2 \\ &= \alpha_1\lambda_1^k\boldsymbol{x}_1 + \alpha_2\lambda_2^k\boldsymbol{x}_2 \\ &= \alpha_1\boldsymbol{x}_1 + \alpha_2(-1)^k\boldsymbol{x}_2. \end{aligned}$$

For example, if $\boldsymbol{z}' = (2,2)$,

$$\begin{pmatrix} 2 \\ 2 \end{pmatrix} = \alpha_1 \begin{pmatrix} 2 \\ 1 \end{pmatrix} + \alpha_2 \begin{pmatrix} -2 \\ 1 \end{pmatrix}.$$

Solving, $\alpha_1 = \frac{3}{2}$, $\alpha_2 = \frac{1}{2}$. Thus, $\boldsymbol{L}^k\boldsymbol{z} = \frac{3}{2}\boldsymbol{x}_1 + \frac{1}{2}(-1)^k\boldsymbol{x}_2$. □

REMARK 15.11: Recall that the first row of $\boldsymbol{L}$ is $(f_1, f_2, \ldots, f_n)$. Let $\{i \mid f_i > 0\}$ be the set of age groups that can have offspring. The greatest common divisor of this set is the largest integer that divides (factors) each age in this set and is denoted gcd $\{i \mid f_i > 0\}$. In the example in Remark 15.10, $\{i \mid f_i > 0\} = 2$, so that $2 = \text{gcd}\,\{i \mid f_i > 0\}$. □

In the case where gcd $\{i \mid f_i > 0\} = 1$, there is a $\bar{k}$ such that $k \geq \bar{k}$ implies that $\boldsymbol{L}^k$ is a strictly positive matrix (has all entries strictly positive): $\boldsymbol{L}^k \gg 0$.

EXAMPLE 15.9: Consider the Leslie matrix:

$$\boldsymbol{L} = \begin{pmatrix} 0 & 2 & 2 \\ 0.4 & 0 & 0 \\ 0 & 0.4 & 0 \end{pmatrix}.$$

This matrix has three eigenvalues. Only one is positive and real and has largest absolute value: $\lambda_1 = 1.051$, with corresponding eigenvector $x_1' = (0.926, 0.352, 0.134)$. The other two eigenvalues are $\lambda_2 = -0.526 + 0.168i$ and $\lambda_2 = -0.526 - 0.168i$, so that $|\lambda_2| = |\lambda_3| = \sqrt{(-0.525)^2 + (0.168)^2} = 0.551$. Considering iterates of $\boldsymbol{L}$,

$$\boldsymbol{L}^{50} = \begin{pmatrix} 5.271 & 13.849 & 10.031 \\ 2.006 & 5.271 & 3.818 \\ 0.764 & 2.006 & 1.453 \end{pmatrix}.$$

All three eigenvalues are distinct, so the eigenvectors are linearly independent. For any initial population distribution, $\boldsymbol{z}_0$, this may be written:

$$\boldsymbol{z}_0 = c_1\boldsymbol{x}_1 + c_2\boldsymbol{x}_2 + c_3\boldsymbol{x}_3,$$

where the c_i may be real or complex numbers and will depend on $\boldsymbol{z}_0$. Therefore,

$$\begin{aligned} \boldsymbol{L}^k\boldsymbol{z}_0 &= \boldsymbol{L}^k(c_1\boldsymbol{x}_1 + c_2\boldsymbol{x}_2 + c_3\boldsymbol{x}_3) \\ &= c_1\boldsymbol{L}^k\boldsymbol{x}_1 + c_2\boldsymbol{L}^k\boldsymbol{x}_2 + c_3\boldsymbol{L}^k\boldsymbol{x}_3 \\ &= c_1\lambda_1^k\boldsymbol{x}_1 + c_2\lambda_2^k\boldsymbol{x}_2 + c_3\lambda_3^k\boldsymbol{x}_3 \\ &= \lambda_1^k\left[c_1\boldsymbol{x}_1 + c_2\left(\frac{\lambda_2}{\lambda_1}\right)^k\boldsymbol{x}_2 + c_3\left(\frac{\lambda_3}{\lambda_1}\right)^k\boldsymbol{x}_3\right]. \end{aligned}$$

Noting that $\left|\frac{\lambda_2}{\lambda_1}\right| = \left|\frac{\lambda_3}{\lambda_1}\right| = \frac{0.551}{1.051}$, and so $\left|\frac{\lambda_2}{\lambda_1}\right|^{50} \approx 0$, so that

$$\boldsymbol{L}^k\boldsymbol{z}_0 \approx \lambda_1^k c_1\boldsymbol{x}_1.$$

For example, let $\boldsymbol{e}_1 = (1,0,0)$ and recall that the eigenvector associated with λ_1 is $\boldsymbol{x}_1 = (0.926, 0.352, 0.134)$. Then

$$\boldsymbol{L}^{50}\boldsymbol{e}_1 = \begin{pmatrix} 5.271 \\ 2.006 \\ 0.764 \end{pmatrix} \approx 5.69 \begin{pmatrix} 0.926 \\ 0.352 \\ 0.134 \end{pmatrix} \approx (1.051)^{50} c_1 \begin{pmatrix} 0.926 \\ 0.352 \\ 0.134 \end{pmatrix} = (1.051)^{50} c_1 \boldsymbol{x}_1.$$

So $5.69 \approx (1.051)^{50} c_1$ or $5.69 \approx 12.026 c_1$ so that $c_1 \approx 0.473$. Let $\boldsymbol{e}_2 = (0,1,0)$ and $\boldsymbol{e}_3 = (0,0,1)$, so that the ith column of $\boldsymbol{L}^{50}$ is $\boldsymbol{L}^{50}\boldsymbol{e}_i$. Comparing $\boldsymbol{L}^{50}$ and $\boldsymbol{x}_i$, with Col_i denoting the ith column of $\boldsymbol{L}^{50}$:

$$\text{Col}_1 = \boldsymbol{L}^{50}\boldsymbol{e}_1 = 5.69\boldsymbol{x}_1, \ \text{Col}_2 = \boldsymbol{L}^{50}\boldsymbol{e}_2 = 14.97\boldsymbol{x}_1, \ \text{Col}_3 = \boldsymbol{L}^{50}\boldsymbol{e}_3 = 10.83\boldsymbol{x}_1. \ \diamondsuit$$

When $\boldsymbol{L}$ has distinct eigenvalues, the eigenvectors are independent and one may form the system

$$\boldsymbol{L}\{x_1, x_2, \ldots, x_n\} = \{x_1, x_2, \ldots, x_n\}\boldsymbol{\Lambda} \quad \text{or} \quad \boldsymbol{L}\boldsymbol{X} = \boldsymbol{X}\boldsymbol{\Lambda},$$

where $\boldsymbol{\Lambda}$ is the diagonal matrix of eigenvalues. Rearranging,

$$\boldsymbol{L} = \boldsymbol{X}\boldsymbol{\Lambda}\boldsymbol{X}^{-1}, \quad \boldsymbol{L}^k = \boldsymbol{X}\boldsymbol{\Lambda}^k\boldsymbol{X}^{-1}.$$

Consider an initial value $\boldsymbol{x}_0$. At time t, the state becomes $\boldsymbol{x}_t$:

$$\boldsymbol{x}_t = \boldsymbol{L}^t\boldsymbol{x}_0 = \boldsymbol{X}\boldsymbol{\Lambda}^t\boldsymbol{X}^{-1}\boldsymbol{x}_0 = \boldsymbol{X}\boldsymbol{\Lambda}^t\boldsymbol{z}_0^*,$$

where $\boldsymbol{z}_0^* = \boldsymbol{X}^{-1}\boldsymbol{z}_0$. Noting that

$$\boldsymbol{\Lambda}^t\boldsymbol{z}_0^* = \begin{pmatrix} \lambda_1^t & 0 & \cdots & 0 \\ 0 & \lambda_2^t & \cdots & 0 \\ \vdots & \vdots & \ddots & \vdots \\ 0 & 0 & \cdots & \lambda_n^t \end{pmatrix} \begin{pmatrix} z_{01}^* \\ z_{02}^* \\ \vdots \\ z_{0n}^* \end{pmatrix} = \begin{pmatrix} \lambda_1^t z_{01}^* \\ \lambda_2^t z_{02}^* \\ \vdots \\ \lambda_n^t z_{0n}^* \end{pmatrix},$$

$$\boldsymbol{z}_t = \boldsymbol{X}\boldsymbol{\Lambda}^t\boldsymbol{z}_0^* = \boldsymbol{x}_1\lambda_1^t\boldsymbol{z}_{01}^* + \boldsymbol{x}_2\lambda_2^t\boldsymbol{z}_{02}^* + \cdots + \boldsymbol{x}_n\lambda_n^t\boldsymbol{z}_{0n}^*.$$

So, if there is a positive eigenvalue, say λ_1, that has norm larger than any other eigenvalue, then this one will dominate and, in the long run, $\boldsymbol{z}_t$ will be proportional to $\boldsymbol{x}_1\lambda_1^t\boldsymbol{z}_{01}^*$. Depending on the magnitude of λ_1, the population will either grow, be constant or decline but, in any case, the relative proportions will be determined by $\boldsymbol{x}_1$.

15.6 Eigenvalues and eigenvectors of symmetric matrices

In some important cases, the matrices considered are symmetric. For example, the Hessian matrix of second partial derivatives arising in optimization is a symmetric matrix, as is any matrix $\boldsymbol{A}$ that may be written $\boldsymbol{A} = \boldsymbol{B}'\boldsymbol{B}$ for some matrix $\boldsymbol{B}$. The eigenvalues and eigenvectors of symmetric matrices satisfy some additional properties, which are outlined next. In what follows, all matrices are assumed to be symmetric.

PROPOSITION 15.1: For an $n \times n$ symmetric matrix,

1. The eigenvalues are all real.
2. If the eigenvalues are distinct, the eigenvectors are orthogonal.
3. If the eigenvalues are not distinct, n eigenvectors, $\boldsymbol{v}_1, \dots, \boldsymbol{v}_n$ can be found, which are linearly independent and orthogonal.

◊

EXAMPLE 15.10:

$$\boldsymbol{A} = \begin{pmatrix} 4 & 2 \\ 2 & 1 \end{pmatrix}, \boldsymbol{A} - \lambda \boldsymbol{I} = \begin{pmatrix} 4-\lambda & 2 \\ 2 & 1-\lambda \end{pmatrix},$$

$$|\boldsymbol{A} - \lambda \boldsymbol{I}| = (4-\lambda)(1-\lambda) - 4 = 4 - \lambda - 4\lambda + \lambda^2 - 4 = \lambda^2 - 5\lambda = 0.$$

The equation $\lambda^2 - 5\lambda = \lambda(1-5\lambda) = 0$ has roots $\lambda_1, \lambda_2 = 0,5$. The eigenvectors are:

$$\boldsymbol{x}_1 = \begin{pmatrix} \frac{2}{\sqrt{5}} \\ \frac{1}{\sqrt{5}} \end{pmatrix}, \boldsymbol{x}_2 = \begin{pmatrix} \frac{1}{\sqrt{5}} \\ \frac{-2}{\sqrt{5}} \end{pmatrix}$$

Confirming:

$$\begin{pmatrix} 4 & 2 \\ 2 & 1 \end{pmatrix} \begin{pmatrix} \frac{2}{\sqrt{5}} \\ \frac{1}{\sqrt{5}} \end{pmatrix} = 5 \begin{pmatrix} \frac{2}{\sqrt{5}} \\ \frac{1}{\sqrt{5}} \end{pmatrix}, \begin{pmatrix} 4 & 2 \\ 2 & 1 \end{pmatrix} \begin{pmatrix} \frac{1}{\sqrt{5}} \\ \frac{-2}{\sqrt{5}} \end{pmatrix} = 0 \begin{pmatrix} \frac{1}{\sqrt{5}} \\ \frac{-2}{\sqrt{5}} \end{pmatrix}.$$

Thus $\boldsymbol{x}_1, \boldsymbol{x}_2, \lambda_1, \lambda_2$ satisfy, $\boldsymbol{A}\boldsymbol{x}_1 = \lambda_1 \boldsymbol{x}_1$ and $\boldsymbol{A}\boldsymbol{x}_2 = \lambda_2 \boldsymbol{x}_2$. Thus, the eigenvalues here are real and distinct, and the eigenvectors are orthogonal. ◊

The next example illustrates the same ideas in the 3×3 case.

EXAMPLE 15.11: Consider the matrix $\boldsymbol{B}$. The eigenvectors are $\boldsymbol{v}_1$, $\boldsymbol{v}_2$ and $\boldsymbol{v}_3$ with corresponding eigenvalues, $\lambda_1 = 5+\sqrt{10}$, $\lambda_2 = 5-\sqrt{10}$ and $\lambda_3 = 4$:

$$\boldsymbol{B} = \begin{pmatrix} 6 & 2 & 1 \\ 2 & 4 & 2 \\ 1 & 2 & 4 \end{pmatrix}, \quad \boldsymbol{v}_1 = \begin{pmatrix} 1+\frac{\sqrt{10}}{5} \\ 2\frac{\sqrt{10}}{5} \\ 1 \end{pmatrix}, \quad \boldsymbol{v}_2 = \begin{pmatrix} 1-\frac{\sqrt{10}}{5} \\ -2\frac{\sqrt{10}}{5} \\ 1 \end{pmatrix}, \quad \boldsymbol{v}_3 = \begin{pmatrix} -2 \\ 1 \\ 2 \end{pmatrix}.$$

Because the eigenvalues are distinct, the eigenvectors are orthogonal. Stacking the $\boldsymbol{v}_i$ vectors in a matrix, $\boldsymbol{C}$:

$$\boldsymbol{C} = \begin{pmatrix} 1+\frac{\sqrt{10}}{5} & 1-\frac{\sqrt{10}}{5} & -2 \\ 2\frac{\sqrt{10}}{5} & -2\frac{\sqrt{10}}{5} & 1 \\ 1 & 1 & 2 \end{pmatrix} = \begin{pmatrix} \boldsymbol{v}_1 & \boldsymbol{v}_2 & \boldsymbol{v}_3 \end{pmatrix},$$

and

$$\boldsymbol{C}'\boldsymbol{C} = \begin{pmatrix} \boldsymbol{v}_1'\boldsymbol{v}_1 & \boldsymbol{v}_1'\boldsymbol{v}_2 & \boldsymbol{v}_1'\boldsymbol{v}_3 \\ \boldsymbol{v}_2'\boldsymbol{v}_1 & \boldsymbol{v}_2'\boldsymbol{v}_2 & \boldsymbol{v}_2'\boldsymbol{v}_3 \\ \boldsymbol{v}_3'\boldsymbol{v}_1 & \boldsymbol{v}_3'\boldsymbol{v}_2 & \boldsymbol{v}_3'\boldsymbol{v}_3 \end{pmatrix} = \begin{pmatrix} 4+2\frac{\sqrt{10}}{5} & 0 & 0 \\ 0 & 4-2\frac{\sqrt{10}}{5} & 0 \\ 0 & 0 & 9 \end{pmatrix}.$$

◇

When the eigenvalues are not distinct, some choices for eigenvectors may not be orthogonal. However, the eigenvectors may always be chosen to form an orthogonal set.

EXAMPLE 15.12: Consider the symmetric matrix:

$$\boldsymbol{A} = \begin{pmatrix} 1 & 2 & 1 \\ 2 & 4 & 2 \\ 1 & 2 & 1 \end{pmatrix}.$$

Then three eigenvectors corresponding to the eigenvalues are:

$$\lambda_1 = 6, \ \boldsymbol{v}_1 = \begin{pmatrix} 1 \\ 2 \\ 1 \end{pmatrix}; \quad \lambda_2 = 0, \ \boldsymbol{v}_2 = \begin{pmatrix} -2 \\ 1 \\ 0 \end{pmatrix}; \quad \lambda_3 = 0, \ \boldsymbol{v}_3 = \begin{pmatrix} -1 \\ 0 \\ 1 \end{pmatrix},$$

so $\boldsymbol{A}\boldsymbol{v}_i = \lambda_i \boldsymbol{v}_i$, $i = 1,2,3$. For example, considering the first eigenvector:

$$\begin{pmatrix} 1 & 2 & 1 \\ 2 & 4 & 2 \\ 1 & 2 & 1 \end{pmatrix} \begin{pmatrix} 1 \\ 2 \\ 1 \end{pmatrix} = 6 \begin{pmatrix} 1 \\ 2 \\ 1 \end{pmatrix}.$$

Note that $\boldsymbol{v}_2 \cdot \boldsymbol{v}_3 = 2 \neq 0$, so that $\boldsymbol{v}_2$ and $\boldsymbol{v}_3$ are not orthogonal.

However, for the repeated root $\lambda_2 = \lambda_3 = 0$, an eigenvector, $\boldsymbol{x}$, must satisfy: $\boldsymbol{A}\boldsymbol{x} = 0$:

$$\begin{pmatrix} 1 & 2 & 1 \\ 2 & 4 & 2 \\ 1 & 2 & 1 \end{pmatrix} \begin{pmatrix} x_1 \\ x_2 \\ x_3 \end{pmatrix} = \begin{pmatrix} 0 \\ 0 \\ 0 \end{pmatrix}.$$

Since the first and third rows are equal, and the second row is twice the first, there is only one independent equation: the system is satisfied by any (x_1, x_2, x_3) with $x_1 + 2x_2 + x_3 = 0$. One solution is to put $x_2 = 0$, $x_1 = 1$, $x_3 = -1$; another has $x_2 = 1$, $x_1 = -1$, $x_3 = -1$. This gives three orthogonal covectors:

$$\boldsymbol{v}_1 = \begin{pmatrix} 1 \\ 2 \\ 1 \end{pmatrix}; \quad \boldsymbol{v}_2 = \begin{pmatrix} -1 \\ 1 \\ -1 \end{pmatrix}; \quad \boldsymbol{v}_3 = \begin{pmatrix} -1 \\ 0 \\ 1 \end{pmatrix}.$$

◊

Definition 15.4: *A square matrix* $\boldsymbol{C}$ *is said to be* orthogonal *if*

$$\boldsymbol{C}'\boldsymbol{C} = \boldsymbol{I}.$$

PROPOSITION 15.2: If $\boldsymbol{C}$ is orthogonal then it is non-singular. ◊

PROOF: Suppose $\exists \alpha_i, i = 1,\ldots,n$, not all zero, such that $\sum \alpha_i \boldsymbol{c}_i = 0$, where $\boldsymbol{c}_i$ is the ith column of $\boldsymbol{C}$. Then for any $j = 1,2,\ldots,n$, $0 = \boldsymbol{c}_j \cdot (\sum \alpha_i \boldsymbol{c}_j) = \sum \alpha_i \boldsymbol{c}_j \cdot \boldsymbol{c}_i = \alpha_j$. This contradicts the assumption that $\alpha_j \neq 0$ for some j.

Alternatively, observe that if $\boldsymbol{C}$ is orthogonal then $|\boldsymbol{C}| = 1$, because $|\boldsymbol{C}'\boldsymbol{C}| = |\boldsymbol{C}||\boldsymbol{C}| = |\boldsymbol{I}| = 1$, so that $|\boldsymbol{C}|^2 = 1$ and therefore $|\boldsymbol{C}| = \pm 1$. And a non-zero determinant implies non-singularity. Thus, if $\boldsymbol{C}$ is orthogonal then $\boldsymbol{C}' = \boldsymbol{C}^{-1}$ and $\boldsymbol{C}'\boldsymbol{C} = \boldsymbol{C}\boldsymbol{C}' = \boldsymbol{I}$. Furthermore, each root of $\boldsymbol{C}$ is either $+1$ or -1. ■

EXAMPLE 15.13: Let

$$\boldsymbol{C} = \begin{pmatrix} \frac{2}{\sqrt{5}} & \frac{1}{\sqrt{5}} \\ \frac{1}{\sqrt{5}} & -\frac{2}{\sqrt{5}} \end{pmatrix}:$$

$\boldsymbol{C}$ is an orthogonal matrix. ◊

PROPOSITION 15.3: If $\boldsymbol{C}$ is orthogonal then its roots are each equal to 0 or 1. ◊

PROOF: To see that the roots are ± 1, let $\lambda_1, \ldots, \lambda_n$ be the roots of $\boldsymbol{C}$, so that for each $\lambda_i, \exists x \neq 0$ such that $\boldsymbol{C}\boldsymbol{x} = \lambda_i \boldsymbol{x}$. This implies that

$$(\boldsymbol{C}\boldsymbol{x})'(\boldsymbol{C}\boldsymbol{x}) = (\boldsymbol{x}'\boldsymbol{C}')(\boldsymbol{C}\boldsymbol{x}) = \boldsymbol{x}'(\boldsymbol{C}'\boldsymbol{C})\boldsymbol{x} = \boldsymbol{x}'\boldsymbol{x} = \lambda_i \boldsymbol{x}'(\lambda_i \boldsymbol{x}) = \lambda_i^2(\boldsymbol{x}'\boldsymbol{x}).$$

Thus $\boldsymbol{x}'\boldsymbol{x} = \lambda_i^2(\boldsymbol{x}'\boldsymbol{x})$ and since $\boldsymbol{x}'\boldsymbol{x} > 0$, this implies that $\lambda_i^2 = 1$ so that $\lambda_i = \pm 1$. ■

PROPOSITION 15.4: If $\boldsymbol{A}$ is a symmetric $n \times n$ matrix, then there is an $n \times n$ orthogonal matrix $\boldsymbol{C}$ such that

1. $\boldsymbol{C}'\boldsymbol{A}\boldsymbol{C}$ is diagonal, and
2. The characteristic roots $\boldsymbol{\Lambda}$ are the elements on the diagonal of $\boldsymbol{C}'\boldsymbol{A}\boldsymbol{C} = \boldsymbol{\Lambda}$. ◊

Some results mentioned earlier are easy to prove in the symmetric case.

PROPOSITION 15.5: If $\boldsymbol{A}$ is symmetric, the rank of $\boldsymbol{A}$ is the number of non-zero diagonal elements of $\boldsymbol{\Lambda}$. ◊

PROOF: $\boldsymbol{C}'\boldsymbol{A}\boldsymbol{C} = \boldsymbol{\Lambda}$ so $r(\boldsymbol{C}'\boldsymbol{A}\boldsymbol{C}) = r(\boldsymbol{A}\boldsymbol{C}) = r(\boldsymbol{A}) = r(\boldsymbol{\Lambda})$ = number of non-zero elements on diagonal of $\boldsymbol{\Lambda}$: $r(\boldsymbol{C}'\boldsymbol{A}\boldsymbol{C}) = r(\boldsymbol{A}\boldsymbol{C}) = r(\boldsymbol{A})$ uses the fact that when $\boldsymbol{B}$ is non-singular, $r(\boldsymbol{B}\boldsymbol{A}) = r(\boldsymbol{A})$ and here $\boldsymbol{C}'$ and $\boldsymbol{C}$ are non-singular. ■

PROPOSITION 15.6: If $\boldsymbol{A}$ is symmetric and idempotent, $\text{tr}(\boldsymbol{A}) = r(\boldsymbol{A})$. ◊

PROOF: From Proposition 15.4, write $\boldsymbol{C}'\boldsymbol{A}\boldsymbol{C} = \boldsymbol{\Lambda}$, with $\boldsymbol{\Lambda}$ the diagonal matrix with 0 or 1 on the diagonal (the roots are 0 or 1 since $\boldsymbol{A}$ is idempotent). Then $\boldsymbol{C}\boldsymbol{C}'\boldsymbol{A}\boldsymbol{C}\boldsymbol{C}' = \boldsymbol{C}\boldsymbol{\Lambda}\boldsymbol{C}'$ or $\boldsymbol{A} = \boldsymbol{C}\boldsymbol{\Lambda}\boldsymbol{C}'$, since $\boldsymbol{C}\boldsymbol{C}' = \boldsymbol{I}$. Thus $\text{tr}(\boldsymbol{A}) = \text{tr}(\boldsymbol{C}\boldsymbol{\Lambda}\boldsymbol{C}') = \text{tr}(\boldsymbol{\Lambda}\boldsymbol{C}'\boldsymbol{C}) = \text{tr}(\boldsymbol{\Lambda}) = r(\boldsymbol{\Lambda})$. Since $\boldsymbol{C}'\boldsymbol{A}\boldsymbol{C} = \boldsymbol{\Lambda}$, $r(\boldsymbol{C}'\boldsymbol{A}\boldsymbol{C}) = r(\boldsymbol{\Lambda}) = \text{tr}(\boldsymbol{\Lambda})$ = number of 1s on the diagonal and since $r(\boldsymbol{A}) = r(\boldsymbol{C}'\boldsymbol{A}\boldsymbol{C}), r(\boldsymbol{A}) = r(\boldsymbol{\Lambda})$. Thus $\text{tr}(\boldsymbol{A}) = \text{tr}(\boldsymbol{\Lambda}) = r(\boldsymbol{\Lambda}) = r(\boldsymbol{A})$. ■

PROPOSITION 15.7: If $\boldsymbol{A}$ is idempotent, the characteristic roots of $\boldsymbol{A}$ are real.◊

PROOF: Suppose that $\boldsymbol{A}$ is idempotent. Given a root λ, there exists some $\boldsymbol{x} \neq 0$ with $\boldsymbol{A}\boldsymbol{x} = \lambda\boldsymbol{x}$. Multiply both sides by $\boldsymbol{A}$: $\boldsymbol{A}(\boldsymbol{A}\boldsymbol{x}) = \lambda\boldsymbol{A}\boldsymbol{x}$ or $\boldsymbol{A}^2\boldsymbol{x} = \lambda\lambda\boldsymbol{x} = \lambda^2\boldsymbol{x}$. Since $\boldsymbol{A}$ is idempotent $\boldsymbol{A}^2 = \boldsymbol{A}$ so $\boldsymbol{A}\boldsymbol{x} = \lambda^2\boldsymbol{x}$ and since $\boldsymbol{A}\boldsymbol{x} = \lambda\boldsymbol{x}$ this implies $\lambda^2\boldsymbol{x} = \lambda\boldsymbol{x}$ and since $\boldsymbol{x} \neq 0, \lambda^2 = \lambda$ so that $\lambda = 0$ or 1. ■

Positive definiteness for a complex matrix is ambiguous since, even if $\boldsymbol{x}$ is real, $\boldsymbol{x}'\boldsymbol{A}\boldsymbol{x}$ may be a complex number. For real matrices, the expression is unambiguous.

PROPOSITION 15.8: Let $\lambda_1,\ldots,\lambda_n$ be the roots of $\boldsymbol{A}$. If $\boldsymbol{A}$ is positive definite then the roots are real and $\lambda_i > 0, \forall i$. If $\boldsymbol{A}$ is positive semi-definite then $\lambda_i \geq 0, \forall i$. ◊

PROOF: Let λ be a root of $\boldsymbol{A}$ so that for some $\boldsymbol{x} \neq 0, \boldsymbol{A}\boldsymbol{x} = \lambda\boldsymbol{x}$. Then $\boldsymbol{x}'\boldsymbol{A}\boldsymbol{x} = \lambda\boldsymbol{x}'\boldsymbol{x} > 0$ so that $\lambda = \frac{\boldsymbol{x}'\boldsymbol{A}\boldsymbol{x}}{\boldsymbol{x}'\boldsymbol{x}}$, a real number. If $\boldsymbol{A}$ is positive definite $0 < \boldsymbol{x}'\boldsymbol{A}\boldsymbol{x}$ so that $\lambda > 0$. If $\boldsymbol{A}$ is positive semi-definite then $\lambda = \frac{\boldsymbol{x}'\boldsymbol{A}\boldsymbol{x}}{\boldsymbol{x}'\boldsymbol{x}} \geq 0$. ■

15.7 Real and complex eigenvalues and eigenvectors

The eigenvalues are the roots of the characteristic equation $\mid \boldsymbol{A} - \lambda\boldsymbol{I} \mid = 0$. In Example 15.2 the roots are real, but this need not be the case. Even if the coefficients, a_{ij}, of the matrix $\boldsymbol{A}$ are real, the roots may be complex, as Example 15.14 shows. (In fact, even if all entries of $\boldsymbol{A}$ are strictly positive, some roots may be complex, as Example 15.6 shows.) So, in general, the eigenvectors and eigenvalues of a real matrix may be complex valued. Section 15.6 considered the case where the matrix is real and symmetric and in that case the eigenvalues and eigenvectors may be chosen. (If $x \neq 0$ satisfies $\boldsymbol{A}x = \lambda x$, then $\boldsymbol{A}(ix) = \lambda(ix)$, so there are always complex eigenvectors. If $x + iy$ is an eigenvector, then $\boldsymbol{A}(x + zy) = \lambda(x + iy)$ so $\boldsymbol{A}x = \lambda x$ and $\boldsymbol{A}_y = \lambda y$ and so the real eigenvalue λ has a real eigenvector.) However, in the study of many problems (such as dynamic systems and Markov processes), real, square, non-symmetric matrices arise naturally, in which case, dealing with complex solutions is unavoidable.

For the calculations to follow, some basic facts regarding the algebra of complex numbers is necessary.

REMARK 15.12: A complex number z may be written $z = a + ib$, where a is the real part and b the imaginary part. Given two complex numbers ($z = a + ib$, $w = c + id$) the rules of addition, subtraction, multiplication and division are as follows:

$$
\begin{aligned}
z+w &= (a+c)+(b+d)i \\
z-w &= (a-c)+(b-d)i \\
zw &= (a+ib)(c+id) = ac+adi+cbi+bdi^2 = ac+(ad+cb)i-bd \\
&= (ac-bd)+(ad+cb)i \\
\frac{1}{z} &= \frac{a-bi}{a^2+b^2}.
\end{aligned}
$$

The complex conjugate of $z = a+bi$ is denoted $\bar{z}$ and defined $\bar{z} = a-bi$. The norm of z is denoted $|z|$ and defined $|z| = \sqrt{z\bar{z}} = (a+bi)(a-bi) = \sqrt{a^2+b^2}$. So, $\frac{1}{z} = z^{-1} = \frac{\bar{z}}{|z|} = \frac{a-bi}{|z|}$. □

REMARK 15.13: To illustrate the use of this algebra in matrix inversion, consider:

$$
\boldsymbol{A} = \begin{pmatrix} 2i & 1+i \\ -2i & 1-3i \end{pmatrix}.
$$

The calculation of the determinant and inverse are the same for a complex matrix as for a real matrix. Thus, a complex matrix $\boldsymbol{A}$ has inverse $\boldsymbol{B}$ if $\boldsymbol{B}$ satisfies $\boldsymbol{BA} = \boldsymbol{AB} = \boldsymbol{I}$. (Given any two matrices $\boldsymbol{C}$ and $\boldsymbol{D}$ (real or complex), if $\boldsymbol{C}$ is an $n \times m$ matrix and $\boldsymbol{D}$ an $m \times r$ matrix, then $\boldsymbol{E} = \boldsymbol{CD}$ has (i,j)-component $e_{ij} = \sum_{s=1}^{m} c_{is} d_{sj}$.) The determinant of A is (recall $i^2 = -1$):

$$
|\boldsymbol{A}| = 2i(1-3i)-(-2i)(1+i) = 2i-6i^2+2i+2i^2 = 4i-4i^2 = 4+4i = 4(1+i).
$$

The minor, cofactor and adjoint matrices are:

$$
\boldsymbol{M} = \begin{pmatrix} 1-3i & -2i \\ 1+i & 2i \end{pmatrix}, \quad \boldsymbol{C} = \begin{pmatrix} 1-3i & +2i \\ -(1+i) & 2i \end{pmatrix}, \quad \text{adj}(\boldsymbol{A}) = \begin{pmatrix} 1-3i & -(1+i) \\ 2i & 2i \end{pmatrix}.
$$

Thus,

$$
\boldsymbol{A}^{-1} = \frac{1}{4(1+i)} \begin{pmatrix} 1-3i & -(1+i) \\ 2i & 2i \end{pmatrix}.
$$

This may be simplified further. Note that $\frac{1}{1+i} = \frac{1}{2}(1-i)$ so $\frac{1}{4(1+i)} = \frac{1}{8}(1-i)$. Dividing through each term in the adjoint:

$$
\begin{aligned}
\frac{1}{4(1+i)}(1-3i) &= \frac{1}{8}(1-i)(1-3i) = \frac{1}{8}[1-4i+3i^2] = \frac{1}{8}[-2-4i] = -\frac{1}{4}(1+2i) \\
-\frac{(1+i)}{4(1+i)} &= -\frac{1}{4} \\
\frac{1}{4(1+i)}(2i) &= \frac{1}{8}(1-i)2i = \frac{1}{8}(2i-2i^2) = 2\frac{1}{8}(i-(-1)) = \frac{1}{4}(1+i).
\end{aligned}
$$

Therefore:

$$A^{-1} = \begin{pmatrix} -\frac{1}{4} - \frac{1}{2}i & -\frac{1}{4} \\ \frac{1}{4} + \frac{1}{4}i & \frac{1}{4} + \frac{1}{4}i \end{pmatrix}. \qquad \square$$

REMARK 15.14: For complex vectors, the definitions of orthogonality and length require modification, using complex conjugates:

- The vector $x' = (1, i) \neq 0$ has $x'x = 1 - 1 = 0$.
- The vectors $x' = (1, i)$ and $y' = (-1, i)$ are orthogonal, but $x'y = (1, i)(-1, i)' = -1 - 1 = -2$.

These examples illustrate the need for a generalized definition of inner product with complex variables. Define the *inner products* of two complex vectors, x and y, according to $x \cdot y = \sum x_i \bar{y}_i$, where $\bar{y}_i$ is the complex conjugate of y_i ($y_i = c + di$, then $\bar{y}_i = c - di$). Note that this is not symmetric: $x \cdot y \neq y \cdot x$. This may be seen in the scalar case with $x = a + bi$ and $y = c + di$ so that $x \cdot y = (a + bi)(c - di) = ac + db - (ad - bc)i$ and $y \cdot x = (c + di)(a - bi) = ac + db + (ad - bc)i$. This definition captures the correct notion of orthogonality. With $x = (1, i)$, $\bar{x} = (1, -i)$ and $x \cdot x = 1 \times 1 + i(-i) = 1 - (-1) = 2$ and $y = (-1, i)$, $\bar{y} = (-1, -i)$ gives $x \cdot y = 1(-1) + i(-i) = -1 - (-1) = 0$. $\square$

REMARK 15.15: Linear independence for complex vectors is defined as follows. Let $\mathscr{C}$ denote the complex numbers and $\mathscr{C}^k = \mathscr{C} \times \cdots \times \mathscr{C}$ the k-product of $\mathscr{C}$. The vectors $x_1, \ldots, x_n$ are said to be linearly independent if $\sum_{i=1}^n c_i x_i = 0$ implies that $c_i = 0$ for all i (here, each c_i may be real or complex).

Let e_j be the real n-vector with 1 in the ith position and 0 elsewhere. Let x be an arbitrary n-vector with jth element x_j. Then, letting $c_j = x_j$, $x = \sum_{j=1}^n c_j e_j$. Therefore, the collection of vectors $\{e_j\}_{j=1}^n$ is a basis for $\mathscr{C}^n$ and $\mathscr{C}^n$ has dimension n. $\square$

Apart from the case of real symmetric matrices, there is limited benefit to assuming that the entries of a matrix are real, since the eigenvalues and eigenvectors may be complex. Example 15.14 shows that matrices with real entries may have complex eigenvalues and eigenvectors.

EXAMPLE 15.14: Consider:

$$A = \begin{pmatrix} 1 & 5 \\ -2 & 3 \end{pmatrix}.$$

Then

$$A - \lambda I = \begin{pmatrix} 1-\lambda & 5 \\ -2 & 3-\lambda \end{pmatrix},$$

The characteristic equation for A is:

$$|A - \lambda I| = (1-\lambda)(3-\lambda) + 10 = \lambda^2 - 4\lambda + 13 = 0,$$

and this has roots or eigenvalues:

$$\lambda_1, \lambda_2 = \frac{4 \pm \sqrt{16 - 4\cdot 13}}{2} = 2 \pm 3i,$$

so the roots are complex. The eigenvectors corresponding to $\lambda_1 = 2+3i$ and $\lambda_2 = 2-3i$ are with normalization of the second component to 1:

$$x_1 = \begin{pmatrix} \frac{1}{2}(1-3i) \\ 1 \end{pmatrix}, \qquad x_2 = \begin{pmatrix} \frac{1}{2}(1+3i) \\ 1 \end{pmatrix}$$

where the computations make use of $(1+3i)(1-3i) = 10$, so that $(1-3i) = \frac{10}{1+3i}$. Recall $z = a + ib$, $z^{-1} = \frac{\bar{z}}{|z|^2}$, $|z| = \sqrt{a^2+b^2}$, or equivalently $\frac{1}{a+bi} = \frac{1}{a^2+b^2}(a-bi)$. ◊

The next example illustrates the same point for a strictly positive matrix, given a real strictly positive matrix with complex eigenvalues.

EXAMPLE 15.15: Consider,

$$A = \begin{pmatrix} 1 & 1 & 20 \\ 8 & 7 & 7 \\ 1 & 20 & 1 \end{pmatrix}, \qquad A - \lambda I = \begin{pmatrix} 1-\lambda & 1 & 20 \\ 8 & 7-\lambda & 7 \\ 1 & 20 & 1-\lambda \end{pmatrix},$$

and $|A - \lambda I| = 2926 + 153\lambda + 9\lambda^2 - \lambda^3 = -(\lambda - 22)(\lambda^2 + 13\lambda + 133)$. Therefore, $\lambda_1 = 22$ is a root of the equation and an eigenvalue. The other two eigenvalues are obtained as the roots:

$$\begin{aligned}\lambda_2, \lambda_2 &= \frac{1}{2}[-13 \pm \sqrt{13^2 - 4(133)}] = \frac{1}{2}[-13 \pm \sqrt{169 - 532)}] = \frac{1}{2}[-13 \pm \sqrt{-363}] \\ &= \frac{1}{2}[-13 \pm \sqrt{-3(11^2)}] \\ &= \frac{1}{2}[-13 \pm (11\sqrt{3})\, i].\end{aligned}$$

◊

REMARK 15.16: One case where complex roots arise naturally is with rotations. Let θ be an angle and define the matrix $\boldsymbol{R}(\theta)$:

$$\boldsymbol{R}(\theta) = \begin{pmatrix} \cos(\theta) & -\sin(\theta) \\ \sin(\theta) & \cos(\theta) \end{pmatrix}.$$

Then, $\boldsymbol{R}(\theta)\boldsymbol{x}$ rotates the vector $\boldsymbol{x}$ counterclockwise θ degrees. A 90° rotation is given by:

$$\boldsymbol{R}(90) = \begin{pmatrix} \cos(90) & -\sin(90) \\ \sin(90) & \cos(90) \end{pmatrix} = \begin{pmatrix} 0 & -1 \\ 1 & 0 \end{pmatrix}.$$

So, for example, the vector $\boldsymbol{x} = (1,1)$ under $\boldsymbol{R}(90)$ (see Figure 15.4) gives:

$$\begin{pmatrix} 0 & -1 \\ 1 & 0 \end{pmatrix}\begin{pmatrix} 1 \\ 1 \end{pmatrix} = \begin{pmatrix} -1 \\ 1 \end{pmatrix} = \boldsymbol{x}'.$$

Every vector is rotated 90° under the transformation $\boldsymbol{R}(90)$. Therefore, for any $\boldsymbol{x}$, $\boldsymbol{x}' = \boldsymbol{R}(90)\boldsymbol{x}$ can never be a real scalar multiple of $\boldsymbol{x}$, since $\boldsymbol{x}$ and $\boldsymbol{x}'$ point in different directions. In such circumstances complex vectors and complex roots arise. □

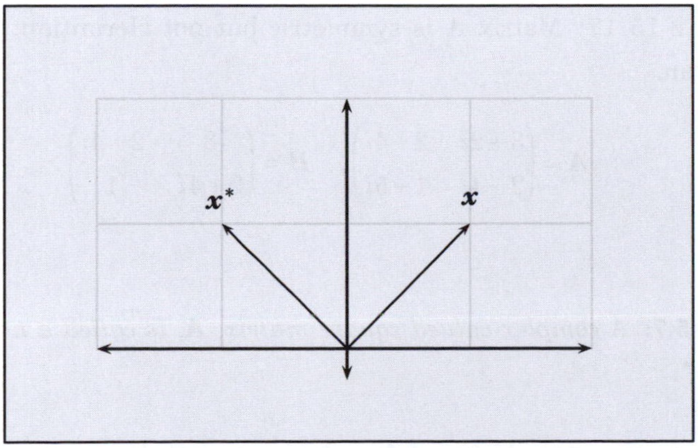

Figure 15.4: Rotation of 90°

Definition 15.5: *Let* **A** *be a square matrix with complex entries (a complex-valued matrix).*

- *The matrix* $\bar{\mathbf{A}}$ *is obtained from* **A** *by taking the complex conjugate of each entry (replace each entry with its complex conjugate).*

- *The conjugate matrix* $\mathbf{A}^* \stackrel{\text{def}}{=} \bar{\mathbf{A}}'$, *is called the conjugate transpose of* $\mathbf{A}$, *obtained by taking the complex conjugate of each entry and then taking the transpose.*

EXAMPLE 15.16: The conjugate transpose:

$$\boldsymbol{A} = \begin{pmatrix} 2i & 1+i \\ -2i & 1-3i \end{pmatrix}, \qquad \bar{\boldsymbol{A}} = \begin{pmatrix} -2i & 1-i \\ 2i & 1+3i \end{pmatrix}, \quad \boldsymbol{A}^* = \begin{pmatrix} -2i & 2i \\ 1-i & 1+3i \end{pmatrix}. \qquad \diamond$$

When $\boldsymbol{A}$ is real, $\boldsymbol{A}^*$ is the transpose of $\boldsymbol{A}$. The conjugate transpose has the following properties: $(\boldsymbol{A}^*)^* = \boldsymbol{A}$; $(\boldsymbol{A}+\boldsymbol{B})^* = \boldsymbol{A}^* + \boldsymbol{B}^*$; $(c\boldsymbol{A})^* = c^*\boldsymbol{A}^*$; $(\boldsymbol{AB})^* = \boldsymbol{B}^*\boldsymbol{A}^*$.

Definition 15.6: *A complex-valued square matrix,* $\mathbf{A}$, *with* $\mathbf{A} = \mathbf{A}^*$ *is called a Hermitian matrix.*

Note that for any matrix $\boldsymbol{A}$, $\boldsymbol{A} = \boldsymbol{A}^*$ implies that the diagonal entries are real (because each diagonal element must equal its conjugate). If a matrix is real, the Hermitian condition is equivalent to symmetry.

EXAMPLE 15.17: Matrix $\boldsymbol{A}$ is symmetric but not Hermitian; matrix $\boldsymbol{B}$ is Hermitian.

$$\boldsymbol{A} = \begin{pmatrix} 3+2i & 2+4i \\ 2-4i & 1+5i \end{pmatrix}, \quad \boldsymbol{B} = \begin{pmatrix} 3 & 2+4i \\ 2+4i & 1 \end{pmatrix}. \qquad \diamond$$

Definition 15.7: *A complex-valued square matrix,* $\mathbf{A}$, *is called a normal matrix if* $\boldsymbol{A}^*\boldsymbol{A} = \boldsymbol{A}\boldsymbol{A}^*$.

If $\boldsymbol{A}$ is Hermitian, then $\boldsymbol{A}^*\boldsymbol{A} = \boldsymbol{AA} = \boldsymbol{AA}^*$, so that Hermitian matrices are normal.

All normal matrices are diagonalizable, but not all diagonalizable matrices are normal.

Definition 15.8: *A complex-valued square matrix,* $\mathbf{A}$, *is called a unitary matrix if* $\boldsymbol{A}^*\boldsymbol{A} = \boldsymbol{AA}^* = \boldsymbol{I}$. *Thus,* $\mathbf{A}$ *is unitary if and only if it has an inverse* $\mathbf{A}^{-1}$ *that is equal to its conjugate transpose* $\mathbf{A}^*$: $\mathbf{A}^{-1} = \mathbf{A}^*$.

A unitary matrix corresponds to orthogonality for a real matrix. From the definitions, both Hermitian and unitary matrices are normal.

EXAMPLE 15.18: Consider:

$$\boldsymbol{A} = \begin{pmatrix} 1-i & 1+i \\ 1+i & 1-i \end{pmatrix}, \quad \boldsymbol{A}\boldsymbol{A}^* = \begin{pmatrix} 1-i & 1+i \\ 1+i & 1-i \end{pmatrix}\begin{pmatrix} 1+i & 1-i \\ 1-i & 1+i \end{pmatrix}$$
$$= \begin{pmatrix} 2(1+i)(1-i) & (1-i)^2+(1+i)^2 \\ (1-i)^2+(1+i)^2 & 2(1+i)(1-i) \end{pmatrix}.$$

Since $(1+i)(1-i) = 1-i^2 = 2$ and $(1-i)^2+(1+i)^2 = 1-2i+i^2+1+2i+i^2 = 2+2i^2 = 0$, $\boldsymbol{A}\boldsymbol{A}^* = 4\boldsymbol{I}$. Similarly, $\boldsymbol{A}^*\boldsymbol{A} = 4\boldsymbol{I}$, so that $\boldsymbol{B} = \frac{1}{2}\boldsymbol{A}$ satisfies $\boldsymbol{B}\boldsymbol{B}^* = \boldsymbol{B}^*\boldsymbol{B} = \boldsymbol{I}$. ◊

15.8 Diagonalization

One important role of eigenvalue theory is the diagonalization of a matrix.

Definition 15.9: *A square matrix* $\mathbf{A}$ *is diagonalizable if there is a non-singular matrix* $\boldsymbol{Q}$ *and a diagonal matrix* $\mathbf{D}$*, such that* $\mathbf{A} = \boldsymbol{QDQ}^{-1}$.

The main theorem on diagonalization is:

Theorem 15.6: *A square* $n \times n$ *matrix* $\boldsymbol{A}$ *is diagonalizable if and only if it has* n *linearly independent eigenvectors.*

If there are n linearly independent eigenvectors, $\{\boldsymbol{x}_i\}_{i=1}^n$, then for each i, $\boldsymbol{A}\boldsymbol{x}_i = \lambda_i \boldsymbol{x}_i$; and forming a matrix $\boldsymbol{X}$ whose columns are the $\boldsymbol{x}_i$ gives $\boldsymbol{AX} = \boldsymbol{X}\boldsymbol{\Lambda}$, where $\boldsymbol{\Lambda}$ is a diagonal matrix with the eigenvalues on the diagonal. Thus $\boldsymbol{X}^{-1}\boldsymbol{AX} = \boldsymbol{\Lambda}$.

A case of particular interest is where the diagonalizing matrix is orthogonal (or unitary, allowing for complex eigenvectors).

Definition 15.10: *A complex-valued square matrix,* $\mathbf{A}$*, is said to be unitarily diagonalizable if there is a unitary matrix* $\mathbf{P}$ *with complex entries such that:*

$$\boldsymbol{P}^{-1}\boldsymbol{AP} = \boldsymbol{P}^*\boldsymbol{AP} = \boldsymbol{\Lambda},$$

where $\boldsymbol{\Lambda}$ *is a diagonal matrix with complex entries.*

Theorem 15.7: *A complex-valued square matrix, **A**, is unitarily diagonalizable if and only if it is normal.*

EXAMPLE 15.19: This theorem leaves open the possibility that a real-valued matrix may be unitarily diagonalizable only with a complex matrix $\boldsymbol{P}$. For example,

$$\boldsymbol{A} = \begin{pmatrix} 0 & 1 \\ -1 & 0 \end{pmatrix}, \quad \boldsymbol{AA}' = \boldsymbol{A}'\boldsymbol{A} = \boldsymbol{I}.$$

This matrix is not Hermitian, but it is normal and hence unitarily diagonalizable. The matrix has eigenvalues $\lambda_1 = i$, $\lambda_2 = -i$ (with corresponding eigenvectors $x_1 = (-i, 1)$ and $x_2 = (i, 1)$). Since $\boldsymbol{A}$ is unitarily diagonalizable, then there is $\boldsymbol{P}$ such that $\boldsymbol{P}^{-1}\boldsymbol{AP} = \boldsymbol{\Lambda}$, a diagonal matrix. Then $\boldsymbol{AP} = \boldsymbol{P\Lambda}$, so that the columns of $\boldsymbol{P}$ are eigenvectors of $\boldsymbol{A}$ and the elements of $\boldsymbol{\Lambda}$ corresponding eigenvalues. (Since the eigenvectors, $\boldsymbol{\Lambda}$, are complex, if $\boldsymbol{P}$ diagonalizes $\boldsymbol{A}$, $\boldsymbol{P}$ must be complex, otherwise $\boldsymbol{AP}$ is real while $\boldsymbol{P\Lambda}$ is complex.) ◊

15.9 Properties of eigenvalues and eigenvectors

The following discussion is concerned with arbitrary matrices, real or complex. As already mentioned, even if a square matrix is real, it may have complex eigenvalues and eigenvectors, so that there is little advantage in assuming the matrix to be real. For complex matrices that are Hermitian, all the eigenvalues are real, but the eigenvalues may be complex. For real symmetric matrices, all eigenvalues and eigenvectors are real. This case is discussed in Section 15.6.

Theorem 15.8: *Let $\lambda_1, \ldots, \lambda_n$ be the roots of **A** (which may be complex). Then*

- *The trace of **A** equals the sum of the eigenvalues of **A**:* $\text{tr}(\boldsymbol{A}) = \sum_{i=1}^{n} \lambda_i$.
- *The determinant of **A** equals the product of its eigenvalues.* $|\boldsymbol{A}| = \prod_{i=1}^{n} \lambda_i$.
- *The trace of $\boldsymbol{A}^k$ is equal to the sum of the roots, each raised to the power of k:* $\text{tr}(\boldsymbol{A}^k) = \sum_i \lambda_i^k$.

For Hermitian matrices:

Theorem 15.9: *If **A** is a Hermitian matrix, all the eigenvalues are real. (Although the eigenvectors must be complex whenever **A** has complex elements.)*

EXAMPLE 15.20:

$$\boldsymbol{A} = \begin{pmatrix} 4 & 1 \\ 2 & 3 \end{pmatrix}, \qquad \lambda_1 = 5,\ \lambda_2 = 2, \quad \text{tr}(\boldsymbol{A}) = \lambda_1 + \lambda_2, \quad |\,\boldsymbol{A}\,| = 10 = \lambda_1 \lambda_2,$$

and, for example,

$$\boldsymbol{A}^3 = \begin{pmatrix} 86 & 39 \\ 78 & 47 \end{pmatrix},$$

with eigenvalues $\hat{\lambda}_1 = 8$, $\hat{\lambda}_2 = 125$, $\text{tr}(\boldsymbol{A}) = 86 + 47 = 133 = 8 + 125$, and $|\,\boldsymbol{A}\,| = 1000 = 8 \times 125$. ◊

Theorem 15.10: *For any square matrix, $\boldsymbol{A}$:*

(i) The roots of $\boldsymbol{A}^*$ are the conjugates of the roots of $\boldsymbol{A}$.

(ii) If $\boldsymbol{A}$ is non-singular, the roots of $\boldsymbol{A}^{-1}$ are $\mu_1, \ldots, \mu_n$ where $\mu_i = \lambda_i^{-1}$ and λ_i is a root of $\boldsymbol{A}$.

(iii) If q is a real number, and λ an eigenvalue of $\boldsymbol{A}$, then $q\lambda$ is an eigenvalue of $q\boldsymbol{A}$.

(iv) If q is a positive integer, and λ an eigenvalue of $\boldsymbol{A}$, then λ^q is an eigenvalue of $\boldsymbol{A}^q$.

PROOF: (i) Recall that $|\bar{\boldsymbol{B}}| = \overline{|\boldsymbol{B}|}$, the determinant of the conjugate is the conjugate of the determinant, and the conjugate of a sum is the sum of the conjugates, $\overline{\boldsymbol{A}+\boldsymbol{B}} = \bar{\boldsymbol{A}} + \bar{\boldsymbol{B}}$. If λ_i is a root of $\boldsymbol{A}$ then $|\,\boldsymbol{A} - \lambda_i \boldsymbol{I}\,| = 0$, so $|\,\bar{\boldsymbol{A}} - \bar{\lambda}_i \boldsymbol{I}\,| = 0$. Therefore, $\bar{\lambda}_i$ is an eigenvalue of $\bar{\boldsymbol{A}}$ and so $\bar{\lambda}_i$ is an eigenvalue of $\bar{\boldsymbol{A}}' = \boldsymbol{A}^*$.
(ii) Let $(\lambda_1, \ldots, \lambda_n)$ be the roots of $\boldsymbol{A}$ so that for each $i, \boldsymbol{A}\boldsymbol{x}_i = \lambda_i \boldsymbol{x}_i$ for some $\boldsymbol{x}_i \neq 0$. Therefore $\boldsymbol{x}_i = \lambda_i \boldsymbol{A}^{-1} \boldsymbol{x}_i$. Since $\lambda_i \neq 0$, dividing by λ_i gives $\boldsymbol{A}^{-1}\boldsymbol{x}_i = \frac{1}{\lambda_i}\boldsymbol{x}_i = \mu_i \boldsymbol{x}_i$, so that $\mu_i = \frac{1}{\lambda_i}$, a root of $\boldsymbol{A}^{-1}$.
(iii) and (iv) follow from the definition. ■

PROPOSITION 15.9: $r(\boldsymbol{A})$ = number of non-zero roots of $\boldsymbol{A}$. ◊

EXAMPLE 15.21: With

$$\boldsymbol{A} = \begin{pmatrix} 1 & 3 \\ 0 & 4 \end{pmatrix} \text{ and } \boldsymbol{B} = \begin{pmatrix} 4 & 2 \\ 2 & 1 \end{pmatrix},$$

$r(\boldsymbol{A}) = 2$, and $\lambda_1 = 1$, $\lambda_2 = 4$; while $r(\boldsymbol{B}) = 1$, and $\lambda_1 = 5$, $\lambda_2 = 0$. So $\boldsymbol{A}$ has rank 2 and two non-zero roots while $\boldsymbol{B}$ has rank 1 and one non-zero root. ◊

Notice that in Example 15.21 the matrix $\boldsymbol{A}$ is not Hermitian (symmetric as a real matrix). Although the eigenvectors are independent, they are not orthogonal.

EXAMPLE 15.22:

$$\boldsymbol{A} = \begin{pmatrix} 1 & 3 \\ 0 & 4 \end{pmatrix}.$$

The characteristic equation is $(1-\lambda)(4-\lambda) = 0$ with roots 1 and 4. Thus, $\boldsymbol{A}\boldsymbol{x}_i = \lambda_i \boldsymbol{x}_i$ or $(\boldsymbol{A} - \lambda_i \boldsymbol{I})\boldsymbol{x}_i = 0$:

$$\begin{pmatrix} 1-\lambda_i & 3 \\ 0 & 4-\lambda_i \end{pmatrix} \begin{pmatrix} x_{i1} \\ x_{i2} \end{pmatrix} = \begin{pmatrix} 0 \\ 0 \end{pmatrix}.$$

Taking $\lambda_i = 1$:

$$\begin{pmatrix} 1-1 & 3 \\ 0 & 4-1 \end{pmatrix} \begin{pmatrix} x_{i1} \\ x_{i2} \end{pmatrix} = \begin{pmatrix} 0 \\ 0 \end{pmatrix}; \quad \begin{pmatrix} 0 & 3 \\ 0 & 3 \end{pmatrix} \begin{pmatrix} x_{i1} \\ x_{i2} \end{pmatrix} = \begin{pmatrix} 0 \\ 0 \end{pmatrix}.$$

So, $\boldsymbol{x}_1 = (s, 0)$ satisfies this equation for any s.

Taking $\lambda_i = 4$:

$$\begin{pmatrix} 1-4 & 3 \\ 0 & 4-4 \end{pmatrix} \begin{pmatrix} x_{21} \\ x_{22} \end{pmatrix} = \begin{pmatrix} 0 \\ 0 \end{pmatrix}; \quad \begin{pmatrix} -3 & 3 \\ 0 & 0 \end{pmatrix} \begin{pmatrix} x_{21} \\ x_{22} \end{pmatrix} = \begin{pmatrix} 0 \\ 0 \end{pmatrix}.$$

So, $\boldsymbol{x}_2 = (t, t)$ satisfies this equation for any t. With the normalizations of $s = t = 1$, the eigenvectors are:

$$\boldsymbol{x}_1 = \begin{pmatrix} 1 \\ 0 \end{pmatrix}, \quad \boldsymbol{x}_2 = \begin{pmatrix} 1 \\ 1 \end{pmatrix}.$$

Whatever values are chosen for s and t, these must be non-zero since the eigenvectors must all be chosen non-zero. Therefore, in this example, if $\tilde{\boldsymbol{x}}_1$ and $\tilde{\boldsymbol{x}}_2$ are eigenvectors, then $\tilde{\boldsymbol{x}}_1 \cdot \tilde{\boldsymbol{x}}_2 = (s,0)\cdot(t,t) = st \neq 0$. So, the eigenvectors cannot be chosen to be orthogonal. ◊

PROPOSITION 15.10: If $\boldsymbol{A}$ is diagonal so that $a_{ij} = 0, i \neq j$, then the roots of $\boldsymbol{A}$ are its diagonal elements. ◊

PROOF: $\boldsymbol{A} = \{a_{ij}\}$, and $a_{ij} = 0$ if $i \neq j$. Then $| \boldsymbol{A} - \lambda \boldsymbol{I} | = 0$ if $\lambda = a_{ii}$ for some i since the ith row (and column) of $\boldsymbol{A} - \lambda \boldsymbol{I}$ contains all zeros and so $\boldsymbol{A} - \lambda \boldsymbol{I}$ must have a zero determinant. ■

In view of Proposition 15.9:

PROPOSITION 15.11: The trace of an idempotent matrix equals its rank. ◊

The following discussion provides insight into the long-run behavior of powers of a matrix. Given the eigenvectors $\boldsymbol{v}_1,\dots,\boldsymbol{v}_n$ and eigenvalues $\lambda_1,\dots,\lambda_n$ of the matrix $\boldsymbol{A}$, for arbitrary numbers $\alpha_1,\dots,\alpha_n$, let

$$\boldsymbol{z} = \alpha_1\boldsymbol{v}_1 + \alpha_2\boldsymbol{v}_2 + \cdots + \alpha_n\boldsymbol{v}_n = \sum_{i=1}^{n} \alpha_i\boldsymbol{v}_i.$$

Then

$$\boldsymbol{A}\boldsymbol{z} = \alpha_1\boldsymbol{A}\boldsymbol{v}_1 + \alpha_2\boldsymbol{A}\boldsymbol{v}_2 + \cdots + \alpha_n\boldsymbol{A}\boldsymbol{v}_n$$
$$\boldsymbol{A}\boldsymbol{z} = \alpha_1\lambda_1\boldsymbol{v}_1 + \alpha_2\lambda_2\boldsymbol{v}_2 + \cdots + \alpha_n\lambda_n\boldsymbol{v}_n.$$

Repeating this,

$$\boldsymbol{A}^2\boldsymbol{z} = \boldsymbol{A}\boldsymbol{A}\boldsymbol{z} = \alpha_1\lambda_1\boldsymbol{A}\boldsymbol{v}_1 + \alpha_2\lambda_2\boldsymbol{A}\boldsymbol{v}_2 + \cdots + \alpha_n\lambda_n\boldsymbol{A}\boldsymbol{v}_n$$
$$\boldsymbol{A}^2\boldsymbol{z} = \boldsymbol{A}\boldsymbol{A}\boldsymbol{z} = \alpha_1\lambda_1^2\boldsymbol{v}_1 + \alpha_2\lambda_2^2\boldsymbol{v}_2 + \cdots + \alpha_n\lambda_n^2\boldsymbol{v}_n.$$

After t iterations,

$$\boldsymbol{A}^t\boldsymbol{z} = \alpha_1\lambda_1^t\boldsymbol{v}_1 + \alpha_2\lambda_2^t\boldsymbol{v}_2 + \cdots + \alpha_n\lambda_n^t\boldsymbol{v}_n.$$

Pick the largest eigenvalue of $\boldsymbol{A}$, say λ_1 is larger than any other eigenvalue: $|\lambda_1| > |\lambda_i|$, $i \neq 1$, and rearrange:

$$\boldsymbol{A}^t\boldsymbol{z} = \lambda_1^t\{\alpha_1\boldsymbol{v}_1 + \alpha_2\left(\frac{\lambda_2}{\lambda_1}\right)^t\boldsymbol{v}_2 + \cdots + \alpha_n\left(\frac{\lambda_n}{\lambda_1}\right)^t\boldsymbol{v}_n\}.$$

Since λ_1 is strictly larger than any other eigenvalue, then for t large, $\boldsymbol{A}^t\boldsymbol{z} \approx \lambda_1^t\alpha_1\boldsymbol{v}_1$ or $(\frac{1}{\lambda_1}\boldsymbol{A})^t\boldsymbol{z} \approx \alpha_1\boldsymbol{v}_1$. Thus, regardless of $\boldsymbol{z}$, considering the sequence $\{\boldsymbol{A}\boldsymbol{z},\boldsymbol{A}^2\boldsymbol{z},\boldsymbol{A}^3\boldsymbol{z},\dots,\boldsymbol{A}^t\boldsymbol{z}\}$, for large t, $\boldsymbol{A}^t\boldsymbol{z}$ is proportional to the eigenvector associated with the largest eigenvalue.

EXAMPLE 15.23: With

$$A = \begin{pmatrix} 1 & 3 \\ 0 & 4 \end{pmatrix}, \quad x_1 = \begin{pmatrix} 1 \\ 0 \end{pmatrix}, \quad x_1 = \begin{pmatrix} 1 \\ 1 \end{pmatrix},$$

A has two roots, $\lambda_1 = 1$, $\lambda_2 = 4$, with corresponding eigenvectors, x_1 and x_2. Taking the largest root, λ_2,

$$\frac{1}{\lambda_2} A = \begin{pmatrix} \frac{1}{4} & \frac{3}{4} \\ 0 & 1 \end{pmatrix}.$$

Then, for example,

$$\left(\frac{1}{\lambda_2} A\right)^5 = \begin{pmatrix} \frac{1}{4} & \frac{3}{4} \\ 0 & 1 \end{pmatrix}^5 = \begin{pmatrix} \frac{1}{1024} & \frac{1023}{1024} \\ 0 & 1 \end{pmatrix} \approx \begin{pmatrix} 0 & 1 \\ 0 & 1 \end{pmatrix},$$

so that

$$\left(\frac{1}{\lambda_2} A\right)^5 \begin{pmatrix} z_1 \\ z_2 \end{pmatrix} = \begin{pmatrix} z_2 \\ z_2 \end{pmatrix} = z_2 \begin{pmatrix} 1 \\ 1 \end{pmatrix} = z_2 \begin{pmatrix} 1 \\ 1 \end{pmatrix},$$

the eigenvector (subject to scaling) of λ_2. ◊

More generally, let $\lambda' = \max_i |\lambda_i| > 0$ and $\hat{\lambda}_i = \frac{\lambda_i}{\lambda'}$ with $|\hat{\lambda}_i| \leq 1$. Assume that $\lambda_1 = \lambda'$ and the expression becomes:

$$A^t z = \lambda_1^t \{\alpha_1 v_1 + \alpha_2 \hat{\lambda}_2^t v_2 + \cdots + \alpha_n \hat{\lambda}_n^t v_n\}.$$

Let $I = \{i \mid |\lambda_i| = |\lambda'|\}$: $j \in I$, if and only if $\lambda_j = \max_i \lambda_i$. Then $|\hat{\lambda}_j| = 1$ if $j \in I$ and $|\hat{\lambda}_j| < 1$ if $j \notin I$, so that for $j \notin I$, $\hat{\lambda}_j^t \to 0$ as $t \to \infty$. In this case,

$$A^t z \to \lambda_1^t \{\sum_{j \in I} .\alpha_j v_j\}.$$

15.9.1 Largest eigenvalues and eigenvalue multiplicity

Let A be a square matrix.

Recall that given a square matrix A, recall that the determinant $|A - \lambda I| = p(\lambda)$ defines a polynomial in λ:

$$p(\lambda) = a_n \lambda^n + a_{n-1} \lambda^{n-1} + \cdots + a_1 \lambda + a_0,$$

with n roots $\lambda_1, \lambda_2, \ldots, \lambda_n$, so that

$$p(\lambda) = a_n(\lambda - \lambda_1)(\lambda - \lambda_2)\cdots(\lambda - \lambda_n).$$

Definition 15.11: *If the root λ_j occurs exactly k times, then it is said to have (algebraic) multiplicity k. In this case, there is a polynomial $f(\lambda)$, $f(\lambda_j) \neq 0$ and $p(\lambda) = (\lambda - \lambda_j)^k f(\lambda)$. If a root has multiplicity 1 it is called a simple root.*

EXAMPLE 15.24: Let

$$\boldsymbol{A} = \begin{pmatrix} 1 & 0 & 0 \\ 0 & 2 & 0 \\ 0 & 0 & 3 \end{pmatrix}.$$

Then the corresponding polynomial is: $p(\lambda) = (1-\lambda)(2-\lambda)(3-\lambda)$. So, each root is simple with multiplicity 1. If

$$\boldsymbol{B} = \begin{pmatrix} 1 & 0 & 0 \\ 0 & 1 & 0 \\ 0 & 0 & 2 \end{pmatrix},$$

then the corresponding polynomial is $p(\lambda) = (1-\lambda)(1-\lambda)(2-\lambda) = (1-\lambda)^2(2-\lambda)$. The root $\lambda = 2$ is simple, the root $\lambda = 1$ has multiplicity 2. ◊

Definition 15.12: *The (geometric) multiplicity of an eigenvalue is the number of linearly independent eigenvectors that have that eigenvalue.*

EXAMPLE 15.25: Taking the matrix $\boldsymbol{B}$ of Example 15.24, the expression $\boldsymbol{Bx} = \lambda\boldsymbol{x}$ or $(\boldsymbol{B} - \lambda I)\boldsymbol{x} = 0$ gives, with $\lambda = 1$:

$$\begin{pmatrix} 1-1 & 0 & 0 \\ 0 & 1-1 & 0 \\ 0 & 0 & 2-1 \end{pmatrix}\begin{pmatrix} x_1 \\ x_2 \\ x_3 \end{pmatrix} = \begin{pmatrix} 0 \\ 0 \\ 0 \end{pmatrix} \quad \text{or} \quad \begin{aligned} 0x_1 + 0x_2 &= 0 \\ 0x_1 + 0x_2 &= 0, \\ 1x_3 &= 0 \end{aligned}$$

so the set of eigenvectors associated with $\lambda = 1$ are vectors of the form:

$$\begin{pmatrix} x_1 \\ x_2 \\ 0 \end{pmatrix}$$

and so the geometric multiplicity of the eigenvector $\lambda = 1$ is equal to 2. The geometric multiplicity of eigenvector $\lambda = 2$ is 1. ◊

If n_i is the algebraic multiplicity and m_i the geometric multiplicity of root λ_i, then $m_i \le n_i$.

EXAMPLE 15.26: Consider the matrix:

$$\boldsymbol{A} = \begin{pmatrix} 1 & 2 \\ 0 & 1 \end{pmatrix}; \quad p(\lambda) = (1-\lambda)^2.$$

Thus, the root $\lambda = 1$ has algebraic multiplicity of 2. With $\lambda = 1$,

$$(\boldsymbol{A} - \lambda \boldsymbol{I})\boldsymbol{x} = 0 \text{ or } \begin{pmatrix} 1-\lambda & 2 \\ 0 & 1-\lambda \end{pmatrix}\begin{pmatrix} x_1 \\ x_2 \end{pmatrix} = \begin{pmatrix} 0 \\ 0 \end{pmatrix}, \text{ or } \begin{pmatrix} 0 & 2 \\ 0 & 0 \end{pmatrix}\begin{pmatrix} x_1 \\ x_2 \end{pmatrix} = \begin{pmatrix} 0 \\ 0 \end{pmatrix}.$$

Any vector with $\boldsymbol{x}_1$ arbitrary and $\boldsymbol{x}_2 = 0$ satisfies this equation. So, the root $\lambda = 1$ has geometric multiplicity of 1. ◊

Exercises for Chapter 15

Exercise 15.1 Let

$$A = \begin{pmatrix} 1 & 2 \\ 4 & 3 \end{pmatrix}.$$

Give the characteristic equation and find the eigenvalues and a corresponding set of eigenvectors.

Exercise 15.2 Let

$$B = \begin{pmatrix} 1 & 0 & 0 \\ 2 & 4 & 2 \\ 4 & 0 & 2 \end{pmatrix}.$$

Give the characteristic equation and find the eigenvalues and a corresponding set of eigenvectors.

Exercise 15.3 Let

$$A = \begin{pmatrix} -\frac{6}{4} & \frac{6}{4} \\ -\frac{12}{4} & \frac{11}{4} \end{pmatrix}.$$

Let λ_1 and λ_2 be the eigenvalues with eigenvectors $\boldsymbol{x}_1$ and $\boldsymbol{x}_2$. Set $\boldsymbol{X}$ as the matrix with columns $\boldsymbol{x}_1$ and $\boldsymbol{x}_2$ and $\boldsymbol{\Lambda}$ the diagonal matrix with λ_1 and λ_2 on the main diagonal. Use the formula $\boldsymbol{A} = \boldsymbol{X}\boldsymbol{\Lambda}\boldsymbol{X}^{-1}$ to show that $\boldsymbol{A}^k \to 0$ as $k \to \infty$.

Exercise 15.4 Let $\boldsymbol{A}$ be an $n \times n$ matrix. Suppose that the sum of the absolute values of the entries in any column is no larger than 1 (for each j, $\sum_{i=1}^{n} | a_{ij} | \le 1$). Show that all eigenvalues are equal to or less than 1 (in absolute value).

Exercise 15.5 Show that a square matrix $\boldsymbol{A}$ is non-singular if and only if all eigenvalues are not zero.

Exercise 15.6 Consider the matrix:

$$\boldsymbol{A} = \begin{pmatrix} 2 & 5 \\ 4 & 3 \end{pmatrix},$$

$$\boldsymbol{A} - \lambda \boldsymbol{I} = \begin{pmatrix} 2-\lambda & 5 \\ 4 & 3-\lambda \end{pmatrix}.$$

Therefore, $| \boldsymbol{A} - \lambda \boldsymbol{I} | = -14 - 5\lambda + \lambda^2$ (which factors to $(\lambda+2)(\lambda-7)$). The characteristic equation is:

$$p(\lambda) = \lambda^2 - 5\lambda - 14 = 0.$$

According to the Cayley–Hamilton theorem, the matrix satisfies the characteristic polynomial

$$p(\boldsymbol{A}) = \boldsymbol{A}^2 - 5\boldsymbol{A} - 14 = 0.$$

Show that this condition is satisfied for this matrix.

Exercise 15.7 Let $p(\lambda) = a_2\lambda^2 + a_1\lambda + a_0 = 0$ be the characteristic equation of a 2×2 matrix $\boldsymbol{A}$. From the Cayley–Hamilton theorem, $p(\boldsymbol{A}) = a_2\boldsymbol{A}^2 + a_1\boldsymbol{A} + a_0\boldsymbol{I} = 0$. This implies that $\boldsymbol{A}^2$ may be computed as $\boldsymbol{A}^2 = -\frac{a_1}{a_2}\boldsymbol{A} - \frac{a_0}{a_2}\boldsymbol{I}$. Illustrate this calculation with the matrix:

$$\boldsymbol{A} = \begin{pmatrix} 4 & 5 \\ 2 & 3 \end{pmatrix}.$$

Show how this procedure may be used to solve for $\boldsymbol{A}^k$ in terms if $\boldsymbol{A}$ and $\boldsymbol{I}$.

Exercise 15.8 Show how the Cayley–Hamilton theorem may be used to solve for $\boldsymbol{A}^k$ in terms of $\boldsymbol{A}$ and $\boldsymbol{I}$ when $\boldsymbol{A}$ is a 2×2 matrix. Show also how it may be used to find the inverse of $\boldsymbol{A}$.

Exercise 15.9 Consider:

$$\boldsymbol{B} = \begin{pmatrix} 1 & 5 \\ 2 & 3 \end{pmatrix}.$$

Find the eigenvalues and eigenvectors of $\boldsymbol{B}$.

Exercise 15.10 Find the eigenvalues of

$$\boldsymbol{A} = \begin{pmatrix} 1 & 2 \\ 3 & 4 \end{pmatrix}.$$

Exercise 15.11 For complex matrices, the inverse is defined in the usual way. If $\boldsymbol{A}$ is a square matrix with complex entries (a complex valued matrix), $\boldsymbol{A}$ is invertible if there is a matrix $\boldsymbol{B}$ with $\boldsymbol{AB} = \boldsymbol{BA} = \boldsymbol{I}$, and $\boldsymbol{B}$ is called the inverse of $\boldsymbol{A}$ and denoted $\boldsymbol{A}^{-1}$. Calculate the determinant and inverse of:

$$\boldsymbol{A} = \begin{pmatrix} 2+2i & 4+i \\ 3+2i & 5+2i \end{pmatrix}.$$

16

Differential equations

16.1 Introduction

In this chapter, dynamic models are considered where some variable varies continuously as a function of some other variable (such as time). If y is a function of x, then assuming differentiability, the derivative is written $\frac{dy}{dx}$ or $y'(x)$. Knowing y, it is easy to determine the derivative $y'(x)$. However, in a variety of problems, the situation is reversed – the derivative, $y'(x)$ is known, but not the actual function $y(x)$. And in this class of problem, the task is to determine $y(x)$ from knowledge of $y'(x)$, called solving the differential equation. Section 16.1.1 provides a few motivating examples to illustrate how such dynamics may arise. Generally, it is difficult to determine $y(x)$ from knowledge of x. For example, given $y(x) = \ln(1+x^2)+xe^x$, differentiation gives $y'(x) = \frac{2x}{1+x^2} + e^x + xe^x = \frac{2x}{1+x^2} + (1+x)e^x$. But if given $y'(x) = \frac{2x}{1+x^2} + (1+x)e^x$, it is more difficult to see that $y(x) = \ln(1+x^2)+xe^x$.

The general approach to this problem is to categorize differential equations into classes of problems of varying difficulty. In a first-order differential equation, the dynamic relation is expressed by a function f with $y' = f(y,x)$. If this can be written in the form $a_1(x)y'(x) + a_0(x)y = b(x)$ it is called a first-order linear differential equation. Similarly, a second-order differential equation has the form $y''(x) = f(y,y',x)$, where y'' is the second derivative of the (unknown) function y. Again, the task is to find the function $y(x)$ that satisfies this equation. Depending on the function f, this problem may be extremely difficult to solve. In Section 16.2, first-order linear differential equations are considered. These arise in a number of economic models – such as growth models. Turning to non-linear differential equations, these are typically difficult to solve but there are some specific functional forms for which a solution may easily be found. Two non-linear first-order differential equations (the logistic and Bernoulli equations) are introduced in Section 16.3. Given a first-order differential equation, $y'(x) = f(y,x)$, a fundamental question concerns the existence of a solution: is there a unique function y that satisfies this equation? Section 16.4 provides sufficient conditions for the existence of a unique solution. Second-order linear differential equations are considered in Section 16.5. Finally, Section 16.6 introduces linear systems of differential equations and uses that framework to study price dynamics in a multimarket system.

16.1.1 Preliminary discussion

A differential equation is a rule governing the change in some variable. For example, suppose there is a stock of wealth invested and growing at the rate r. Thus, if the current stock is $y(t)$, and the value a short time later is $y(t+\Delta t)$ then the percentage increase is given by $\frac{y(t+\Delta t)-y(t)}{y(t)}$. The increase per unit of time over the time period t to $t+\Delta t$ is given by $\frac{1}{\Delta t}\frac{y(t+\Delta t)-y(t)}{y(t)} = \frac{1}{y(t)}\frac{y(t+\Delta t)-y(t)}{\Delta t}$. If $y(t)$ varies smoothly with t, then this is approximately $\frac{1}{y(t)}\frac{dy}{dt}$, as Δt goes to 0. With a constant growth rate of r, if $y(t)$ is the stock of wealth at time t, the percentage change in the stock is given by the equation $\frac{1}{y(t)}\frac{dy}{dt} = r$, so that

$$\frac{dy}{dt} - ry(t) = 0.$$

This is a differential equation where the change in y over time is governed by the current value of y and the parameter r. A solution to the differential equation is a function $y(t)$ satisfying the expression $\frac{dy}{dt} - ry(t) = 0$ for each t. This differential equation is easily solved, recognizing that $\frac{d\ln y(t)}{dt} = \frac{1}{y(t)}\frac{dy}{dt}$, so that

$$\ln y(t) - \ln y(t_0) = \int_{t_0}^{t} \frac{d\ln y(t)}{dt}dt = \int_{t_0}^{t} \frac{1}{y(t)}\frac{dy(t)}{dt}dt = \int_{t_0}^{t} r\,dt = r(t-t_0).$$

Thus, $\ln y(t) = \ln y(t_0) + r(t-t_0)$, so that $y(t) = e^{\ln y(t_0)+r(t-t_0)} = y(t_0)e^{r(t-t_0)}$.

More generally, given $y'(x)$, the function $y(x)$ is an unknown function of the variable x (or t when it is convenient to emphasize time), but the derivative $y'(x)$ is specified in terms of some function f:

$$\frac{dy}{dx} = f(y,x).$$

Such an equation is called an ordinary differential equation of order 1. Writing $y^{(n)}$ for the nth derivative of y with respect to x, an equation of the form

$$y^{(n)} = f(y^{(n-1)}, y^{(n-2)}, \ldots, y^{(1)}, y, x)$$

is called an nth-order ordinary differential equation. A solution is a function $y(x)$ satisfying the equation.

The next example illustrates a simple application (a first-order linear differential equation), determining the flow of payments required to amortize a loan in some fixed period of time.

EXAMPLE 16.1: Suppose that an individual takes out a loan of value P_0 (the principal at time $t = 0$). The interest rate per unit of time is r and the individual agrees to pay D per unit of time to repay the loan. Thus, if the loan has outstanding value $P_{(t)}$ at time t, then the change in the value of the outstanding loan over a short period of time, Δt is

$$\begin{aligned}\Delta P(t) &= P(t)r\Delta t - D\Delta t\\ &= (P(t)r - D)\Delta t.\end{aligned}$$

Dividing by Δt and letting the time increment, Δt go to 0,

$$\frac{dP(t)}{dt} = P(t)r - D$$

or

$$P'(t) - rP(t) = -D.$$

Comparing with the general formula, see Equation 16.6, $P(t)$ is given by:

$$\begin{aligned}P(t) &= \left(P_0 - \frac{D}{r}\right)e^{rt} + \frac{D}{r}\\ &= P_0e^{rt} + \frac{D}{r}(1 - e^{rt}).\end{aligned}$$

With this formula, one may consider such questions as the per unit of time repayment, $D(T)$, required to amortize the loan in some period, T, so that $P(T) = 0$. The condition $P(T) = 0$ implies that $0 = P_0e^{rT} + \frac{D}{r}(1 - e^{rT})$, so that

$$D(T) = -\frac{r}{1 - e^{rT}}P_0e^{rT}.$$

The figure plots $D(t)$ for $P_0 = 100$ and $r = 0.05$.

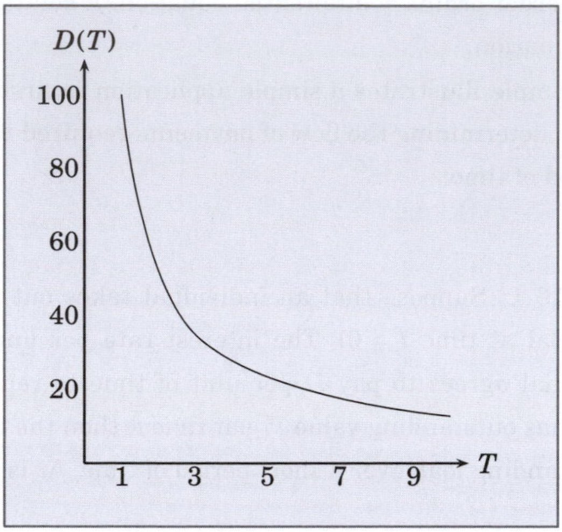

◊

16.2 First-order linear differential equations

A first-order linear differential equation has the form:

$$a_1(x)\frac{dy}{dx} + a_0(x)y = b(x).$$

This expression defines the movement of y, $\frac{dy}{dx}$, in terms of its current value, y, and the value of a variable x. A solution to the differential equation is a function $y(x)$ satisfying this equation on some interval in the domain of x.

16.2.1 Constant coefficients

When the coefficients a_1, a_0 and b do not depend on x and $a_1(x) \neq 0$, $a_1\frac{dy}{dx} + a_0 y = b$ may be rewritten $\frac{dy}{dx} + \frac{a_0}{a_1}y = \frac{b}{a_1}$, giving the simple first-order differential equation:

$$\frac{dy}{dx} + \alpha y = \beta. \tag{16.1}$$

A *particular solution*, y, is any function satisfying Equation 16.1. If y^a and y^b satisfy Equation 16.1, then $y^c = y^a - y^b$ satisfies the *homogeneous* Equation:

$$\frac{dy}{dx} + \alpha y = 0. \tag{16.2}$$

A solution to the homogeneous equation is called a *complementary* solution. Therefore, any two particular solutions differ by a complementary solution. So, given any particular solution, y^p, to Equation 16.1, every particular solution, y, to Equation 16.1 is equal to y^p plus a complementary solution, y^c: $y = y^p + y^c$. The family of particular solutions is called the *general* solution.

Therefore, every solution to Equation 16.1 may be found by finding a particular solution and adding a complementary solution. Using this observation, the general solution may be obtained by finding a particular solution and then all complementary solutions.

For a particular solution, with constant coefficients, this is easy to find: set $\frac{dy}{dx} = 0$ and pick the constant y to solve $\alpha y = \beta$: $y^p(x) = \frac{\beta}{\alpha}$. Next, consider the homogeneous case where $\beta = 0$, $\frac{dy}{dx} + \alpha y = 0$, and rearrange to give $\frac{1}{y}\frac{dy}{dx} = -\alpha$. Integrating,

$$\int_{x_0}^{x} \frac{1}{y}\frac{dy}{dx}dx = \int_{x_0}^{x} \frac{1}{y(\tilde{x})}y'(\tilde{x})d\tilde{x} = \ln y(x) - \ln y(x_0) = -\int_{x_0}^{x} \alpha d\tilde{x} = -\alpha(x - x_0).$$

Therefore, $\ln y(x) = [\ln y(x_0) + \alpha x_0] - \alpha x$. So, the complementary solution is:

$$y^c(x) = [e^{\ln y(x_0) + \alpha x_0}]e^{-\alpha x} = c_0 e^{-\alpha x}.$$

Note that c_0 is an indeterminate constant of integration, so this expression is a family of functions. Combining the particular and complementary solutions gives:

$$y(x) = y^p(x) + y^c(x) \tag{16.3}$$

$$= \frac{\beta}{\alpha} + c_0 e^{-\alpha x}. \tag{16.4}$$

If the value of y is given at x_0, $y_0 = y(x_0)$, then $y_0 = \frac{\beta}{\alpha} + c_0 e^{-\alpha x_0}$ or $y_0 - \frac{\beta}{\alpha} = c_0 e^{-\alpha x_0}$ or $c_0 = \left(y_0 - \frac{\beta}{\alpha}\right)e^{\alpha x_0}$. Therefore,

$$y(x) = y^p(x) + y^c(x) \tag{16.5}$$

$$= \frac{\beta}{\alpha} + \left(y_0 - \frac{\beta}{\alpha}\right)e^{-\alpha(x - x_0)}. \tag{16.6}$$

REMARK 16.1: What happens to the solution function $y(x)$ as x becomes very large? From the solution 16.4, it is clear that the answer depends on α. If $\alpha > 0$, then $e^{-\alpha x}$ goes to 0 as x becomes large, and if $\alpha < 0$, then $e^{-\alpha x}$ goes to infinity (or minus infinity, depending on c_0) as x becomes large. Call the first case "stable" and the second "unstable". Considering the complementary equation, $\frac{dy}{dx} + \alpha y = 0$, a trial solution for y is $y(x) = e^{rx}$, so that the equation becomes

$0 = re^{rx} + \alpha e^{rx} = (r+\alpha)e^{rx}$ giving $(r+\alpha) = 0$. This is called the characteristic polynomial of the homogeneous equation and has root $r = -\alpha$. Therefore, the dynamic system is stable if the characteristic equation has a negative root. □

EXAMPLE 16.2: The Harrod–Domar model of economic growth postulates that output and capital are related according the formula $K = vY$, where K is the aggregate amount of capital, Y national output and v a coefficient determining the amount of capital required per unit of output produced. Savings, S, is related to output according to the formula, $S = sY$, so that savings is proportional to Y. Investment I equals saving S, $I = S$. Capital changes over time owing to two effects: investment raises capital and depreciation reduces capital, where depreciation is assumed to be at rate δ. Therefore, over an increment of time Δt, the change in capital is

$$\begin{aligned}\Delta K &= (I - \delta K)\Delta t \\ \Delta K &= (sY - \delta v Y)\Delta t,\end{aligned}$$

using $I = sY$ and $K = vY$. Since $K = vY$, $\Delta K = v\Delta Y$ and so,

$$\begin{aligned}v\Delta Y &= (I - \delta K)\Delta t \\ \Delta Y &= \left(\frac{s}{v} - \delta\right) Y \Delta t \\ \frac{\Delta Y}{\Delta t} &= \left(\frac{s}{v} - \delta\right) Y.\end{aligned}$$

In continuous time, this is written: $\frac{dY}{dt} = \left(\frac{s}{v} - \delta\right)Y$, or $\frac{dY}{dt} - \left(\frac{s}{v} - \delta\right)Y = 0$. From the earlier discussion, this has solution:

$$Y(t) = Y(0) \times e^{(\frac{s}{v} - \delta)t}.$$

◊

The following example gives the rate of price adjustment as dependent on the gap between supply and demand.

EXAMPLE 16.3: The linear demand and supply model is given by the equations:

$$\begin{aligned}Q_d &= \alpha - \beta p \\ Q_s &= \gamma + \delta p,\end{aligned}$$

where $\alpha > \gamma > 0$ and $\beta, \delta > 0$. Define a dynamic model by setting $\frac{dp}{dt} = \omega(Q_d - Q_s)$, where $\omega > 0$ is a constant. Thus, $\frac{dp}{dt} = \omega[\alpha - \gamma - (\beta + \delta)p]$ or

$$\frac{dp}{dt} + \omega(\beta + \delta)p = \omega(\alpha - \gamma).$$

Solving,

$$p(t) = c_0 e^{-\omega(\beta+\delta)t} + \frac{\alpha - \gamma}{\beta + \delta}.$$

At $t = 0$, $p(0) = c_0 + \frac{\alpha-\gamma}{\beta+\delta}$, so that $c_0 = p(0) - \frac{\alpha-\gamma}{\beta+\delta}$, and substituting this gives,

$$p(t) = \left[p(0) - \frac{\alpha - \gamma}{\beta + \delta}\right] e^{-\omega(\beta+\delta)t} + \frac{\alpha - \gamma}{\beta + \delta}. \qquad \diamond$$

The next example, taken from a physics application, illustrates the calculation of terminal velocity – where a falling object reaches a constant speed when the resistance force from the air equals the force of gravity.

EXAMPLE 16.4: According to Newton's second law, $F = ma$, force equals mass times acceleration. If s denotes speed or velocity, $\frac{ds}{dt} = a(t)$, acceleration is equal to change in speed. Thus $F = m\frac{ds}{dt}$. Consider an object falling from the sky. In a vacuum, gravity creates force proportional to mass: mg where g is a gravity parameter measured in terms of acceleration (g is roughly 9.8 m/s^2). So, for an object falling in a vacuum, $mg = m\frac{ds}{dt}$. However, in the air, an object with mass encounters resistance proportional to speed, tending to offset gravity. Suppose the proportionality factor is k, so that the acceleration formula becomes $m\frac{ds}{dt} = mg - ks(t)$. Or, $\frac{ds}{dt} + \frac{k}{m}s(t) = g$. Terminal velocity $\left(\frac{ds}{dt} = 0\right)$ occurs when speed satisfies $mg - ks(t) = 0$. A natural question to ask is: how many seconds does it take for the speed of the object to approach terminal velocity?

Consider the differential equation describing a falling object: $\frac{ds}{dt} + \frac{k}{m}s(t) = g$. Suppose that at time $t_0 = 0$, the object begins falling so that, initially, the speed is 0: $s_0 = s(t_0) = 0$. According to the formula:

$$s(t) = \left(s_0 - \frac{gm}{k}\right)e^{-kt} + \frac{gm}{k} = \frac{gm}{k}(1 - e^{-kt}).$$

Thus, the speed converges to $s(\infty) = \frac{gm}{k}$, so that the speed at time t as a fraction of terminal velocity is $f = \frac{s(t)}{s(\infty)} = (1 - e^{-kt})$. Therefore, $e^{-kt} = (1 - f)$

and $-kt = \ln(1-f)$ so that $t = -\frac{1}{k}\ln(1-f)$. For a skydiver in an outstretched position, terminal velocity is in the approximate range of 120–130 miles per hour. ◊

16.2.2 Variable coefficients

A general first-order (linear) differential equation has the form:

$$a_1(x)\frac{dy}{dx} + a_0(x)y = b(x), \tag{16.7}$$

where $a_1(x) \neq 0$, $\forall x$. Equation 16.7 is called the *inhomogeneous equation*. The *homogeneous equation* is obtained when $b(x) = 0$, $\forall x$:

$$a_1(x)\frac{dy}{dx} + a_0(x)y = 0. \tag{16.8}$$

To solve Equation 16.7 write it in the form:

$$\frac{dy}{dx} + \frac{a_0(x)}{a_1(x)}y = \frac{b(x)}{a_1(x)},$$

or, with $\alpha(x) = \frac{a_0(x)}{a_1(x)}$, and $\beta(x) = \frac{b(x)}{a_1(x)}$,

$$\frac{dy}{dx} + \alpha(x)y = \beta(x). \tag{16.9}$$

The corresponding homogeneous equation is:

$$\frac{dy}{dx} + \alpha(x)y = 0.$$

As noted earlier, a solution to Equation 16.9, y^p, is called a *particular solution*. The family of all particular solutions is called the *general solution*. If y^c is a solution of the homogeneous equation (the complementary solution) and y^p is a particular solution, then $y^p + y^c$ is also a particular solution. Note that if y^c is a solution to the homogeneous equation, then so is ky^c for any constant k (since $\frac{dy^c}{dx} + \alpha(x)y^c = 0$ implies $k\frac{dy^c}{dx} + \alpha(x)ky^c = 0$). Therefore, if there is a solution, there will be an infinite number of solutions, requiring some initial condition to provide a unique solution.

Considering the homogeneous component, $\frac{dy}{dx} + \alpha(x)y = 0$, rewrite as:

$$\frac{1}{y}\frac{dy}{dx} = -\alpha(x).$$

Integrating, $\ln y(x) - \ln y(x_0) = -\int_{x_0}^{x} \alpha(\tilde{x})d\tilde{x}$, so that

$$y(x) = y(x_0)e^{-\int_{x_0}^{x} \alpha(\tilde{x})d\tilde{x}} \tag{16.10}$$

is a solution of the homogeneous equation. We can check this by differentiating:

$$\begin{aligned}\frac{dy(x)}{dx} &= y(x_0)[-\alpha(x)]e^{-\int_{x_0}^{x} \alpha(\tilde{x})d\tilde{x}} \\ &= [-\alpha(x)]\left\{y(x_0)e^{-\int_{x_0}^{x} \alpha(\tilde{x})d\tilde{x}}\right\} \\ &= [-\alpha(x)]y(x).\end{aligned}$$

Solving the general (inhomogeneous) case, given in Equation 16.9, to obtain a particular solution makes use of integrating factors, described next.

16.2.2.1 Integrating factors

Returning to the general equation, Equation 16.9,

$$\frac{dy(x)}{dx} + \alpha(x)y(x) = \beta(x),$$

observe that the left side is similar in form to $q(x)y'(x) + q'(x)y(x)$, which is the derivative of $q(x)y(x)$. If Equation 16.9 is expressed in this way, recovery of the function $y(x)$ is simplified. To carry out this procedure, multiply the equation by some function $q(x)$ (called an integrating factor) to give:

$$q(x)\frac{dy(x)}{dx} + q(x)\alpha(x)y(x) = q(x)\beta(x).$$

If $q'(x) = q(x)\alpha(x)$, then the expression is

$$q(x)\frac{dy(x)}{dx} + q'(x)y(x) = \frac{d}{dx}\{q(x)y(x)\} = q(x)\beta(x). \tag{16.11}$$

In this case, integrating both sides of Equation 16.11 gives:

$$\int_{x_0}^{x} \frac{d}{dx}\{q(x)y(x)\} = q(x)y(x)\,|_{x_0}^{x} = q(x)y(x) - q(x_0)y(x_0) = \int_{x_0}^{x} q(\tilde{x})\beta(\tilde{x})d\tilde{x}, \tag{16.12}$$

or

$$q(x)y(x) = q(x_0)y(x_0) + \int_{x_0}^{x} q(\tilde{x})\beta(\tilde{x})d\tilde{x}, \tag{16.13}$$

so that a solution for $y(x)$ is given:

$$y(x) = \frac{1}{q(x)}\left[q(x_0)y(x_0) + \int_{x_0}^{x} q(\tilde{x})\beta(\tilde{x})d\tilde{x}\right]. \tag{16.14}$$

What is q? Since $q'(x) = q(x)\alpha(x)$, $\frac{q'(x)}{q(x)} = \alpha(x)$ so that $\ln q(x) |_{x_0}^{x} = \int_{x_0}^{x} \alpha(\tilde{x})d\tilde{x}$, or $\ln\left\{\frac{q(x)}{q(x_0)}\right\} = \int_{x_0}^{x} \alpha(\tilde{x})d\tilde{x}$, so that

$$\frac{q(x)}{q(x_0)} = e^{\int_{x_0}^{x} \alpha(\tilde{x})d\tilde{x}} \text{ or } q(x) = q(x_0)e^{\int_{x_0}^{x} \alpha(\tilde{x})d\tilde{x}}.$$

Therefore,

$$\begin{aligned} y(x) &= \frac{1}{q(x)}\left[q(x_0)y(x_0) + \int_{x_0}^{x} q(x')\beta(x')dx'\right] \\ &= \frac{1}{q(x)}\left[q(x_0)y(x_0) + q(x_0)\int_{x_0}^{x} e^{\int_{x_0}^{x'} \alpha(\tilde{x})d\tilde{x}}\beta(x')dx'\right] \\ &= \frac{y(x_0) + \int_{x_0}^{x} e^{\int_{x_0}^{x'} \alpha(\tilde{x})d\tilde{x}}\beta(x')dx'}{e^{\int_{x_0}^{x} \alpha(\tilde{x})d\tilde{x}}}. \end{aligned} \tag{16.15}$$

Notice that when $\beta(x) = 0$ for all x, this expression reduces to Equation 16.10, the solution to the homogeneous equation.

EXAMPLE 16.5: Consider the linear demand and supply model of Example 16.3 with dynamic adjustment modified to $\frac{dp}{dt} = \omega(t)(Q_d - Q_s)$, where ω now depends on time and with $Q_d = \alpha - \beta p$ and $Q_s = \gamma + \delta p$. Dynamic adjustment is now given by:

$$\frac{dp}{dt} + \omega(t)(\beta + \delta)p = \omega(t)(\alpha - \gamma).$$

From Equation 16.15,

$$p(t) = \frac{\left[p(t_0) + \int_{t_0}^{t} e^{\int_{t_0}^{t'} \omega(\tilde{t})(\beta+\delta)d\tilde{t}}\omega(t')(\alpha - \gamma)dt'\right]}{e^{\int_{t_0}^{t} \omega(\tilde{t})(\beta+\delta)d\tilde{t}}}.$$

To simplify this expression, let $t_0 = 0$ and set $\omega(t) = \omega \cdot t$. Then

$$\int_{t_0}^{t'} \omega(\tilde{t})(\beta + \delta)d\tilde{t} = \omega(\beta + \delta)\frac{1}{2}t'^2 = kt'^2,$$

where $k = \omega(\beta+\delta)\frac{1}{2}$. Next,

$$\int_{t_0}^{t} e^{\int_{t_0}^{t'} \omega(\tilde{t})(\beta+\delta)d\tilde{t}} \omega(t')(\alpha-\gamma)dt' = \omega(\alpha-\gamma)\int_{t_0}^{t} e^{kt'^2} t' dt'$$

$$= \omega(\alpha-\gamma)\frac{1}{2}\frac{e^{kt^2}-1}{k}$$

$$= \frac{(\alpha-\gamma)}{\beta+\delta}[e^{kt^2}-1].$$

Thus,

$$p(t) = \frac{p(0) + \frac{(\alpha-\gamma)}{\beta+\delta}\left[e^{kt^2}-1\right]}{e^{kt^2}}$$

$$= p(0)e^{-kt^2} + \frac{(\alpha-\gamma)}{\beta+\delta}[1-e^{-kt^2}].$$

Writing $p_e = \frac{(\alpha-\gamma)}{\beta+\delta}$, the market clearing price $p(t)$ is a weighted average of the initial price $p(0)$ and the market clearing price p_e: $p(t) = p(0)e^{-kt^2} + p_e[1-e^{-kt^2}]$. ◊

EXAMPLE 16.6:

$$\frac{dy}{dx} + axy = x.$$

Here, $\alpha(x) = ax$ so that $\int_{x_0}^{x} \alpha(\tilde{x})d\tilde{x} = \frac{a}{2}[x^2 - x_0^2]$. Thus,

$$e^{\int_{x_0}^{x} \alpha(\tilde{x})d\tilde{x}} = e^{\frac{a}{2}x^2} e^{-\frac{a}{2}x_0^2} = k(x_0)e^{\frac{a}{2}x^2}$$

so that $q(x) = q(x_0)k(x_0)e^{\frac{a}{2}x^2} = c_0 e^{\frac{a}{2}x^2}$. Multiplying both sides of $\frac{dy}{dx} + axy = x$ by $c_0 e^{\frac{a}{2}x^2}$ gives

$$\left[\frac{dy}{dx} + axy\right] e^{\frac{a}{2}x^2} = xe^{\frac{a}{2}x^2}$$

or

$$\frac{d}{dx}\left\{ye^{\frac{a}{2}x^2}\right\} = xe^{\frac{a}{2}x^2}$$

so that

$$\int_{x_0}^{x'} \frac{d}{dx}\left\{ye^{\frac{a}{2}x^2}\right\}dx = \int_{x_0}^{x'} xe^{\frac{a}{2}x^2}dx$$

or

$$\begin{aligned} y(x')e^{\frac{a}{2}x'^2} - y(x_0)e^{\frac{a}{2}x_0{}^2} &= \int_{x_0}^{x'} xe^{\frac{a}{2}x^2}dx \\ &= \frac{1}{a}\left[e^{\frac{1}{2}ax'^2} - e^{\frac{1}{2}ax_0^2}\right] \end{aligned}$$

or

$$y(x')e^{\frac{a}{2}x'^2} = c_0 + \frac{1}{a}e^{\frac{1}{2}ax'^2}.$$

Therefore,

$$y(x') = c_0e^{-\frac{a}{2}x'^2} + \frac{1}{a}.$$

◊

16.3 Some non-linear first-order differential equations

This section considers two well-known non-linear first-order differential equations that have exact solutions: the logistic and Bernoulli differential equations.

16.3.1 The logistic model

The linear model $\frac{dy}{dx} + ay = 0$ has solution $y(x) = y_0e^{-ax}$ so that as x becomes large $y(x)$ tends to 0 or to ∞, depending on whether a is positive or negative. Models of population growth where the habitat is finite cannot be modeled by this equation over long periods of time. The logistic model provides a growth model that asymptotes to a finite limit. Although the differential equation is non-linear, it is easily solved.

Consider

$$\frac{y'}{y} = \alpha\left[1 - \frac{y}{k}\right] = \alpha\left[\frac{k-y}{k}\right] = \frac{\alpha}{k}[k-y], \quad \alpha > 0, k > 0$$

From this expression, when $k-y>0$ or $y<k$, $y'>0$ and the value of y grows. Similarly, when $k-y<0$ the value of y decreases. At $y=k$, $y'=0$, and k is the steady state value for y. The equation may be rearranged as,

$$\frac{1}{y}\frac{k}{k-y}y'=\alpha.$$

Observe that $\frac{1}{y}+\frac{1}{k-y}=\frac{1(k-y)}{y(k-y)}+\frac{1y}{y(k-y)}=\frac{k}{y(k-y)}$. Consequently,

$$\frac{1}{y}\frac{k}{k-y}y'=\frac{1}{y}y'+\frac{1}{k-y}y'=\alpha,$$

and so

$$\int_{x_0}^{x^*}\frac{1}{y(x)}\frac{k}{k-y(x)}y'(x)dx=\int_{x_0}^{x^*}\frac{1}{y(x)}y'(x)dx+\int_{x_0}^{x^*}\frac{1}{k-y(x)}y'(x)dx=\alpha[x^*-x_0],$$

or

$$\ln y(x^*)-\ln(x_0)-[\ln(k-y(x^*))-\ln(k-y(x_0))]=\alpha[x^*-x_0]$$

$$\ln\frac{y(x^*)}{k-y(x^*)}-\ln\frac{y(x_0)}{k-y(x_0)}=\alpha[x^*-x_0]$$

$$\ln\frac{y(x^*)}{k-y(x^*)}=\ln\frac{y(x_0)}{k-y(x_0)}+\alpha[x^*-x_0]$$

$$\ln\frac{y(x^*)}{k-y(x^*)}=c_0+\alpha x^*$$

$$\frac{y(x^*)}{k-y(x^*)}=e^{c_0+\alpha x^*}$$

$$=ce^{\alpha x^*}.$$

Therefore,

$$y(x^*)=kce^{\alpha x^*}-y(x^*)ce^{\alpha x^*},$$

so that

$$y(x^*)=\frac{kce^{\alpha x^*}}{1+ce^{\alpha x^*}}=\frac{k}{1+c'e^{-\alpha x^*}},\quad c'=\frac{1}{c}.$$

The value of c' is determined by the initial condition $y(x_0)=y_0$: $y_0=\frac{k}{1+c'e^{-\alpha x_0}}$ or $1+c'e^{-\alpha x_0}=\frac{k}{y_0}$ or $c'=\left(\frac{k}{y_0}-1\right)e^{\alpha x_0}$. The function $y(x)=\frac{k}{1+ce^{-\alpha x}}$ is called the *logistic* function (Figure 16.1).

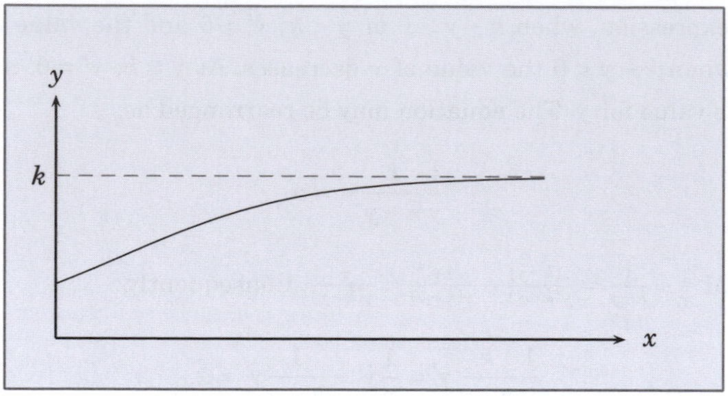

Figure 16.1: The logistic curve

16.3.2 The Bernoulli equation

One particular non-linear differential equation that has appeared in economic applications is the Bernoulli equation. This is a differential equation of the form:

$$\frac{dy}{dx} + \alpha(x)y = \beta(x)y^{\alpha}, \quad \alpha \neq 0, 1.$$

Dividing by y^{α},

$$\frac{1}{y^{\alpha}}\frac{dy}{dx} + \alpha(x)y^{1-\alpha} = \beta(x).$$

Define $w = y^{1-\alpha}$, so that $w' = (1-\alpha)y^{-\alpha}y'$ or $\frac{dw}{dx} = (1-\alpha)\frac{1}{y^{\alpha}}\frac{dy}{dx}$. With this change of variables:

$$\frac{1}{(1-\alpha)}\frac{dw}{dx} + \alpha(x)w = \beta(x).$$

This is a first-order linear differential equation, which may be solved for $w(x)$ as described previously. The function $y(x)$ is then obtained as $y(x) = w(x)^{\frac{1}{1-\alpha}}$.

EXAMPLE 16.7: The Solow growth model provides a differential equation governing the long-run dynamics of per capita capital. A constant returns to scale production function $Y = f(K, L)$ is assumed, so that $y = \frac{Y}{L} = \frac{1}{L}f(K, L) = f(\frac{K}{L}, \frac{L}{L}) = f(k)$, where $k = \frac{K}{L}$. For example, $f(K, L) = K^{\alpha}L^{1-\alpha}$ gives $f(k) = k^{\alpha}$, with $0 < \alpha < 1$. Savings, S, is a fixed proportion, s, of income, Y, and investment I is equal to saving: $I = S = sY$. Capital growth per unit of time is equal to investment less depreciation: $I - \delta K$: $\Delta K = (sY - \delta K)\Delta t$ or, in continuous time, $\frac{dK}{dt} = (sY - \delta K)$. Therefore, $\frac{1}{K}\frac{dK}{dt} = \left(s\frac{Y}{K} - \delta\right)$. Since $\frac{Y}{K} = \frac{Y/L}{K/L} = \frac{y}{k}$,

$$\frac{1}{K}\frac{dK}{dt} = \left(s\frac{y}{k} - \delta\right).$$

To convert to per capita terms, observe that

$$\frac{dk}{dt} = \frac{L\frac{dK}{dt} - K\frac{dL}{dt}}{L^2} = \frac{1}{L}\frac{dK}{dt} - \frac{1}{L}k\frac{dL}{dt}.$$

So, with $\frac{1}{k} = \frac{L}{K}$,

$$\frac{1}{k}\frac{dk}{dt} = \frac{1}{K}\frac{dK}{dt} - \frac{1}{L}\frac{dL}{dt}.$$

Let $n = \frac{1}{L}\frac{dL}{dt}$ be the continuous time growth rate of the labor supply, so that

$$\frac{1}{k}\frac{dk}{dt} = \frac{1}{K}\frac{dK}{dt} - n.$$

Collecting terms,

$$\frac{1}{K}\frac{dK}{dt} = \frac{1}{k}\frac{dk}{dt} + n = \left(s\frac{y}{k} - \delta\right).$$

Therefore,

$$\frac{dk}{dt} = sy - (\delta + n)k = sf(k) - (\delta + n)k. \tag{16.16}$$

From this expression, there are two points where k is constant: at 0, and at $\hat{k}$, defined by $sf(\hat{k}) - (\delta + n)\hat{k} = 0$.

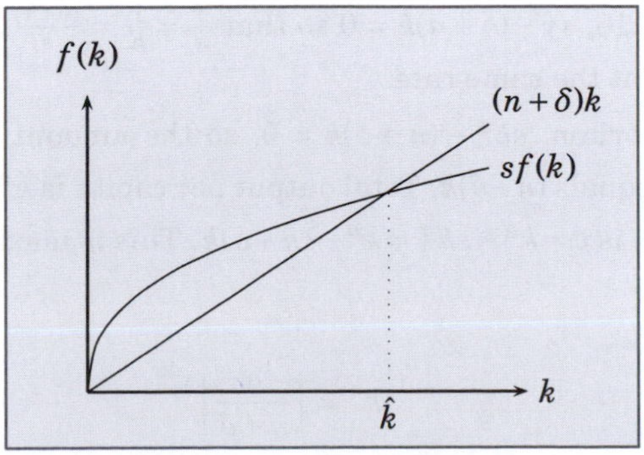

With $f(k) = k^\alpha$, the differential equation becomes

$$\frac{dk}{dt} = sk^\alpha - (\delta + n)k$$

or $\frac{dk}{dt} + (\delta + n)k = sk^\alpha$, which is a simple Bernoulli equation. Rearrange to get:

$$\frac{1}{k^\alpha}\frac{dk}{dt} + (\delta + n)k^{1-\alpha} = s$$

and define $w = k^{1-\alpha}$, so that

$$\frac{dw}{dt} = (1-\alpha)k^{-\alpha}\frac{dk}{dt} = (1-\alpha)\frac{1}{k^\alpha}\frac{dk}{dt}, \quad \text{or} \quad \frac{1}{1-\alpha}\frac{dw}{dt} = \frac{1}{k^\alpha}\frac{dk}{dt}.$$

With this change of variable, the differential equation is:

$$\frac{1}{1-\alpha}\frac{dw}{dt} + (\delta + n)w = s \quad \text{or} \quad \frac{dw}{dt} + (1-\alpha)(\delta + n)w = (1-\alpha)s.$$

The solution to a differential equation of the form $\frac{dw}{dt} + aw = b$ has the form:

$$w(t) = w(0)e^{-at} + \frac{b}{a} = w(0)e^{-[(1-\alpha)(\delta+n)]t} + \frac{s}{\delta + n}.$$

Since $k = w^{\frac{1}{1-\alpha}}$

$$k(t) = \left[k(0)^{1-\alpha}e^{-[(1-\alpha)(\delta+n)]t} + \frac{s}{\delta + n}\right]^{\frac{1}{1-\alpha}}.$$

At $t \to \infty$, $k(t) \to \left[\frac{s}{\delta+n}\right]^{\frac{1}{1-\alpha}}$. From Equation 16.16, $\frac{dk}{dt} = 0$ when $sk^\alpha - (\delta+n)k = 0$ or $k = \left[\frac{s}{\delta+n}\right]^{\frac{1}{1-\alpha}}$, so the solution gives the same steady state. (Note that $k = 0$ is also a solution but one in which there is no output.) Also, from Equation 16.16, $sy - (\delta + n)k = 0$ so that $\frac{y}{k} = \frac{Y}{K} = \frac{(\delta+n)}{s}$ is constant and so Y and K grow at the same rate.

In equilibrium, $sk^\alpha - (n + \delta)k = 0$, so the amount of output per capita saved, sk^α, equals $(n+\delta)k$. Total output per capita is k^α and so consumption per capita, c, is $c = k^\alpha - sk^\alpha = k^\alpha - (n+\delta)k$. This is maximized when $\alpha k^{\alpha-1} - (n + \delta) = 0$, or

$$k^* = \left[\frac{\alpha}{n+\delta}\right]^{\frac{1}{1-\alpha}}.$$

The solution is called the *golden rule* level of capital, the level that maximizes per capita consumption. Comparing this with the long-run steady state level of $k = \left[\frac{s}{n+\delta}\right]^{\frac{1}{1-\alpha}}$; these are equal when $s = \alpha$. The savings rate that maximizes the long-run per capita consumption is called the *golden rule* savings rate. ◊

16.4 First-order differential equations: existence of solutions

Consider the (differential) equation

$$\frac{dy}{dx} = f(x,y),$$

associating the change in y to the current values of x and y. When $f(x,y) = \beta(x) - \alpha(x)y$, this gives the first-order differential equation $\frac{dy}{dx} + \alpha(x)y = \beta(x)$. A solution to the equation on the interval (a,b) is a function $y(x)$ satisfying $\frac{dy(x)}{dx} = f(x,y(x))$ for $x \in (a,b)$. Two natural questions arise. Does the differential equation have a solution? And is the solution unique?

Definition 16.1: *Let Q be a region in $\mathcal{R}^2$. Say that the function $f(x,y)$ satisfies a uniform Lipschitz condition in Q if there is some constant, K, such that for any pairs (x,y_1), $(x,y_2) \in Q$:*

$$|f(x,y_1) - f(x,y_2)| \leq K|y_1 - y_2|.$$

With this definition, a key theorem on the existence of a solution to a differential equation is the following.

Theorem 16.1: *Given the differential equation:*

$$\frac{dy}{dx} = f(x,y),$$

with initial condition $y(x_0) = y_0$, where f is a continuous function on $Q = [x_0 - a, x_0 + a] \times [y_0 - b, y_0 + b]$. Then for some $a' > 0$, there is a function y defined on $[x_0 - a', x_0 + a']$ with $y(x_0) = y_0$ and $y'(x) = f(x,y(x)), x \in [x_0 - a', x_0 + a']$. If f satisfies a uniform Lipschitz condition on Q, the solution is unique.

In the case of a linear differential equation, $f(x,y) = \beta(x) - \alpha(x)y$, with $\alpha(x)$ continuous,

$$f(x,y_1) - f(x,y_2) = \mid \alpha(x)y_1 - \alpha(x)y_2 \mid \leq \mid \alpha(x) \mid \mid y_1 - y_2 \mid \leq K \mid y_1 - y_2 \mid,$$

where $K = \max_{x \in [x_0 - a', x_0 + a']} \mid \alpha(x) \mid$, and so the uniform Lipschitz condition is satisfied.

16.5 Second- and higher-order differential equations

An equation is an nth-order linear differential equation if it has the form:

$$a_n(x)\frac{d^n y}{dx^n} + a_{n-1}(x)\frac{d^{n-1}y}{dx^{n-1}} + \cdots + a_1(x)\frac{dy}{dx} + a_0(x)y = b(x).$$

A second-order differential equation has the form:

$$a_2(x)\frac{d^2 y}{dx^2} + a_1(x)\frac{dy}{dx} + a_0(x)y = b(x).$$

In general, this is difficult to solve for the function y. The remainder of the discussion focuses on the second-order differential equation with constant coefficients.

16.5.1 Constant coefficients

When the coefficients are constant, the second-order linear differential equation has the form:

$$a_2\frac{d^2 y}{dx^2} + a_1\frac{dy}{dx} + a_0 y = b.$$

This equation has a particular solution $\bar{y} = \frac{b}{a_0}$. Consider the homogeneous equation,

$$a_2\frac{d^2 y}{dx^2} + a_1\frac{dy}{dx} + a_0 y = 0.$$

Observe that if y_1 and y_2 are solutions to the homogeneous equation, then so is $y^* = c_1 y_1 + c_2 y_2$ for any c_1 and c_2.

A candidate solution $y(x) = ke^{rx}$ has the property that $y'(x) = rke^{rx} = ry(x)$ and $y''(x) = r^2ke^{rx} = r^2y(x)$, so that if this function satisfies the homogeneous equation then

$$a_2r^2y(x) + a_1ry(x) + a_0y(x) = (a_2r^2 + a_1r^1 + a_0)y(x) = 0.$$

Since $y(x) = ke^{rx} \neq 0$, this requires $a_2r^2 + a_1r^1 + a_0 = 0$. This quadratic expression has two roots:

$$r_1, r_2 = \frac{-a_1 \pm \sqrt{a_1^2 - 4a_2a_0}}{2a_2}.$$

The solution to the differential equation depends on the roots, which may be real and distinct, real and repeated or complex.

$$y(x) = \begin{cases} c_1e^{r_1x} + c_2e^{r_2x} + \frac{b}{a_0}, & \text{roots real and distinct} \\ c_1e^{rx} + c_2xe^{rx} + \frac{b}{a_0}, & \text{roots real and equal} \\ c_1e^{\alpha x}\cos(\beta x) + c_2e^{\alpha x}\sin(\beta x) + \frac{b}{a_0}, & \text{roots complex.} \end{cases}$$

The first two cases are easy to confirm, the third involves some computation. These issues are developed in Section 16.5.2.

REMARK 16.2: Considering stability, in the first case it is necessary for r_1 and r_2 to be negative, and in the second case for r to be negative. In the complex case, the roots are repeated and α is the real part of both roots, so in that case, it is necessary for α to be negative. In sum, a sufficient condition for stability of the solution is that the real parts of the roots be negative. □

16.5.2 Derivation of solution

Regarding these roots, there are three distinct regions of the parameter space:

1. Roots real and distinct: $a_1^2 - 4a_2a_0 > 0$. The general solution of the homogeneous equation has the form: $y(x) = c_1e^{r_1x} + c_2e^{r_2x}$.

2. Roots real and equal: $a_1^2 - 4a_2a_0 = 0$, so that $r_1 = r_2 = r$. The general solution of the homogeneous equation has the form: $y(x) = c_1e^{rx} + c_2xe^{rx}$.

3. Roots complex: $a_1^2 - 4a_2a_0 < 0$, so that $r_1 = \alpha + \beta i$ and $r_2 = \alpha - \beta i$, where $\alpha = -\frac{a_1}{2a_2}$ and $\beta = \left(\sqrt{4a_2a_0 - a_1^2}\right)/2a_2$. The general solution of the homogeneous equation has the form: $y(x) = c_1e^{r_1x} + c_2e^{r_2x}$.

According to Euler's formula, for any real number x, $e^{ix} = \cos(x) + i\sin(x)$ (Figure 16.2).

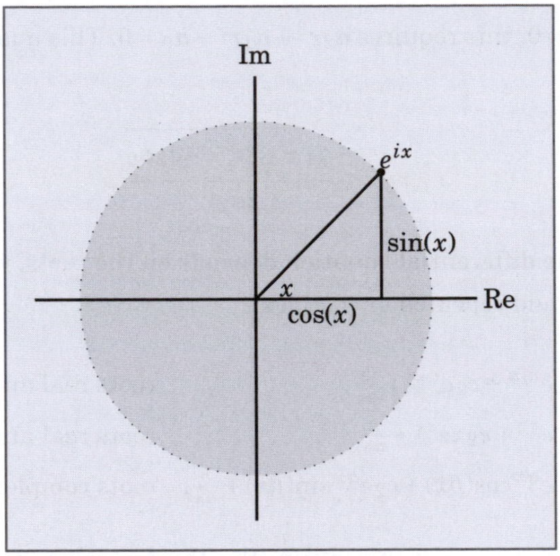

Figure 16.2: Euler's formula

Thus,

$$e^{r_1 x} = e^{\alpha x + i\beta x} = e^{\alpha x}e^{i\beta x} = e^{\alpha x}[\cos(\beta x) + i\sin(\beta x)]$$

and

$$e^{r_2 x} = e^{\alpha x - i\beta x} = e^{\alpha x}e^{i(-\beta x)} = e^{\alpha x}[\cos(-\beta x) + i\sin(-\beta x)] = e^{\alpha x}[\cos(\beta x) - i\sin(\beta x)].$$

Therefore,

$$\begin{aligned} y(x) &= c_1 e^{r_1 x} + c_2 e^{r_2 x} \\ &= c_1 e^{\alpha x}[\cos(\beta x) + i\sin(\beta x)] + c_2 e^{\alpha x}[\cos(\beta x) - i\sin(\beta x)] \\ &= e^{\alpha x}[(c_1 + c_2)\cos(\beta x) + i(c_1 - c_2)\sin(\beta x)]. \end{aligned}$$

Since y is a real-valued function, setting $c_1 - c_2 = 0$ and letting $c_1 + c_2 = c$, gives

$$y_a(x) = ce^{\alpha x}\cos(\beta x)$$

Similarly, letting $c_1 = -i\frac{c}{2}$ and $c_2 = i\frac{c}{2}$ gives $c_1 + c_2 = 0$ and $i(c_1 - c_2) = i[-i - i]\frac{c}{2} = -2i^2\frac{c}{2} = c$. This gives

$$y_b(x) = ce^{\alpha x}\sin(\beta x)$$

as a solution. Since any combination of solutions is a solution, solutions of the homogeneous equation have the form:

$$\begin{aligned} y(x) &= c_1 e^{\alpha x}\cos(\beta x) + c_2 e^{\alpha x}\sin(\beta x) \\ &= e^{\alpha x}[c_1\cos(\beta x) + c_2\sin(\beta x)], \end{aligned}$$

where c_1 and c_2 are real numbers. This may be rearranged. Consider a right-angled triangle with adjacent c_1, opposite c_2 and hypotenuse $h = \sqrt{c_1^2 + c_2^2}$. The angle, θ, formed by the hypotenuse and adjacent has cosine, $\cos(\theta) = \frac{c_1}{h}$ and sine, $\sin(\theta) = \frac{c_2}{h}$. Thus, $c_1 = h\cos(\theta)$ and $c_2 = h\sin(\theta)$. Then

$$\begin{aligned} y(x) &= c_1 e^{\alpha x}\cos(\beta x) + c_2 e^{\alpha x}\sin(\beta x) \\ &= he^{\alpha x}[\cos(\theta)\cos(\beta x) + \sin(\theta)\sin(\beta x)] \\ &= he^{\alpha x}\cos(\beta x - \theta), \end{aligned}$$

using the formula $\cos(a)\cos(b) + \sin(a)\sin(b) = \cos(a - b)$.

Summarizing,

$$y(x) = \begin{cases} c_1 e^{r_1 x} + c_2 e^{r_2 x} + \frac{b}{a_0}, & \text{roots real and distinct} \\ c_1 e^{rx} + c_2 x e^{rx} + \frac{b}{a_0}, & \text{roots real and equal} \\ c_1 e^{\alpha x}\cos(\beta x) + c_2 e^{\alpha x}\sin(\beta x) + \frac{b}{a_0}, & \text{roots complex.} \end{cases}$$

EXAMPLE 16.8: According to Hooke's law, the force acting on an object is proportional to its displacement from equilibrium. If the equilibrium value of y is 0, then when the displacement is $y < 0$ the force acts to increase y, and when $y > 0$ the force acts to decrease y. With a proportionality factor k, this may be written $F = -ky$, with force proportional to displacement. According to Newton's second law, force equals mass times acceleration: $F = ma = m\frac{d^2y}{dt^2}$. Hence

$$m\frac{d^2y}{dt} + ky = 0.$$

Rearranging, $\frac{d^2y}{dt} + \frac{k}{m}y = 0$. The associated roots are

$$r_1, r_2 = \frac{\pm\sqrt{-4(1)(\frac{k}{m})}}{2} = \pm i\sqrt{\frac{k}{m}}.$$

Thus, the roots are complex and according to the previous discussion,

$$y(t) = h\cos\left(\sqrt{\frac{k}{m}}\cdot t - \theta\right)$$

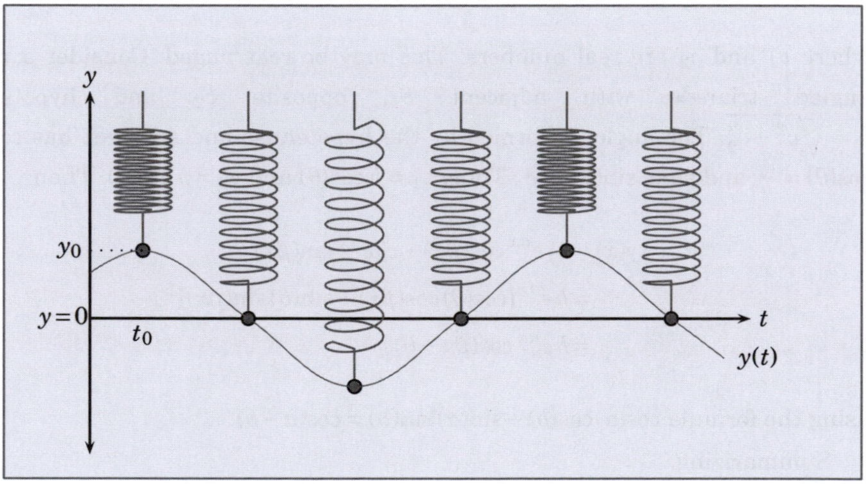

At t_0, observe that $y(t_0) = y_0$ and $y'(t_0) = 0$. At t_0, $\cos\left(\sqrt{\frac{k}{m}}\cdot t - \theta\right)$ is maximized so that $\sqrt{\frac{k}{m}}\cdot t_0 - \theta = 0$ or $\sqrt{\frac{k}{m}}\cdot t_0 = \theta$, and so $\sqrt{\frac{k}{m}}\cdot t - \theta = \sqrt{\frac{k}{m}}\cdot(t - t_0)$. Since $\cos(0) = 1$, $y_0 = h\cos\left(\sqrt{\frac{k}{m}}\cdot(t_0 - t_0)\right) = h\cos(0) = h$. So, the differential equation is given by:

$$y(t) = y_0\cos\left(\sqrt{\frac{k}{m}}\cdot(t - t_0)\right).$$

This example may be generalized when the spring is damped: $m\frac{d^2y}{dt^2} + q\frac{dy}{dt} + ky = 0$ (with real roots if and only if $q^2 - 4mk > 0$). ◊

Finally, the following result confirms the existence and uniqueness of a solution in the linear case.

Theorem 16.2: *The* nth*-order linear differential equation:*

$$a_n(x)\frac{d^n y}{dx^n} + a_{n-1}(x)\frac{d^{n-1}y}{dx^{n-1}} + \cdots + a_1(x)\frac{dy}{dx} + a_0(x)y = b(x)$$

with initial conditions $y(x_0) = y_0$, $y'(x_0) = y_1$, $y''(x_0) = y_2, \ldots, y^{(n-1)}(x_0) = y_{n-1}$ *has a unique solution.*

16.6 Systems of differential equations

Let $\boldsymbol{A}$ be an $n \times n$ matrix and $\boldsymbol{\beta}$ an $n \times 1$ vector of constants. A system of first-order linear differential equations is given by:

$$\frac{d\boldsymbol{y}}{dx}(x) + \boldsymbol{A}\boldsymbol{y}(x) = \boldsymbol{\beta}. \tag{16.17}$$

To simplify the discussion, assume that $\boldsymbol{A}$ has n linearly independent eigenvectors $\boldsymbol{v}_i$ and corresponding eigenvalues λ_i, then for each i, $\boldsymbol{A}\boldsymbol{v}_i = \lambda_i \boldsymbol{v}_i$. Let $\boldsymbol{\Lambda}$ be the diagonal matrix with λ_i in the ith row and column and let $\boldsymbol{P}$ be the matrix with $\boldsymbol{v}_i$ in the ith column. Then this collection of n equations may be written as $\boldsymbol{AP} = \boldsymbol{P\Lambda}$. Therefore, $\boldsymbol{P}^{-1}\boldsymbol{AP} = \boldsymbol{P}^{-1}\boldsymbol{P\Lambda} = \boldsymbol{\Lambda}$. Considering Equation 16.17, define $\boldsymbol{z} = \boldsymbol{P}^{-1}\boldsymbol{y}$, so that $\boldsymbol{y} = \boldsymbol{Pz}$ and substitute into Equation 16.17 to obtain:

$$\boldsymbol{P}\frac{d\boldsymbol{z}}{dx}(x) + \boldsymbol{AP}\boldsymbol{z}(x) = \boldsymbol{\beta}.$$

Multiplying by $\boldsymbol{P}^{-1}$:

$$\frac{d\boldsymbol{z}}{dx}(x) + \boldsymbol{P}^{-1}\boldsymbol{AP}\boldsymbol{z}(x) = \boldsymbol{P}^{-1}\boldsymbol{\beta}$$

or

$$\frac{d\boldsymbol{z}}{dx}(x) + \boldsymbol{\Lambda}\boldsymbol{z}(x) = \boldsymbol{P}^{-1}\boldsymbol{\beta} = \bar{\boldsymbol{\beta}}.$$

Because $\boldsymbol{\Lambda}$ is diagonal, this system has the form:

$$\frac{dz_i}{dx}(x) + \lambda_i z_i(x) = \bar{\boldsymbol{\beta}}_i,\ i = 1, \ldots, n, \tag{16.18}$$

a collection of n first-order linear differential equations with solutions $z_i(x) = c_i e^{-\lambda_i x} + \delta_i$. Given initial values $y_i(x_0)$, values $z_i(x_0)$ are determined according to $\boldsymbol{z}(0) = \boldsymbol{P}^{-1}\boldsymbol{y}(0)$, and this determines c_i and δ_i, as in Equations 16.1 and 16.5.

Since $\boldsymbol{y} = \boldsymbol{P}\boldsymbol{z}$,

$$y_i(x) = \sum_{j=1}^{n} P_{ij} z_j = \sum_{j=1}^{n} P_{ij}[c_j e^{-\lambda_j x} + \delta_j] = \sum_{j=1}^{n} \gamma_{ij} e^{-\lambda_j x} + \bar{\delta}_j, \tag{16.19}$$

where $\gamma_{ij} = P_{ij} c_j$ and $\bar{\delta}_i = \sum_{j=1}^{n} P_{ij}\delta_j$. Note that the roots λ_i appearing in Equation 16.18 may be complex. Considering the characteristic polynomial $\boldsymbol{p}_A(\lambda) = |\boldsymbol{A} - \lambda \boldsymbol{I}|$, this is an nth-order polynomial with real coefficients. Such a polynomial has exactly n roots, including multiplicities (repeated roots), and because the polynomial coefficients are real, complex roots appear in conjugate pairs. Because $\boldsymbol{P}^{-1}\boldsymbol{A}\boldsymbol{P} = \boldsymbol{\Lambda}$, $|\boldsymbol{P}^{-1}\boldsymbol{A}\boldsymbol{P}| = |\boldsymbol{A}| = |\boldsymbol{\Lambda}|$ and $\mathrm{tr}(\boldsymbol{P}^{-1}\boldsymbol{A}\boldsymbol{P}) = \mathrm{tr}(\boldsymbol{A}\boldsymbol{P}\boldsymbol{P}^{-1}) = \mathrm{tr}(\boldsymbol{A}) = \mathrm{tr}(\boldsymbol{\Lambda})$. That is, $|\boldsymbol{A}| = \times_{i=1}^{n} \lambda_i$ and $\sum_{i=1}^{n} a_{ii} = \sum_{i=1}^{n} \lambda_i$.

Writing $\boldsymbol{P} = (\boldsymbol{v}_1, \boldsymbol{v}_2, \ldots, \boldsymbol{v}_n)$ where $\boldsymbol{v}_i$ is the ith eigenvector, Equation 16.19 may be written as

$$y(x) = \sum_{j=1}^{n} \boldsymbol{v}_j c_j e^{-\lambda_j x} + \sum_{j=1}^{n} \boldsymbol{v}_j \delta_j.$$

The homogeneous equation $\boldsymbol{y}' + \boldsymbol{A}\boldsymbol{y} = 0$ solves to give $y_c(x) = \sum_{j=1}^{n} v_j c_j e^{-\lambda_j x}$ as the complementary solution of the homogeneous equation and a particular solution is given as the solution to $\boldsymbol{A}\boldsymbol{y} = \boldsymbol{\beta}$ or $\boldsymbol{y}_p = \boldsymbol{A}^{-1}\boldsymbol{\beta}$. A general solution of the homogeneous equation is then:

$$\boldsymbol{y}(x) = \sum_{j=1}^{n} \boldsymbol{v}_j c_j e^{-\lambda_j x} + \boldsymbol{A}^{-1}\boldsymbol{\beta}.$$

EXAMPLE 16.9: Consider the two-good market given by:

$$\begin{aligned} \text{Market 1}: \quad D_1 = S_1 \quad &\text{or} \quad a_0 + a_1 p_1 + a_2 p_2 = b_0 + b_1 p_1 + b_2 p_2 \\ \text{Market 2}: \quad D_2 = S_2 \quad &\text{or} \quad \alpha_0 + \alpha_1 p_1 + \alpha_2 p_2 = \beta_0 + \beta_1 p_1 + \beta_2 p_2. \end{aligned}$$

Suppose that price dynamics are linear, with price adjusting to eliminate excess demand over time: $\frac{dp_i}{dt} = (D_i(t) - S_i(t))$. Thus:

$$\begin{aligned} \frac{dp_1}{dt} &= (a_0 - b_0) + (a_1 - b_1)p_1 + (a_2 - b_2)p_2 \\ \frac{dp_2}{dt} &= (\alpha_0 - \beta_0) + (\alpha_1 - \beta_1)p_1 + (\alpha_2 - \beta_2)p_2. \end{aligned}$$

Letting

$$\boldsymbol{A} = \begin{pmatrix} (a_1 - b_1) & (a_2 - b_2) \\ (\alpha_1 - \beta_1) & (\alpha_2 - \beta_2) \end{pmatrix}, \quad \boldsymbol{\beta} = \begin{pmatrix} (a_0 - b_0) \\ (\alpha_0 - \beta_0) \end{pmatrix}, \quad \boldsymbol{p} = \begin{pmatrix} p_1 \\ p_2 \end{pmatrix},$$

the system may be written:

$$\frac{d\boldsymbol{p}}{dt} = \boldsymbol{A}\boldsymbol{p} + \boldsymbol{\beta} \quad \text{or} \quad \frac{d\boldsymbol{p}}{dt} - \boldsymbol{A}\boldsymbol{p} = \boldsymbol{\beta}.$$

Suppose that $(a_1, b_1) = \left(-\frac{5}{20}, \frac{6}{20}\right)$, $(a_2, b_2) = \left(\frac{3}{20}, \frac{0}{20}\right)$, $(\alpha_1, \beta_1) = \left(\frac{1}{20}, \frac{3}{20}\right)$, and $(\alpha_2, \beta_2) = \left(\frac{2}{20}, \frac{6}{20}\right)$. Then

$$\boldsymbol{A} = \begin{pmatrix} -\frac{11}{20} & \frac{3}{20} \\ -\frac{2}{20} & -\frac{4}{20} \end{pmatrix}.$$

Noting that the characteristic equation is $| \boldsymbol{A} - \lambda \boldsymbol{I} | = \lambda^2 + \frac{3}{4}\lambda + \frac{1}{8}$, the matrix $\boldsymbol{A}$ has eigenvalues: $\lambda_1 = -\frac{1}{4}$, $\lambda_2 = -\frac{1}{2}$, with corresponding eigenvectors, $\boldsymbol{v}_1' = (1,2)$ and $\boldsymbol{v}_2' = (3,1)$. Write $\boldsymbol{V} = (\boldsymbol{v}_1, \boldsymbol{v}_2)$, a matrix whose columns are the eigenvectors of $\boldsymbol{A}$ and $\boldsymbol{\Lambda}$ for the diagonal matrix with λ_1 and λ_2 on the main diagonal.

Considering the system $\frac{d\boldsymbol{p}}{dt} - \boldsymbol{A}\boldsymbol{p} = \boldsymbol{\beta}$, multiplying by $\boldsymbol{V}^{-1}$ and setting $\boldsymbol{z} = \boldsymbol{V}^{-1}\boldsymbol{p}$, so that $\boldsymbol{p} = \boldsymbol{V}\boldsymbol{z}$ gives:

$$\frac{d\boldsymbol{z}}{dt} - \boldsymbol{V}^{-1}\boldsymbol{A}\boldsymbol{V}\boldsymbol{z} = \frac{d\boldsymbol{z}}{dt} - \boldsymbol{\Lambda}\boldsymbol{z} = \boldsymbol{V}^{-1}\boldsymbol{\beta} = \boldsymbol{\beta}^*.$$

So $\frac{dz_i}{dt} - \lambda_i z_i = \beta_i^*$, with solution $z_i = c_i e^{\lambda_i t} + \delta_i$. Thus, as in Equation 16.19,

$$\boldsymbol{p}(t) = \sum_{j=1}^{n} \boldsymbol{v}_j c_j e^{\lambda_j t} - \boldsymbol{A}^{-1}\boldsymbol{\beta}.$$

Because the roots are negative between 0 and -1, the system converges to $\boldsymbol{p}_e = -\boldsymbol{A}^{-1}\boldsymbol{\beta}$. For example, if $(a_0, b_0) = (3,45)$ and $(\alpha_0, \beta_0) = (15,20)$, then

$$\boldsymbol{p}_e = \begin{pmatrix} p_{1e} \\ p_{2e} \end{pmatrix} = -\begin{pmatrix} -\frac{11}{20} & \frac{3}{20} \\ -\frac{2}{20} & -\frac{4}{20} \end{pmatrix}^{-1} \begin{pmatrix} -15 \\ -5 \end{pmatrix} = \begin{pmatrix} 30 \\ 10 \end{pmatrix}.$$

Fixing an initial value $p(0)$,

$$\boldsymbol{p}(0) = \boldsymbol{v}_1 c_1 e^{\lambda_1 0} + \boldsymbol{v}_2 c_2 e^{\lambda_2 0} + \boldsymbol{p}_e = \boldsymbol{v}_1 c_1 + \boldsymbol{v}_2 c_2 + \boldsymbol{p}_e = \boldsymbol{V}\boldsymbol{c} + \boldsymbol{p}_e,$$

where $\boldsymbol{c} = (c_1, c_2)$. Thus, $\boldsymbol{p}(0) - \boldsymbol{p}_e = \boldsymbol{V}\boldsymbol{c}$ so that $\boldsymbol{c} = \boldsymbol{V}^{-1}(\boldsymbol{p}(0) - \boldsymbol{p}_e)$. Writing $\boldsymbol{\Lambda}^*(t)$ for the diagonal matrix with $e^{\lambda_i t}$ the ith diagonal term, the price dynamics may be written:

$$\boldsymbol{p}(t) = \boldsymbol{V}\boldsymbol{\Lambda}^*(t)\boldsymbol{V}^{-1}(\boldsymbol{p}(0) - \boldsymbol{p}_e) + \boldsymbol{p}_e.$$

Expanding the term $\boldsymbol{V}\boldsymbol{\Lambda}^*(t)\boldsymbol{V}^{-1}$ gives:

$$\boldsymbol{p}(t) = \begin{pmatrix} -\frac{1}{5}e^{\lambda_1 t} + \frac{6}{5}e^{\lambda_2 t} & \frac{3}{5}e^{\lambda_1 t} - \frac{3}{5}e^{\lambda_2 t} \\ -\frac{2}{5}e^{\lambda_1 t} + \frac{2}{5}e^{\lambda_2 t} & \frac{6}{5}e^{\lambda_1 t} - \frac{1}{5}e^{\lambda_2 t} \end{pmatrix} (\boldsymbol{p}(0) - \boldsymbol{p}_e) + \boldsymbol{p}_e.$$

◊

EXAMPLE 16.10: Let

$$\begin{pmatrix} y_1' \\ y_2' \end{pmatrix} = \begin{pmatrix} \frac{1}{5} & 0 \\ \frac{2}{5} & \frac{4}{5} \end{pmatrix} \begin{pmatrix} y_1 \\ y_2 \end{pmatrix} + \begin{pmatrix} 2 \\ 4 \end{pmatrix}.$$

The coefficient matrix, $\boldsymbol{A} = \begin{pmatrix} \frac{1}{5} & 0 \\ \frac{2}{5} & \frac{4}{5} \end{pmatrix}$ has eigenvalues $\lambda_1 = \frac{1}{5}$ and $\lambda_2 = \frac{4}{5}$. The corresponding eigenvectors are $\boldsymbol{v}_1 = (-\frac{3}{2}, 1)$ and $\boldsymbol{v}_2 = (0, 1)$. Therefore, the general solution is:

$$\begin{pmatrix} y_1(x) \\ y_2(x) \end{pmatrix} = \begin{pmatrix} -\frac{3}{2} \\ 1 \end{pmatrix} c_1 e^{-\frac{1}{5}x} + \begin{pmatrix} 0 \\ 1 \end{pmatrix} c_2 e^{-\frac{4}{5}} + \begin{pmatrix} 5 & 0 \\ -\frac{5}{2} & \frac{5}{4} \end{pmatrix} \begin{pmatrix} 2 \\ 4 \end{pmatrix}$$
$$= \begin{pmatrix} -\frac{3}{2}c_1 e^{-\frac{1}{5}x} + 10 \\ c_1 e^{-\frac{1}{5}x} + c_2 e^{-\frac{4}{5}} \end{pmatrix}.$$

◊

16.7 Stability

The long-run behavior of a dynamic system depends on the roots of the characteristic equation. A sufficient condition for stability of the dynamic system is that the roots of the characteristic polynomial have negative real parts. Consider the polynomial:

$$p(\lambda) = a_o\lambda^n + a_1\lambda^{n-1} + a_2\lambda^{n-2} + \cdots + a_{n-1}\lambda + a_n,$$

where $a_0 > 0$ (a_0 can always be chosen positive, if necessary by multiplying a polynomial with negative coefficient on λ^n by -1).

Definition 16.2: *The polynomial $p(\lambda)$ is called stable if all roots of p have negative real part. (If a root λ^* is real, it is negative, if complex with $\lambda^* = a^* + ib^*$, then $a^* < 0$.)*

The Routh–Hurwicz theorem provides a means to determine the stability of a polynomial. Define the $n \times n$ (Hurwicz) matrix, $\boldsymbol{H}$:

$$\boldsymbol{H} = \begin{pmatrix} a_1 & a_3 & a_5 & a_7 & \cdots & a_k & 0 & \cdots & \cdots & 0 \\ a_0 & a_2 & a_4 & a_6 & \cdots & a_{k-1} & a_{k+1} & \cdots & \cdots & 0 \\ 0 & a_1 & a_3 & a_5 & \cdots & a_{k-2} & a_k & a_{k+2} & \cdots & 0 \\ 0 & a_0 & a_2 & a_4 & \cdots & a_{k-3} & a_{k-1} & \cdots & \ddots & \vdots \\ 0 & 0 & a_1 & a_3 & \ddots & a_{k-4} & \vdots & \cdots & \cdots & \vdots \\ 0 & 0 & a_0 & a_2 & \cdots & \ddots & & & \cdots & \\ \vdots & & & & & & \ddots & & \cdots & \\ \vdots & & & & & & & \ddots & \cdots & \\ \vdots & & & & & & & \cdots & a_{n-1} & 0 \\ 0 & & & & & & & & \cdots & a_n \end{pmatrix}, \qquad (16.20)$$

where $k = n$ if n is odd, and $k = n-1$ if k is even, and for $j > n$ define $a_j = 0$. For example, the cases $n = 3$, $n = 5$ and $n = 6$ are:

$$\boldsymbol{H} = \begin{pmatrix} a_1 & a_3 & 0 \\ a_0 & a_2 & 0 \\ 0 & a_1 & a_3 \end{pmatrix}, \quad \boldsymbol{H} = \begin{pmatrix} a_1 & a_3 & a_5 & 0 & 0 \\ a_0 & a_2 & a_4 & 0 & 0 \\ 0 & a_1 & a_3 & a_5 & 0 \\ 0 & a_0 & a_2 & a_4 & 0 \\ 0 & 0 & a_1 & a_3 & a_5 \end{pmatrix}, \quad \boldsymbol{H} = \begin{pmatrix} a_1 & a_3 & a_5 & 0 & 0 & 0 \\ a_0 & a_2 & a_4 & a_6 & 0 & 0 \\ 0 & a_1 & a_3 & a_5 & 0 & 0 \\ 0 & a_0 & a_2 & a_4 & a_6 & 0 \\ 0 & 0 & a_1 & a_3 & a_5 & 0 \\ 0 & 0 & a_0 & a_2 & a_4 & a_6 \end{pmatrix}.$$

Theorem 16.3: *The polynomial $p(\lambda)$ is stable if and only if all the leading principal minors of $\boldsymbol{H}$ are positive. More formally, letting $\boldsymbol{H}_j$ denote the matrix obtained from $\boldsymbol{H}$ by deleting the last $n-j$ rows and columns for $j = 1,\dots,n$, the polynomial is stable if and only if $|\boldsymbol{H}_j| > 0$ for all j.*

In the case of a second-order polynomial,

$$\boldsymbol{H} = \begin{pmatrix} a_1 & 0 \\ a_0 & a_2 \end{pmatrix}, \quad \boldsymbol{H}_1 = \{a_1\}, \quad \boldsymbol{H}_2 = \boldsymbol{H}.$$

so that $|\boldsymbol{H}_1| = a_1$ and $|\boldsymbol{H}_2| = a_1 a_2$. These are positive if and only if a_1 and a_2 are positive. And, since $a_0 > 0$, stability requires that all coefficients in the characteristic polynomial $p(\lambda) = a_0\lambda^2 + a_1\lambda + a_2$ are positive.

Consider Example 16.9 with coefficient matrix:

$$\boldsymbol{A} = \begin{pmatrix} -\frac{11}{20} & \frac{3}{20} \\ -\frac{2}{20} & -\frac{4}{20} \end{pmatrix}$$

The corresponding characteristic equation is $\mid \boldsymbol{A} - \lambda \boldsymbol{I} \mid = \lambda^2 + \frac{3}{4}\lambda + \frac{1}{8}$, so that all coefficients are positive. Note that the matrix has eigenvalues: $\lambda_1 = -\frac{1}{4}$, $\lambda_2 = -\frac{1}{2}$, leading to stable dynamics of the system.

Exercises for Chapter 16

Exercise 16.1 Consider a loan with outstanding value $P(t)$ at time t. The growth of the debt at time t, over a small instant of time, Δt, is $P(t)r\Delta t$, where r is the continuous time rate of interest. Suppose that repayments are divided into two periods: an early period with no interest – from time $t = 0$ to t^* and then on a continuous basis at the rate d or a repayment of $d\Delta t$ over the period of time Δt. Find the value of d that amortizes the loan over the length of time $T > t^*$.

Exercise 16.2 Suppose that:

$$\frac{dy}{dt} - \frac{3}{4}y(t) = 5$$

and y satisfies the initial condition, $y(0) = 0$. Find $y(t)$.

Exercise 16.3 The elasticity of demand is:

$$\epsilon(p) = \frac{p}{Q}\frac{dQ(p)}{dp} = f(p).$$

Find the demand function when the elasticity of demand is:

(a) $f(p) = c$

(b) $f(p) = p^c$

(c) $f(p) = cp$.

The parameter c may be positive or negative depending on context.

Exercise 16.4 Suppose that:

$$\frac{dy}{dx} + \alpha(x)y(x) = 0.$$

Find $y(x)$ by direct integration.

Exercise 16.5 Suppose that:

$$\frac{dy}{dx} + \frac{1}{2}y(x) = x$$

and y satisfies the initial condition, $y(0) = 2$. Find $y(x)$.

Exercise 16.6 Suppose that:

$$\frac{dy}{dt} - \alpha y(t) = t,$$

where α is a constant and y satisfies the initial condition, $y(t) = y(t_0)$ at $t = t_0$. Find $y(t)$.

Exercise 16.7 The theory of population growth from Malthus suggested that as the population grows, competition for resources will work to inhibit that population growth. One model that captures this feature is the logistic growth model:

$$\frac{dP}{dt} = aP\left(1 - \frac{P}{P*}\right).$$

Find the growth rate of the population as a function of t.

Exercise 16.8 Consider a production function $y = f(k) = k^{\frac{1}{2}}$. Suppose that k evolves over time according to the formula:

$$\frac{dk}{dt} = sk^{\frac{1}{2}} - (\delta + n)k = \frac{1}{20}k^{\frac{1}{2}} - \left(\frac{1}{10} + \frac{1}{30}\right)k,$$

where $s = \frac{1}{20}$, $\delta = \frac{1}{10}$ and $n = \frac{1}{30}$ are the savings, depreciation and population growth rates. Find $k(t)$.

Exercise 16.9 Suppose that:

$$\frac{d^2y}{dt^2} - \frac{3}{4}\frac{dy(t)}{dt} + \frac{1}{8}y(t) = 5$$

and y satisfies the initial conditions, $y(0) = 0$, $y(1) = 1$. Find $y(t)$.

Exercise 16.10 Consider the system:

$$\begin{pmatrix} \frac{dy_1(t)}{dt} \\ \frac{dy_2(t)}{dt} \end{pmatrix} + \begin{pmatrix} \frac{7}{4} & -\frac{3}{4} \\ \frac{5}{2} & -1 \end{pmatrix} \begin{pmatrix} y_1(t) \\ y_2(t) \end{pmatrix} = \begin{pmatrix} 2 \\ 3 \end{pmatrix}.$$

(a) Find the eigenvalues and eigenvectors of the coefficient matrix $\boldsymbol{A}$, showing that $\boldsymbol{v}_1' = (3,5)$ and $\boldsymbol{v}_2' = (1,2)$ are eigenvectors of $\boldsymbol{A}$.

(b) Let $\boldsymbol{P}$ be the matrix of eigenvectors; define $\boldsymbol{z} = \boldsymbol{P}^{-1}\boldsymbol{y}$. Express the system in terms of $\boldsymbol{z}$.

(c) Solve for $\boldsymbol{z}$.

17

Linear difference equations

17.1 Introduction

In the study of many problems, the change in a variable is most naturally modeled in discrete units of time. For many bank loans, interest is computed on a monthly basis so that the level of debt is updated on a monthly basis. Government data are routinely released on a quarterly or annual basis, firms report profit each quarter. Thus, unemployment data or profit data are seen as sequences $u_t, u_{t-1}, \ldots$ or $\pi_t, \pi_{t-1}, \ldots$. When the data evolve according to a systematic pattern, difference equations may be used to study long-run behavior.

Section 17.2 provides some introductory examples to motivate the use of difference equations. Difference equations are categorized by the order of the difference equation. In a first-order difference equation the current value of the variable of interest depends on its value in the previous period; in an nth-order difference equation, the current value depends on the value in the previous n periods. The entire discussion here considers only first- and second-order difference equations, developed in Sections 17.3 and 17.4, respectively. Section 17.3.3 describes the well-known cobweb model of cycles in the demand and price of a single good and illustrates the process of solving a first-order difference equation. A generalization of the first-order difference equation is considered in Section 17.3.4. In Section 17.4.5 some examples of second-order difference equations are considered, in particular the multiplier-accelerator model from macroeconomic theory. Finally, Section 17.5 considers vector difference equations.

17.2 Motivating examples

Suppose that you obtain a loan from a bank, to be repaid over time. At the end of each period, interest is computed on the outstanding balance and a payment is made against the loan. Suppose that at the beginning of period t the outstanding balance is y_t. At the end of the period, the outstanding balance has grown to $(1+r)y_t$ when the interest rate is r. Then a payment, m, is made to the bank. So, at the beginning of period $t+1$, the outstanding balance is $y_{t+1}=(1+r)y_t-m$. This may be rewritten as

$$y_{t+1}-(1+r)y_t=-m. \tag{17.1}$$

What sequence $\{y_t\}$ might satisfy this rule? Observe that if y_t were constant over time, $y_t=\bar{y}$, then $\bar{y}-(1+r)\bar{y}=-m$ or $\bar{y}=\frac{m}{r}$. In this case, the outstanding debt does not change and the repayment just covers interest on the principal. From the lender's perspective, the loan is an asset, generating m dollars per period in perpetuity. This may be seen in present value terms.

Consider an asset that pays m dollars at the end of each year. In present-value terms, the payment of m one year from now is currently worth $\frac{1}{1+r}m$, the payment of m two years from now is currently worth $\frac{1}{(1+r)^2}m$ and the payment of m in t years from now is currently worth $\frac{1}{(1+r)^t}m$. So, the present value of the cash flow is:

$$\begin{aligned}
PV &= \frac{1}{(1+r)}m+\frac{1}{(1+r)^2}m+\frac{1}{(1+r)^3}m+\cdots \\
&= m\frac{1}{(1+r)}\left[1+\frac{1}{(1+r)}+\frac{1}{(1+r)^2}+\cdots\right] \\
&= m\frac{1}{(1+r)}\left[\frac{1+r}{r}\right] \\
&= \frac{m}{r}.
\end{aligned} \tag{17.2}$$

The loan of $\bar{y}$ with repayment only of interest generates a yearly cash flow of m dollars. The value, $\bar{y}$, satisfies $r\bar{y}=m$; m is the repayment each period that exactly covers interest accrued on the outstanding amount $\bar{y}$, so the principal $\bar{y}$ is unchanged from period to period and equals PV.

Notice that if there were some sequence $\{\tilde{y}_t\}$ satisfying $\tilde{y}_{t+1}-(1+r)\tilde{y}_t=0$, then

$$[\bar{y}-(1+r)\bar{y}]+[\tilde{y}_{t+1}-(1+r)\tilde{y}_t]=-m+0=-m,$$

or $(\bar{y}+\tilde{y}_{t+1})-(1+r)(\bar{y}+\tilde{y}_t)=-m$, so that the sequence $\{y_t^*\}=\{\bar{y}+\tilde{y}_t\}$ also satisfies Equation 17.1. Considering $\tilde{y}_{t+1}-(1+r)\tilde{y}_t=0$, this expression asserts that

$\tilde{y}_{t+1} = (1+r)\tilde{y}_t$, so that starting from the initial period the sequence increments over time by multiplying by $(1+r)$: $\tilde{y}_{t+1} = A(1+r)^t$ for $t \geq 1$, with A arbitrary in this expression. Thus, $y_t^* = \tilde{y}_{t+1} + \bar{y} = A(1+r)^t + \frac{m}{r}$. Suppose that some terminal date T is set such that the loan is fully repaid: $0 = y_T^* = A(1+r)^T + \frac{m}{r}$. Therefore, $A(1+r)^T = -\frac{m}{r}$ or $A = -\frac{m}{r}\frac{1}{(1+r)^T}$. Thus,

$$\begin{aligned} y_t^* &= A(1+r)^t + \frac{m}{r} \\ &= \left[-\frac{m}{r}\frac{1}{(1+r)^T}\right](1+r)^t + \frac{m}{r} \\ &= \left[\frac{m}{r}\right]\left[1 - \frac{(1+r)^t}{(1+r)^T}\right]. \end{aligned} \tag{17.3}$$

If the repayment per period is m and the loan is repaid in T periods, the amount borrowed must have been: $y_0^* = \left[\frac{m}{r}\right]\left[1 - \frac{(1+r)^0}{(1+r)^T}\right] = \left[\frac{m}{r}\right]\left[1 - \frac{1}{(1+r)^T}\right]$.

As a final motivating example, consider an employment model where workers transition in and out of employment. Let $p(e \mid e)$ be the probability of being employed next period given a person is currently employed, and let $p(u \mid e)$ be the probability of being unemployed next period given a person is currently employed. $p(e \mid u)$ and $p(u \mid u)$ denote the corresponding probabilities for an individual currently unemployed. From the definitions, $p(e \mid e) + p(u \mid e) = 1$ and $p(e \mid u) + p(u \mid u) = 1$. If l is the labor force, then at any point in time current employment e_t and unemployment u_t equal the labor force: $l = e_t + u_t$. Of those that are currently employed, $e_t p(e \mid e)$ remain employed and $e_t p(u \mid e)$ become unemployed. Of those that are currently unemployed, $u_t p(e \mid u)$ become employed and $u_t p(u \mid u)$ remain unemployed. Thus, unemployment next period is

$$\begin{aligned} u_{t+1} &= e_t p(u \mid e) + u_t p(u \mid u) \\ &= (l - u_t) p(u \mid e) + u_t p(u \mid u) \\ &= l p(u \mid e) + u_t[p(u \mid u) - p(u \mid e)] \\ &= k + \delta u_t, \end{aligned}$$

where $k = l p(u \mid e)$ and $\delta = [p(u \mid u) - p(u \mid e)]$. In the steady state, $u_{t+1} = u_t = \bar{u}$, and so $\bar{u} = k + \delta\bar{u}$, $\bar{u}(1-\delta) = k$ or $\bar{u} = \frac{k}{1-\delta}$. Also, $u_{t+1} - \bar{u} = \delta(u_t - \bar{u})$, so that $u_{t+j} = \delta^j(u_t - \bar{u})$. So, for example, if there is a 25% chance of an unemployed worker finding a job, and a 5% chance of an employed worker losing a job, then $\delta = 0.25 - 0.05 = 0.2$.

This discussion illustrates the solution of a first-order difference equation. The next section formalized the analysis of first-order difference equations.

17.3 First-order linear difference equations

A *first-order linear difference equation* describes the movement of some variable y_t over time according to an equation of the form:

$$a_1 y_t + a_0 y_{t-1} = f(t).$$

Assume $a_1 \neq 0$, since otherwise the equation gives $y_{t-1} = \frac{f(t)}{a_0}$, in which case y_t evolves deterministically according to the function f and the behavior of y_t is completely described. When $a_1 \neq 0$, the equation can be rearranged:

$$y_t + a y_{t-1} = \varphi(t), \tag{17.4}$$

where $a = \frac{a_0}{a_1}$ and $\varphi(t) = \frac{1}{a_1} f(t)$. This is the general form of the first-order linear difference equation: $y_t + a y_{t-1} = \varphi(t)$.

A solution to this dynamic equation is a sequence $\{y_t^*\}$ satisfying the equation for each t: $y_t^* + a y_{t-1}^* = \varphi(t)$.

Definition 17.1: *A function y_t^p is called a particular solution of $y_t + a y_{t-1} = \varphi(t)$ if*

$$y_t^p + a y_{t-1}^p = \varphi(t), \ \forall t.$$

The collection of all particular solutions of $y_t + a y_{t-1} = \varphi(t)$ is called the general solution.

In general, there are many particular solutions. A unique particular solution is selected by imposing an initial condition. For example, if the value of y_0 is set, then $y_1 = \varphi(t) - a y_0$, and so on: the sequence $\{y_t\}$ is fully determined. If y_t and x_t are two particular solutions, then

$$(y_t + a y_{t-1}) - (x_t + a x_{t-1}) = (y_t - x_t) + a(y_{t-1} - x_{t-1}) = z_t + a z_{t-1} = \varphi(t) - \varphi(t) = 0.$$

So, if y_t and x_t are two particular solutions then $z_t = y_t - x_t$ satisfies:

$$z_t + a z_{t-1} = 0.$$

Definition 17.2: *A function y_t^c is called a complementary solution if*

$$y_t^c + a y_{t-1}^c = 0, \ \forall t.$$

In view of the preceding discussion, if y_t and x_t are particular solutions, then $z_t = y_t - x_t$ is a complementary solution. Consequently, given any particular solution x_t, any other particular solution, y_t, may be written $y_t = x_t + z_t$, for some complementary solution z_t.

REMARK 17.1: Note, for example, that multiplying any complementary solution, y_t^c by a constant, k, gives a complementary solution ky_t^c. And, if x and y are complementary solutions, then, for any numbers α and β, $\alpha x + \beta y$ is a complementary solution. There are many complementary and (hence) many particular solutions. □

17.3.1 Solving first-order linear difference equations

These observations suggest a general method of solution for the first-order linear difference equation, $y_t + ay_{t-1} = \varphi(t)$:

1. Find a complementary solution, y_t^c, to the *homogeneous equation*: $y_t + ay_{t-1} = 0$.
2. Find a particular solution, y_t^p, to the *inhomogeneous equation*: $y_t + ay_{t-1} = \varphi(t)$.
3. Combine the complementary and particular solutions to give the *general solution*: $y_t = y_t^c + y_t^p$.
4. Use the initial condition to select a particular solution.

To simplify the discussion, the case where $\varphi(t)$ is constant and the case where it is variable are discussed separately.

17.3.2 The constant forcing function

The function $\varphi(t)$ is sometimes called the forcing function of the difference equation. When this is constant, $\varphi(t) = c$ for all t, the difference equation is:

$$y_t + ay_{t-1} = c.$$

The homogeneous equation is $y_t + ay_{t-1} = 0$. Finding a complementary solution is straightforward: $y_t = -ay_{t-1} = (-a)^2 y_{t-2} = (-a)^3 y_{t-3} = \cdots$, so it is clear that $y_t = k(-a)^t$ satisfies this expression for any constant k. When $a = -1$, $k(-a)^t = k$.

The inhomogeneous equation $y_t + ay_{t-1} = c$ has a constant particular solution if $a \neq -1$: pick $y_t = \bar{y}$ where $\bar{y} + a\bar{y} = c$, so $y_t = \bar{y} = \frac{c}{1+a}$ is a solution. If $a = -1$, the equation is $y_t - y_{t-1} = c$, so that $y_t, y_{t-1}, \ldots$ lie on a line with slope c: $y_t = ct + k'$ for any number k'.

Combining the complementary and particular solutions,

$$y_t = \begin{cases} k(-a)^t + \frac{c}{1+a} & a \neq -1 \\ (k+k') + ct & a = -1. \end{cases}$$

Imposing an initial condition eliminates the arbitrary parameter. In the case where $a \neq -1$, the initial condition that the solution equal y_0 at $t = 0$ gives: $y_0 = k(-a)^0 + \frac{c}{1+a} = k + \frac{c}{1+a}$, so that $k = y_0 - \frac{c}{1+a}$. In the case where $a = -1$, $y_0 = k + k'$ and $y_t = y_0 + ct$. Thus, with initial condition y_0, the solution (depicted in Figure 17.1 for $y_0 = 6$, $a = \frac{1}{2}$ and $c = \frac{1}{2}$) is:

$$y_t = \begin{cases} \left[y_0 - \frac{c}{1+a}\right](-a)^t + \frac{c}{1+a}, & a \neq -1 \\ y_0 + ct, & a = -1. \end{cases}$$

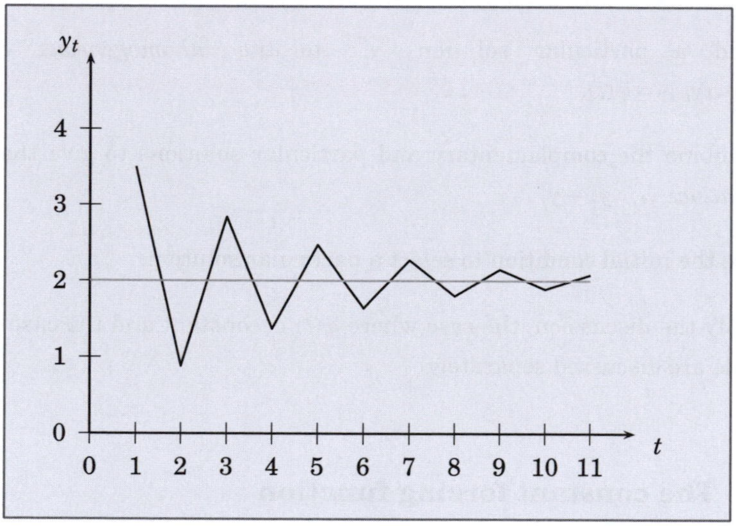

Figure 17.1: Plot for $y_t + \left(\frac{3}{4}\right)y_{t-1}$, $c = \frac{7}{2}$, $y_0 = \frac{7}{2}$.

EXAMPLE 17.1: $y_t - \frac{1}{2}y_{t-1} = 4$, $y_0 = 1$. The particular solution is: $y_t^p = y = \frac{4}{1-\frac{1}{2}} = 8$. The complementary solution is: $y_t^c = A\left(\frac{1}{2}\right)^t$. This gives $y_t = A\left(\frac{1}{2}\right)^t + 8$ and using $y_0 = 1$ gives $1 = y_0 = A + 8 \Rightarrow A = (-7)$. Therefore, $y_t = (-7)\left(\frac{1}{2}\right)^t + 8$. ◊

EXAMPLE 17.2: $y_t - y_{t-1} = 4$, $y_0 = 1$. The particular solution is $y_t^p = 4t$ and the complementary solution is $y_t^c = A(1)^t = A$. This gives $y_t = A(1)^t + 4t = A + 4t$, so $1 = y_0 = A$. Thus, $y_t = 1 + 4t$. ◊

The cobweb model of price dynamics described in Section 17.3.3 illustrates a first-order linear difference equation with constant forcing function.

17.3.3 The cobweb model

This is a supply and demand model with the key feature that supply in the current period depends on the previous period's price. Thus, in any given period, price alone adjusts to clear the market:

$$D_t = a + bp_t,\ a > 0,\ b < 0$$
$$S_t = c + dp_{t-1},\ 0 < c < a,\ d > 0.$$

Market clearing each period gives:

$$D_t = a + bp_t = c + dp_{t-1} = S_t.$$

In a stationary equilibrium (price constant over time): $p_t = p_{t-1} = p_e$.

$$a + bp_e = c + dp_e \Rightarrow p_e = \frac{a-c}{d-b}.$$

17.3.3.1 Dynamic behavior of price

With market clearing each period, $D_t = S_t$, current price adjusts to clear the market: $a + bp_t = c + dp_{t-1}$. Rearranging, $bp_t - dp_{t-1} = c - a$, so that:

$$p_t - \left(\frac{d}{b}\right)p_{t-1} = \frac{c-a}{b},$$

or

$$p_t + \alpha p_{t-1} = \delta,$$

where $\alpha = -\frac{d}{b}$ and $\delta = \frac{c-a}{b}$. Note that $\alpha = -\frac{d}{b} > 0$ since $d > 0$, $b < 0$. The general solution is:

$$p_t = k\left(\frac{d}{b}\right)^t + \frac{\delta}{1+\alpha}.$$

Observe that $\frac{\delta}{1+\alpha} = \frac{(c-a)}{b\left(1-\frac{d}{b}\right)} = \frac{c-a}{b-d} = p_e > 0$, so the solution can be written

$$p_t = k\left(\frac{d}{b}\right)^t + p_e.$$

With an initial price p_0, k is determined: $k = p_0 - p_e$ and this gives the particular solution:

$$p_t = (p_0 - p_e)\left(\frac{d}{b}\right)^t + p_e.$$

This is stable ($p_t \to p_e$) provided $\left|\frac{d}{b}\right| < 1$.

17.3.3.2 Stability

A different perspective can be given in terms of elasticities. The supply elasticity is:

$$\frac{p_{t-1}}{S_t}\frac{dS_t}{dp_{t-1}} = \frac{p_{t-1}}{S_t}d = e_{sp-1}(p_{t-1}),$$

and the demand elasticity is:

$$\frac{p_t}{D_t}\frac{dD_t}{dp_t} = \frac{p_t}{D_t}b = e_{Dp}(p_t).$$

At $p_e = p_{t-1} = p_t$, $S_t = D_t$, so that $\frac{e_{sp-1}(p_e)}{e_{Dp}(p_e)} = \left(\frac{d}{b}\right)$ = ratio of supply to demand elasticity at the equilibrium point. Since $d > 0$ and $b < 0$, $\frac{d}{b} < 0$. There are three cases to consider.

1. $\frac{d}{b} < -1$, ($\left|\frac{d}{b}\right| > 1$). In this case, supply is more responsive to price changes than demand giving, unstable dynamics.

2. $\frac{d}{b} = -1$, ($\left|\frac{d}{b}\right| = 1$). When demand and supply have slopes of equal magnitude, cycles occur.

3. $\frac{d}{b} > -1$, ($\left|\frac{d}{b}\right| < 1$). When supply is less responsive to price changes than demand, stable dynamics result and the price converges to p_e.

Figure 17.2 illustrates.

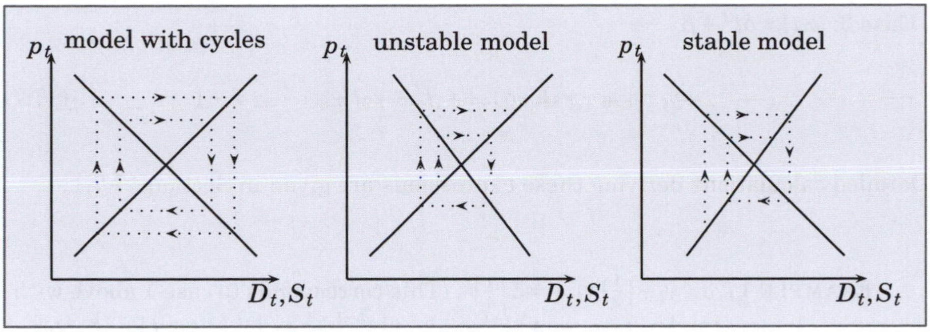

Figure 17.2: Cobweb dynamics

17.3.4 Equations with variable forcing function

Considering again the equation

$$y_t + ay_{t-1} = \varphi(t).$$

The homogeneous part is the same, $y_t + ay_{t-1} = 0$, regardless of $\varphi(t)$, and the complementary solution is unchanged: $y_t = k(-a)^t$. Finding a particular solution is more complex. The usual practice is to postulate a functional form for the particular solution that is similar to the functional form of the forcing function φ. For example, $\varphi(t) = t^2 - 3$ suggests a quadratic particular solution y_t: $y_t = \alpha t^2 + \beta t + \gamma$, whereas $\varphi(t) = 2^t$ suggests a solution of the form $y_t = g2^t$. The discussion here focuses on three cases:

1. $\varphi(t) = gd^t$
2. $\varphi(t) = \alpha t + \beta$.
3. $\varphi(t) = \alpha t^2 + \beta$.

Case 1: $\varphi(t) = gd^t$

$$y_t = \begin{cases} \left[y_0 - \frac{gd}{a+d}\right](-a)^t + \frac{gd^{t+1}}{a+d}, & a+d \neq 0 \\ y_0(-a)^t - \left(\frac{g}{a}\right)td^{t+1}, & a+d = 0. \end{cases} \tag{17.5}$$

Case 2: $\varphi(t) = \delta t + \beta$

$$y_t = \left\{y_o - \frac{1}{1+a}\left[\beta + \alpha\frac{\delta}{1+a}\right]\right\}(-a)^t + \left\{\frac{\delta}{1+a}t + \frac{1}{1+a}\left[\beta + a\frac{\delta}{1+a}\right]\right\}, \quad a \neq -1. \tag{17.6}$$

Case 3: $\varphi(t) = \delta t^2 + \beta$

$$y_t = [y_0 - \gamma_3](-a)^t + \gamma_1 t^2 + \gamma_2 t + \gamma_3, \quad a \neq -1. \tag{17.7}$$

Detailed calculations deriving these expressions are given in Section 17.7.1.

EXAMPLE 17.3: $y_t - \left(\frac{1}{2}\right) y_{t-1} = 5\left(\frac{1}{4}\right)^t$. (This corresponds to case 1 above with $a = -\frac{1}{2}$ and $d = \frac{1}{4}$.) So $y_t = A\left(\frac{1}{2}\right)^t + \frac{5}{\left(-\frac{1}{4}\right)}\left(\frac{1}{4}\right)^{t+1} = A\left(\frac{1}{2}\right)^t - (20)\left(\frac{1}{4}\right)^{t+1}$. Next, $1 = y_0 = A - 20\left(\frac{1}{4}\right) = A - 5 \Rightarrow A = 6$. Thus, $y_t = 6\left(\frac{1}{2}\right)^t - 20\left(\frac{1}{4}\right)^{t+1}$. ◊

EXAMPLE 17.4: $y_t - \frac{1}{2} y_{t-1} = 5\left(\frac{1}{2}\right)^t$. Therefore $y_t = A\left(\frac{1}{2}\right)^t - \frac{5}{\frac{1}{2}} t\left(\frac{1}{2}\right)^{t+1} = A\left(\frac{1}{2}\right)^t - 10t\left(\frac{1}{2}\right)^{t+1}$. Finally, $1 = y_0 = A - 10\left(\frac{1}{2}\right) = A - 5 \Rightarrow A = 6$, so $y_t = 6\left(\frac{1}{2}\right)^t - 10t\left(\frac{1}{2}\right)^{t+1}$. ◊

17.4 Second-order linear difference equations

The simplest form of second-order difference equation is:

$$y_t + a_1 y_{t-1} + a_2 y_{t-2} = c.$$

As in the case of first-order difference equations, the problem of finding a solution is dealt with as follows:

- Solve the inhomogeneous equation: $y_t + a_1 y_{t-1} + a_2 y_{t-2} = c$ to obtain the particular solution.
- Solve the homogeneous equation: $y_t + a_1 y_{t-1} + a_2 y_{t-2} = 0$ to obtain the complementary solution.
- Obtain the general solution as the sum of the particular and complementary solutions.
- Use initial conditions to identify a unique solution.

17.4.1 The inhomogeneous equation

The form of solution for the inhomogeneous equation depends on the values of a_1 and a_2. The simplest candidate particular solution is a constant solution $y_t = y^*$ and if this satisfies the inhomogeneous equation: $y^*(1+a_1+a_2) = c$, y^* is determined as $y^* = \frac{c}{1+a_1+a_2}$.

However, $y^*(1+a_1+a_2) = c$ is insoluble if $(1+a_1+a_2) = 0$ (and $c \neq 0$). In this case, try a particular solution $y = \rho t$, so that $\rho t + a_1\rho(t-1) + a_2\rho(t-2) = c$. Rearranging, $\rho(1+a_1+a_2)t - (a_1+2a_2)\rho = c$ or $-(a_1+2a_2)\rho = c$ and this solves for $\rho = -\frac{c}{(a_1+2a_2)}$, provided $a_1+2a_2 \neq 0$.

Finally, if both $(1+a_1+a_2)$ and (a_1+2a_2) equal 0, try as a particular solution $y = \rho t^2$. Then $\rho t^2 + a_1\rho(t-1)^2 + a_2\rho(t-2)^2 = c$, or $\rho t^2(1+a_1+a_2) - 2\rho t(a_1+2a_2) + (a_1+4a_2)\rho = c$, so that $\rho = \frac{c}{a_1+4a_2}$. The equations $(1+a_1+a_2) = 0$ and $(a_1+2a_2) = 0$ imply $a_2 = 1$, $a_1 = -1$, so that $\rho = \frac{c}{2}$. So, depending on the parameters, a particular solution is given by:

$$y_t^p = \begin{cases} \frac{c}{1+a_1+a_2}, & a_1+a_2 \neq -1 \\ \frac{c}{a_1+2a_2}t, & a_1+a_2 = -1, \quad a_1+2a_2 \neq 0 \\ \frac{c}{2}t^2, & a_1+a_2 = -1, \quad a_1+2a_2 = 0, \quad (a_1 = -2, \text{ and } a_2 = 1). \end{cases} \tag{17.8}$$

17.4.2 The homogeneous equation

Now, consider the homogeneous equation, $y_t + a_1y_{t-1} + a_2y_{t-2} = 0$. As a trial complementary solution consider $y_t^c = kb^t$. This gives the equation:

$$kb^t + a_1kb^{t-1} + a_2kb^{t-2} = 0.$$

Canceling k gives the quadratic equation:

$$b^2 + a_1b + a_2 = 0.$$

Use the usual formula to find the roots, λ_1 and λ_2:

$$\lambda_1, \lambda_2 = \frac{-a_1 \pm \sqrt{a_1^2 - 4a_2}}{2}.$$

So, for $i = 1,2$, $y_t = k\lambda_i^t$ satisfies the complementary equation for any value of k. More generally, if y_t and z_t are solutions of the homogeneous equation, then so is $k_1y_t + k_2z_t$ for any numbers k_1 and k_2. So any y_t of the form:

$$y_t = k_1\lambda_1^t + k_2\lambda_2^t$$

satisfies the complementary equation. The dynamic behavior of this equation depends critically on the roots, and concerning the roots there are three possibilities:

1. Both roots are real and distinct ($a_1^2 - 4a_2 > 0$).

2. Both root are real but repeated ($\lambda_1 = \lambda_2$, when $a_1^2 - 4a_2 = 0$).

3. Both roots are complex and distinct ($a_1^2 - 4a_2 < 0$).

Depending on which of these three cases applies, the complementary solution is:

$$y_t^c = \begin{cases} k_1\lambda_1^t + k_2\lambda_2^t, & \text{roots real and distinct} \\ k_1\lambda_1^t + k_2 t\lambda_2^t, & \text{roots real and equal} \\ r^t[B_1\cos(\theta t) + B_2\sin(\theta t)], & \text{roots complex.} \end{cases} \tag{17.9}$$

The derivation of these equations is given in Section 17.7.2.

17.4.3 The general solution

Together, Equations 17.8 and 17.9 define the solution:

$$y_t = y_t^p + y_t^c.$$

For example, if the roots are real and distinct, and $a_1 + a_2 \neq 1$, then

$$y_t = k_1\lambda_1^t + k_2\lambda_2^t + \frac{c}{1+a_1+a_2}.$$

Suppose now that y_0 and y_1 are known. Then,

$$y_0 = k_1 + k_2 + \frac{c}{1+a_1+a_2}$$
$$y_1 = k_1\lambda_1 + k_2\lambda_2 + \frac{c}{1+a_1+a_2}.$$

Rearranging:

$$\begin{pmatrix} y_0 - \frac{c}{1+a_1+a_2} \\ y_1 - \frac{c}{1+a_1+a_2} \end{pmatrix} = \begin{pmatrix} 1 & 1 \\ \lambda_1 & \lambda_2 \end{pmatrix} \begin{pmatrix} k_1 \\ k_2 \end{pmatrix},$$

and, since the roots are not equal, this determines k_1 and k_2.

How does y_t behave in the long run as t becomes large? This is discussed next.

17.4.4 Stability

From the particular solution in Equation 17.8, it is clear that if $a_1 + a_2 = -1$, then the system diverges, since $|y_t^p| \to \infty$. So, a necessary condition for stability is that $a_1 + a_2 \neq -1$. Considering the complementary solution, in the case where the roots λ_1 and λ_2 are real and distinct, the roots are less than 1 in absolute value if $-2 < a_1 < 2$, $1 + a_1 + a_2 > 0$ and $1 - a_1 + a_2 > 0$. These conditions, along with the condition that the root be real, $a_1^2 - 4a_2 > 0$, define the stability region for real distinct roots. For stability in the case where the roots are real and equal $a_1^2 - 4a_2 = 0$, this reduces to the conditions $a_1 + a_2 \neq -1$ and $-2 < a_1 < 2$. When the roots λ_1 and λ_2 are complex (and hence conjugate), then it is necessary that they have absolute value less than 1: $|\lambda_1| = |\lambda_2| < -1$ and this translates to the condition $a_2 < 1$. Assuming that $a_1 + a_2 \neq 1$, the regions of stability determined by the parameters is depicted in Figure 17.3. Detailed calculations are given in Section 17.7.3.

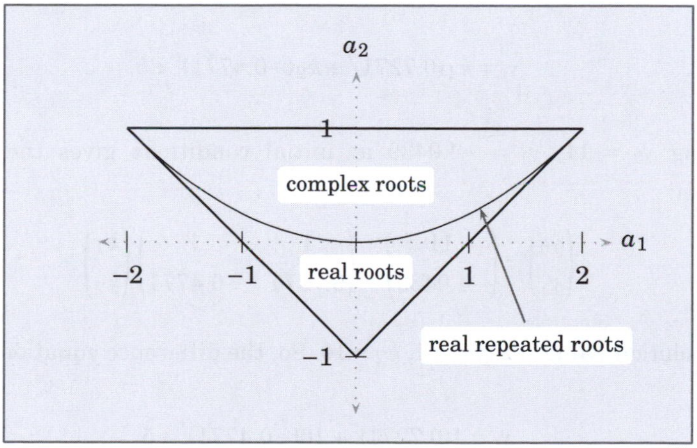

Figure 17.3: Stability of second-order system

17.4.5 Examples

> EXAMPLE 17.5: In the equation $y_t + a_1 y_{t-1} + a_2 y_{t-2} = 9$, suppose that $a_1 = -0.5$ and $a_2 = -0.3$, so that
>
> $$y_t - 0.5y_{t-1} - 0.3y_{t-2} = 9.$$

The roots are real since $a_1^2 - 4a_2 = 0.25 - 4(-0.3) = 1.45 > 0$. The stability conditions are satisfied, since:

$$1 + a_1 + a_2 = 1 - 0.8 > 0$$

$$|\, a_2 \,| < 2$$

$$1 - a_1 + a_2 = 1.2 > 0.$$

The general form of the solution is:

$$y_t = k_1 \lambda_1^t + k_2 \lambda_2^t + \frac{c}{1 + a_1 + a_2}.$$

The roots are $\lambda_1 = \frac{1}{2}[(0.25) + \sqrt{1.45}] \approx 0.7271$ and $\lambda_2 = \frac{1}{2}[(0.25) - \sqrt{1.45}] \approx -0.4771$, and $\frac{c}{1+a_1+a_2} = \frac{9}{1.8} = 5$. Substituting these numbers:

$$y_t = k_1(0.7271)^t + k_2(-0.4771)^t + 5.$$

Picking $y_0 = 11$, $y_1 = -4.0439$ as initial conditions gives the equation system:

$$\begin{pmatrix} y_0 \\ y_1 \end{pmatrix} = \begin{pmatrix} 11 \\ -4.0439 \end{pmatrix} = \begin{pmatrix} 1 & 1 \\ 0.7271 & -0.4771 \end{pmatrix} \begin{pmatrix} k_1 \\ k_2 \end{pmatrix}.$$

The solution values are $k_1 = 1$, $k_2 = 10$. So, the difference equation is:

$$y_t = 1(0.7271)^t + 10(-0.4771)^t + 5.$$

Figure 17.4 depicts the time path of the process. ◊

17.4.6 The Samuelson multiplier-accelerator model

The next example is the Samuelson multiplier-accelerator model, where aggregate household consumption is a fraction of aggregate income and investment responds to increases in household consumption.

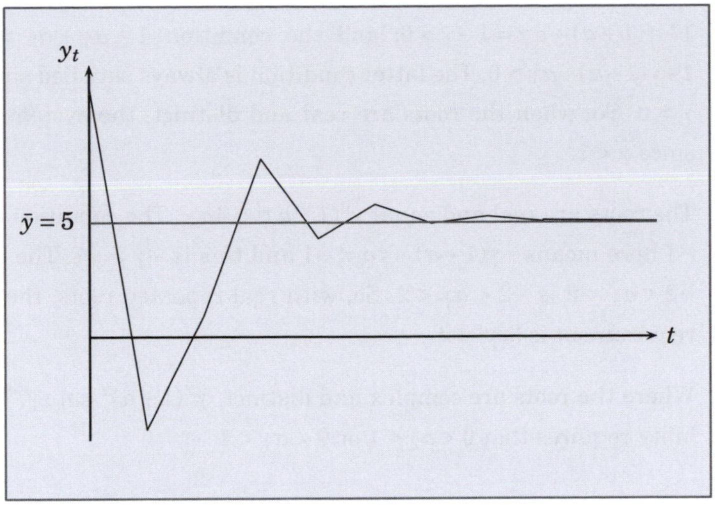

Figure 17.4: A converging process

EXAMPLE 17.6: Aggregate output equals aggregate consumption:

$$Y_t = C_t + I_t + G$$
$$C_t = \gamma Y_{t-1},\ 0 < \gamma < 1$$
$$I_t = \alpha(C_t - C_{t-1}),\ \alpha > 0.$$

Substitution gives a second-order difference equation:

$$Y_t - \gamma Y_{t-1} - \alpha(\gamma Y_{t-1} - \gamma Y_{t-2}) = G$$
$$Y_t - \gamma(1+\alpha)Y_{t-1} + \alpha\gamma Y_{t-2} = G.$$

Solving for the particular solution:

$$Y_t^p = \frac{G}{1-\gamma(1+\alpha)+\alpha\gamma} = \frac{G}{1-\gamma}.$$

To find the complementary solution, we need to find the roots of: $\lambda^2 - \gamma(1+\alpha)\lambda + \alpha\gamma = 0$:

$$\lambda_1, \lambda_2 = \frac{\gamma(1+\alpha) \pm \sqrt{[\gamma(1+\alpha)]^2 - 4\alpha\gamma}}{2}.$$

There are three cases to consider:

1. The roots are real and distinct: $\gamma^2(1+\alpha)^2 > 4\alpha\gamma$ (or $\gamma > \frac{4\alpha}{(1+\alpha)^2}$). With $a_1 = -\gamma(1+\alpha)$ and $a_2 = \alpha\gamma$, the condition $1 + a_1 + a_2 > 0$ gives

$1-\gamma(1+\alpha)+\alpha\gamma = 1-\gamma > 0$; and the condition $1-a_1+a_2 > 0$ gives $1+\gamma(1+\alpha)+\alpha\gamma > 0$. The latter condition is always satisfied since α and $\gamma > 0$. So, when the roots are real and distinct, the system is stable since $\gamma < 1$.

2. The roots are real and equal $\gamma^2(1+\alpha)^2 = 4\alpha\gamma$. The condition $a_1+a_2 \neq -1$ here means $-\gamma(1+\alpha)+\gamma\alpha \neq -1$ and this is $-\gamma \neq -1$. The condition $-2 < a_1 < 2$ is $-2 < \alpha\gamma < 2$. So, with real repeated roots, the stability requirement is $|\alpha\gamma| < 2$.

3. Where the roots are complex and distinct, $\gamma^2(1+\alpha)^2 < 4\alpha\gamma$. Then stability requires that $0 < a_2 < 1$ or $0 < \alpha\gamma < 1$. ◊

This multiplier-accelerator model may be augmented with the addition of a monetary sector.

EXAMPLE 17.7: Taking the previous example, modify by adding a monetary sector. Let the money supply be $\bar{m}$ and money demand be $m_t = m_0 + m_1 Y_{t-1} + m_2 r_t$. Assume money market equilibrium in each period: $\bar{m} = m_t$. Also modify the investment equation: $I_t = \alpha(C_t - C_{t-1}) + i_0 r_t$.

From the money market equilibrium, $\bar{m} = m_t = m_0 + m_1 Y_{t-1} + m_2 r_t$, we can solve for $r_t : r_t = \frac{\bar{m}}{m_2} - \frac{m_0}{m_2} - \frac{m_1}{m_2} Y_{t-1}$. So,

$$\begin{aligned}
Y_t &= C_t + I_t + G_0 \\
&= \gamma Y_{t-1} + \alpha(\gamma Y_{t-1} - \gamma Y_{t-2}) + \frac{i_0(\bar{m}-m_0)}{m_2} - i_0 \frac{m_1}{m_2} Y_{t-1} + G \\
&= \gamma(1+\alpha) Y_{t-1} - \alpha\gamma Y_{t-2} - i_0 \frac{m_1}{m_2} Y_{t-1} + \frac{i_0(\bar{m}-m_0)}{m_2} + G,
\end{aligned}$$

so

$$Y_t - \left[\gamma(1+\alpha) - \frac{i_0 m_1}{m_2}\right] Y_{t-1} + \alpha\gamma Y_{t-2} + \frac{i_0(\bar{m}-m_0)}{m_2} + G = 0$$

gives a second-order difference equation. Thus, the characteristic equation is:

$$\lambda^2 - \left[\gamma(1+\alpha) - \frac{i_0 m_1}{m_2}\right] \lambda + \alpha\gamma = 0.$$

In the form $y_t + a_1 y_{t-1} + a_2 y_{t-2} = 0$, the conditions for stability are: $1+a_1+a_2 > 0$, $1-a_1+a_2 > 0$, and $a_2 < 1$. Here, a_1 corresponds to $-[\gamma(1+\alpha) - \frac{i_0 m_1}{m_2}]$

and a_2 corresponds to $\alpha\gamma$. So, the conditions for stability are:

$$1-\left[\gamma(1+\alpha)-\frac{i_0 m_1}{m_2}\right]+\alpha\gamma>0$$

$$1+\left[\gamma(1+\alpha)-\frac{i_0 m_1}{m_2}\right]+\alpha\gamma>0$$

$$1-\alpha\gamma>0.$$

◊

17.5 Vector difference equations

Consider a dynamic system where the movement of two variables, x_t and y_t, over time is described:

$$a_{11}x_t+a_{12}y_t+b_{11}x_{t-1}+b_{12}y_{t-1}=c_1$$

$$a_{21}x_t+a_{22}y_t+b_{21}x_{t-1}+b_{22}y_{t-2}=c_2.$$

Let

$$\boldsymbol{z}_t=\begin{bmatrix}x_t\\ y_t\end{bmatrix},\ \boldsymbol{A}=\begin{bmatrix}a_{11} & a_{21}\\ a_{12} & a_{22}\end{bmatrix},\ \boldsymbol{B}=\begin{bmatrix}b_{11} & b_{21}\\ b_{12} & b_{22}\end{bmatrix},\ \boldsymbol{C}=\begin{bmatrix}c_1\\ c_2\end{bmatrix},$$

so that

$$\boldsymbol{A}\boldsymbol{z}_t+\boldsymbol{B}\boldsymbol{z}_{t-1}=\boldsymbol{C}.$$

As before define complementary, particular and general solutions. A complementary solution solves: $\boldsymbol{A}\boldsymbol{z}_t^c+\boldsymbol{B}\boldsymbol{z}_{t-1}^c=0$; the particular solution solves: $\boldsymbol{A}\boldsymbol{z}_t^p+\boldsymbol{B}\boldsymbol{z}_t^p=\boldsymbol{C}$; and the general solution is the sum of both: $\boldsymbol{z}_t=\boldsymbol{z}_t^c+\boldsymbol{z}_t^p$.

17.5.1 The particular solution

To find the particular solution, $\boldsymbol{z}_t^p$, set:

$$\boldsymbol{z}_t^p=\bar{\boldsymbol{z}}=\begin{bmatrix}\bar{x}\\ \bar{y}\end{bmatrix}$$

Then $\boldsymbol{A}\bar{\boldsymbol{z}}+\boldsymbol{B}\bar{\boldsymbol{z}}=\boldsymbol{C}$ or $(\boldsymbol{A}+\boldsymbol{B})\bar{\boldsymbol{z}}=\boldsymbol{C}$. Assume $\boldsymbol{A}+\boldsymbol{B}$ is non-singular, so that $\boldsymbol{z}_t^p=\bar{\boldsymbol{z}}=(\boldsymbol{A}+\boldsymbol{B})^{-1}\boldsymbol{C}$.

17.5.2 The complementary solution

The complementary solution satisfies $\boldsymbol{A}\boldsymbol{z}_t^c + \boldsymbol{B}\boldsymbol{z}_t^c = 0$. To solve this, try a solution of the form:

$$\boldsymbol{z}_t^c = \begin{bmatrix} sb^t \\ ub^t \end{bmatrix} \quad \text{so that} \quad \boldsymbol{A}\begin{bmatrix} sb^t \\ ub^t \end{bmatrix} + \boldsymbol{B}\begin{bmatrix} sb^{t-1} \\ ub^{t-1} \end{bmatrix} = 0.$$

Divide by b^{t-1} to get:

$$(\boldsymbol{A}b + \boldsymbol{B})\begin{bmatrix} s \\ u \end{bmatrix} = 0.$$

If the vector (s,u) is not equal to 0, then $|\boldsymbol{A}b + \boldsymbol{B}| = 0$, so

$$|\boldsymbol{A}b + \boldsymbol{B}| = \begin{vmatrix} a_{11}b + b_{11} & a_{12}b + b_{12} \\ a_{21}b + b_{21} & a_{22}b + b_{22} \end{vmatrix} = 0.$$

This gives:

$$0 = (a_{11}b + b_{11})(a_{22}b + b_{22}) - (a_{21}b + b_{21})(a_{12}b + b_{12})$$

$$= (a_{11}a_{22} - a_{21}a_{12})b^2 + (a_{11}b_{22} + a_{22}b_{11} - a_{21}b_{12} - a_{12}b_{21})b + (b_{11}b_{22} - b_{21}b_{12})$$

$$= k_1 b^2 + k_2 b + k_3 = 0.$$

Denote the roots of this quadratic b_1, b_2. The system

$$[\boldsymbol{A}b_i + \boldsymbol{B}]\begin{bmatrix} s \\ u \end{bmatrix} = 0$$

has solution (s_i, u_i) and since $(\boldsymbol{A}b_i + \boldsymbol{B})$ is singular, this implies that (s,u) is determined up to a scalar multiple: if (s_i, u_i) satisfies this equation, so does $\lambda(s_i, u_i)$ for any real number, λ. Fixing the value of one (s_i or u_i), determines the value of the other. This gives the complementary solution:

$$\boldsymbol{z}_t^c = \begin{pmatrix} x_t^c \\ y_t^c \end{pmatrix} = \begin{pmatrix} s_1 b_1^t + s_2 b_2^t \\ u_1 b_1^t + u_2 b_2^t \end{pmatrix}.$$

The general solution is

$$\boldsymbol{z}_t = \boldsymbol{z}_t^c + \boldsymbol{z}_t^p = \begin{pmatrix} x_t^c \\ y_t^c \end{pmatrix} = \begin{pmatrix} s_1 b_1^t + s_2 b_2^t \\ u_1 b_1^t + u_2 b_2^t \end{pmatrix} + (\boldsymbol{A} + \boldsymbol{B})^{-1}\begin{pmatrix} c_1 \\ c_2 \end{pmatrix}.$$

If the roots are real and repeated, then for the complementary solution try:

$$\boldsymbol{z}_t^c = \begin{bmatrix} s_1 b_1^t + s_2 t b_2^t \\ u_1 b_1^t + u_2 t b_2^t \end{bmatrix}.$$

The particular solution is unchanged as long as $|\boldsymbol{A}+\boldsymbol{B}| \neq 0$.

EXAMPLE 17.8: Consider an oligopoly of two firms, where each expects the other to produce last year's output. In the usual case:

$$\pi_1(Q_1,Q_2) = Q_1(a-b[Q_1+Q_2]) - cQ_1.$$

In the present circumstances:

$$\pi_1(Q_{1t},Q_{2t-1}) = Q_{1t}(a-b(Q_{1t}+Q_{2t-1})) - cQ_{1t}.$$

This has first-order condition: $a-2bQ_{1t}-bQ_{2t-1}-c=0$. Therefore,

$$Q_{1t} + \frac{1}{2}Q_{2t-1} = \frac{a-c}{2b}.$$

Similarly,

$$Q_{2t} + \frac{1}{2}Q_{1t-1} = \frac{a-c}{2b}.$$

These equations give the system:

$$\begin{bmatrix} 1 & 0 \\ 0 & 1 \end{bmatrix} \begin{bmatrix} Q_{1t} \\ Q_{2t} \end{bmatrix} + \begin{bmatrix} \frac{1}{2} & 0 \\ 0 & \frac{1}{2} \end{bmatrix} \begin{bmatrix} Q_{1t-1} \\ Q_{2t-1} \end{bmatrix} = \begin{bmatrix} \frac{(a-c)}{2b} \\ \frac{(a-c)}{2b} \end{bmatrix}.$$

For a particular solution $Q_{it} = \bar{Q}_i$:

$$1\frac{1}{2}\bar{Q}_1 = \frac{a-c}{2b} \Rightarrow \bar{Q}_1 = \frac{a-c}{3b}$$

$$1\frac{1}{2}\bar{Q}_2 = \frac{a-c}{2b} \Rightarrow \bar{Q}_2 = \frac{a-c}{3b}.$$

For the general solution:

$$\begin{vmatrix} b+\frac{1}{2} & 0 \\ 0 & b+\frac{1}{2} \end{vmatrix} = b^2 + b + \frac{1}{4} = 0,$$

the roots are

$$b_1, b_2 = \frac{-1 \pm (1-4(\frac{1}{4}))^{\frac{1}{2}}}{2} = -\frac{1}{2}, -\frac{1}{2}.$$

Thus the roots are repeated and real. The solution is:

$$Q_{1t} = s_1\left(-\frac{1}{2}\right)^t + s_2\left(-\frac{1}{2}\right)^t t + \frac{a-c}{3b}$$

$$Q_{2t} = u_1\left(-\frac{1}{2}\right)^t + u_2\left(-\frac{1}{2}\right)^t t + \frac{a-c}{3b}.$$

These reaction functions are depicted in Figure 17.5. The model is stable since $\left(-\frac{1}{2}\right)^t \to 0$ and $t\left(-\frac{1}{2}\right)^t \to 0$. ◊

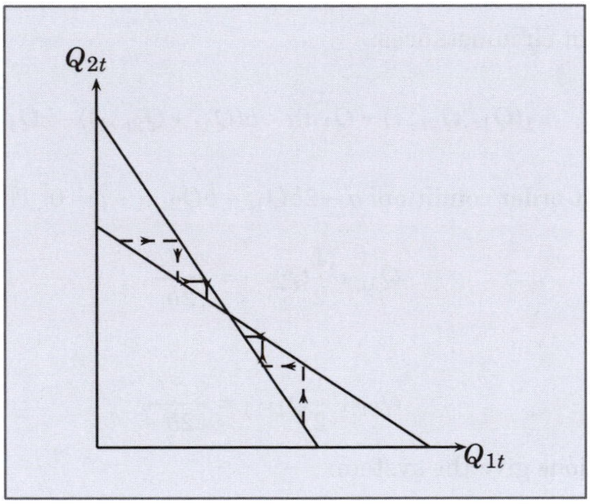

Figure 17.5: Dynamic reaction functions

17.6 The n-variable case

Consider a system of n equations governing the evolution of the n-vector $\boldsymbol{x}$ over time:

$$\boldsymbol{x}_t - \boldsymbol{A}\boldsymbol{x}_{t-1} = c, \tag{17.10}$$

then provided $(\boldsymbol{I}-\boldsymbol{A})$ is invertible, $\bar{\boldsymbol{x}} = (\boldsymbol{I}-\boldsymbol{A})^{-1}c$ is a particular solution of Equation 17.10. Provided the eigenvalues of $\boldsymbol{A}$ are distinct, finding the complementary solution is relatively straightforward. When the eigenvectors are all distinct, $\boldsymbol{A}$ may be written $\boldsymbol{A} = \boldsymbol{U}\boldsymbol{\Lambda}\boldsymbol{U}^{-1}$ where $\boldsymbol{\Lambda}$ is a diagonal matrix with the eigenvalues on the diagonal and where $\boldsymbol{U}$ is a matrix composed of the n linearly independent eigenvectors. Since $\bar{\boldsymbol{x}} - \boldsymbol{A}\bar{\boldsymbol{x}} = c$,

$$\begin{aligned}(\boldsymbol{x}_t - \bar{\boldsymbol{x}}) - \boldsymbol{A}(\boldsymbol{x}_{t-1} - \bar{\boldsymbol{x}}) &= 0 \qquad (17.11)\\ (\boldsymbol{x}_t - \bar{\boldsymbol{x}}) - \boldsymbol{U}\boldsymbol{\Lambda}\boldsymbol{U}^{-1}(\boldsymbol{x}_{t-1} - \bar{\boldsymbol{x}}) &= 0\\ (\boldsymbol{x}_t - \bar{\boldsymbol{x}}) - \boldsymbol{U}\boldsymbol{\Lambda}\boldsymbol{U}^{-1}\boldsymbol{U}\boldsymbol{\Lambda}\boldsymbol{U}^{-1}(\boldsymbol{x}_{t-2} - \bar{\boldsymbol{x}}) &= 0\\ (\boldsymbol{x}_t - \bar{\boldsymbol{x}}) - \boldsymbol{U}\boldsymbol{\Lambda}^2\boldsymbol{U}^{-1}(\boldsymbol{x}_{t-2} - \bar{\boldsymbol{x}}) &= 0.\end{aligned}$$

Thus,

$$(\boldsymbol{x}_t - \bar{\boldsymbol{x}}) - \boldsymbol{U}\boldsymbol{\Lambda}^t\boldsymbol{U}^{-1}(\boldsymbol{x}_0 - \bar{\boldsymbol{x}}) = 0$$

or

$$\boldsymbol{x}_t = \boldsymbol{U}\boldsymbol{\Lambda}^t\boldsymbol{U}^{-1}(\boldsymbol{x}_0 - \bar{\boldsymbol{x}}) + \bar{\boldsymbol{x}}.$$

Provided all roots of $\boldsymbol{A}$ are less than 1, $\boldsymbol{\Lambda}^t$ converges to the $\boldsymbol{0}$ matrix and $\boldsymbol{x}_t$ converges to $\bar{\boldsymbol{x}}$.

17.7 Miscellaneous calculations

The following sections provide detailed computations underlying the results stated in Sections 17.3.4 and 17.4.2.

17.7.1 Various forcing functions: solutions

The following sections solve for the forcing functions given in Equations 17.5, 17.6 and 17.7.

17.7.1.1 The case $\varphi(t) = gd^t$

Assume that the particular solution has the form, $y_t = \rho d^t$. Substituting into the inhomogeneous equation gives $\rho d^t + a\rho d^{t-1} = gd^t$. Factoring d^{t-1},

$$d^{t-1}(\rho d + a\rho - gd) = 0, \quad \text{or} \quad \rho d + a\rho - gd = 0.$$

This equation has no solution in ρ if $a+d=0$. If $(a+d)\neq 0$, rearranging gives $\rho=\frac{gd}{a+d}$ and a particular solution

$$y_t=\frac{gd}{a+d}d^t=\frac{gd^{t+1}}{a+d}.$$

Then, the general solution is $y_t=k(-a)^t+\frac{gd^{t+1}}{a+d}$. With initial condition $y_0=k(-a)^0+\frac{gd^{0+1}}{a+d}=k+\frac{gd}{a+d}$, $k=y_0-\frac{gd}{a+d}$ and

$$y_t=\left[y_0-\frac{gd}{a+d}\right](-a)^t+\frac{gd^{t+1}}{a+d}.$$

If $a+d=0$, try as an alternative candidate for a particular solution: $y_t=\rho t d^t$. Then $\rho t d^t+a\rho(t-1)d^{t-1}=gd^t$. Factoring d^{t-1} gives

$$\begin{aligned}&d^{t-1}(\rho td+a\rho(t-1)-gd)=0, \quad \text{or} \quad \rho td+a\rho(t-1)-gd)=0\\ &\text{or} \quad \rho(d+a)t-\rho a-gd=0.\end{aligned}$$

Since $d+a=0, \rho(-a)=gd$ or $\rho=-\frac{gd}{a}$. Thus $y_t=-\left(\frac{gd}{a}\right)td^t=-\left(\frac{g}{a}\right)td^{t+1}$ is a particular solution. Now the general solution has the form, $y_t=k(-a)^t-\left(\frac{g}{a}\right)td^{t+1}$. With initial condition y_0, $y_0=k$ and $y_t=y_0(-a)^t-\left(\frac{g}{a}\right)td^{t+1}$. Summarizing:

$$y_t=\begin{cases}\left[y_0-\frac{gd}{a+d}\right](-a)^t+\frac{gd^{t+1}}{a+d}, & a+d\neq 0\\ y_0(-a)^t-\left(\frac{g}{a}\right)td^{t+1}, & a+d=0.\end{cases}$$

17.7.1.2 The case $\varphi(t)=\delta t+\beta$

In this case, try $y_t=\gamma_1 t+\gamma_2$ as a particular solution. Then,

$$\begin{aligned}\gamma_1 t+\gamma_2+a[\gamma_1(t-1)+\gamma_2]&=\delta t+\beta\\ \gamma_1 t+\gamma_2+a\gamma_1(t-1)+a\gamma_2&=\delta t+\beta\\ \gamma_1 t+\gamma_2+a\gamma_1 t-a\gamma_1+a\gamma_2&=\delta t+\beta\\ (\gamma_1+a\gamma_1)t+(1+a)\gamma_2-a\gamma_1&=\delta t+\beta.\end{aligned}$$

Equating coefficients, $(\gamma_1+a\gamma_1)=\delta$ and $(1+a)\gamma_2-a\gamma_1=\beta$. So, assuming $(1+a)\neq 0$, $\gamma_1=\frac{\delta}{1+a}$ and $\gamma_2=\frac{1}{1+a}\left[\beta+a\frac{\delta}{1+a}\right]$. Thus, the general solution is:

$$y_t=k(-a)^t+\frac{\delta}{1+a}t+\frac{1}{1+a}\left[\beta+a\frac{\delta}{1+a}\right].$$

The value of k is determined given an initial value y_0: $y_0 = k + \frac{1}{1+a}\left[\beta + a\frac{\delta}{1+a}\right]$, so that $k = y_0 - \frac{1}{1+a}\left[\beta + a\frac{\delta}{1+a}\right]$.

17.7.1.3 The case $\varphi(t) = \delta t^2 + \beta$

In this case, try $y_t = \gamma_1 t^2 + \gamma_2 t + \gamma_3$ as a particular solution. Then,

$$\begin{aligned}
\gamma_1 t^2 + \gamma_2 t + \gamma_3 + a[\gamma_1(t-1)^2 + \gamma_2(t-1) + \gamma_3] &= \delta t^2 + \beta \\
\gamma_1 t^2 + \gamma_2 t + \gamma_3 + a\gamma_1(t-1)^2 + a\gamma_2(t-1) + a\gamma_3 &= \delta t^2 + \beta \\
\gamma_1 t^2 + \gamma_2 t + \gamma_3 + a\gamma_1 t^2 - 2a\gamma_1 t + a\gamma_1 + a\gamma_2 t - a\gamma_2 + a\gamma_3 &= \delta t^2 + \beta \\
(\gamma_1 + a\gamma_1)t^2 + (\gamma_2 - 2a\gamma_1 + a\gamma_2)t + (\gamma_3 + a\gamma_1 - a\gamma_2 + a\gamma_3) &= \delta t^2 + \beta \\
(1+a)\gamma_1 t^2 + [(1+a)\gamma_2 - 2a\gamma_1]t + [(1+a)\gamma_3 + a(\gamma_1 - \gamma_2)] &= \delta t^2 + \beta.
\end{aligned}$$

This has no solution if $a = -1$ because in that case t^2 does not appear on the left side, but does on the right side, so assume that $a \neq -1$. Then, comparing terms:

$$\begin{aligned}
(1+a)\gamma_1 t^2 &= \delta t^2 \\
\left[(1+a)\gamma_2 - 2a\gamma_1\right]t &= 0 \\
(1+a)\gamma_3 + a\frac{1-a}{a+1} &= \beta.
\end{aligned}$$

Therefore $\gamma_1 = \frac{\delta}{1+a}$, and $[(1+a)\gamma_2 - 2a\gamma_1] = 0$ gives $\gamma_2 = \frac{2a}{1+a}\gamma_1 = \frac{2a}{1+a}\frac{\delta}{1+a} = \frac{2a}{(1+a)^2}\delta$. Finally, $[(1+a)\gamma_3 + a(\gamma_1 - \gamma_2)] = \beta$. Since $\gamma_2 - \gamma_1 = \left(\frac{2a}{1+a} - 1\right) = \frac{a-1}{a+1}$, this gives $(1+a)\gamma_3 + a\frac{1-a}{a+1} = \beta$, so $\gamma_3 = a\frac{a-1}{(a+1)^2} + \frac{\beta}{1+a}$. With initial condition y_0, $y_0 = k(-a)^0 + \gamma_1(0)^2 + \gamma_2(0) + \gamma_3 = k + \gamma_3$ and so $k = y_0 - \gamma_3$. Therefore,

$$y_t = [y_0 - \gamma_3](-a)^t + \gamma_1 t^2 + \gamma_2 t + \gamma_3, \quad a \neq -1.$$

17.7.2 Roots of the characteristic equation

In Section 17.4.2, the solution for the homogeneous equation of the second-order difference equation is given. The following sections elaborate on that result. The three possible cases are considered in turn.

17.7.2.1 Roots of the characteristic equation: real and distinct

In this case, $a_1^2 - 4a_2 > 0$, and a complementary solution is of the form

$$y_t = k_1\lambda_1^t + k_2\lambda_2^t.$$

The constants k_1 and k_2 are arbitrary real numbers. If $|\lambda_1|, |\lambda_2| < 1$ then the complementary solution converges to 0 for any values of k_1 and k_2: $\lim_{t\to\infty} y_t = 0$.

The general solution is: $y_t = y_t^c + y_t^p$, where y_t^c is a complementary solution and y_t^p a particular solution. Depending on the values of a_1 and a_2, the general solution is:

$$y_t = \begin{cases} k_1\lambda_1^t + k_2\lambda_2^t + \frac{c}{1+a_1+a_2}, & a_1 + a_2 \neq -1 \\ k_1\lambda_1^t + k_2\lambda_2^t + \frac{c}{a_1+2a_2}t, & a_1 + a_2 = -1, a_1 + 2a_2 \neq 0. \end{cases} \tag{17.12}$$

Note that a third possibility: $a_1 + a_2 = -1$, $a_1 + 2a_2 = 0$ does not arise since these two equations imply that $a_1 = -2$, $a_2 = 1$, which would imply that $a_1^2 - 4a_2 = 0$. When the roots are real and distinct, the equation pair $a_1 + a_2 = -1$, $a_1 + 2a_2 = 0$ cannot be satisfied.

Initial conditions In either of these general solutions, the constants k_1 and k_2 are as yet undetermined. These can be determined uniquely with additional information about y_t. If the values of y_t at $t = 0$ and $t = 1$, y_0 and y_1, are known, this is sufficient to determine k_1 and k_2, as described next.

Suppose that $a - 1 + a_2 \neq -1$, so that $y_t = k_1\lambda_1^t + k_2\lambda_2^t + \frac{c}{1+a_1+a_2}$, and let $c^* = \frac{c}{1+a_1+a_2}$. Let the two initial conditions be the values of y_t at $t = 0$ and $t = 1$: y_0, y_1. Then $y_0 = k_1 + k_2 + c^*$ and $y_1 = k_1\lambda_1 + k_2\lambda_2 + c^*$. Rearranging, and solving for k_1 and k_2:

$$\begin{bmatrix} y_0 - c^* \\ y_1 - c^* \end{bmatrix} = \begin{bmatrix} 1 & 1 \\ \lambda_1 & \lambda_2 \end{bmatrix} \begin{bmatrix} k_1 \\ k_2 \end{bmatrix} \implies \begin{bmatrix} k_1 \\ k_2 \end{bmatrix} = \frac{1}{\lambda_2 - \lambda_1} \begin{bmatrix} -\lambda_2 & \lambda_1 \\ 1 & 1 \end{bmatrix} \begin{bmatrix} y_0 - c^* \\ y_1 - c^* \end{bmatrix}.$$

In the second case, $a_1 + a_2 = -1$, $a_1 + 2a_2 \neq 0$. Now, $y_0 = k_1 + k_2$ and $y_1 = k_1\lambda_1 + k_2\lambda_2 + \bar{c}$, where $\bar{c} = \frac{c}{a_1+2a_2}$. Solving for k_1, k_2:

$$\begin{bmatrix} k_1 \\ k_2 \end{bmatrix} = \frac{1}{\lambda_2 - \lambda_1} \begin{bmatrix} -\lambda_2 & \lambda_1 \\ 1 & 1 \end{bmatrix} \begin{bmatrix} y_0 \\ y_1 - \bar{c} \end{bmatrix}.$$

17.7.2.2 Roots of the characteristic equation: real and repeated

With repeated roots, $a_1^2 - 4a_2 = 0$. In this case $\lambda_1 = \lambda_2 = \lambda = -\frac{a_1}{2}$ and the complementary solution is:

$$y_t = k_1\lambda^t + k_2 t\lambda^t.$$

This may be checked by substitution into the homogeneous equation: $y_t + a_1 y_{t-1} + a_2 y_{t-2} = 0$. So,

$$\begin{aligned}(k_1\lambda^t + k_2 t\lambda^t) + a_1(k_1\lambda^{t-1} + k_2(t-1)\lambda^{t-1}) + a_2(k_1\lambda^{t-2} + k_2(t-2)\lambda^{t-2}) &= 0\\ \lambda^{t-2}[(k_1\lambda^2 + k_2 t\lambda^2) + a_1(k_1\lambda + k_2(t-1)\lambda) + a_2(k_1 + k_2(t-2))] &= 0\\ (k_1\lambda^2 + a_1 k_1\lambda + a_2 k_1) + (k_2 t\lambda^2 + k_2 a_1(t-1)\lambda + k_2 a_2(t-2)) &= 0\\ k_1(\lambda^2 + a_1\lambda + a_2) + k_2(t\lambda^2 + a_1(t-1)\lambda + a_2(t-2)) &= 0\end{aligned}$$

Now $\lambda = -\frac{a_1}{2}$ and $a_1^2 - 4a_2 = 0$ in this case. Therefore

$$\lambda^2 + a_1\lambda + a_2 = \frac{a_1^2}{4} - \frac{a_1^2}{2} + \frac{a_1^2}{4} = 0.$$

Also,

$$\begin{aligned}(t\lambda^2 + a_1(t-1)\lambda + a_2(t-2)) &= t\frac{a_1^2}{4} - (t-1)\frac{a_1^2}{2} + (t-2)\frac{a_1^2}{4}\\ &= (t-2)\left[\frac{a_1^2}{4} - \frac{a_1^2}{2} + \frac{a_1^2}{4}\right] + 2\left(\frac{a_1^2}{4}\right) - \left(\frac{a_1^2}{2}\right)\\ &= 0.\end{aligned}$$

This confirms that y_t is a complementary solution.

The curve $a_1^2 - 4a_2 = 0$ is tangent to the line $a_1 + a_2 = -1$ at $a_1 = -2$, $a_2 = 1$, the only point common to both. At this point, $a_1 + 2a_2 = 0$. So, together, $a_1^2 - 4a_2 = 0$ and $a_1 + a_2 = -1$ eliminate the possibility that $a_1 + 2a_2 \neq 0$ (see Figure 17.6).

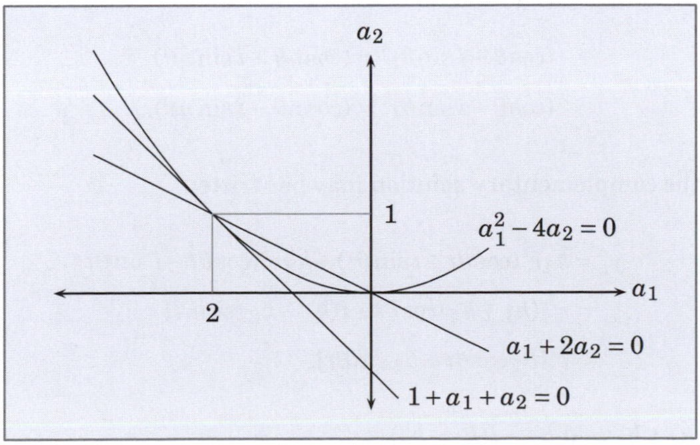

Figure 17.6: Plot of $a_1^2 - 4a_2 = 0$, $1 + a_1 + a_2 = 0$ and $a_1 + 2a_2 = 0$

So, when $a_1^2 - 4a_2 = 0$, there are two possibilities for the particular solution: $y_t^p = \frac{c}{1+a_1+a_2}$ if $a_1 + a_2 \neq -1$ and $y_t^p = \frac{c}{2}t$ if $a_1 + a_2 = -1$. Therefore, when $a_1^2 - 4a_2 = 0$, the general solution is:

$$y_t = \begin{cases} k_1\lambda^t + k_2 t\lambda^t + \frac{c}{1+a_1+a_2}, & a_1 + a_2 \neq -1 \\ k_1\lambda^t + k_2 t\lambda^t - \frac{c}{2}t^2, & a_1 + a_2 = -1, a_1 + 2a_2 = 0. \end{cases}$$

If $|\lambda| < 1$, then both λ^t and $t\lambda^t$ go to 0. The solution converges to $\frac{c}{1+a_1+a_2}$ if $a_1 + a_2 \neq -1$. If $a_1 + a_2 = -1$, then the solution diverges because $\frac{c}{2}t^2$ diverges.

Initial conditions Choosing values for some pair $y_{t'}, y_{t''}$ determine k_1 and k_2.

17.7.2.3 Roots of the characteristic equation: complex and distinct

With complex roots, $a_1^2 - 4a_2 < 0$. Let $\alpha = -\frac{a_1}{2}, \beta = \frac{\sqrt{4a_2 - a_1^2}}{2}$, then $\lambda_1, \lambda_2 = \alpha \pm i\beta$, where $i = \sqrt{-1}$, give the roots of $b^2 + a_1 b + a_2 = 0$. Thus, the complementary solution has the form:

$$y_t^c = k_1\lambda_1^t + k_2\lambda_2^t.$$

Considering λ_1, this has value α in the real plane (x-axis) and β in the imaginary plane (y-axis). The angle formed by the ray from the origin is θ given by $\theta = \cos^{-1}(\frac{\alpha}{r}) = \sin^{-1}(\frac{\beta}{r})$, so we can write $\alpha = r\cos\theta$ and $\beta = r\sin\theta$, where $r = |\lambda_1| = |\lambda_2| = \sqrt{\alpha^2 + \beta^2}$ (the length of the hypotenuse). Thus, $\lambda_1 = r(\cos\theta + i\sin\theta)$ and $\lambda_2 = r(\cos\theta - i\sin\theta)$. According to de Moivre's theorem

$$(\cos\theta + i\sin\theta)^n = (\cos n\theta + i\sin n\theta)$$
$$(\cos\theta - i\sin\theta)^n = (\cos n\theta - i\sin n\theta).$$

Therefore, the complementary solution may be written:

$$\begin{aligned} y_t^c &= k_1 r^t(\cos\theta t + i\sin\theta t) + k_2 r^t(\cos\theta t - i\sin\theta t) \\ &= r^t[(k_1 + k_2)\cos\theta t + i(k_1 - k_2)\sin\theta t] \\ &= r^t[b_1\cos\theta t + b_2\sin\theta t], \end{aligned}$$

where $b_1 = k_1 + k_2$ and $b_2 = i(k_1 - k_2)$.

Since the solution is real, b_1 and b_2 must be real. Setting $k_1 = k_2 = \frac{k}{2}$, $b_1 = k$, $b_2 = 0$ and $y_t^c = r^t c\cos(\theta t)$. Similarly, setting $k_1 = -i\frac{b}{2}$, $k_2 = i\frac{b}{2}$, $b_1 = k_1 + k_2 = 0$ and

$b_2 = i(k_1 - k_2) = i(-i)c = c$, and $y_t^c = r^t c\sin(\theta t)$. Thus, the complementary solution is of the form $y_t^c = r^t[c_1\cos(\theta t) + c_2\sin(\theta t)]$.

For the particular solution, observe that if $(1 + a_1 + a_2) = 0$ then $a_1 = -(1 + a_2)$ so that $a_1^2 - 4a_2 = [-(1+a_2)]^2 - 4a_2 = 1 + 2a_2 + a_2^2 - 4a_2^2 = 1 - 2a_2 + a_2^2 = (1-a_2)^2 \geq 0$. Thus, $1 + a_1 + a_2 = 0$ implies $a_1^2 - 4a_2 \geq 0$. With complex roots, $a_1^2 - 4a_2 < 0$, it must be that $1 + a_1 + a_2 \neq 0$. Consequently, considering

$$y_t + a_1 y_{t-1} + a_2 y_{t-2} = c,$$

the particular solution is given by: $y_t^p = \frac{c}{1+a_1+a_2}$. So, the general solution is:

$$y_t = r^t[k_1\cos(\theta t) + k_2\sin(\theta t)] + \frac{c}{1 + a_1 + a_2}.$$

Note that $r = \sqrt{\frac{a_1^2}{4} + \frac{4a_2 - a_1^2}{4}} = \sqrt{a_2}$ and $a_2 > 0$ since $a_1^2 - 4a_2 < 0$. Also, $|\cos(\theta t)| \leq 1$ and $|\sin(\theta t)| \leq 1$.

Summarizing the discussion of the second-order difference equation, $y_t + a_1 y_{t-1} + a_2 y_{t-2} = c$: the particular and complementary solutions have the form:

$$y_t^p = \begin{cases} \frac{c}{1+a_1+a_2}, & \textit{if } a_1 + a_2 \neq -1 \\ \frac{c}{a_1+2a_2}t, & \textit{if } a_1 + a_2 = -1, a_1 + 2a_2 \neq 0 \\ \frac{c}{2}t, & \textit{if } a_1 + a_2 = -1, a_1 + 2a_2 = 0 \end{cases}$$

$$y_t^c = \begin{cases} k_1\lambda_1^t + k_2\lambda_2^t, & \text{roots real and distinct} \\ k_1\lambda_1^t + k_2 t\lambda_2^t, & \text{roots real and equal} \\ r^t[B_1\cos(\theta t) + B_2\sin(\theta t)], & \text{roots complex,} \end{cases}$$

with general solution $y_t = y_t^p + y_t^c$.

17.7.3 Stability

An important question concerns the dynamic behavior of the solution over time: how does the general solution, y_t, vary over time? To study this, the three cases must be considered separately.

17.7.3.1 Stability: roots real and distinct

In this case $a_1^2 - 4a_2 > 0$. For stability, the roots must have absolute value less than 1. Therefore

$$(a) \quad -1 < \lambda_1 = \frac{-a_1 + \sqrt{a_1^2 - 4a_2}}{2} < 1, \text{ and } (b) \quad -1 < \lambda_2 = \frac{-a_1 - \sqrt{a_1^2 - 4a_2}}{2} < 1.$$

These conditions may be written:

$$(a) \quad -2 + a_1 < \sqrt{a_1^2 - 4a_2} < 2 + a_1 \text{ and } (b) \quad -2 + a_1 < -\sqrt{a_1^2 - 4a_2} < 2 + a_1.$$

Since $-\sqrt{a_1^2 - 4a_2} < 0 < \sqrt{a_1^2 - 4a_2}$, we can combine these equalities:

$$-2 + a_1 < -\sqrt{a_1^2 - 4a_2} < 0 < \sqrt{a_1^2 - 4a_2} < 2 + a_1.$$

Squaring the right side, $\left(\sqrt{a_1^2 - 4a_2}\,\right)^2 < (2 + a_1)^2$ gives:

$$a_1^2 - 4a_2 < 4 + 4a_1 + a_1^2, \quad \text{or} \quad 0 < 4 + 4a_1 + 4a_2 \quad \text{or} \quad 0 < 1 + a_1 + a_2$$

Considering the left side, $-2 + a_1 < -\sqrt{a_1^2 - 4a_2} < 0$, implies $(-2 + a_1)^2 > a_1^2 - 4a_2$ (if x, y are two negative numbers, then $x < y$ implies that $x^2 > y^2$):

$$(-2 + a_1)^2 > a_1^2 - 4a_2 \quad \text{or} \quad 4 - 4a_1 + a_1^2 > a_1^2 - 4a_2 \quad \text{or} \quad 4 - 4a_1 + 4a_2 > 0$$
$$\text{or} \quad 1 - a_1 + a_2 > 0.$$

Also, note that $0 < 2 + a_1$ and $-2 + a_1 < 0$, so that $-2 < a_1 < 2$ (and since $a_1^2 - 4a_2 > 0$, $a_2 < 1$). To summarize, with $a_1^2 - 4a_2 > 0$, the model is stable if:

- $-2 < a_1 < 2$
- $1 + a_1 + a_2 > 0$
- $1 - a_1 + a_2 > 0$.

17.7.3.2 Real and equal roots

In this case, $a_1^2 - 4a_2 = 0$. In this case we require $a_1 + a_2 \neq -1$ (so the particular solution is constant) and $-1 < \frac{a_1}{2} < 1$, so the complementary solution converges.

The condition $|\frac{a_1}{2}| < 1$ is equivalent to $-2 < a_1 < 2$. Therefore, in the case $a_1^2 - 4a_2 = 0$, stability requires:

- $a_1 + a_2 \neq -1$
- $-2 < a_1 < 2$.

17.7.3.3 Complex distinct roots

Now $a_1^2 - 4a_2 < 0$. This implies that $a_1 + a_2 \neq -1$, so the particular solution is constant. The complementary solution converges if $a_2 < 1$ (and $a_2 > 0$ since $a_1^2 - 4a_2 < 0$). So, in the case where $a_1^2 - 4a_2 < 0$, stability requires that

- $a_2 < 1$.

17.7.3.4 Stability regions

These three cases determine three regions within the set $\{(a_1, a_2) \mid -2 \leq a_1 \leq 2, -1 \leq a_2 \leq 1\}$.

1. Real distinct roots: (i) $a_1^2 - 4a_2 > 0$, (ii) $1 + a_1 + a_2 > 0$, (iii) $1 - a_1 + a_2 > 0$.
2. Real equal roots: (i) $a_1^2 - 4a_2 = 0$, (ii) $-2 < a_1 < 2$.
3. Complex distinct roots: (i) $a_1^2 - 4a_2 < 0$, (ii) $a_2 < 1$.

A graph of these regions is given in Figure 17.7.

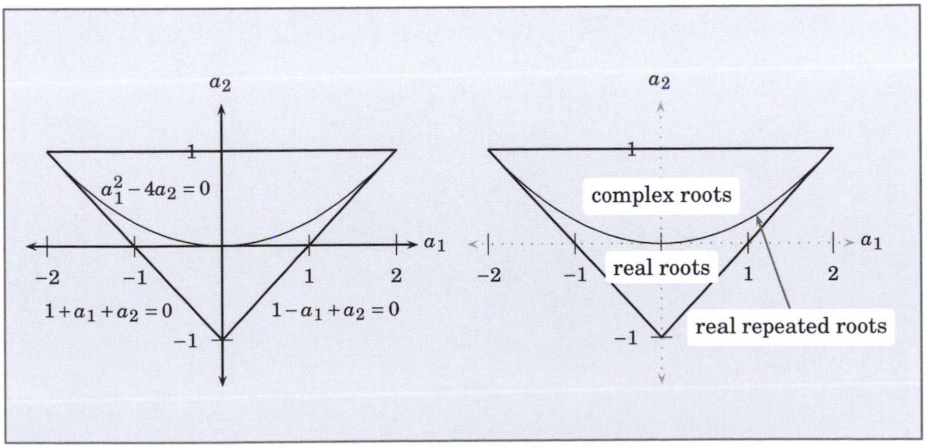

Figure 17.7: Stable regions: real, repeated and complex roots

Exercises for Chapter 17

Exercise 17.1 Let y_0 be the initial value of y. Find the long-run dynamics of the following four systems:

$$y_t = \frac{1}{2}y_{t-1} + 2 \tag{17.13}$$

$$y_t = -\frac{1}{2}y_{t-1} + 2 \tag{17.14}$$

$$y_t = 2y_{t-1} + 2 \tag{17.15}$$

$$y_t = -2y_{t-1} + 2. \tag{17.16}$$

Exercise 17.2 Suppose that an amount P_0 is borrowed at time $t = 0$. The interest rate is r so after one period the debt grows to $P_0(1+r)$. Each period a repayment d is made, so that the end of period 1, the outstanding debt is $P_1 = P_0(1+r) - d$. The value of the loan proceeds over time in this way.

(a) Find the equation governing the evolution of P_t.

(b) Given a period T, find the value of d that amortizes (pays off) the loan exactly at the end of period T.

Exercise 17.3 In a labor-market model, suppose that $p(u \mid e)$ is the fraction of employed people who become unemployed next period and $p(u \mid u)$ the fraction of unemployed who remain unemployed. Let $p(u \mid e) = 0.05$ and $p(u \mid u) = 0.3$. If u_t

is the unemployment rate (and $1-u_t$ the unemployment rate), then u_t evolves according to the equation:

$$u_{t+1} = p(u \mid e)(1-u_t) + p(u \mid u)u_t.$$

Find the long-run dynamic behavior of u_t.

Exercise 17.4 Consider the dynamic model

$$y_{t+1} = ay_t + c.$$

(a) Plot the equation $y_{t+1} = ay_t + c$ and the steady-state line $y_{t+1} = y_t$ in (y_{t+1}, y_t) space, for:

(i) $a = \frac{1}{2}$ and $c = 4$.

(ii) $a = -\frac{1}{2}$ and $c = 4$.

(iii) $a = 2$ and $c = -1$.

(b) In each case, trace the dynamics from the initial value of $y_0 = 1$.

Exercise 17.5 Let y_0 be the initial value of y. Find the long-run dynamics of the following four systems:

$$y_t = \frac{1}{2}y_{t-1} + 3t \tag{17.17}$$

$$y_t = -\frac{1}{2}y_{t-1} + 3t \tag{17.18}$$

$$y_t = 2y_{t-1} + 3t \tag{17.19}$$

$$y_t = -2y_{t-1} + 3t. \tag{17.20}$$

Exercise 17.6 Let y_0 be the initial value of y. Find the long-run dynamics of the following four systems:

$$y_t = \frac{1}{2}y_{t-1} + 3^t \tag{17.21}$$

$$y_t = -\frac{1}{2}y_{t-1} + 2^t \tag{17.22}$$

$$y_t = 2y_{t-1} + \left(\frac{1}{2}\right)^t \tag{17.23}$$

$$y_t = -2y_{t-1} + 2^t. \tag{17.24}$$

18

Probability and distributions

18.1 Introduction

A probability distribution gives the frequency of occurrence of some variable. For example, the income distribution is summarized by a function F, where $F(y)$ gives the fraction of the population with income at or below y. If a person is selected at random from the population, then the probability that the person will have income below y is $F(y)$. Similarly, one may consider the age distribution, represented by G, so that $G(a)$ is the fraction of people with age no larger than a, or a wage distribution where $H(w)$ gives the proportion of people earning a wage no larger than w. Thus, a distribution provides information regarding the frequency of a certain characteristic or property in a population. Considering the income distribution, suppose a person is selected at random from the population and income recorded. At the selection process, let Y be the as yet to be determined income. This unknown number is called a random variable. The probability that the person selected at random will have an income below y is denoted $P(Y \leq y)$ and this is equal to $F(y)$. In Section 18.2, random variables are introduced and briefly described. Intuitively, every distribution function describes the behavior of a random variable. Section 18.3 introduces the distribution function and its density.

In some cases, a random variable may only have a finite number of possible values. For example, if age is recorded in years, then a person selected at random will have an age in the set $\{0,1,\dots,\bar{a}\}$ where $\bar{a}$ is the age of the oldest person alive. Or there may be an infinite number of possibilities – consider picking a positive integer, so the possible outcomes are given by the set $\{1,2,\dots\}$. In this case there is a (countable) infinity of possible outcomes. Finally, the outcome possibilities may be infinite: if a number is chosen at random from the unit interval, then there is a continuum of possible outcomes (any number in the interval [0,1]). In the case where the random variable has a finite or countable number of possible outcomes, say that the random variable is discrete, if there are a continuum of possible outcomes, say the random variable is a continuous variable. These categorizations lead to an analogous categorization of distributions into those which attach positive probability to a finite or countable set of points, called discrete distributions, and another group where it is not possible to describe the distribution in terms of the probability of each individual outcome. For a certain category, these distributions are represented by a continuous distribution function (and this may be generalized to allow for discontinuities). Section 18.4 considers a few important discrete distributions (the Bernoulli, binomial and Poisson distributions). Finally, Section 18.5 describes the primary distributions used in hypothesis testing in economics: the uniform, normal, t, chi-square and F distributions.

18.2 Random variables

Consider the experiment of tossing a six-sided die with each side (labeled 1 to 6) equally likely. For the experiment, there are six possible outcomes $\{1,2,3,4,5,6\}$ and each outcome is equally likely with probability $\frac{1}{6}$. We can formalize this by denoting the outcome of the die toss by X, where X can take on any one of the six possible values. Write $P(X = i)$ to denote the probability that X takes the value i (the toss results in side i facing up). With the probability of each outcome assigned a number, we can determine the probability that the number drawn is less than or equal to 3 ($P(X \leq 3) = P(X = 1) + P(X = 2) + P(X = 3)$), that the probability is greater than 3 ($P(X > 3) = P(X = 4) + P(X = 5) + P(X = 6) = 1 - P(X \leq 3)$), and so on. The notion of a random variable gives a formal way of examining random outcomes with associated probabilities. Any experiment with a finite number of outcomes $\{x_1, x_2, \ldots, x_n\}$ can be represented in this way: if outcome x_i has probability p_i, the associated random variable has corresponding probability $P(X = x_i) = p_i$. With this formalism in place one can identify defining features of the random process. The mean is denoted μ and defined $\mu = \sum_{i=1}^{n} p_i x_i$, and the variance, σ^2, is defined $\sigma^2 = \sum_{i=1}^{n} p_i (x_i - \mu)^2$. The mean provides one measure of "central tendency". For example, if the die were tossed very many times we would expect 1 to be drawn roughly one-sixth of the time, and likewise for the other numbers – so that the average would be approximately equal to the mean

$\mu = \sum_i p_i x_i = \frac{1}{6}1 + \frac{1}{6}2 + \cdots + \frac{1}{6}6 = 3\frac{1}{2}$. While the mean provides a measure of "central tendency", variance measures dispersion. In the die-tossing case, the variance is:

$$\sigma^2 = \left(1-3\tfrac{1}{2}\right)^2 + \left(2-3\tfrac{1}{2}\right)^2 + \left(3-3\tfrac{1}{2}\right)^2 + \left(4-3\tfrac{1}{2}\right)^2 + \left(5-3\tfrac{1}{2}\right)^2 + \left(6-3\tfrac{1}{2}\right)^2 = 17\tfrac{1}{2}.$$

This formalism allows one to study or represent arbitrary situations where outcomes are unknown. For example, we could let X denote the number of inches (or centimeters) of rainfall in a given city for next year, so that $P(X = i) = p_i$ denotes the probability that there will be i inches of rainfall. In this case, we can say that the probability that there will be i inches of rainfall or less next year is $P(X \le i) = \sum_{j \le i} p_j$. From this perspective, the numbers p_i are given – laws of nature generate the rainfall according to the same distribution from year to year. We may attempt to estimate the distribution using historical data: let $\hat{p}_i$ denote the fraction of recorded years in which rainfall was i. Then, $\hat{p}_i$ provides an estimate of the value of p_i. Recall that a random variable is called discrete if it can have only a finite or countable number (distinct values may be labeled with the integers) of possible values. In the finite case, the random variable is described by the outcomes $\{x_1, x_2, \ldots, x_n\}$ and probabilities $\{p_1, p_2, \ldots, p_n\}$. For example, with eight possible outcomes, the distribution may be depicted as in Figure 18.1.

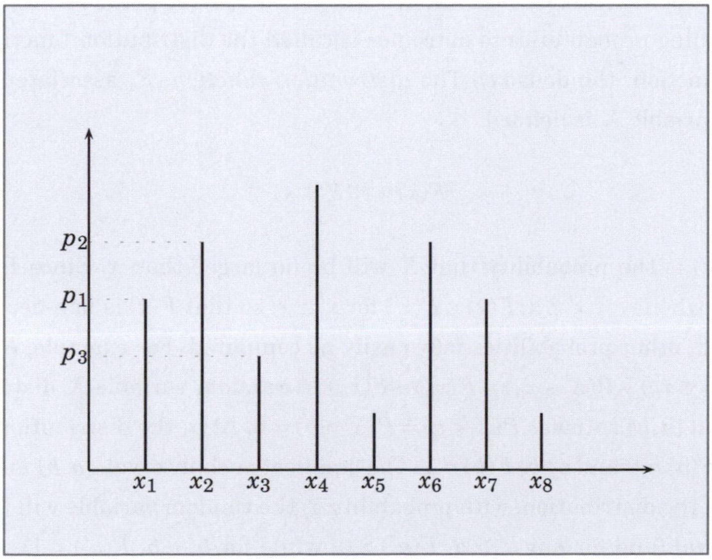

Figure 18.1: Probability density

As an example of the countable case, let the possible outcomes be $\{x_1, x_2, \ldots, x_n, \ldots\}$, where $P(X = x_i) = (1-\delta)\delta^{i-1}$. Thus, $P(X = x_1) = (1-\delta)$, $P(X = x_2) = (1-\delta)\delta$,

$P(X = x_3) = (1-\delta)\delta^2$, and so on. Thus, $\sum_i P(X = x_i) = (1-\delta)[1+\delta+\delta^2+\cdots] = 1$, so the probabilities add up to 1. For a discrete random variable X that has outcome x_i with probability p_i, the mean $\mu = \sum x_i p_i$ is often called the expected value of X and written $E\{X\}$. Similarly, the variance, $\sigma^2 = \sum(x_i-\mu)^2 p_i$ is written $E\{(X-\mu)^2\}$.

If the random variable is *continuous*, there is a continuum of possible outcomes. Some random events are most suitably modeled using continuous variables. For example, consider a device for filling cans of oil. Design specification requires that each can be filled to, say, 330 ml. In operation, few cans will be filled exactly; although the average fill may be 330 ml, a can selected at random may have, for example, 328.755 ml. However, most cans will be filled to a level close to 330 ml. Describing the dispersion around 330 ml; is most easily modeled with a continuous random variable. In this case, for example, consider a random variable X with the property that $P(X < 320) = 0$ and $P(X > 340) = 0$, and for $x \in [320, 340]$, $P(X < x) = \frac{x-320}{340-320} = \frac{x-320}{20}$. Thus, for example, $P(X \le 330) = \frac{10}{20}$.

18.3 The distribution function and density

A random variable takes on different values with specific probabilities. The function attaching probabilities to outcomes is called the distribution function (or the related function, the density). The *distribution function*, F, associated with the random variable X is defined

$$F(x) = P(X \le x). \tag{18.1}$$

Thus, $F(x)$ is the probability that X will be no larger than x. Since $P(X \le x) \le P(X \le x')$ whenever $x' \ge x$, $F(x) \le F(x')$ for $x' \ge x$, so that $F(x)$ is non-decreasing in x. Given F, other probabilities may easily be computed. For example, $P(x_a < X \le x_b) = P(X \le x_b) - P(X \le x_a) = F(x_b) - F(x_a)$. A random variable X distributed on an interval $[a,b]$ satisfies $P(X < a) = P(X > b) = 0$. At a, the distribution function satisfies $F(a) = 0$ and at b, $F(b) = 1$. The smallest such interval, $[a,b]$ is called the *support* of the distribution: with probability 1, the random variable will fall within this interval (and for any $a' > a$, $F(a') > 0$, while for $b' < b$, $F(b') < 1$). A random variable X is non-negative if $P(X \ge 0) = 1$. Figure 18.2 illustrates two distribution functions with supports $[0,1]$ and $[0,4]$, respectively.

In the examples depicted, the distribution functions are continuous. In the case of discrete random variables, the distribution function is defined in the same way: $F(x) = P(X \le x)$. Since X takes values from a set $\{x_1, x_2, \ldots, x_k, \ldots\}$, with

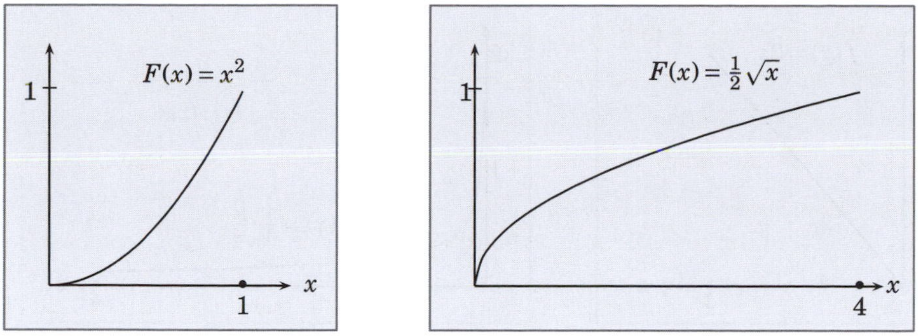

Figure 18.2: Cumulative distribution functions

$x_i < x_{i+1}$ for all i, $F(x) = \sum_{x_i \le x} P(X = x_i)$, so that the distribution is a step function increasing by the amount $P(X = x_i) = p_i$ at $x = x_i$.

In general, a distribution function may have features of continuous and discrete distributions. For example, $F(x) = \frac{1}{2}x$ for $x < \frac{1}{2}$ and $F(x) = x$ for $x \ge \frac{1}{2}$ is a distribution function of a random variable with support $[0,1]$ that takes the value $\frac{1}{2}$ with probability $\frac{1}{4}$.

For a continuous distribution, F, for any x, $F(x)-F(x-\Delta x) \to 0$ as $\Delta x \to 0$. However, such a distribution may not be differentiable. For example, $F(x) = 2x^2$, $x < \frac{1}{2}$ and $F(x) = x$, $\frac{1}{2} \le x \le 1$ is a distribution function with support $[0,1]$ that is not differentiable at $\frac{1}{2}$. When F is continuous and differentiable, denote the derivative of F by $f(x) = F'(x)$; this is called the *density* of F. From the definition of a derivative, $f(x) = \lim_{\Delta x \to 0} \frac{F(x+\Delta x)-F(x)}{\Delta x}$. Thus, when $\Delta x > 0$ is small $f(x) \approx \frac{1}{\Delta x}[F(x+\Delta x) - F(x)]$ or:

$$f(x)\Delta x \approx F(x+\Delta x) - F(x) = P(X \le x + \Delta x) - P(X \le x) = P(x < X \le x + \Delta x).$$

Therefore, $f(x)\Delta x$ is approximately the probability that the random variable X will fall in the interval $[x, x+\Delta x]$. From the definition of f,

$$\int_a^b f(x)dx = \int_a^b F'(x)dx = F(b) - F(a) = P(a < X \le b),$$

so the distribution function F can be obtained from the density f. Considering the two distribution functions plotted in Figure 18.2, $F(x) = x^2$ has density $f(x) = 2x$, and $F(x) = \frac{1}{2}\sqrt{x}$ has density $f(x) = \frac{1}{4}\frac{1}{\sqrt{x}}$ and these are depicted in Figure 18.3.

While the distribution function has values between 0 and 1, the density function can have arbitrarily large positive values. If F is a distribution function with

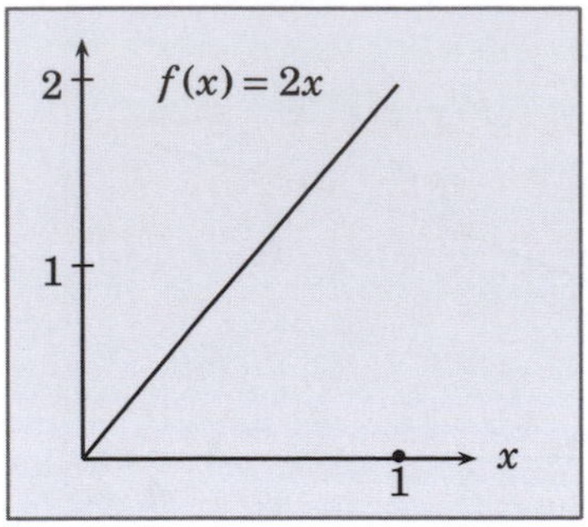

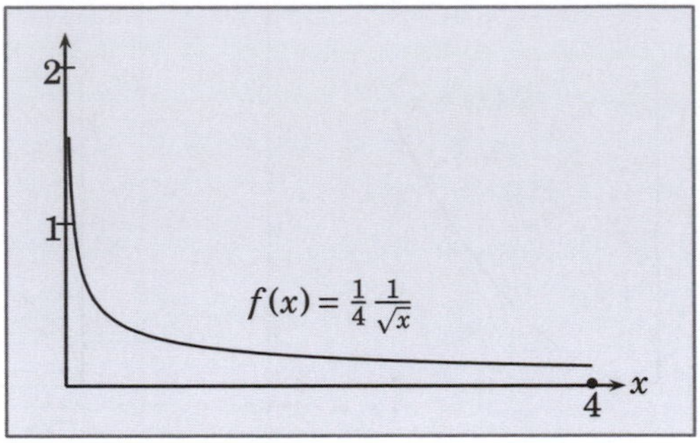

Figure 18.3: Density functions

$F(a) = 0$, $F(b) = 1$ and density f, then $\int_a^b f(x)dx = F(b) - F(a) = 1$, so that the area under f is 1. (Here it may be that $a = -\infty$ or $b = \infty$, or both.)

18.3.1 The conditional distribution

The probability that the random variable falls between a and b ($a < b$) is given by $F(b) - F(a)$. If b is the largest possible value for the random variable, $F(b) = 1$ then the probability that the random variable is larger than a is given by $1 - F(a)$. The conditional probability that the random variable X lies between a and $a' > a$ is given by:

$$P(X \leq a' \mid X > a) = \frac{F(a') - F(a)}{1 - F(a)}.$$

The logic for this expression is simple. The probability that X lies in the interval $[a, a']$ is $F(a') - F(a)$ and the probability that X falls above a' is $1 - F(a')$. The total probability of these two outcomes is $1 - F(a) = (F(a') - F(a)) + (1 - F(a'))$, so the relative likelihood of each outcome adds up to 1 with

$$1 = \frac{F(a') - F(a)}{1 - F(a)} + \frac{1 - F(a')}{1 - F(a)}.$$

Given that X is above a, the probability that it is in the interval $[a, a']$ is $\frac{F(a')-F(a)}{1-F(a)}$ and the probability that it is on or above a' is $\frac{1-F(a')}{1-F(a)}$. Hence, the conditional probability of X falling in $[a, a']$ given that $X \geq a$ is $P(X \leq a' \mid X > a) = \frac{F(a')-F(a)}{1-F(a)} = F(X \leq a' \mid x \geq a)$.

18.3.2 Joint distributions

Given two random variables, X and Y, let the probability that X is in a set (called an event) A and Y in a set B be $P(X \in A, Y \in B)$. When A and B are two half intervals, $A = [-\infty, \bar{x}]$ and $B = [-\infty, \bar{y}]$, this becomes

$$P(X \in A,\ Y \in B) = P(X \in (-\infty, x],\ Y \in (-\infty, y]) = P(X \le x,\ Y \le y). \tag{18.2}$$

Definition 18.1: *The joint distribution of two random variables X, Y is a function $F(x, y)$ giving the probability that X is no larger than x and Y is no larger than y, and is defined:*

$$F(x, y) = P(X \le x \text{ and } Y \le y). \tag{18.3}$$

Assuming sufficient differentiability, the density $f(x, y)$ associated with F is defined as $f(x, y) = \frac{\partial^2 F}{\partial x \partial y}$.

EXAMPLE 18.1:

$$F(x, y) = \frac{xy}{(2 - xy)}, \quad 0 \le x \le 1,\ 0 \le y \le 1 \tag{18.4}$$

(Figure 18.4) with corresponding density

$$F_{xy}(x, y) = f(x, y) = \frac{1}{(2 - xy)} + \frac{3yx}{(2 - xy)^2} + \frac{2x^2y^2}{(2 - xy)^3}.$$

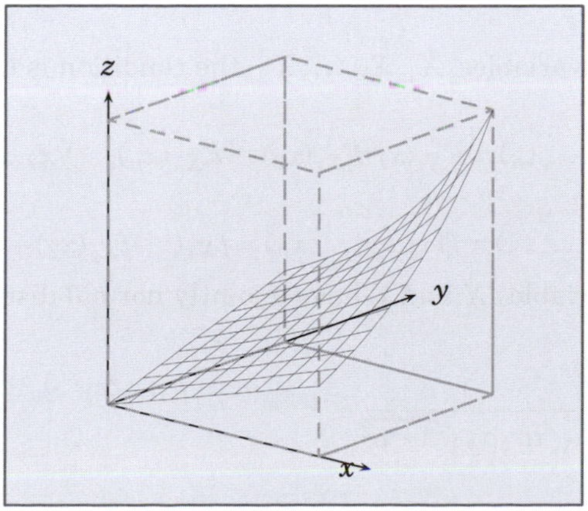

Figure 18.4: The function $F(x, y) = \frac{xy}{2 - xy}$

Because X and Y are random variables, we may also consider the probability that X or Y lie in specific sets, such as $P(X \in A)$. In this case, the distribution of X is $F_X(x) = P(X \leq x)$ and the distribution of Y, $F_Y(y) = P(Y \leq y)$.

18.3.2.1 Independence

Definition 18.2: *The random variables X and Y are said to be independent if for any events A and B,*

$$P(X \in A, Y \in B) = P(X \in A)P(Y \in B). \tag{18.5}$$

An equivalent condition is that X and Y are independent if

$$P(X \leq x, Y \leq y) = P(X \leq x)P(Y \leq y), \quad \forall x, \forall y. \tag{18.6}$$

In terms of the distribution function, two random variables X and Y are independent if the joint distribution $F(x,y)$ may be written $F(x,y) = F_X(x)F_Y(y)$ for all x,y, where F_X (F_Y) is the distribution of X (Y). Equivalently, X and Y are independent if the density factors satisfy $f(x,y) = f_X(x)f_Y(y)$, where f_X (f_Y) is the density of X (Y).

> EXAMPLE 18.2: If $F(x,y) = xy$, then with $F_X(x) = x$ and $F_Y(y) = y$, $F(x,y) = F_X(x)F_Y(y)$, and X and Y are independent. ◊

With n random variables, $X_1, X_2, \ldots, X_n$, the condition is unchanged:

$$F(x_1, x_2, \ldots, x_n) = F_{X_1}(x_1)F_{X_2}(x_2)\cdots F_{X_n}(x_n), \quad \forall x_2, x_2, \ldots, x_n,$$

and $F_{x_1x_2\cdots x_n}(x_1, x_2, \ldots, x_n) = f(x_1, x_2, \ldots, x_n) = f_{X_1}(x_1)f_{X_2}(x_2)\cdots f_{X_n}(x_n)$.

Two random variables X and Y have a jointly normal distribution if

$$f(x,y) = \frac{1}{\sqrt{\pi}\sigma_x\sigma_y\sqrt{1-\rho^2}} e^{-\frac{1}{2(1-\rho)}[(\frac{x-\mu_x}{\sigma_x})^2 + (\frac{y-\mu_y}{\sigma_y})^2 - 2\rho\frac{(x-\mu_x)(y-\mu_y)}{\sigma_x\sigma_y}]},$$

where $\rho = \frac{\mathrm{Cov}(X,Y)}{\sigma_x\sigma_y} = \frac{E\{(X-\mu_x)(Y-\mu_y)\}}{\sigma_x\sigma_y}$ is the correlation coefficient between X and Y.

In the special case where $\rho = 0$,

$$
\begin{aligned}
f(x,y) &= \frac{1}{\pi\sigma_x\sigma_y} e^{-\frac{1}{2}[(\frac{x-\mu_x}{\sigma_x})^2+(\frac{y-\mu_y}{\sigma_y})^2]} \\
&= \frac{1}{\sqrt{\pi}\sigma_x} e^{-\frac{1}{2}(\frac{x-\mu_x}{\sigma_x})^2} \frac{1}{\sqrt{\pi}\sigma_y} e^{\frac{1}{2}(\frac{y-\mu_y}{\sigma_y})^2} \\
&= f(x)f(y),
\end{aligned}
$$

where $f(x) = \frac{1}{\sqrt{\pi}\sigma_x} e^{-\frac{1}{2}(\frac{x-\mu_x}{\sigma_x})^2}$.

REMARK 18.1: Note, therefore, that for the normal distribution, 0-correlation implies independence: if two normally distributed random variables are uncorrelated, then they are independent. □

18.4 Discrete distributions

A discrete distribution is characterized by a set of possible outcomes $\{x_1, x_2, \ldots\}$ and corresponding probabilities $\{p_1, p_2, \ldots\}$. Thus, the random variable X takes the value x_i with probability p_i. The expectation of X is $\mu = \sum_i p_i x_i$; the variance $\sigma^2 = \sum_i p_i (x_i - \mu)^2$. More generally, given any function h, one can define the expectation of h as $E\{h(X)\} = \sum_i p_i h(x_i)$. The following section describes three distributions, the Bernoulli, binomial and Poisson distributions.

18.4.1 The Bernoulli distribution

Consider a random variable X with two possible outcomes, 0 or 1. Suppose the probability of the outcome 1 is given by p (and $1-p$ for the outcome 0): $P(X=1) = p$, $P(X=0) = 1-p$. Then, the expected value of X is $E\{X\} = p \cdot 1 + (1-p) \cdot 0 = p$; and the variance $\sigma^2 = p(1-p)^2 + (1-p)(0-p)^2 = p(1-p)^2 + (1-p)p^2 = p(1-p)$ $[(1-p)+p] = p(1-p)$.

18.4.2 The binomial distribution

Suppose that there is a large number of identical coins (a population of coins), labeled 0 on one side and 1 on the other. Each coin when tossed has probability p of side 1 coming up, and probability $(1-p)$ of 0 coming up. Suppose that

n coins are drawn randomly from this population and each coin tossed in turn with the outcome recorded. The resulting outcome is a list of n entries such as $(0,1,1,0,\cdots,1)$ or $(1,1,0,1,\cdots,0)$. In each position, there are two possible entries: 0 or 1, so there are 2^n possible lists. The probability of each such list is easy to determine. For example, the probability of the list $(0,1,1,0,\cdots,1)$ is the probability that 0 is drawn first, followed by 1, followed by 1, followed by 0, and so on, giving the probability $(1-p)\cdot p\cdot p\cdot(1-p)\cdot\cdots\cdot p$. If there are k 1s and $n-k$ 0s, then this number is $p^k(1-p)^{n-k}$. Any other list with exactly k 1s will have the same probability. How many lists are there with exactly k 1s? This is discussed next.

18.4.2.1 Combinations and permutations

Suppose that there are n distinct objects, labeled 1 to n. One could imagine putting these objects side by side. How many distinct arrangements are there? First, observe that in placing an object in the first position, there are n choices – for any one of the n objects. With that done, there are $n-1$ possible choices left for the next (second) position, then $n-2$ choices for the third position, and so on. Thus, there are $n\times(n-1)\times(n-2)\times\cdots\times 1$ possible arrangements. This expression is denoted $n!$. Consider next the case where there are k objects to be placed in n possible positions. There are n possible positions for the first object, $(n-1)$ positions for the second, and $n-k$ positions for the kth object. Thus, there are $n\times(n-1)\cdots\times(n-k+1)$ possible arrangements of the k objects in the n positions. Observe that $n\times(n-1)\times\cdots\times(n-k+1)=\frac{n\times(n-1)\times\cdots\times 2\times 1}{(n-k)\times(n-k-1)\times\cdots\times 1}=\frac{n!}{(n-k)!}$. This expression is called *the number of permutations of n elements taking k at a time* and is denoted $P_{n,k}$. For example, with two objects labeled 1 and 2, and three possible positions, these objects can be arranged as follows: $(1,2,\star)$, $(1,\star,2)$, $(2,1,\star)$, $(2,\star,1)$, $(\star,1,2)$, $(\star,2,1)$, and here $3\times 2=\frac{3!}{(3-2)!}=6$.

Notice that in the example, comparing the first and third arrangement, you see that the same positions are occupied, and likewise comparing the second and fourth, and the fifth and sixth arrangements. If instead, we were just interested in the positions occupied by the set of k items, then to fill any given k positions with k items, there are k possible choices for the first entry, $k-1$ for the second, and so forth. Thus, there are $k!$ ways in which we can place the k objects to make an arrangement. In the example, the first and third arrangement have the same locations occupied by interchanging the items. Since any given k positions in an arrangement can be occupied in $k!$ different ways (by assigning the objects), the number of distinct ways in which the n positions can be filled by the k objects is $\frac{n!}{(n-k)!}\times\frac{1}{k!}=\frac{n!}{(n-k)!k!}$. In the example, denoting a position filled by "+", gives $(+,+,\star)$,

$(+,\star,+)$, $(\star,+,+)$. There are $3 = \frac{3!}{(3-2)!2!}$ ways to fill two of the three positions. This observation answers the question of how many ways we can arrange k 1s in n possible positions: $\frac{n}{(n-k)!k!}$. The formula is *the number of combinations of n objects taking k at a time*. It is common to write $\binom{n}{k}$ for this expression: $\binom{n}{k} = \frac{n}{(n-k)!k!}$.

More generally, suppose that the k items can be categorized into r distinct groups. Say that there are k_i items of category i with $k = k_1 + k_2 + \cdots + k_r$. Again, the formula $\frac{n!}{(n-k)!}$ counts as distinct all possible arrangements of a given category. If the items in category i are allocated to a fixed set of positions in a list, there are $k_i!$ ways to do this. Hence, if the items in a category are not distinguished in a list, there are $\frac{n!}{(n-k)!k_1!k_2!\cdots k_r!}$ such lists. For example, if there are k_g green balls, k_b blue balls, k_r red balls and k_y yellow balls with $k = k_g + k_b + k_r + k_y < n$, then there are $\frac{n!}{(n-k)!k_g!k_b!k_r!k_y!}$ color patterns possible (consider the $n-k$ positions white).

18.4.2.2 The binomial distribution continued

From the previous discussion, the number of ways in which we can have k 1s in n possible positions or locations is $\frac{n!}{(n-k)!k!}$. Thus, the probability of having a profile with $n-k$ 0s and k 1s is $\binom{n}{k} p^k(1-p)^{(n-k)}$. According to the *binomial theorem*, for any numbers x and y and any positive integer n,

$$(x+y)^n = \sum_{k=0}^{n} \binom{n}{k} x^k y^{(n-k)}. \tag{18.7}$$

In the particular case where $x = p$ any $y = 1-p$,

$$1 = (p+1-p)^n = \sum_{k=0}^{n} \binom{n}{k} p^k (1-p)^{(n-k)}. \tag{18.8}$$

So, the probabilities add up to 1 (as they should).

Suppose that we have n identical coins with 0 on one face and 1 on the other. Suppose also that the probability that any one of these coins when tossed has probability p of falling with face 1 up. Let X_i be the random variable representing the ith coin: $P(X_i = 1) = p$ and $P(X_i = 0) = 1-p$. Consider the random variable $X = X_1 + X_2 + \cdots + X_n$. Tossing each coin, we can write down the outcome – to produce a list such as $(1,0,0,1,\ldots,1)$, in which case the first and fourth coin fell with 1 up, the second and third coin with 0 up, and so on. The outcome on any coin is independent of the outcome on any other. From the earlier discussion, the probability of exactly k 1s is:

$$P(X = k) = \binom{n}{k} p^k (1-p)^{(n-k)}, \quad k = 0,\dots,n. \tag{18.9}$$

So, the sum of n independent Bernoulli variables has the binomial distribution (Figure 18.5).

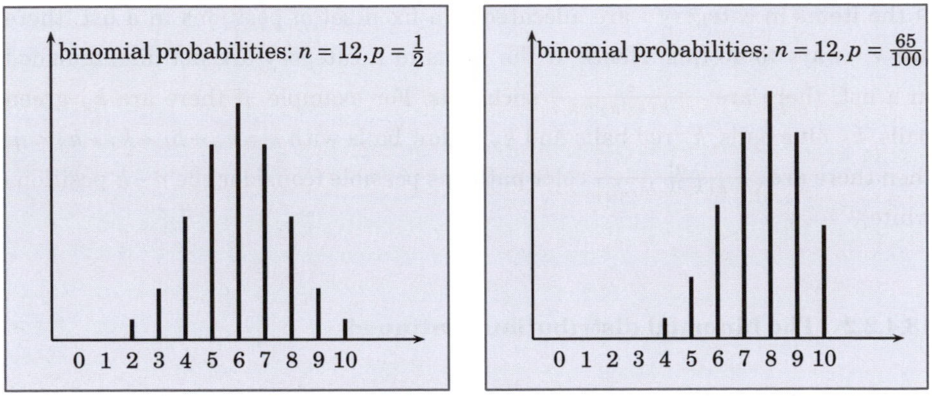

Figure 18.5: The binomial distribution

18.4.3 The Poisson distribution

Let X be a random variable denoting the number of occurrences in one unit of time. Let $P(X = k)$ be the probability of k occurrences, where k is a non-negative integer. The Poisson distribution assigns probability:

$$P(X = k) = \frac{e^{-\lambda}\lambda^k}{k!}, \quad k = 0,1,\dots \tag{18.10}$$

to k occurrences, where $\lambda > 0$ is a given parameter. A mathematical fact is that for any real number α, $e^\alpha = \sum_{k=0}^{\infty} \frac{\alpha^k}{k!}$. Thus, $\sum_{k=0}^{\infty} \frac{e^{-\lambda}\lambda^k}{k!} = e^{-\lambda} \sum_{k=0}^{\infty} \frac{\lambda^k}{k!} = e^{-\lambda} e^{\lambda} = 1$, so that $\{P(X = k)\}_{k=0}^{\infty}$ does indeed define a distribution (Figure 18.6).

When the unit of time is not necessarily 1 but, say, Δ, so that $P(X = k)$ denotes the probability of k occurrences in the period Δ is given by:

$$P(X = k) = \frac{e^{-\lambda\Delta}(\Delta\lambda)^k}{k!}, \quad k = 0,1,\dots \tag{18.11}$$

So, for example, $P(X = 0) = e^{-\lambda\Delta}$ and $P(X = 1) = e^{-\lambda\Delta}(\Delta\lambda)$.

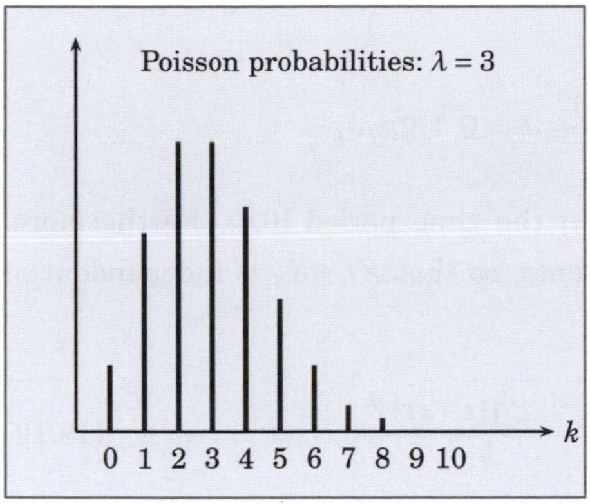

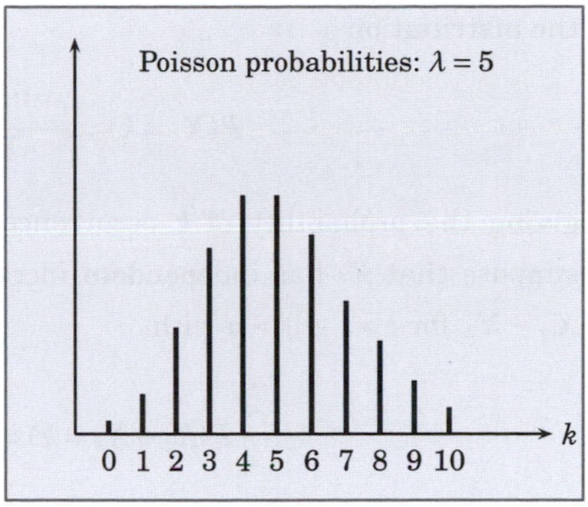

Figure 18.6: The Poisson distribution: $P(X=k)=\frac{e^{-\lambda}\lambda^k}{k!}$

Comparing the probability of k and $k+1$ occurrences: $P(X=k)=\frac{e^{-\lambda\Delta}(\Delta\lambda)^k}{k!}$ and

$$\begin{aligned}P(X=k+1) &= \frac{e^{-\lambda\Delta}(\Delta\lambda)^{k+1}}{k+1!}\\ &= \frac{e^{-\lambda\Delta}(\Delta\lambda)^k}{k!}\cdot\frac{(\Delta\lambda)}{k+1}\\ &= P(X=k)\cdot\frac{(\Delta\lambda)}{k+1}.\end{aligned}$$

Thus, $P(X=k+1)=P(X=k)\times(\frac{\lambda}{k+1})\Delta$. The expected value of X is

$$E\{X\}=\sum_{k=0}^{\infty}P(X=k)k=\sum_{k=0}^{\infty}\frac{e^{-\lambda\Delta}(\Delta\lambda)^k}{k!}k=\Delta\lambda.$$

The variance is:

$$\text{Var}(X)=\sum_{k=0}^{\infty}P(X=k)(k-\Delta\lambda)^2=\Delta\lambda,$$

so that the mean and variance are equal. The Poisson distribution is sometimes used to model queuing or arrival of customers at a service point.

18.4.4 The Poisson process

For each $t \geq 0$, let X_t be a random variable denoting the number of occurrences of an event in the time interval $[0,t]$. For example, X_t might denote the number of customers entering a store in the time period from 0 up to time t. Let X_t have the distribution

$$P(X_t = k) = \frac{e^{-\lambda t}(t\lambda)^k}{k!},\ k = 0,1,2,\dots,$$

giving the probability of k occurrences in the time period $[0,t)$. Furthermore, suppose that X_t has independent increments, so that $X_t - X_s$ is independent of $X_q - X_p$ for $t > s \geq q > p$ with

$$P(X_t - X_s = k) = \frac{e^{-\lambda(t-s)}[(t-s)\lambda]^k}{k!}. \tag{18.12}$$

(Two random variables, Y and Z, are said to be independent if for any pair of intervals (a,b) and (c,d), $P(Y \in (a,b) \text{ and } Z \in (c,d)) = P(Y \in (a,b)) \times P(Z \in (c,d))$.) From Equation 18.12, $X_t - X_s$ has a Poisson distribution, giving the probability of k occurrences in the time interval $[s,t]$. So, for any interval of time, Δ, the distribution over occurrences of the event is $P(X_{s+\Delta} - X_s = k) = \frac{e^{-\lambda\Delta}(\Delta\lambda)^k}{k!}$, for any $s \geq 0$.

EXAMPLE 18.3: To illustrate this, suppose that the number of customers arriving in a store is 20 per hour. From earlier calculations, λ gives the mean number of occurrences per unit of time (1 hour), so that $\lambda = 20$: $P(X_{t+1} - X_t = k) = \frac{e^{-\lambda 1}[1\lambda]^k}{k!}$. The probability of k occurrences in a ten-minute period ($\frac{1}{6}$ of an hour) is, for any t, $P(X_{t+\frac{1}{6}} - X_t = k) = \frac{e^{-\lambda\frac{1}{6}}(\frac{1}{6}\lambda)^k}{k!}$. ◊

REMARK 18.2: The Poisson process is characterized by a few simple postulates.

1. For a short length of time Δ, for all t in the family of random variables X_t, $t \geq 0$ satisfies:

   (a) $P(X_{t+\Delta} - X_t = 0) = 1 - \lambda\Delta + o(\Delta)$

   (b) $P(X_{t+\Delta} - X_t = 1) = \lambda\Delta + o(\Delta)$

   (c) $P(X_{t+\Delta} - X_t \geq 2) = o(\Delta)$,

   where $o(\Delta)$ is a term of smaller order than Δ (so that $\frac{o(\Delta)}{\Delta} \to 0$ as $\Delta \to 0$).

2. If $t_1 < t_2 < t_3 < t_4$, then $X_{t_4} - X_{t_3}$ and $X_{t_2} - X_{t_1}$ are independent random variables.

In this case, the family of random variables X_t, $t \geq 0$, satisfy Equation 18.12. □

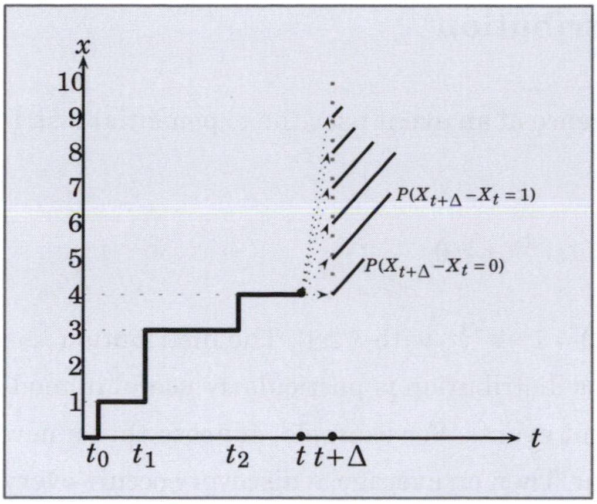

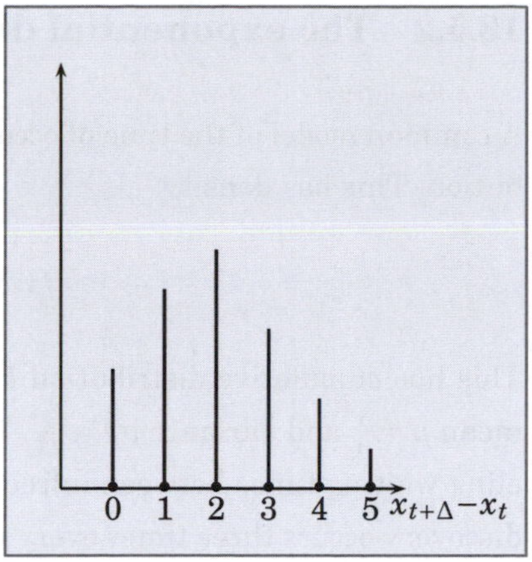

Figure 18.7: Poisson process: density at t

18.5 Continuous distributions

In this section, continuous distributions are considered, characterized by a cumulative distribution F and density $f = F'$. In this case, the mean is $E\{X\} = \mu = \int xf(x)dx$ and variance $E\{(X-\mu)^2\} = \sigma^2 = \int(x-\mu)^2 f(x)dx$. Similarly, given a function $h(x)$, $E\{h(X)\} = \int h(x)f(x)dx$.

The following discussion introduces the primary distributions used in hypothesis testing in economics: the uniform, normal, t, chi-square and F distributions.

18.5.1 The uniform distribution

Consider the task of selecting a number at random from the interval $[a,b]$. With each number "equally likely", this random variable is said to be uniformly distributed on $[a,b]$ with distribution given by

$$F(x) = \frac{1}{b-a}(x-a),\ x \in [a,b].$$

The corresponding density is $f(x) = \frac{1}{b-a}$. The uniform distribution has mean

$$\mu = \int_a^b xf(x)dx = \int_a^b x\frac{1}{b-a}dx = \frac{1}{b-a}\int_a^b xdx = \frac{1}{2}\frac{1}{b-a}x^2 \,|_a^b = \frac{1}{2}\frac{b^2-a^2}{b-a}$$
$$= \frac{1}{2}\frac{(b+a)(b-a)}{b-a} = \frac{1}{2}(b+a).$$

The variance is $\sigma^2 = \frac{1}{12}(b-a)^2$.

18.5.2 The exponential distribution

A common model of the time of occurrence of an event uses the exponential distribution. This has density:

$$f(x) = \lambda e^{-\lambda x}, x \geq 0.$$

This has cumulative distribution $F(x) = 1 - e^{-\lambda x}$ with $x \geq 0$. The distribution has mean $\mu = \frac{1}{\lambda}$ and variance $\sigma^2 = \frac{1}{\lambda^2}$. This distribution is particularly useful in modeling waiting times between infrequent events. For example, suppose that a new discovery occurs three times every year. Then on average, a discovery occurs every four months or $\frac{1}{3}$ of a year. The expected time between innovations is $\frac{1}{3}$. Model the random waiting time to the arrival of an innovation as having density $f(x) = 3e^{-3x}$. So, for example, the probability that an innovation arrives in 2 months or $\frac{1}{6}$ of a year is then $F\left(\frac{1}{6}\right) = 1 - e^{-3\frac{1}{6}} = 1 - e^{-\frac{1}{2}} = 1 - 0.6065 = 0.3935$; and the probability that an innovation arrives within a year is $F(1) = 1 - e^{-3 \cdot 1} = 1 - e^{-3} = 0.9502$. One interesting feature of the exponential distribution is that the conditional distribution and the unconditional distribution are the same. That the event has not occurred does not imply that it will occur relatively soon:

$$\begin{aligned}
P(X \in [a, a'] \mid X \geq a) &= \frac{F(a') - F(a)}{1 - F(a)} = \frac{e^{-\lambda a} - e^{-\lambda a'}}{e^{-\lambda a}} = \frac{e^{-\lambda a}[1 - e^{-\lambda a'} e^{\lambda a}]}{e^{-\lambda a}} \\
&= \frac{-e^{\lambda a}[1 - e^{-\lambda(a'-a)}]}{e^{\lambda a}} \\
&= 1 - e^{\lambda(a'-a)} \\
&= F(X < a' \mid x > a) \\
&= 1 - e^{\lambda s}, \quad s = a' - a \\
&= F(s).
\end{aligned}$$

The distribution of the wait time is the same at time $x = a$, given no arrival, as it is at time $x = 0$.

18.5.3 The normal distribution

The normal distribution is often used to model the spread of a characteristic in a population. For example, IQ in a population is commonly believed to be normally distributed. The density of the normal distribution with mean μ and variance σ^2 is depicted in Figure 18.8.

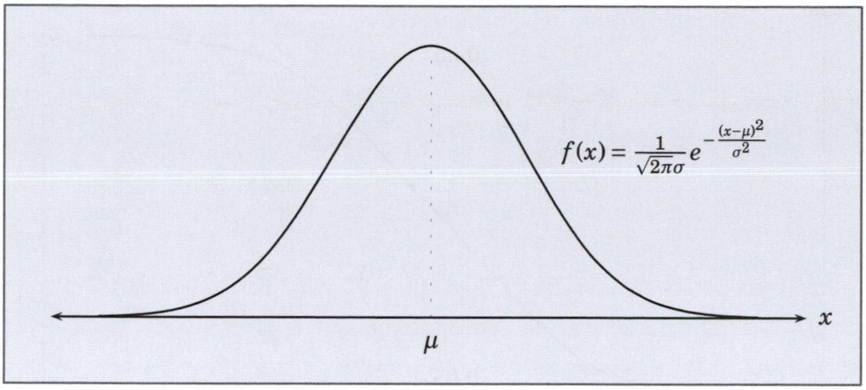

Figure 18.8: The normal density

The normal distribution has density:

$$f(x)=\frac{1}{\sqrt{2\pi}\sigma}e^{-\frac{(x-\mu)^2}{2\sigma^2}}, \quad -\infty<x<\infty, \tag{18.13}$$

with corresponding cumulative distribution:

$$F(x)=\int_{-\infty}^{x} f(\xi)d\xi=\int_{-\infty}^{x}\frac{1}{\sqrt{2\pi}\sigma}e^{-\frac{(\xi-\mu)^2}{2\sigma^2}}d\xi. \tag{18.14}$$

This does not integrate to a closed form solution. In the special case where $\mu=0$ and $\sigma^2=1$, the distribution is called the standard normal distribution, with density $\varphi(x)=\frac{1}{\sqrt{2\pi}}e^{-\frac{1}{2}x^2}$ and cumulative distribution $\Phi(x)=\int_{-\infty}^{x}\varphi(\xi)d\xi=\int_{-\infty}^{x}\frac{1}{\sqrt{2\pi}}e^{-\frac{1}{2}\xi^2}d\xi$. The cumulative distribution of the standard normal distribution is depicted in Figure 18.9.

Remark 18.3: Because $\Phi(x)\to 1$ as $x\to\infty$, $1=\int_{-\infty}^{\infty}\frac{1}{\sqrt{2\pi}}e^{-\frac{1}{2}\xi^2}d\xi$, and this implies that $\sqrt{2\pi}=\int_{-\infty}^{\infty}e^{-\frac{1}{2}\xi^2}d\xi$. □

Notice that $\Phi(-1.645)=0.05$ and $\Phi(1.645)=0.95$, so that a random variable drawn from a standard normal distribution will fall below -1.645 with probability 0.05 and above 1.645 with probability 0.05. Therefore, with probability equal to 0.9, the random variable will fall in the range $[-1.645, 1.646]$. For reference:

x	−3.090	−2.576	−2.326	−1.960	−1.645	−1.282	1.282	1.645	1.960	2.326	2.576	3.090
$\Phi(x)$	0.001	0.005	0.010	0.025	0.050	0.100	0.900	0.950	0.975	0.990	0.995	0.999

A useful fact about normally distributed random variables is:

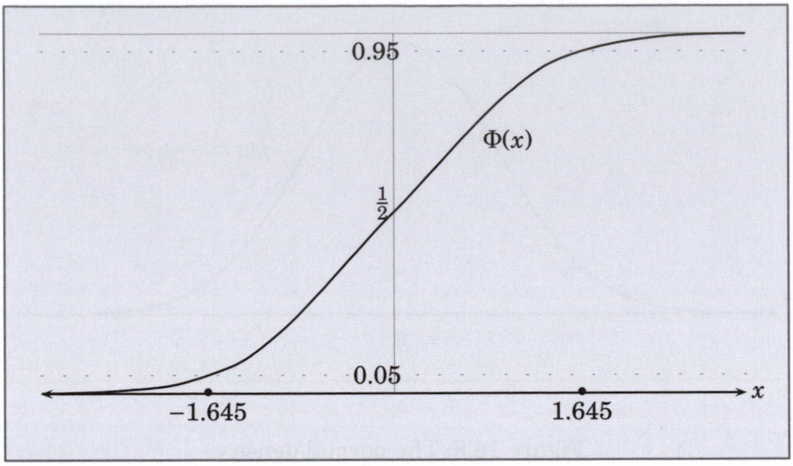

Figure 18.9: Distribution function of standard normal random variable

Theorem 18.1: *If X is a normally distributed random variable with mean μ and variance σ^2, then $Z = a + bX$ has a normal distribution with mean $a + b\mu$ and variance $b^2\sigma^2$.*

Given a normal random variable X with mean μ and variance σ^2, consider the transformation $X' = -\frac{\mu}{\sigma} + \frac{1}{\sigma}X$. Then, according to the theorem, X' has mean $-\frac{\mu}{\sigma} + \frac{\mu}{\sigma} = 0$ and variance $\frac{1}{\sigma^2}\sigma^2 = 1$, so that $X' = -\frac{\mu}{\sigma} + \frac{1}{\sigma}X = \frac{X-\mu}{\sigma}$ has the standard normal distribution. Since $\frac{X-\mu}{\sigma} \leq k$ whenever $X - \mu \leq k\sigma$ or $X \leq \mu + k\sigma$, so:

$$P\left(\frac{X-\mu}{\sigma} \leq k\right) = P(X \leq \mu + k\sigma).$$

For example, $P\left(\frac{X-\mu}{\sigma} \leq 1.960\right) = 0.975$. Since $\frac{X-\mu}{\sigma} \leq 1.960$ whenever $X - \mu \leq 1.960\sigma$ or $X \leq \mu + 1.960\sigma$,

$$P\left(\frac{X-\mu}{\sigma} \leq 1.960\right) = P(X \leq \mu + 1.960\sigma) = 0.975.$$

Thus, for example, if a random variable is drawn from a normal distribution, there is a 95% probability that it will lie within 1.96 standard deviations of the mean. A second useful fact regarding normally distributed random variables is:

Theorem 18.2: *Suppose that $X_1, X_2, \ldots, X_n$ are normally distributed random variables. Then*

- *$X = \sum_{i=1}^{n} a_i X_i$ is normally distributed with mean $\mu = \sum_{i=1}^{n} a_i \mu_i$, where μ_i is the mean of X_i.*
- *If the random variables are independent then the variance of X is given by: $\sigma^2 = \sum_{i=1}^{n} a_i^2 \sigma_i^2$, where σ_i^2 is the variance of X_i.*

18.5.4 Moment-generating functions

The expectation of a function $h(x)$ is $E\{h(X)\} = \int h(x)f(x)dx$. One special case occurs when $h(x) = e^{tx}$. The expectation is called the *moment-generating function* (assuming the integral exists):

$$m_X(t) = E\{e^{tX}\} = \int e^{tx} f(x)dx.$$

(In the discrete case, $m(t) = \sum e^{tx_i} p(x_i)$.)

Different distributions determine different moment generating functions.

Theorem 18.3: *Suppose that $m_X(t)$ is defined on an interval $(-\epsilon,\epsilon)$, $\epsilon > 0$. Then m_X uniquely determines the distribution. Thus, if $m_X = m_Y$ on $(-\epsilon,\epsilon)$ then X and Y have the same distribution.*

One implication of this theorem is that finding the moment-generating function of a random variable identifies the distribution of the random variable.

One important property of moment-generating functions is the following. Suppose that $X_1,\ldots,X_n$ are n independent random variables and let $Y = X_1 + X_2 + \cdots + X_n$. Then

$$m_Y(t) = E\{e^{t(X_1+X_2+\cdots+X_n)}\} = E\{e^{tX_1}\}E\{e^{tX_2}\}\cdots E\{e^{tX_n}\} = m_{X_1}(t)\cdot m_{X_2}(t)\cdots m_{X_n}(t).$$

Thus, the moment-generating function of a sum of independent random variables is equal to the product of the moment-generating functions of the individual random variables. In the case where the independent variables have the same distribution, then they have the same moment-generating function, m_X so that

$$m_Y(t) = m_{X_1}(t)\cdot m_{X_2}(t)\cdots m_{X_n}(t) = m_X(t)^n.$$

In the case of the normal distribution, the moment-generating function is: $m(t) = e^{\mu t+\frac{1}{2}t^2\sigma^2}$, where μ is the mean and σ^2 the variance. Thus, for example, the sum of n independent identically distributed random variables has moment-generating function $m(t)^n = e^{n\mu t+\frac{1}{2}t^2 n\sigma^2}$, a normal distribution with mean $n\mu$ and variance $n\sigma^2$.

18.5.5 The chi-square distribution

A chi-square distribution with k degrees of freedom (k a positive integer) is defined by density:

$$f(x) = \frac{1}{\Gamma\left(\frac{k}{2}\right)}\left(\frac{1}{2}\right)^{\frac{k}{2}} x^{\frac{k}{2}-1}e^{-\frac{1}{2}x}, \quad x > 0, \tag{18.15}$$

where $\Gamma(\cdot)$ is the gamma function $\Gamma(q) = \int_0^\infty \xi^{q-1}e^{-\xi}d\xi$. The chi-square distribution with n degrees of freedom has mean n and variance $2n$.

REMARK 18.4: This gamma function is defined for any $q > 0$. If $q > 1$, then $\Gamma(q) = (q-1)\Gamma(q-1)$. To see this, integrating by parts,

$$\begin{aligned}\int_0^\infty \xi^{q-1}e^{-\xi}d\xi &= -\int_0^\infty \xi^{q-1}d\{e^{-\xi}\}\\ &= -([\xi^{q-1}e^{-\xi}]_0^\infty - \int_0^\infty e^{-\xi}d\{\xi^{q-1}\}) = \int_0^\infty e^{-\xi}d\{\xi^{q-1}\}.\end{aligned}$$

The latter integral is $\int_0^\infty e^{-x}d\{\xi^{q-1}\} = (q-1)\int_0^\infty e^{-\xi}\xi^{q-2}d\xi = (q-1)\Gamma(q-1)$.

Furthermore, from the definition $\Gamma(1) = 1$, and at $q = \frac{1}{2}$, $\Gamma(\frac{1}{2}) = \sqrt{\pi}$. Because $\Gamma(x+1) = x\Gamma(x)$, if x is an integer then $\Gamma(x+1) = x!$. And, if $x = n$ is an integer,

$$\begin{aligned}\Gamma\left(n+\frac{1}{2}\right) &= \left(n-\frac{1}{2}\right)\Gamma\left(n-\frac{1}{2}\right) = \left(n-\frac{1}{2}\right)\left(n-\frac{3}{2}\right)\Gamma\left(n-\frac{3}{2}\right)\\ &= \left(n-\frac{1}{2}\right)\left(n-\frac{1}{2}\right)\left(n-\frac{3}{2}\right)\cdots\left(\frac{3}{2}\right)\left(\frac{1}{2}\right)\Gamma\left(\frac{1}{2}\right)\\ &= \left(\frac{2n-1}{2}\right)\left(\frac{2n-3}{2}\right)\cdots\left(\frac{3}{2}\right)\left(\frac{1}{2}\right)\sqrt{\pi}\\ &= \left(\frac{1}{2^n}\right)1\cdot 3\cdot 5\cdots(2n-1)\sqrt{\pi}.\end{aligned}$$

□

REMARK 18.5: The chi-square distribution is a special case of the gamma distribution, which has the density:

$$f(x) = \frac{\beta^\alpha}{\Gamma(\alpha)}x^{\alpha-1}e^{-\beta x},$$

where $\alpha > 0$ and $\beta > 0$ and where $x \geq 0$. If $\beta = \frac{1}{2}$ and $\alpha = \frac{k}{2}$, the distribution is the chi-square distribution.

Because $\int_0^\infty f(x)dx = 1$, $\int_0^\infty \frac{\beta^\alpha}{\Gamma(\alpha)}x^{\alpha-1}e^{-\beta x}dx = \frac{\beta^\alpha}{\Gamma(\alpha)}\int_0^\infty x^{\alpha-1}e^{-\beta x}dx = 1$, which implies the following useful fact:

$$\int_0^\infty x^{\alpha-1}e^{-\beta x}dx = \frac{\Gamma(\alpha)}{\beta^\alpha}. \tag{18.16}$$

□

The chi-square distribution has a close relation to the normal distribution.

Theorem 18.4:

- *If X has a standard normal distribution, then X^2 has a chi-square distribution with one degree of freedom.*

- *If $X_1, X_2, \ldots, X_n$ are independent standard normal random variables, then $Z = X_1^2 + X_2^2 + \cdots + X_n^2$ has a chi-square distribution with n degrees of freedom.*

PROOF: For the first part, let X be a standard normal random variable and let $Z = X^2$. Then for $z > 0$,

$$\text{Prob}(Z \le z) = \text{Prob}(X^2 \le z) = \text{Prob}(-\sqrt{z} \le X \le \sqrt{z}),$$

and with $F(z) = \text{Prob}(Z \le z)$,

$$F(z) = \int_{-\sqrt{z}}^{\sqrt{z}} \frac{1}{\sqrt{2\pi}} e^{-\frac{1}{2}x^2}\,dx = \frac{1}{\sqrt{2\pi}} \int_{-\sqrt{z}}^{\sqrt{z}} e^{-\frac{1}{2}x^2}\,dx.$$

Differentiating,

$$\begin{aligned}
F'(z) = f(z) &= \frac{1}{\sqrt{2\pi}} \left[\left(\frac{1}{2} z^{-\frac{1}{2}} \right) e^{-\frac{1}{2}[\sqrt{z}]^2} - \left(-\frac{1}{2} z^{-\frac{1}{2}} \right) e^{-\frac{1}{2}[-\sqrt{z}]^2} \right] \\
&= \frac{1}{\sqrt{2\pi}} \left(\frac{1}{2} z^{-\frac{1}{2}} \right) [e^{-\frac{1}{2}z} + e^{-\frac{1}{2}z}] \\
&= \frac{1}{\sqrt{2\pi}} z^{-\frac{1}{2}} e^{-\frac{1}{2}z} \\
&= \frac{1}{\sqrt{\pi}} \left(\frac{1}{2} \right)^{\frac{1}{2}} z^{\frac{1}{2}-1} e^{-\frac{1}{2}z},
\end{aligned}$$

which is the chi-square distribution with one degree of freedom ($\Gamma\left(\frac{1}{2}\right) = \sqrt{\pi}$).

For the second part, $Z = X_1^2 + X_2^2 + \cdots + X_n^2 = Z_1 + \cdots + Z_n$ so that Z is the sum of independent identically distributed chi-square random variables each with one degree of freedom. Take as given that a chi-square distribution with k degrees of freedom has moment-generating function $m_Z(t) = (1-2t)^{-\frac{k}{2}}$. Therefore, each Z_i has moment-generating function $m_{Z_i}(t) = (1-2t)^{-\frac{1}{2}}$, and since

$$m_Z(t) = \Pi_{i=1}^n m_{Z_i}(t) = \left[(1-2t)^{-\frac{1}{2}} \right]^n = (1-2t)^{-\frac{n}{2}},$$

the moment-generating function of Z is that of a chi-square distribution with n degrees of freedom (Figure 18.10). ■

18.5.6 The *t* distribution

The random variable X is said to have a t distribution with k degrees of freedom if the density of X is given by:

$$f(x) = \frac{\Gamma(\frac{1}{2}(k+1))}{\Gamma(\frac{k}{2})} \frac{1}{\sqrt{k\pi}} \frac{1}{(1+\frac{x^2}{k})^{\frac{k+1}{2}}}, \quad -\infty < x < \infty. \tag{18.17}$$

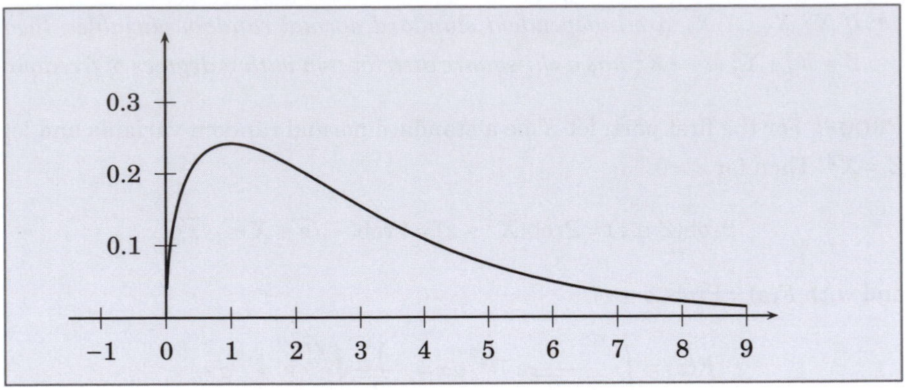

Figure 18.10: The chi-square distribution with three degrees of freedom

The parameter $k > 0$, the degrees of freedom, is usually taken to be an integer, although this is not necessary ($f(x)$ is a density for any $k > 0$). The t distribution with k degrees of freedom has mean 0 ($k > 1$) and variance $\frac{k}{k-2}$ ($k > 2$). One important fact regarding the t distribution is:

Theorem 18.5: *Suppose that Y has a standard normal distribution and suppose that Z has a chi-square distribution with k degrees of freedom. Suppose also that Y and Z are independently distributed. Then the random variable*

$$X = \frac{Y}{\sqrt{\frac{Z}{k}}}$$

has a t distribution with k degrees of freedom.

PROOF: Consider

$$\text{Prob}(X \le t) = \text{Prob}\left(\frac{Y}{\sqrt{\frac{Z}{k}}} \le t\right) = \text{Prob}\left(Y \le t\sqrt{\frac{Z}{k}}\right),$$

where Y is a standard normal with density φ and Z a chi-square variable with density ψ. For any Z, the set of Y satisfies $Y \le t\sqrt{\frac{Z}{k}}$ (Figure 18.11). Thus, with Y having normal density φ and Z having chi-square density ψ with k degrees of freedom:

$$\begin{aligned} F(t) = \text{Prob}(X \le t) &= \text{Prob}\left(\left\{(y,z) \mid y \le t\sqrt{\frac{z}{k}}\right\}\right) \\ &= \int_0^{\infty}\int_{-\infty}^{t\sqrt{\frac{z}{k}}} \varphi(y)\psi(z)dydz \\ &= \int_0^{\infty}\left\{\int_{-\infty}^{t\sqrt{\frac{z}{k}}} \varphi(y)dy\right\}\psi(z)dz. \end{aligned}$$

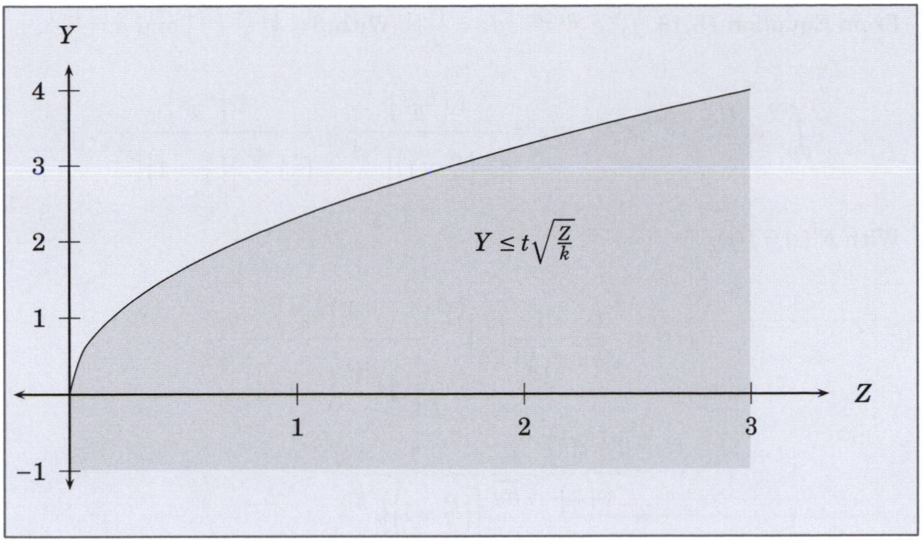

Figure 18.11: The region $\{(y,z) \mid y \le t\sqrt{\frac{z}{k}}\}$

But

$$\frac{d}{dt}\left\{\int_{-\infty}^{t\sqrt{\frac{z}{k}}} \varphi(y)dy\right\} = \varphi\left(t\sqrt{\frac{z}{k}}\right)\sqrt{\frac{z}{k}}$$

and

$$\psi(z) = \frac{1}{\Gamma(\frac{k}{2})}\left(\frac{1}{2}\right)^{\frac{k}{2}} z^{\frac{k}{2}-1}e^{-\frac{1}{2}z},$$

so that

$$\begin{aligned}
F'(t) &= \int_0^\infty \left[\varphi\left(t\sqrt{\frac{z}{k}}\right)\cdot\sqrt{\frac{z}{k}}\right]\frac{1}{\Gamma\left(\frac{k}{2}\right)}\left(\frac{1}{2}\right)^{\frac{k}{2}} z^{\frac{k}{2}-1}e^{-\frac{1}{2}z}dz \\
&= \int_0^\infty \left[\frac{1}{\sqrt{2\pi}}e^{-\frac{1}{2}\left(t\sqrt{\frac{z}{k}}\right)^2}\cdot\sqrt{\frac{z}{k}}\right]\frac{1}{\Gamma\left(\frac{k}{2}\right)}\left(\frac{1}{2}\right)^{\frac{k}{2}} z^{\frac{k}{2}-1}e^{-\frac{1}{2}z}dz \\
&= \frac{1}{\sqrt{2\pi}}\int_0^\infty \left[e^{-\frac{1}{2}(t^2\frac{z}{k})}\cdot z^{\frac{1}{2}}\frac{1}{\sqrt{k}}\right]\frac{1}{\Gamma\left(\frac{k}{2}\right)}\left(\frac{1}{2}\right)^{\frac{k}{2}} z^{\frac{k}{2}-1}e^{-\frac{1}{2}z}dz \\
&= \frac{1}{\sqrt{2\pi}}\frac{1}{\sqrt{k}}\frac{1}{\Gamma\left(\frac{k}{2}\right)}\left(\frac{1}{2}\right)^{\frac{k}{2}}\int_0^\infty \left[e^{-\frac{1}{2}(t^2\frac{z}{k})}\cdot z^{\frac{1}{2}}\right] z^{\frac{k}{2}-1}e^{-\frac{1}{2}z}dz \\
&= \frac{1}{\sqrt{k\pi}}\frac{1}{\Gamma\left(\frac{k}{2}\right)}\left(\frac{1}{2}\right)^{\frac{k+1}{2}}\int_0^\infty e^{-\frac{1}{2}\left(\frac{t^2}{k}+1\right)z}\cdot z^{\frac{k+1}{2}-1}dz.
\end{aligned}$$

From Equation 18.16, $\int_0^\infty e^{-\beta z} z^{\alpha-1} dz = \frac{\Gamma(\alpha)}{\beta^\alpha}$. With $\beta = \frac{1}{2}\left(\frac{t^2}{k}+1\right)$ and $\alpha = \frac{k+1}{2}$:

$$\int_0^\infty e^{-\frac{1}{2}\left(\frac{t^2}{k}+1\right)z} \cdot z^{\frac{k+1}{2}-1} dz = \frac{\Gamma\left(\frac{k+1}{2}\right)}{\left[\frac{1}{2}\left(\frac{t^2}{k}+1\right)\right]^{\frac{k+1}{2}}} = \frac{\Gamma\left(\frac{k+1}{2}\right)}{\left(\frac{1}{2}\right)^{\frac{k+1}{2}}\left[\left(\frac{t^2}{k}+1\right)\right]^{\frac{k+1}{2}}}.$$

With $F'(t) = f(t)$,

$$\begin{aligned} f(t) &= \frac{1}{\sqrt{k\pi}} \frac{1}{\Gamma\left(\frac{k}{2}\right)} \left(\frac{1}{2}\right)^{\frac{k+1}{2}} \frac{\Gamma\left(\frac{k+1}{2}\right)}{\left[\left(\frac{1}{2}\right)^{\frac{k+1}{2}}\left(\frac{t^2}{k}+1\right)\right]^{\frac{k+1}{2}}} \\ &= \frac{\Gamma\left(\frac{k+1}{2}\right)}{\Gamma\left(\frac{k}{2}\right)} \frac{1}{\sqrt{k\pi}} \frac{1}{\left[\left(\frac{t^2}{k}+1\right)\right]^{\frac{k+1}{2}}}, \end{aligned}$$

which is the density of the t distribution. ∎

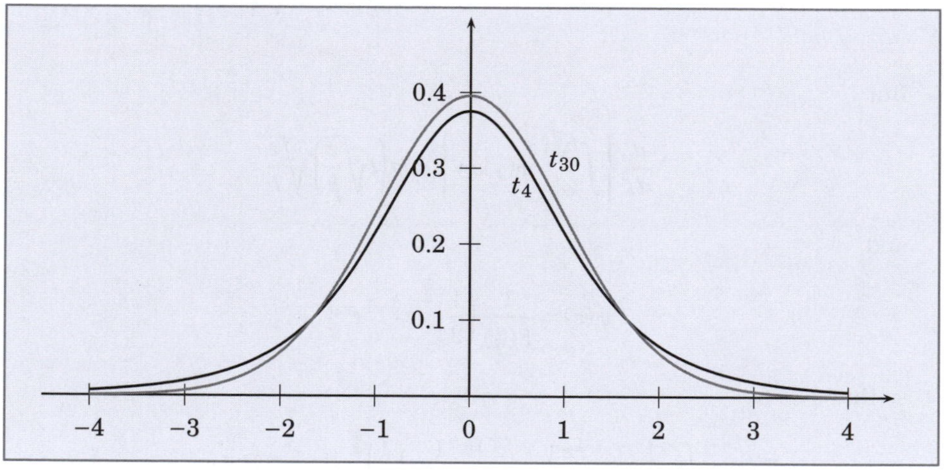

Figure 18.12: The t distribution with 4 and 30 degrees of freedom

REMARK 18.6: Given a collection of independently identically distributed standard normal random variables with mean 0 and variance 1, let $\bar{x} = \frac{1}{n}\sum_i x_i$ and $s^2 = \frac{1}{n-1}\sum_{i=1}^n (x_i - \bar{x})^2$. The following discussion shows that $\bar{x}$ and s^2 are independently distributed, $\bar{x}$ has a normal distribution with mean 0 and variance $\frac{1}{n}$ and $(n-1)s^2$ has a chi-square distribution with $n-1$ degrees of freedom. Therefore $\frac{\bar{x}}{s}$ has a t distribution with $n-1$ degrees of freedom.

Note that $\text{Var}\left(\frac{1}{n}\sum_i X_i\right) = \frac{1}{n^2}\sum_i \text{Var}(X_i) = \frac{1}{n^2} n \cdot 1 = \frac{1}{n}$. and $E\{\bar{X}\} = E\left\{\frac{1}{n}\sum_i X_i\right\} = \frac{1}{n}\sum_i E\{X_i\} = 0$, so that $\bar{x}$ has mean 0 and variance $\frac{1}{n}$. Finally, $\sum_i X_i$, the sum of normally distributed random variables, is normally distributed.

To consider the distribution of s^2, let $\boldsymbol{\iota}' = (1, \cdots, 1)$ be a vector of 1s and define the matrix $\boldsymbol{M}$:

$$\boldsymbol{M} = \boldsymbol{I} - \boldsymbol{\iota}(\boldsymbol{\iota}'\boldsymbol{\iota})^{-1}\boldsymbol{\iota}' = \begin{pmatrix} 1 & 0 & \cdots & 0 \\ 0 & 1 & \cdots & 0 \\ \vdots & \vdots & \ddots & 0 \\ 0 & 0 & \cdots & 1 \end{pmatrix} - \begin{pmatrix} 1 \\ 1 \\ \vdots \\ 1 \end{pmatrix} \frac{1}{n} \begin{pmatrix} 1 & 1 & \cdots & 1 \end{pmatrix}$$

$$= \begin{pmatrix} 1 & 0 & \cdots & 0 \\ 0 & 1 & \cdots & 0 \\ \vdots & \vdots & \ddots & 0 \\ 0 & 0 & \cdots & 1 \end{pmatrix} - \frac{1}{n} \begin{pmatrix} 1 & 1 & \cdots & 1 \\ 1 & 1 & \cdots & 1 \\ \vdots & \vdots & \ddots & 1 \\ 1 & 1 & \cdots & 1 \end{pmatrix},$$

since $(\boldsymbol{\iota}'\boldsymbol{\iota}) = n$, $(\boldsymbol{\iota}'\boldsymbol{\iota})^{-1} = \frac{1}{n}$. Notice that $\boldsymbol{M}$ is symmetric and idempotent; $\boldsymbol{MM} = \boldsymbol{M}$. Also

$$\begin{pmatrix} x_1 - \bar{x} \\ x_2 - \bar{x} \\ \vdots \\ x_n - \bar{x} \end{pmatrix} = \left[\begin{pmatrix} 1 & 0 & \cdots & 0 \\ 0 & 1 & \cdots & 0 \\ \vdots & \vdots & \ddots & 0 \\ 0 & 0 & \cdots & 1 \end{pmatrix} - \frac{1}{n} \begin{pmatrix} 1 & 1 & \cdots & 1 \\ 1 & 1 & \cdots & 1 \\ \vdots & \vdots & \ddots & 1 \\ 1 & 1 & \cdots & 1 \end{pmatrix} \right] \begin{pmatrix} x_1 \\ x_2 \\ \vdots \\ x_n \end{pmatrix}.$$

Let $\boldsymbol{x}' = (x_1, x_2, \ldots, x_n)$ so that $\boldsymbol{x}$ is the column vector of the x_i and $\bar{\boldsymbol{x}}' = (\bar{x}, \bar{x}, \ldots, \bar{x})$, with some abuse of notation. Then

$$\boldsymbol{x} - \bar{\boldsymbol{x}} = \boldsymbol{Mx},$$

and so

$$(n-1)\boldsymbol{s}^2 = \sum_i (x_i - \bar{x})^2 = (x - \bar{x})'(x - \bar{x}) = \boldsymbol{x}'\boldsymbol{Mx}.$$

Because $\boldsymbol{x} - \bar{\boldsymbol{x}} = \boldsymbol{Mx}$, $\boldsymbol{x} - \bar{\boldsymbol{x}}$ is a vector of normal (but not independent) random variables. The expected value of the vector $\boldsymbol{x} - \bar{\boldsymbol{x}}$ is $\boldsymbol{E}\{\boldsymbol{x} - \bar{\boldsymbol{x}}\} = 0$. The variance–covariance matrix of $\boldsymbol{x} - \bar{\boldsymbol{x}}$ is

$$\boldsymbol{E}\{(\boldsymbol{x} - \bar{\boldsymbol{x}})(\boldsymbol{x} - \bar{\boldsymbol{x}})'\} = \boldsymbol{E}\{\boldsymbol{Mxx}'\boldsymbol{M}'\} = \boldsymbol{ME}\{\boldsymbol{xx}'\}\boldsymbol{M}' = \boldsymbol{MIM}' = \boldsymbol{MM} = \boldsymbol{M}.$$

The symmetry of $\boldsymbol{M}$ implies that there is an orthogonal matrix $\boldsymbol{P}$ and eigenvalue matrix $\boldsymbol{\Lambda}$, such that $\boldsymbol{M}$ may be written as $\boldsymbol{M} = \boldsymbol{P\Lambda P}'$. Considering $(n-1)\boldsymbol{s}^2 = \boldsymbol{x}'\boldsymbol{Mx}$,

$$\boldsymbol{x}'\boldsymbol{Mx} = \boldsymbol{x}'\boldsymbol{P\Lambda P}'\boldsymbol{x} = \boldsymbol{y}'\boldsymbol{\Lambda y},$$

where $\boldsymbol{y} = \boldsymbol{P}'\boldsymbol{x}$. Because $\boldsymbol{y}$ is a linear combination of normal random variables, $\boldsymbol{y}$ is a vector of normally distributed random variables. The vector $\boldsymbol{y}$ has mean 0:

$$\boldsymbol{E}\{\boldsymbol{y}\} = \boldsymbol{E}\{\boldsymbol{P}'\boldsymbol{x}\} = \boldsymbol{P}'\boldsymbol{E}\{\boldsymbol{x}\} = \boldsymbol{M}\boldsymbol{0} = \boldsymbol{0}.$$

The variance of $\boldsymbol{y}$ is:

$$\boldsymbol{E}\{\boldsymbol{y}\boldsymbol{y}'\} = \boldsymbol{E}\{\boldsymbol{P}'\boldsymbol{x}\boldsymbol{x}'\boldsymbol{P}\} = \boldsymbol{P}'\boldsymbol{E}\{\boldsymbol{x}\boldsymbol{x}'\}\boldsymbol{P} = \boldsymbol{P}'\boldsymbol{I}\boldsymbol{P} = \boldsymbol{P}'\boldsymbol{P} = \boldsymbol{I}.$$

So, the vector $\boldsymbol{y}' = (y_1, y_2, \ldots, y_n)$ are n uncorrelated random bariables and, since normal, they are independent, each with mean 0 and variance 1.

Notice that if $(\lambda, \boldsymbol{x}_v)$ is an eigenvalue–eigenvector pair associated with $\boldsymbol{M}$, then $\boldsymbol{M}\boldsymbol{x}_v = \lambda\boldsymbol{x}_v$, since $\boldsymbol{M}\boldsymbol{M} = \boldsymbol{M}$, multiplying this equation gives $\boldsymbol{M}\boldsymbol{M}\boldsymbol{x}_v = \lambda\boldsymbol{M}\boldsymbol{x}_v$ or $\boldsymbol{M}\boldsymbol{x}_v = \lambda^2\boldsymbol{x}_v$, so that $\lambda\boldsymbol{x}_v = \lambda^2\boldsymbol{x}_v$. Because $\boldsymbol{x}_v \neq 0$, λ_v must equal 0 or 1. Hence, all eigenvalues of $\boldsymbol{M}$ are either 0 or 1. Recall that the sum of diagonal elements of $\boldsymbol{M}$ is equal to the sum of the eigenvalues. And the sum of the diagonal elements of $\boldsymbol{M}$ is the trace of the matrix:

$$\mathrm{tr}(\boldsymbol{M}) = \mathrm{tr}(\boldsymbol{I} - \boldsymbol{\iota}(\boldsymbol{\iota}'\boldsymbol{\iota})^{-1}\boldsymbol{\iota}') = \mathrm{tr}(\boldsymbol{I}) - \mathrm{tr}(\boldsymbol{\iota}(\boldsymbol{\iota}'\boldsymbol{\iota})^{-1}\boldsymbol{\iota}')) = n - \mathrm{tr}((\boldsymbol{\iota}'\boldsymbol{\iota})^{-1}\boldsymbol{\iota}'\boldsymbol{\iota})) = n - 1.$$

Since all eigenvalues are 0 or 1 and the sum of the eigenvalues is $n-1$, there are $n-1$ eigenvalues each equal to 1. Consequently, $\boldsymbol{y}'\boldsymbol{\Lambda}\boldsymbol{y}$ is the sum of squares of $n-1$ normal independent random variables and so it has a chi-square distribution with $n-1$ degrees of freedom. Since $\boldsymbol{x}'\boldsymbol{M}\boldsymbol{x} = \boldsymbol{y}'\boldsymbol{\Lambda}\boldsymbol{y}$, $\boldsymbol{x}'\boldsymbol{M}\boldsymbol{x}$ has a chi-square distribution with $n-1$ degrees of freedom. And therefore, $(n-1)\boldsymbol{s}^2$ has a chi-square distribution with $n-1$ degrees of freedom.

Considering $\bar{\boldsymbol{x}} = \frac{1}{n}\boldsymbol{\iota}'\boldsymbol{x}$ and $(n-1)\boldsymbol{s}^2 = \boldsymbol{x}'\boldsymbol{M}\boldsymbol{x} = \boldsymbol{x}'\boldsymbol{M}'\boldsymbol{M}\boldsymbol{x} = (\boldsymbol{M}\boldsymbol{x})'\boldsymbol{M}\boldsymbol{x}$, observe that the covariance of $\bar{\boldsymbol{x}}$ and $\boldsymbol{M}\boldsymbol{x}$ is 0: $\boldsymbol{E}\{\boldsymbol{M}\boldsymbol{x}\boldsymbol{x}'\boldsymbol{\iota}\} = \boldsymbol{M}\boldsymbol{E}\{\boldsymbol{x}\boldsymbol{x}'\}\boldsymbol{\iota} = \boldsymbol{M}\boldsymbol{\iota} = [\boldsymbol{I} - \boldsymbol{\iota}(\boldsymbol{\iota}'\boldsymbol{\iota})^{-1}\boldsymbol{\iota}']$ $\boldsymbol{\iota} = [\boldsymbol{\iota} - \boldsymbol{\iota}(\boldsymbol{\iota}'\boldsymbol{\iota})^{-1}\boldsymbol{\iota}'\boldsymbol{\iota}] = \boldsymbol{\iota} - \boldsymbol{\iota} = 0$. Therefore, the random variables $\bar{\boldsymbol{x}}$ and $\boldsymbol{M}\boldsymbol{x}$ are independent, because $\boldsymbol{\iota}'\boldsymbol{x}$ and $\boldsymbol{M}\boldsymbol{x}$ are normally distributed. Since $\boldsymbol{s}^2$ is a function of $\boldsymbol{M}\boldsymbol{x}$, $\boldsymbol{s}^2$ and $\bar{\boldsymbol{x}}$ are independent.

Consequently:

$$\frac{\bar{X}}{\sqrt{\frac{(n-1)s^2}{n-1}}} = \frac{\bar{X}}{s} = \frac{\bar{X}}{\sqrt{\frac{1}{n-1}\sum_i (x_i - \bar{x})^2}}$$

has t distribution with $n-1$ degrees of freedom (Figure 18.12). □

18.5.7 The *F* distribution

The random variable X is said to have the F distribution with degrees of freedom m and n (Figure 18.3) if the density of X is given by:

$$f(x) = \frac{\Gamma\left(\frac{m+n}{2}\right)}{\Gamma\left(\frac{m}{2}\right)\Gamma\left(\frac{n}{2}\right)}\left(\frac{m}{n}\right)^{\frac{m}{2}} \frac{x^{\frac{m-2}{2}}}{\left(1+\left(\frac{m}{n}\right)x\right)^{\frac{m+n}{2}}}, \quad x > 0. \tag{18.18}$$

Usually, m and n are taken to be positive integers, although $f(x)$ is a density for $m, n > 0$.

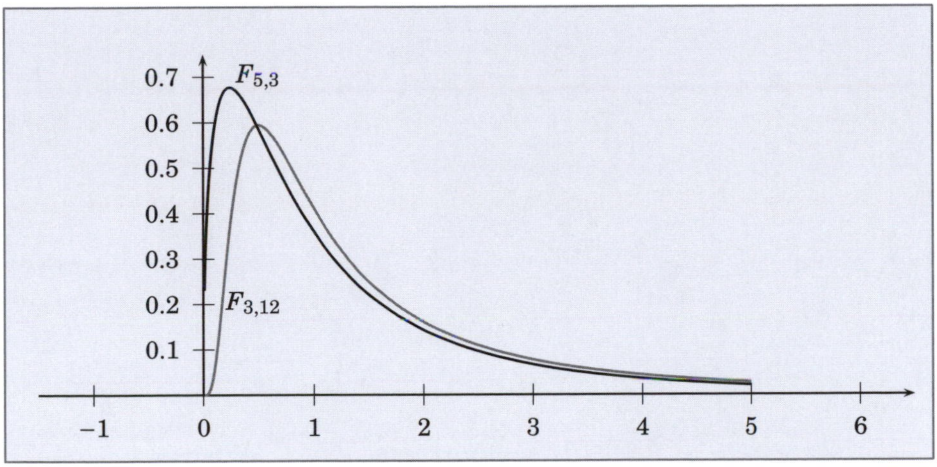

Figure 18.13: The F distribution: $F_{3,12}$ and $F_{5,3}$ distributions

Theorem 18.6: *Suppose that Y and Z are independent chi-square distributions with m and n degrees of freedom, respectively. Then the random variable*

$$X = \frac{\frac{Y}{m}}{\frac{Z}{n}}$$

has an F distribution with degrees of freedom m and n.

The F distribution with degrees of freedom m and n has mean $\frac{n}{n-2}$ $(n > 2)$ and variance $\frac{2n^2(m+n-2)}{m(n-2)^2(n-4)}$ $(n > 2)$. The proof of this theorem uses techniques similar to the proof for the t distribution. If φ_m and φ_n are independent chi-square distributions of the random variables Y and Z, then

$$F(s) = \text{Prob}\left(\frac{\frac{Y}{m}}{\frac{Z}{n}} \leq s\right) = \text{Prob}\left(Y \leq s\left(\frac{m}{n}\right)Z\right) = \int_0^\infty \int_0^{s\left(\frac{m}{n}\right)z} \varphi_m(y)\varphi_n(z)\,dy\,dz$$

and

$$\begin{aligned} F'(s) &= \int_0^\infty \frac{d}{ds}\left\{\int_0^{s\left(\frac{m}{n}\right)z} \varphi_m(y)dy\right\}\varphi_n(z)dz \\ &= \int_0^\infty \left(\frac{m}{n}\right) z\varphi_m\left(s\left(\frac{m}{n}\right)z\right)\varphi_n(z)dz. \end{aligned}$$

Exercises for Chapter 18

Exercise 18.1 Suppose that an instructor has 100 questions available and must write an exam consisting of four questions. A question may be asked only once in the exam. How many different exams can the instructor set? (Two exams are different if there is at least one question on one exam that is not on the other.)

Exercise 18.2 Suppose that an instructor has 100 questions available and must write an exam consisting of four questions. Furthermore, each question must come from a distinct category: A, B, C, D. The instructor has 15 category A questions, 25 category B questions, 40 category C questions and 30 category D questions. A question may be asked only once in the exam.

(a) How many different exams can the instructor set? (Two exams are different if there is at least one question on one exam that is not on the other.)

(b) Suppose instead that the exam consists of eight questions, and two must be drawn from each section. How many distinct exams are possible?

Exercise 18.3 Let

$$f(x) = \begin{cases} a(1-x^4) & \text{for } 0 < x < 1 \\ 0 & \text{otherwise} \end{cases}$$

be a probability density function. Find a.

Exercise 18.4 Let

$$f(x) = \begin{cases} a(1-x^n) & \text{for } 0 < x < 1 \\ 0 & \text{otherwise} \end{cases}$$

be a probability density function where n is a positive integer. Find a.

Exercise 18.5 Let X be a random variable with density

$$f(x) = \begin{cases} 4x^3 & \text{for } 0 < x < 1 \\ 0 & \text{otherwise} \end{cases}$$

Find

$$\text{Prob}\left(\frac{1}{3} < X < \frac{2}{3}\right).$$

Exercise 18.6 For a random variable X, the moments of the distribution are defined:

$$\mu_1 = E\{X\},\ \mu_2 = E\{X^2\}, \ldots, \mu_k = E\{X^k\}$$

with the mean $\mu = \mu_1$. Show that the variance, σ^2, satisfies $\sigma^2 = \mu_2 - \mu_1^2$.

Exercise 18.7 The moment-generating function for a discrete random variable X is defined:

$$m(t) = \sum_x e^{tx} p(x),$$

where $p(x) = \text{Prob}(X = x)$. Let the set of possible values of x be $\{x_1, \ldots, x_n\}$, so the moment-generating function is:

$$m(t) = \sum_{i=1}^{n} e^{tx_i} p(x_i).$$

Show that $m'(0) = \sum_{i=1}^{n} x_i p(x_i)$ and $m''(0) = \sum_{i=1}^{n} x_i^2 p(x_i)$.

Exercise 18.8 The Bernoulli distribution is:

$$f(x) \quad = \quad \begin{cases} p^x(1-p)^{1-x} & \text{for } x = 0,1 \\ 0 & \text{otherwise.} \end{cases}$$

(a) Find the mean, μ, and variance, σ^2, of a Bernoulli distributed random variable.

(b) Find the moment-generating function of f.

(c) Use the moment-generating function to find μ and σ^2.

Exercise 18.9 The exponential distribution is:

$$f(x) = \begin{cases} \theta e^{-\theta x} & \text{for } x > 0 \\ 0 & \text{for } x \leq 0. \end{cases}$$

(a) Find the mean and variance of an exponentially distributed random variable.

(b) Find the moment-generating function of f.

Exercise 18.10 The exponential distribution is:

$$f(x) = \begin{cases} \theta e^{-\theta x} & \text{for } x > 0 \\ 0 & \text{for } x \leq 0. \end{cases}$$

The hazard function is defined:

$$h(x) = \frac{f(x)}{1 - F(x)}.$$

Show that, for the exponential function, the hazard function is constant.

Exercise 18.11 The Poisson distribution is:

$$f(x) = \begin{cases} \frac{e^{-\lambda}\lambda^x}{x!} & \text{for } x = 0, 1, 2, \ldots \\ 0 & \text{otherwise.} \end{cases}$$

If a random event has a mean number of occurrences in a given time period, then in that time period the number of occurrences has the Poisson distribution. Suppose that a coffee shop has an average of 24 customers per hour.

(a) What is the probability that the shop will have six or fewer customers in any given hour?

(b) Considering two successive hours, what is the probability that the shop will have six or fewer customers in the first hour and more than six customers in the second hour?

Exercise 18.12 Suppose that X and Y are two independent random variables with Poisson distributions: $\frac{e^{-\lambda_x}\lambda_x^x}{x!}$ and $\frac{e^{-\lambda_y}\lambda_y^y}{y!}$. Then $Z = X + Y$ has a Poisson distribution with parameter $\lambda_z = \lambda_x + \lambda_y$.

Suppose that two shops A and B have an average number of customers per hour of 5 and 10, respectively. What is the probability that, together, the two stores will have fewer than 20 customers in any given hour?

Exercise 18.13 Use the fact that, for any real number γ,

$$e^{\gamma} = \sum_{x=0}^{\infty} \frac{\gamma^x}{x!}$$

to show that the moment-generating function for the Poisson distribution is

$$m(t) = e^{-\lambda} e^{\lambda e^t}.$$

From this conclude that the Poisson distribution has mean λ.

Exercise 18.14 Suppose that a ball will fall onto the real number line. The distribution of the point of impact is given by a standard normal distribution. Suppose you can place an interval of length $k > 0$ on the line. How should the interval be located to maximize the probability that the ball will fall within the interval?

Exercise 18.15 Let X be a standard normal random variable. Define the random variable $Z = \delta X^2$ where $\delta > 0$. Find the distribution of Z.

Exercise 18.16 The random variable defined:

$$T_n = \frac{X}{\sqrt{\frac{Z}{n}}},$$

where X is a standard normal variable and Z a chi-square variable with n degrees of freedom, has a t distribution with n degrees of freedom.

Considering the denominator, a chi-square variable with n degrees of freedom is equal to the sum of squares on n independent standard normal variables: $Z = \sum_{i=1}^{n} X_i^2$. Let $V_i = X_i^2$ and since X_i is a standard normal, $E\{V_i\} = 1$. Thus, $\frac{Z}{n} = \frac{1}{n}\sum_{i=1}^{n} V_i$. According to the law of large numbers, the average of n independent identically distributed random variables converges to the average. So, $\frac{Z}{n} = \frac{1}{n}\sum_{i=1}^{n} V_i \to 1$, and so $\sqrt{\frac{Z}{n}} \to 1$. Use this observation to argue that, as n becomes large, the distribution of a T variable converges to that of a standard normal distribution.

Exercise 18.17 The random variable defined:

$$F_n = \frac{\frac{Y}{m}}{\frac{Z}{n}},$$

where Y, Z are independent and where Y is a chi-square distributed random variable with m degrees of freedom and Z a chi-square variable with n degrees of freedom, has an F distribution with m, n degrees of freedom.

Considering the denominator, a chi-square variable with n degrees of freedom is equal to the sum of squares on n independent standard normal variables: $Z = \sum_{i=1}^{n} X_i^2$. Let $V_i = X_i^2$ and since X_i is a standard normal, $E\{V_i\} = 1$. Thus, $\frac{Z}{n} = \frac{1}{n}\sum_{i=1}^{n} V_i$. According to the law of large numbers, the average of n independent identically distributed random variables converges to the average. So, $\frac{Z}{n} = \frac{1}{n}\sum_{i=1}^{n} V_i \to 1$. Use this observation to argue that, as n becomes large, the distribution of an F variable converges to that of a chi-square (with m degrees of freedom) divided by its degrees of freedom.

19

Estimation and hypothesis testing

19.1 Introduction

This chapter summarizes some of the key concepts in estimation and hypothesis testing that arise in statistical and econometric work in economics. Section 19.2 discusses estimation, unbiasedness and efficiency. The behavior of sequences of random variables in terms of long-run averages and the distribution of such averages is considered in Sections 19.3.1 and 19.3.2. The fact that the mean of a sequence of independently and identically distributed random variables has a statistical distribution approaching a normal distribution is an important observation used in hypothesis testing. Hypothesis testing is discussed in Section 19.4. Finally, these techniques are applied in the context of econometric models in Section 19.5.

19.2 Estimation

Consider a sample of n random variables, $X_1, X_2, \ldots, X_n$, all drawn from the same distribution, F, with density f. For example, flipping the same coin n times and writing 1 to denote heads and 0 to denote tails after each flip gives n such random variables X_i – where the ith random variable X_i is 1 or 0, depending on the outcome of the ith flip. The mean $\mu = \int x f(x) dx$ and variance $\sigma^2 = \int (x-\mu)^2 f(x) dx$ are important summary statistics describing the distribution. Typically, such statistics are not known and must be estimated from sample data. The average of the n variables is $\bar{X}_n = \frac{1}{n}\sum_{i=1}^{n} X_i$ and provides an estimate of the mean, μ. Similarly, $s_n^2 = \frac{1}{n-1}\sum_{i=1}^{n}(X_i - \bar{X}_n)$ gives an estimate of the variance σ^2. Since

$$E\{\bar{X}_n\} = E\left\{\frac{1}{n}\sum_{i=1}^{n} X_i\right\} = \frac{1}{n}\sum_{i=1}^{n} E\{X_i\} = \frac{1}{n}\sum_{i=1}^{n} \mu = \mu, \tag{19.1}$$

the mean $\bar{X}_n$ is called an *unbiased estimator* of μ. Provided the X_i are independent with common mean, μ, and variance, σ^2, s_n^2 is an unbiased estimator of σ^2: $E\{s_n^2\} = \sigma^2$. Observe that $\bar{X}_n$ and s_n^2 are functions of the data observed, $X_1, X_2, \ldots, X_n$. This is what an estimator is – a function of the observed data estimating underlying parameters that generated the data.

19.2.1 Unbiasedness and efficiency

In general, we assume that the data are generated by some given distribution F (or density f) and this distribution may be characterized in terms of its "moments" (the mean μ, variance σ^2, and higher moments $k_j = \int (x-\mu)^k f(x)dx$). For the normal distribution, the mean and variance fully define the distribution (Equation 18.13); for the t distribution (Equation 18.17), the degrees of freedom k fully define the distribution, and so on.

Given a distribution F where θ is some parameter of the distribution, $F(\cdot;\theta)$, let $h(\theta)$ be a function of that parameter. Finally, let $g_n(X_1, X_2, \ldots, X_n)$ be a function of the data that provides an estimate of $h(\theta)$. For example, if θ is the mean and $h(\theta) = \theta$, $g_n(X_1, \ldots, X_n)$ provides an estimate of the mean. How good is an estimator? Two primary criteria are unbiasedness and efficiency.

Definition 19.1: *The estimator g_n is an unbiased estimator of $h(\theta)$ if:*

$$E\{g_n\} = h(\theta) \quad \textit{for all} \quad \theta.$$

Note that the expectation $E\{g_n\}$ is computed using the distribution function $F(\cdot;\theta)$ and so depends on θ. This may be made explicit by writing $E_\theta\{\cdot\}$. If g_n is an estimator of μ, the bias of the estimator is given by $\text{bias}(g_n) = E\{g_n\} - \mu$.

Definition 19.2: *The mean squared error of an estimator g_n of $h(\theta)$ is given by:*

$$E_\theta\{[g_n - h(\theta)]^2\}. \tag{19.2}$$

The mean squared error (MSE) may be decomposed into the variance and bias of the estimator, as follows:

$$\begin{aligned}
E_\theta\{[g_n-h(\theta)]^2\} &= E_\theta\left\{\Big[(g_n-E_\theta\{g_n\})+(E_\theta\{g_n\}-h(\theta))\Big]^2\right\} \qquad (19.3)\\
&= E_\theta\Big\{\Big[(g_n-E_\theta\{g_n\})^2+(E_\theta\{g_n\}-h(\theta))^2\\
&\quad +2(g_n-E_\theta\{g_n\})(E_\theta\{g_n\}-h(\theta))\Big]\Big\}\\
&= E_\theta\left\{(g_n-E_\theta\{g_n\})^2\right\}+(E_\theta\{g_n\}-h(\theta))^2\\
&\quad +2E_\theta\left\{(g_n-E_\theta\{g_n\})(E_\theta\{g_n\}-h(\theta))\right\}\\
&= E_\theta\{(g_n-E_\theta\{g_n\})^2\}+(E_\theta\{g_n\}-h(\theta))^2\\
&= \mathrm{Var}(g_n)+(E_\theta\{g_n\}-h(\theta))^2\\
&= \mathrm{Var}(g_n)+[\mathrm{bias}(g_n)]^2. \qquad (19.4)
\end{aligned}$$

Unbiasedness is one desirable property of an estimator. A second is precision of the estimator in the sense that it is "likely" to be close to the value of the parameter to be estimated. For example, let $J \subset \{1,2,\dots,n\}$ containing $j<n$ elements. Then consider the estimator of the mean: $\tilde{X}_n^J = \frac{1}{j}\sum_{i\in J}X_i$. $E\{\tilde{X}_n^J\} = \frac{1}{j}\sum_{i\in J}E\{X_i\} = \frac{1}{j}\sum_{i\in J}\mu = \mu$, so $\tilde{X}_n^J$ is an unbiased estimator of μ. However, the variance of $\tilde{X}_n^J$ is $\frac{1}{j}\sigma^2$. Thus,

$$\mathrm{Var}(\tilde{X}_n^J) = \frac{1}{j}\sigma^2 > \frac{1}{n}\sigma^2 = \mathrm{Var}(\bar{X}_n), \qquad (19.5)$$

so that while both estimators are unbiased, one has a higher variance and so is a "less-accurate" estimator.

Definition 19.3: *An estimator g_n of $h(\theta)$ is efficient if it is unbiased and there is no other unbiased estimator $\hat{g}_n$ with smaller variance.*

REMARK 19.1: Such an estimator may not minimize the mean square error. For example, consider the two estimates of the variance:

$$s^2 = \frac{1}{n}\sum(X_i-\bar{X}_i)^2, \quad \text{and} \quad \hat{s}^2 = \frac{1}{n-1}\sum(X_i-\bar{X}_i)^2 = \frac{n}{n-1}s^2.$$

Since $\frac{1}{\sigma^2}\hat{s}^2$ satisfies $E\left\{\frac{n-1}{\sigma^2}\hat{s}^2\right\} = n-1$, $E\{\hat{s}^2\} = \sigma^2$, $\hat{s}^2$ is unbiased. Next observe that $\mathrm{Var}\left\{\frac{n-1}{\sigma^2}\hat{s}^2\right\} = 2(n-1)$ so that $\mathrm{Var}\{\hat{s}^2\} = 2(n-1)\frac{(\sigma^2)^2}{(n-1)^2} = \frac{2\sigma^4}{n-1}$. Thus, $\hat{s}^2$ is unbiased with variance $\frac{2\sigma^4}{n-1}$.

By contrast, $E\{s^2\} = \frac{n-1}{n}E\{\hat{s}^2\} = \frac{n-1}{n}\sigma^2$, so the bias is $\left(1-\frac{1}{n}\right)\sigma^2-\sigma^2 = -\frac{1}{n}\sigma^2$. The variance of s^2 equals, $\mathrm{Var}\left\{\frac{n-1}{n}\hat{s}^2\right\} = \left(\frac{n-1}{n}\right)^2\mathrm{Var}\{\hat{s}^2\} = \frac{(n-1)^2}{n^2}\frac{2\sigma^4}{n-1} = \frac{2(n-1)}{n^2}\sigma^4$.

Therefore:

$$
\begin{aligned}
MSE(s^2) &= \mathrm{Var}(s^2) + \mathrm{bias}(s^2) \\
&= \frac{2(n-1)}{n^2}\sigma^4 + \left(-\frac{1}{n}\sigma^2\right)^2 \\
&= \left[\frac{2(n-1)}{n^2} + \frac{1}{n^2}\right]\sigma^4 \\
&= \frac{2n-1}{n^2}\sigma^4.
\end{aligned}
$$

But $MSE(\hat{s^2}) = \mathrm{Var}(\hat{s}^2) = \frac{2\sigma^4}{n-1}$. Thus,

$$
\begin{aligned}
MSE(\hat{s}^2) - MSE(s^2) &= \frac{2\sigma^4}{n-1} - \frac{2n-1}{n^2}\sigma^4 \\
&= \frac{2n^2 - (n-1)(2n-1)}{(n-1)n^2}\sigma^4 \\
&= \frac{3n-1}{(n-1)n^2}\sigma^4.
\end{aligned}
$$

Since $n > 1$, $(3n-1) > 0$ and $MSE(\hat{s}^2) - MSE(s^2) > 0$. The biased estimator of σ^2, s^2, has smaller MSE than the unbiased estimator $\hat{s}^2$. □

19.3 Large-sample behavior

Let $g_n(X_1,\ldots,X_n)$ be an estimator of some parameter of a distribution F. As the sample size increases, the estimator varies. How does this behave as the sample size becomes large? The law of large numbers concerns the limiting behavior of the estimator. What does $g_n(X_1,\ldots,X_n)$ converge to as n becomes large (assuming it converges)? Because $g_n(X_1,\ldots,X_n)$ is a random variable, it has a distribution. The central limit theorem studies how this distribution behaves as n becomes large.

19.3.1 Law of large numbers

Given n random variables, $X_1, X_2, \ldots, X_n$ suppose they are independently distributed with common distribution F. Suppose also, that the mean of each X_i is μ and variance σ^2. Consider the average: $\bar{X}_n = \frac{1}{n}\sum_{i=1}^n X_i$. Under what circumstances does $\bar{X}_n$ converge to μ as n becomes large (in the sense of being close to μ

with high probability)? The expectation of $\bar{X}_n$ is

$$E\{\bar{X}_n\} = E\left\{\frac{1}{n}\sum_{i=1}^{n} X_i\right\} = \frac{1}{n}\sum_{i=1}^{n} E\{X_i\} = \frac{1}{n}\sum_{i=1}^{n}\mu = \frac{1}{n}n\mu = \mu.$$

So, the mean of $\bar{X}$ is μ. The variance of $\bar{X}_n$ is

$$\sigma^2_{\bar{X}_n} = E\{(\bar{X}_n - \mu)^2\} = E\left\{\left(\frac{1}{n}\sum_{i=1}^{n} X_i - \mu\right)^2\right\}. \qquad (19.6)$$

Rearranging:

$$E\left\{\left(\frac{1}{n}\sum_{i=1}^{n} X_i - \mu\right)^2\right\} = E\left\{\left(\frac{1}{n}\sum_{i=1}^{n} X_i - \frac{1}{n}n\mu\right)^2\right\} = E\left\{\left(\frac{1}{n}[\sum_{i=1}^{n} X_i - n\mu]\right)^2\right\}$$
$$= E\left\{\left(\frac{1}{n}\sum_{i=1}^{n}(X_i - \mu)\right)^2\right\},$$

and this equals

$$E\left\{\left(\frac{1}{n}\right)^2\left(\sum_{i=1}^{n}(X_i - \mu)\right)^2\right\} = \left(\frac{1}{n}\right)^2 E\left\{\left(\sum_{i=1}^{n}(X_i - \mu)\right)^2\right\} = \left(\frac{1}{n}\right)^2 E\left\{\left(\sum_{i=1}^{n}(X_i - \mu)\right)^2\right\}.$$

Rearranging the latter term:

$$\left(\frac{1}{n}\right)^2 E\left\{\sum_{i=1}^{n}\sum_{j=1}^{n}(X_i - \mu)(X_j - \mu)\right\} = \left(\frac{1}{n}\right)^2 \sum_{i=1}^{n}\sum_{j=1}^{n} E\{(X_i - \mu)(X_j - \mu)\} = \left(\frac{1}{n}\right)^2 \sum_{i=1}^{n}\sum_{j=1}^{n}\sigma_{ij},$$

where $\sigma_{ij} = E\{(X_i - \mu)(X_j - \mu)\}$. In the case where the X_i are independent, $\sigma_{ij} = 0$ whenever $i \neq j$, then $\sum_{i=1}^{n}\sum_{j=1}^{n}\sigma_{ij} = \sum_{i=1}^{n}\sigma_{ii}$, and since the variance is the same for each X: $\sigma_{ii} = \sigma^2$;

$$\sigma^2_{\bar{X}_n} = E\{(\bar{X} - \mu)^2\} = \left(\frac{1}{n}\right)^2 n\sigma^2 = \frac{1}{n}\sigma^2. \qquad (19.7)$$

Thus, as n increases, the variance of $\bar{X}_n$ declines, so that the dispersion of $\bar{X}_n$ around the mean μ declines as n increases – with large n, $\bar{X}_n$ is close to X with high probability. The following discussion presents this in a somewhat more precise way: that for any $\epsilon > 0$, $P(|\bar{X}_n - \mu| \geq \epsilon)$ goes to 0 as n becomes large. The proof of the law of large numbers utilizes the Chebyshev inequality.

Theorem 19.1 (Chebyshev inequality): *If X is a random variable with mean μ and variance σ^2, then*

$$1 - P(-\epsilon < X - \mu < \epsilon) = P(|X - \mu| \geq \epsilon) \leq \frac{\sigma^2}{\epsilon^2}, \quad \epsilon > 0. \tag{19.8}$$

PROOF: Let $g(\cdot)$ be a non-negative function defined on the real number line.

$$\begin{aligned}
E\{g(X)\} &= \int_{-\infty}^{\infty} g(x)f(x)dx \\
&= \int_{\{x|g(x)\geq\delta\}} g(x)f(x)dx + \int_{\{x|g(x)<\delta\}} g(x)f(x)dx \\
&\geq \int_{\{x|g(x)\geq\delta\}} g(x)f(x)dx \\
&\geq \int_{\{x|g(x)\geq\delta\}} \delta f(x)dx = \delta \int_{\{x|g(x)\geq\delta\}} f(x)dx \\
&\geq \delta P(g(X) \geq \delta).
\end{aligned} \tag{19.9}$$

Consequently,

$$\frac{E\{g(X)\}}{\delta} \geq P(g(X) \geq \delta).$$

Taking $g(x) = (x-\mu)^2$, $E\{g(X)\} = E\{(x-\mu)^2\} = \sigma^2$, gives

$$\frac{\sigma^2}{\delta} \geq P((X-\mu)^2 \geq \delta) = P(|X - \mu| \geq \sqrt{\delta}).$$

Letting $\delta = \epsilon^2$, gives

$$\frac{\sigma^2}{\epsilon^2} \geq P(|X - \mu| \geq \epsilon). \tag{19.10}$$

Alternatively, letting $\delta = \epsilon^2\sigma^2$, gives

$$\frac{1}{\epsilon^2} \geq P(|X - \mu| \geq \epsilon\sigma). \tag{19.11}$$

This completes the proof. ■

The weak law of large numbers is discussed next.

Theorem 19.2 (Weak law of large numbers): *Let $X_1, X_2, \ldots, X_n$, be n independent random variables with common distribution having mean μ and variance σ^2. Then:*

$$\lim_{n\to\infty} P(-\epsilon < \bar{X}_n - \mu < \epsilon) = 1.$$

PROOF: Applying the Chebyshev inequality to the random variable $\bar{X}_n = \frac{1}{n}\sum_{i=1}^{n} X_i$ gives:

$$\frac{1}{n}\frac{\sigma^2}{\epsilon^2} = \frac{\sigma^2_{\bar{X}_n}}{\epsilon^2} \geq P(|\bar{X}_n - \mu| \geq \epsilon) = 1 - P(-\epsilon < \bar{X}_n - \mu < \epsilon). \tag{19.12}$$

Therefore,

$$1 \geq P(-\epsilon < \bar{X}_n - \mu < \epsilon) \geq 1 - \frac{1}{n}\frac{\sigma^2}{\epsilon^2}. \tag{19.13}$$

Since $\lim_{n\to\infty}\left[1 - \frac{1}{n}\frac{\sigma^2}{\epsilon^2}\right] = 1$,

$$\lim_{n\to\infty} P(-\epsilon < \bar{X}_n - \mu < \epsilon) = 1. \tag{19.14}$$

■

As the sample size increases the distribution of the $\bar{X}_n$ concentrates around the mean μ. From earlier computations, the variance of $\bar{X}_n$ is $\sigma^2_{\bar{X}_n} = \frac{1}{n}\sigma^2$, which goes to 0 as n becomes large. However, with a suitable rescaling of $\bar{X}_n$, $(\sqrt{n}\bar{X}_n)$, the distribution does not collapse as n becomes large, but instead converges to a normal distribution.

19.3.2 The central limit theorem

For any constant α, and any random variable Y with variance σ^2_Y, the variance of αY is $\alpha^2\sigma^2_Y$. In the present context, the variance of $\sqrt{n}\bar{X}_n$ is $(\sqrt{n})^2\sigma^2_{\bar{X}_n} = (\sqrt{n})^2\left(\frac{1}{n}\sigma^2\right) = \sigma^2$. Thus, multiplying $\bar{X}_n$ by $\sqrt{n}$ gives a random variable, $\sqrt{n}\bar{X}_n$, whose variance does not collapse. Furthermore, with $E\{X_i\} = \mu = E\{\bar{X}_n\}$, $\sqrt{n}(\bar{X}_n - \mu)$ has mean 0 and variance σ^2. Going one step further, the random variable $\sqrt{n}(\bar{X}_n - \mu)/\sigma$ has mean 0 and variance 1. The central limit theorem asserts that this random variable is approximately normally distributed as n becomes large.

Theorem 19.3 (Central limit theorem): *Let $X_1, X_2, \ldots, X_n$ be a collection of n independently and identically distributed random variables with mean μ and variance σ^2. Then, as n becomes large,*

$$\frac{\sqrt{n}(\bar{X}_n - \mu)}{\sigma} \tag{19.15}$$

is approximately normally distributed. Specifically, for each x,

$$\lim_{n\to\infty} \text{Prob}\left(\frac{\sqrt{n}(\bar{X}_n - \mu)}{\sigma} \leq x\right) = \Phi(x). \tag{19.16}$$

So, for n large, $\text{Prob}\left(\frac{\sqrt{n}(\bar{X}_n-\mu)}{\sigma} \le x\right) \approx \Phi(x)$. Therefore, when n is large,

$$\begin{aligned}\text{Prob}\left(-x \le \frac{\sqrt{n}(\bar{X}_n-\mu)}{\sigma} \le x\right) &= \text{Prob}\left(\frac{\sqrt{n}(\bar{X}_n-\mu)}{\sigma} \le x\right) - \text{Prob}\left(\frac{\sqrt{n}(\bar{X}_n-\mu)}{\sigma} \le -x\right) \\ &\approx \Phi(x) - \Phi(-x). \end{aligned} \tag{19.17}$$

If $x = 1.96$ then $\Phi(-x) = 0.025$ and $\Phi(x) = 0.975$ and $\Phi(x) - \Phi(-x) = 0.95$. Thus, for n large,

$$\text{Prob}\left(-1.96 \le \frac{\sqrt{n}(\bar{X}_n-\mu)}{\sigma} \le 1.96\right) \approx 0.95. \tag{19.18}$$

Thus, for large n, there is approximately 0.95% probability that $\sqrt{n}(\bar{X}_n-\mu)$ will lie in the interval $[-1.96\sigma, 1.96\sigma]$. Alternatively, for large n, there is approximately 0.95% probability that $(\bar{X}_n-\mu)$ will lie in the interval $\left[-1.96\frac{\sigma}{\sqrt{n}}, 1.96\frac{\sigma}{\sqrt{n}}\right]$ or that $\bar{X}_n$ will lie in the interval $\left[\mu - 1.96\frac{\sigma}{\sqrt{n}}, \mu + 1.96\frac{\sigma}{\sqrt{n}}\right]$.

19.3.2.1 Confidence intervals

The expression

$$\text{Prob}\left(-x \le \frac{\sqrt{n}(\bar{X}_n-\mu)}{\sigma} \le x\right) \approx \Phi(x) - \Phi(-x) \tag{19.19}$$

may be rearranged as

$$\text{Prob}\left(x \ge \frac{\sqrt{n}(-\bar{X}_n+\mu)}{\sigma} \ge -x\right) \approx \Phi(x) - \Phi(-x) \tag{19.20}$$

or

$$\text{Prob}\left(x\frac{\sigma}{\sqrt{n}} \ge (-\bar{X}_n+\mu) \ge -x\frac{\sigma}{\sqrt{n}}\right) \approx \Phi(x) - \Phi(-x) \tag{19.21}$$

or

$$\text{Prob}\left(\bar{X}_n + x\frac{\sigma}{\sqrt{n}} \ge \mu \ge \bar{X}_n - x\frac{\sigma}{\sqrt{n}}\right) \approx \Phi(x) - \Phi(-x). \tag{19.22}$$

or

$$\text{Prob}\left(\bar{X}_n - x\frac{\sigma}{\sqrt{n}} \le \mu \le \bar{X}_n + x\frac{\sigma}{\sqrt{n}}\right) \approx \Phi(x) - \Phi(-x). \tag{19.23}$$

This expression asserts that given n independent identically distributed random variables with mean μ and variance σ^2, if n is large then the probability that the interval $\left[\bar{X}_n - x\frac{\sigma}{\sqrt{n}}, \bar{X}_n + x\frac{\sigma}{\sqrt{n}}\right]$ contains μ is approximately $\Phi(x) - \Phi(-x)$. In the case where the X_i are normally distributed, the approximation is replaced by equality. When $x = 1.96$ this gives a 95% *confidence interval*.

REMARK 19.2: If for each x,

$$\lim_{n\to\infty} \text{Prob}\left(\frac{\sqrt{n}(\bar{X}_n - \mu)}{\sigma} \leq x\right) = \Phi(x), \tag{19.24}$$

then $Z_n = \frac{\sqrt{n}(\bar{X}_n - \mu)}{\sigma}$ is said to *converge in distribution* to the standard normal distribution. A common notation for this is:

$$Z_n \xrightarrow{\mathscr{D}} \mathscr{N}(0,1). \tag{19.25}$$

Because Z_n is a random variable, it has a distribution, say F_n, so that $P(Z_n \leq z) = F_n(z)$ for each z. As n increases, there is a corresponding sequence of random variables, Z_n, and related distribution functions F_n. The expression $Z_n \xrightarrow{\mathscr{D}} \mathscr{N}(0,1)$ means that for each x, $F_n(x) \to \Phi(x)$ for each x.

In the previous discussion, σ was assumed known both for the central limit theorem and the construction of the confidence interval for μ. In general, σ would not be known and must be replaced by an estimate, s^2. The estimate of the variance is routinely estimated as $s_n^2 = \frac{1}{n-1}\sum_{i=1}^n (X_i - \bar{X}_n)^2$. If the X_i are independent with common mean, μ, and variance σ^2, then $E\{s_n^2\} = \sigma^2$. As n becomes large, the random variable s_n^2 should provide a more "accurate" estimate of σ^2. Suppose that $\lim_{n\to\infty} \mathrm{P}(\mid s_n^2 - \sigma^2 \mid \geq \epsilon) \to 0$ for each $\epsilon > 0$, then s_n^2 is said to converge in probability to σ^2. If this condition is satisfied, then with $s_n = \sqrt{s_n^2}$, $\hat{Z}_n = \frac{\sqrt{n}(\bar{X}_n - \mu)}{s_n}$ converges in distribution to a standard normal distribution, $\mathscr{N}(0,1)$. □

19.4 Hypothesis testing

Consider a data sample $\underline{X} = (X_1, X_2, \ldots, X_n)$ drawn independently from or according to some statistical distribution $f(x)$. Thus, the joint density is given by $f(\underline{x}) = f(x_1, x_2, \ldots, x_n) = f(x_1)f(x_2)\cdots f(x_n)$. To make parameter dependence clear, write $f(x \mid \theta)$. The parameter θ is unknown, but belongs to some set Θ. Suppose that Θ is partitioned into two sets, Θ_1 and Θ_2, and one wishes to know whether or not $\theta \in \Theta_1$. In general, with a finite set of data, $X_1, \ldots X_n$, this cannot be answered with certainty, but statistical procedures exist for providing an assessment of whether $\theta \in \Theta_1$, or not: hypothesis testing. Suppose that we have some hypothesis H_0 (that $\theta \in \Theta_1$) called the *null* hypothesis (the hypothesis to be tested), regarding this distribution. The (unknown) parameter θ is the subject of the hypothesis – that θ lies in some set Θ_0, or is equal to some specific value θ_0. The alternative hypothesis is that this is not the case, that $\theta \notin \Theta_0$. Label this hypothesis H_1.

With the parameter space, Ω, partitioned into two regions, Ω_1 and Ω_2, a hypothesis may be formulated:

$$H_0: \quad \theta \in \Omega_0$$
$$H_1: \quad \theta \in \Omega_1.$$

19.4.1 A hypothesis test

A procedure for testing the null hypothesis, a test $\mathscr{T}$, is a procedure for making a judgement as to whether the hypothesis, H_0, is true or not. This decision is based on the data observed.

The n data observations $(X_1,\dots,X_n)$ lie in the sample space, the set of possible values for this vector. For example, if a coin is tossed three times with outcome H or T each time, then the set of possible outcomes is $S = \{(H,H,H),(H,H,T),(H,T,H),(H,T,T),(T,H,H),(T,H,T),(T,T,H),(T,T,T)\}$. With n tosses of the coin, an element of the sample space is a list of n possible values – such as $(H,H,T,H,T,T,\dots,T)$, a list with n elements.

To test a hypothesis, the sample space S is partitioned into two regions, the critical region C, and its complement C^c: $S = C \cup C^c$. If the sample $X_1,X_2,\dots,X_n$ lies in the region C, the hypothesis H_0 is rejected, and accepted otherwise. Note that the value of the underlying parameter or parameters, θ, determines the distribution of the sample, so the likelihood of the sample lying on C or C^c depends on θ. Depending on θ, some sample values will be more or less likely than others. In this way, the value of the observed sample provides information on whether a particular value of θ is plausible, given the data. So, C is chosen so that the observed sample would be unlikely if the maintained or null hypothesis were true: one rejects the null hypothesis because the sample data would be unlikely to be observed were the null hypothesis true.

19.4.2 Types of error

In making a decision regarding the truth of a hypothesis (testing the hypothesis), two possible errors may be made: deciding that the hypothesis is false when in fact it is true, or alternatively, not rejecting the hypothesis when in fact it is false. A *type I error* involves rejecting the null hypothesis when it is true (for a simple given hypothesis). The probability of rejecting the null hypothesis when it is true is called the *size* or *level of significance* of the test. A *type II error* involves not rejecting the null hypothesis when it is false.

	H_0 *true*	H_0 *false*
Accept H_0	Correct	Type II error
Reject H_0	Type I error	Correct

Definition 19.4: *The power function, π, associated with a test is defined:*

$$\pi(\theta) = \text{Prob}(\text{Reject}\, H_0 \mid \theta), \quad \theta \in \Omega.$$

In terms of the critical region, C, $\pi(\theta) = \text{Prob}((X_1,\ldots,X_n) \in C \mid \theta)$.

Definition 19.5: *The size of a test, denoted α, is the largest probability of rejecting the null hypothesis when it is true:*

$$\alpha = \sup_{\theta \in \Omega_0} \pi(\theta).$$

19.4.2.1 Error trade-offs

Ideally, the probability of an error should be made as small as possible. However, in selecting the critical region, C, there is trade-off between the probability of a type I and a type II error. Recall, the decision rule is reject H_0 if $(X_1,\ldots,X_n) \in C$. Consider the following two choices for C:

1. Set $C = S$. In this case always reject H_0.

2. Set $C = \{\emptyset\}$, the empty set. In this case never reject H_0.

In the first case, H_0 is rejected, regardless of the data, so in particular, H_0 is rejected when H_0 is true; and since H_0 is never accepted, it is not accepted when false. Therefore, in case 1, P(Type I error) = 1 and P(Type II error) = 0. Conversely, in the second case, P(Type I error) = 0 and P(Type II error) = 1. Thus there is a trade-off when attempting to reduce *both* types of error. The next example illustrates this trade-off.

EXAMPLE 19.1: Suppose that $(X_1,\dots,X_n)$ are random variables drawn from a normal distribution with known variance, $\sigma^2 = 4$, and unknown mean μ, which is either 0 or 1. Consider the following null and alternative hypotheses:

$$\begin{aligned} H_0 &: \quad \mu = 0 \\ H_1 &: \quad \mu = 1. \end{aligned}$$

Suppose the critical region is chosen as $C = C(k) = \{(X_1,\dots,X_n) \mid \bar{X} > k\}$, for some number k. Then,

$$\alpha(C(k)) = P(\bar{X} > k \mid \mu = 0) = 1 - P(\bar{X} \le k \mid \mu = 0) = 1 - \Phi\left(\frac{k-0}{\frac{2}{\sqrt{n}}}\right)$$

$$\beta(C(k)) = 1 - P(\bar{X} \ge k \mid \mu = 1) = 1 - \left[1 - \Phi\left(\frac{k-1}{\frac{2}{\sqrt{n}}}\right)\right] = \Phi\left(\frac{k-1}{\frac{2}{\sqrt{n}}}\right).$$

For any $k \in (-\infty,\infty)$, there is a pair $(\alpha(C(k)), \beta(C(k)))$. A plot of this pair as k varies shows the trade-off between the probability of a type I and a type II error (Figure 19.1). ◊

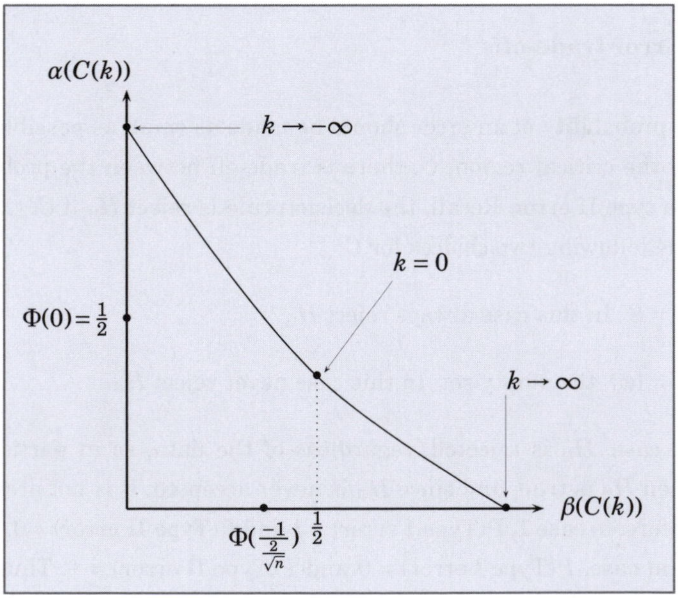

Figure 19.1: Trade-off between Type I and Type II error

19.4.2.2 Simple hypotheses and minimizing error probabilities

A simple hypothesis with simple alternative has the form:

$$H_0: \quad \theta = \theta_0$$
$$H_1: \quad \theta = \theta_1.$$

Let C be the critical region: reject H_0 is $(X_1,\ldots,X_n) \in C$. Define

$$\alpha(C) = \text{Prob}((X_1,\ldots,X_n) \in C \mid H_0) = \text{Prob(Type I error)}$$
$$\beta(C) = 1 - \text{Prob}((X_1,\ldots,X_n) \in C \mid H_1) = \text{Prob(Type II error)}.$$

In this context, we envisage that the sample data are drawn either from a distribution with density $f(\underline{x} \mid \theta_0) = f(x_1 \mid \theta_0)\cdots f(x_n \mid \theta_0) = \prod_{i=1}^n f(x_i \mid \theta_0)$ or from a distribution with density $f(x \mid \theta_1) = \prod_{i=1}^n f(x_i \mid \theta_1)$. Consider an average of the error probabilities:

$$\begin{aligned} L(C) = a\alpha(C) + b\beta(C) &= a\int_C f(\underline{x} \mid \theta_0)d\underline{x} + b\left[1 - \int_C f(\underline{x} \mid \theta_1)d\underline{x}\right] \\ &= b + \left[a\int_C f(\underline{x} \mid \theta_0)d\underline{x} - b\int_C f(\underline{x} \mid \theta_1)d\underline{x}\right] \\ &= b + \int_C [af(\underline{x} \mid \theta_0) - bf(\underline{x} \mid \theta_1)]d\underline{x}. \end{aligned}$$

To make this as small as possible, pick $C = C^* = \{\underline{x} \mid af(\underline{x} \mid \theta_0) - bf(\underline{x} \mid \theta_1) < 0\}$, to be the set of $\underline{x}$ where $af(\underline{x} \mid \theta_0) < bf(\underline{x} \mid \theta_1)$. With this approach, we reject H_0 when $\frac{f(\underline{x}|\theta_1)}{f(\underline{x}|\theta_0)} > \frac{a}{b}$.

REMARK 19.3: In view of the construction of C^*, for any other test procedure, $\hat{C}$:

$$\alpha(C) \le \alpha(C^*) \Rightarrow \beta(C) \ge \beta(C^*)$$

and

$$\alpha(C) < \alpha(C^*) \Rightarrow \beta(C) > \beta(C^*).$$

(This is a version of the Neyman–Pearson lemma.) □

Given the size of the test, the power is as large as possible on the alternative hypothesis.

EXAMPLE 19.2: Suppose that $(X_1,\ldots,X_n)$ are random variables drawn from a normal distribution with known variance, $\sigma^2 = 1$, and unknown mean μ. Consider the following null and alternative hypotheses:

$$H_0: \quad \mu = \mu_0$$
$$H_1: \quad \mu = \mu_1.$$

Assume that $\mu_1 > \mu_0$. Consider the set $C = \{\underline{x} = (x_1, \ldots, x_n) \mid \frac{f(\underline{x}|\mu_1)}{f(\underline{x}|\mu_0)} \geq k\}$. The density of n normal independent random variables with mean μ and variance σ^2 is:

$$f(\underline{x} \mid \mu, \sigma^2) = \left(\frac{1}{\sqrt{2\pi\sigma^2}}\right)^n e^{-\frac{1}{2}\frac{1}{\sigma^2}\sum(x_i-\mu)^2}.$$

Therefore,

$$\begin{aligned}\frac{f(\underline{x} \mid \mu_1, \sigma^2)}{f(\underline{x} \mid \mu_0, \sigma^2)} &= \frac{\left(\frac{1}{\sqrt{2\pi\sigma^2}}\right)^n e^{-\frac{1}{2}\frac{1}{\sigma^2}\sum(x_i-\mu_1)^2}}{\left(\frac{1}{\sqrt{2\pi\sigma^2}}\right)^n e^{-\frac{1}{2}\frac{1}{\sigma^2}\sum(x_i-\mu_0)^2}} \\ &= e^{-\frac{1}{2}\frac{1}{\sigma^2}\sum(x_i-\mu_1)^2} e^{+\frac{1}{2}\frac{1}{\sigma^2}\sum(x_i-\mu_0)^2} \\ &= e^{-\frac{1}{2}\frac{1}{\sigma^2}[\sum(x_i-\mu_1)^2 - \sum(x_i-\mu_0)^2]}.\end{aligned}$$

Since $\sum(x_i - \mu)^2 = \sum x_i^2 - 2\mu\sum x_i + n\mu^2 = \sum x_i^2 - 2n\mu\bar{x} + n\mu^2$, $\sum(x_i - \mu_1)^2 - \sum(x_i - \mu_0)^2 = (-2n\mu_1\bar{x} + n\mu_1^2) - (-2n\mu_0\bar{x} + n\mu_0^2) = 2n\bar{x}(\mu_0 - \mu_1) + n(\mu_1^2 - \mu_0^2)$. Thus, with $\sigma^2 = 1$,

$$\begin{aligned}\frac{f(\underline{x} \mid \mu_1, \sigma^2)}{f(\underline{x} \mid \mu_0, \sigma^2)} &= e^{-\frac{1}{2\sigma^2}[2n\bar{x}(\mu_0-\mu_1) + n(\mu_1^2-\mu_0^2)]} \\ &= e^{-[n\bar{x}(\mu_0-\mu_1) + \frac{1}{2}n(\mu_1^2-\mu_0^2)]}.\end{aligned}$$

The condition $\frac{f(\underline{x}|\mu_1,\sigma^2)}{f(\underline{x}|\mu_0,\sigma^2)} \geq k$ implies $e^{-[n\bar{x}(\mu_0-\mu_1)+\frac{1}{2}n(\mu_1^2-\mu_0^2)]} \geq k$ or $-\left[n\bar{x}(\mu_0 - \mu_1) + \frac{1}{2}n(\mu_1^2 - \mu_0^2)\right] \geq \ln k$. Rearranging: $n\bar{x}(\mu_1 - \mu_0) \geq \ln k + \frac{1}{2}n(\mu_1^2 - \mu_0^2)$ or $\bar{x} \geq \frac{\ln k + \frac{1}{2}n(\mu_1^2-\mu_0^2)}{n(\mu_1-\mu_0)} = \hat{k}(k)$. Finally, the size of the test, α, may be fixed by choosing k appropriately. $P(\bar{x} \geq \hat{k}(k) \mid \mu = \mu_0) = \alpha$.

So, $P(\bar{x} - \mu_0 \geq \hat{k}(k) - \mu_0 \mid \mu_0) = \alpha$, or $P(\sqrt{n}(\bar{x} - \mu_0) \geq \sqrt{n}(\hat{k}(k) - \mu_0) \mid \mu_0) = \alpha$. Since $\sqrt{n}(\bar{x} - \mu_0)$ is a standard normal variable, this is $P(Z \geq \sqrt{n}(\hat{k}(k) - \mu_0) \mid \mu_0) = \alpha$. If z_α satisfies $P(Z \geq z_\alpha) = \alpha$, then set $\sqrt{n}(\hat{k}(k) - \mu_0) = z_\alpha$ or $\hat{k}(k) = \frac{1}{\sqrt{n}}z_\alpha + \mu_0$. Then, $\ln k = \left[\frac{1}{\sqrt{n}}z_\alpha + \mu_0\right](\mu_1 - \mu_0) - \frac{1}{2}(\mu_1^2 - \mu_0^2)$. Finally,

$$k = e^{\left[\frac{1}{\sqrt{n}}z_\alpha + \mu_0\right](\mu_1-\mu_0) - \frac{1}{2}(\mu_1^2-\mu_0^2)}.$$

◇

19.4.3 Simple null and alternative hypotheses

Recall that a hypothesis test consists of the null hypothesis H_0 (the hypothesis to be tested) and the alternative hypothesis H_1 (the hypothesis assumed to be true if the null hypothesis is false). For example, suppose that data come from one of two distributions – the sample $X_1,\ldots,X_n$ is drawn from the distribution $f(x \mid \theta_0)$, or else from $f(x \mid \theta_1)$. Suppose that we hypothesize that the true value of θ is θ_0, the null hypothesis is $H_0 : \theta = \theta_0$ and the alternative hypothesis is $H_1 : \theta = \theta_1$. Sensible intuition suggests that we should compare how likely the data is to have come from $f(\cdot \mid \theta_0)$ with the likelihood of the data having come from $f(\cdot \mid \theta_1)$. Define the likelihood of the sample under the parameter θ as $\mathscr{L}(X_1,\ldots,X_n \mid \theta) = \Pi_{i=1}^n f(X_i \mid \theta)$. This is the density associated with the sample at parameter θ. The relative likelihood of the sample (likelihood ratio) is then given by:

$$LR(X_1,\ldots,X_n) = \frac{\mathscr{L}(X_1,\ldots,X_n \mid \theta_0)}{\mathscr{L}(X_1,\ldots,X_n \mid \theta_1)} = \frac{\Pi_{i=1}^n f(X_i \mid \theta_0)}{\Pi_{i=1}^n f(X_i \mid \theta_1)}.$$

If $LR(X_1,\ldots,X_n)$ is "large", it seems plausible that the sample is drawn according to the parameter θ_0. In this case, the decision rule is: accept H_0 if $LR(X_1,\ldots,X_n) \geq l$ and reject H_0 otherwise, where l is some appropriately chosen threshold. The standard way to choose l is to set the size of the test to some specific value.

Suppose that we reject H_0 whenever $LR(X_1,\ldots,X_n) \leq \lambda$; then the rejection region for the test is $C_\lambda = \{(X_1,\ldots,X_n) \mid LR(X_1,\ldots,X_n) \leq \lambda\}$, giving the sample region on which H_0 is rejected. Choosing λ small means that few samples will satisfy the condition; choosing λ large leads to most samples satisfying the condition. The probability of the sample being drawn in C_λ when the value of θ is θ_0 is given by $\text{Prob}\{(X_1,\ldots,X_n) \in C_\lambda \mid \theta_0\}$. If λ is chosen so that $\text{Prob}\{(X_1,\ldots,X_n) \in C_\lambda \mid \theta_0\} = \alpha$ then the probability of rejecting the null hypothesis, H_0, when it is true is α: the size of the test is α. Similarly, we can consider $\text{Prob}\{(X_1,\ldots,X_n) \in C_\lambda \mid \theta_1\}$, the probability of rejecting H_0 when it is false. Let $\pi_\lambda(\theta) = \text{Prob}\{(X_1,\ldots,X_n) \in C_\lambda \mid \theta\}$ be the power function, so that $\pi_\lambda(\theta_0) = \alpha$.

A key result, the *Neyman–Pearson* lemma asserts that defining the rejection region using the likelihood ratio as described above is optimal in the sense that any alternative test region with no larger size will have a smaller power at the alternative hypothesis. For any alternative test with power function $\pi^*(\theta)$, if $\pi^*(\theta_0) = \pi_\lambda(\theta_0) = \alpha$, then $\pi^*(\theta_1) \leq \pi_\lambda(\theta_1)$.

EXAMPLE 19.3: The exponential distribution has the form $f(x) = \theta e^{-\theta x}$, where $\theta > 0$ and $0 < x < \infty$. Write $f(x \mid \theta)$ to highlight the parameter. Given a sample $(X_1,\ldots,X_n)$,

$$\mathscr{L}(X_1,\ldots,X_n \mid \theta) = \Pi_{i=1}^n f(X_i \mid \theta) = \Pi_{i=1}^n \theta e^{-\theta X_i} = \theta^n e^{-\theta \sum_{i=1}^n X_i}, \qquad (19.26)$$

so that the likelihood ratio is:

$$LR(X_1,\ldots,X_n) = \frac{\mathscr{L}(X_1,\ldots,X_n \mid \theta_0)}{\mathscr{L}(X_1,\ldots,X_n \mid \theta_1)} = \frac{\theta_0^n e^{-\theta_0 \sum_{i=1}^n X_i}}{\theta_1^n e^{-\theta_1 \sum_{i=1}^n X_i}} = \left(\frac{\theta_0}{\theta_1}\right)^n e^{-(\theta_0-\theta_1)\sum_{i=1}^n X_i}.$$

So, $LR(X_1,\ldots,X_n) \le \lambda$ corresponds to $\left(\frac{\theta_0}{\theta_1}\right)^n e^{-(\theta_0-\theta_1)\sum_{i=1}^n X_i} \le \lambda$ or $e^{-(\theta_0-\theta_1)\sum_{i=1}^n X_i} \le \left(\frac{\theta_1}{\theta_0}\right)^n \lambda$. Taking logarithms, this is equivalent to $-(\theta_0-\theta_1)\sum_{i=1}^n X_i \le \ln\left[\left(\frac{\theta_1}{\theta_0}\right)^n \lambda\right]$ or $\sum_{i=1}^n X_i \le \frac{1}{(\theta_1-\theta_0)} \ln\left[\left(\frac{\theta_1}{\theta_0}\right)^n \lambda\right]$, giving the condition under which H_0 is rejected. When $\theta = \theta_0$, we want this to have probability α. ◊

19.4.4 Uniformly most powerful tests

With simple null and alternative hypothesis, and with fixed size α, the test that rejects H_0 when $LR(X_1,\ldots,X_n) = \frac{\mathscr{L}(X_1,\ldots,X_n|\theta_0)}{\mathscr{L}(X_1,\ldots,X_n|\theta_1)} \le \lambda$ is the most powerful test on the alternative H_1: no test of equal size has larger power on the alternative hypothesis $H_1 : \theta = \theta_1$. But this is true for any $\theta > \theta_0$. So, the test "reject H_0 if $\bar{x} \le k'$ some k'", is a *uniformly most powerful test* of the null hypothesis, $H_0 : \theta = \theta_0$, against the alternative hypothesis, $H_1 : \theta > \theta_0$. (Uniformly most powerful because it is most powerful at *each* $\theta_1 > \theta_0$.) Considering Example 19.3, let the null hypothesis be $H_0 : \theta \in \Omega_0 = \{\theta \le \theta_0\}$ and $H_1 : \theta \in \Omega_1 = \{\theta > \theta_0\}$. Fix a size α and choose k so that $\sup_{\theta\in\Omega_0} \pi_C(\theta) = \alpha$, where π_C is the power function determined by k, $C = \{(x_1,\ldots,x_n) \mid \bar{x} \le k\}$. Then, for any other critical region C', with $\pi_{C'}(\theta) \le \alpha$, $\pi_{C'}(\theta) \le \pi_C(\theta)$ for all $\theta \in \Omega_1$.

Definition 19.6: *Let $f(\underline{x} \mid \theta) = \prod_{i=1}^n f(x_i \mid \theta)$ be the joint distribution on n independent identically distributed random variables. Let $T = r(\underline{x})$ be a statistic (a function of the observed data). Say that $f(\underline{x} \mid \theta)$ has a monotone likelihood ratio in θ if, for $\theta_2 > \theta_1$,*

$$\frac{f(\underline{x} \mid \theta_2)}{f(\underline{x} \mid \theta_1)}$$

depends only on $\underline{x}$ through T and the ratio is a non-increasing or non-decreasing function of T over the range of T.

In Example 19.3, putting $T = r(\underline{x}) = \sum_i x_i$,

$$\frac{f(\underline{x} \mid \theta_2)}{f(\underline{x} \mid \theta_1)} = \left(\frac{\theta_2}{\theta_1}\right)^n e^{-(\theta_2-\theta_1)\sum_{i=1}^n X_i} = \left(\frac{\theta_2}{\theta_1}\right)^n e^{-(\theta_2-\theta_1)T},$$

and with $\theta_2 > \theta_1$, the ratio is monotonically decreasing in T.

Theorem 19.4: *Suppose that $f(\underline{x} \mid \theta)$ has a monotone likelihood ratio in $T = r(\underline{x})$, and consider the null and alternative hypotheses:*

$$H_0 : \theta \leq \theta_0$$
$$H_1 : \theta > \theta_0 \tag{19.27}$$

1. *If $\frac{f(\underline{x}|\theta')}{f(\underline{x}|\theta'')}$ is non-decreasing in $r(\underline{x})$, $\forall \theta' > \theta''$) then a test of the form: reject H_0 if $r(\underline{x}) \geq c$ with $P(r(\underline{x}) \geq c \mid \theta = \theta_0) = \alpha$ is a uniformly most powerful (UMP) test of the hypothesis 19.27 at level of significance α.*

2. *If $\frac{f(\underline{x}|\theta')}{f(\underline{x}|\theta'')}$ is non-increasing in $r(\underline{x})$, $\forall \theta' > \theta''$) then a test of the form: reject H_0 if $r(\underline{x}) \leq c$ with $P(r(\underline{x}) \leq c \mid \theta = \theta_0) = \alpha$ is a uniformly most powerful test of the hypothesis 19.27 at level of significance α.*

REMARK 19.4: If the hypotheses are specified:

$$H_0 : \theta \geq \theta_0$$
$$H_1 : \theta < \theta_0,$$

then uniformly most powerful tests are possible. For example, reject H_0 when the likelihood ratio $\frac{f(\underline{x}|\theta')}{f(\underline{x}|\theta'')}$ is non-decreasing and in $r(\underline{x})$, $\forall \theta' > \theta''$) and $r(\underline{x}) \leq c$ with c chosen so that $P(r(\underline{x}) \leq c \mid \theta = \theta_0) = \alpha$ is a uniformly most powerful (UMP) test at significance level α. Conversely, when the likelihood ratio is increasing, the UMP test has the form "reject H_0 when $r(\underline{x}) \geq c$". □

Thus, for one-sided hypotheses, uniformly most powerful tests exist.

19.4.4.1 A detailed example

Suppose that in the past a compact car has averaged 30 mpg ($\mu = 30$) with a standard deviation of 5 mpg ($\sigma = 5$).The manufacturer wishes to test the claim that mpg has increased as a result of engine redesign, against the hypothesis of no improvement. It is known that the redesign has left the standard deviation unchanged. Thus

$$H_0 : \mu \leq 30$$
$$H_1 : \mu > 30$$

Suppose that the sample size is 50 ($n = 50$); data are available on the mpg performance of 50 cars. Consider:

$$Z_{50} = \frac{\bar{x}_{50} - 30}{\frac{5}{\sqrt{50}}}.$$

19.4.4.2 Distribution of the test statistic

If the distribution of car mpg is normal, Z_{50} is normally distributed; if the distribution of car mpg is unknown, then, appealing to the central limit theorem, Z_{50} is approximately normally distributed. If Z_{50} is normally distributed, the procedure we will follow is exact; if not we will appeal to the central limit theorem and the procedure will be approximately correct.

19.4.4.3 Choice of significance level

Next we decide on a level of significance for the test: the probability of rejecting the null hypothesis when it is true. It is usual to choose a level of significance of 1%, 5% or 10%. On the null hypothesis, Z_{50} has a standard normal (or approximately a standard normal) distribution. Pick a level of significance of 5%: then

$$P(Z_{50} \geq 1.65) = P\left(\frac{\bar{x}_{50} - 30}{\frac{5}{\sqrt{50}}} > 1.65\right) = 0.05. \tag{19.28}$$

This gives

$$P\left(\bar{x} \geq 30 + \frac{1}{\sqrt{2}}1.65\right) = P(\bar{x} \geq 31.2) = 0.05.$$

If engine redesign has not increased mpg, then in a sample of 50 new cars there would be a 5% chance that their average mpg would exceed 31.2 (Figure 19.2). In other words: on H_0, $P(\bar{x} \geq 31.2) = 0.05$. Note that this probability is correct only when the underlying mean is unchanged at 30 mpg. Our decision rule is to reject H_0 if $\bar{x} \geq 31.2$, because it is very unlikely that we would have a sample average above 31.2 if H_0 were true – there is only a 5% chance. However, there is a 5% chance that we will reject H_0 when H_0 is true. With this test $P(\text{reject } H_0 | H_0 \text{ true}) = 0.05$. The choice of significance level fixes the probability of a type I error (at the level of significance).

19.4.4.4 Significance level: Type I and II errors

One could go further and reduce the probability of a type I error to 0 as follows. With Z_{50} a standard normal variable (or approximately so), on H_0,

$$P\left(\frac{\bar{x}_{50}-30}{\frac{5}{\sqrt{50}}} \geq 6\right) = 0, \quad \text{or} \quad P(\bar{x}_{50} \geq 34.24) = 0. \tag{19.29}$$

If we reject H_0 when $\bar{x}_{50} > 34.24$, there is 0 probability that we will reject H_0 when it is true.

Suppose that we adopt this rule: reject H_0 if $\bar{x} > 34.24$ and suppose also that the true mean is 31. Then $\frac{\bar{x}_{50}-31}{\frac{5}{\sqrt{50}}}$ is a standard normal variable. Consider the probability of $\bar{x}$ being greater than 34.24 when the true mean is 31. This probability is

$$P(\bar{x} > 34.24) = P\left(\frac{\bar{x}_{50}-31}{\frac{5}{\sqrt{50}}} > \frac{34.24-31}{\frac{5}{\sqrt{50}}}\right) \tag{19.30}$$

$$= P(Z > 4.54). \tag{19.31}$$

Since, for a standard normal random variable Z, $P(Z < 4.58) = 1$, there is 0 probability of rejecting the null hypothesis H_0, even when $\mu = 31$. So, though there is a 0 probability of rejecting the null hypothesis when it is true, there is also 0 probability of rejecting when it is false (and $\mu = 31$). Reducing the probability of rejection when the null is true also reduces the probability of rejection when the null is false (increase the probability of accepting the null hypothesis when it is false). Reducing the probability of a type I error increases the probability of a type II error.

Returning to the decision rule (reject H_0 if $x > 31.2$), if the true mean were 31, then

$$P(\bar{x} > 31.2) = P\left(\frac{\bar{x}_{50}-31}{\frac{5}{\sqrt{50}}} > \frac{31.2-31}{\frac{5}{\sqrt{50}}}\right) = P(Z > 0.28) = 0.39$$

$$P(\bar{x}_{50} > 31.2) = P\left(\frac{\bar{x}-31}{\frac{5}{\sqrt{50}}} > \frac{31.2-31}{\frac{5}{\sqrt{50}}}\right) = P(Z \geq 0.28) = 0.39. \tag{19.32}$$

Therefore, $P(\text{reject } H_0 \mid H_0 \text{ false}) = 0.39$ and $P(\text{accept } H_0 \mid H_0 \text{ false}) = 0.61$, when the true mean is 31 (Figure 19.3). To summarize, while the old rule (reject H_0 if $\bar{x} > 31.2$) has a 5% chance of rejecting H_0 when H_0 is true (Type I error) and a 61% chance of accepting H_0 when H_0 is false and $\mu = 31$ (Type II error),

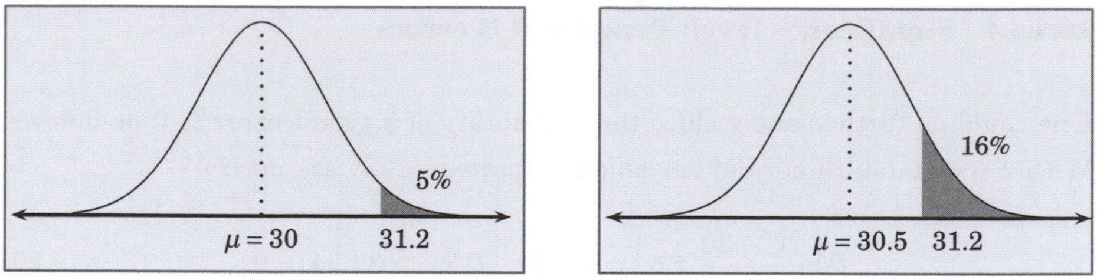

Figure 19.2: Distributions at $\mu = 30$ and $\mu = 30.5$

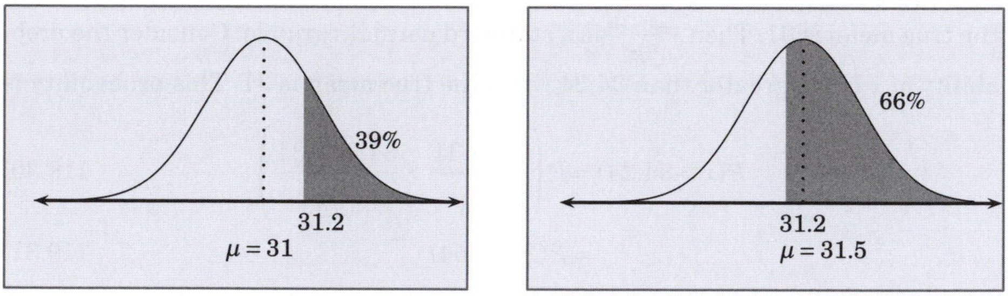

Figure 19.3: Distributions at $\mu = 31$ and $\mu = 31.5$

the decision rule to reject H_0 if $\bar{x} > 34.24$ has a 0% chance of rejecting H_0 when it is true and a 100% chance of accepting H_0 when it is false and $\mu = 31$. The point of these observations is that there is a trade-off between the probabilities of Type I and Type II errors. Reducing the probability of a Type I error increases the probability of a Type II error.

19.4.4.5 Power of a test

Ideally we would like to have a test that always rejects H_0 when it is false and always accepts H_0 when it is true. The power of a test when the true mean is μ is defined to be the probability of rejecting H_0 (when the true mean is μ). Thus an ideal test would have probability 0 of rejecting H_0 when it is true and probability 1 of rejecting H_0 when it is false. In the present case, H_0 is true when $\mu < 30$ and H_0 is false when $\mu > 30$. Denote by $\pi(\mu)$ the power of the test when the true mean is μ. To calculate the power function, we calculate the probability of rejecting H_0 when the true mean is varied. Then, for our example, π would look like Figure 19.4. This function is drawn in Figure 19.5.

$$\mu = 30.0: \qquad P(\bar{x} > 31.2) = P\left(\frac{\bar{x}-30.0}{\frac{5}{\sqrt{50}}} > \frac{31.2-30.0}{\frac{5}{\sqrt{50}}}\right) = P(Z > 1.697) = 0.04$$

$$\mu = 30.5: \qquad P(\bar{x} > 31.2) = P\left(\frac{\bar{x}-30.5}{\frac{5}{\sqrt{50}}} > \frac{31.2-30.5}{\frac{5}{\sqrt{50}}}\right) = P(Z > 0.99) = 0.16$$

$$\mu = 31.0: \qquad P(\bar{x} > 31.2) = P\left(\frac{\bar{x}-31.0}{\frac{5}{\sqrt{50}}} > \frac{31.2-31.0}{\frac{5}{\sqrt{50}}}\right) = P(Z > 0.2828) = 0.39$$

$$\mu = 31.5: \qquad P(\bar{x} > 31.2) = P\left(\frac{\bar{x}-31.5}{\frac{5}{\sqrt{50}}} > \frac{31.2-31.5}{\frac{5}{\sqrt{50}}}\right) = P(Z > -0.4243) = 0.66$$

$$\mu = 32.0: \qquad P(\bar{x} > 31.2) = P\left(\frac{\bar{x}-32.0}{\frac{5}{\sqrt{50}}} > \frac{31.2-32.0}{\frac{5}{\sqrt{50}}}\right) = P(Z > -1.13) = 0.87$$

Figure 19.4: Power function; one-sided test

19.4.4.6 Increasing the sample size

Increasing the sample size changes the power function. Suppose the sample size is 225 instead of 50 ($\sqrt{225} = 15$). Again, under the null hypothesis, $P\left(\frac{\bar{X}-30}{\frac{5}{15}} > z_\alpha\right) = 0.05$ requires $z_\alpha = 1.65$, so setting the size to 5% requires rejection of the null hypothesis when $\bar{X} > 30 + \frac{5}{15}1.65 = 30.55$.

With this rejection threshold, when the true mean is 30.5:

$$P(\bar{x} \geq 30.55) = P\left(\frac{\bar{X}-30.5}{\frac{5}{15}} > \frac{30.55-30.5}{\frac{5}{15}}\right) = P(Z > 0.15) = 0.44.$$

And when the true mean is 31:

$$P(\bar{x} \geq 30.55) = P\left(\frac{\bar{X}-31}{\frac{5}{15}} > \frac{30.55-31}{\frac{5}{15}}\right) = P(Z > -1.35) = 0.91.$$

Similarly, when the true mean is 31.5, $P(\bar{X} > 30.55) = 99.78$.

Increasing the sample size raises the power on the alternative hypothesis.

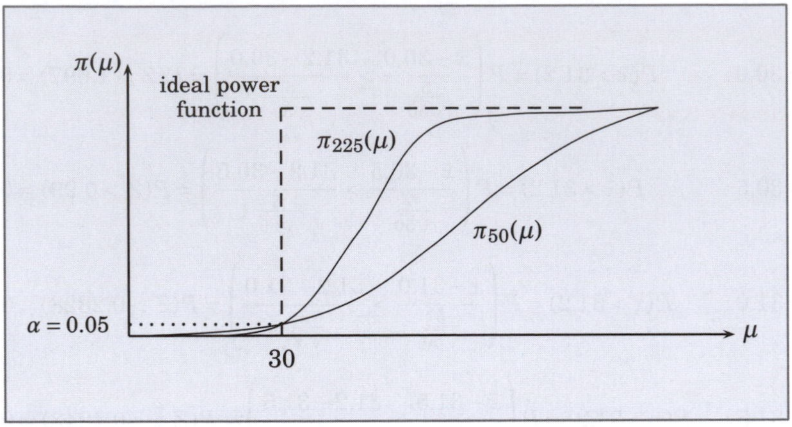

Figure 19.5: Power functions for sample sizes of 50 and 225

19.4.5 Two-sided tests

Consider now a (null) hypothesis that some parameter, say the mean, is equal to a specific value against the (alternative) hypothesis that the parameter is greater than or less than that specific value. In this case, one expects to reject the null hypothesis when the sample average is either "high" or "low" relative to the null hypothesis.

EXAMPLE 19.4:

$$H_0 : \mu = 30$$
$$H_1 : \mu \neq 30.$$

For a standard normal random variable, Z, $P(-1.96 < Z < 1.96) = 0.05$. With a sample size of $n = 50$ and population standard deviation, $\sigma = 5$, $\frac{\bar{x}-30}{\frac{5}{\sqrt{50}}} = \frac{\sqrt{50}}{5}(\bar{x}-30)$ has a standard normal distribution so that

$$P\left(-1.96 < \frac{\bar{x}-30}{\frac{5}{\sqrt{50}}} < 1.96\right) = 0.05$$
$$P\left(30 - \left(\frac{5}{\sqrt{50}}\right)1.96 < \bar{x} < 30 + \left(\frac{5}{\sqrt{50}}\right)1.96\right) = 0.05$$
$$P(28.6 < \bar{x} < 31.4) = 0.05.$$

Thus, when the null hypothesis is true, there is a 5% chance that $\bar{x}$ will fall outside the range [28.6,31.4]. Next, consider the power function of this test. Suppose that $\mu = 30.5$. Then

$$\begin{aligned} P(28.6 < \bar{x} < 31.4 \mid \mu = 30.5) &= P(28.6 - 30.5 < \bar{x} - 30.5 < 31.4 - 30.5) \\ &= P\left(\left(\frac{\sqrt{50}}{5}\right)[28.6 - 30.5] < \left(\frac{\sqrt{50}}{5}\right)[\bar{x} - 30.5] \right. \\ &\qquad \left. < \left(\frac{\sqrt{50}}{5}\right)[31.4 - 30.5]\right) \\ &= P(-2.69 < Z < 1.27) \\ &= P(-2.69 < Z \leq 0) + P(0 < Z < 1.27) \\ &= 0.4964 + 0.3980 \\ &= 0.8944. \end{aligned}$$

Thus, $\pi(30.5) = 0.1056$. For $\mu = 31$:

$$\begin{aligned} P(28.6 < \bar{x} < 31.4 \mid \mu = 31) &= P(28.6 - 31 < \bar{x} - 31 < 31.4 - 31) \\ &= P\left(\left(\frac{\sqrt{50}}{5}\right)[28.6 - 31] < \left(\frac{\sqrt{50}}{5}\right)[\bar{x} - 31] \right. \\ &\qquad \left. < \left(\frac{\sqrt{50}}{5}\right)[31.4 - 31]\right) \\ &= P(-3.39 < Z < 0.56) \\ &= 0.4997 + 0.2123 \\ &= 0.712. \end{aligned}$$

Thus, $\pi(31) = 0.288$. For $\mu = 31.5$:

$$\begin{aligned} P(28.6 < \bar{x} < 31.4 \mid \mu = 31.5) &= P(28.6 - 31.5 < \bar{x} - 31.5 < 31.4 - 31.5) \\ &= P(-4.20 < Z < -0.14) \\ &= 0.4443. \end{aligned}$$

So, $\pi(31.5) = 0.5557$. Similar calculations give the power function at various points for the two-sided test as:

$\pi(28)$	$\pi(28.5)$	$\pi(29)$	$\pi(29.5)$	$\pi(30)$	$\pi(30.5)$	$\pi(31)$	$\pi(31.5)$	$\pi(32)$
0.7995	0.5557	0.288	0.1056	0.05	0.1056	0.288	0.5557	0.7995

The power function for the one-sided alternative (see Figure 19.4) is:

$\pi(28)$	$\pi(28.5)$	$\pi(29)$	$\pi(29.5)$	$\pi(30)$	$\pi(30.5)$	$\pi(31)$	$\pi(31.5)$	$\pi(32)$
0.0	0.0	0.0	0.0072	0.05	0.16	0.39	0.66	0.87

◊

Denote the power functions for the one and two-sided tests $\pi_1(\mu)$ and $\pi_2(\mu)$. These are plotted in Figure 19.6:

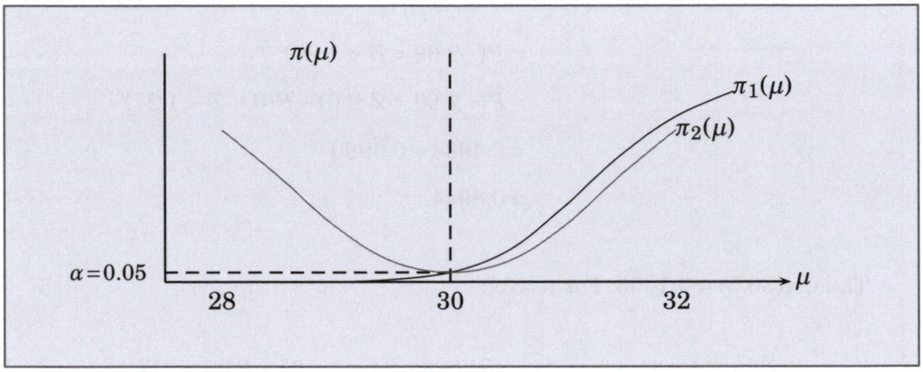

Figure 19.6: Power functions for sample sizes of 50 and 225

19.4.6 Unbiased tests

In the previous example, notice that $\pi_2(\mu) > \pi_1(\mu)$, $\mu > 30$, so that the power function associated with the one-sided test has greater power on the region $\{\mu' \mid \mu' > 30\}$ and less power on the region $\{\mu' \mid \mu' < 30\}$. So, the two-sided test is not uniformly most powerful on the alternative hypothesis.

Definition 19.7: *A test satisfying:*

$$\pi(\theta) \leq \alpha, \quad \theta \in \Omega_0$$
$$\pi(\theta) \geq \alpha, \quad \theta \in \Omega_1$$

is an unbiased test of significance level α where $\max_{\theta \in \Omega_0} = \alpha$.

In the case where $\Omega_0 = \{\mu_0\}$, a single value, and where the power function is differentiable, since π is minimized at $\mu = \mu_0$, it must be that $\pi'(\mu) = 0$. The test given in Example 19.4 is a uniformly most powerful unbiased test.

19.4.7 Use of chi-square and F tests

19.4.7.1 A chi-square test for the variance

Let $X_1,\ldots,X_n$ be independent random variables, normally distributed with mean μ and variance σ^2. Consider:

$$H_0 : \sigma^2 = \sigma_0^2$$
$$H_1 : \sigma^2 \neq \sigma_0^2.$$

Then, on the null hypothesis,

$$\frac{1}{\sigma_0^2}\sum(X_i - \bar{x})^2 = \frac{(n-1)s^2}{\sigma_0^2}$$

has a chi-square distribution with $n-1$ degrees of freedom. Suppose the significance level is set at α. For a random variable Z that has a chi-square distribution with $n-1$ degrees of freedom, let $\chi_{\frac{\alpha}{2}L}$ and $\chi_{\frac{\alpha}{2}U}$ satisfy:

$$P(Z \leq \chi_{\frac{\alpha}{2}L}) = 1 - P(Z \leq \chi_{\frac{\alpha}{2}U}) = P(Z > \chi_{\frac{\alpha}{2}U}) = \frac{\alpha}{2}.$$

Then, on the null hypothesis,

$$P\left(\chi_{\frac{\alpha}{2}L} \leq \frac{(n-1)s^2}{\sigma_0^2} \leq \chi_{\frac{\alpha}{2}U}\right) = \alpha.$$

This gives a test of H_0: reject the null hypothesis if $\frac{(n-1)s^2}{\sigma_0^2}$ is outside the interval $[\chi_{\frac{\alpha}{2}L}, \chi_{\frac{\alpha}{2}U}]$.

For example, if $n = 20$ then $\frac{(n-1)s^2}{\sigma_0^2}$ has a chi-square distribution with $n-1 = 19$ degrees of freedom. A chi-square random variable, Z, with 19 degrees of freedom satisfies: $P(Z \leq 8.907) = 0.025$ and $P(Z \leq 32.852) = 0.975$. So, on the null hypothesis,

$$P\left(8.907 \leq \frac{(n-1)s^2}{\sigma_0^2} \leq 32.852\right) = 0.05.$$

So, reject H_0 when $\frac{(n-1)s^2}{\sigma_0^2}$ lies outside the interval $[8.907, 32.852]$.

19.4.7.2 An F test for difference in variances

Let $X_1,\ldots,X_n$ be n independently normally distributed random variables with mean μ_X and variance σ_X^2. Similarly, let $Y_1,\ldots,Y_n$ be m independently normally distributed random variables with mean μ_Y and variance σ_Y^2. Suppose also that the X_i and Y_j are independent. Consider the hypotheses:

$$H_0 : \sigma_X^2 = \sigma_Y^2$$
$$H_1 : \sigma_X^2 \neq \sigma_Y^2.$$

Let $S_X^2 = \frac{1}{n-1}\sum(X_i - \bar{X})^2$ and $S_Y^2 = \frac{1}{m-1}\sum(Y_j - \bar{Y})^2$. Then $\frac{(n-1)s_X^2}{\sigma_X^2}$ and $\frac{(m-1)s_Y^2}{\sigma_Y^2}$ have, respectively, chi-square distributions with $n-1$ and $m-1$ degrees of freedom. Consider:

$$\frac{\left(\frac{(n-1)s_X^2}{\sigma_X^2}\right)/(n-1)}{\left(\frac{(m-1)s_Y^2}{\sigma_Y^2}\right)/(m-1)} = \frac{\left(\frac{s_X^2}{\sigma_X^2}\right)}{\left(\frac{s_Y^2}{\sigma_Y^2}\right)} = \left(\frac{s_X^2}{s_Y^2}\right)\left(\frac{\sigma_Y^2}{\sigma_X^2}\right).$$

This has an F distribution with $(n-1, m-1)$ degrees of freedom, and on H_0, $\sigma_Y^2 = \sigma_X^2$, so that on H_0, $\frac{s_X^2}{s_Y^2}$ has an F distribution with $(n-1, m-1)$ degrees of freedom. For a random variable Z with an F distribution having these degrees of freedom, define $F_{n-1,m-1}^L$ and $F_{n-1,m-1}^U$ so that $P(Z < F_{n-1,m-1}^L) = P(Z > F_{n-1,m-1}^U) = \frac{\alpha}{2}$. Then,

$$P\left(F_{n-1,m-1}^L < \frac{s_X^2}{s_Y^2} < F_{n-1,m-1}^U\right) = \alpha.$$

This yields the test criterion: reject H_0 if $\frac{s_X^2}{s_Y^2}$ is outside the interval $\left[F_{n-1,m-1}^L, F_{n-1,m-1}^U\right]$. For example, with $\alpha = 0.05$, if $n = 25$ and $m = 21$, $F_{24,20}^L = 0.430$ and $F_{24,20}^U = 2.41$.

19.5 Econometric applications

19.5.1 Matrix computations

The mean or expectation of a random variable X with density $f(x)$ is defined $\mu = \int x f(x) dx$. This is often written $\boldsymbol{E}\{X\}$, the expected value of X. The same notation may be applied to vectors and matrices of random variables. Thus, if $\boldsymbol{X} = (X_1, X_2, \ldots, X_n)$ is a vector of random variables with means μ_i, then $\boldsymbol{E}\{\boldsymbol{X}\} = (E\{X_1\}, E\{X_2\}, \ldots, E\{X_n\}) = (\mu_1, \ldots, \mu_n)$. Similarly, if $\boldsymbol{M}$ is a matrix of random variables, X_{ij}, then expectation of $\boldsymbol{M}$ is defined coordinate-wise:

$$\boldsymbol{M} = \begin{pmatrix} X_{11} & X_{12} & \cdots & X_{1n} \\ X_{21} & X_{22} & \cdots & X_{2n} \\ \vdots & \vdots & \ddots & \vdots \\ X_{n1} & X_{n2} & \cdots & X_{nn} \end{pmatrix}, \quad \boldsymbol{E}\{\boldsymbol{M}\} = \begin{pmatrix} E\{X_{11}\} & E\{X_{12}\} & \cdots & E\{X_{1n}\} \\ E\{X_{21}\} & E\{X_{22}\} & \cdots & E\{X_{2n}\} \\ \vdots & \vdots & \ddots & \vdots \\ E\{X_{n1}\} & E\{X_{n2}\} & \cdots & E\{X_{nn}\} \end{pmatrix}. \tag{19.33}$$

Certain rules of manipulation are clear. If $\boldsymbol{P}$ is an $m \times k$ matrix of constants and $\boldsymbol{Q}$ is an $r \times n$ matrix of constants and $\boldsymbol{M}$ is a $k \times r$ matrix of random variables, then:

$$\boldsymbol{E}\{\boldsymbol{PM}\} = \boldsymbol{PE}\{\boldsymbol{M}\} \quad \text{and} \quad \boldsymbol{E}\{\boldsymbol{PMQ}\} = \boldsymbol{PE}\{\boldsymbol{M}\}\boldsymbol{Q}. \tag{19.34}$$

Note, however, that if both $\boldsymbol{P}$ and $\boldsymbol{M}$ are matrices of random variables, then in general $\boldsymbol{E}\{\boldsymbol{PM}\} \neq \boldsymbol{E}\{\boldsymbol{P}\}\boldsymbol{E}\{\boldsymbol{M}\}$. Given a vector of random variables $\boldsymbol{X} = (X_1, X_2, \ldots, X_n)$, the corresponding vector of means is $\boldsymbol{\mu} = (\mu_1, \ldots, \mu_n)$. Then the vector of deviations around the means is

$$\boldsymbol{X} - \boldsymbol{\mu} = \begin{pmatrix} X_1 - \mu_1 \\ X_2 - \mu_2 \\ \vdots \\ X_n - \mu_n \end{pmatrix}, \tag{19.35}$$

and

$$(\boldsymbol{X} - \boldsymbol{\mu})(\boldsymbol{X} - \boldsymbol{\mu})' = \begin{pmatrix} X_1 - \mu_1 \\ X_2 - \mu_2 \\ \vdots \\ X_n - \mu_n \end{pmatrix} (X_1 - \mu_1, X_2 - \mu_2, \cdots, X_n - \mu_n) \tag{19.36}$$

$$= \begin{pmatrix} (X_1 - \mu_1)(X_1 - \mu_1) & (X_1 - \mu_1)(X_2 - \mu_2) & \cdots & (X_1 - \mu_1)(X_n - \mu_n) \\ (X_2 - \mu_2)(X_1 - \mu_1) & (X_2 - \mu_2)(X_2 - \mu_2) & \cdots & (X_2 - \mu_2)(X_n - \mu_n) \\ \vdots & \vdots & \ddots & \vdots \\ (X_n - \mu_n)(X_1 - \mu_1) & (X_n - \mu_n)(X_2 - \mu_2) & \cdots & (X_n - \mu_n)(X_n - \mu_n) \end{pmatrix}.$$

Let $\sigma_{ij} = \boldsymbol{E}\{(X_i - \mu_i)(X_j - \mu_j)\}$. Then

$$\boldsymbol{E}\{(X - \mu)(X - \mu)'\} = \boldsymbol{\Omega} = \begin{pmatrix} \sigma_{11} & \sigma_{12} & \cdots & \sigma_{1n} \\ \sigma_{21} & \sigma_{22} & \cdots & \sigma_{2n} \\ \vdots & \vdots & \ddots & \vdots \\ \sigma_{n1} & \sigma_{n2} & \cdots & \sigma_{nn} \end{pmatrix}. \tag{19.37}$$

The matrix $\boldsymbol{\Omega}$ is called the variance–covariance matrix of the collection of random variables $(X_1,\ldots,X_n)$, with σ_{ij} the covariance between X_i and X_j. When the random variables are uncorrelated, $\sigma_{ij}=0$. Furthermore, if the random variables have the same variance $\sigma_{ii}=\sigma_{jj}$, write σ^2 for the common variance. Thus, when the random variables are uncorrelated with common variance σ^2, $\boldsymbol{\Omega}=\sigma^2\boldsymbol{I}$, where $\boldsymbol{I}$ is the identity matrix.

19.5.2 Ordinary least squares

A linear model relating Y_t to $X_{1t},X_{2t},\ldots,X_{kt}$ is given by:

$$Y_t=\beta_0+\beta_1X_{it}+\beta_2X_{2t}+\cdots+X_{kt}+\epsilon_t. \tag{19.38}$$

Suppose that data are available on T time periods: $t=1,\ldots,T$. Writing

$$\boldsymbol{Y}=\begin{pmatrix}Y_1\\Y_2\\\vdots\\Y_T\end{pmatrix},\quad \boldsymbol{X}=\begin{pmatrix}1&X_{11}&X_{21}&\cdots&X_{k1}\\1&X_{12}&X_{22}&\cdots&X_{k2}\\\vdots&\vdots&\vdots&\ddots&\vdots\\1&X_{1T}&X_{2T}&\cdots&X_{kT}\end{pmatrix},\quad \boldsymbol{\epsilon}=\begin{pmatrix}\epsilon_1\\\epsilon_2\\\vdots\\\epsilon_T\end{pmatrix},\quad \boldsymbol{\beta}=\begin{pmatrix}\beta_0\\\beta_1\\\vdots\\\beta_k\end{pmatrix}, \tag{19.39}$$

the T observations in Equation 19.38 may be written

$$\boldsymbol{Y}=\boldsymbol{X}\boldsymbol{\beta}+\boldsymbol{\epsilon}. \tag{19.40}$$

19.5.2.1 The OLS estimator

Assuming that $\boldsymbol{X}$ has rank of at least k, the OLS (ordinary least squares) estimator is given by:

$$\hat{\boldsymbol{\beta}}=(\boldsymbol{X}'\boldsymbol{X})^{-1}\boldsymbol{X}'\boldsymbol{Y}. \tag{19.41}$$

This may be expanded:

$$\begin{aligned}\hat{\boldsymbol{\beta}}&=(\boldsymbol{X}'\boldsymbol{X})^{-1}\boldsymbol{X}'\boldsymbol{Y}\\&=(\boldsymbol{X}'\boldsymbol{X})^{-1}\boldsymbol{X}'[\boldsymbol{X}\boldsymbol{\beta}+\boldsymbol{\epsilon}].\\&=(\boldsymbol{X}'\boldsymbol{X})^{-1}\boldsymbol{X}'\boldsymbol{X}\boldsymbol{\beta}+(\boldsymbol{X}'\boldsymbol{X})^{-1}\boldsymbol{X}'\boldsymbol{\epsilon}.\\&=\boldsymbol{\beta}+(\boldsymbol{X}'\boldsymbol{X})^{-1}\boldsymbol{X}'\boldsymbol{\epsilon}.\end{aligned} \tag{19.42}$$

The difference between the estimator, $\hat{\boldsymbol{\beta}}$, and the true value, $\boldsymbol{\beta}$, is:

$$\hat{\boldsymbol{\beta}} - \boldsymbol{\beta} = (\boldsymbol{X}'\boldsymbol{X})^{-1}\boldsymbol{X}'\boldsymbol{\epsilon}, \tag{19.43}$$

a linear combination of the error terms. Assuming $\boldsymbol{X}$ is non-stochastic (non-random), so that it treated as a constant,

$$\boldsymbol{E}\{\hat{\boldsymbol{\beta}} - \boldsymbol{\beta}\} = \boldsymbol{E}\{(\boldsymbol{X}'\boldsymbol{X})^{-1}\boldsymbol{X}'\boldsymbol{\epsilon}\} = (\boldsymbol{X}'\boldsymbol{X})^{-1}\boldsymbol{X}'\boldsymbol{E}\{\boldsymbol{\epsilon}\} = (\boldsymbol{X}'\boldsymbol{X})^{-1}\boldsymbol{X}'\boldsymbol{0} = \boldsymbol{0}. \tag{19.44}$$

So $\boldsymbol{E}\{\hat{\boldsymbol{\beta}} - \boldsymbol{\beta}\} = 0$, and $\hat{\boldsymbol{\beta}}$ is an unbiased estimator of $\boldsymbol{\beta}$. With $\boldsymbol{X}$ non-stochastic, the variance of the estimator is:

$$\begin{aligned}\boldsymbol{E}\{(\hat{\boldsymbol{\beta}} - \boldsymbol{\beta})(\hat{\boldsymbol{\beta}} - \boldsymbol{\beta})'\} &= \boldsymbol{E}\{[(\boldsymbol{X}'\boldsymbol{X})^{-1}\boldsymbol{X}'\boldsymbol{\epsilon}][(\boldsymbol{X}'\boldsymbol{X})^{-1}\boldsymbol{X}'\boldsymbol{\epsilon}]'\} \\ &= \boldsymbol{E}\{[(\boldsymbol{X}'\boldsymbol{X})^{-1}\boldsymbol{X}'\boldsymbol{\epsilon}]\boldsymbol{\epsilon}'\boldsymbol{X}(\boldsymbol{X}'\boldsymbol{X})^{-1}\} \\ &= (\boldsymbol{X}'\boldsymbol{X})^{-1}\boldsymbol{X}'\boldsymbol{E}\{\boldsymbol{\epsilon}\boldsymbol{\epsilon}'\}\boldsymbol{X}(\boldsymbol{X}'\boldsymbol{X})^{-1} \\ &= (\boldsymbol{X}'\boldsymbol{X})^{-1}\boldsymbol{X}'\boldsymbol{\Omega}\boldsymbol{X}(\boldsymbol{X}'\boldsymbol{X})^{-1},\end{aligned} \tag{19.45}$$

using $(\boldsymbol{ABC})' = \boldsymbol{C}'\boldsymbol{B}'\boldsymbol{A}'$, $(\boldsymbol{A}')' = \boldsymbol{A}$, $((\boldsymbol{A}'\boldsymbol{A})^{-1})' = (\boldsymbol{A}'\boldsymbol{A})^{-1}$, since $\boldsymbol{A}'\boldsymbol{A}$ is symmetric and therefore so is $(\boldsymbol{A}'\boldsymbol{A})^{-1}$. When $\boldsymbol{\Omega} = \boldsymbol{\sigma}^2\boldsymbol{I}$,

$$\begin{aligned}(\boldsymbol{X}'\boldsymbol{X})^{-1}\boldsymbol{X}'\boldsymbol{\Omega}\boldsymbol{X}(\boldsymbol{X}'\boldsymbol{X})^{-1} &= \boldsymbol{\sigma}^2(\boldsymbol{X}'\boldsymbol{X})^{-1}\boldsymbol{X}'\boldsymbol{I}\boldsymbol{X}(\boldsymbol{X}'\boldsymbol{X})^{-1} \\ &= \boldsymbol{\sigma}^2(\boldsymbol{X}'\boldsymbol{X})^{-1}\boldsymbol{X}'\boldsymbol{X}(\boldsymbol{X}'\boldsymbol{X})^{-1} = \boldsymbol{\sigma}^2(\boldsymbol{X}'\boldsymbol{X})^{-1}.\end{aligned}$$

In this case,

$$\boldsymbol{E}\{(\hat{\boldsymbol{\beta}} - \boldsymbol{\beta})(\hat{\boldsymbol{\beta}} - \boldsymbol{\beta})'\} = \boldsymbol{\sigma}^2(\boldsymbol{X}'\boldsymbol{X})^{-1}. \tag{19.46}$$

This discussion may be summarized:

Theorem 19.5: *Given the linear model $\boldsymbol{Y} = \boldsymbol{X}\boldsymbol{\beta} + \boldsymbol{\epsilon}$, if $\boldsymbol{X}$ is non-stochastic and the ϵ are uncorrelated with 0 mean and common variance, then $\hat{\boldsymbol{\beta}}$ has mean $\boldsymbol{\beta}$ and variance $\boldsymbol{\sigma}^2(\boldsymbol{X}'\boldsymbol{X})^{-1}$.*

19.5.2.2 Errors and residuals

Let $\hat{\boldsymbol{Y}} = \boldsymbol{X}\hat{\boldsymbol{\beta}}$ and define the vector of residuals:

$$\boldsymbol{e} = \boldsymbol{Y} - \hat{\boldsymbol{Y}}. \tag{19.47}$$

This vector may be be viewed as the variation in $\boldsymbol{y}$ unexplained by $\boldsymbol{X}\hat{\boldsymbol{\beta}}$. Let $\bar{\boldsymbol{y}} = \sum_{t=1}^{n} y_t$ and $\bar{\boldsymbol{Y}} = (\bar{y}, \dots, \bar{y})'$ be an $n \times 1$ column vector with $\bar{y}$ in each location.

Then $(\boldsymbol{Y}-\bar{\boldsymbol{Y}}) = (\boldsymbol{Y}-\hat{\boldsymbol{Y}})+(\hat{\boldsymbol{Y}}-\bar{\boldsymbol{Y}})$. This breaks the deviation around the mean into two components; the part "explained" by the estimated value, $\hat{\boldsymbol{Y}}-\bar{\boldsymbol{Y}}$, and the part that us "unexplained" by $\hat{\boldsymbol{Y}}$, $\boldsymbol{Y}-\hat{\boldsymbol{Y}}$, the residuals. Squaring,

$$(\boldsymbol{Y}-\bar{\boldsymbol{Y}})'(\boldsymbol{Y}-\bar{\boldsymbol{Y}}) = (\boldsymbol{Y}-\hat{\boldsymbol{Y}})'(\boldsymbol{Y}-\hat{\boldsymbol{Y}})+(\hat{\boldsymbol{Y}}-\bar{\boldsymbol{Y}})'(\hat{\boldsymbol{Y}}-\bar{\boldsymbol{Y}})+2(\boldsymbol{Y}-\hat{\boldsymbol{Y}})'(\hat{\boldsymbol{Y}}-\bar{\boldsymbol{Y}}). \quad (19.48)$$

Considering the last term, note that $(\boldsymbol{Y}-\hat{\boldsymbol{Y}}) = \boldsymbol{Y}-\boldsymbol{X}\hat{\boldsymbol{\beta}} = \boldsymbol{e}$, so that $(\boldsymbol{Y}-\hat{\boldsymbol{Y}})'(\hat{\boldsymbol{Y}}-\bar{\boldsymbol{Y}}) = \boldsymbol{e}'(\hat{\boldsymbol{Y}}-\bar{\boldsymbol{Y}}) = \boldsymbol{e}'\hat{\boldsymbol{Y}}-\boldsymbol{e}'\bar{\boldsymbol{Y}}$. From the definition of $\hat{\boldsymbol{\beta}}$, $0 = \boldsymbol{X}'(\boldsymbol{Y}-\boldsymbol{X}\hat{\boldsymbol{\beta}}) = \boldsymbol{X}'\boldsymbol{e}$ so $\boldsymbol{X}'\boldsymbol{e} = 0$. Then $\boldsymbol{e}'\hat{\boldsymbol{Y}} = \boldsymbol{e}'\boldsymbol{X}\hat{\boldsymbol{\beta}} = 0$, since $\boldsymbol{e}'\boldsymbol{X} = 0$. Finally, $\boldsymbol{e}'\bar{\boldsymbol{Y}} = \boldsymbol{e}'\boldsymbol{\iota}\bar{\boldsymbol{y}}$. If the first column of $\boldsymbol{X}$ is $\boldsymbol{\iota}$, then $\boldsymbol{X}'\boldsymbol{e} = 0$ implies that $\boldsymbol{e}'\boldsymbol{\iota} = 0$. Provided the model has an intercept (a column of $\boldsymbol{X}$ is $\boldsymbol{\iota}$), $\boldsymbol{e}'\bar{\boldsymbol{Y}}$. In this case:

$$\begin{aligned}(\boldsymbol{Y}-\bar{\boldsymbol{Y}})'(\boldsymbol{Y}-\bar{\boldsymbol{Y}}) &= (\boldsymbol{Y}-\hat{\boldsymbol{Y}})'(\boldsymbol{Y}-\hat{\boldsymbol{Y}})+(\hat{\boldsymbol{Y}}-\bar{\boldsymbol{Y}})'(\hat{\boldsymbol{Y}}-\bar{\boldsymbol{Y}}) \\ SST &= SSE+SSR.\end{aligned}$$

This decomposes the sum of squared deviations of y around the mean into a variation around the mean that is explained by the regression (SSR) and the remainder that is unexplained (SSE).

Definition 19.8: *The R^2 statistic is a measure of goodness of fit of the regression:*

$$R^2 = 1-\frac{SSE}{SST} = \frac{SST-SSE}{SST}. \quad (19.49)$$

When there is a constant in the $\boldsymbol{X}$ matrix, $R^2 = \frac{SSR}{SST}$, and equals the fraction of total variation in the dependent variable explained by the regression.

The residual vector $\boldsymbol{e} = \boldsymbol{Y}-\hat{\boldsymbol{Y}} = \boldsymbol{Y}-\boldsymbol{X}\hat{\boldsymbol{\beta}}$, and substituting for $\hat{\boldsymbol{\beta}}$

$$\boldsymbol{e} = \boldsymbol{Y}-\boldsymbol{X}\hat{\boldsymbol{\beta}} = \boldsymbol{Y}-\boldsymbol{X}(\boldsymbol{X}'\boldsymbol{X})^{-1}\boldsymbol{X}'\boldsymbol{Y} = \boldsymbol{I}\boldsymbol{Y}-\boldsymbol{X}(\boldsymbol{X}'\boldsymbol{X})^{-1}\boldsymbol{X}'\boldsymbol{Y} = (\boldsymbol{I}-\boldsymbol{X}(\boldsymbol{X}'\boldsymbol{X})^{-1}\boldsymbol{X}')\boldsymbol{Y}. \quad (19.50)$$

Let $\boldsymbol{M} = (\boldsymbol{I}-\boldsymbol{X}(\boldsymbol{X}'\boldsymbol{X})^{-1}\boldsymbol{X}')$, then $\boldsymbol{e} = \boldsymbol{M}\boldsymbol{Y}$. Observe that $\boldsymbol{M}' = \boldsymbol{M}$ and that $\boldsymbol{M}\boldsymbol{M} = \boldsymbol{M}$; the matrix is symmetric and idempotent. One important property of $\boldsymbol{M}$ is that $\boldsymbol{M}\boldsymbol{X} = 0$ since $\boldsymbol{M}\boldsymbol{X} = [\boldsymbol{I}-\boldsymbol{X}(\boldsymbol{X}'\boldsymbol{X})^{-1}\boldsymbol{X}']\boldsymbol{X} = \boldsymbol{X}-\boldsymbol{X}(\boldsymbol{X}'\boldsymbol{X})^{-1}\boldsymbol{X}'\boldsymbol{X} = \boldsymbol{X}-\boldsymbol{X} = 0$. Therefore, $\boldsymbol{e} = \boldsymbol{M}\boldsymbol{Y} = \boldsymbol{M}[\boldsymbol{X}\hat{\boldsymbol{\beta}}+\boldsymbol{\epsilon}] = \boldsymbol{M}\boldsymbol{\epsilon}$, so the vector of residuals is a linear transformation of $\boldsymbol{\epsilon}$.

The sum of squared residuals is $SSE = \boldsymbol{e}'\boldsymbol{e}$; observe $\boldsymbol{e}'\boldsymbol{e} = \boldsymbol{\epsilon}'\boldsymbol{M}'\boldsymbol{M}\boldsymbol{\epsilon} = \boldsymbol{\epsilon}'\boldsymbol{M}\boldsymbol{\epsilon}$. The expected value of the sum of squared residuals $\boldsymbol{E}\{\boldsymbol{e}'\boldsymbol{e}\} = \boldsymbol{E}\{\boldsymbol{e}'\boldsymbol{M}\boldsymbol{e}\}$. Recall that for two matrices $\boldsymbol{A}-p\times q$ and $\boldsymbol{B}-q\times p$, $\mathrm{tr}\boldsymbol{A}\boldsymbol{B} = \mathrm{tr}\boldsymbol{B}\boldsymbol{A}$. In the present context,

$\boldsymbol{\epsilon}'\boldsymbol{M}\boldsymbol{\epsilon} = \text{tr}(\boldsymbol{\epsilon}'\boldsymbol{M}\boldsymbol{\epsilon}) = \text{tr}(\boldsymbol{M}\boldsymbol{\epsilon}\boldsymbol{\epsilon}')$. Taking expectations,

$$\boldsymbol{E}\{\boldsymbol{\epsilon}'\boldsymbol{M}\boldsymbol{\epsilon}\} = \boldsymbol{E}\{\text{tr}(\boldsymbol{M}\boldsymbol{\epsilon}\boldsymbol{\epsilon}')\} = \text{tr}(\boldsymbol{M}\boldsymbol{E}\{\boldsymbol{\epsilon}\boldsymbol{\epsilon}'\}) = \text{tr}(\boldsymbol{M}\boldsymbol{\sigma}^2\boldsymbol{I}) = \boldsymbol{\sigma}^2\text{tr}\boldsymbol{M}. \tag{19.51}$$

Finally, observe that $\boldsymbol{M} = \boldsymbol{I} - \boldsymbol{X}(\boldsymbol{X}'\boldsymbol{X})^{-1}\boldsymbol{X}'$ where $\boldsymbol{I}$ is an $n \times n$ identity matrix. To emphasis this, write $\boldsymbol{I}_n$ for an $n \times n$ identity matrix. Then

$$\begin{aligned}\text{tr}(\boldsymbol{M}) &= \text{tr}(\boldsymbol{I}_n - \boldsymbol{X}(\boldsymbol{X}'\boldsymbol{X})^{-1}\boldsymbol{X}') = \text{tr}\boldsymbol{I}_n - \text{tr}\boldsymbol{X}(\boldsymbol{X}'\boldsymbol{X})^{-1}\boldsymbol{X}' = \text{tr}\boldsymbol{I}_n - \text{tr}(\boldsymbol{X}'\boldsymbol{X})^{-1}\boldsymbol{X}'\boldsymbol{X} \\ &= \text{tr}\boldsymbol{I}_n - \text{tr}\boldsymbol{I}_k \\ &= n - k \end{aligned} \tag{19.52}$$

Therefore, $\boldsymbol{E}\{SSE\} = \boldsymbol{E}\{\boldsymbol{e}'\boldsymbol{e}\} = \boldsymbol{E}\{\boldsymbol{\epsilon}'\boldsymbol{M}\boldsymbol{\epsilon}\} = \boldsymbol{\sigma}^2(n-k)$. This implies:

Theorem 19.6: *An unbiased estimator of σ^2 is given by $s^2 = \frac{e'e}{n-k}$.*

19.5.3 Distributions and hypothesis testing

In what follows, assume that $\boldsymbol{\epsilon}$ is a vector of normal independently distributed random variables.

19.5.3.1 The distribution of $\hat{\beta}$

Recall that $\hat{\boldsymbol{\beta}} - \boldsymbol{\beta} = (\boldsymbol{X}'\boldsymbol{X})^{-1}\boldsymbol{X}'\boldsymbol{\epsilon}$, so that if $\boldsymbol{\epsilon}$ is normally distributed, then so is $\hat{\boldsymbol{\beta}} - \boldsymbol{\beta}$. In this case, from earlier computations, $\hat{\boldsymbol{\beta}} - \boldsymbol{\beta}$ is normally distributed with mean 0 and variance $(\boldsymbol{X}'\boldsymbol{X})^{-1}\boldsymbol{\sigma}^2$.

REMARK 19.5 (properties of quadratic forms): Because $\boldsymbol{M}$ is symmetric, there is an orthogonal matrix $\boldsymbol{C}$ such that $\boldsymbol{C}'\boldsymbol{M}\boldsymbol{C} = \boldsymbol{\Lambda}$, where $\boldsymbol{\Lambda}$ is a diagonal matrix with the characteristic roots of $\boldsymbol{M}$ on the diagonal. Since $\boldsymbol{M}$ is an idempotent matrix, these roots are each either 0 or 1. The number of non-zero roots is equal to the rank of $\boldsymbol{M}$ and this is equal to the trace of $\boldsymbol{M}$, which is $n-k$. If Λ_{jj} is the entry of $\boldsymbol{\Lambda}$ in the jth row and column, then for each j, $\Lambda_{jj} \in \{0,1\}$. Thus, $\sum_{j=1}^{n} \Lambda_{jj} = \sum_{j \in J} \Lambda_{jj} = n-k$.

From $\boldsymbol{C}'\boldsymbol{M}\boldsymbol{C} = \boldsymbol{\Lambda}$, $\boldsymbol{M} = \boldsymbol{C}\boldsymbol{C}'\boldsymbol{M}\boldsymbol{C}\boldsymbol{C} = \boldsymbol{C}\boldsymbol{\Lambda}\boldsymbol{C}'$ so that $\boldsymbol{\epsilon}'\boldsymbol{M}\boldsymbol{\epsilon} = \boldsymbol{\epsilon}'\boldsymbol{C}\boldsymbol{\Lambda}\boldsymbol{C}'\boldsymbol{\epsilon} = \boldsymbol{z}'\boldsymbol{\Lambda}\boldsymbol{z}$, where $\boldsymbol{z} = \boldsymbol{C}'\boldsymbol{\epsilon}$. If $\boldsymbol{\epsilon}$ is normally distributed with mean 0 and variance $\boldsymbol{\sigma}^2\boldsymbol{I}$, then $\boldsymbol{z}$, as a linear combination of normal variables, is also normally distributed with $\boldsymbol{E}\{\boldsymbol{z}\} = \boldsymbol{E}\{\boldsymbol{C}'\boldsymbol{\epsilon}\} = \boldsymbol{C}'\boldsymbol{E}\{\boldsymbol{\epsilon}\} = 0$ and with $\text{Var}(\boldsymbol{z}) = \boldsymbol{E}\{\boldsymbol{z}\boldsymbol{z}'\} = \boldsymbol{E}\{\boldsymbol{C}'\boldsymbol{\epsilon}\boldsymbol{\epsilon}'\boldsymbol{C}\} = \boldsymbol{C}'\boldsymbol{E}\{\boldsymbol{\epsilon}\boldsymbol{\epsilon}'\}\boldsymbol{C} = \boldsymbol{C}'(\boldsymbol{\sigma}^2\boldsymbol{I})\boldsymbol{C} = \boldsymbol{\sigma}^2\boldsymbol{C}'\boldsymbol{C} = \boldsymbol{\sigma}^2\boldsymbol{I}$. Therefore, $\boldsymbol{z}$ is normally distributed with mean 0 and variance $\boldsymbol{\sigma}^2\boldsymbol{I}$, so that $\left(\frac{z}{\sigma}\right)$ is normally distributed with mean 0 and variance $\boldsymbol{I}$, so that $\boldsymbol{z}$ is a vector of independent normal variables with mean 0 and variance 1. Since $\boldsymbol{z}'\boldsymbol{\Lambda z} = \sum_{j=1}^{n} \Lambda_{jj} z_j^2 = \sum_{j\in J} \Lambda_{jj} z_j^2 = \sum_{j\in J} z_j^2$, $\boldsymbol{z}'\boldsymbol{\Lambda z}$ is the sum of $n-k$ independent squared normal variables with mean 0 and variance σ^2. Therefore, $\left(\frac{\boldsymbol{z}}{\sigma}\right)\boldsymbol{\Lambda}\left(\frac{\boldsymbol{z}}{\sigma}\right)$ is the sum of $n-k$ squared standard normal variables and so has a chi-square distribution having $n-k$ degrees of freedom. □

19.5.3.2 The distribution of s^2

The variance is estimated by $s^2 = \frac{1}{n-k}e'e = \frac{1}{n-k}SSE$. Observe that $\left(\frac{e}{\sigma}\right)'\left(\frac{e}{\sigma}\right) = \left(\frac{\epsilon}{\sigma}\right)\boldsymbol{M}\left(\frac{\epsilon}{\sigma}\right) = \left(\frac{z}{\sigma}\right)'\boldsymbol{\Lambda}\left(\frac{z}{\sigma}\right)$, so that $\left(\frac{e}{\sigma}\right)'\left(\frac{e}{\sigma}\right)$ has a chi-square distribution with $n-k$ degrees of freedom. Since the mean and variance of a chi-square distribution with r degrees of freedom are r and $2r$, respectively, the mean and variance of $\frac{SSE}{\sigma^2} = \left(\frac{e}{\sigma}\right)'\left(\frac{e}{\sigma}\right) = \frac{1}{\sigma^2}(e'e)$ is $n-k$ and $2(n-k)$. Hence the mean and variance of SSE are $\sigma^2(n-k)$ and $(\sigma^2)^2 2(n-k) = \sigma^4 2(n-k)$. Since $s^2 = \frac{1}{n-k}SSE$, the mean of s^2 is σ^2, the variance is $\sigma^4 2\frac{1}{n-k}$ and $(n-k)\frac{s^2}{\sigma^2}$ has a chi-square distribution with $n-k$ degrees of freedom.

REMARK 19.6: Observe that $\boldsymbol{e} = \boldsymbol{M\epsilon}$, so that $\boldsymbol{e}$ is a linear combination of normal random variables with $\boldsymbol{E}\{\boldsymbol{e}\} = \boldsymbol{E}\{\boldsymbol{M\epsilon}\} = 0$ and $\mathrm{Var}(\boldsymbol{e}) = \boldsymbol{E}\{\boldsymbol{M\epsilon\epsilon}'\boldsymbol{M}\} = \boldsymbol{ME}\{\boldsymbol{\epsilon\epsilon}'\}\boldsymbol{M} = \boldsymbol{M\sigma}^2\boldsymbol{IM} = \boldsymbol{\sigma}^2\boldsymbol{M}$. So that the residuals are normally distributed with mean 0 and variance $\boldsymbol{\sigma}^2\boldsymbol{M}$. Note that the variance–covariance matrix $\boldsymbol{\sigma}^2\boldsymbol{M}$ is singular, since $\boldsymbol{M}$ has rank $n-k<n$. □

The following theorem is useful.

Theorem 19.7: *Let $\boldsymbol{M}$ be a symmetric idempotent $n\times n$ matrix and let $\boldsymbol{B}$ be a $k\times n$ matrix such that $\boldsymbol{BM} = 0$. If $\boldsymbol{\epsilon}$ is normally distributed with mean 0 and variance $\boldsymbol{\sigma}^2\boldsymbol{I}$, then $\boldsymbol{A\epsilon}$ and $\boldsymbol{B\epsilon}$ are independently distributed.*

PROOF: The variables $\boldsymbol{v} = \boldsymbol{M\epsilon}$ and $\boldsymbol{w} = \boldsymbol{B\epsilon}$ are normally distributed with mean 0. The covariance is $\boldsymbol{E}\{\boldsymbol{vw}'\} = \boldsymbol{E}\{\boldsymbol{M\epsilon\epsilon B}'\} = \boldsymbol{\sigma}^2\boldsymbol{MMB}' = 0$. Normally distributed variables that are uncorrelated are independent. ■

Furthermore, if $\boldsymbol{v}$ and $\boldsymbol{w}$ are independent, then functions of $\boldsymbol{v}$ and $\boldsymbol{w}$ must also be independent.

Theorem 19.8: *The estimators $\hat{\boldsymbol{\beta}}$ and $\mathbf{s}^2$ are independent.*

To see this, observe that $\hat{\boldsymbol{\beta}} - \boldsymbol{\beta} = (\boldsymbol{X}'\boldsymbol{X})^{-1}\boldsymbol{X}'\boldsymbol{\epsilon}$ and $(\boldsymbol{n}-\boldsymbol{k})\mathbf{s}^2 = \boldsymbol{\epsilon M\epsilon} = \boldsymbol{\epsilon MM\epsilon}$. Since $\boldsymbol{MX} = (\boldsymbol{I} - \boldsymbol{X}(\boldsymbol{X}'\boldsymbol{X})^{-1}\boldsymbol{X}')\boldsymbol{X} = \boldsymbol{X} - \boldsymbol{X} = 0$, $\hat{\boldsymbol{\beta}}$ and $\mathbf{s}^2$ are independent.

19.5.4 Test statistics

The estimator $\hat{\boldsymbol{\beta}} = (\boldsymbol{X}'\boldsymbol{X})^{-1}\boldsymbol{X}'\boldsymbol{Y} = \boldsymbol{\beta} + (\boldsymbol{X}'\boldsymbol{X})^{-1}\boldsymbol{X}'\boldsymbol{\epsilon}$ is normally distributed with mean $\boldsymbol{\beta}$ and variance $\boldsymbol{\sigma}^2(\boldsymbol{X}'\boldsymbol{X})^{-1}$; $\hat{\boldsymbol{\beta}} - \boldsymbol{\beta}$ is normally distributed with mean 0 and variance $\boldsymbol{\sigma}^2(\boldsymbol{X}'\boldsymbol{X})^{-1}$. Letting $(X'X)^{-1}_{kk}$ be the entry in the kth row and column of $(\boldsymbol{X}'\boldsymbol{X})^{-1}$, $\hat{\beta}_k - \beta_k$ is normally distributed with mean 0 and variance $\sigma^2(X'X)^{-1}_{kk}$. Therefore,

$$\frac{\hat{\beta}_k - \beta_k}{\sqrt{\sigma^2(X'X)^{-1}_{kk}}} \tag{19.53}$$

is normally distributed with mean 0 and variance 1. However, this cannot be used to test hypotheses regarding β_k since σ^2 is not observed. However, $(n-k)\frac{s^2}{\sigma^2}$ had a chi-square distribution with $n-k$ degrees of freedom and is independent of $\hat{\beta}_k$. Therefore,

$$\frac{\frac{\hat{\beta}_k-\beta_k}{\sqrt{\sigma^2(X'X)^{-1}_{kk}}}}{\sqrt{\frac{\frac{(n-k)s^2}{\sigma^2}}{n-k}}} = \frac{\frac{\hat{\beta}_k-\beta_k}{\sigma\sqrt{(X'X)^{-1}_{kk}}}}{\frac{s}{\sigma}} = \frac{\hat{\beta}_k - \beta_k}{s\sqrt{(X'X)^{-1}_{kk}}} \tag{19.54}$$

has a t distribution with $n-k$ degrees of freedom. With this, one may, for example, test the hypothesis that $\hat{\boldsymbol{\beta}} - \boldsymbol{\beta} = 0$.

More generally, one may test the hypothesis that the parameter vector satisfies a set of linear restrictions: $\boldsymbol{R\beta} = \boldsymbol{p}$. Suppose that $\boldsymbol{R}$ is an $r \times k$ matrix of full rank and recall that $(\hat{\boldsymbol{\beta}} - \boldsymbol{\beta}) = (\boldsymbol{X}'\boldsymbol{X})^{-1}\boldsymbol{X}'\boldsymbol{\epsilon}$, so that $\boldsymbol{R}(\hat{\boldsymbol{\beta}} - \boldsymbol{\beta}) = \boldsymbol{R}(\boldsymbol{X}'\boldsymbol{X})^{-1}\boldsymbol{X}'\boldsymbol{\epsilon} = \boldsymbol{Q\epsilon}$. This is a linear combination of normally distributed random variables, with mean 0 and variance $\boldsymbol{E}\{\boldsymbol{R}(\boldsymbol{X}'\boldsymbol{X})^{-1}\boldsymbol{X}'\boldsymbol{\epsilon\epsilon}\boldsymbol{X}(\boldsymbol{X}'\boldsymbol{X})^{-1}\boldsymbol{R}'\} = \boldsymbol{\sigma}^2\boldsymbol{R}(\boldsymbol{X}'\boldsymbol{X})^{-1}\boldsymbol{X}'\boldsymbol{X}(\boldsymbol{X}'\boldsymbol{X})^{-1}\boldsymbol{R}' = \boldsymbol{\sigma}^2\boldsymbol{R}(\boldsymbol{X}'\boldsymbol{X})^{-1}\boldsymbol{R}'$. Recalling that $(n-k)s^2 = \boldsymbol{\epsilon MM\epsilon}$, since $\boldsymbol{QM} = 0$, $\boldsymbol{R}(\hat{\boldsymbol{\beta}} - \boldsymbol{\beta})$ is independent of s^2.

Consider $\boldsymbol{\epsilon}'\left[\boldsymbol{X}(\boldsymbol{X}'\boldsymbol{X})^{-1}\boldsymbol{R}'[\boldsymbol{R}(\boldsymbol{X}'\boldsymbol{X})^{-1}\boldsymbol{R}']^{-1}\boldsymbol{R}(\boldsymbol{X}'\boldsymbol{X})^{-1}\boldsymbol{X}'\right]\boldsymbol{\epsilon} = \boldsymbol{\epsilon}'\boldsymbol{H\epsilon}$. Observe that $\boldsymbol{H}$ is symmetric and idempotent. Considering $\left(\frac{\epsilon}{\sigma}\right)'\boldsymbol{H}\left(\frac{\epsilon}{\sigma}\right)$, this has a chi-square distribution with degrees of freedom equal to the trace of $\boldsymbol{H}$:

$$\begin{aligned}
&\mathrm{tr}(\boldsymbol{X}(\boldsymbol{X}'\boldsymbol{X})^{-1}\boldsymbol{R}'[\boldsymbol{R}(\boldsymbol{X}'\boldsymbol{X})^{-1}\boldsymbol{R}']^{-1}\boldsymbol{R}(\boldsymbol{X}'\boldsymbol{X})^{-1}\boldsymbol{X}') \\
&\quad = \mathrm{tr}([\boldsymbol{R}(\boldsymbol{X}'\boldsymbol{X})^{-1}\boldsymbol{R}']^{-1}\boldsymbol{R}(\boldsymbol{X}'\boldsymbol{X})^{-1}\boldsymbol{X}'\boldsymbol{X}(\boldsymbol{X}'\boldsymbol{X})^{-1}\boldsymbol{R}') \\
&\quad = \mathrm{tr}([\boldsymbol{R}(\boldsymbol{X}'\boldsymbol{X})^{-1}\boldsymbol{R}']^{-1}\boldsymbol{R}(\boldsymbol{X}'\boldsymbol{X})^{-1}\boldsymbol{X}'\boldsymbol{X}(\boldsymbol{X}'\boldsymbol{X})^{-1}\boldsymbol{R}') \\
&\quad = \mathrm{tr}(\boldsymbol{I}_r) \\
&\quad = r.
\end{aligned} \tag{19.55}$$

Consider the hypothesis that $\boldsymbol{R\beta}=\boldsymbol{p}$, so that $\boldsymbol{\beta}$ satisfies $\boldsymbol{p}$ equality restrictions. In this case, $\boldsymbol{R}(\hat{\boldsymbol{\beta}}-\boldsymbol{\beta})=(\boldsymbol{R}\hat{\boldsymbol{\beta}}-\boldsymbol{p})$, $\boldsymbol{R}\hat{\boldsymbol{\beta}}-\boldsymbol{p}=\boldsymbol{R}(\boldsymbol{X}'\boldsymbol{X})^{-1}\boldsymbol{X}'\boldsymbol{\epsilon}$ and $(\boldsymbol{R}\hat{\boldsymbol{\beta}}-\boldsymbol{p})'[\boldsymbol{R}(\boldsymbol{X}'\boldsymbol{X})^{-1}\boldsymbol{R}']^{-1}$ $(\boldsymbol{R}\hat{\boldsymbol{\beta}}-\boldsymbol{p})=\boldsymbol{\epsilon}'\boldsymbol{H}\boldsymbol{\epsilon}$ so that $\frac{1}{\sigma^2}(\boldsymbol{R}\hat{\boldsymbol{\beta}}-\boldsymbol{p})'[\boldsymbol{R}(\boldsymbol{X}'\boldsymbol{X})^{-1}\boldsymbol{R}']^{-1}(\boldsymbol{R}\hat{\boldsymbol{\beta}}-\boldsymbol{p})$ has a chi-square distribution with p degrees of freedom and is independent of s^2. Therefore:

$$\frac{\frac{(R\hat{\beta}-p)'[R(X'X)^{-1}R']^{-1}(R\hat{\beta}-p)}{\sigma^2}\left(\frac{1}{p}\right)}{\frac{(n-k)s^2}{\sigma^2}(\frac{1}{n-k})} = \frac{\frac{(R\hat{\beta}-p)'[R(X'X)^{-1}R']^{-1}(R\hat{\beta}-p)}{p}}{s^2}$$
$$= \frac{\frac{(R\hat{\beta}-p)'[R(X'X)^{-1}R']^{-1}(R\hat{\beta}-p)}{p}}{\frac{e'e}{n-k}} \qquad (19.56)$$

is distributed as an F distribution with p and $n-k$ degrees of freedom.

Theorem 19.9: *Let $\boldsymbol{M}$ and $\boldsymbol{M}^*$ be symmetric idempotent matrices with $\boldsymbol{MM}^*=0$. If $\xi\sim\mathcal{N}(0,\sigma^2\boldsymbol{I})$, then $\boldsymbol{\xi}'\boldsymbol{M}\boldsymbol{\xi}$ and $\boldsymbol{\xi}'\boldsymbol{M}^*\boldsymbol{\xi}$ are independently distributed.*

REMARK 19.7: Recall that if $\boldsymbol{M}$ is a symmetric idempotent matrix then

$$\frac{1}{\sigma^2}\boldsymbol{\xi}'\boldsymbol{M}\boldsymbol{\xi}\sim\chi^2_{trM},\quad \xi\sim\mathcal{N}(0,\sigma^2\boldsymbol{I}).$$

□

REMARK 19.8: Let $\boldsymbol{M}$ and $\boldsymbol{M}^*$ be symmetric idempotent matrices with $\boldsymbol{MM}^*=\boldsymbol{M}^*$. Then $\boldsymbol{M}-\boldsymbol{M}^*$ is symmetric and idempotent, and $\boldsymbol{M}^*(\boldsymbol{M}-\boldsymbol{M}^*)=0$. To see this note that $(\boldsymbol{M}-\boldsymbol{M}^*)(\boldsymbol{M}-\boldsymbol{M}^*)=\boldsymbol{M}-\boldsymbol{M}^*\boldsymbol{M}-\boldsymbol{MM}^*+\boldsymbol{M}^*=\boldsymbol{M}-\boldsymbol{M}^*-\boldsymbol{M}^*+\boldsymbol{M}^*=\boldsymbol{MM}^*$. □

Recall that if $X\sim\chi^2_{n_1}$ and $Y\sim\chi^2_{n_2}$ are independent, then $\frac{X/n_1}{Y/n_2}\sim F_{n_1,n_2}$.

19.5.5 Hypothesis testing for the linear model

1. Consider the linear model:

$$\boldsymbol{y}=\boldsymbol{X}\boldsymbol{\beta}_1+\boldsymbol{W}\boldsymbol{\beta}_2+\epsilon,\quad \epsilon\sim\mathcal{N}(0,\sigma^2\boldsymbol{I}), \qquad (19.57)$$

where $\boldsymbol{X}-(n\times k_1)$, $\boldsymbol{W}-(n\times k_2)$, and let $k=k_1+k_2$. Letting $\boldsymbol{X}^*=(\boldsymbol{X}:\boldsymbol{W})$ and $\boldsymbol{\beta}'=(\boldsymbol{\beta}_1,\boldsymbol{\beta}_2)$, so that $\boldsymbol{X}^*-(n\times k)$, Equation 19.57 may be expressed as $\boldsymbol{y}=\boldsymbol{X}^*\boldsymbol{\beta}+\boldsymbol{\epsilon}$. Consider the hypothesis:

$$H_0:\boldsymbol{\beta}_2=0$$
$$H_1:\boldsymbol{\beta}_2\neq 0$$

$$X = (X : W)\begin{pmatrix} I \\ 0 \end{pmatrix} = X^* A, \quad \text{where} \quad X^* = (X : W), \quad A = \begin{pmatrix} I \\ 0 \end{pmatrix}.$$

Let $M = I - X(X'X)^{-1}X'$ and $M^* = I - X^*(X^{*\prime}X^*)^{-1}X^{*\prime}$. Note that $MX = 0$ and $M^*X = M^*X^*A = 0$.

(a) On H_0, $y = X\beta_1 + \epsilon = (X^*A)\beta_1 + \epsilon$. Regressing y on X, let the estimated coefficient be $\hat{\beta}_1$ and the residuals $e = y - X\hat{\beta}_1 = y - X(X'X)^{-1}X'y = My$. On H_0, $e = My = M(X\beta_1 + \epsilon) = M\epsilon$.

(b) Regress y on X^* and let β^* be the estimated coefficient, giving residuals $e^* = y - X\beta^* = y - X^*(X^{*\prime}X)^{-1}X^{*\prime}y = (I - X^*(X^{*\prime}X)^{-1}X^{*\prime})y = M^*y$.

Consider:

$$\frac{\frac{e'e - e^{*\prime}e^*}{\text{tr}(M - M^*)}}{\frac{e^{*\prime}e^*}{\text{tr}M^*}} = \frac{\frac{\epsilon'(M - M^*)\epsilon}{\text{tr}(M - M^*)}}{\frac{\epsilon' M^* \epsilon}{\text{tr}M^*}} \sim F_{\text{tr}(M - M^*), \text{tr}M^*} = F_{k - k_1, n - k}$$

Since $(M - M^*)M^* = 0$, $\text{tr}(M - M^*) = \text{tr}M - \text{tr}M^* = (n - k_1) - (n - k) = k - k_1$ and $\text{tr}M^* = n - k$.

2. Suppose that $y_i = X_i\beta_i + \epsilon_i$, $i = 1, 2$, where $X_i - n_i \times k$ and $n = n_1 + n_2$ and $n_i \geq k$. Consider

$$H_0 : \beta_1 = \beta_2$$
$$H_1 : \beta_1 \neq \beta_2.$$

Let

$$y = \begin{pmatrix} y_1 \\ y_2 \end{pmatrix} = \begin{pmatrix} X_1 & 0 \\ 0 & X_2 \end{pmatrix}\begin{pmatrix} \beta_1 \\ \beta_2 \end{pmatrix} + \begin{pmatrix} \epsilon_1 \\ \epsilon_2 \end{pmatrix}$$
$$y = X^*\beta + \epsilon.$$

Let $X = \begin{pmatrix} X_1 \\ X_2 \end{pmatrix} = X^*\begin{pmatrix} I \\ I \end{pmatrix}$. Observe that $M^*X = M^*X^*\begin{pmatrix} I \\ I \end{pmatrix} = 0$.

(a) Regress y on X^* giving residuals e^*.

(b) Regress y on X giving residuals e.

Consider:

$$\frac{\frac{e'e-e^{*\prime}e^*}{\text{tr}(M-M^*)}}{\frac{e^{*\prime}e^*}{\text{tr}M^*}} = \frac{\frac{\frac{1}{\sigma^2}\epsilon'(M-M^*)\epsilon}{\text{tr}(M-M^*)}}{\frac{\frac{1}{\sigma^2}\epsilon' M^*\epsilon}{\text{tr}M^*}} \sim F_{\text{tr}(M-M^*),\text{tr}M^*} = F_{k-k_1,n-k} = F_{k,n-2k}.$$

(Note that $\text{tr}(M-M^*) = n-k-(n-2k) = k$ and $\text{tr}M^* = n-2k$.)

3. Consider

$$y_i = (X_i\ Z_i)\begin{pmatrix}\beta_i\\ \gamma_i\end{pmatrix} + \epsilon_i.$$

This may be written:

$$y = \begin{pmatrix}y_1\\ y_2\end{pmatrix} = \begin{pmatrix}X_1 & 0 & Z_1 & 0\\ 0 & X_2 & 0 & Z_2\end{pmatrix}\begin{pmatrix}\beta_1\\ \beta_2\\ \gamma_1\\ \gamma_2\end{pmatrix} + \begin{pmatrix}\epsilon_1\\ \epsilon_2\end{pmatrix}$$

$$= X^*\alpha + \epsilon.$$

Consider the hypotheses:

$$H_0 : \beta_1 = \beta_2$$
$$H_1 : \beta_1 \neq \beta_2.$$

Let

$$X = \begin{pmatrix}X_1 & Z_1 & 0\\ X_2 & 0 & Z_2\end{pmatrix} = X^*\begin{pmatrix}I & 0 & 0\\ 0 & I & 0\\ 0 & 0 & I\\ 0 & 0 & I\end{pmatrix}.$$

Let e be the residuals from the regression of y on X, and e^* be the residuals from the regression of y on X^*. Noting that $M^*X = 0$, it follows that

$$\frac{\frac{e'e-e^{*\prime}e^*}{\text{tr}(M-M^*)}}{\frac{e^{*\prime}e^*}{\text{tr}M^*}} \sim F_{\text{tr}(M-M^*),\text{tr}M^*}$$

4. Let $y_i = X_i\beta_i + \epsilon_i$, where $X_1 - n_1 \times k$, $X_2 - n_2 \times k$, $n_2 < k$ and $n = n_1 + n_2$. This is the same model as discussed previously by the assumption that $n_2 < k$ implies that $X^{*\prime}X^*$ is not invertible, where $X^* = \begin{pmatrix}X_1 & 0\\ 0 & X_2\end{pmatrix}$.

As before, the hypothesis is:

$$H_0 : \beta_1 = \beta_2$$
$$H_1 : \beta_1 \neq \beta_2.$$

Let $\boldsymbol{X} = \begin{pmatrix} X_1 \\ X_2 \end{pmatrix}$, $\boldsymbol{M} = \boldsymbol{I} - \boldsymbol{X}(\boldsymbol{X}'\boldsymbol{X})^{-1}\boldsymbol{X}'$, $\boldsymbol{M}_1 = \boldsymbol{I} - \boldsymbol{X}_1(\boldsymbol{X}_1'\boldsymbol{X}_1)^{-1}\boldsymbol{X}_1'$ and $\boldsymbol{M}^* = \begin{pmatrix} \boldsymbol{M}_1 & 0 \\ 0 & 0 \end{pmatrix}$. Note that $\boldsymbol{X}'\boldsymbol{M} = (\boldsymbol{X}_1', \boldsymbol{X}_2')\boldsymbol{M}^* = (0,0)$, so $\boldsymbol{M}\boldsymbol{M}^* = (\boldsymbol{I} - \boldsymbol{X}(\boldsymbol{X}'\boldsymbol{X})^{-1}\boldsymbol{X}')\boldsymbol{M}^* = \boldsymbol{M}^*$. Consider:

(a) Regress $\boldsymbol{y}_1$ on $\boldsymbol{X}_1$, giving residuals $\boldsymbol{y}_1 - \boldsymbol{X}_1\hat{\boldsymbol{\beta}}_1 = \boldsymbol{M}_1\boldsymbol{y}_1$,

(b) Regress $\boldsymbol{y}$ on $\boldsymbol{X}$, giving residuals $\boldsymbol{y} - \boldsymbol{X}\hat{\boldsymbol{\beta}} = \boldsymbol{M}y$.

Then $\begin{pmatrix} e_1 \\ 0 \end{pmatrix} = \boldsymbol{M}^*\boldsymbol{y} = \boldsymbol{M}^*\left[\begin{pmatrix} \boldsymbol{X}_1 \\ \boldsymbol{X}_2 \end{pmatrix}\boldsymbol{\beta} + \boldsymbol{\epsilon}\right] = \boldsymbol{M}^*\boldsymbol{\epsilon}$, on H_0.

Therefore,

$$\frac{\frac{\boldsymbol{e}'\boldsymbol{e} - \boldsymbol{e}^{*\prime}\boldsymbol{e}^*}{\text{tr}(\boldsymbol{M}-\boldsymbol{M}^*)}}{\frac{\boldsymbol{e}^{*\prime}\boldsymbol{e}^*}{\text{tr}\boldsymbol{M}^*}} \sim \boldsymbol{F}_{\text{tr}(\boldsymbol{M}-\boldsymbol{M}^*),\text{tr}\boldsymbol{M}^*} = \boldsymbol{F}_{n_2,n_1-k}.$$

Since $\text{tr}(\boldsymbol{M} - \boldsymbol{M}^*) = n - k - (n_1 - k) = n_2$.

In each of these cases, on H_0 there is a small probability that the F statistic is large. When H_0 is true, $\boldsymbol{e}'\boldsymbol{e}$, the sum of squared residuals with the restriction, should be close to $\boldsymbol{e}^{*\prime}\boldsymbol{e}^*$, the sum of squared residuals without the restriction – so the numerator in the F statistic is small.

Exercises for Chapter 19

Exercise 19.1 In a firm, the mean number of errors in random batches of 100 units is 14. It is known that the distribution of errors is normal with standard deviation $\sigma = 2$. An error-reduction policy is implemented and then a sample of 16 batches of 100 items is taken. Give a test with 5% significance level of the hypothesis that errors have been reduced (against the null hypothesis that there is no improvement).

Exercise 19.2 Recall that the power, $\pi(\mu)$, of a test is the probability of rejecting the null hypothesis when the true parameter value is μ. Suppose that a test of the hypothesis $H_0 : \mu \geq 14$ against the alternative $H_1 : \mu < 14$ at the 5% significance level had critical region $\{\bar{X} \mid \bar{X} < 13.175\}$, so that H_0 is rejected if $\bar{X}$ is less than 13.175. Calculate the power function of the test.

Exercise 19.3 The P value associated with a test is the probability of observing a text statistic value that deviates from the null hypothesis by more than the actual statistic, given the null hypotheses is true. Suppose a mean of $\bar{X} = 13$ is observed from a sample of size 16 from a distribution that is assumed to be normal with standard deviation of 2 under the null hypothesis that $\mu = 14$. Calculate the P value.

Exercise 19.4 A store sells pet food to a large number of customers. The typical customer purchases 80 kilos of pet food each year. The standard deviation in purchase quantity is $\sigma = 0.5$ kilos and the distribution of consumption is assumed to be normal.

A sample of 16 customer purchases is taken at the end of each year to detect drift in consumption patterns from the mean consumption level of 80 kilos. Determine a test for the hypothesis $H_0 : \mu = 80$ against the alternative $H_1 : \mu \neq 80$ at the 5% significance level.

Exercise 19.5 For Exercise 19.4, calculate the power function.

Exercise 19.6 A production process makes carpets with a variance in length of 9 cm. However, the manufacturer does not know the mean length of carpets produced and wishes to estimate the mean length of carpets produced, such that there is a 95% probability of the estimate being within 1 cm of the true mean, μ. How large a sample should be chosen?

Exercise 19.7 Suppose that the advertised operating life of a battery is 50 hours. As a user, you suspect that the battery life is shorter. Assuming that battery life is normally distributed, a sample of 12 batteries produces an average battery life of $\bar{X} = 53.8$ and a sample standard deviation $s = 7.5$. What do you conclude? Test the hypothesis at a 5% significance level. What value of $\bar{X}$ would lead you to conclude that battery lifetime was shorter than 50?

Exercise 19.8 Consider the test in the Exercise 19.7. Calculate the power function associated with this test.

Exercise 19.9 An advertising agency has designed an advertising program on the assumption that 50% of the audience are over 30 years of age. The agency wishes to determine if this assumption is correct. To test the hypothesis, the agency surveyed 400 audience members. Of these, 210 were over 30 and 190 were 30 years or younger. Test the hypothesis that the audience population has 50% over 30.

Exercise 19.10 A manufacturing process produces car tires with the characteristics that the distribution of tread life is normal with a mean of 30 thousand miles and a standard deviation of 3 thousand miles. If 4 tires are selected at random, what is the probability that the sample mean tread life will be less than 26 thousand miles?

Exercise 19.11 Continuing with the problem in Exercise 19.10, suppose the company selects 16 tires at random. What mean tread life has a 99% probability of being exceeded?

Exercise 19.12 The heights of women and men in a population are normally and independently distributed with $\mu_w = 65$, $\sigma_w = 1$, $\mu_m = 68$, $\sigma_m = 2$, using the subscript w for women and m for men. Suppose that a woman and a man are selected from the population. What is the probability that the woman is taller than the man?

Exercise 19.13 According to the central limit theorem, if $X_1, X_2, \ldots, X_n$ is a random sample from some distribution, with $E\{X_i\} = \mu$ and $\mathrm{Var}(X_i) = \sigma^2$ for all i, then with $\bar{X}_n = \frac{1}{n}\sum_{i=1}^{n} X_i$,

$$Z_n = \frac{\sqrt{n}(\bar{X}_n - \mu)}{\sigma}$$

converges in distribution to a standard normal random variable.

Recall that if X is a random variable with distribution function F, and $\{X_n\}$ a sequence of random variables with corresponding distribution functions $\{F_n\}$, then the sequence of random variables is said to converge in distribution to the random variable X if $F_n(x) \to F(x)$ for each x at which F is continuous. Use this result to answer the following question.

Suppose that the average yearly consumption of sugar by people frequenting a store is 175 ounces, with a standard deviation of 20 ounces. What is the probability that 16 people selected at random would have a combined consumption exceeding 3000 ounces?

Exercise 19.14 A firm regularly receives a large shipment of widgets. A random sample of 400 is taken each time and each widget tested for quality. A shipment is rejected if more than 50 widgets in the sample are defective. Furthermore, it is known that in any shipment, 90% of the widgets are good (and 10% defective). What is the probability that a shipment will be rejected?

Exercise 19.15 A population has 50% green items and 50% red. A random sample of 100 items is drawn from the population. What is the probability (approximately) that at least 55 items drawn are green?

Exercise 19.16 Suppose a random sample of size 75 is drawn from a uniform distribution on the unit interval. Then $E\{X_i\} = 0.5$ and $\mathrm{Var}(X_i) = \frac{1}{12}$. What is the probability (approximately) that the mean of the sample will be between 0.45 and 0.55?

Exercise 19.17 A lightbulb manufacturer makes lightbulbs with mean life of 500 hours and with standard deviation of 500 hours. What is the probability that a sample of 25 lightbulbs will have a mean life less than 450 hours? What is the probability that a sample of 400 will have a mean life less than 450 hours?

Bibliography

Apostol, T. M., *Mathematical Analysis*, 2nd edition, Addison Wesley Longman, Reading, MA, 1974.

Bunday, B., *Basic Optimization Methods*, Edward Arnold, London, 1984.

Cull, P., Flahive, M. E. and Robson, R. O., *Difference Equations: From Rabbits to Chaos*, Springer-Verlag, New York, 2005.

Debreu, D. (1952), "Definite and Semidefinite Quadratic Forms" *Econometrica*, Vol. 20, No. 2, 295–300.

Gilbert, G. T. (1991), "Positive Definite Matrices and Sylvester's Criterion" *The American Mathematical Monthly*, Vol. 98, No. 1, 44–46.

Hestenes, M., *Optimization Theory: The Finite Dimensional Case*, Krieger Publishing Company, New York, 1981.

Horn, R. A. and Johnson, C. R., *Matrix Analysis*, Cambridge University Press, Cambridge, 1985.

Silverman, R. A., *Modern Calculus and Analytic Geometry*, MacMillan, New York, 1969.

Vajda, S. (1974), "Tests of Optimality in Constrained Optimization" *Journal of the Institute of Mathematics and its Applications*, Vol. 13, No. 2, 187–200.

Väliaho, H. (1982), "On the Definity of Quadratic Forms Subject to Linear Constraints" *Journal of Optimization Theory and Applications*, Vol. 38, No. 1, 143–145.

Index